Student Solutions Manual to accompany Calculus

Robert Ellis | Denny Gulick

AF903843

Cengage

Australia • Brazil • Canada • Mexico • Singapore • United Kingdom • United States

Cengage

Student Solutions Manual to accompany Calculus

Robert Ellis | Denny Gulick

Executive Editors:
Michele Baird
Maureen Staudt
Michael Stranz

Project Development Manager:
Linda deStefano

Senior Marketing Coordinators:
Sara Mercurio
Lindsay Shapiro

Senior Production / Manufacturing Manager:
Donna M. Brown

PreMedia Services Supervisor:
Rebecca A. Walker

Rights & Permissions Specialist:
Kalina Hintz

Cover Image:
Getty Images*

* Unless otherwise noted, all cover images used by Custom Solutions, a part of Cengage Learning, have been supplied courtesy of Getty Images with the exception of the Earthview cover image, which has been supplied by the National Aeronautics and Space Administration (NASA).

© 2004 Cengage Learning

ALL RIGHTS RESERVED. No part of this work covered by the copyright herein may be reproduced, transmitted, stored or used in any form or by any means graphic, electronic, or mechanical, including but not limited to photocopying, recording, scanning, digitizing, taping, Web distribution, information networks, or information storage and retrieval systems, except as permitted under Section 107 or 108 of the 1976 United States Copyright Act, without the prior written permission of the publisher.

For product information and technology assistance, contact us at
Cengage Learning Customer & Sales Support, 1-800-354-9706
For permission to use material from this text or product,
submit all requests online at **cengage.com/permissions**
Further permissions questions can be emailed to
permissionrequest@cengage.com

ISBN-13: 979-8-214-04775-1

Cengage Learning
5191 Natorp Boulevard
Mason, Ohio 45040
USA

Cengage Learning is a leading provider of customized learning solutions with office locations around the globe, including Singapore, the United Kingdom, Australia, Mexico, Brazil, and Japan. Locate your local office at:
international.cengage.com/region

Cengage Learning products are represented in Canada by Nelson Education, Ltd.

For your lifelong learning solutions, visit **custom.cengage.com**

Visit our corporate website at **cengage.com**

Printed in the United States of America

Contents

Chapter 1

Functions

1.1 The Real Numbers

1. Notice that $4 \cdot 16 = 64 > 63 = 9 \cdot 7$. Multiplying by the positive number $1/(9 \cdot 16)$, we have $(4 \cdot 16)/(9 \cdot 16) > (9 \cdot 7)/(9 \cdot 16)$ or $\frac{4}{9} > \frac{7}{16}$. Therefore, $a > b$.

3. Since $\pi^2 > (3.14)^2 = 9.8596$, we have $a > b$.

5. $(1.41)^2 = 1.9881 < 2$, so $\sqrt{2} > 1.41$.

7. Closed, bounded: 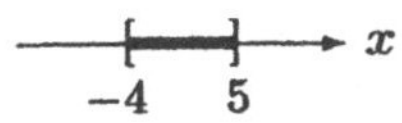

9. Open, unbounded: 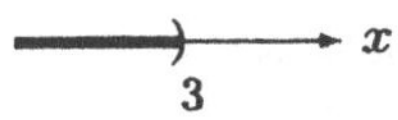

11. Closed, unbounded: 

13. Closed, unbounded: 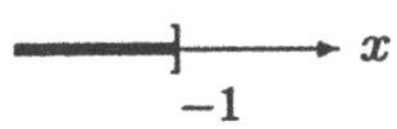

15. $(-3, 4)$

17. $(1, \infty)$

19. If $-6x - 2 > 5$, then $-6x > 7$, so $x < -\frac{7}{6}$. Thus the solution is $(-\infty, -\frac{7}{6})$.

21. If $-1 \le 2x - 3 < 4$, then $2 \le 2x < 7$, so $1 \le x < \frac{7}{2}$. Thus the solution is $[1, \frac{7}{2})$.

23. From the diagram we see that the solution is the union of $(-\infty, -\frac{1}{2}]$ and $[1, \infty)$.

$x - 1$	−	−	−	−	−	−	−	0	+	+	+
$x + \frac{1}{2}$	−	−	0	+	+	+	+	+	+	+	+
$(x-1)(x+\frac{1}{2})$	+	+	0	−	−	−	−	0	+	+	+

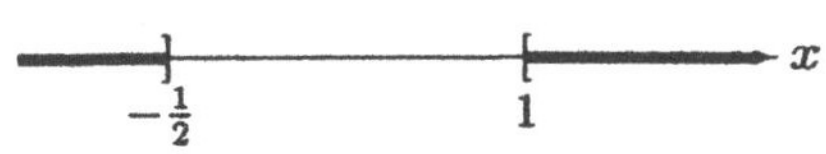

25. From the diagram we see that the solution is the union of $(-\infty, -\frac{1}{3})$ and $(0, \frac{2}{3})$.

x	− − − − 0 + + + + + +
$x-\frac{2}{3}$	− − − − − − − − − − 0 + +
$x+\frac{1}{3}$	− − 0 + + + + + + + +
$x(x-\frac{2}{3})(x+\frac{1}{3})$	− − 0 + 0 − − − 0 + +

27. From the diagram we see that the solution is the union of $(-\infty, -3)$ and $(-1, \infty)$.

$(2x-1)^2$	+ + + + + + + + + + 0 +
$x+1$	− − − − − − − 0 + + + +
$x+3$	− − 0 + + + + + + + + +
$\dfrac{(2x-1)^2}{(x+1)(x+3)}$	+ + − − − + + 0 +

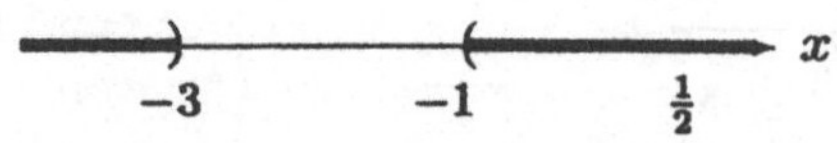

29. The given inequality is equivalent to $2x^2(2x-3) \leq 0$. From the diagram we see that the solution is $(-\infty, \frac{3}{2}]$.

x^2	+ + 0 + + + + + + + + +
$2x-3$	− − − − − − − − 0 + + +
$2x^2(2x-3)$	− − 0 − − − − 0 + + +

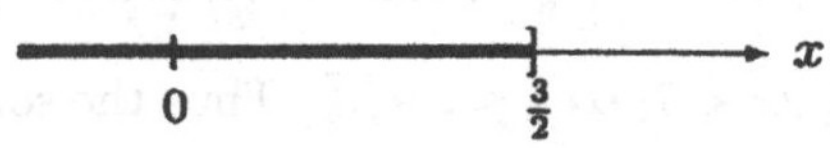

31. The given inequality is equivalent to $(8x^3-1)/x^2 > 0$. From the diagram we see that the solution is $(\frac{1}{2}, \infty)$.

$8x^3-1$	− − − − − − − − 0 + + +
x^2	+ + 0 + + + + + + + + +
$\dfrac{8x^3-1}{x^2}$	− − − − − − 0 + + +

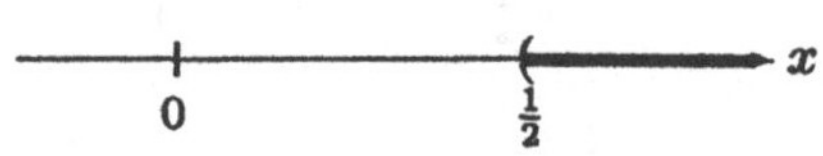

33. The given inequality is equivalent to

$$\frac{4x(x+\sqrt{6})(x-\sqrt{6})}{(x+2)(x-2)} < 0.$$

From the diagram we see that the solution is the union of $(-\infty, -\sqrt{6})$, $(-2, 0)$, and $(2, \sqrt{6})$.

x	$- - - - - 0 + + + + +$
$x+\sqrt{6}$	$- 0 + + + + + + + + +$
$x-\sqrt{6}$	$- - - - - - - - - - 0 +$
$x+2$	$- - - 0 + + + + + + +$
$x-2$	$- - - - - - - - 0 + + +$
$\dfrac{4x(x+\sqrt{6})(x-\sqrt{6})}{(x+2)(x-2)}$	$- 0 + \quad - 0 + \quad - 0 +$

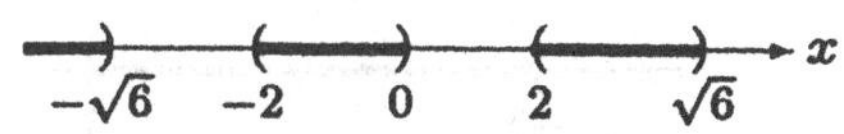

35. The given inequality is equivalent to

$$\frac{(t+2)(t-1)}{(t+1)^3(t-1)^3} \geq 0 \quad \text{or} \quad \frac{t+2}{(t+1)^3(t-1)^2} \geq 0.$$

From the diagram we see that the solution is the union of $(-\infty, -2]$, $(-1, 1)$, and $(1, \infty)$.

$t+2$	$- - 0 + + + + + + + + +$
$(t+1)^3$	$- - - - 0 + + + + + + +$
$(t-1)^2$	$+ + + + + + + + + 0 + +$
$\dfrac{t+2}{(t+1)^3(t-1)^2}$	$+ + 0 - \quad + + + \quad + +$

37. Observe that $\sqrt{9-6x}$ is defined only for $x \leq \frac{9}{6} = \frac{3}{2}$ and that $\sqrt{9-6x} > 0$ for $x < \frac{3}{2}$. Thus the given inequality is equivalent to the pair of inequalities $x < \frac{3}{2}$ and $2 - x > 0$ (that is, $x < 2$). Thus the solution is $(-\infty, \frac{3}{2})$.

39. The given inequality is equivalent to

$$\frac{1}{x+1} - \frac{3}{2} > 0 \quad \text{or} \quad \frac{-3(x+\frac{1}{3})}{2(x+1)} > 0.$$

From the diagram we see that the solution is $(-1, -\frac{1}{3})$.

$x+\frac{1}{3}$	$- - - - - - - - - 0 + +$
$x+1$	$- - 0 + + + + + + + + +$
$\dfrac{-3\,(x+\frac{1}{3})}{2(x+1)}$	$- - \quad + + + + + 0 - -$

41. The given inequality is equivalent to

$$\frac{x+1}{x-1} - \frac{1}{2} \le 0, \quad \text{or} \quad \frac{x+3}{2(x-1)} \le 0.$$

From the diagram we see that the solution is $[-3, 1)$.

$x+3$	$- \ - \ 0 \ + \ + \ + \ + \ + \ + \ + \ +$
$x-1$	$- \ - \ - \ - \ - \ - \ - \ - \ - \ 0 \ + \ +$
$\dfrac{x+3}{2(x-1)}$	$+ \ + \ 0 \ - \ - \ - \ - \ - \ - \quad + \ +$

43. $-|-3| = -3$

45. $|-5| + |5| = 5 + 5 = 10$

47. $|x| = 1$ if $x = 1$ or $-x = 1$; the solution is $-1, 1$.

49. $|x-1| = 2$ if $x - 1 = 2$ (so that $x = 3$), or $-(x-1) = 2$ (so that $-x+1 = 2$, or $x = -1$); the solution is -1, 3.

51. $|6x+5| = 0$ if $6x + 5 = 0$, or $x = -\frac{5}{6}$; the solution is $-\frac{5}{6}$.

53. If $|x| = |x|^2$, then either $|x| = 0$ or we may divide by $|x|$ to obtain $1 = |x|$ (so that $x = -1$ or $x = 1$). The solution is $-1, 0, 1$.

55. If $|x+1|^2 + 3|x+1| - 4 = 0$, then $(|x+1|+4)(|x+1|-1) = 0$. Since $|x+1| + 4 \neq 0$ it follows that $|x+1| - 1 = 0$, or $|x+1| = 1$. Thus either $x + 1 = 1$ (so that $x = 0$), or $-(x+1) = 1$ (so that $-x = 2$, or $x = -2$). The solution is $0, -2$.

57. If $|x+4| = |x-4|$, then either $x + 4 = x - 4$ (so that $4 = -4$, which is impossible), or $x + 4 = -(x-4)$ (so that $2x = 0$, or $x = 0$). The solution is 0.

59. If $|x-2| < 1$, then $-1 < x - 2 < 1$, or $1 < x < 3$. The solution is $(1, 3)$.

61. If $|x+1| < 0.01$, then $-0.01 < x + 1 < 0.01$, or $-1.01 < x < -0.99$. The solution is $(-1.01, -0.99)$.

63. If $|x+3| \ge 3$, then $x + 3 \ge 3$ (so that $x \ge 0$), or $-(x+3) \ge 3$ (so that $x \le -6$). The solution is the union of $(-\infty, -6]$ and $[0, \infty)$.

65. If $|2x+1| \ge 1$, then either $2x + 1 \ge 1$ (so that $x \ge 0$), or $-(2x+1) \ge 1$, (so that $2x \le -2$, or $x \le -1$). The solution is the union of $(-\infty, -1]$ and $[0, \infty)$.

67. If $|2x - \frac{1}{3}| > \frac{2}{3}$, then either $2x - \frac{1}{3} > \frac{2}{3}$ (so that $2x > 1$, or $x > \frac{1}{2}$), or $-(2x - \frac{1}{3}) > \frac{2}{3}$ (so that $-2x > \frac{1}{3}$, or $x < -\frac{1}{6}$). The solution is the union of $(-\infty, -\frac{1}{6})$ and $(\frac{1}{2}, \infty)$.

69. Since $|4 - 2x| \geq 0 > -1$ for all x, the given inequality is equivalent to $|4 - 2x| < 1$, so that $-1 < 4 - 2x < 1$, or $-5 < -2x < -3$, or $\frac{5}{2} > x > \frac{3}{2}$. The solution is $(\frac{3}{2}, \frac{5}{2})$.

71. $\dfrac{69^{800}}{59^{800}} = \left(\dfrac{69}{59}\right)^{800} \approx 2.498407507 \times 10^{54}$

73. $\dfrac{(0.123)^{9000}}{(0.125)^{9000}} = \left(\dfrac{0.123}{0.125}\right)^{9000} \approx 9.034120564 \times 10^{-64}$

75. We desire all x such that $|x-12|+|x-13| > 4$. If $x \geq 13$, then the inequality becomes $x-12+x-13 > 4$, that is, $2x > 4+25 = 29$, so $x > 14.5$. If $x \leq 12$, then the inequality becomes $-(x-12)-(x-13) > 4$, that is, $-2x > 4 - 25 = -21$, so $x < 10.5$. Finally, if $12 < x < 13$, then $|x - 12| \leq 1$ and $|x - 13| \leq 1$, so $|x - 12| + |x - 13| < 4$. Consequently the solution is the union of $(-\infty, 10.5)$ and $(14.5, \infty)$.

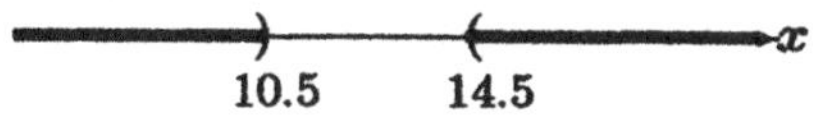

77. Yes, because if $x > 5$, then $x^2 > 25$.

79. No, because if $0 < x \leq 1$, then $1/x \geq 1 \geq x$.

81. a. If $a \geq 0$ and $b \geq 0$ (or $a \leq 0$ and $b \leq 0$), then $|ab| = ab = |a|\,|b|$. If $a \geq 0$ and $b \leq 0$ (or $a \leq 0$ and $b \geq 0$), then $|ab| = -(ab) = |a|\,|b|$. In either case, $|ab| = |a|\,|b|$.

    b. If $b \geq 0$, then $b = |b| \geq -|b|$. If $b < 0$, then $-b = |b| > -|b|$, so $b = -|b| < |b|$. In either case, $-|b| \leq b \leq |b|$.

    c. $|a - b| = |(-1)(b - a)| = |-1|\,|b - a| = |b - a|$.

83. $(|a| + |b|)^2 = |a|^2 + 2|a|\,|b| + |b|^2 = a^2 + 2|ab| + b^2 \geq a^2 + 2ab + b^2 = (a + b)^2 = |a + b|^2$, with equality holding if and only if $|ab| = ab$. But $|ab| = ab$ if and only if $ab \geq 0$. Thus $(|a| + |b|)^2 = |a + b|^2$, and hence $|a| + |b| = |a + b|$, if and only if $ab \geq 0$.

85. If $0 < a < b$, then $0 < (\sqrt{b/2} - \sqrt{a/2})^2 = (b/2 - 2\sqrt{ab/4} + a/2) = [(a+b)/2 - \sqrt{ab}]$, so $\sqrt{ab} < (a+b)/2$. Also $a = \sqrt{a^2} < \sqrt{ab}$.

87. If $0 < a < b$, then $a = \sqrt{aa} < \sqrt{ab}$, so $-2a > -2\sqrt{ab}$, and thus $(\sqrt{b} - \sqrt{a})^2 = b - 2\sqrt{ab} + a < b - 2a + a = b - a$. Consequently $\sqrt{b} - \sqrt{a} < \sqrt{b - a}$.

89. Assume $\sqrt{3} = p/q$, where p and q are integers such that at most one of them is divisible by 3. Then $3 = p^2/q^2$, or $p^2 = 3q^2$. Thus 3 divides p^2, so 3 divides p—say $p = 3a$. Then $3q^2 = p^2 = 9a^2$, so we have $q^2 = 3a^2$. Thus 3 divides q^2 and hence q. Therefore 3 divides both p and q, contradicting our assumption. Consequently $\sqrt{3}$ is irrational.

91. Let a and b be adjacent sides of the rectangle. Then $P = 2(a+b)$. By Exercise 85, $\sqrt{ab} \leq (a+b)/2$, so $ab \leq [(a+b)/2]^2 = (P/4)^2$. But ab is the area of the rectangle, whereas $(P/4)^2$ is the area of a square with perimeter P.

93. **If A_R is the area of a rectangle of perimeter P, and A_S the area of a square of perimeter P, then by Exercise 91, $A_R \le A_S$. If A_C is the area of a circle of circumference (perimeter) P, then by Exercise 92, $A_S < A_C$. Thus $A_R < A_C$.**

1.2 Points and Lines in the Plane

1.

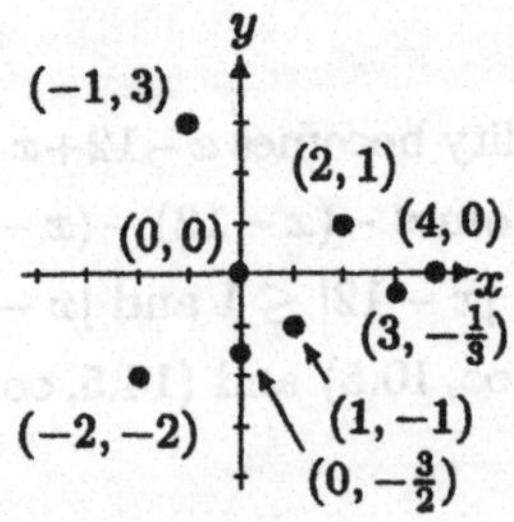

3. Distance $= \sqrt{(-2-3)^2 + (0-0)^2} = 5$

5. Distance $= \sqrt{(6-2)^2 + (-3-1)^2} = \sqrt{32} = 4\sqrt{2}$

7. Distance $= \sqrt{(-3-6)^2 + (-4-5)^2} = \sqrt{162} = 9\sqrt{2}$

9. Distance $= \sqrt{(\sqrt{3}-\sqrt{2})^2 + (2-1)^2} = \sqrt{6 - 2\sqrt{6}}$

11. Distance $= \sqrt{(b-a)^2 + (b-a)^2} = \sqrt{2(b-a)^2} = \sqrt{2}\,|b-a|$

13. By (3),
$y - (-1) = 3(x-2)$,
or $y = 3x - 7$.

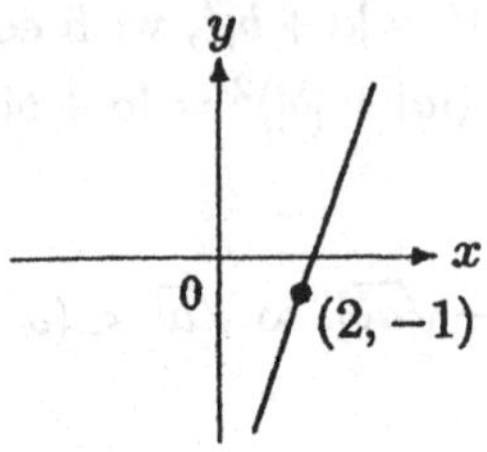

15. By (3),
$y - \frac{1}{2} = -1(x - \frac{1}{2})$,
or $y = -x + 1$.

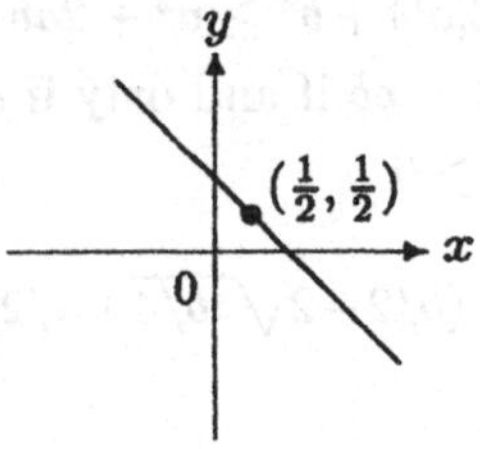

17. By (4),
$y = (-1)x + 0$,
or $y = -x$.

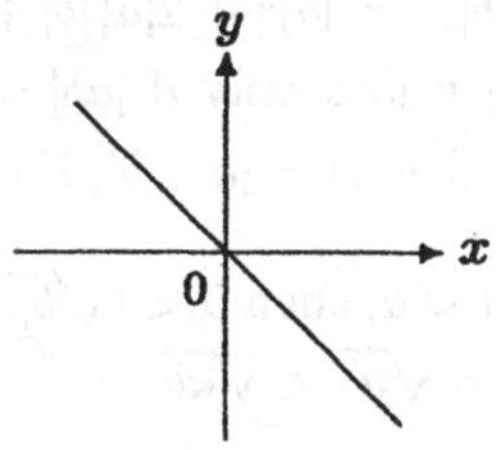

19. By (4),
$y = 3x - 3$.

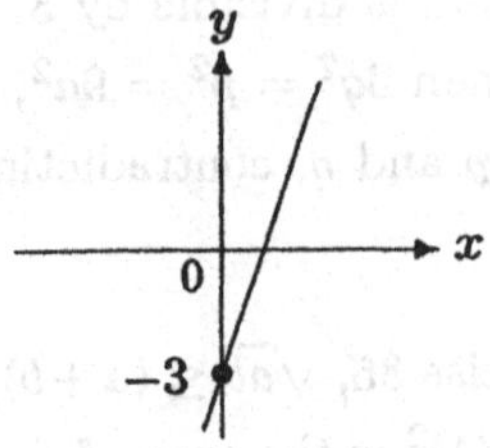

21. By (4),
$m = -1$ and $b = 0$.

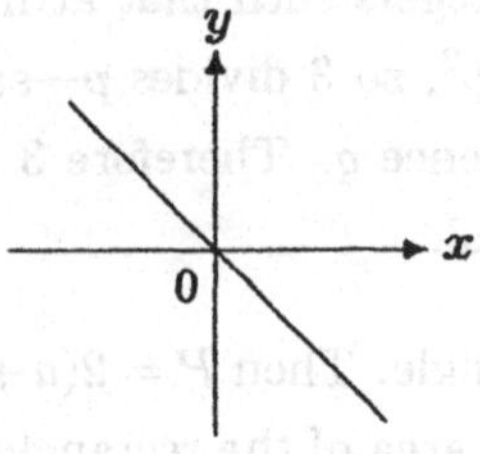

23. The slope-intercept equation of the line is $y = 2x + 7$. By (4), $m = 2$ and $b = 7$.

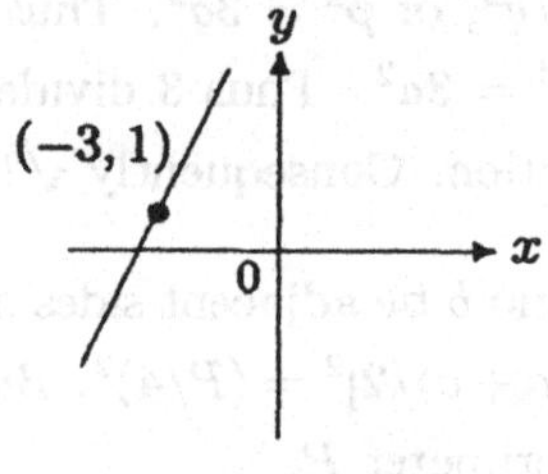

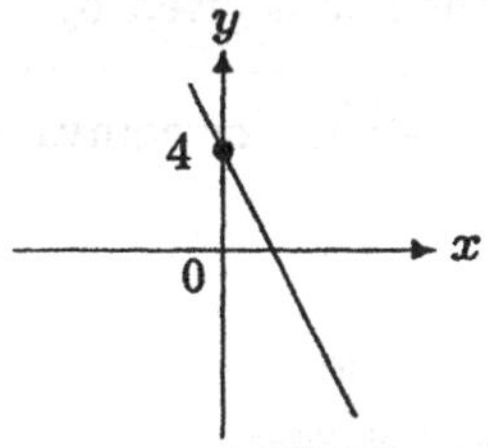

Exercise 25

25. The slope-intercept equation of the line is $y = -2x + 4$. By (4), $m = -2$ and $b = 4$.

27. $m_1 = 2$; $m_2 = -\frac{1}{2}$. Since $m_1 m_2 = -1$, the lines are perpendicular. If (x, y) is the point of intersection, then $y = 2x + 3$ and $y = -\frac{1}{2}x - 1$, so that $2x + 3 = -\frac{1}{2}x - 1$, or $\frac{5}{2}x = -4$, or $x = -\frac{8}{5}$. Then $y = 2(-\frac{8}{5}) + 3 = -\frac{1}{5}$. The point of intersection is $(-\frac{8}{5}, -\frac{1}{5})$.

29. $m_1 = 1$; $m_2 = 1$. Since $m_1 = m_2$, the lines are parallel.

31. $m_1 = -\frac{2}{3}$; $m_2 = \frac{3}{2}$. Since $m_1 m_2 = -1$, the lines are perpendicular. If (x, y) is the point of intersection, then $y = -\frac{2}{3}x - \frac{1}{3}$ and $y = \frac{3}{2}x - \frac{5}{2}$, so $-\frac{2}{3}x - \frac{1}{3} = \frac{3}{2}x - \frac{5}{2}$, or $-\frac{13}{6}x = -\frac{13}{6}$, or $x = 1$. Then $y = -\frac{2}{3}(1) - \frac{1}{3} = -1$. The point of intersection is $(1, -1)$.

33. Both lines are vertical. Thus the lines are parallel.

35. $m_1 = -2$; $m_2 = -2$. Since $m_1 = m_2$, the lines are parallel.

37. The slope of l is 3; the desired line has the same slope. From the point-slope equation, we get $y - (-1) = 3(x - 2)$, or $y = 3x - 7$.

39. The slope of l is -1; the desired line has the same slope. From the point-slope equation, we get $y - 0 = -1(x - 0)$, or $y = -x$.

41. The slope of l is $\frac{2}{3}$; the desired line has the same slope. From the point-slope equation, we get $y - 1 = \frac{2}{3}(x - 2)$, or $y = \frac{2}{3}x - \frac{1}{3}$.

43. The slope of l is 2; from Theorem 1.3, the desired line has slope $-\frac{1}{2}$. By (3), an equation is $y - (-3) = -\frac{1}{2}(x - (-1))$, or $y = -\frac{1}{2}x - \frac{7}{2}$.

45. The slope of l is $-\frac{2}{3}$; from Theorem 1.3, the desired line has slope $\frac{3}{2}$. By (3), an equation is $y - 3 = \frac{3}{2}(x - 2)$, or $y = \frac{3}{2}x$.

47. The slope of l is 2; from Theorem 1.3, the desired line has slope $-\frac{1}{2}$. By (3), an equation is $y - (-5) = -\frac{1}{2}(x - 4)$, or $y = -\frac{1}{2}x - 3$.

49. By (1), the slope m is given by $m = (y_2 - y_1)/(x_2 - x_1)$. By (3), a point-slope equation of l is given by

$$y - y_1 = \frac{y_2 - y_1}{x_2 - x_1}(x - x_1).$$

51. By Exercise 49 with $(x_1, y_1) = (-2, 4)$, the line is given by

$$y - 4 = \frac{3-4}{-1-(-2)}(x-(-2)), \quad \text{or equivalently,} \quad y - 4 = -(x+2).$$

53.

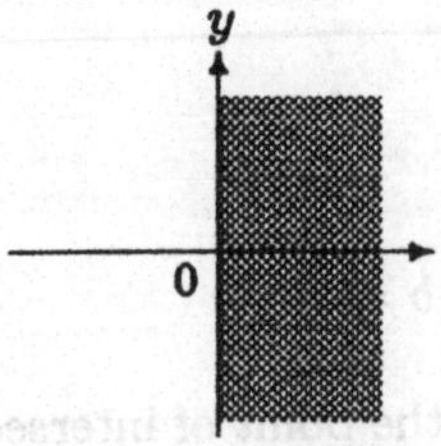

55.

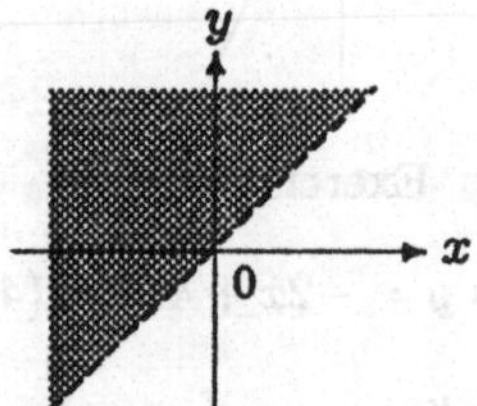

57.

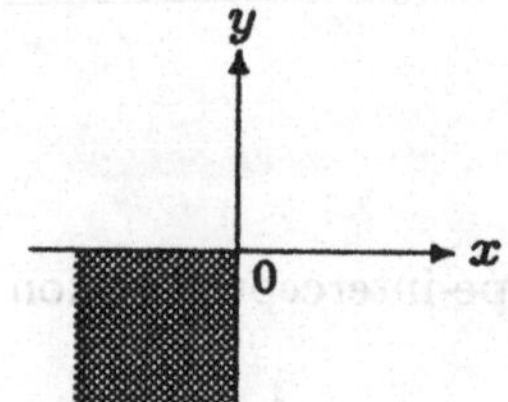

59.

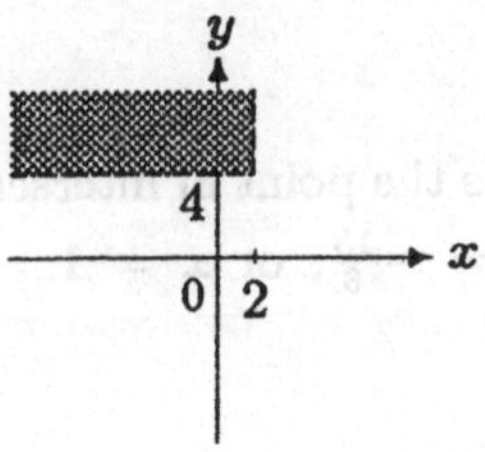

61.

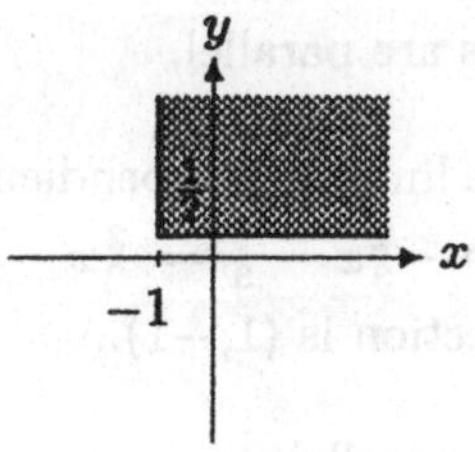

63. a. The requirement $(x, y) = (x, -y)$ is satisfied only when $y = -y$, or $y = 0$. Thus the region is the x axis.

b. The requirement $(x, y) = (-x, y)$ is satisfied only when $x = -x$, or $x = 0$. Thus the region is the y axis.

65. By the figure we deduce that the other vertices of the square are $(-3, 7)$, $(-7, 7)$, and $(-7, 3)$.

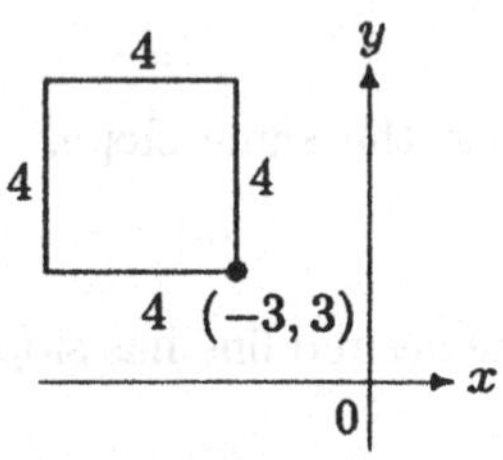

Exercise 65

67. Let d_1, d_2, and d_3 be the lengths of the three sides. Then

$$d_1 = \sqrt{\left(\sqrt{3}-1-(-1)\right)^2 + (3-2)^2} = 2$$

$$d_2 = \sqrt{\left(-1-(\sqrt{3}-1)\right)^2 + (4-3)^2} = 2$$

$$d_3 = \sqrt{\left(-1-(-1)\right)^2 + (4-2)^2} = 2$$

Thus the triangle is equilateral.

69. Using the figure, we find that the midpoints are $(a/2, 0)$, $((a+b)/2, c/2)$, and $(b/2, c/2)$. The sum of the squares of the lengths of the sides is $a^2 + [(b-a)^2 + c^2] + (b^2 + c^2) = 2a^2 + 2b^2 + 2c^2 - 2ab$. The sum of the squares of the lengths of the medians is

$$\left[\left(b-\frac{a}{2}\right)^2 + c^2\right] + \left[\left(\frac{a+b}{2}\right)^2 + \left(\frac{c}{2}\right)^2\right] + \left[\left(\frac{b}{2}-a\right)^2 + \left(\frac{c}{2}\right)^2\right]$$

$$= b^2 - ab + \frac{a^2}{4} + c^2 + \frac{a^2}{4} + \frac{ab}{2} + \frac{b^2}{4} + \frac{c^2}{4} + \frac{b^2}{4} - ab + a^2 + \frac{c^2}{4} = \frac{3}{4}(2a^2 + 2b^2 + 2c^2 - 2ab).$$

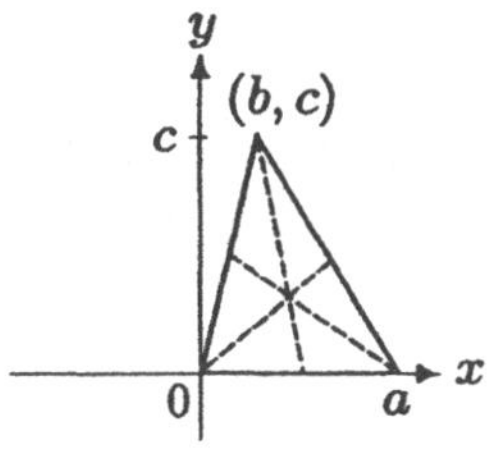

Exercise 69

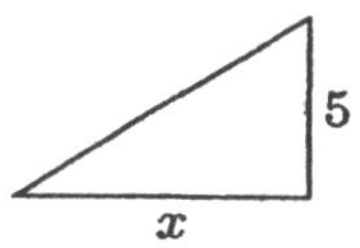

Exercise 71

71. Let x denote the distance between the base of the ramp and the wall. By hypothesis, $5/x = 0.28$, so that $x = 5/0.28 \approx 17.85714286$. Therefore the distance is approximately 17.9 feet.

1.3 Functions

1. $f(\sqrt{5}) = \sqrt{3}$; $f(\pi) = \sqrt{3}$

3. $f(0) = 1 - 0 + 0^3 = 1$; $f(-1) = 1 - (-1) + (-1)^3 = 1$

5. $g(\sqrt{2}) = \dfrac{1}{2(\sqrt{2})^2} = \dfrac{1}{4}$

7. $g(27) = \sqrt[3]{27} = 3$; $g(-\frac{1}{8}) = \sqrt[3]{-\frac{1}{8}} = -\frac{1}{2}$

9. $f(2) = \dfrac{2-1}{2^2+4} = \dfrac{1}{8}$

11. $f(3.2) = \frac{-2}{169}(3.2)^2 + \frac{4}{13}(3.2) + 3 \approx 3.863431953$; $f(25.5) = \frac{-2}{169}(25.5)^2 + \frac{4}{13}(25.5) + 3 \approx 3.150887574$

13. $g(0.5) = \dfrac{100.24(0.5)^6}{(0.24)(0.5)^6 - 1} \approx -1.572145546$; $g(-7.31) = \dfrac{100.24(-7.31)^6}{(0.24)(-7.31)^6 - 1} \approx 417.6780725$

15. All real numbers

17. $[-2, 8]$

19. $[-2, \infty)$

21. Since $x(x-1) \geq 0$ for $x \leq 0$ and for $x \geq 1$, the domain is the union of $(-\infty, 0]$ and $[1, \infty)$.

23. Since $3 - 1/t^2 \geq 0$ if $3 \geq 1/t^2$, or equivalently, $t^2 \geq \frac{1}{3}$, so that $|t| \geq 1/\sqrt{3} = \sqrt{3}/3$, the domain is the union of $(-\infty, -\sqrt{3}/3]$ and $[\sqrt{3}/3, \infty)$.

25. Since the cube root function is defined for all real numbers, the domain is all real numbers.

27. Since $x - 1 = 0$ if $x = 1$, the domain is all real numbers except 1.

29. Since $w^2 - 16 = 0$ if $w = -4$ or 4, the domain is all real numbers except -4 and 4.

31. Since $x^2 + 4 > 0$ for all x, the domain is all real numbers.

33. Union of $[-4, -1]$ and $(0, 6)$

35. x is in the domain if $1 - \sqrt{9 - x^2} \geq 0$ (so that $1 \geq \sqrt{9 - x^2}$, or $1 \geq 9 - x^2$, or $x^2 \geq 8$) and $9 - x^2 \geq 0$ (so that $x^2 \leq 9$). Thus $8 \leq x^2 \leq 9$, so the domain is the union of $[-3, -2\sqrt{2}]$ and $[2\sqrt{2}, 3]$.

37. The set consisting of the number -1.

39. If $x < 4$, then $f(x) = 3x - 2 < 3 \cdot 4 - 2 = 10$. Thus the range is $(\infty, 10)$.

41. y is in the range of f if there is an x such that $y = \dfrac{1}{x - 1}$. This is equivalent to $x - 1 = \dfrac{1}{y}$, or $x = \dfrac{1}{y} + 1$. This equation has a solution x unless $y = 0$. Thus the range consists of all real numbers except 0.

43. a. A function is described.

    b. A function is not described because two values are assigned to the number -2.

    c. f is not a function because f assigns two values to every positive number.

    d. f is not a function because f assigns two values to every number.

    e. g is a function.

    f. g is a function.

    g. g is a function. (Note that $g(2) = 9$.)

    h. g is not a function because $g(1) = 2 - 3 = -1$ from the first line of the formula, whereas $g(1) = 3 - 3 = 0$ from the second line of the formula.

    i. f is a function.

    j. f is not a function because it assigns two values to numbers such as $\sqrt{2}$: $(\sqrt{2})^2 = 2$, which is rational, so that $f(\sqrt{2}) = 2$; but $\sqrt{2}$ is irrational, so $f(\sqrt{2}) = \sqrt{2}$.

45. $f_1 = f_5$ and $f_2 = f_3$.

47. a. x is in the domain of f if and only if $x^2 - 1 \geq 0$, or $x^2 \geq 1$, so the domain of f is the union of $(-\infty, -1]$ and $[1, \infty)$. x is in the domain of g if and only if $x^2 - 1 \geq 0$ and $x + \sqrt{x^2 - 1} \neq 0$. The first condition requires x to be in $(-\infty, -1]$ or $[1, \infty)$. Since under this condition, $0 \leq x^2 - 1 < x^2$, we have $\sqrt{x^2 - 1} < |x|$. Therefore $\sqrt{x^2 - 1} \neq -x$, so $x + \sqrt{x^2 - 1} \neq x + (-x) = 0$. Thus the domain of f and the domain of g are equal to the union of $(-\infty, -1]$ and $[1, \infty)$.

b. For all x in the domain of f,

$$f(x) = (x - \sqrt{x^2-1})\frac{x+\sqrt{x^2-1}}{x+\sqrt{x^2-1}} = \frac{x^2-(x^2-1)}{x+\sqrt{x^2-1}} = \frac{1}{x+\sqrt{x^2-1}} = g(x).$$

Since f and g have the same domain and assign the same value to each x in that domain, $f = g$.

49. $f(x) = \frac{1}{2}x^2\sqrt[3]{x/5}$ for $x \geq 0$

51. Let x denote the length of a side. Thus

$$A(x) = \frac{1}{2}x\left(\frac{\sqrt{3}}{2}x\right) = \frac{\sqrt{3}}{4}x^2 \quad \text{for } x \geq 0.$$

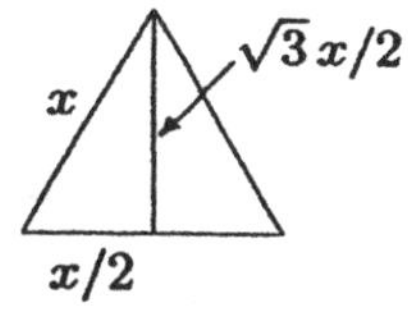

53. a. If the initial height is 832 feet, then by the conversion table the height in meters is approximately $832/3.28 \approx 254$. Similarly, the initial speed of 144 feet per second corresponds to $144/3.28 \approx 43.9$ meters. Therefore the height of the ball in Example 1 is given (approximately) by $h(t) = -4.9t^2 - 43.9t + 254$.

b. Using the result of part (a), we find that $h(4) = -4.9(16) - 43.9(4) + 253.6 = -.4$. Thus the ball hits the ground in approximately 4 seconds (as before).

55. From Exercise 54, $h(t) = -4.9t^2 + v_0t + 30.625$ and $h(2) = 0$. Thus $-4.9(2)^2 + v_0(2) + 30.625 = 0$, so that $v_0 = \frac{1}{2}[(4.9)(4) - 30.625] = -5.5125$. Thus the ball was thrown downward at 5.5125 meters per second.

57. a. By (1), $h(t) = -4.9t^2 - 5t + 30$, so that $h(\frac{1}{2}) = -4.9(\frac{1}{2})^2 - 5(\frac{1}{2}) + 30 = 26.275$ and $h(1) = -4.9(1)^2 - 5(1) + 30 = 20.1$. Thus after $\frac{1}{2}$ second and after 1 second the ball is 26.275 meters and 20.1 meters, respectively, above the ground.

b. We need to find t such that $10 = h(t) = -4.9t^2 - 5t + 30$, or equivalently, $4.9t^2 + 5t - 20 = 0$. By the quadratic formula,

$$t = \frac{-5 \pm \sqrt{25 - 4(4.9)(-20)}}{2(4.9)}.$$

Since $t \geq 0$, we find that $t \approx 1.573528353$. Therefore it takes approximately 1.6 seconds for the ball to reach the window.

59. a. If $T = 2\pi\sqrt{L/g}$, then $T^2 = 4\pi^2L/g$, so that $L = gT^2/4\pi^2$.

b. Letting $g = 9.8$ and $L = 21.8$ in the formula for T, we have

$$T = 2\pi\sqrt{\frac{21.8}{9.8}} \approx 9.371197208.$$

Thus the period is approximately 9.4 seconds.

61. If t represents time in hours starting at noon and D distance in miles, then

$$D(t) = \begin{cases} 400t & \text{for } 0 \leq t < 2 \\ |400t - 800(t-2)| = |1600 - 400t| & \text{for } 2 \leq t \leq 5 \end{cases}$$

63. a. $R(0) = \dfrac{0}{c+0} = 0$ and $R(2) = \dfrac{2}{c+2d}$.

$R(0) = 0$ indicates no response in the absence of the drug.

b. Since $R(x)(c + dx) = x$, or equivalently, $cR(x) + dxR(x) = x$, we have $x - dxR(x) = cR(x)$, so that $x = cR(x)/[1 - dR(x)]$.

1.4 Graphs

1.

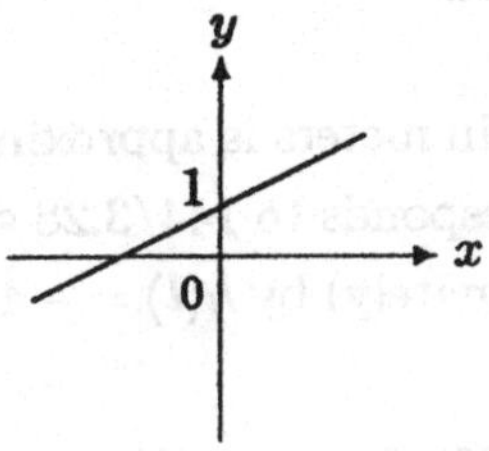

3.

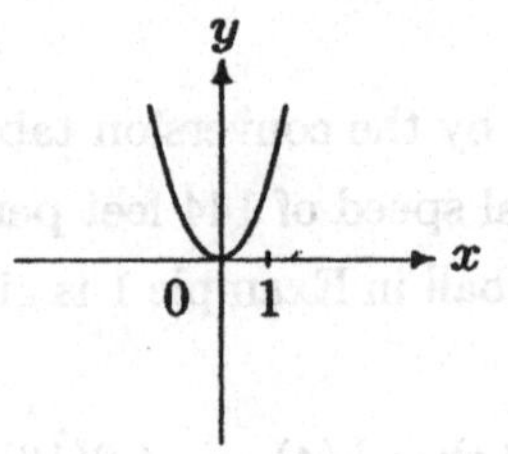

5.

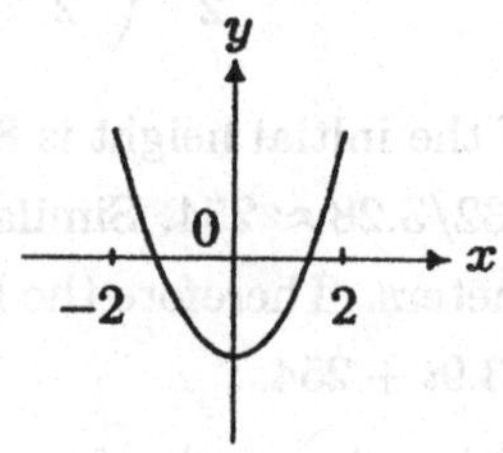

7.

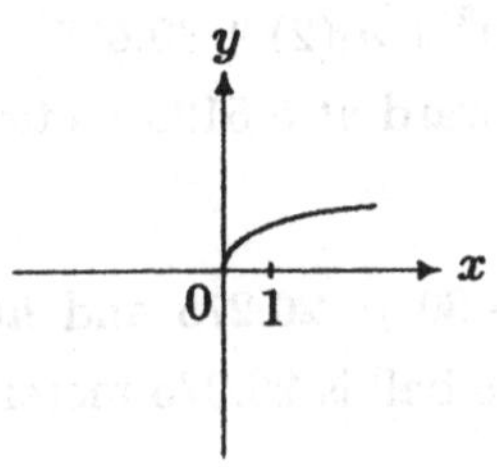

9.

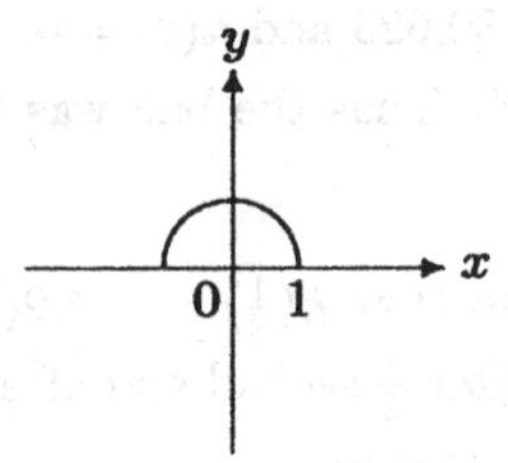

11.

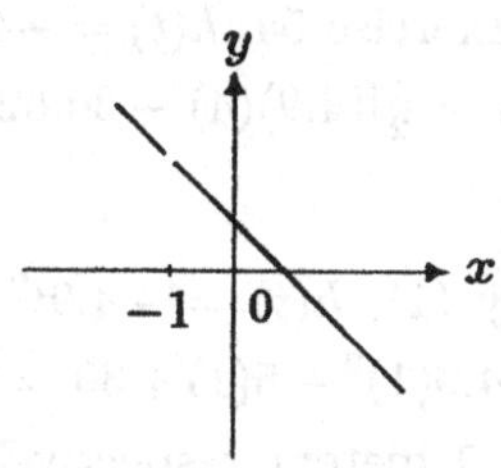

13.

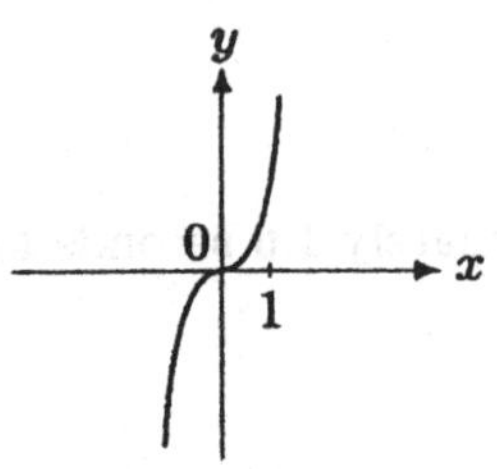

15.

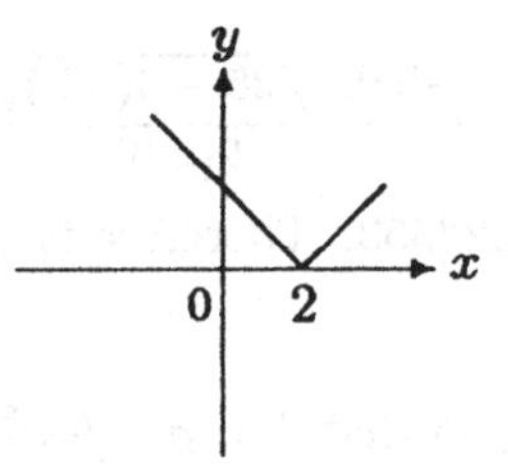

17.

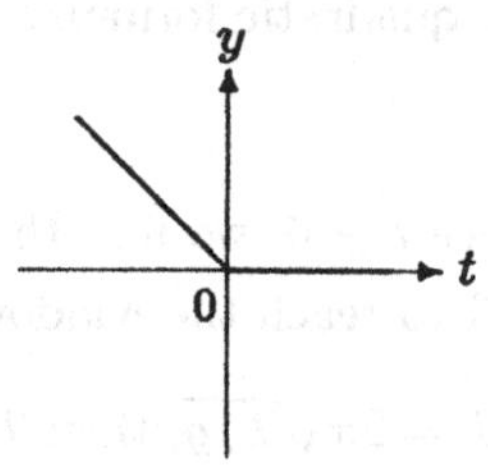

19.

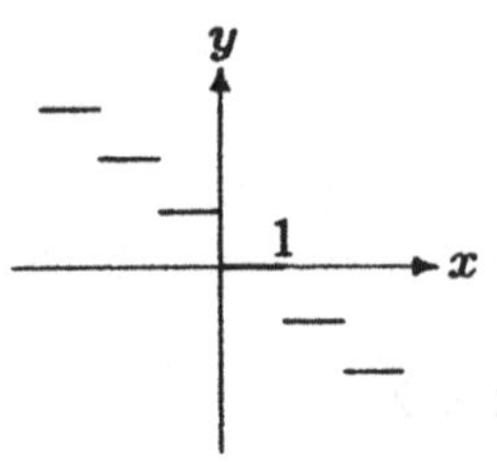

21.

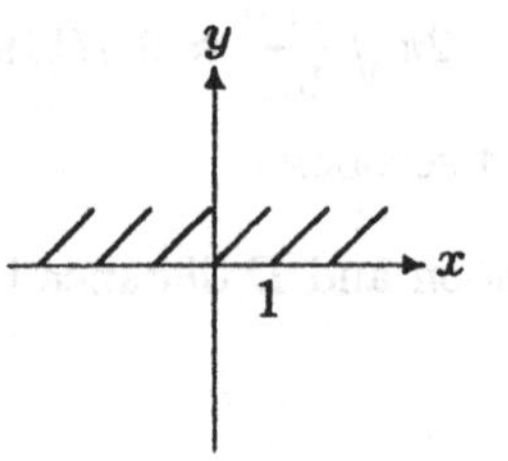

23.

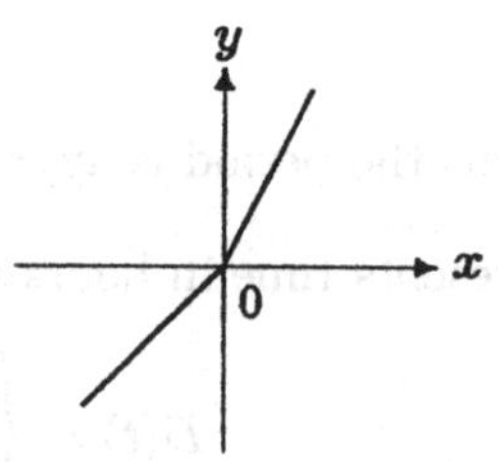

25.

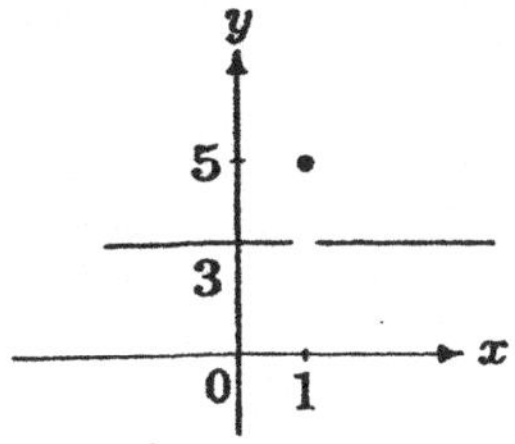

27.

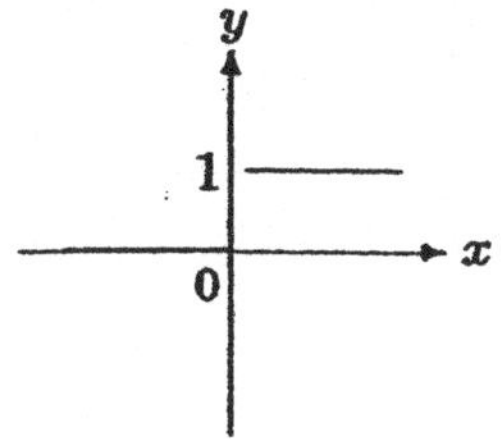

29. By a graphics calculator, the smallest value is approximately -0.1232.

31. By a graphics calculator, the smallest value is approximately -1.82494.

33. By a graphics calculator, the smallest value is approximately -.0175242.

35. not graph of function

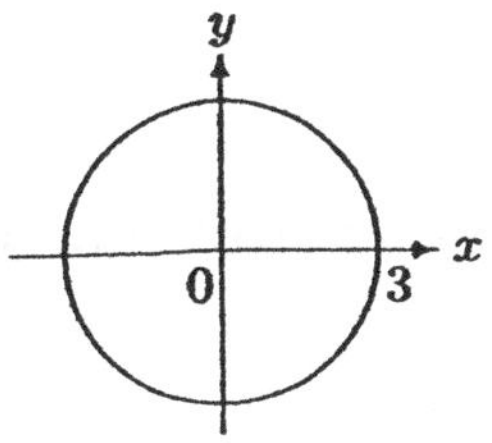

37. graph of function

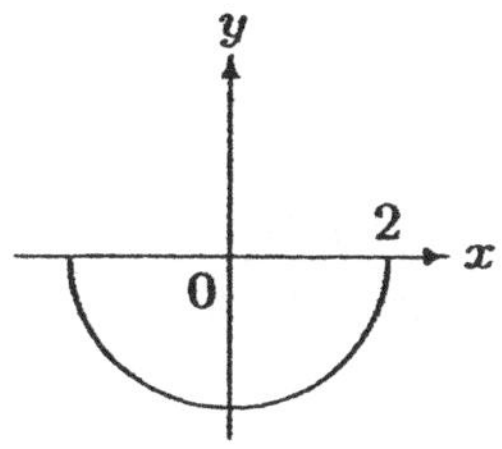

39. not graph of function

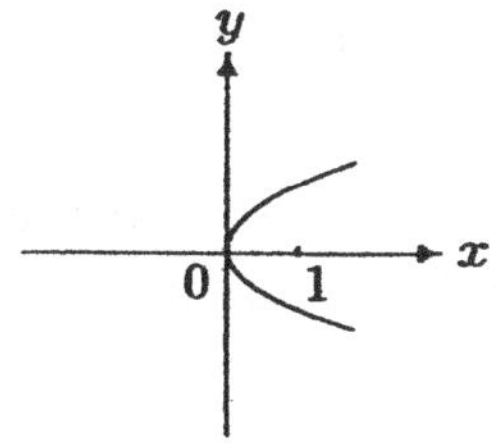

41. not graph of function

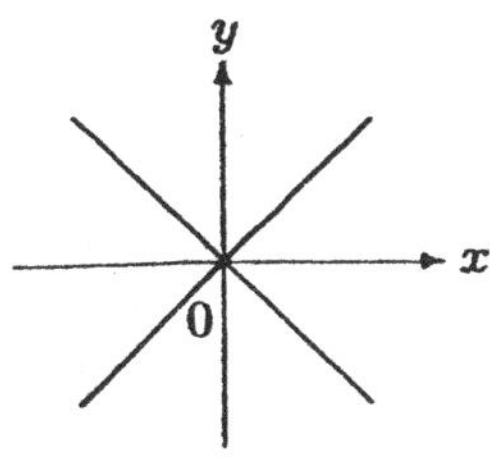

43. graph consists of the two axes; not graph of function

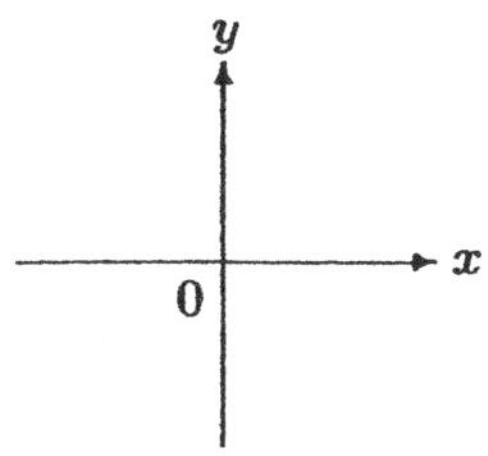

45. not graph of function

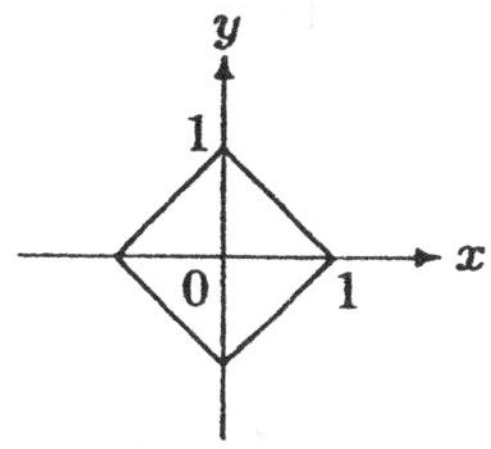

47. c. If $c > 0$, then the graph of $y = f(x + c)$ is obtained from the graph of f by shifting to the left c units. If $c < 0$, then the graph of $y = f(x + c)$ is obtained from the graph of f by shifting to the right $|c|$ units.

49. a. $-2, -1, 1$ and 3

b. The union of $[-4, -2)$, $(-1, 1)$ and $(3, 4]$

c. The union of $(-2, -1)$ and $(1, 3)$

d. None

e. $[-4, 4]$

51. $p = 2$, $q = 1$, $r = -2$

53. f is a possible revenue function, g a cost function, and h a profit function.

55.

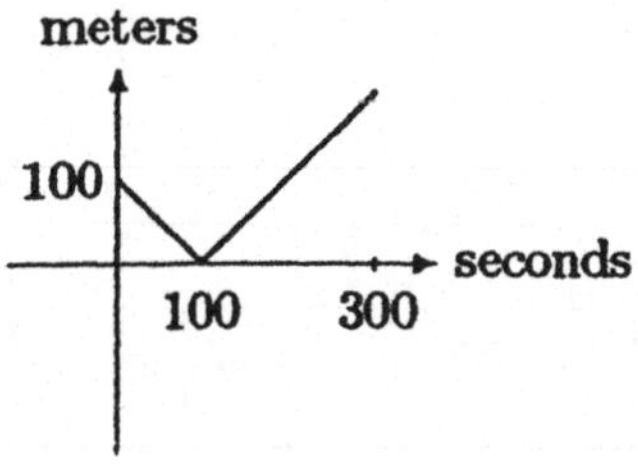

57. a.

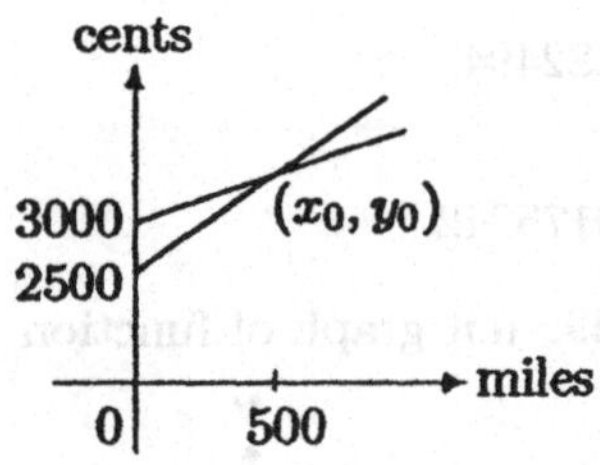

b. If $P = (x_0, y_0)$ is the intersection, then if you expect to travel fewer than x_0 miles, you should rent from Econ Agency; otherwise rent from Budge Agency.

59.

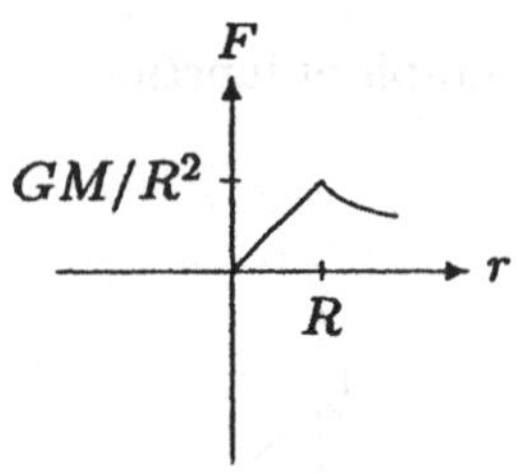

1.5 Aids to Graphing

			Symmetry		
	y intercepts	x intercepts	x axis	y axis	origin
1.	$-\sqrt{\frac{2}{3}}, \sqrt{\frac{2}{3}}$	-2	yes	no	no
3.	none	$-1, 1$	yes	yes	yes
5.	0	0	no	yes	no
7.	none	none	yes	yes	yes
9.	none	$-1, 1$	no	no	yes
11.	0	$[0, 1)$	no	no	no
13.	3	$-3, 3$	no	yes	no
15.	1	1	no	no	no

17. y intercept: 0; x intercept: 0;
symmetry with respect to origin

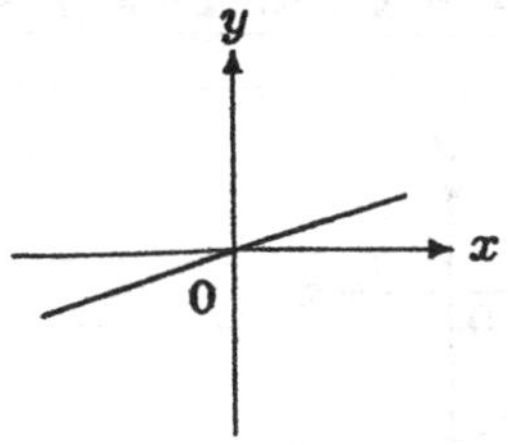

19. y intercept: -3; x intercepts: $-\sqrt{3}$, $\sqrt{3}$;
symmetry with respect to y axis

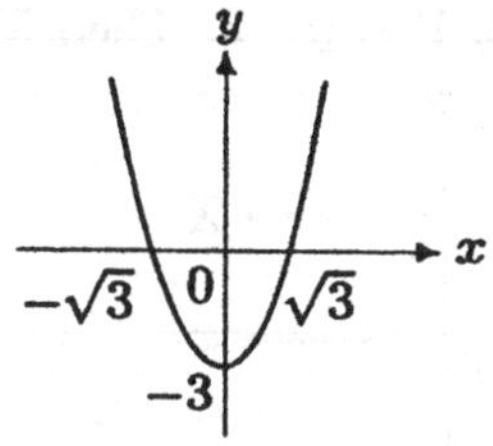

21. y intercepts: -1, 1; symmetry with respect to x axis, y axis, origin

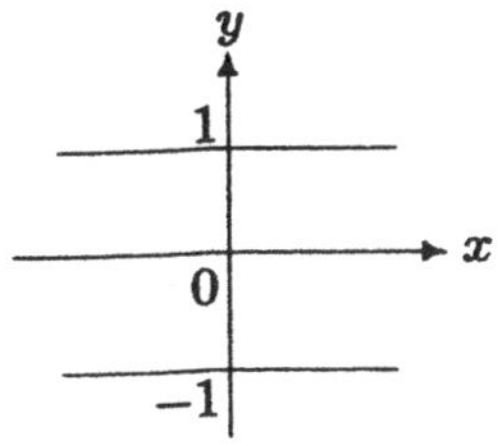

23. y intercepts: -2, 2; x intercept: 2;
symmetry with respect to x axis

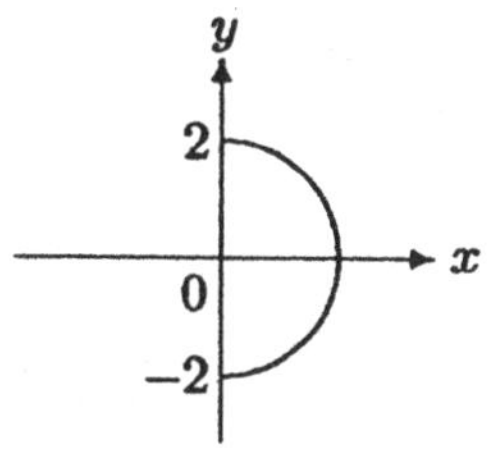

25. y intercept: 0; x intercept: 0;
symmetry with respect to x axis, y axis, origin

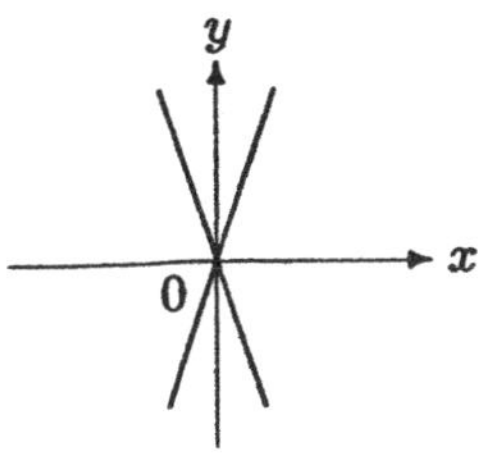

27. $(x-1)^2 + (y-3)^2 = 4$
Let $X = x - 1$, $Y = y - 3$. Then $X^2 + Y^2 = 4$.

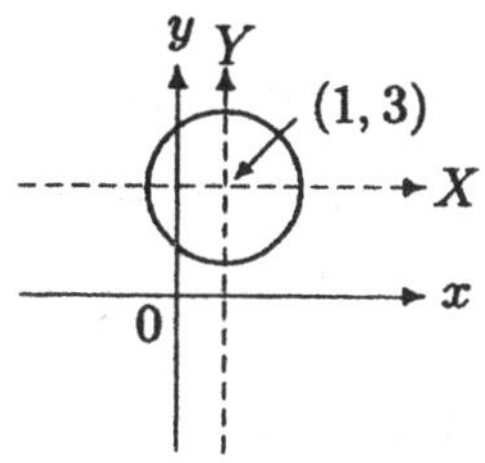

29. $x^2 - 2x + y^2 = 3$, so $(x-1)^2 + y^2 = 4$
Let $X = x - 1$, $Y = y$. Then $X^2 + Y^2 = 4$.

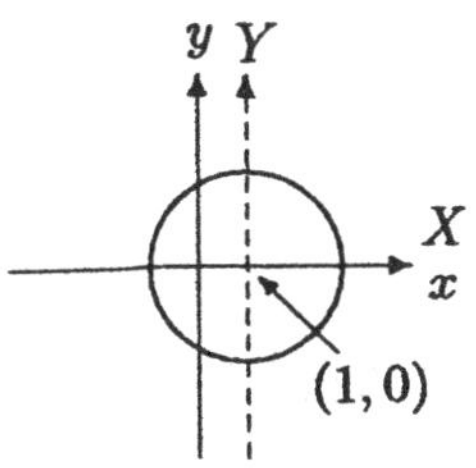

31. $x^2 + y - 3 = 0$, so $y - 3 = -x^2$
Let $X = x$, $Y = y - 3$. Then $Y = -X^2$.

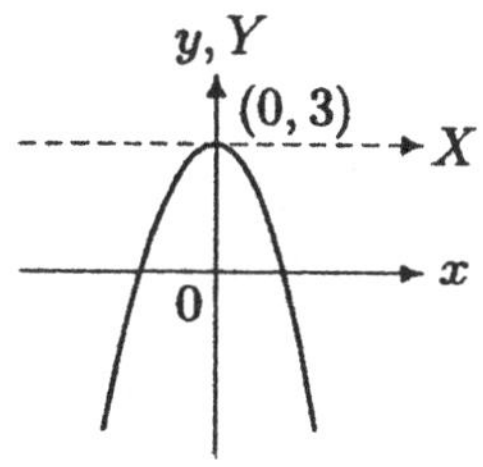

33. $x^2 + y^2 + 4x - 6y + 13 = 0$,
so $(x+2)^2 + (y-3)^2 = 0$
Let $X = x + 2$, $Y = y - 3$. Then $X^2 + Y^2 = 0$.

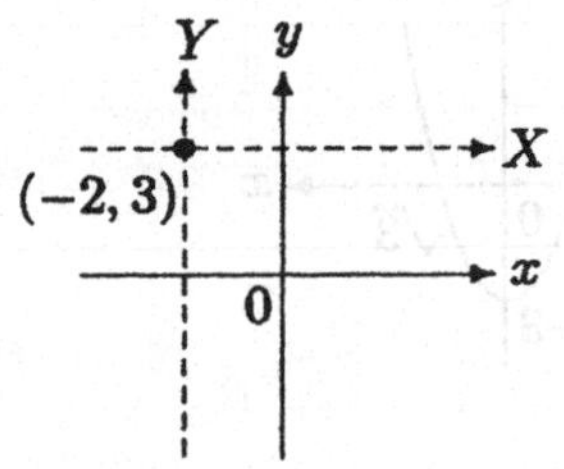

35. $y + 4 = \dfrac{1}{x+2}$
Let $X = x + 2$, $Y = y + 4$. Then $Y = 1/X$.

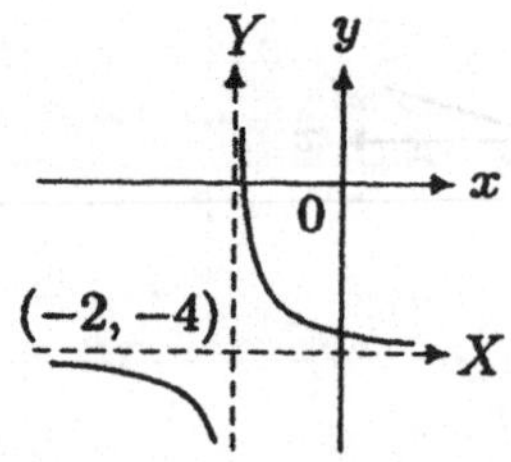

37. $x - 2 = |y - 2|$
Let $X = x - 2$, $Y = y - 2$. Then $X = |Y|$.

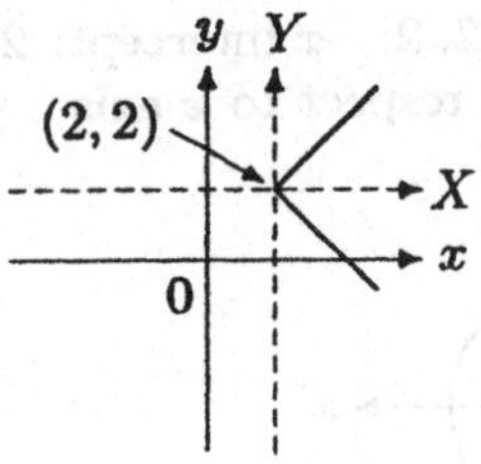

39. a. $f(-x) = -(-x) = x = -f(x)$, so f is odd.

b. $f(-x) = 5(-x)^2 - 3 = 5x^2 - 3 = f(x)$, so f is even.

c. $f(-x) = (-x)^3 + 1 = -x^3 + 1 \neq (x^3 + 1)$ or $-(x^3 + 1)$ if $x \neq 0$, so f is neither even nor odd.

d. $f(-x) = (-x-2)^2 = \big(-(x+2)\big)^2 = (x+2)^2 \neq (x-2)^2$ or $-(x-2)^2$ if $x \neq 0$, so f is neither even nor odd.

e. $f(-x) = \big((-x)^2 + 3\big)^3 = (x^2+3)^3 = f(x)$, so f is even.

f. $f(-x) = -x\big((-x)^2 + 1\big)^2 = -x(x^2+1)^2 = -f(x)$, so f is odd.

g. $\dfrac{-x}{(-x)^2 + 4} = -\dfrac{x}{x^2+4}$, so the function is odd.

h. $|-x| = |x|$, so the function is even.

i. $\dfrac{|-x|}{-x} = \dfrac{|x|}{-x} = -\dfrac{|x|}{x}$, so the function is odd.

41. By Exercise 40 (or (1)), $x = \dfrac{-(-3) \pm \sqrt{(-3)^2 - 4(1)(1)}}{2(1)} = \dfrac{3 \pm \sqrt{5}}{2} = \dfrac{3}{2} + \dfrac{1}{2}\sqrt{5}$ or $\dfrac{3}{2} - \dfrac{1}{2}\sqrt{5}$.

43. Since $b^2 - 4ac = 7^2 - 4(2)(7) = -7$, g has no zero.

45. By (1), $x = \dfrac{-5.1 \pm \sqrt{(5.1)^2 - 4(-4.9)(1.2)}}{2(-4.9)} \approx -0.20$ or 1.24.

47. f has no real zeros if and only if the discriminant $b^2 - 4ac < 0$. But then $4ac > 0$. Thus the discriminant of g is $b^2 - 4a(-c) = b^2 + 4ac > 0$. Therefore g has two real zeros.

49. $(-1, 0)$ and $(1.54, 4.67)$

51. $(-1.21, \infty)$

53. a. The graph of g is 3 units to the left of the graph of f.

$g(x) = f(x+3) = (x+3)^2$ $\qquad$ $g(x) = f(x-2) = |x-2|$

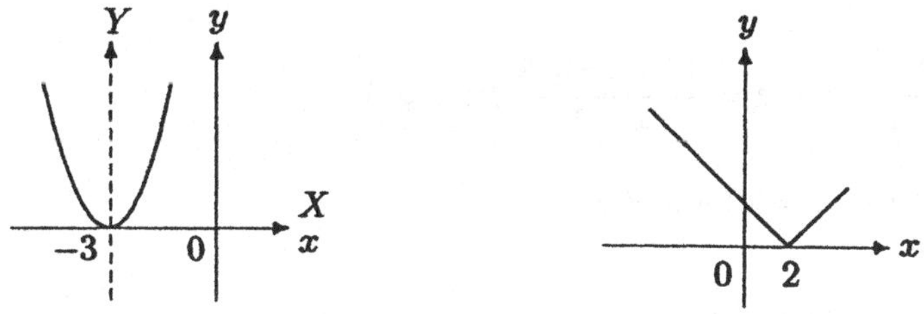

(a) $\qquad$ (b)

b. The graph of g is 2 units to the right of the graph of f.

55. The graph of g is d units above the graph of f if $d \geq 0$, and is $-d$ units below the graph of f if $d < 0$.

57. Let (x, y) be on the graph. Since the graph is symmetric with respect to the x axis, $(x, -y)$ is on the graph. But then symmetry with respect to the y axis implies that $(-x, -y)$ is on the graph. Thus $(-x, -y)$ is on the graph whenever (x, y) is on the graph, so the graph is symmetric with respect to the origin. The converse is not true. For example, the graph of $xy = 1$ is symmetric with respect to the origin but not with respect to either axis.

59. Since $f(c-x) = f(c+x)$, the point $(c-x, y)$ is on the graph of f if and only if the point $(c+x, y)$ is on the graph of f. Thus the graph of f is symmetric with respect to the line $x = c$.

61. The wave is moving to the right.

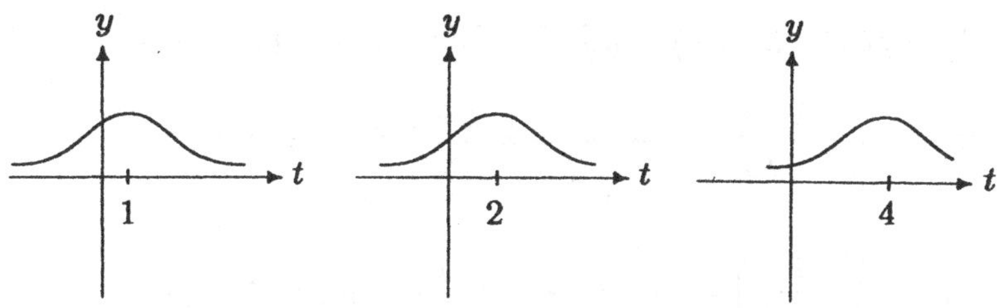

1.6 Combining Functions

1. $(f+g)(-1) = f(-1) + g(-1) = (-3) + (2) = -1$

3. $(fg)(\frac{1}{2}) = f(\frac{1}{2})g(\frac{1}{2}) = (-3)(\frac{11}{4}) = -\frac{33}{4}$

5. $f(g(1)) = f(2) = 6$

7. $\dfrac{f(x) - f(2)}{x-2} = \dfrac{2x^2 + x - 10}{x-2} = \dfrac{(x-2)(2x+5)}{x-2} = 2x + 5$

9. $(f+g)(16) = f(16) + g(16) = \dfrac{15}{257} + 2 = \dfrac{529}{257}$

11. $(fg)(9) = f(9)g(9) = \dfrac{8}{82}\sqrt{3} = \dfrac{4\sqrt{3}}{41}$

13. $f(g(1)) = f(1) = 0$

15. $(f+g)(x) = \dfrac{2}{x-1} + x - 1 = \dfrac{x^2-2x+3}{x-1}$ for $x \neq 1$

$(fg)(x) = \dfrac{2}{x-1}(x-1) = 2$ for $x \neq 1$

$\left(\dfrac{f}{g}\right)(x) = \dfrac{2}{(x-1)^2}$ for $x \neq 1$

17. $(f+g)(t) = t^{3/4} + t^2 + 3$ for $t \geq 0$

$(fg)(t) = t^{3/4}(t^2+3) = t^{11/4} + 3t^{3/4}$ for $t \geq 0$

$\left(\dfrac{f}{g}\right)(t) = \dfrac{t^{3/4}}{t^2+3}$ for $t \geq 0$

19. $(g \circ f)(x) = g(f(x)) = g(1-x) = 2(1-x) + 5 = -2x + 7$ for all x

$(f \circ g)(x) = f(g(x)) = f(2x+5) = 1 - (2x+5) = -2x - 4$ for all x

21. $(g \circ f)(x) = g(f(x)) = g(x^2) = \sqrt{x^2} = |x|$ for all x

$(f \circ g)(x) = f(g(x)) = f(\sqrt{x}) = (\sqrt{x})^2 = x$ for $x \geq 0$

23. $(g \circ f)(x) = g(f(x)) = g(\sqrt{x}) = (\sqrt{x})^2 - 5\sqrt{x} + 6 = x - 5\sqrt{x} + 6$ for $x \geq 0$

$(f \circ g)(x) = f(g(x)) = f(x^2 - 5x + 6) = \sqrt{x^2 - 5x + 6}$ for $x \leq 2$ or $x \geq 3$

25. $(g \circ f)(x) = g(f(x)) = g\left(\dfrac{1}{x-1}\right) = \dfrac{1}{\dfrac{1}{x-1} + 1} = \dfrac{x-1}{x}$ for $x \neq 0, 1$

$(f \circ g)(x) = f(g(x)) = f\left(\dfrac{1}{x+1}\right) = \dfrac{1}{\dfrac{1}{x+1} - 1} = -\dfrac{x+1}{x}$ for $x \neq -1, 0$

27. Let $f(x) = x - 3$, $g(x) = \sqrt{x}$. Then $h(x) = g(x-3) = g(f(x)) = (g \circ f)(x)$.

29. Let $f(x) = 3x^2 - 5\sqrt{x}$, $g(x) = x^{1/3}$. Then $h(x) = g(3x^2 - 5\sqrt{x}) = g(f(x)) = (g \circ f)(x)$.

31. Let $f(x) = x + 3$, $g(x) = 1/(x^2+1)$. Then $h(x) = g(x+3) = g(f(x)) = (g \circ f)(x)$. Alternatively, let $f(x) = (x+3)^2 + 1$ and $g(x) = 1/x$.

33. Let $f(x) = \sqrt{x} - 1$, $g(x) = \sqrt{x}$. Then $h(x) = g(\sqrt{x} - 1) = g(f(x)) = (g \circ f)(x)$. Alternatively, let $f(x) = \sqrt{x}$ and $g(x) = \sqrt{x-1}$.

35. $g(x) = -|x-2|$

37. If $0 \le x+3 \le 4$, then $-3 \le x \le 1$; the domain of g is $[-3,1]$.

39. f is red; g is green; $f+g$ is blue; and fg is black.

41. a. $h = f$ if $c = -1$

    b. Since $2 = h(1) = g(1+c)$, and $g(.4) \approx 2$, c must satisfy the equation $1 + c \approx .4$, so $c \approx -.6$.

43. The graph of h is so short because $\sqrt{x+1}$ is defined for $x \ge -1$, and $\sqrt{2-x}$ is defined for $x \le 2$. Thus the domain of h is $[-1,2]$.

45. All functions

47. Assume that $f(x) \ge 0$ for all x in the domain of f. Then let $g(x) = (f(x))^2 - 1$ for all such x. It follows that $(f(x))^2 = 1 + g(x)$ and thus $f(x) = \sqrt{1+g(x)}$. Thus all functions f such that $f(x) \ge 0$ for all x in either domain of f allow such a g.

49. Since f is even and g is odd, $-x$ is in the domain of f and g and hence fg if x is. For such x, $(fg)(-x) = f(-x)g(-x) = f(x)(-g(x)) = -f(x)g(x) = -(fg)(x)$. Thus fg is odd.

51. $f(f(x)) = f(a-x) = a-(a-x) = x$

53. $g(x) = f(x+p) - f(x) = [a(x+p)+b] - (ax+b) = ap$

55. a. Since $g(-x) = \frac{1}{2}[f(-x)+f(x)] = g(x)$, g is an even function.

    b. Since $h(-x) = \frac{1}{2}[f(-x)-f(x)] = -\frac{1}{2}[f(x)-f(-x)] = -h(x)$, h is an odd function.

    c. $g(x) + h(x) = \frac{1}{2}[f(x)+f(-x)] + \frac{1}{2}[f(x)-f(-x)] = f(x)$

57. The nth iterate is close to 0.

59. For large values of n, the nth iterate oscillates between numbers close to 0.80 and numbers close to 0.51.

61. For large values of n, the nth iterate oscillates between numbers close to .16, .50, and .96.

63. $P(x) = R(x) - C(x) = 5x^2 - \frac{1}{10}x^4 - (4x^2 - 24x + 38) = -\frac{1}{10}x^4 + x^2 + 24x - 38$. Thus $P(1) = -\frac{1}{10} + 1 + 24 - 38 = -13.1$; $P(2) = -\frac{16}{10} + 4 + 48 - 38 = 12.4$. Since $P(1) = -13.1$, there is a loss when $x = 1$. Since $P(2) = 12.4$, there is a profit when $x = 2$.

65. a. $V(r(s)) = \frac{4}{3}\pi\left(\frac{1}{2}\sqrt{\frac{s}{\pi}}\right)^3 = \frac{1}{6\sqrt{\pi}}s^{3/2}$ for $s \ge 0$

    b. $V(r(6)) = \frac{1}{6}\cdot 6\sqrt{\frac{6}{\pi}} = \sqrt{\frac{6}{\pi}}$

1.7 Trigonometric Functions

1. a. $210° = \left(\frac{\pi}{180} \cdot 210\right)$ radians $= \frac{7\pi}{6}$ radians

   b. $-405° = -\left(\frac{\pi}{180} \cdot 405\right)$ radians $= -\frac{9\pi}{4}$ radians

   c. $1° = \left(\frac{\pi}{180} \cdot 1\right)$ radian $= \frac{\pi}{180}$ radian

3. a. $\sin \frac{11\pi}{6} = \sin\left(-\frac{\pi}{6} + 2\pi\right) = \sin\left(-\frac{\pi}{6}\right) = -\sin\frac{\pi}{6} = -\frac{1}{2}$

   b. $\sin\left(-\frac{2\pi}{3}\right) = -\sin\frac{2\pi}{3} = -\frac{\sqrt{3}}{2}$

   c. $\cos\frac{5\pi}{4} = \cos\left(\pi + \frac{\pi}{4}\right) = -\cos\frac{\pi}{4} = -\frac{\sqrt{2}}{2}$

   d. $\cos\left(-\frac{7\pi}{6}\right) = \cos\frac{7\pi}{6} = \cos\left(\pi + \frac{\pi}{6}\right) = -\cos\frac{\pi}{6} = -\frac{\sqrt{3}}{2}$

   e. $\tan\frac{4\pi}{3} = \frac{\sin(4\pi/3)}{\cos(4\pi/3)} = \frac{\sin[\pi + (\pi/3)]}{\cos[\pi + (\pi/3)]} = \frac{-\sin(\pi/3)}{-\cos(\pi/3)} = \frac{-\sqrt{3}/2}{-1/2} = \sqrt{3}$

   f. $\tan\left(-\frac{\pi}{4}\right) = \frac{\sin(-\pi/4)}{\cos(-\pi/4)} = \frac{-\sin(\pi/4)}{\cos(\pi/4)} = \frac{-\sqrt{2}/2}{\sqrt{2}/2} = -1$

   g. $\cot\frac{\pi}{6} = \frac{\cos(\pi/6)}{\sin(\pi/6)} = \frac{\sqrt{3}/2}{1/2} = \sqrt{3}$

   h. $\cot\left(-\frac{17\pi}{3}\right) = \frac{\cos[-(17\pi)/3]}{\sin[-(17\pi)/3]} = \frac{\cos[-6\pi + (\pi/3)]}{\sin[-6\pi + (\pi/3)]} = \frac{\cos(\pi/3)}{\sin(\pi/3)} = \frac{1/2}{\sqrt{3}/2} = \frac{1}{\sqrt{3}} = \frac{\sqrt{3}}{3}$

   i. $\sec 3\pi = \frac{1}{\cos 3\pi} = \frac{1}{\cos(2\pi + \pi)} = \frac{1}{\cos\pi} = \frac{1}{-1} = -1$

   j. $\sec\left(-\frac{\pi}{3}\right) = \frac{1}{\cos(-\pi/3)} = \frac{1}{\cos(\pi/3)} = \frac{1}{1/2} = 2$

   k. $\csc\frac{\pi}{2} = \frac{1}{\sin(\pi/2)} = \frac{1}{1} = 1$

   l. $\csc[-(5\pi)/3] = \frac{1}{\sin[-(5\pi)/3]} = \frac{1}{\sin[-2\pi + (\pi/3)]} = \frac{1}{\sin(\pi/3)} = \frac{1}{\sqrt{3}/2} = \frac{2}{\sqrt{3}} = \frac{2\sqrt{3}}{3}$

5. $\tan x = \frac{\sin x}{\cos x} = -\frac{4}{3}$; $\cot x = \frac{1}{\tan x} = -\frac{3}{4}$; $\sec x = \frac{1}{\cos x} = -\frac{5}{3}$; $\csc x = \frac{1}{\sin x} = \frac{5}{4}$

7. $7\pi/6$, $11\pi/6$

9. $\sin x = \sin 2x = 2\sin x \cos x$, so $\sin x\,(2\cos x - 1) = 0$. But $\sin x = 0$ for $x = 0, \pi$; $2\cos x - 1 = 0$ if $\cos x = \frac{1}{2}$, which happens for $x = \pi/3, 5\pi/3$. Solutions: $0, \pi/3, \pi, 5\pi/3$.

11. The union of $[0, 7\pi/6)$ and $(11\pi/6, 2\pi)$

13. The union of $[\pi/4, \pi/2)$ and $[5\pi/4, 3\pi/2)$

15. The union of $(0, \pi/4]$, $(\pi/2, 3\pi/4]$, $(\pi, 5\pi/4]$, and $(3\pi/2, 7\pi/4]$

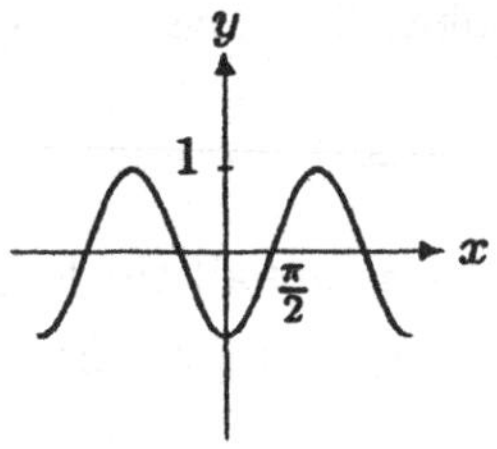

Exercise 17

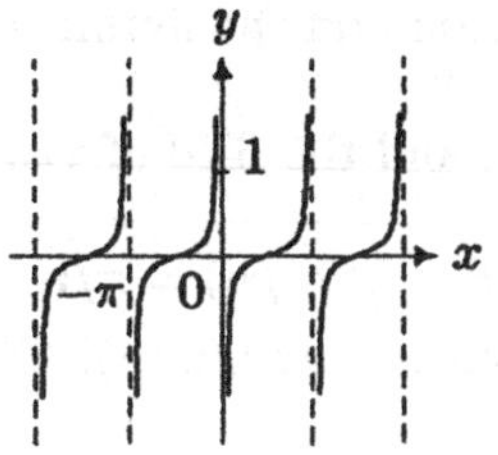

Exercise 19

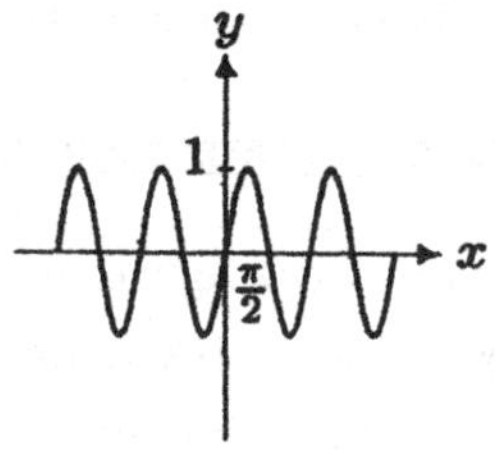

Exercise 23

17. $\cos(\pi - x) = \cos\pi\,\cos x + \sin\pi\,\sin x = -\cos x$; y intercept: -1; x intercepts: $\pi/2 + n\pi$ for any integer n; symmetric with respect to the y axis; an even function.

19. y intercepts: none; x intercepts: $\pi/2 + n\pi$ for any integer n; symmetric with respect to origin; an odd function. (Let $X = x + \pi/2$, $Y = y$; equation becomes $Y = \tan X$.)

21. y intercept: 1; x intercepts: none; symmetric with respect to y axis; an even function $\sec(2\pi - x) = \sec(-x) = \sec x$; see Figure 1.70(c).

23. y intercept: 0; x intercepts: $n\pi/2$ for any integer n; symmetric about origin; an odd function.

25. $\sin\dfrac{7\pi}{12} = \sin\left(\dfrac{\pi}{3} + \dfrac{\pi}{4}\right) = \sin\dfrac{\pi}{3}\cos\dfrac{\pi}{4} + \cos\dfrac{\pi}{3}\sin\dfrac{\pi}{4} = \dfrac{\sqrt{3}}{2}\cdot\dfrac{\sqrt{2}}{2} + \dfrac{1}{2}\cdot\dfrac{\sqrt{2}}{2} = \dfrac{\sqrt{2}}{4}(\sqrt{3}+1)$

27. $2\sin^2 x + \sin x - 1 = 0$ if and only if $(2\sin x - 1)(\sin x + 1) = 0$, so the solution consists of all x for which $\sin x = \frac{1}{2}$ or $\sin x = -1$. Solution: $\pi/6 + 2n\pi$, $5\pi/6 + 2n\pi$, $3\pi/2 + 2n\pi$ for any integer n.

29. If $\cos x \neq 0$, then

$$\frac{\sin^2 x + \cos^2 x}{\cos^2 x} = \frac{1}{\cos^2 x} \quad \text{or} \quad \left(\frac{\sin x}{\cos x}\right)^2 + 1 = \left(\frac{1}{\cos x}\right)^2.$$

Thus $1 + \tan^2 x = \sec^2 x$ whenever $\tan x$ and $\sec x$ are defined.

31. a. By (10), $\sin(\pi - x) = \sin\pi\,\cos x - \cos\pi\,\sin x = \sin x$.

b. By (10), $\sin[(3\pi/2) - x] = \sin(3\pi/2)\cos x - \cos(3\pi/2)\sin x = -\cos x$.

c. By (11), $\cos(\pi - x) = \cos\pi\,\cos x + \sin\pi\,\sin x = -\cos x$.

d. By (11), $\cos[(3\pi/2) - x] = \cos(3\pi/2)\cos x + \sin(3\pi/2)\sin x = -\sin x$.

33. $m_1 = 4$, $m_2 = \dfrac{-2}{3}$; $\tan\theta = \dfrac{m_2 - m_1}{1 + m_1 m_2} = \dfrac{(-2/3) - 4}{1 + (-2/3)(4)} = \dfrac{14}{5}$.

35. a. π b. $2\pi/3$ c. π d. π

37. Since $\beta = \dfrac{\pi}{2} - \alpha$, we have

$$\cos(\alpha - \beta) = \cos\left[\alpha - \left(\frac{\pi}{2} - \alpha\right)\right] = \cos\left(2\alpha - \frac{\pi}{2}\right) = \cos\left(\frac{\pi}{2} - 2\alpha\right) = \sin 2\alpha.$$

39. In each case, the iterates approach a number approximately 0.739, so we conjecture that the same is true for any real number a.

41. Let d be the distance between the beacon and the illuminated point. Then $\sec\theta = d/2$, so $d = 2\sec\theta$.

43. $\pi/4$; the angle grows larger and larger, and the field of vision eventually becomes blocked.

45. a. Let θ be as in the figure, so that $\theta = \frac{1}{2}(2\pi/n) = \pi/n$. Next let x be half the length of one side of the polygon, as in the figure. Then $x = r\sin(\pi/n)$. Thus the perimeter of the n-sided polygon is given by $p_n(r) = 2nr\sin(\pi/n)$.

b. From part (a), $2x = 2r\sin(\pi/n)$ or $r = 2x/[2\sin(\pi/n)]$. For the Pentagon, $2x = 921$ and $n = 5$, so $r = 921/[2\sin(\pi/5)] \approx 783.4$ (feet).

47. b. Since $g < 0$ and since $\cos\theta$ is the largest for $\theta = 0$ and smallest for $\theta = \pi$, it follows that T is greatest when $\cos\theta$ is smallest, that is, for $\theta = \pi$. Analogously, T is smallest when $\cos\theta$ is largest, that is, for $\theta = 0$.

49. a. Since $\frac{2\pi}{5}t = 2\pi$ for $t = 5$, the period of R and hence of the cycle is 5 seconds.

b. By part (a), there are 12 cycles in a minute.

c.

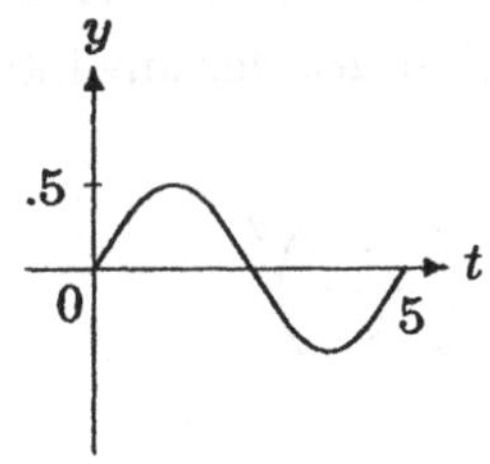

d. Positive values of R correspond to a flow into the lungs; negative values of R correspond to a flow out of the lungs.

e. When $t = 3$, $R = 0.5\sin\frac{6\pi}{5} \approx -0.29$.

1.8 Exponential and Logarithmic Functions

1. By (8), $\ln e^3 = 3$.

3. By (8), $e^{\ln 3x} = 3x$.

5. By (8), $\ln(e^{\ln e}) = \ln e = 1$.

7. By (7), $7^{\log_7 2x} = 2x$.

9. By (6), $\log_9 3 = \log_9 9^{1/2} = \frac{1}{2}$.

11. By (6) and the Law of Logarithms (*iv*), $\log_{\frac{1}{4}} 2^x = x\log_{\frac{1}{4}} 2 = x\log_{\frac{1}{4}}(\frac{1}{4})^{-1/2} = x(-\frac{1}{2}) = -\frac{1}{2}x$.

13. $e^{-1.24} \approx 0.2893842179$

15. $2^{7/2} \approx 11.31370850$

17. By (11), $\log_3 5 = \dfrac{\ln 5}{\ln 3} \approx 1.464973521$.

19. By (11), $\log_\pi e = \dfrac{\ln e}{\ln \pi} = \dfrac{1}{\ln \pi} \approx 0.8735685268$.

21. $(x\ln x)^2 = x^2(\ln x)^2 \neq x^2\ln(x^2)$ because $(\ln x)^2 \neq \ln(x^2)$.

23. $f(-x) = \dfrac{e^{-x}}{e^{-2x}+1} = \dfrac{e^{-x}}{e^{-2x}+1}\dfrac{e^{2x}}{e^{2x}} = \dfrac{e^x}{1+e^{2x}} = f(x)$, so that f is an even function.

25. $f(x) = e^{2+x}$

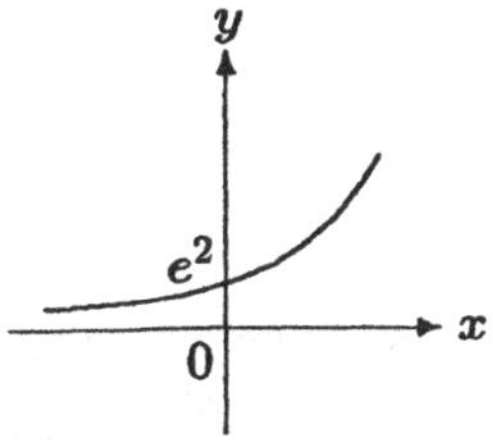

27. $f(x) = \ln(x+1)$

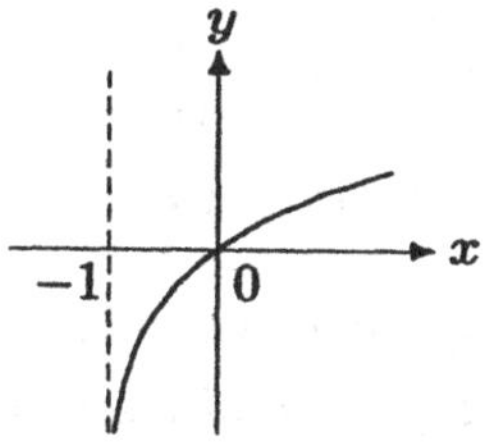

29. The graphs intersect at (x, y) if and only if $e^x = e^{1-x}$, so by (3), $x = 1 - x$ and thus $x = \frac{1}{2}$. Notice that $f(\frac{1}{2}) = e^{1/2} = g(\frac{1}{2})$. Thus the graphs intersect at $(\frac{1}{2}, e^{1/2})$.

31. The graphs intersect at (x, y) if and only if $e^{3x} = 3e^x$, so by (8) and the Law of Logarithms (*i*), $3x = \ln(e^{3x}) = \ln(3e^x) = \ln 3 + \ln e^x = \ln 3 + x$. Thus $x = \frac{1}{2}\ln 3$. Notice that $f(\frac{1}{2}\ln 3) = e^{3(\ln 3)/2} = (e^{\ln 3})^{3/2} = 3^{3/2} = g(\frac{1}{2}\ln 3)$. Thus the graphs intersect at $(\frac{1}{2}\ln 3, 3^{3/2})$.

33. The graphs intersect at (x, y) if and only if $\log_3 x = \log_2 x$, so by (11),

$$\frac{\ln x}{\ln 3} = \frac{\ln x}{\ln 2}.$$

Since $\ln 3 \neq \ln 2$, this can occur only for $x = 1$. Notice that $f(1) = \log_3 1 = 0 = \log_2 1 = g(1)$. Thus the graphs intersect at $(1, 0)$.

35. By the Law of Logarithms (*i*), $\ln x + \ln(3x-1) = 0$ if $\ln[x(3x-1)] = 0$, which by (5) is equivalent to $x(3x-1) = 1$, or $3x^2 - x - 1 = 0$. By the quadratic formula,

$$x = \frac{-(-1) \pm \sqrt{(-1)^2 - 4(3)(-1)}}{2(3)} = \frac{1 \pm \sqrt{13}}{6}.$$

Since x must be positive, $x = (1 + \sqrt{13})/6$.

37. Since
$$x+\sqrt{x^2-1}=(x+\sqrt{x^2-1})\frac{x-\sqrt{x^2-1}}{x-\sqrt{x^2-1}}=\frac{x^2-(x^2-1)}{x-\sqrt{x^2-1}}=\frac{1}{x-\sqrt{x^2-1}}$$
it follows by the Law of Logarithms (*ii*) that
$$\ln(x+\sqrt{x^2-1})=\ln\frac{1}{x-\sqrt{x^2-1}}=-\ln(x-\sqrt{x^2-1}).$$

39. By two applications of the Law of Exponents (*i*), $a^{b+c+d}=a^{b+(c+d)}=a^ba^{c+d}=a^ba^ca^d$.

41. By the Law of Logarithms,
$$f(x)=\ln(4x)-\ln x^3+\ln x^2=\ln\frac{(4x)x^2}{x^3}=\ln 4.$$
This is the reason that the graph of f is a horizontal line.

43. a. $f(x+1)-f(x)=[a(x+1)+b]-(ax+b)=ax+a-ax=a$

b. $\dfrac{g(x+1)}{g(x)}=\dfrac{ba^{x+1}}{ba^x}=\dfrac{ba^xa}{ba^x}=a$

45. Approximately 0.2591711018.

47. Let x be the amplitude of the earthquake's largest wave 100 kilometers from the epicenter, and a the corresponding amplitude of a zero-level earthquake. By (13), $\log(x/a)=2$, so $x/a=10^2=100$. But $a=0.001$. Therefore $x=100a=0.1$ (millimeters).

49. Let x_1 be the maximal amplitude of an earthquake of amplitude 8.5, and x_2 the maximal amplitude of an earthquake of amplitude 8.4. By (13), $8.5=\log(x_1/a)$ and $8.4=\log(x_2/a)$. Thus
$$\log\frac{x_2}{x_1}=\log\frac{x_2/a}{x_1/a}=\log\frac{x_2}{a}-\log\frac{x_1}{a}=8.4-8.5=-0.1.$$
Therefore $x_2/x_1=10^{-0.1}\approx 0.794328$.

51. Since 90 kilotons releases $90(10^{20})$ ergs, let $E=90(10^{20})$. Then
$$11.4+1.5M=\log E=\log[90(10^{20})]=\log 90+20.$$
Therefore $M=\frac{1}{1.5}(\log 90+8.6)\approx 7.036$. Consequently the magnitude would be approximately 7.

53. a. $L(10^{-12})=10\log\dfrac{10^{-12}}{10^{-16}}=10\log 10^4=10(4)=40$; thus the threshold is at 40 decibels (approximately).

b. $L(10^{-11})=10\log\dfrac{10^{-11}}{10^{-16}}=10\log 10^5=10(5)=50$; thus leaves rustle at 50 decibels (approximately).

c. $L(10^{-2})=10\log\dfrac{10^{-2}}{10^{-16}}=10\log 10^{14}=10(14)=140$; thus a power mower has 140 decibels (approximately).

d. $L(10) = 10 \log \dfrac{10^1}{10^{-16}} = 10 \log 10^{17} = 10(17) = 170$; thus a jackhammer has 170 decibels (approximately).

55. If $L(y) = L(x) + 100$, then $10 \log(y/I_0) = 10 \log(x/I_0) + 100$, so $\log(x/I_0) - \log(y/I_0) = -10$. Thus

$$\log \frac{x}{y} = \log \frac{x/I_0}{y/I_0} = \log \frac{x}{I_0} - \log \frac{y}{I_0} = -10$$

so that $x/y = 10^{-10}$. Therefore the ratio of the intensities is 10^{-10}.

57. a. $p(0) \approx 29.92$ (inches of mercury)

b. $p(5) \approx (29.92)e^{(-0.2)(5)} \approx 11.00695288$ (inches of mercury)

c. $p(10) \approx (29.92)e^{(-0.2)(10)} \approx 4.049231674$ (inches of mercury)

59. Let I_1 denote the intensity of the X-ray beam with wavelength 5×10^{-11}, and I_2 that of the X-ray beam with wavelength 10^{-10}. Then

$$I_{1,\,\text{exit}} = I_{1,\,\text{entry}}\, e^{-\sigma_1(.002)} \quad \text{and} \quad I_{2,\,\text{exit}} = I_{2,\,\text{entry}}\, e^{-\sigma_2(.002)}.$$

Since $I_{1,\,\text{entry}} = I_{2,\,\text{entry}}$ by hypothesis,

$$\frac{I_{1,\,\text{exit}}}{I_{2,\,\text{exit}}} = \frac{e^{-\sigma_1(.002)}}{e^{-\sigma_2(.002)}} = \frac{e^{-(5.4\times10^2)(.002)}}{e^{-(4.1\times10^3)(.002)}} = \frac{e^{-1.08}}{e^{-8.2}} \approx 1236.450433.$$

Thus the X-ray beam with wavelength 5×10^{-11} exits with approximately 1200 times the intensity of the other X-ray beam.

Chapter 1 Review

1. From the diagram we see that the solution is the union of $(-\infty, -\frac{3}{2})$ and $(4, \infty)$.

$2x+3$	− − 0 + + + + + + + +
$(x-4)^3$	− − − − − − − − − 0 + +
$\dfrac{2x+3}{(x-4)^3}$	+ + 0 − − − − − + +

3. The given inequality is equivalent to

$$\frac{2}{3-x} - 4 \geq 0, \quad \text{or} \quad \frac{4(x-\frac{5}{2})}{3-x} \geq 0.$$

From the diagram we see that the solution is $[\frac{5}{2}, 3)$.

$x - \frac{5}{2}$	$- \; - \; 0 \; + \; + \; + \; + \; + \; + \; + \; +$
$3 - x$	$+ \; + \; + \; + \; + \; + \; + \; + \; 0 \; - \; -$
$\dfrac{4(x - \frac{5}{2})}{3 - x}$	$- \; - \; 0 \; + \; + \; + \; + \; + \quad - \; -$

5. If $|4 - 6x| < \frac{1}{2}$, then $-\frac{1}{2} < 4 - 6x < \frac{1}{2}$, or $-\frac{9}{2} < -6x < -\frac{7}{2}$, or $\frac{7}{12} < x < \frac{3}{4}$. The solution is $(\frac{7}{12}, \frac{3}{4})$.

7. a. For $a \geq 0$ we have $|a| = a$, so $(|a| + a)/2 = (a + a)/2 = a$.

   b. For $a < 0$ we have $|a| = -a$, so $(|a| + a)/2 = (-a + a)/2 = 0$.

9. The slope is $\frac{1-(-3)}{0-1} = -4$, so a point-slope equation is $y - 1 = -4(x - 0)$.

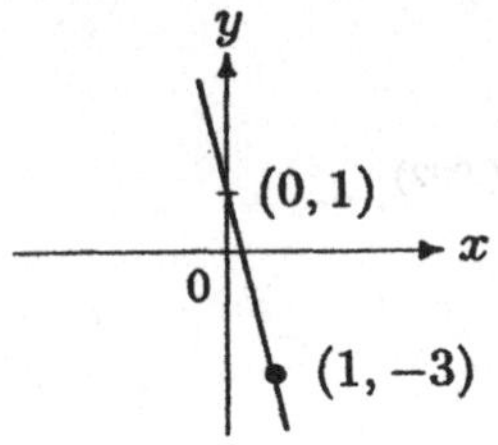

11. A slope-intercept equation is $y = -\frac{1}{3}x + 6$.

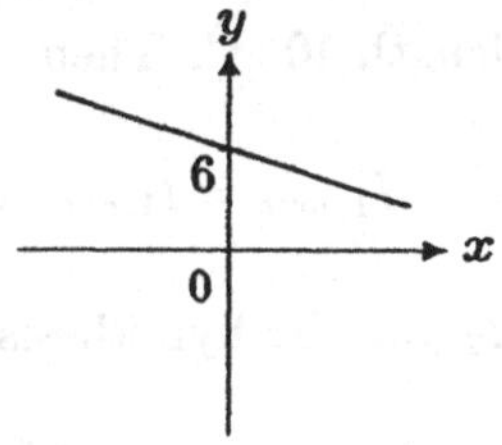

13. $m_1 = 2$; $m_2 = -\frac{1}{2}$. Since $m_1 m_2 = -1$, the lines are perpendicular.

15. The slope of l is $-\frac{1}{2}$.

   a. The desired line has slope $-\frac{1}{2}$ also, so an equation is $y - (-2) = -\frac{1}{2}(x - 1)$, or $y = -\frac{1}{2}x - \frac{3}{2}$.

   b. The desired line has slope 2, so an equation is $y - (-2) = 2(x - 1)$, or $y = 2x - 4$.

17. The first and fourth lines are parallel, with slope 1; the second and third lines are parallel, with slope -1. Thus the lines form a rectangle. The points of intersection are $(1, 1)$, $(5, 5)$, $(-3, 5)$, and $(1, 9)$. The lengths of the sides of the figure are

$$\sqrt{(5-1)^2 + (5-1)^2} = \sqrt{(5-1)^2 + (5-9)^2}$$
$$= \sqrt{(1-(-3))^2 + (9-5)^2} = \sqrt{(-3-1)^2 + (5-1)^2} = 4\sqrt{2}.$$

Since the sides all have the same length, the rectangle is a square.

19. Since $x(x^2 + 4x + 3) = x(x + 3)(x + 1) = 0$ only if $x = -3, -1$ or 0, the domain is all numbers except $-3, -1$ and 0.

21. The domain consists of all x such that $e^x/(e^x - 1) > 0$. Since $e^x > 0$ for all x, this means that $e^x - 1 > 0$, or $e^x > 1$, or $x > 0$. Thus the domain is $(0, \infty)$.

23.

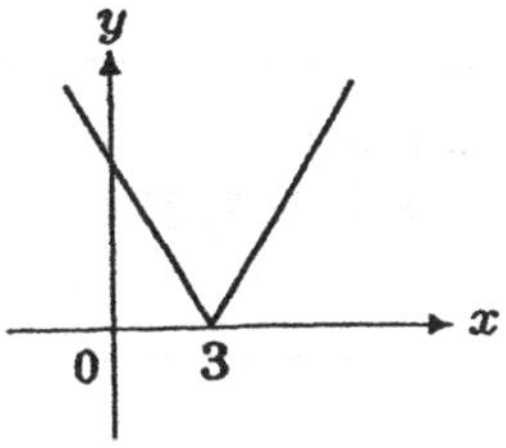

25.

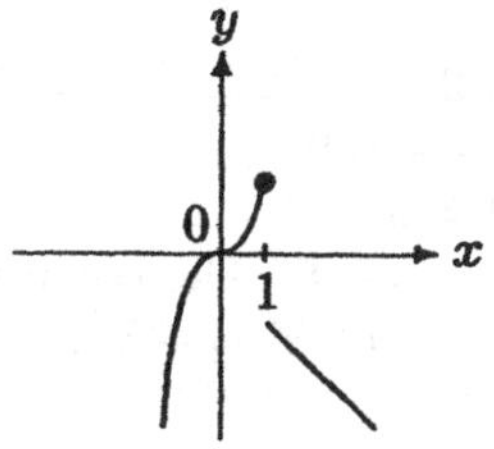

27. y intercept: 1; x intercept: 1;
not symmetric with respect to either axis or origin

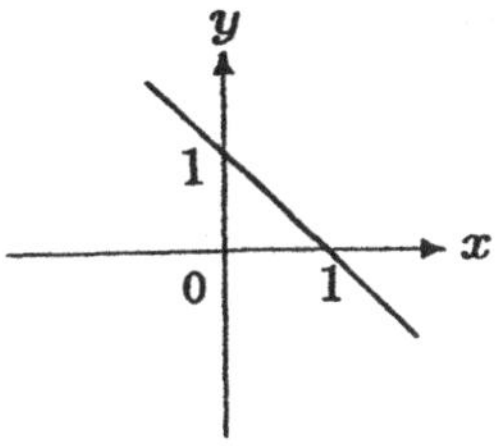

29. y intercepts: $-\sqrt{3}$, $\sqrt{3}$; x intercepts: $-\sqrt{3}$, $\sqrt{3}$;
symmetric with respect to the x axis, y axis and origin

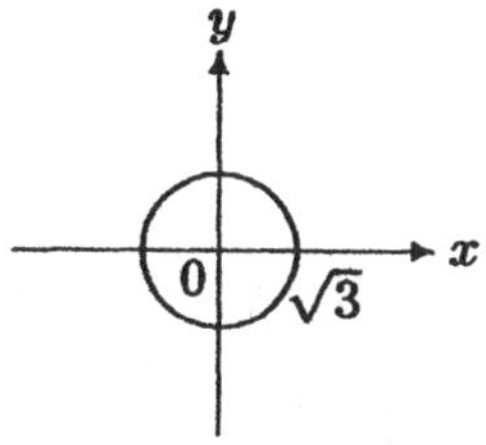

31. Notice that if $X = x$ and $Y = y - 1$, then $y = 1 - \sin x$ becomes $Y = -\sin X$.
y intercept: 1 x intercepts: $\pi/2 + 2n\pi$ for any integer n; not symmetric with respect to either axis or origin

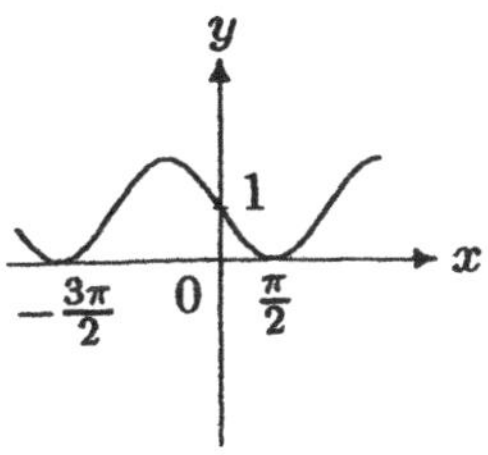

33. Notice that if $X = x - (\pi/4)$ and $Y = y$, then $y = \tan[x - (\pi/4)]$ becomes $Y = \tan X$.
y intercept: -1; x intercepts: $(\pi/4) + n\pi$ for any integer n; not symmetric with respect to either axis or origin

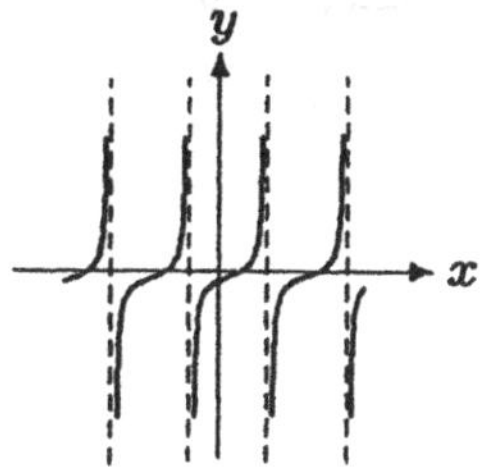

35. The domain is $(-1, 1)$; x and y intercepts: 0;
symmetric with respect to the y axis:
an even function.

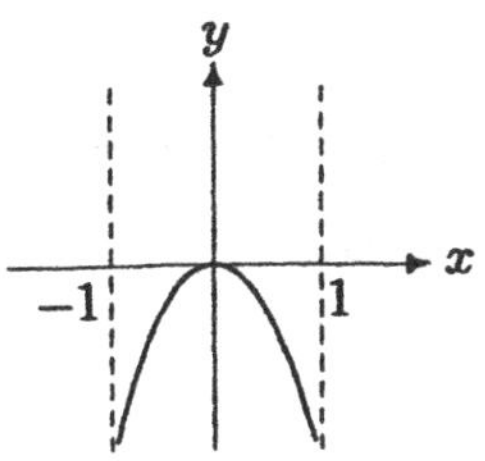

37. $f(-x) = (-x)^2 - \cos(-x) = x^2 - \cos x = f(x)$, so f is an even function.

39. $f(-x) = e^{-x} + e^{-(-x)} = e^{-x} + e^{x} = f(x)$, so f is an even function.

41. $x^2 + 6x + y + 4 = 0$
$(x^2 + 6x + 9) + y + 4 = 9$
$(x + 3)^2 + (y - 5) = 0$
Let $X = x + 3$, $Y = y - 5$. The equation becomes $X^2 + Y = 0$, or $Y = -X^2$.

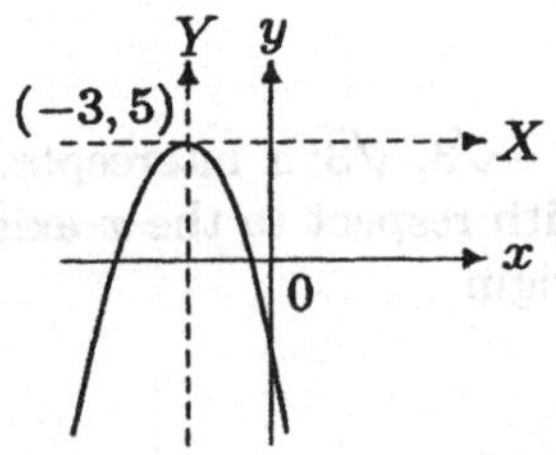

43. $x - 2 = 1/(y + 3)$
Let $X = x - 2$, $Y = y + 3$.
The equation becomes $X = 1/Y$.

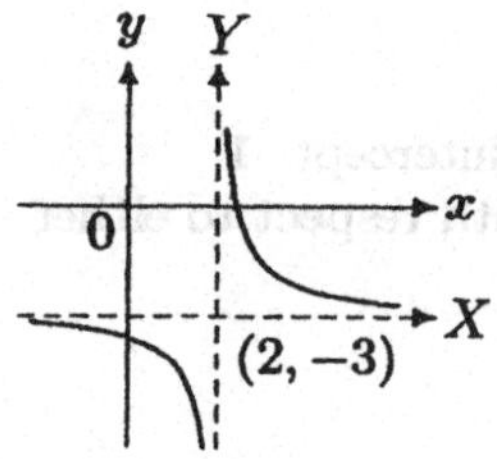

45. a. $f(x) = \begin{cases} -(x-2) - (x-3) = -2x + 5 & \text{for } x \le 2 \\ (x-2) - (x-3) = 1 & \text{for } 2 < x < 3 \\ (x-2) + (x-3) = 2x - 5 & \text{for } x \ge 3 \end{cases}$

$g(x) = \begin{cases} -(x-2) + (x-3) = -1 & \text{for } x \le 2 \\ (x-2) + (x-3) = 2x - 5 & \text{for } 2 < x < 3 \\ (x-2) - (x-3) = 1 & \text{for } x \ge 3 \end{cases}$

b. 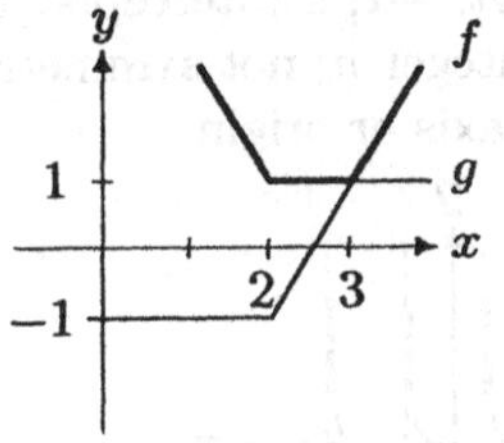

47. $(f - g)(x) = \dfrac{x+2}{x^2 - 4x + 3} - \dfrac{x+1}{x^2 - 2x - 3} = \dfrac{x+2}{(x-1)(x-3)} - \dfrac{x+1}{(x-3)(x+1)}$
$= \dfrac{x+2}{(x-1)(x-3)} - \dfrac{1}{x-3} = \dfrac{(x+2) - (x-1)}{(x-1)(x-3)} = \dfrac{3}{(x-1)(x-3)}$ for $x \neq -1$, 1, and 3

$\left(\dfrac{f}{g}\right) = \dfrac{(x+2)/(x^2 - 4x + 3)}{(x+1)/(x^2 - 2x - 3)} = \dfrac{(x+2)(x-3)(x+1)}{(x-1)(x-3)(x+1)} = \dfrac{x+2}{x-1}$ for $x \neq -1$, 1, and 3

49. a. Domain of f: $[-1, \infty)$; domain of g: $[-2, \infty)$; domain of h: the union of $(-\infty, -2]$ and $[-1, \infty)$.

b. Domain of fg: $[-1, \infty)$. The domain of fg is smaller than the domain of h (although the rules of fg and h are the same).

51. If $x \neq 1$, then $x/(x-1) \neq 1$, so x is in the domain of $f \circ f$.

a. $f(f(x)) = \dfrac{x/(x-1)}{[x/(x-1)] - 1} = \dfrac{x}{x - (x-1)} = x$

b. By part (a), $f\Big(f\big(f(f(f(x)))\big)\Big) = f(f(x)) = x$ for $x \neq 1$.

53. The domain of f is $(-\infty, \infty)$ if $c = 0$ and $a \neq 0$, and in this case, $f(x) = -x - (b/a)$, so $f(f(x)) = -[-x - (b/a)] - b/a = x$. Next, the domain consists of all $x \neq a/c$ if $c \neq 0$. In this case, if $f(x) = a/c$, then $(ax+b)/(cx-a) = a/c$, so that $acx + bc = acx - a^2$, and hence $a^2 + bc = 0$. By hypothesis $a^2 + bc \neq 0$, so $f(x) \neq a/c$ for all x in the domain of f. Consequently the domain of $f \circ f$ is the domain of f. Finally,

$$f(f(x)) = f\left(\frac{ax+b}{cx-a}\right) = \frac{a[(ax+b)/(cx-a)]+b}{c[(ax+b)/(cx-a)]-a} = \frac{a^2x+ab+bcx-ab}{acx+bc-acx+a^2} = \frac{(a^2+bc)x}{a^2+bc} = x.$$

55. $\cos x = \dfrac{\sin x}{\tan x} = \dfrac{-2/3}{-2\sqrt{5}/5} = \dfrac{\sqrt{5}}{3}$; $\cot x = \dfrac{1}{\tan x} = -\dfrac{5}{2\sqrt{5}} = -\dfrac{\sqrt{5}}{2}$; $\sec x = \dfrac{1}{\cos x} = \dfrac{3}{\sqrt{5}} = \dfrac{3\sqrt{5}}{5}$;

$\csc x = \dfrac{1}{\sin x} = -\dfrac{3}{2}$

57. $|\sin x| = |\cos x|$ for $x = \pi/4$, $3\pi/4$, $5\pi/4$, and $7\pi/4$. By comparing values in the intervals $[0, \pi/4)$, $(\pi/4, 3\pi/4)$, $(3\pi/4, 5\pi/4)$, $(5\pi/4, 7\pi/4)$, and $(7\pi/4, 2\pi)$, we see that $|\sin x| \geq |\cos x|$ for x in the intervals $[\pi/4, 3\pi/4]$ and $[5\pi/4, 7\pi/4]$.

59. a. $a\sin(x+b) = a\sin x\cos b + a\cos x\sin b = \sin x + \sqrt{3}\cos x$ if $a\cos b = 1$ and $a\sin b = \sqrt{3}$. Thus $\tan b = (a\sin b)/(a\cos b) = \sqrt{3}/1 = \sqrt{3}$, so $b = (\pi/3) + n\pi$ for any integer n. If $b = \pi/3 + 2n\pi$, then $a = 1/(\cos b) = 1/[\cos(\pi/3)] = 1/\frac{1}{2} = 2$. If $b = 4\pi/3 + 2n\pi$, then $a = 1/(\cos b) = 1/[\cos(4\pi/3)] = 1/(-\frac{1}{2}) = -2$. Thus the solutions are $a = 2$, $b = \pi/3 + 2n\pi$ for any integer n, and $a = -2$, $b = 4\pi/3 + 2n\pi$ for any integer n.

b. $a\cos(x+b) = a\cos x\cos b - a\sin x\sin b = \sin x + \sqrt{3}\cos x$ if $a\cos b = \sqrt{3}$ and $-a\sin b = 1$. Thus $\tan b = (a\sin b)/(a\cos b) = -1/\sqrt{3}$, so $b = -\pi/6 + n\pi$ for any integer n. If $b = -\pi/6 + 2n\pi$, then

$$a = \frac{\sqrt{3}}{\cos b} = \frac{\sqrt{3}}{\cos(-\pi/6)} = \frac{\sqrt{3}}{\sqrt{3}/2} = 2.$$

If $b = 5\pi/6 + 2n\pi$, then $a = \sqrt{3}/(\cos b) = \sqrt{3}/\cos(5\pi/6) = \sqrt{3}/(-\sqrt{3}/2) = -2$. Thus the solutions are $a = 2$, $b = -\pi/6 + 2n\pi$ for any integer n, and $a = -2$, $b = 5\pi/6 + 2n\pi$ for any integer n.

61. $2e^x \geq e^{2x}$ if and only if $2e^x - e^{2x} \geq 0$, that is, $e^x(2 - e^x) \geq 0$. Since $e^x > 0$ for all x, the inequality holds if and only if $2 - e^x \geq 0$, or equivalently, $e^x \leq 2$. Thus $x \leq \ln 2$. Therefore $2e^x \geq e^{2x}$ for x in $(-\infty, \ln 2]$.

63. a. red in (a) b. blue in (b) c. blue in (a) d. red in (b) e. green in (a) f. green in (b)

65. If $0 \leq x \leq \pi/3$, then $\frac{1}{2} \leq \cos x \leq 1$, and thus $1 \leq 1/(\cos x) \leq 2$. Since $\tan x = (\sin x)/(\cos x)$, it follows that for such x, $\sin x \leq (\sin x)/(\cos x) \leq 2\sin x$, or equivalently, $\sin x \leq \tan x \leq 2\sin x$.

67. Let P be the perimeter of the square and the equilateral triangle. Then each side of the square has length $\frac{1}{4}P$, so the area of the square is $\frac{1}{16}P^2$. Each side of the triangle has length $\frac{1}{3}P$, so by Exercise 51 of Section 1.3, the area of the triangle is $\frac{1}{4}\sqrt{3}(\frac{1}{3}P)^2 = (\sqrt{3}/36)P^2$. Since $\sqrt{3}/36 < \frac{1}{16}$, the square has the larger area. However, one can make the area of the rectangle with perimeter P as small as one wishes by making the length of one side of the rectangle small enough. Thus an equilateral triangle can have a larger area or smaller area than a rectangle with the same perimeter.

69. The fourth vertex could be $(1,3)(\text{opposite}(2,0))$, $(5,1)(\text{opposite}(0,1))$, or $(-1,-1)(\text{opposite}(3,2))$.

71. By the Pythagorean Theorem, $|AB| = \sqrt{24^2 + 18^2} = \sqrt{900} = 30$, so that if you pass through B, then you drive $30 + 14 + 16 = 60$ kilometers. Since

$$|AC| = \sqrt{24^2 + (18+14)^2} = \sqrt{1600} = 40$$

if you do not pass through B, then you drive $40 + 16 = 56$ kilometers. At a speed of 60 kilometers per hour, the trip through B takes 1 hour, whereas the trip that avoids B takes 56 minutes. Since your trip lasts 1 hour, you pass through B.

73. Using (1) in Section 1.7 and the notation in the diagram, we have $800 = 600\theta$, so $\theta = \frac{4}{3}$ radians. Thus $d = 600 \sin \frac{4}{3} \approx 583$ feet.

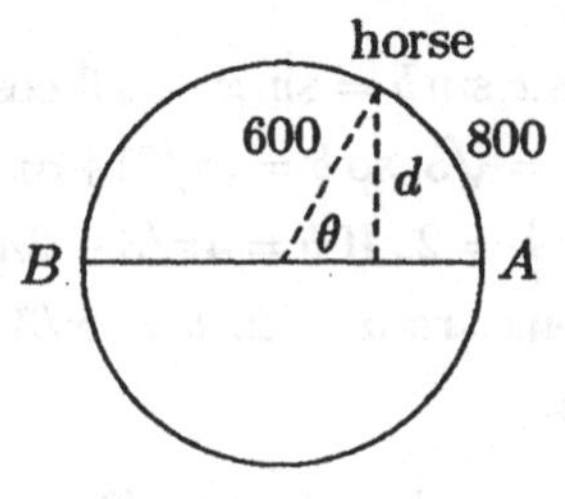

75. a. By hypothesis, $34{,}703 = P(0) = b$ and $44{,}485 = P(10) = 10a + 34{,}703$, so that $a = 978.2$. Thus the predicted population in the year 2000 is $P(30) = (978.2)(30) + 34{,}703 = 64{,}049$.

b. By hypothesis, $34{,}703 = P(0) = ae^0 = a$ and $44{,}485 = P(10) = 34{,}703e^{10b}$, so that $e^{10b} = \frac{44{,}485}{34{,}703}$. Thus $b = \frac{1}{10} \ln \frac{44{,}485}{34{,}703} \approx 0.0248325915$. Thus the predicted population in the year 2000 is $P(30) = 34{,}703e^{30(0.0248325915)} \approx 73{,}098$.

Chapter 2

Limits and Continuity

2.1 Informal Discussion of Limits

1. 3 3. 0 5. $-\frac{1}{11}$ 7. $\frac{2}{5}$

9. $\lim_{x\to-2}\frac{x^2-4}{x+2}=\lim_{x\to-2}\frac{(x+2)(x-2)}{x+2}=\lim_{x\to-2}(x-2)=-4$

11. $\lim_{x\to-5}\frac{x^2+4x-5}{x+5}=\lim_{x\to-5}\frac{(x-1)(x+5)}{x+5}=\lim_{x\to-5}(x-1)=-6$

13. $\lim_{x\to1}\frac{x^3-1}{x-1}=\lim_{x\to1}\frac{(x-1)(x^2+x+1)}{x-1}=\lim_{x\to1}(x^2+x+1)=3$

15. $\lim_{x\to0}3(\sin^2 x+\cos^2 x)=\lim_{x\to0}3=3$

17. $\lim_{x\to-1}-\frac{|x|}{x}=\lim_{x\to-1}-\frac{-x}{x}=\lim_{x\to-1}1=1$

19. 1.003346721, 1.000033335, 1.000000333, 1.003346721, 1.000033335, 1.000000333; $\lim_{x\to0}f(x)=1$

21. 0.0499583472, 0.0049999583, 0.0005, −0.0499583472, −0.0049999583, −0.0005; $\lim_{x\to0}f(x)=0$

23. 0.6164048071, 0.6001600395, 0.6000016, 0.6164048071, 0.6001600395, 0.6000016;

$$\lim_{x\to0}\frac{\sin 3x}{\sin 5x}=0.6$$

25. 0.0500417084, 0.0050000417, 0.0004999999, −0.0500417084, −0.0050000417, −0.0004999999;

$$\lim_{x\to0}(\csc x-\cot x)=0$$

27. 3.498588076, 3.045453395, 3.004504503, 2.591817793, 2.955446645, 2.995504497;

$$\lim_{x\to0}\frac{e^{3x}-1}{x}=3$$

29. 1.6

31. The limit equals 0.

33. The limit does not exist.

35. b. 1

37. $(f(2) - f(1))/(2 - 1)$ is larger.

39. $f(3) = 9$ and

$$\frac{f(x) - f(3)}{x - 3} = \frac{x^2 - 9}{x - 3} = \frac{(x + 3)(x - 3)}{x - 3} = x + 3.$$

The slope of the line tangent to the graph of f at $(3, 9)$ is

$$\lim_{x \to 3} \frac{f(x) - f(3)}{x - 3} = \lim_{x \to 3} (x + 3) = 6.$$

41. $f(4) = 33$ and

$$\frac{f(x) - f(4)}{x - 4} = \frac{(2x^2 + 1) - 33}{x - 4} = \frac{2(x^2 - 16)}{x - 4} = \frac{2(x + 4)(x - 4)}{x - 4} = 2(x + 4).$$

The slope of the line tangent to the graph of f at $(4, 33)$ is

$$\lim_{x \to 4} \frac{f(x) - f(4)}{x - 4} = \lim_{x \to 4} 2(x + 4) = 16.$$

43. .5129329439, .5012541824, .5001250418, .4879016417, .4987541511, .4998750416; the slope is .5.

45. 1.583844403, 1.570925532, 1.570797619, 1.583844403, 1.570925532, 1.570797619; the slope is approximately 1.57.

47. average velocity $= \dfrac{f(t) - f(\frac{1}{2})}{t - \frac{1}{2}} = \dfrac{t^2 - \frac{1}{4}}{t - \frac{1}{2}} = \dfrac{(t - \frac{1}{2})(t + \frac{1}{2})}{t - \frac{1}{2}} = t + \dfrac{1}{2}$; $v(\dfrac{1}{2}) =$
$\lim_{t \to 1/2} (t + \frac{1}{2}) = 1$

49. a. You should use as small a time interval containing $t = 50$ minutes as possible. Thus you should use the distance at $t = 50$ and the distance at either $t = 49$ or $t = 51$. The distances at $t = 48$ and $t = 52$ would be ignored.

b. Using the average velocity over the time interval $[50, 51]$ to approximate $v(50)$, we find that

$$v(50) \approx \frac{29 - 28.4}{51 - 50} = 0.6 \text{ (miles per hour)}$$

Using the average velocity over the time interval $[49, 50]$, we find that

$$v(50) \approx \frac{28.4 - 27.8}{50 - 49} = 0.6 \text{ (miles per hour)}$$

Since we obtained the same answer both ways, our estimate of the velocity at $t = 50$ is 0.6 (miles per hour).

15. Let $y = (e^x - 1)/x$. Then $\lim_{x\to 0} y = \lim_{x\to 0}(e^x - 1)/x = 1$ by (5) in Section 2.1. By the Substitution Rule,

$$\lim_{x\to 0}\ln\left(\frac{e^x - 1}{x}\right) = \lim_{y\to 1}\ln y = \ln 1 = 0$$

17. By Example 8 and the Substitution Rule with $y = 4x$,

$$\lim_{x\to 0}\frac{\cos 4x - 1}{x} = \lim_{y\to 0}\frac{\cos y - 1}{y/4} = 4\lim_{y\to 0}\frac{\cos y - 1}{y} = 4\cdot 0 = 0$$

19. Using Example 8, we have

$$\lim_{t\to 0}\frac{\cos t - 1}{t^{1/3}} = \lim_{t\to 0}\left(\frac{\cos t - 1}{t^{1/3}}\cdot\frac{t^{2/3}}{t^{2/3}}\right) = \lim_{t\to 0}\frac{\cos t - 1}{t}\cdot\lim_{t\to 0}t^{2/3} = 0\cdot 0 = 0$$

21. Using Example 7 and the Substitution Rule with $y = 2x$, we have

$$\lim_{x\to 0}\frac{\sin x}{\sin 2x} = \lim_{x\to 0}\frac{(\sin x)/(2x)}{(\sin 2x)/(2x)} = \frac{1}{2}\frac{\lim_{x\to 0}[(\sin x)/x]}{\lim_{x\to 0}[(\sin 2x)/(2x)]} = \frac{1}{2}\frac{1}{\lim_{y\to 0}[(\sin y)/y]} = \frac{1}{2}$$

23. Using the comment following Example 6, we have

$$\lim_{x\to\pi}\frac{\tan^2 x}{1+\sec x} = \lim_{x\to\pi}\frac{\sec^2 x - 1}{1+\sec x} = \lim_{x\to\pi}\frac{(\sec x + 1)(\sec x - 1)}{1+\sec x} = \lim_{x\to\pi}(\sec x - 1) = \sec\pi - 1 = -2$$

25. $$\lim_{x\to -1}\frac{x^2 - 1}{x+1} = \lim_{x\to -1}\frac{(x-1)(x+1)}{x+1} = \lim_{x\to -1}(x-1) = -1-1 = -2$$

27. $$\lim_{x\to -2}\frac{x^4 - 16}{4 - x^2} = -\lim_{x\to -2}\frac{x^4 - 16}{x^2 - 4} = -\lim_{x\to -2}\frac{(x^2-4)(x^2+4)}{x^2 - 4} = -\lim_{x\to -2}(x^2+4) = -((-2)^2+4) = -8$$

29. $$\lim_{x\to 3}\frac{x^2 - x - 6}{x^3 - 3x^2 + x - 3} = \lim_{x\to 3}\frac{(x+2)(x-3)}{(x-3)(x^2+1)} = \lim_{x\to 3}\frac{x+2}{x^2+1} = \frac{3+2}{3^2+1} = \frac{1}{2}$$

31. $$\lim_{y\to 2}\frac{\sqrt{y}-\sqrt{2}}{y^2 - 2y} = \lim_{y\to 2}\frac{\sqrt{y}-\sqrt{2}}{y(y-2)} = \lim_{y\to 2}\frac{\sqrt{y}-\sqrt{2}}{y(\sqrt{y}-\sqrt{2})(\sqrt{y}+\sqrt{2})} = \lim_{y\to 2}\frac{1}{y(\sqrt{y}+\sqrt{2})}$$

$$= \frac{1}{2(\sqrt{2}+\sqrt{2})} = \frac{1}{4\sqrt{2}} = \frac{\sqrt{2}}{8}$$

33. $$\lim_{x\to -2}\left(\frac{x^2}{x+2} - \frac{4}{x+2}\right) = \lim_{x\to -2}\frac{x^2 - 4}{x+2} = \lim_{x\to -2}\frac{(x-2)(x+2)}{x+2} = \lim_{x\to -2}(x-2) = -2-2 = -4$$

35. $\lim_{x\to 0}\frac{e^{2x}-1}{e^x-1}=\lim_{x\to 0}\frac{(e^x)^2-1}{e^x-1}=\lim_{x\to 0}\frac{(e^x+1)(e^x-1)}{e^x-1}=\lim_{x\to 0}(e^x+1)=e^0+1=1+1=2$

37. Since $-1\le\cos(1/x^2)\le 1$, we have $-|x|\le x\cos(1/x^2)\le |x|$. Since $\lim_{x\to 0}(-|x|)=0$ and $\lim_{x\to 0}|x|$ 0, it follows from the Squeezing Theorem that $\lim_{x\to 0}x\cos(1/x^2)=0$.

39. Since $-1\le\sin[1/(x-1)]\le 1$ and $-1\le\cos x\le 1$, we have

$$-1\le\sin\frac{1}{x-1}\cos x\le 1\quad\text{and hence}\quad -|\ln x|\le(\ln x)\sin\frac{1}{x-1}\cos x\le|\ln x|.$$

By the Substitution Rule with $y=\ln x$, $\lim_{x\to 1}|\ln x|=\lim_{y\to 0}|y|=|0|=0$ and hence $\lim_{x\to 1}-|\ln x|=-0=0$. It follows from the Squeezing Theorem that

$$\lim_{x\to 1}(\ln x)\sin\frac{1}{x-1}\cos x=0.$$

41. Since $x^4\le f(x)\le 2x^4$ for all x in $[-1,1]$, and since $\lim_{x\to 0}x^4=0$ and $\lim_{x\to 0}2x^4=0$, it follows from the Squeezing Theorem that $\lim_{x\to 0}f(x)=0$.

43. $\lim_{x\to 1}\frac{f(x)-f(1)}{x-1}=\lim_{x\to 1}\frac{(x^6-1)-0}{x-1}=\lim_{x\to 1}\frac{(x-1)(x^5+x^4+x^3+x^2+x+1)}{x-1}$

$=\lim_{x\to 1}(x^5+x^4+x^3+x^2+x+1)=6$

l: $y-0=6(x-1)$, or $y=6(x-1)$

45. $\lim_{x\to -1}\frac{f(x)-f(-1)}{x-(-1)}=\lim_{x\to -1}\frac{1/(x+3)-1/2}{x+1}=\lim_{x\to -1}\frac{\frac{-1-x}{2(x+3)}}{x+1}=\lim_{x\to -1}\frac{-1}{2(x+3)}=-\frac{1}{4}$

47. By Example 8, $\lim_{x\to 0}\frac{f(x)-f(0)}{x-0}=\lim_{x\to 0}\frac{\cos x-1}{x}=0$; l: $y-1=0(x-0)$, or $y=1$.

49. Using the Substitution Rule with $y=4-x^2$, we have

$$\lim_{x\to 1}\frac{f(x)-f(1)}{x-1}=\lim_{x\to 1}\frac{\sqrt{4-x^2}-\sqrt{3}}{x-1}=\lim_{x\to 1}\frac{(\sqrt{4-x^2}-\sqrt{3})(\sqrt{4-x^2}+\sqrt{3})}{(x-1)(\sqrt{4-x^2}+\sqrt{3})}$$

$$=\lim_{x\to 1}\frac{4-x^2-3}{(x-1)(\sqrt{4-x^2}+\sqrt{3})}=\lim_{x\to 1}\frac{1-x^2}{(x-1)(\sqrt{4-x^2}+\sqrt{3})}=\lim_{x\to 1}\frac{(1-x)(1+x)}{(x-1)(\sqrt{4-x^2}+\sqrt{3})}$$

$$=\lim_{x\to 1}-\frac{1+x}{\sqrt{4-x^2}+\sqrt{3}}=-\frac{\lim_{x\to 1}(1+x)}{\lim_{y\to 3}(\sqrt{y}+\sqrt{3})}=\frac{-2}{2\sqrt{3}}=-\frac{\sqrt{3}}{3};$$

l: $y-\sqrt{3}=-\frac{\sqrt{3}}{3}(x-1)$.

51. Yes

53. f is a polynomial function, so f is continuous at 2.

55. f is a rational function and 0 is in the domain of f, so f is continuous at 0.

57. a. Since $\lim_{x\to 2} f(x) = 1.5$ and $\lim_{x\to 2} g(x) = 2$, it follows from the Sum Rule that $\lim_{x\to 2}[f(x)+g(x)] = 1.5 + 2 = 3.5$.

b. Since $\lim_{x\to 2} f(x) = 1.5$ and $\lim_{x\to 2} g(x) = 2$, it follows from the Product Rule that $\lim_{x\to 2}[f(x)\,g(x)] = (2)(1.5) = 3$.

c. Since $\lim_{x\to 2}(x-4) = -2$ and $\lim_{x\to 2} g(x) = 2$, it follows from the Product Rule that $\lim_{x\to 2}(x-4)g(x) = (-2)(2) = -4$.

59. a. $\lim\limits_{x\to 0} \dfrac{f(x)-f(0)}{x-0} = \lim\limits_{x\to 0} \dfrac{x^2-0}{x-0} = \lim\limits_{x\to 0} x = 0$; $\lim\limits_{x\to 0} \dfrac{g(x)-g(0)}{x-0} = \lim\limits_{x\to 0} \dfrac{x^3-0}{x-0} = \lim\limits_{x\to 0} x^2 = 0$.

Since the tangent lines through $(0,0)$ have the same slope, they are the same.

b. $\lim\limits_{x\to 0} \dfrac{f(x)-f(0)}{x-0} = \lim\limits_{x\to 0} \dfrac{(x^2+1)-1}{x-0} = \lim\limits_{x\to 0} x = 0$;

$\lim\limits_{x\to 0} \dfrac{g(x)-g(0)}{x-0} = \lim\limits_{x\to 0} \dfrac{(-x^2+1)-1}{x-0} = \lim\limits_{x\to 0}(-x) = 0$.

Since the tangent lines through $(0,1)$ have the same slope, they are the same.

61. Since $g(x) = f(x) - [f(x) - g(x)]$, and $\lim_{x\to -1} f(x)$ and $\lim_{x\to -1}[f(x)-g(x)]$ exist, it follows from the Difference Rule that $\lim_{x\to -1} g(x)$ exists and $\lim_{x\to -1} g(x) = \lim_{x\to -1} f(x) - \lim_{x\to -1}[f(x)-g(x)] = 4 - 6 = -2$.

63. $\lim_{x\to 0}[f(x)+g(x)] = \lim_{x\to 0}(1/x - 1/x) = \lim_{x\to 0} 0 = 0$, but by Example 3, $\lim_{x\to 0}(1/x)$ does not exist and $\lim_{x\to 0}(-1/x) = -\lim_{x\to 0}(1/x)$ does not exist.

65. a. One example is defined by

$$f(x) = \begin{cases} -1/x & \text{for } -1 \le x < 0 \\ 0 & \text{for } x = 0 \\ 1/x & \text{for } 0 < x \le 1 \end{cases}$$

b. One example is defined by

$$f(x) = \begin{cases} 100/x & \text{for } 0 < |x| \le 1 \\ 0 & \text{for } x = 0 \end{cases}$$

67. $\dfrac{x\sin^2(1/x)}{1+\sin^2(1/x)} = f(x)g(x)$, where $f(x) = x$ and $g(x) = \dfrac{\sin^2(1/x)}{1+\sin^2(1/x)}$.

Since $\lim_{x\to 0} f(x) = \lim_{x\to 0} x = 0$ and $|g(x)| \le 1$ for all $x \ne 0$, it follows from Exercise 65 with $M = 1$ that

$$\lim_{x\to 0} \frac{x\sin^2(1/x)}{1+\sin^2(1/x)} = 0.$$

69. a. Since $f(x) \leq g(x)$ for all $x \neq a$, we have $L \leq M$.

b. Since $f(x) < g(x)$ for all $x \neq a$, we have $L \leq M$. Notice that we cannot conclude that $L < M$. For example, if $f(x) = 0$ and $g(x) = x^2$, then $f(x) < g(x)$ for $x \neq 0$. Nevertheless $\lim_{x\to 0} f(x) = 0 = \lim_{x\to 0} f(x)$.

71. Let $y = x^n$. Since x^n approaches 0 as x approaches 0, it follows that y approaches 0, so by the

73. a. $\lim_{x\to 0} 1/(x^2 + x)$ does not exist, so we cannot use the Product Rule to conclude that

$$\lim_{x\to 0}\left[x\left(\frac{1}{x^2+x}\right)\right] = \left(\lim_{x\to 0} x\right)\left(\lim_{x\to 0}\frac{1}{x^2+x}\right)$$

b. $\lim_{x\to 0}\left[x\left(\frac{1}{x^2+x}\right)\right] = \lim_{x\to 0}\frac{1}{x+1} = \frac{1}{\lim_{x\to 0}(x+1)} = 1$

75. Since v is a rational function whose denominator has no roots, v is continuous at every number in $(0, \infty)$, i.e., $\lim_{t\to t_0} v(t) = v(t_0)$ for every $t_0 > 0$. By the Product and Constant Multiple Rules, we conclude that for any $t_0 > 0$,

$$\lim_{t\to t_0} K(t) = \lim_{t\to t_0}\frac{1}{2}m[v(t)]^2 = \frac{1}{2}m[\lim_{t\to t_0} v(t)]^2 = \frac{1}{2}m[v(t_0)]^2 = K(t_0)$$

Therefore K is continuous at any positive number t_0.

2.4 One-Sided and Infinite Limits

1. $\lim_{x\to -2^+}(x^3+3x-5) = \lim_{x\to -2}(x^3+3x-5) = (-2)^3+3(-2)-5 = -19$

3. $\lim_{x\to 2^-} \dfrac{x^2-4}{x-2} = \lim_{x\to 2^-} \dfrac{(x-2)(x+2)}{x-2} = \lim_{x\to 2^-}(x+2) = \lim_{x\to 2}(x+2) = 4$

5. $\lim_{x\to 1^+} \dfrac{x^2+3x-4}{x^2-1} = \lim_{x\to 1^+} \dfrac{(x-1)(x+4)}{(x-1)(x+1)} = \lim_{x\to 1^+} \dfrac{x+4}{x+1} = \lim_{x\to 1} \dfrac{x+4}{x+1} = \dfrac{5}{2}$

7. Since $t-5>0$ for $t>5$, $\lim_{t\to 5^+}|t-5|/(5-t) = \lim_{t\to 5^+}(t-5)/(5-t) = \lim_{t\to 5^+}(-1) = -1$.

9. If $x<0$, then $x^3<0$. Thus $4/x^3<0$, so $\lim_{x\to 0^-}(4/x^3) = -\infty$.

11. For all x except $x=0$, we have $-1/x^2<0$. Thus $\lim_{x\to 0^+}(-1/x^2) = \lim_{x\to 0^-}(-1/x^2) = -\infty$. Therefore $\lim_{x\to 0}(-1/x^2) = -\infty$.

13. If $y<-1$, then $y+1<0$, so $\pi/(y+1)<0$. Therefore $\lim_{y\to -1^-}[\pi/(y+1)] = -\infty$.

15. For $0<z<\pi/2$, $\tan z>0$. Since $\tan z = (\sin z)/(\cos z)$, $\lim_{z\to\pi/2}\cos z = 0$, and $\lim_{z\to\pi/2}\sin z = 1$, we have $\lim_{z\to\pi/2^-}\tan z = \infty$.

17. If $-1<x<1$, then $0<1-|x|<1$. Therefore the substitution $y=1-|x|$ yields $\lim_{x\to 1^-}\ln(1-|x|) = \lim_{y\to 0^+}\ln y = -\infty$.

19. If $x>0$, then $\sqrt{x}>0$, so that $1-e^{\sqrt{x}}<0$. By the Substitution Rule with $y=\sqrt{x}$, $\lim_{x\to 0^+}(1-e^{\sqrt{x}}) = \lim_{y\to 0^+}(1-e^y) = 1-e^0 = 0$. By the version of the Substitution Rule for one-sided limits with $y=1-e^{\sqrt{x}}$,

$$\lim_{x\to 0^+} \frac{1}{1-e^{\sqrt{x}}} = \lim_{y\to 0^-} \frac{1}{y} = -\infty.$$

21. If $x>5$ then $x-5>0$, so $\sqrt{x-5}$ is defined. By the version of the Substitution Rule for one-sided limits, $\lim_{x\to 5^+}\sqrt{x-5} = \lim_{y\to 0^+}\sqrt{y} = 0$. Thus $\lim_{x\to 5^+} 1/(x\sqrt{x-5}) = \infty$.

23. By the versions of the Product and Substitution Rules for one-sided limits, with $y=2x$ and $z=\cos y$,

$$\lim_{x\to 0^+}\sqrt{x\cos 2x} = \lim_{x\to 0^+}\sqrt{x}\,\lim_{x\to 0^+}\sqrt{\cos 2x} = \lim_{x\to 0^+}\sqrt{x}\,\lim_{y\to 0^+}\sqrt{\cos y} = \lim_{x\to 0^+}\sqrt{x}\,\lim_{z\to 1^-}\sqrt{z} = 0\cdot 1 = 0.$$

25. $\lim_{x\to 0}\sqrt{\dfrac{1}{x^2}} = \lim_{x\to 0}\dfrac{1}{\sqrt{x^2}} = \lim_{x\to 0}\dfrac{1}{|x|} = \infty$

27. If $x<-\frac{1}{2}$, then $x+\frac{1}{2}<0$ and $4x-7<0$. Thus $(4x-7)/(x+\frac{1}{2})>0$, so

$$\lim_{x\to -1/2^-} \frac{4x-7}{x+\frac{1}{2}} = \lim_{x\to -1/2^-}(4x-7)\cdot\frac{1}{x+\frac{1}{2}} = \infty.$$

29. If $0 < x < 2$, then $x + 1 > 0$ and $x^2 - 4 < 0$, so $(x+1)/[2(x^2-4)] < 0$, and thus

$$\lim_{x\to 2^-} \frac{x+1}{2(x^2-4)} = \lim_{x\to 2^-} \left[\frac{x+1}{2} \cdot \frac{1}{x^2-4}\right] = -\infty.$$

31. If $y < 3$, then $3 - y > 0$ and $\lim_{y\to 3^-} \sqrt{3-y} = 0$. Thus $\lim_{y\to 3^-}(-1/\sqrt{3-y}) = -\infty$.

33. For $y < 0$, $\sqrt{y}$ is undefined. Therefore $\lim_{y\to 0^-}[(5+\sqrt{1+y^2})/\sqrt{y}]$ does not exist.

35. $$\lim_{t\to 0} \frac{1-\cos t}{t^2} = \lim_{t\to 0}\left[\frac{1-\cos t}{t^2} \cdot \frac{1+\cos t}{1+\cos t}\right] = \lim_{t\to 0}\left[\frac{\sin^2 t}{t^2} \cdot \frac{1}{1+\cos t}\right]$$
$$= \lim_{t\to 0} \frac{\sin t}{t} \lim_{t\to 0} \frac{\sin t}{t} \lim_{t\to 0} \frac{1}{1+\cos t} = 1 \cdot 1 \cdot \frac{1}{2} = \frac{1}{2}$$

37. For $x \neq 3, -3$, we have

39. $\lim_{x\to 0^+} e^{2+\ln x} = \lim_{x\to 0^+} e^2 e^{\ln x} = \lim_{x\to 0^+} e^2 x = e^2 \lim_{x\to 0^+} x = e^2 \cdot 0 = 0$.

41. If $x < 0$, then $e^x < 1$, so $e^x - 1 < 0$, which implies that $\sqrt{e^x - 1}$ is not defined for $x < 0$. Therefore $\lim_{x\to 0^-} \sqrt{e^x - 1}$ does not exist.

43. If $x > 1$, then $\ln x > 0$. If $y = \ln x$, then by the version of the Substitution Rule for one-sided limits, $\lim_{x\to 1^+} \ln(\ln x) = \lim_{y\to 0^+} \ln y = -\infty$.

45. $\lim_{x\to 0^-} f(x) = \lim_{x\to 0^-}(2x-4) = \lim_{x\to 0}(2x-4) = -4$;

$\lim_{x\to 0^+} f(x) = \lim_{x\to 0^+} -(x+2)^2 = \lim_{x\to 0} -(x+2)^2 = -4$.

Therefore $\lim_{x\to 0} f(x) = -4$.

47. $\lim_{x\to 3^+}(x+2)/(x-3) = \infty$, so $x = 3$ is a vertical asymptote.

49. $\lim_{x\to 1^+} f(x) = \infty = \lim_{x\to 1^-} f(x)$, so $x = 1$ and $x = -1$ are vertical asymptotes. (Notice that

$$\lim_{x\to 0} f(x) = \lim_{x\to 0}\left[\frac{\sin x}{x} \cdot \frac{1}{x^2-1}\right] = \lim_{x\to 0} \frac{\sin x}{x} \lim_{x\to 0} \frac{1}{x^2-1} = (1)(-1) = -1$$

so $x = 0$ is *not* a vertical asymptote.)

51. If $x \neq -5, -2, 0$, then

$$\frac{x^2+2x-15}{x^3+7x^2+10x} = \frac{(x-3)(x+5)}{x(x+2)(x+5)} = \frac{x-3}{x(x+2)}.$$

Thus $\lim_{x\to 0^-} f(x) = \infty = \lim_{x\to -2^+} f(x)$, so $x = -2$ and $x = 0$ are vertical asymptotes.

53. $\dfrac{x+1/x}{x^4+1} = \dfrac{x^2+1}{x(x^4+1)}$, so $\lim_{x\to 0^+} f(x) = \infty$, and thus $x = 0$ is a vertical asymptote.

55. $\lim_{x\to 0}(\sin x)/x = 1$, so there are no vertical asymptotes.

57. $\lim_{x\to(\pi/2+n\pi)^-} \tan x = \infty$ for any integer n, so $x = \pi/2 + n\pi$ is a vertical asymptote for any integer n.

59. $$\lim_{x\to 0} \frac{f(x)-f(0)}{x-0} = \lim_{x\to 0} \frac{x^{1/5}-0}{x-0} = \lim_{x\to 0} \frac{1}{x^{4/5}} = \infty$$

Thus there is a vertical tangent line l at $(0,0)$; l: $x = 0$.

61. $$\lim_{x\to 0} \frac{f(x)-f(0)}{x-0} = \lim_{x\to 0} \frac{(1-5x^{3/5})-1}{x-0} = \lim_{x\to 0} \frac{-5}{x^{2/5}} = -\infty$$

Thus there is a vertical tangent line l at $(0,1)$; l: $x = 0$.

63.

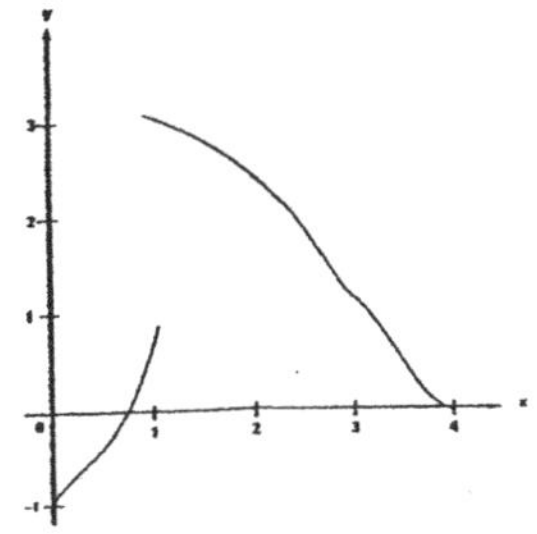

65. Let n be an arbitrary integer. If $n-1 < x < n$, then $[x] = n-1$, so that $\lim_{x\to n^-}[x] = \lim_{x\to n^-}(n-1) = n-1$. If $n < x < n+1$, then $[x] = n$, so that $\lim_{x\to n^+}[x] = \lim_{x\to n^+} n = n$.

67. $$\lim_{x\to 0} \frac{f(x)-f(0)}{x-0} = \lim_{x\to 0} \frac{x^{1/n}-0}{x-0} = \lim_{x\to 0} \frac{1}{x^{1-1/n}} = \lim_{x\to 0} \frac{1}{x^{(n-1)/n}} = \infty \text{ since } n-1 \text{ is even.}$$

Thus there is a vertical tangent line at $(0,0)$.

69. a. Let $t = 0$ correspond to the time a water droplet reaches the top of the falls. By (15) of Section 2.2 the height of a descending water droplet is given by $h(t) = -4.9t^2 - 2t + 40$. A water droplet reaches bottom at the positive time t_0 for which $h(t_0) = 0$, or $-4.9t_0^2 - 2t_0 + 40 = 0$. By the quadratic formula,

$$t_0 = \frac{2 - \sqrt{(-2)^2 - 4(-4.9)(40)}}{2(-4.9)} = \frac{-2+\sqrt{788}}{9.8} \approx 2.7 \,(\text{seconds}).$$

The velocity at time t_0 is

$$v(t_0) = \lim_{t\to t_0^-} \frac{h(t)-h(t_0)}{t-t_0} = \lim_{t\to t_0^-} \frac{(-4.9t^2-2t+40)-(-4.9t_0^2-2t_0+40)}{t-t_0}$$

$$= \lim_{t\to t_0} \frac{-4.9(t^2-t_0^2)-2(t-t_0)}{t-t_0} = \lim_{t\to t_0} [-4.9(t+t_0)-2]$$

$$= -9.8t_0 - 2 = -\sqrt{788} = -2\sqrt{197} \,(\text{meters per second}).$$

b. Since the final velocity is $-2\sqrt{197}$ by part (a), we will find the time t_1 at which $v(t_1)$ is half the terminal velocity, that is,

$$\frac{1}{2}(-2\sqrt{197}) = -\sqrt{197}.$$

As in the solution of part (a),

$$v(t_1) = \lim_{t\to t_1} \frac{h(t) - h(t_1)}{t - t_1} = \lim_{t\to t_1} \frac{-4.9(t^2 - t_1^2) - 2(t - t_1)}{t - t_1}$$

$$= \lim_{t\to t_1} [-4.9(t + t_1) - 2] = -9.8t_1 - 2.$$

Thus $-9.8t_1 - 2 = -\sqrt{197}$, so

$$t_1 = \frac{\sqrt{197} - 2}{9.8} \approx 1.2 \text{ (seconds)}.$$

By calculator we find that $h_{1/2} = h(t_1) \approx 30.2$ (meters).

71. $$\lim_{x\to 0^-} E(x) = \lim_{x\to 0^-} c\left[\frac{x}{-x} - \frac{x}{(x^2 + r^2)^{1/2}}\right] = \lim_{x\to 0} c\left[-1 - \frac{x}{(x^2 + r^2)^{1/2}}\right] = -c;$$

$$\lim_{x\to 0^+} E(x) = \lim_{x\to 0^+} c\left[\frac{x}{x} - \frac{x}{(x^2 + r^2)^{1/2}}\right] = \lim_{x\to 0} c\left[1 - \frac{x}{(x^2 + r^2)^{1/2}}\right] = c$$

Since $c \neq 0$ by hypothesis, the two one-sided limits are not equal, so that $\lim_{x\to 0} E(x)$ does not exist.

2.5 Continuity, the Intermediate Value Theorem, and the Bisection Method

1. Since $\sqrt{x}$ and $\cos x$ are continuous at $\pi/4$, so is their product, by Theorem 2.9.

3. Since $\lim_{x\to 0} f(x) = \lim_{x\to 0} (\sin x)/x = 1 = f(0)$, f is continuous at 0, by Definition 2.8.

5. The functions in (b) and (f) are continuous. For the function in (a), $\lim_{x\to 2}$ exists but does not equal $f(2)$, so f is not continuous at 2. For the functions in (c), (d) and (e), $\lim_{x\to 2}$ does not exist, so none of those functions is continuous at 2.

7. As x approaches 2 from the right, $x-2$ approaches 0 from the right. Thus f is continuous from the right at 2. Note that if $x < 2$, then $x - 2 < 0$, so $\sqrt{x-2}$ is not defined for $x < 2$ and hence f is not continuous from the left at 2.

9. As x approaches 0 from the left, $1 - e^x$ approaches 0 from the right. Thus $\sqrt{1-e^x}$ is continuous from the left at 0. Note that if $x > 0$, then $1 - e^x < 0$, so $\sqrt{1-e^x}$ is not defined for $x > 0$, and hence f is not continuous from the right at 0.

11. Notice that $e^{4-4} = e^0 = 1$, and

$$\frac{|x-4|}{x-4} = \begin{cases} 1 & \text{if } x > 4 \\ -1 & \text{if } x < 4. \end{cases}$$

Thus f is continuous from the right but is not continuous from the left at 4. Since $f(x) = -1$ for $x < 4$, it follows that f is continuous at 0.

13. Since $f(t) = t^2\sqrt{t^2 - t^4} = t^2\sqrt{t^2(1-t^2)}$, the domain is $[-1, 1]$. Next,

$$\lim_{t\to 0} t^2\sqrt{t^2 - t^4} = \left(\lim_{t\to 0} t^2\right)\left(\lim_{t\to 0} \sqrt{t^2 - t^4}\right) = 0 \cdot 0 = 0 = f(0).$$

Thus f is continuous at 0. As t approaches 1 from the left, $t^2 - t^4 = t^2(1-t^2)$ approaches 0 from the right, so $\lim_{t\to 1^-} t^2\sqrt{t^2-t^4} = (\lim_{t\to 1^-} t^2)(\lim_{t\to 1^-} \sqrt{t^2-t^4}) = 1 \cdot 0 = 0 = f(1)$. Thus f is continuous from the left at 1. If $t > 1$, then $t^2 - t^4 < 0$, so $\sqrt{t^2-t^4}$ and hence $t^2\sqrt{t^2-t^4}$ are not defined. Thus f is not continuous from the right at 1.

15. a. For f to be continuous at 3, we need $\lim_{x\to 3} f(x) = f(3)$. Since $f(x) = x$ for $x \neq 3$, we have $\lim_{x\to 3} f(x) = \lim_{x\to 3} x = 3$. Thus if we redefine $f(3)$ to be 3, then the resulting function will be continuous at 3.

b. For f to be continuous at 1, we need $\lim_{x\to 1} f(x) = f(1)$. Thus there must be a $\delta > 0$ such that if $0 < |x-1| < \delta$, then $|f(x) - f(1)| < \frac{1}{2}$. However, for x close to 1 and $x < 1$, $f(x) = 2x - 3$, so $f(x)$ is close to $2 \cdot 1 - 3 = -1$. By contrast, for x close to 1 and $x > 1$, we have $f(x) = 3x - 2$, so $f(x)$ is close to $3 \cdot 1 - 2 = 1$. Therefore there is no way of redefining $f(1)$ so that f is continuous at 1.

17. Since $\sin x$ and x are both continuous on $(-\infty,\infty)$ and $x \neq 0$ on $(-\infty,0)$ and $(0,\infty)$, Theorem 2.9 implies that f is continuous on $(\infty, 0)$ and $(0,\infty)$.

19. Since $x+3$ is continuous on $(-\infty,\infty)$, the square root function is continuous on $[0,\infty)$, and $x+3 \geq 0$ for x in $[-3,\infty)$, Theorem 2.10 implies that f is continuous on $[-3,\infty)$.

21. Since x^2-x^4 is continuous on $(-\infty,\infty)$, the square root function is continuous on $[0,\infty)$, and $x^2-x^4 > 0$ for x in $(0,1)$, Theorem 2.10 implies that $\sqrt{x^2-x^4}$ is continuous on $(0,1)$. Finally, the constant function 1 is continuous on $(-\infty,\infty)$. Therefore Theorem 2.9 implies that f is continuous on $(0,1)$.

23. Observe that $1-x \geq 0$ and $3x-2 > 0$ for $\frac{2}{3} < x \leq 1$, so that $(1-x)/(3x-2) \geq 0$ for $\frac{2}{3} < x \leq 1$. Since $(1-x)/(3x-2)$, being a rational function, is conti ious on $(\frac{2}{3}, 1]$, and since the square root function is continuous on $[0,\infty)$, it follows from Theorem 2.10 that f is continuous on $(\frac{2}{3}, 1]$.

25. This follows from (14) in Section 2.3. Alternatively, one could apply Theorem 2.9 to the quotient of the continuous functions $\sin x$ and $\cos x$, noting that $\cos x \neq 0$ for x in $(-\pi/2, \pi/2)$.

27. Since $1/x$ is continuous on $(0,\infty)$, $\cot x = (\cos x)/(\sin x)$ is continuous on $(0,\pi)$, and $0 < 1/x < \pi$ for x in $(1/\pi,\infty)$, Theorem 2.10 implies that f is continuous on $(1/\pi,\infty)$.

29. Since e^t is continuous on $(-\infty,\infty)$, with $e^t > 0$ for all t, and since the square root function is continuous on $[0,\infty)$, Theorem 2.10 implies that $\sqrt{e^t}$ is continuous on $(-\infty,\infty)$.

31. Since e^t is continuous on $(-\infty,\infty)$ and $1-e^t > 0$ for $t < 0$, and since the natural logarithm function is continuous on $(0,\infty)$, it follows from Theorem 2.10 that $\ln(1-e^t)$ is continuous on $(-\infty,0)$.

33. Let $f(x) = x^4 - x - 1$. Then f is continuous. Since $f(-1) = 1$ and $f(1) = -1$, the Intermediate Value Theorem implies that there is a c in $[-1,1]$ such that $f(c) = 0$, that is, $c^4 - c - 1 = 0$.

35. Let $f(x) = x^2 + 1/x$. Then f is continuous. Since $f(-2) = \frac{7}{2}$ and $f(-\frac{1}{2}) = -\frac{7}{4}$, the Intermediate Value Theorem implies that there is a c in $[-2, -\frac{1}{2}]$ such that $f(c) = 1$, that is, $c^2 + (1/c) = 1$.

37. Let $f(x) = x^3 + x^2 + x - 2$. Then f is continuous. Since $f(-1) = -3$ and $f(1) = 1$, the Intermediate Value Theorem implies that there is a c in $[-1,1]$ such that $f(c) = 0$, that is, $c^3 + c^2 + c - 2 = 0$.

39. Let $f(x) = \cos x - x$. Then f is continuous. Since $f(0) = 1$ and $f(\pi/2) = -\pi/2$, the Intermediate Value Theorem implies that there is a c in $[0, \pi/2]$ such that $f(c) = 0$, that is, $\cos c = c$.

41. Let $f(x) = e^{-x} - x$. Then f is continuous. Since $f(0) = 1$ and $f(1) = e^{-1} - 1 < 0$, the Intermediate Value Theorem implies that there is a c in $(0,1)$ such that $e^{-c} - c = 0$, that is, $e^{-c} = c$.

43. Let $f(x) = (2-x)^2(40-8x)$. Then $f(x) = 8(2-x)^2(5-x) = 0$ if $x = 2$ or 5.

Interval	c	$f(c)$	Sign of $f(x)$ on interval
$(-\infty, 2)$	0	160	+
$(2, 5)$	3	16	+
$(5, \infty)$	6	-128	$-$

Therefore $(2-x)^2(40-8x) > 0$ on the union of $(-\infty, 2)$ and $(2, 5)$.

45. Let $f(x) = (x^4+x)(x+3)$. Then $f(x) = x(x^3+1)(x+3) = 0$ if $x = -3, -1$, or 0.

Interval	c	$f(c)$	Sign of $f(x)$ on interval
$(-\infty, -3)$	-4	-252	$-$
$(-3, -1)$	-2	14	+
$(-1, 0)$	$-\frac{1}{2}$	$-\frac{35}{32}$	$-$
$(0, \infty)$	1	8	+

Therefore $(x^4+x)(x+3) \leq 0$ on the union of $(-\infty, -3]$ and $[-1, 0]$.

47. Let $f(x) = (x-1)(x-3)/[(2x+1)(2x-1)]$. Then $(2x+1)(2x-1) = 0$ if $x = -\frac{1}{2}$ or $\frac{1}{2}$, so the domain of f is the union of $(-\infty, -\frac{1}{2})$, $(-\frac{1}{2}, \frac{1}{2})$, and $(\frac{1}{2}, \infty)$. Next, $f(x) = 0$ if $x = 1$ or 3.

Interval	c	$f(c)$	Sign of $f(x)$ on interval
$(-\infty, -\frac{1}{2})$	-1	$\frac{8}{3}$	+
$(-\frac{1}{2}, \frac{1}{2})$	0	-3	$-$
$(\frac{1}{2}, 1)$	$\frac{3}{4}$	$\frac{9}{20}$	+
$(1, 3)$	2	$-\frac{1}{15}$	$-$
$(3, \infty)$	4	$\frac{1}{21}$	+

Therefore $(x-1)(x-3)/[(2x+1)(2x-1)] \geq 0$ on the union of $(-\infty, -\frac{1}{2})$, $(\frac{1}{2}, 1]$ and $[3, \infty)$.

49. Let $f(x) = -2(x-1)(x^2+2x+4)/(27-x^3)$. Then $27 - x^3 = 0$ if $x = 3$, so the domain of f is the union of $(-\infty, 3)$ and $(3, \infty)$. Next, since $x^2+2x+4 = (x+1)^2+3 > 0$ for all x, $f(x) = 0$ if $x = 1$.

Interval	c	$f(c)$	Sign of $f(x)$ on interval
$(-\infty, 1)$	0	$\frac{8}{27}$	+
$(1, 3)$	2	$-\frac{24}{19}$	$-$
$(3, \infty)$	4	$\frac{168}{37}$	+

Therefore $\dfrac{-2(x-1)(x^2+2x+4)}{27-x^3} < 0$ on $(1, 3)$.

51. Since $f(1) < 0$ and $f(2) > 0$, we let $a = 1$ and $b = 2$, and assemble the following table:

Interval	Length	Midpoint c	$f(c)$
$[1, 2]$	1	$\frac{3}{2}$	$\frac{1}{4}$
$[1, \frac{3}{2}]$	$\frac{1}{2}$	$\frac{5}{4}$	$-\frac{7}{16}$
$[\frac{5}{4}, \frac{3}{2}]$	$\frac{1}{4}$	$\frac{11}{8}$	$-\frac{7}{64}$
$[\frac{11}{8}, \frac{3}{2}]$	$\frac{1}{8}$		

Since the length of $[\frac{11}{8}, \frac{3}{2}]$ is $\frac{1}{8}$, and neither $\frac{11}{8}$ nor $\frac{3}{2}$ is a zero of f, the midpoint $\frac{23}{16}$ of $[\frac{11}{8}, \frac{3}{2}]$ is less than $\frac{1}{16}$ from a zero of f.

53. Since $f(1) < 0$ and $f(2) > 0$, we let $a = 1$ and $b = 2$, and assemble the following table:

Interval	Length	Midpoint c	$f(c)$
$[1, 2]$	1	$\frac{3}{2}$	$\frac{3}{8}$
$[1, \frac{3}{2}]$	$\frac{1}{2}$	$\frac{5}{4}$	$-\frac{67}{64}$
$[\frac{5}{4}, \frac{3}{2}]$	$\frac{1}{4}$	$\frac{11}{8}$	$-\frac{205}{512}$
$[\frac{11}{8}, \frac{3}{2}]$	$\frac{1}{8}$		

Since the length of $[\frac{11}{8}, \frac{3}{2}]$ is $\frac{1}{8}$,and neither $\frac{11}{8}$ nor $\frac{3}{2}$ is a zero of f; the midpoint $\frac{23}{16}$ of $[\frac{11}{8}, \frac{3}{2}]$ is less than $\frac{1}{16}$ from a zero of f.

55. Let $f(x) = x^2 - 5$, so that $\sqrt{5}$ is the zero of f. Since $f(2) < 0$ and $f(3) > 0$, the zero lies in $(2, 3)$. We let $a = 2$ and $b = 3$, and assemble the following table:

Interval	Length	Midpoint c	$f(c)$
$[2, 3]$	1	$\frac{5}{2}$	$\frac{5}{4}$
$[2, \frac{5}{2}]$	$\frac{1}{2}$	$\frac{9}{4}$	$\frac{1}{16}$
$[2, \frac{9}{4}]$	$\frac{1}{4}$	$\frac{17}{8}$	$-\frac{31}{64}$
$[\frac{17}{8}, \frac{9}{4}]$	$\frac{1}{8}$		

Since the length of $[\frac{17}{8}, \frac{9}{4}]$ is $\frac{1}{8}$, and neither $\frac{17}{8}$ nor $\frac{9}{4}$ is a zero of f, the midpoint $\frac{35}{16}$ of $[\frac{17}{8}, \frac{9}{4}]$ is less than $\frac{1}{16}$ from $\sqrt{5}$.

57. Let $f(x) = x^2 - 0.7$, so that $\sqrt{0.7}$ is the zero of f. Since $f(0) < 0$ and $f(1) > 0$, the zero lies in $(0, 1)$. We let $a = 0$ and $b = 1$, and assemble the following table:

Interval	Length	Midpoint c	$f(c)$
$[0, 1]$	1	$\frac{1}{2}$	$-\frac{9}{20}$
$[\frac{1}{2}, 1]$	$\frac{1}{2}$	$\frac{3}{4}$	$-\frac{11}{80}$
$[\frac{3}{4}, 1]$	$\frac{1}{4}$	$\frac{7}{8}$	$\frac{21}{320}$
$[\frac{3}{4}, \frac{7}{8}]$	$\frac{1}{8}$		

Since the length of $[\frac{3}{4}, \frac{7}{8}]$ is $\frac{1}{8}$, and neither $\frac{3}{4}$ nor $\frac{7}{8}$ is a zero of f, the midpoint $\frac{13}{16}$ of $[\frac{3}{4}, \frac{7}{8}]$ is less than $\frac{1}{16}$ from $\sqrt{0.7}$.

59. .738...

61. .567...

63. Let p be any number, and let $f(x) = x^3$. If $p = 0$ and $c = 0$, then $c^3 = p$. If $p > 0$, then $f(0) = 0$, whereas $f(p+1) = (p+1)^3 \geq p+1 > p > 0$: The Intermediate Value Theorem implies that there exists a number c in $[0, p+1]$ such that $c^3 = f(c) = p$. If $p < 0$, then $f(0) = 0$ and $f(p-1) = (p-1)^3 < p-1 < p < 0$. The Intermediate Value Theorem implies that there is a number c in $[p-1, 0]$ such that $c^3 = f(c) = p$. In any case, c is the desired cube root of p.

65. Let $A(0) = \pi x^2$. Notice that A is continuous, with $A(0) = 0$ and $A(10) = 100\pi > 300$. By the Intermediate Value Theorem there is a c in $[0, 10]$ such that $A(c) = 200$.

67. Let n be a given positive integer, and let $f(x) = x^n + 1 - x$. Then $c = c^n + 1$ if and only if $f(c) = 0$. Thus we only need to discover those positive integers n for which there is a number c such that $f(c) = 0$, that is, f has a zero. If $n = 1$, then $f(x) = x + 1 - x = 1 > 0$, so f does not have a zero. Next, if n is an odd integer greater than 1, then $f(0) = 1 > 0$ and $f(-2) = (-2)^n + 3 < -8 + 3 < 0$. Since f is continuous on $(-\infty, \infty)$ and has both positive and negative values, the Intermediate Value Theorem implies that f has a zero (even in the interval $(-2, 0)$). Finally, suppose that n is an even positive number. On the one hand, if $x < 1$, then $f(x) = x^n + 1 - x > 0$ since $x^n \geq 0$. On the other hand, if $x \geq 1$, then $x^n \geq x$, so that $f(x) = x^n + 1 - x \geq 1 > 0$. Consequently if n is a positive even integer, then f has no zero. We conclude that the only positive integers n for which there is a number c such that $c = c^n + 1$ are the odd integers greater than 1.

69. a. Continuous b. Continuous c. Has discontinuities d. Continuous e. Has discontinuities

71. F is a polynomial function, so F is continuous at every point in its domain.

73. a. From the diagram, the radius R of the cylinder satisfies

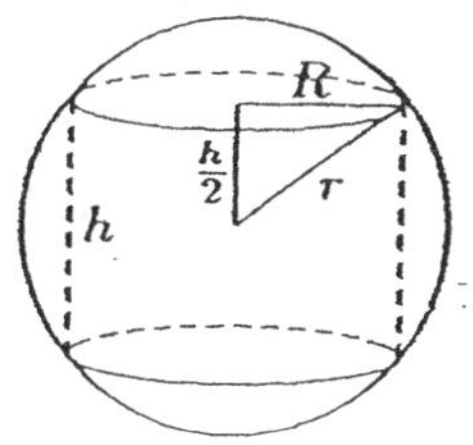

$$R^2 = r^2 - \left(\frac{h}{2}\right)^2 = r^2 - \frac{1}{4}h^2.$$

Therefore the volume $V(h)$ is given by

$$V(h) = \pi R^2 h = \pi\left(r^2 - \frac{1}{4}h^2\right)h = \pi r^2 h - \frac{\pi}{4}h^3.$$

Since the height of the cylinder cannot exceed the diameter $2r$ of the sphere, the domain of V is $[0, 2r]$.

b. Since V is a polynomial function, it is continuous on $[0, 2r]$.

75. Since 2θ and the sine function are continuous on $(-\infty, \infty)$, so is $\sin 2\theta$ by Theorem 2.10 Therefore $(v_0^2/g)\sin 2\theta$ is continuous on $(-\infty, \infty)$ by Theorem 2.9. It follows that R is continuous on $[0, \pi/2]$.

77. Since G, M, and R are constants, the function GMr/R^3 is a polynomial in r and the function GM/r^2 is a rational function in r. Therefore F is continuous on $(0, R)$ and on (R, ∞). To investigate continuity at R, we examine the one-sided limits of F at R:

$$\lim_{r\to R^-} F(r) = \lim_{r\to R^-} \frac{GMr}{R^3} = \lim_{r\to R} \frac{GMr}{R^3} = \frac{GMR}{R^3} = \frac{GM}{R^2};$$

$$\lim_{r\to R^+} F(r) = \lim_{r\to R^+} \frac{GM}{r^2} = \lim_{r\to R} \frac{GM}{r^2} = \frac{GM}{R^2}.$$

Since the two one-sided limits are equal, we have $\lim_{r\to R} F(r) = GM/R^2 = F(R)$. Therefore F is continuous at R. Thus F is continuous on $(0,\infty)$.

Chapter 2 Review

1. By (5) of Section 2.3, $\lim\limits_{x\to 2} \dfrac{-4x+3}{x^2-1} = \dfrac{-4(2)+3}{2^2-1} = -\dfrac{5}{3}$.

3. By the Product Rule, $\lim_{x\to 1} x\sqrt{x+3} = (\lim_{x\to 1} x)(\lim_{x\to 1} \sqrt{x+3}) = \lim_{x\to 1} \sqrt{x+3}$. Let $y = x+3$, so that y approaches 4 as x approaches 1. By the Substitution Rule, $\lim_{x\to 1} \sqrt{x+3} = \lim_{y\to 4} \sqrt{y} = \sqrt{4} = 2$. Thus $\lim_{x\to 1} x\sqrt{x+3} = \lim_{x\to 1} \sqrt{x+3} = 2$.

5. Let $y = x + 3\pi/4$. Since y approaches π as x approaches $\pi/4$, the Substitution Rule implies that $\lim_{x\to\pi/4} \sec^3(x+3\pi/4) = \lim_{y\to\pi} \sec^3 y = \sec^3 \pi = (-1)^3 = -1$.

7. Let $y = 6v$. Since y approaches 0 as v approaches 0, it follows from the Substitution Rule, the Constant Multiple Rule, and Example 6 of Section 2.3 that

$$\lim_{v\to 0} \frac{5\sin 6v}{4v} = \lim_{y\to 0} \frac{5\sin y}{4(y/6)} = \frac{15}{2} \lim_{y\to 0} \frac{\sin y}{y} = \frac{15}{2}.$$

9. Using (9) of Section 2.3, along with the Quotient, Sum, and Difference Rules, we have

$$\lim_{x\to 0} \frac{1-e^x}{1+e^x} = \frac{1-e^0}{1+e^0} = \frac{1-1}{1+1} = 0.$$

11. If $x \neq -10$, then $(x+10)^2 > 0$ and $\lim_{x\to -10}(x+10)^2 = 0$. Thus

$$\lim_{x\to -10} \frac{4x}{(x+10)^2} = -\infty.$$

13. If $x < 0$, then $-x > 0$, so $\sqrt{-x} > 0$ and thus $9 + \sqrt{-x} > 9$, so $\sqrt{9+\sqrt{-x}} > 0$. Using the Substitution Rule twice, we find that

$$\lim_{x\to 0^-} \sqrt{9+\sqrt{-x}} = \lim_{y\to 0^+} \sqrt{9+y} = \lim_{z\to 9^+} \sqrt{z} = \sqrt{9} = 3.$$

15. First notice that

$$\frac{\sqrt{2+3w}-\sqrt{2-3w}}{w} = \frac{\sqrt{2+3w}-\sqrt{2-3w}}{w}\,\frac{\sqrt{2+3w}+\sqrt{2-3w}}{\sqrt{2+3w}+\sqrt{2-3w}}$$
$$= \frac{(2+3w)-(2-3w)}{w(\sqrt{2+3w}+\sqrt{2-3w})} = \frac{6}{\sqrt{2+3w}+\sqrt{2-3w}}.$$

By the Substitution Rule, $\lim_{w\to 0}\sqrt{2+3w} = \lim_{y\to 2}\sqrt{y} = \sqrt{2}$ and $\lim_{w\to 0}\sqrt{2-3w} = \lim_{y\to 2}\sqrt{y} = \sqrt{2}$, so it follows that

$$\lim_{w\to 0}\frac{\sqrt{2+3w}-\sqrt{2-3w}}{w} = \lim_{w\to 0}\frac{6}{\sqrt{2+3w}+\sqrt{2-3w}} = \frac{6}{\sqrt{2}+\sqrt{2}} = \frac{3}{2}\sqrt{2}.$$

17. Let $f(x) = \cos(1/w)$. If $w = 1/2n\pi$ for any positive integer n, then $0 < w < 1/n$ and $f(w) = 1$. If $w = 1/(2n\pi + \pi)$ for any positive integer n, then $0 < w < 1/n$ and $f(w) = -1$. Thus $\lim_{w\to 0^+} f(w)$ does not exist.

it follows that $\lim_{x\to 0^+} e^x/\ln x = 0$.

21. Since $\lim_{x\to 4^+}(3-2|x|)/(x-4) = -\infty$, $x = 4$ is a vertical asymptote.

23. We have
$$\frac{x^2+3x-4}{x^2-5x-14} = \frac{(x+4)(x-1)}{(x-7)(x+2)}.$$
Since $\lim_{x\to 7^+} f(x) = \infty = \lim_{x\to -2^+} f(x)$, $x = -2$ and $x = 7$ are vertical asymptotes.

25. Since
$$\lim_{x\to -2^+} f(x) = \lim_{x\to -2^+}\frac{(x-2)^2}{(x-2)(x+2)} = \lim_{x\to -2^+}\frac{x-2}{x+2} = -\infty$$
$x = -2$ is a vertical asymptote. (Since $\lim_{x\to 2} f(x) = \lim_{x\to 2}(x-2)/(x+2) = 0$, $x = 2$ is *not* a vertical asymptote.)

27. Notice that $-1/x > 0$ for $x < 0$. Thus $\lim_{x\to 0^-}(-1/x) = \infty$ and hence $\lim_{x\to 0^-} e^{-1/x} = \infty$. Therefore $x = 0$ is a vertical asymptote.

29. Since $\lim_{x\to 0^+}\ln x = -\infty$ and $\lim_{x\to 0^+} e^x = 1$, it follows that $\lim_{x\to 0^+}(\ln x)/e^x = -\infty$. Therefore $x = 0$ is a vertical asymptote.

31. $f(x)$ is defined for all x in $(-\infty, -\sqrt{13}]$ and for all x in $[\sqrt{13}, \infty)$. Thus f is continuous from the left at $-\sqrt{13}$.

33. Since $\lim_{x\to 2^+} f(x) = \lim_{x\to 2^+}(3x+4) = 10 = f(2)$ but $\lim_{x\to 2^-} f(x) = \lim_{x\to 2^-}(3x-4) = 2 \neq f(2)$, f is continuous from the right at 2 but is not continuous at 2.

35. Since f is a rational function whose denominator is not zero for x in the interval $(2,\infty)$, f is continuous on $(2,\infty)$.

37. If x is in $[\sqrt{2}/3,4)$, then $3x-\sqrt{2}\geq 0$ and $4-x>0$. Thus for such x, $(3x-\sqrt{2})/(4-x)$ is nonnegative, so on $[\sqrt{2}/3,4)$, the rational function $(3x-\sqrt{2})/(4-x)$ is continuous. Since the square root function is continuous for x in $[0,\infty)$, Theorem 2.10 implies that f is continuous on $[\sqrt{2}/3,4)$.

39. Since x^2 and e^x are continuous on $(-\infty,\infty)$, Theorem 2.10 implies that $e^{(x^2)}$ is continuous on $(-\infty,\infty)$. Since $e^{(x^2)}>0$ for all x and since the square root function is continuous on $(0,\infty)$, it follows from Theorem 2.10 that f is continuous on $(-\infty,\infty)$.

41. Since $\lim_{x\to 2} f(x)=\lim_{x\to 2}[(x+8)(x-2)]/(x-2)=\lim_{x\to 2}(x+8)=10$, we can define $f(2)=10$ to make f continuous at 2.

43. a. Since $\lim_{x\to 0^+} f(x)$ exists and is positive, and since $\lim_{x\to 0^+} g(x)=\infty$, it follows that $\lim_{x\to 0^+} f(x)/g(x)=0$.

    b. Since $\lim_{x\to a} f(x)=f(a)=g(a)=\lim_{x\to a} g(x)\neq 0$, it follows that $\lim_{x\to a} f(x)/g(x)=f(a)/g(a)=1$.

    c. Since $g(x)>0$ for $x<b$, $\lim_{x\to b^-} g(x)=0$, and $\lim_{x\to b^-} f(x)=f(b)>0$, it follows that $\lim_{x\to b^-} f(x)/g(x)=\infty$.

    d. Since $g(x)<0$ for $x>b$, $\lim_{x\to b^+} g(x)=0$, and $\lim_{x\to b^+} f(x)=f(b)>0$, it follows that $\lim_{x\to b^+} f(x)/g(x)=-\infty$.

45. Let $f(x)=(2x-3)(4-x)/(1+x)^2$. Then $(1+x)^2=0$ if $x=-1$, so the domain of f is the union of $(-\infty,-1)$ and $(-1,\infty)$. Next, $(2x-3)(4-x)=0$ if $x=\frac{3}{2}$ or 4.

Interval	c	$f(c)$	Sign of $f(x)$ on interval
$(-\infty,-1)$	-2	-42	$-$
$(-1,\frac{3}{2})$	0	-12	$-$
$(\frac{3}{2},4)$	2	$\frac{2}{9}$	$+$
$(4,\infty)$	5	$-\frac{7}{36}$	$-$

Therefore $(2x-3)(4-x)/(1+x)^2<0$ on the union of $(-\infty,-1)$, $(-1,\frac{3}{2})$ and $(4,\infty)$.

47. Let $f(x)=x^3+3x-3$. Since f is continuous on $[-1,1]$, $f(-1)<0$ and $f(1)>0$, the Intermediate Value Theorem implies that there is a c in $[-1,1]$ such that $f(c)=0$, that is, a solution of the given equation. For the bisection method we let $a=-1$ and $b=1$, and assemble the following table:

Interval	Length	Midpoint c	$f(c)$
$[-1,1]$	2	0	-3
$[0,1]$	1	$\frac{1}{2}$	$-\frac{11}{8}$
$[\frac{1}{2},1]$	$\frac{1}{2}$	$\frac{3}{4}$	$-\frac{21}{64}$
$[\frac{3}{4},1]$	$\frac{1}{4}$	$\frac{7}{8}$	$\frac{151}{512}$
$[\frac{3}{4},\frac{7}{8}]$	$\frac{1}{8}$		

Since the length of $[\frac{3}{4},\frac{7}{8}]$ is $\frac{1}{8}$, and neither $\frac{3}{4}$ nor $\frac{7}{8}$ is a zero of f, the midpoint $\frac{13}{16}$ of $[\frac{3}{4},\frac{7}{8}]$ is less than $\frac{1}{16}$ from a zero of f, and hence from a solution of the equation $x^3+3x-3=0$.

49. Let $f(x) = x + \sin x - \cos x$. Then the zeros of f are the solutions of the given equation. By plotting f and zooming in, we find that the only solution of the equation is approximately .456.

51. $$\lim_{x\to 0}\frac{e^{-ax}-e^{-bx}}{x} = \lim_{x\to 0}\frac{(e^{-ax}-1)-(e^{-bx}-1)}{x} = \lim_{x\to 0}\frac{e^{-ax}-1}{x} - \lim_{x\to 0}\frac{e^{-bx}-1}{x},$$

provided each of these limits exists. Let $y=-ax$. Then by the Substitution Rule, and (8) of Section 2.1,

$$\lim_{x\to 0}\frac{e^{-ax}-1}{x} = \lim_{y\to 0}\frac{e^y-1}{y/(-a)} = -a\lim_{y\to 0}\frac{e^y-1}{y} = (-a)1 = -a.$$

Similarly

$$\lim_{x\to 0}\frac{e^{-bx}-1}{x} = -b.$$

Consequently

$$\lim_{x\to 0}\frac{e^{-ax}-e^{-bx}}{x} = (-a)-(-b) = b-a.$$

53. It does not include the phrase “provided the limits of the two functions both exist.”

55. a. **Since the absolute value function is continuous, the result follows from Theorem 2.10since $|f|$ is a composite of f and the absolute value function.**

b. $|f(x)| = 1$ for all x, so $|f|$ is continuous at 2. Since $\lim_{x\to 2^+} f(x) = 1 \neq f(2)$, f is not continuous at 2.

57. a. $$m_1 = \lim_{x\to 0}\frac{f(x)-f(0)}{x-0} = \lim_{x\to 0}\frac{2\sin x - 0}{x-0} = 2$$

$$m_2 = \lim_{x\to 1}\frac{g(x)-g(0)}{x-1} = \lim_{x\to 1}\frac{(-1/x^2)-(-1)}{x-1} = \lim_{x\to 1}\frac{x^2-1}{x^2(x-1)} = \lim_{x\to 1}\frac{x+1}{x^2} = 2$$

Since $m_1 = m_2$, the tangent lines are parallel.

b. $$m_1 = \lim_{x\to -1}\frac{f(x)-f(-1)}{x-(-1)} = \lim_{x\to -1}\frac{x^3-(-1)}{x+1} = \lim_{x\to -1}(x^2-x+1) = 3$$

$$m_2 = \lim_{x\to 4}\frac{g(x)-g(4)}{x-4} = \lim_{x\to 4}\frac{(3/8)x^2-6}{x-4} = \lim_{x\to 4}(3/8)(x+4) = 3$$

Since $m_1 = m_2$, the tangent lines are parallel.

c. $m_1 = \lim_{x\to-4} \frac{f(x)-f(-4)}{x-(-4)} = \lim_{x\to-4} \frac{(x^2+5x+1)-(-3)}{x+4} = \lim_{x\to-4} \frac{x^2+5x+4}{x+4} = \lim_{x\to-4}(x+1) = -3$

$m_2 = \lim_{x\to3} \frac{g(x)-g(3)}{x-3} = \lim_{x\to3} \frac{(-3x-7)-(-16)}{x-3} = \lim_{x\to3} \frac{-3x+9}{x-3} = -3$

Since $m_1 = m_2$, the tangent lines are parallel.

59. First we find the point of intersection of the line and the graph of f:

$$\sqrt{x} = \frac{x+4}{4}$$
$$4\sqrt{x} = x+4$$
$$16x = x^2+8x+16$$
$$x^2-8x+16 = 0$$
$$(x-4)^2 = 0, \text{ or } x = 4$$

Since

$$\lim_{x\to4} \frac{f(x)-f(4)}{x-4} = \lim_{x\to4} \frac{\sqrt{x}-2}{x-4} = \lim_{x\to4} \frac{1}{\sqrt{x}+2} = \frac{1}{4}$$

the tangent line at $(4,2)$ is $y-2 = \frac{1}{4}(x-4)$, or $4y = x+4$. The point of tangency is $(4,2)$.

61. a. By (15) of Section 2.2, the height of the drip above the basin is given by $h(t) = -4.9t^2 + h_0$, where h_0 is the height of the faucet above the basin. Since $h(\frac{1}{4}) = 0$, we have $-4.9(\frac{1}{4})^2 + h_0 = 0$, so that $h_0 = \frac{4.9}{16} = .30625$ meters. Thus the faucet is approximately .3 meters (or approximately 1 foot) high.

b. As the drip hits the basin, its velocity is

$$\lim_{t\to1/4^-} \frac{h(t)-h(1/4)}{t-1/4} = \lim_{t\to1/4^-} \frac{(-4.9t^2+h_0)-0}{t-1/4} = \lim_{t\to1/4} \frac{-4.9t^2+4.9/16}{t-1/4}$$

$$= \lim_{t\to1/4} \frac{-4.9(t+1/4)(t-1/4)}{t-1/4} = \lim_{t\to1/4}[-4.9(t+1/4)] = -4.9(1/4+1/4) = -2.45.$$

Thus the drip hits the basin with a speed of $|-2.45| = 2.45$ meters per second.

Chapter 3

Derivatives

3.1 The Derivative

1. $f'(4) = \lim_{x\to4} \dfrac{f(x) - f(4)}{x - 4} = \lim_{x\to4} \dfrac{5-5}{x-4} = \lim_{x\to4} 0 = 0$

3. $f'(1) = \lim_{x\to1} \dfrac{f(x) - f(1)}{x - 1} = \lim_{x\to1} \dfrac{(2x+3) - 5}{x-1} = \lim_{x\to1} \dfrac{2x-2}{x-1} = \lim_{x\to1} 2 = 2$

5. $f'(0) = \lim_{x\to0} \dfrac{f(x) - f(0)}{x - 0} = \lim_{x\to0} \dfrac{x^3 - 0}{x} = \lim_{x\to0} x^2 = 0$

7. $f'(1) = \lim_{x\to1} \dfrac{f(x) - f(1)}{x - 1} = \lim_{x\to1} \dfrac{(2x - 3/x) - (-1)}{x-1} = \lim_{x\to1} \dfrac{2x^2 + x - 3}{x(x-1)}$

$= \lim_{x\to1} \dfrac{(2x+3)(x-1)}{x(x-1)} = \lim_{x\to1} \dfrac{2x+3}{x} = 5$

9. $f'(4) = \lim_{x\to4} \dfrac{f(x) - f(4)}{x - 4} = \lim_{x\to4} \dfrac{(1/\sqrt{x} - (1/2)}{x-4} = \lim_{x\to4} \dfrac{2-\sqrt{x}}{2\sqrt{x}\,(\sqrt{x}-2)(\sqrt{x}+2)}$

$= \lim_{x\to4} \dfrac{-1}{2\sqrt{x}\,(\sqrt{x}+2)} = -\dfrac{1}{16}$

11. $\lim_{x\to2^-} \dfrac{f(x) - f(2)}{x - 2} = \lim_{x\to2^-} \dfrac{x^2-4}{x-2} = \lim_{x\to2^-} \dfrac{(x-2)(x+2)}{x-2} = \lim_{x\to2^-} (x+2) = 4;$

$\lim_{x\to2^+} \dfrac{f(x) - f(2)}{x - 2} = \lim_{x\to2^+} \dfrac{(4x-4)-4}{x-2} = \lim_{x\to2^+} 4 = 4.$ Thus $f'(2) = \lim_{x\to2} \dfrac{f(x) - f(2)}{x-2} = 4.$

13. $f'(x) = \lim_{t\to x} \dfrac{f(t) - f(x)}{t - x} = \lim_{t\to x} \dfrac{-\pi - (-\pi)}{t-x} = \lim_{t\to x} 0 = 0$

15. $f'(x) = \lim_{t\to x} \dfrac{f(t) - f(x)}{t - x} = \lim_{t\to x} \dfrac{-5t^2 - (-5x^2)}{t-x} = \lim_{t\to x} \dfrac{-5(t-x)(t+x)}{t-x}$

$= \lim_{t\to x} -5(t+x) = -5(2x) = -10x$

17. $g'(x) = \lim_{t\to x} \dfrac{g(t) - g(x)}{t - x} = \lim_{t\to x} \dfrac{t^3 - x^3}{t - x} = \lim_{t\to x} \dfrac{(t - x)(t^2 + tx + t^2)}{t - x}$

$= \lim_{t\to x}(t^2 + tx + x^2) = (x^2 + x^2 + x^2) = 3x^2$

19. $k'(x) = \lim_{t\to x} \dfrac{k(t) - k(x)}{t - x} = \lim_{t\to x} \dfrac{[(1/t^2) - \sqrt{7}] - [(1/x^2) - \sqrt{7}]}{t - x} = \lim_{t\to x} \dfrac{x^2 - t^2}{t^2x^2(t - x)}$

$= \lim_{t\to x} \dfrac{(x - t)(x + t)}{t^2x^2(t - x)} = \lim_{t\to x} \dfrac{-(t + x)}{t^2x^2} = \dfrac{-2x}{x^4} = -\dfrac{2}{x^3}$

21. $\dfrac{dy}{dx} = \lim_{t\to x} \dfrac{\frac{7}{3} - \frac{7}{3}}{t - x} = \lim_{t\to x} 0 = 0$

23. $\dfrac{dy}{dx} = \lim_{t\to x} \dfrac{(3t^2 + 1) - (3x^2 + 1)}{t - x} = \lim_{t\to x} \dfrac{3(t - x)(t + x)}{t - x} = \lim_{t\to x} 3(t + x) = 3(2x) = 6x$

25. $\left.\dfrac{dy}{dx}\right|_{x=2} = \lim_{x\to 2} \dfrac{0.25 - 0.25}{x - 2} = \lim_{x\to 2} 0 = 0$

27. $\left.\dfrac{dy}{dx}\right|_{x=2} = \lim_{x\to 2} \dfrac{(x^2 - 3) - 1}{x - 2} = \lim_{x\to 2} \dfrac{(x - 2)(x + 2)}{x - 2} = \lim_{x\to 2}(x + 2) = 4$

29. $\lim_{x\to 0} \dfrac{f(x) - f(0)}{x - 0} = \lim_{x\to 0} \dfrac{x^{1/3} - 0}{x - 0} = \lim_{x\to 0} \dfrac{1}{x^{2/3}} = \infty$; no derivative at 0.

31. $\lim_{x\to 0^-} \dfrac{f(x) - f(0)}{x - 0} = \lim_{x\to 0^-} \dfrac{(-x - x) - 0}{x - 0} = \lim_{x\to 0^-} (-2) = -2$;

$\lim_{x\to 0^+} \dfrac{f(x) - f(0)}{x - 0} = \lim_{x\to 0^+} \dfrac{(x - x) - 0}{x - 0} = \lim_{x\to 0^+} 0 = 0$; no derivative at 0.

33. $\lim_{x\to 3} \dfrac{g(x) - g(3)}{x - 3} = \lim_{x\to 3} \dfrac{(x + 3) - 6}{x - 3} = \lim_{x\to 3} 1 = 1$; $g'(3) = 1$.

35. $\lim_{x\to 0^-} \dfrac{k(x) - k(0)}{x - 0} = \lim_{x\to 0^-} \dfrac{(-x^2 + 4x) - (-1)}{x - 0} = \lim_{x\to 0^-} \left(-x + 4 + \dfrac{1}{x}\right) = -\infty$; no derivative at 0.

37. $f'(-2) = 2(-2) = -4$ from Example 4. Thus l: $y - 4 = -4(x - (-2))$, or $y = -4x - 4$.

39. $f'(4) = \frac{1}{2} \cdot 4^{-1/2} = \frac{1}{4}$ from Example 5. Thus l: $y - 2 = \frac{1}{4}(x - 4)$, or $y = \frac{1}{4}x + 1$.

41. It appears that $f'(0) \approx -2$, $f'(1) \approx -1.5$, $f'(2) \approx -.5$ and $f'(3) \approx .7$.

43. f is decreasing on $[-10, 2]$ and is increasing on $[2, 10]$.

45. f is decreasing on $[-10, -1]$ and on $[0, 2]$, and is increasing on $[-1, 0]$ and on $[2, 10]$.

47. f is increasing on $[-10, 10]$.

49. $m_2 = \dfrac{f(2.1) - f(2)}{0.1} = \dfrac{(2.1)^2 - 4}{0.1} = \dfrac{0.41}{0.1} = 4.1;$

$$f'(2) = \lim_{x\to 2} \frac{f(x) - f(2)}{x - 2} = \lim_{x\to 2} \frac{x^2 - 4}{x - 2} = \lim_{x\to 2} \frac{(x-2)(x+2)}{x-2} = \lim_{x\to 2}(x+2) = 4;$$

$|f'(2) - m_2| = |4 - 4.1| = 0.1$

51. It appears that $f'(x) = x$.

53. It appears that $f'(x) = -\sin x$.

55. $f'(-2) \approx \dfrac{1.9536874 - 2.0457932}{-2.006579 - (-1.993421)} \approx 6.9999848$ (Thus we conjecture that $f'(-2) = 7$.)

57. $f'(\frac{1}{2}) \approx \dfrac{1.6514352 - 1.6460118}{.50164474 - .49835526} \approx 1.648710434$

59. a. $f(x) = x^4$, $a = 2$;

$$f'(2) = \lim_{x\to 2} \frac{x^4 - 16}{x - 2} = \lim_{x\to 2} \frac{(x-2)(x+2)(x^2+4)}{x-2} = \lim_{x\to 2}(x+2)(x^2+4) = 32$$

b. $x^3 + 2x^2 + 4x + 8 = \dfrac{(x^3 + 2x^2 + 4x + 8)(x-2)}{x-2} = \dfrac{x^4 - 16}{x - 2}$; $f(x) = x^4$, $a = 2$; $f'(2) = 32$

61. By Exercise 60, $f'(a) = \lim_{h\to 0} \dfrac{f(a+h) - f(a)}{h}$. Using the Substitution Theorem with $y = -h$, we rewrite the last limit as $\lim_{y\to 0} \dfrac{f(a-y) - f(a)}{-y}$, which equals $-\lim_{y\to 0} \dfrac{f(a-y) - f(a)}{y}$. Replacing y by h, we conclude that $f'(a) = -\lim_{h\to 0} \dfrac{f(a-h) - f(a)}{h}$, or equivalently, $\lim_{h\to 0} \dfrac{f(a-h) - f(a)}{h} = -f'(a)$.

63. a. Since $f(-a) = f(a)$, we use the solution of Exercise 61 to deduce that

$$f'(-a) = \lim_{h\to 0} \frac{f(-a+h) - f(-a)}{h} = \lim_{h\to 0} \frac{f(a-h) - f(a)}{h} = -f'(a) = -2.$$

b. Since $f(-a) = -f(a)$, we use the solution of Exercise 61 to deduce that

$$f'(-a) = \lim_{h\to 0} \frac{f(-a+h) - f(-a)}{h} = \lim_{h\to 0} \frac{-f(a-h) + f(a)}{h} = -\lim_{h\to 0} \frac{f(a-h) - f(a)}{h} = f'(a) = 2.$$

65. No. Let $f(x) = 1$ and $g(x) = x$ for all x. Then $f(1) = 1 = g(1)$, but $f'(1) = 0$ and $g'(1) = 1$.

67. $\lim_{x\to a} g(x) = \lim_{x\to a} \dfrac{f(x) - f(a)}{x - a} = f'(a) = g(a)$. Thus g is continuous at a.

69. Although $|f(x)|$ is small when x is near 0, the graph oscillates with slopes ranging from -1 to 1 on any interval about 0. Thus there could not be any tangent line. Consequently $f'(0)$ cannot exist.

71. The velocity if given by

$$v(1/2) = \lim_{t\to 1/2} \frac{f(t) - f(1/2)}{t - 1/2} = \lim_{t\to 1/2} \frac{2t^2 - 2(1/2)^2}{t - 1/2}$$

$$= \lim_{t\to 1/2} \frac{2(t^2 - (1/2)^2)}{t - 1/2} = \lim_{t\to 1/2} \frac{2(t + 1/2)(t - 1/2)}{t - 1/2} = \lim_{t\to 1/2} 2(t + 1/2) = 2.$$

73. $$m_C(50) = C'(50) = \lim_{x\to 50} \frac{C(x) - C(50)}{x - 50} = \lim_{x\to 50} \frac{(10{,}000 + 3/x) - (10{,}000 + \frac{3}{50})}{x - 50}$$

$$= \lim_{x\to 50} \frac{3(50 - x)}{50x(x - 50)} = \lim_{x\to 50} \frac{-3}{50x} = -\frac{3}{2500}$$

75. a. $$m_C(1) = \lim_{x\to 1} \frac{C(x) - C(1)}{x - 1} = \lim_{x\to 1} \frac{(3 + 12x - 2x^2) - (3 + 12 - 2)}{x - 1}$$

$$= \lim_{x\to 1} \left[\frac{12(x - 1)}{x - 1} - \frac{2(x^2 - 1)}{x - 1}\right] = \lim_{x\to 1} [12 - 2(x + 1)]$$

$= 8$ (thousand dollars per thousand gallons)

b. $$m_C(a) = \lim_{x\to a} \frac{C(x) - C(a)}{x - a} = \lim_{x\to a} \frac{(3 + 12x - 2x^2) - (3 + 12a - 2a^2)}{x - a}$$

$$= \lim_{x\to a} \left[\frac{12(x - a)}{x - a} - \frac{2(x^2 - a^2)}{x - a}\right] = \lim_{x\to a} [12 - 2(x + a)] = 12 - 4a$$

For $m_C(a) = \frac{1}{2}C(1)$, we must have $12 - 4a = \frac{1}{2}(13)$, or $a = \frac{11}{8}$.

77. Since the cost has been increased by 1 thousand dollars, the cost function is given by

$$C(x) = 1 + (3 + 12x - 2x^2) = 4 + 12x - 2x^2 \quad \text{for } 0 \le x \le 3.$$

Thus

$$m_C(1) = \lim_{x\to 1} \frac{C(x) - C(1)}{x - 1} = \lim_{x\to 1} \frac{(4 + 12x - 2x^2) - (4 + 12 - 2)}{x - 1}$$

$$= \lim_{x\to 1} \left[\frac{12(x - 1) - 2(x^2 - 1)}{x - 1}\right] = \lim_{x\to 1} [12 - 2(x + 1)] = 8.$$

Thus the marginal cost has remained the same.

79. a. By hypothesis, after t minutes one boat is at $(2t, 0)$. The other boat is $2t$ units from the origin on the line $y = \sqrt{3}\,x$, so

$$2t = \sqrt{x^2 + y^2} = \sqrt{x^2 + 3x^2} = 2x;$$

thus $x = t$, so $y = \sqrt{3}\,t$, and the second boat is at $(t, \sqrt{3}\,t)$. The distance $D(t)$ between the boats t minutes after starting is given by

$$D(t) = \sqrt{(t - 2t)^2 + (\sqrt{3}t - 0)^2} = \sqrt{t^2 + 3t^2} = 2t.$$

Then $D'(t) = 2$, so the rate of the increase is 2 meters per minute.

b. Let the speed of the boats be b meters per minute. After t minutes one boat is at $(bt, 0)$. The other boat is bt units from the origin on the line $y = \sqrt{3}\,x$, and is traveling at b meters per minute. Thus

$$bt = \sqrt{x^2 + (\sqrt{3}\,x)^2} = 2x,$$

so that $x = bt/2$ and $y = \sqrt{3}\,bt/2$. Thus the second boat is at $(bt/2, \sqrt{3}\,bt/2)$. The distance $D(t)$ between the boats is given by

$$D(t) = \sqrt{(bt/2 - bt)^2 + (\sqrt{3}\,bt/2)^2} = bt\sqrt{\frac{1}{4} + \frac{3}{4}} = bt.$$

Since $D'(t) = b$, and since $D'(t) = 3$ by hypothesis, it follows that $b = 3$. Consequently the speed of the boats is 3 meters per minute.

3.2 Differentiable Functions

1. By Example 1, $f'(x) = 0$ for all x, so $f'(1) = 0$.

3. By (1), $f'(x) = 2x$ for all x, so $f'(3/2) = 3$ and $f'(0) = 0$.

5. By (1), $f'(x) = 4x^3$, so $f'(\sqrt[3]{2}) = 4(\sqrt[3]{2})^3 = 4 \cdot 2 = 8$.

7. By (1), $f'(x) = 10x^9$, so $f'(1) = 10$.

9. By (4), $f'(t) = -\sin t$, so $f'(0) = 0$ and $f'(-\pi/3) = \sqrt{3}/2$.

11. $f'(x) = \lim\limits_{t\to x} \dfrac{f(t) - f(x)}{t - x} = \lim\limits_{t\to x} \dfrac{(-2t - 1) - (-2x - 1)}{t - x} = \lim\limits_{t\to x}(-2) = -2$

13. $f'(x) = \lim\limits_{t\to x} \dfrac{f(t) - f(x)}{t - x} = \lim\limits_{t\to x} \dfrac{t^5 - x^5}{t - x} = \lim\limits_{t\to x} \dfrac{(t - x)(t^4 + t^3x + t^2x^2 + tx^3 + x^4)}{t - x} = 5x^4$

15. $f'(x) = \lim\limits_{t\to x} \dfrac{f(t) - f(x)}{t - x} = \lim\limits_{t\to x} \dfrac{\dfrac{t}{t+1} - \dfrac{x}{x+1}}{t - x} = \lim\limits_{t\to x} \dfrac{t - x}{(t+1)(x+1)(t-x)}$

$$= \lim_{t\to x} \frac{1}{(t+1)(x+1)} = \frac{1}{(x+1)^2}$$

17. $f'(x) = \lim\limits_{t\to x} \dfrac{f(t) - f(x)}{t - x} = \lim\limits_{t\to x} \dfrac{[(t^2 - 1)/(t^2 + 1)] - [(x^2 - 1)/(x^2 + 1)]}{t - x}$

$$= \lim_{t\to x} \frac{(t^2 - 1)(x^2 + 1) - (x^2 - 1)(t^2 + 1)}{(t^2 + 1)(x^2 + 1)(t - x)} = \lim_{t\to x} \frac{2(t - x)(t + x)}{(t^2 + 1)(x^2 + 1)(t - x)}$$

$$= \lim_{t\to x} \frac{2(t + x)}{(t^2 + 1)(x^2 + 1)} = \frac{4x}{(x^2 + 1)^2}$$

19. $\dfrac{dy}{dx} = \lim\limits_{h\to 0} \dfrac{-3\cos(x + h) - (-3\cos x)}{h} = -3 \lim\limits_{h\to 0} \dfrac{\cos(x + h) - \cos x}{h}$

$= (-3)(-\sin x) = 3\sin x$ by the discussion preceding (4).

21. If $x \neq 0$, then

$$\frac{dy}{dx} = \lim_{t\to x} \frac{t^{2/3} - x^{2/3}}{t - x} = \lim_{t\to x} \frac{(t^{1/3} - x^{1/3})(t^{1/3} + x^{1/3})}{(t^{1/3} - x^{1/3})(t^{2/3} + t^{1/3}x^{1/3} + x^{2/3})}$$

$$= \lim_{t\to x} \frac{t^{1/3} + x^{1/3}}{t^{2/3} + t^{1/3}x^{1/3} + x^{2/3}} = \frac{2x^{1/3}}{3x^{2/3}} = \frac{2}{3x^{1/3}} = \frac{2}{3}x^{-1/3}.$$

23. If $x > 1$, then

$$\frac{dy}{dx} = \lim_{t\to x} \frac{\sqrt{t-1} - \sqrt{x-1}}{t - x} = \lim_{t\to x} \frac{\sqrt{t-1} - \sqrt{x-1}}{t - x} \cdot \frac{\sqrt{t-1} + \sqrt{x-1}}{\sqrt{t-1} + \sqrt{x-1}}$$

$$= \lim_{t\to x} \frac{(t - 1) - (x - 1)}{(t - x)(\sqrt{t-1} + \sqrt{x-1})} = \lim_{t\to x} \frac{1}{\sqrt{t-1} + \sqrt{x-1}} = \frac{1}{2\sqrt{x-1}}.$$

25. $f'(x) = \lim_{h\to 0} \dfrac{f(x+h)-f(x)}{h} = \lim_{h\to 0} \dfrac{e^{2(x+h)} - e^{2x}}{h} = \lim_{h\to 0} \dfrac{e^{2x}e^{2h} - e^{2x}}{h} = e^{2x} \lim_{h\to 0} \dfrac{e^{2h}-1}{h}$

Let $y = 2h$. Then

$$\lim_{h\to 0} \frac{e^{2h}-1}{h} = 2 \lim_{y\to 0} \frac{e^y - 1}{y} = 2$$

by the Substitution Rule. Therefore $f'(x) = e^{2x}(2) = 2e^{2x}$.

27. For all x,

$$\lim_{t\to x} \frac{f(t)-f(x)}{t-x} = \lim_{t\to x} \frac{(t^2+t)-(x^2+x)}{t-x} = \lim_{t\to x} \frac{(t^2-x^2)+(t-x)}{t-x}$$

$$= \lim_{t\to x} \frac{(t-x)(t+x)+(t-x)}{t-x} = \lim_{t\to x}(t+x+1) = 2x+1.$$

Thus f is differentiable on $(-\infty, \infty)$.

29. For $x > 4$,

$$\lim_{t\to x} \frac{f(t)-f(x)}{t-x} = \lim_{t\to x} \frac{[1/(4-t)]-[1/(4-x)]}{t-x}$$

$$= \lim_{t\to x} \frac{(4-x)-(4-t)}{(4-t)(4-x)(t-x)} = \lim_{t\to x} \frac{1}{(4-t)(4-x)} = \frac{1}{(4-x)^2}.$$

Therefore f is differentiable on $(4, \infty)$.

31. For $x \geq 1$, $f(x) = x - 1$, and for $x > 1$,

$$\lim_{t\to x} \frac{f(t)-f(x)}{t-x} = \lim_{t\to x} \frac{(t-1)-(x-1)}{t-x} = \lim_{t\to x} 1 = 1.$$

Therefore f is differentiable on $(1, \infty)$. Also,

$$\lim_{t\to 1^+} \frac{f(t)-f(1)}{t-1} = \lim_{t\to 1^+} \frac{t-1-0}{t-1} = 1.$$

Thus f is differentiable on $[1, \infty)$.

33. a. $f'(x) = \lim_{t\to x} \dfrac{-2t^2+2x^2}{t-x} = \lim_{t\to x}[-2(t+x)] = -4x$

Since $f'(a) = 12$, we have $-4a = 12$, so $a = -3$.

b. $f'(x) = \lim_{t\to x} \dfrac{(3t+t^2)-(3x+x^2)}{t-x} = \lim_{t\to x}[3+(t+x)] = 3+2x$

Since $f'(a) = 13$, we have $3 + 2a = 13$, so $a = 5$.

c. $f'(x) = \lim_{t\to x} \dfrac{(1/t)-(1/x)}{t-x} = \lim_{t\to x} \dfrac{-(t-x)}{tx(t-x)} = \lim_{t\to x} \dfrac{-1}{tx} = -\dfrac{1}{x^2}$

Since $f'(a) = -\frac{1}{9}$, we have $-1/a^2 = -\frac{1}{9}$. Thus $a = 3$ or $a = -3$.

d. $f'(x) = \cos x$ by (3). Since $f'(a) = \sqrt{3}/2$, we have $\cos a = \sqrt{3}/2$. Thus $a = -\pi/6 + 2n\pi$ or $a = \pi/6 + 2n\pi$, where n is any integer.

35. The graphs are not virtually coincident because $((x+.1)^{1/2} - x^{1/2})/0.1$ is continuous on $[0, 1]$ and $\frac{1}{2}x^{-1/2}$ is not.

37. Since $f'(x) > 0$ for $x > 1$, f is increasing on $(1, \infty)$. Therefore the graph of f is in Figure 3.12(b).

39. Observe from Figure 3.14 that $f'(x) < 0$ for $x < a$ and $f'(x) > 0$ for $x > a$, for some a in $(0, 1)$. By inspection $h(x) < 0$ for $x < a$ and $h(x) > 0$ for $x > a$, whereas k does not have these properties. Thus it appears that $h = f'$. Similarly you can show that $k = g'$.

41. a. By (1), $f'(x) = 2x$, so that $f'(-1) = -2$. Thus the slope of the tangent line at $(-1, 1)$ is -2, so that the slope of the normal line at $(-1, 1)$ is $\frac{1}{2}$. An equation of the normal line is $y - 1 = \frac{1}{2}(x + 1)$.

b. $$f'(-1) = \lim_{x \to -1} \frac{f(x) - f(-1)}{x - (-1)} = \lim_{x \to -1} \frac{(2x - 3) - (-5)}{x + 1} = \lim_{x \to -1} \frac{2(x + 1)}{x + 1} = \lim_{x \to -1} 2 = 2$$

Thus the slope of the tangent line at $(-1, -5)$ is 2, so that the slope of the normal line is $-\frac{1}{2}$. An equation of the normal line is $y - (-5) = -\frac{1}{2}(x - (-1))$ or $y + 5 = -\frac{1}{2}(x + 1)$.

c. By (3), $f'(x) = \cos x$, so that $f'(0) = 1$. Thus the slope of the tangent line at $(0, 0)$ is 1, so that the slope of the normal line is -1. An equation of the normal line is $y - 0 = -1(x - 0)$, or $y = -x$.

43. $$v(t) = f'(t) = \lim_{h \to 0} \frac{f(t + h) - f(t)}{h} = \lim_{h \to 0} \frac{-3\sin(t + h) - (-3\sin t)}{h}$$

$$= -3 \lim_{h \to 0} \frac{\sin(t + h) - \sin t}{h} = -3\cos t$$

by the discussion preceding (3). In particular, $v(\pi/6) = -3\cos(\pi/6) = -\frac{3}{2}\sqrt{3}$.

45. a. We have

$$\lim_{r \to R^-} \frac{F(r) - F(R)}{r - R} = \lim_{r \to R^-} \frac{\dfrac{GMr}{R^3} - \dfrac{GM}{R^2}}{r - R} = \frac{GM}{R^3} \lim_{r \to R^-} \frac{r - R}{r - R} = \frac{GM}{R^3}$$

and

$$\lim_{r \to R^+} \frac{F(r) - F(R)}{r - R} = \lim_{r \to R^+} \frac{\dfrac{GM}{r^2} - \dfrac{GM}{R^2}}{r - R} = GM \lim_{r \to R^+} \frac{R^2 - r^2}{r^2 R^2 (r - R)}$$

$$= GM \lim_{r \to R^+} \frac{(R - r)(R + r)}{r^2 R^2 (r - R)} = -GM \lim_{r \to R^+} \frac{R + r}{r^2 R^2} = -\frac{2GM}{R^3}.$$

Thus F is not differentiable at R, and hence F is not differentiable on $(0, \infty)$.

b. We have

$$\lim_{r \to b^-} \frac{f(r) - f(b)}{r - b} = \lim_{r \to b^-} \frac{\dfrac{ar}{b} - \dfrac{c}{b^2}}{r - b} = \frac{1}{b^2} \lim_{r \to b^-} \frac{abr - c}{r - b}.$$

Notice that this limit only exists if $c = ab^2$, for then we would have

$$\frac{1}{b^2} \lim_{r \to b^-} \frac{abr - c}{r - b} = \frac{1}{b^2} \lim_{r \to b^-} \frac{abr - ab^2}{r - b} = \frac{a}{b} \lim_{r \to b^-} \frac{r - b}{r - b} = \frac{a}{b}.$$

Thus if $f'(b)$ exists, then $f'(b) = a/b > 0$. Next we see that

$$\lim_{r\to b^+} \frac{f(r) - f(b)}{r - b} = \lim_{r\to b^+} \frac{\dfrac{c}{r^2} - \dfrac{c}{b^2}}{r - b} = c \lim_{r\to b^+} \frac{b^2 - r^2}{r^2b^2(r - b)}$$

$$= c \lim_{r\to b^+} \frac{(b - r)(b + r)}{r^2b^2(r - b)} = -c \lim_{r\to b^+} \frac{b + r}{r^2b^2} = -\frac{2c}{b^3}.$$

Since c is positive by hypothesis, if $f'(b)$ exists, then $f'(b) = -2c/b^3 < 0$. Consequently $f'(b)$ cannot exist, so f cannot be differentiable on $(0, \infty)$.

47. If $0 < x < 6$, then $R'(x) = \lim_{t\to x}(1432t - 1432x)/(t - x) = 1432$, so R is differentiable on $(0, 6)$. If $x > 6$, then $R'(x) = \lim_{t\to x}(8592 - 8592)/(t - x) = 0$, so R is differentiable on $(6, \infty)$. Since

$$\lim_{x\to 6^-} \frac{R(x) - R(6)}{x - 6} = \lim_{x\to 6^-} \frac{1432x - 8592}{x - 6} = \lim_{x\to 6^-} 1432 = 1432$$

and

$$\lim_{x\to 6^+} \frac{R(x) - R(6)}{x - 6} = \lim_{x\to 6^+} \frac{8592 - 8592}{x - 6} = 0$$

R is not differentiable at 6.

49. The area of a circle of radius r is given by $A(r) = \pi r^2$ for $r > 0$. Thus

$$A'(r) = \lim_{t\to r} \frac{A(t) - A(r)}{t - r} = \lim_{t\to r} \frac{\pi t^2 - \pi r^2}{t - r} = \lim_{t\to r} \pi(t + r) = 2\pi r$$

(which is the circumference of the circle).

51. a. The altitude of the triangle with side of length x is $(\sqrt{3}/2)x$, so the area of the triangle is given by $A(x) = \frac{1}{2}[(\sqrt{3}/2)x](x) = (\sqrt{3}/4)x^2$. Then

$$A'(x) = \lim_{t\to x} \frac{(\sqrt{3}/4)t^2 - (\sqrt{3}/4)x^2}{t - x} = \lim_{t\to x} \frac{\sqrt{3}}{4}(t + x) = \frac{\sqrt{3}}{2}x \quad \text{for } x > 0.$$

b. If $A'(x) = A(x)$ and $x > 0$, then $(\sqrt{3}/2)x = (\sqrt{3}/4)x^2$, or $2x = x^2$, so that $x = 2$.

53. a. 7 minutes

b. 2 minutes (from 5 to 7 minutes after leaving home)

c. From 0 to 5 minutes after leaving home.

3.3 Derivatives of Combinations of Functions

1. $f'(x) = -12x^2$

3. $f'(x) = 16x^3 + 9x^2 + 4x + 1$

5. $f'(t) = \dfrac{36}{t^{10}}$

7. $g'(x) = 2(x+5) + (2x-3)\cdot 1 = 4x + 7$

9. $g'(x) = \left(-\dfrac{1}{x^2}\right)\left(2 - \dfrac{1}{x}\right) + \left(1 + \dfrac{1}{x}\right)\left(\dfrac{1}{x^2}\right) = -\dfrac{1}{x^2} + \dfrac{2}{x^3}$

11. $g'(x) = 12x^{-4} - 2\sin x$

13. $f'(z) = -6z^2 + 4\sec z\,\tan z$

15. $f'(z) = 2z\sin z + z^2\cos z$

17. $f(x) = \sin x\,\sin x$, so $f'(x) = \cos x\,\sin x + \sin x\,\cos x = 2\sin x\,\cos x = \sin 2x$

19. $f'(x) = \dfrac{2(4x-1) - (2x+3)4}{(4x-1)^2} = -\dfrac{14}{(4x-1)^2}$

21. $f'(t) = \dfrac{1(t^2+4t+4) - (t+2)(2t+4)}{(t^2+4t+4)^2} = -\dfrac{1}{t^2+4t+4}$

23. $f'(t) = \dfrac{(2t+5)(t^2+t-20) - (t^2+5t+4)(2t+1)}{(t^2+t-20)^2} = \dfrac{-4(t^2+12t+26)}{(t^2+t-20)^2}$

25. $f'(x) = \dfrac{(-\sin x)\sin x - \cos x\,\cos x}{(\sin x)^2} = \dfrac{-1}{\sin^2 x} = -\csc^2 x$

27. $f'(y) = \dfrac{1}{2\sqrt{y}}\sec y + \sqrt{y}\,\sec y\,\tan y$

29. $f'(x) = e^x + \dfrac{e^x}{(e^x)^2} = e^x + \dfrac{1}{e^x}$

31. $f'(t) = \dfrac{2e^t - 2te^t}{(e^t)^2} = \dfrac{2-2t}{e^t}$

33. $\dfrac{dy}{dx} = 3(2x^2-5x) + (3x+1)(4x-5) = 18x^2 - 26x - 5$

35. $\dfrac{dy}{dx} = 2x - \dfrac{2}{x^3}$

37. $\dfrac{dy}{dx} = \dfrac{3x^2(x^4+1) - (x^3-1)(4x^3)}{(x^4+1)^2} = \dfrac{-x^6+4x^3+3x^2}{(x^4+1)^2}$

39. $\dfrac{dy}{dx} = (-\csc x \cot x)\sec x + \csc x\,(\sec x \tan x) = -\csc^2 x + \sec^2 x$

41. $\dfrac{dy}{dx} = \dfrac{(\sin x + x\cos x)(x^2+1) - (x\sin x)(2x)}{(x^2+1)^2} = \dfrac{(x^3+x)\cos x + (1-x^2)\sin x}{(x^2+1)^2}$

43. $\dfrac{dy}{dx} = \dfrac{(2\sin x + 2x\cos x)e^x - (2x\sin x)e^x}{(e^x)^2} = \dfrac{2\sin x + 2x\cos x - 2x\sin x}{e^x}$

45. $f'(x) = 63/x^{10}$; $f'(1) = 63$

47. $f'(x) = (1/\pi)(6x-4)$; $f'(-2) = -16/\pi$

49. $f'(x) = e^x + xe^x$; $f'(1) = e^1 + 1(e^1) = 2e$

51. $f'(x) = 2x - 3$, so $f'(2) = 4 - 3 = 1$. Thus l: $y - (-6) = 1(x-2)$, or $y = x - 8$.

53. $f'(x) = \cos x - (-\sin x) = \cos x + \sin x$, so $f'(\pi/2) = 1$. Thus l: $y - 1 = 1(x - (\pi/2))$, or $y = x + 1 - (\pi/2)$.

55. $f'(x) = 6x^2 - 18x + 12 = 6(x^2 - 3x + 2) = 6(x-1)(x-2)$, and the tangent is horizontal if $f'(x) = 0$. Now $f'(x) = 0$ for $x = 1$ and $x = 2$. If $x = 1$ then $f(x) = 6$, and if $x = 2$ then $f(x) = 5$. Thus the points on the graph of f at which the tangent line is horizontal are $(1,6)$ and $(2,5)$.

57. a. Since $f'(x) = 2x$, an equation of the line tangent at any point $(c, 1+c^2)$ on the graph of f is $y - (1+c^2) = 2c(x-c)$, or $y = 2cx + 1 - c^2$. Similarly, since $g'(x) = -2x$, an equation of the line tangent at any point $(d, -1-d^2)$ on the graph of g is $y - (-1-d^2) = -2d(x-d)$, or $y = -2dx - 1 + d^2$. Two such lines will be the same if and only if their slopes are the same, i.e., if $2c = -2d$, and their y intercepts are the same, i.e., $1 - c^2 = -1 + d^2$. Therefore $d = -c$ and hence $1 - c^2 = -1 + c^2$, so $c = 1$ and $d = -1$, or $c = -1$ and $d = 1$. The points of tangency to the graphs of f and g for the line corresponding to $c = 1$ are $(1,2)$ and $(-1,-2)$, respectively. Similarly, the points of tangency for the line corresponding to $c = -1$ are $(-1,2)$ and $(1,-2)$, respectively.

b. We proceed as in part (a). Since $f'(x) = 2x$, an equation of the line tangent at any point $(c, a+c^2)$ on the graph of f is $y - (a+c^2) = 2c(x-c)$, or $y = 2cx + a - c^2$. Similarly, since $g'(x) = -2x$, an equation of the line tangent at any point $(d, -b-d^2)$ on the graph of g is $y - (-b-d^2) = -2d(x-d)$, or $y = -2dx - b + d^2$. Two such lines will be the same if and only if their slopes are the same, i.e., if $2c = -2d$, and their y intercepts are the same, i.e., $a - c^2 = -b + d^2$. Therefore $d = -c$ and hence $a - c^2 = -b + c^2$, so $c = (a+b)/2$ and $d = -(a+b)/2$, or $c = -(a+b)/2$ and $d = (a+b)/2$. The points of tangency to the graphs of f and g for the line corresponding to $c = (a+b)/2$ are $((a+b)/2, a+(a+b)^2/4)$ and $(-(a+b)/2, -b-(a+b)^2/4)$, respectively. Similarly, the points of tangency for the line corresponding to $c = -(a+b)/2$ are $(-(a+b)/2, a+(a+b)^2/4)$ and $((a+b)/2, -b-(a+b)^2/4)$, respectively.

59. b. Let g denote the desired polynomial. Then

$$g(x) = 1 - \frac{x^2}{2!} + \frac{x^4}{4!} - \frac{x^6}{6!} + \frac{x^8}{8!} - \frac{x^{10}}{10!} + \frac{x^{12}}{12!} - \frac{x^{14}}{14!}.$$

61. $(f+g)(x) = |x| - |x| = 0$, so $(f+g)'(x) = 0$ for all x. By Example 3 in Section 3.1, $f'(x)$ exists only for $x \neq 0$, and the Constant Multiple Rule therefore implies that $g'(x)$ also exists only for $x \neq 0$.

63. Since $f(x) = [(fg)(x)]/g(x)$ for all x in the domain of fg for which $g(x) \neq 0$, and since $(fg)'(a)$ and $g'(a)$ exist with $g(a) \neq 0$, it follows from Theorem 3.7 that $f'(a)$ exists.

65. $$\frac{f(x)g(x) - f(a)g(a)}{x-a} = \frac{f(x)g(x) - f(a)g(x) + f(a)g(x) - f(a)g(a)}{x-a}$$
$$= \frac{f(x)-f(a)}{x-a}g(x) + f(a)\frac{g(x)-g(a)}{x-a} = \frac{f(x)-f(a)}{x-a}g(x)$$

because $f(a) = 0$ by hypothesis. Since $f'(a)$ exists and g is continuous at a, the above calculation implies that

$$(fg)'(a) = \lim_{x\to a} \frac{f(x)g(x) - f(a)g(a)}{x-a} = \lim_{x\to a}\left[\frac{f(x)-f(a)}{x-a}g(x)\right]$$
$$= \lim_{x\to a}\frac{f(x)-f(a)}{x-a}\lim_{x\to a} g(x) = f'(a)g(a).$$

67. By (6) we have $(1/g)'(a) = [-g'(a)]/[g(a)]^2$. Thus

$$\left(\frac{f}{g}\right)'(a) = \left(f\frac{1}{g}\right)'(a) = f'(a)\frac{1}{g(a)} + f(a)\left\{\frac{-g'(a)}{[g(a)]^2}\right\} = \frac{f'(a)g(a) - f(a)g'(a)}{[g(a)]^2}.$$

69. The height of the ball at any time t before the ball hits the ground is given by $h(t) = -16t^2 + 128t + 8$. Thus $h(t) = 8$ if $-16t^2 + 128t = 0$. But $-16t^2 + 128t = -16(t-8)$, so the ball returns to the 8-foot level after 8 seconds. Since $v(t) = h'(t) = -32t + 128$, we conclude that $v(8) = -32(8) + 128 = -128$ (feet per second).

71. a. Let $f(x) = ax^2 + bx + c$, so that $f'(x) = 2ax + b$. From Figure 3.18(a), $f(1) = 0$, so $a + b + c = 0$. Also, $f(-1) = 0$, so $a - b + c = 0$. Since $a + b + c = 0$ and $a - b + c = 0$, it follows that $b = 0$ and that $a = -c$. Next, $f'(-1) = -1$, so that $-1 = -2a + b = -2a$. Thus $a = \frac{1}{2}$ and hence $c = -\frac{1}{2}$. Consequently there is a unique polynomial of degree 2, namely $f(x) = \frac{1}{2}x^2 - \frac{1}{2}$, that has the desired properties.

b. Let $f(x) = ax^2 + bx + c$, so that $f'(x) = 2ax + b$. From Figure 3.18(b), $f(-1) = 0$, so $a - b + c = 0$. Also $f(0) = -1$, so $c = -1$. Thus $a - b = 1$. Next $f'(0) = 0$, so $2a \cdot 0 + b = 0$, and thus $b = 0$. Therefore $a = 1 + b = 1$. Finally, $f'(-1) = -1$, so that $-1 = 2a(-1) + b = 2(1)(-1) + 0 = -2$. This is a contradiction. Therefore there is no polynomial of degree 2 with the desired properties.

73. Since $\mu > 0$, we have $\mu \sin x + \cos x > 0$ for $0 \le x \le \pi/2$. Since the sine and cosine functions are differentiable on $(-\infty, \infty)$, so is the function $\mu \sin x + \cos x$. By the Quotient Rule, the function $50\mu/(\mu \sin x + \cos x)$ is differentiable at every point in $[0, \pi/2]$. Thus F is differentiable on $[0, \pi/2]$; $F'(x) = [-50\mu(\mu \cos x - \sin x)]/(\mu \sin x + \cos x)^2$.

75. $$f'(x) = \frac{100nkx^{n-1}(1+kx^n) - (100kx^n)(nkx^{n-1})}{(1+kx^n)^2} = \frac{100nkx^{n-1}}{(1+kx^n)^2} \quad \text{for } x > 0.$$

77. $$P(x) = R(x) - C(x) = \sqrt{x} - \frac{x+3}{\sqrt{x}+1} = \frac{x + \sqrt{x} - (x+3)}{\sqrt{x}+1} = \frac{\sqrt{x}-3}{\sqrt{x}+1} \quad \text{for } 1 \le x \le 15.$$

Thus $P(x) = 0$ if $\sqrt{x} - 3 = 0$, that is, if $x = 9$. Since the function $(\sqrt{x}-3)/(\sqrt{x}+1)$ is differentiable on $(0, \infty)$, it follows that P is differentiable on $[1, 15]$, and

$$m_P(x) = P'(x) = \frac{[1/(2\sqrt{x})](\sqrt{x}+1) - (\sqrt{x}-3)[1/(2\sqrt{x})]}{(\sqrt{x}+1)^2}$$

$$= \frac{(\sqrt{x}+1) - (\sqrt{x}-3)}{2\sqrt{x}(\sqrt{x}+1)^2} = \frac{2}{\sqrt{x}(\sqrt{x}+1)^2} \ne 0 \quad \text{for } 1 \le x \le 15.$$

Therefore for no value of x is the marginal profit equal to 0.

3.4 The Chain Rule

1. $f'(x) = \frac{9}{4}x^{5/4}$

3. $f'(x) = \frac{3}{2}(1-3x)^{1/2}(-3) = -\frac{9}{2}(1-3x)^{1/2}$

5. $f'(x) = \sqrt{2-7x^2} + x\dfrac{1}{2\sqrt{2-7x^2}}(-14x) = \dfrac{2-14x^2}{\sqrt{2-7x^2}}$

7. $f'(t) = (\cos 5t)(5) = 5\cos 5t$

9. $f'(t) = 4\sin^3 t\,\cos t + 4\cos^3 t\,(-\sin t) = 4\sin^3 t\,\cos t - 4\cos^3 t\,\sin t$

11. $g'(x) = \frac{1}{3}(1-\sin x)^{-2/3}(-\cos x) = (-\frac{1}{3}\cos x)(1-\sin x)^{-2/3}$

13. $f'(x) = [-\sin(\sin x)]\cos x$

15. $f'(x) = 3\left(\dfrac{x-1}{x+1}\right)^2\left(\dfrac{(x+1)-(x-1)}{(x+1)^2}\right) = \dfrac{6(x-1)^2}{(x+1)^4}$

17. $f'(x) = \dfrac{-1}{(x\sqrt{5-2x})^2}\left[\sqrt{5-2x} + x\dfrac{-2}{2\sqrt{5-2x}}\right] = -\dfrac{1}{x^2(5-2x)}\left[\dfrac{(5-2x)-x}{\sqrt{5-2x}}\right] = \dfrac{3x-5}{x^2(5-2x)^{3/2}}$

19. $f'(x) = \cos\dfrac{1}{x} + x\left(-\sin\dfrac{1}{x}\right)\left(-\dfrac{1}{x^2}\right) = \cos\dfrac{1}{x} + \dfrac{1}{x}\sin\dfrac{1}{x}$

21. $g'(z) = \frac{1}{2}[2z-(2z)^{1/3}]^{-1/2}[2-\frac{1}{3}(2z)^{-2/3}(2)] = [2z-(2z)^{1/3}]^{-1/2}[1-\frac{1}{3}(2z)^{-2/3}]$

23. $g'(z) = [2\cos(3z^6)][-\sin(3z^6)][18z^5] = -36z^5\cos(3z^6)\sin(3z^6)$

25. $f'(x) = [-\sin(1+\tan 2x)](\sec^2 2x)(2) = -2[\sin(1+\tan 2x)]\sec^2 2x$

27. By (4), $f'(x) = 5x^4e^{(x^5)}$.

29. By (4), $f'(t) = (\sec^2(e^{3t}))(3e^{3t}) = 3e^{3t}(\sec^2(e^{3t}))$.

31. By (7), $g'(t) = \dfrac{1}{t^2+1}(2t) = \dfrac{2t}{t^2+1}$.

33. By (7), $g'(t) = (2\ln t)\dfrac{1}{t} = \dfrac{2\ln t}{t}$.

35. By (3), $f'(x) = (\ln 3)(3^{5x-7})(5) = 5(\ln 3)3^{5x-7}$.

37. By (11) of Section 1.8, $\log_3(x^2+4) = \dfrac{\ln(x^2+4)}{\ln 3}$. With (8) of this present section, this implies that

$$f'(x) = \frac{1}{\ln 3}\frac{1}{x^2+4}(2x) = \frac{2x}{(\ln 3)(x^2+4)}.$$

39. $\dfrac{dy}{dx} = 3(-\frac{2}{3})x^{-5/3} = -2x^{-5/3}$

41. $\dfrac{dy}{dx} = -\sqrt{1+3x^2} - x \cdot \dfrac{1}{2}(1+3x^2)^{-1/2}(6x) = -\dfrac{1+6x^2}{\sqrt{1+3x^2}}$

43. $\dfrac{dy}{dx} = \dfrac{2}{3}\left(\dfrac{1}{x\sin x}\right)^{-1/3}\left[\dfrac{-1}{(x\sin x)^2}\right][\sin x + x\cos x] = -\dfrac{2}{3}\dfrac{\sin x + x\cos x}{(x\sin x)^{5/3}}$

45. $\dfrac{dy}{dx} = [3\tan^2(\frac{1}{2}x)][\sec^2(\frac{1}{2}x)](\frac{1}{2}) = \frac{3}{2}\tan^2(\frac{1}{2}x)\sec^2(\frac{1}{2}x)$

47. $\dfrac{dy}{dx} = e^{\cos x}(-\sin x) = -(\sin x)e^{\cos x}$

49. By (3), $\dfrac{dy}{dx} = (\ln a)(a^x)\cos bx - a^x(\sin bx)(b) = a^x[(\ln a)(\cos bx) - b\sin bx]$.

51. $\dfrac{d}{dx}(y^5) = 5y^4\dfrac{dy}{dx}$

53. $\dfrac{d}{dx}\left(\dfrac{2}{y}\right) = \dfrac{-2}{y^2}\dfrac{dy}{dx}$

55. $\dfrac{d}{dx}(\sin\sqrt{y}) = (\cos\sqrt{y})\dfrac{1}{2\sqrt{y}}\dfrac{dy}{dx} = \left(\dfrac{1}{2\sqrt{y}}\cos\sqrt{y}\right)\dfrac{dy}{dx}$

57. $\dfrac{d}{dx}(x^3y^2) = 3x^2y^2 + x^3\left(2y\dfrac{dy}{dx}\right) = 3x^2y^2 + 2x^3y\dfrac{dy}{dx}$

59. $\dfrac{d}{dx}(\sqrt{x^2+y^2}) = \dfrac{1}{2\sqrt{x^2+y^2}}\left[2x + 2y\dfrac{dy}{dx}\right] = \dfrac{x + y(dy/dx)}{\sqrt{x^2+y^2}}$

61. $f'(x) = \dfrac{1}{x+\sqrt{x^2-1}}\left[1 + \dfrac{1}{2\sqrt{x^2-1}}(2x)\right] = \dfrac{\sqrt{x^2-1}+x}{(x+\sqrt{x^2-1})\sqrt{x^2-1}} = \dfrac{1}{\sqrt{x^2-1}}$

63. $f'(x) = -2(x+1)^{-3}(1)$, so $f'(0) = -2$. Thus l: $y - 1 = -2(x-0)$, or $y = -2x+1$.

65. $f'(x) = (-2)(-\sin 3x)(3) = 6\sin 3x$, so $f'(\pi/3) = 6\sin\pi = 0$. Thus l: $y = 2$.

67. $f'(x) = 2e^{-3x}(-3) = -6e^{-3x}$, so $f'(0) = -6e^{-3(0)} = -6$. Thus l: $y - 2 = -6(x-0)$, or $y = -6x+2$.

69. Yes, because $(\ln(x+0.1) - \ln x)/(0.1)$ should be approximately $(d/dx)(\ln x)$, which is $1/x$ for $x > 0$.

71. a. $s'(x) = \left[\cos\left(\dfrac{\pi}{180}x\right)\right]\dfrac{\pi}{180} = \dfrac{\pi}{180}c(x)$; $c'(x) = \left[-\sin\left(\dfrac{\pi}{180}x\right)\right]\dfrac{\pi}{180} = -\dfrac{\pi}{180}s(x)$

73. a. Since $g(x) = f(-x)$, the Chain Rule implies that

$$g'(x) = \frac{d}{dx}[f(-x)] = [f'(-x)](-1) = -f'(-x)$$

b. Let f be an even function. Then $f(-x) = f(x)$. Differentiating both sides using part (a), we find that $-f'(-x) = f'(x)$, so that $f'(-x) = -f'(x)$. Therefore f' is an odd function.

c. Let f be an odd function. Then $f(-x) = -f(x)$. Differentiating both sides using part (a), we find that $-f'(-x) = -f'(x)$, so that $f'(-x) = f'(x)$. Therefore f' is an even function.

75. Since the cosine function and the polynomial $\pi t/24$ are differentiable on $(-\infty, \infty)$, $\cos(\pi t/24)$ and hence F are differentiable on $[0, 24]$. Also

$$F'(t) = \frac{336{,}000}{\pi}\left(\sin\frac{\pi t}{24}\right)\left(\frac{\pi}{24}\right) = 14{,}000 \sin\frac{\pi t}{24}.$$

77. a. $\dfrac{dE}{dv} = 100(.4)\left(1 - \dfrac{v}{V}\right)^{-0.6}\left(\dfrac{-1}{V}\right) = \dfrac{-40}{V}\left(1 - \dfrac{v}{V}\right)^{-0.6}$

b. $\dfrac{dE}{dV} = 100(.4)\left(1 - \dfrac{v}{V}\right)^{-0.6}\left(\dfrac{v}{V^2}\right) = \dfrac{40v}{V^2}\left(1 - \dfrac{v}{V}\right)^{-0.6}$

79. a. $v'(r) = \dfrac{1}{2}\left(\dfrac{192{,}000}{r} + v_0^2 - 48\right)^{-1/2}\left(\dfrac{-192{,}000}{r^2}\right) = -\dfrac{96{,}000}{r^2}\left(\dfrac{192{,}000}{r} + v_0^2 - 48\right)^{-1/2}$

b. If $v_0 = 8$, then

$$v'(24{,}000) = -\frac{96{,}000}{(24{,}000)^2}\left(\frac{192{,}000}{24{,}000} + 16\right)^{-1/2} = -\frac{\sqrt{6}}{72{,}000} \text{ (miles per second per mile).}$$

81. $\dfrac{dV}{dt} = \dfrac{dV}{dr}\dfrac{dr}{dt} = [\tfrac{4}{3}\pi(3r^2)](10) = 40\pi r^2$.

83. $\dfrac{dA}{dh} = \dfrac{dA}{dx}\dfrac{dx}{dh} = \left(\dfrac{\sqrt{3}}{2}x\right)\left(\dfrac{2\sqrt{3}}{3}\right) = x = \tfrac{2}{3}\sqrt{3}\,h$, so that $\left.\dfrac{dA}{dh}\right|_{h=\sqrt{3}} = \dfrac{2\sqrt{3}}{3}(\sqrt{3}) = 2$

85. Since $D'(x) = \frac{1}{2}(3 - 2x)^{-1/2}(-2) = -(3 - 2x)^{-1/2}$, we have $D'(x) < 0$ for $0 < x < \frac{3}{2}$.

3.5 Higher Derivatives

1. $f'(x) = 5$; $f''(x) = 0$

3. $f'(x) = -60x^4 + 2x^3 - \frac{1}{2}(1-x)^{-1/2}(-1) = -60x^4 + 2x^3 + \frac{1}{2}(1-x)^{-1/2}$

   $f''(x) = -240x^3 + 6x^2 + \frac{1}{2}(-\frac{1}{2})(1-x)^{-3/2}(-1) = -240x^3 + 6x^2 + \frac{1}{4}(1-x)^{-3/2}$

5. $f'(x) = 2(-2)(1-4x)^{-3}(-4) = 16(1-4x)^{-3}$; $f''(x) = 16(-3)(1-4x)^{-4}(-4) = 192(1-4x)^{-4}$

7. $f'(x) = a(-n)x^{-n-1}$; $f''(x) = (-an)(-n-1)x^{-n-2} = an(n+1)x^{-n-2}$

9. $f'(x) = \dfrac{-3x^2}{(x^3-1)^2}$

   $$f''(x) = \frac{-6x(x^3-1)^2 + 3x^2(2)(x^3-1)(3x^2)}{(x^3-1)^4} = \frac{6x(x^3-1)(-x^3+1+3x^3)}{(x^3-1)^4} = \frac{6x(2x^3+1)}{(x^3-1)^3}$$

11. $f'(x) = \sec^3 x$; $f''(x) = 2\sec x(\sec x \tan x) = 2\sec^2 x \tan x$

13. $f'(x) = 3(x^2 + \sin x)^2(2x + \cos x)$

    $$\begin{aligned} f''(x) &= 6(x^2 + \sin x)(2x + \cos x)^2 + 3(x^2 + \sin x)^2(2 - \sin x) \\ &= 3(x^2 + \sin x)[2(2x + \cos x)^2 + (x^2 + \sin x)(2 - \sin x)] \end{aligned}$$

[illegible] $(1 + \sin x)^{\text{[illegible]}}(\cos x)^2 + \frac{1}{2}(1 + \sin x)^{-1/2}(-\sin x)$

$= -\frac{1}{4}(1 + \sin x)^{-3/2}\cos^2 x - \frac{1}{2}(1 + \sin x)^{-1/2}\sin x$

17. $f'(x) = -\dfrac{1}{x^2}e^{1/x}$; $f''(x) = \dfrac{2}{x^3}e^{1/x} + \dfrac{1}{x^4}e^{1/x}$

19. $f'(x) = \dfrac{2x}{1+x^2}$; $f''(x) = \dfrac{2(1+x^2) - 2x(2x)}{(1+x^2)^2} = \dfrac{2-2x^2}{(1+x^2)^2}$

21. $\dfrac{dy}{dx} = \dfrac{3}{2}x^{1/2}$; $\dfrac{d^2y}{dx^2} = \dfrac{3}{4}x^{-1/2}$

23. $\dfrac{dy}{dx} = 3(x^4 - \tan x)^2(4x^3 - \sec^2 x)$

    $\dfrac{d^2y}{dx^2} = 3[2(x^4 - \tan x)(4x^3 - \sec^2 x)^2 + (x^4 - \tan x)^2 \cdot (12x^2 - 2\sec^2 x \tan x)]$

25. $\dfrac{dy}{dx} = 2ax + b$; $\dfrac{d^2y}{dx^2} = 2a$

27. $\dfrac{dy}{dx} = \dfrac{-1}{(3-x)^2}(-1) = \dfrac{1}{(3-x)^2}$; $\dfrac{d^2y}{dx^2} = (-2)\dfrac{1}{(3-x)^3}(-1) = \dfrac{2}{(3-x)^3}$

29. $\dfrac{dy}{dx} = -\csc x \cot x$; $\dfrac{d^2y}{dx^2} = -(-\csc x \cot x)\cot x - \csc x(-\csc^2 x) = \csc x \cot^2 x + \csc^3 x$

31. $\dfrac{dy}{dx} = e^x \sin x + e^x \cos x;\ \dfrac{d^2y}{dx^2} = (e^x \sin x + e^x \cos x) + (e^x \cos x - e^x \sin x) = 2e^x \cos x$

33. $\dfrac{dy}{dx} = 2x \ln x + x^2\left(\dfrac{1}{x}\right) = 2x \ln x + x;\ \dfrac{d^2y}{dx^2} = 2 \ln x + 2x\left(\dfrac{1}{x}\right) + 1 = 2 \ln x + 3$

35. $f'(x) = -8x;\ f''(x) = -8;\ f^{(3)}(x) = 0$

37. $f'(x) = 2x \cos x^2;\ f''(x) = 2 \cos x^2 - 4x^2 \sin x^2$

$f^{(3)}(x) = -4x \sin x^2 - 8x \sin x^2 - 8x^3 \cos x^2 = -12x \sin x^2 - 8x^3 \cos x^2$

39. $f'(x) = -1/x^2;\ f''(x) = 2/x^3;\ f^{(3)}(x) = -6/x^4$

41. $f'(x) = \dfrac{3(4x+5) - (3x)(4)}{(4x+5)^2} = \dfrac{15}{(4x+5)^2};\ f''(x) = \dfrac{15(-2)(4)}{(4x+5)^3} = \dfrac{-120}{(4x+5)^3}$

$f^{(3)}(x) = \dfrac{(-120)(-3)(4)}{(4x+5)^4} = \dfrac{1440}{(4x+5)^4}$

43. $f'(x) = 2xe^{(x^2)};\ f''(x) = 2e^{(x^2)} + 4x^2e^{(x^2)};\ f^{(3)}(x) = 4xe^{(x^2)} + 8xe^{(x^2)} + 8x^3e^{(x^2)} = 12e^{(x^2)} + 8x^3e^{(x^2)}$

45. $\dfrac{dy}{dx} = \dfrac{7}{2}x^{5/2} - 5x^{3/2};\ \dfrac{d^2y}{dx^2} = \dfrac{35}{4}x^{3/2} - \dfrac{15}{2}x^{1/2};\ \dfrac{d^3y}{dx^3} = \dfrac{105}{8}x^{1/2} - \dfrac{15}{4}x^{-1/2}$

47. $\dfrac{dy}{dx} = -\csc x \cot x;\ \dfrac{d^2y}{dx^2} = -(-\csc x \cot x)\cot x - \csc x(-\csc^2 x) = \csc x \cot^2 x + \csc^3 x$

$\dfrac{d^3y}{dx^3} = (-\csc x \cot x)\cot^2 x + \csc x\,[2 \cot x(-\csc^2 x)] + 3\csc^2 x(-\csc x \cot x)$

$= -\csc x \cot^3 x - 5\csc^3 x \cot x$

49. $\dfrac{dy}{dx} = 3ax^2 + 2bx + c;\ \dfrac{d^2y}{dx^2} = 6ax + 2b;\ \dfrac{d^3y}{dx^3} = 6a$

51. $f'(x) = 24x^7 + \frac{9}{2}x^5 - 3x^{-1/4} - 2x^{-2}$

$f''(x) = 168x^6 + \frac{45}{2}x^4 + \frac{3}{4}x^{-5/4} + 4x^{-3}$

$f^{(3)}(x) = 1008x^5 + 90x^3 - \frac{15}{16}x^{-9/4} - 12x^{-4}$

$f^{(4)}(x) = 5040x^4 + 270x^2 + \frac{135}{64}x^{-13/4} + 48x^{-5}$

53. $f'(x) = \pi \cos \pi x;\ f''(x) = -\pi^2 \sin \pi x;\ f^{(3)}(x) = -\pi^3 \cos \pi x;\ f^{(4)}(x) = \pi^4 \sin \pi x$

55. $f'(x) = -\sqrt{2}\,e^{-\sqrt{2}\,x};\ f''(x) = (-\sqrt{2})^2e^{-\sqrt{2}\,x} = 2e^{-\sqrt{2}\,x};$

$f^{(3)}(x) = -2\sqrt{2}\,e^{-\sqrt{2}\,x};\ f^{(4)}(x) = (-\sqrt{2})(-2\sqrt{2}\,e^{-\sqrt{2}\,x}) = 4e^{-\sqrt{2}\,x}$

57. $v(t) = f'(t) = -32t + 3;\ a(t) = v'(t) = f''(t) = -32$

59. $v(t) = f'(t) = 2\cos t + 3\sin t$; $a(t) = v'(t) = f''(t) = -2\sin t + 3\cos t$

61. Let $f(x) = c_n x^n + c_{n-1}x^{n-1} + \cdots + c_1 x + c_0$, where $c_n \neq 0$. Then $f'(x) = nc_n x^{n-1} + (n-1)c_{n-1}x^{n-2} + \cdots + c_1$. Thus f' is a polynomial of degree $n-1$. In the same way we find that f'' is a polynomial of degree $n-2$. Continuing, we see that $f^{(n)}$ is a polynomial of degree 0, that is $f^{(n)}$ is a constant function. Thus $f^{(n+1)} = 0$, and consequently $f^{(n+2)} = 0$ also.

63. $f'(x) = e^x = f(x)$, so $f^{(n)}(x) = e^x$ for any $n \geq 1$.

65. $f'(x) = e^{-x} - xe^{-x} = e^{-x} - f(x)$; $f''(x) = -e^{-x} - e^{-x} + xe^{-x} = -2e^{-x} + xe^{-x} = -2e^{-x} + f(x)$

$f^{(3)}(x) = 2e^{-x} + e^{-x} - xe^{-x} = 3e^{-x} - f(x)$;

in general, $f^{(n)}(x) = (-1)^n(xe^{-x} - ne^{-x}) = (-1)^n e^{-x}(x-n)$ for any $n \geq 1$.

67. $f'(x) = -\dfrac{1}{(1-x)^2}(-1) = \dfrac{1}{(1-x)^2}$; $f''(x) = \dfrac{2!}{(1-x)^3}$; $f^{(3)}(x) = \dfrac{3!}{(1-x)^4}$;

in general, $f^{(n)}(x) = \dfrac{n!}{(1-x)^{n+1}}$ for any $n \geq 1$.

69. Notice from Figure 3.24 that g is negative on the intervals on which h is decreasing, and g is positive on the intervals on which h is increasing. Similarly, k is negative on the interval on which g is decreasing, and k is positive on the intervals on which g is increasing. We conclude that $g = h'$ and $k = g'$, and hence that $h = f$, $g = f'$, and $k = f''$.

71. a. $f''(x) = 2xf(x) + x^2 f'(x) = 2xf(x) + x^2(x^2 f(x)) = 2xf(x) + x^4 f(x) = (2x + x^4)f(x)$

b. $f''(3) = (2 \cdot 3 + 3^4)f(3) = (6 + 81)\pi = 87\pi$

73. $h'(x) = f'(x)g(x) + f(x)g'(x)$;

$h''(x) = f''(x)g(x) + f'(x)g'(x) + f'(x)g'(x) + f(x)g''(x) = f''(x)g(x) + 2f'(x)g'(x) + f(x)g''(x)$

75. Angular velocity $= f'(t) = -2\pi\omega a \sin 2\pi\omega t$; angular acceleration $= f''(t) = -4\pi^2\omega^2 a \cos 2\pi\omega t$.

77. $\dfrac{ds}{dt} = kae^{kt} - bke^{-kt}$; $\dfrac{d^2s}{dt^2} = k^2ae^{kt} + bk^2e^{-kt} = k^2(ae^{kt} + be^{-kt}) = k^2 s$, so acceleration is k^2 times the distance traveled.

79. a. $f'(x) = \dfrac{a^2bke^{-kax}}{(1+be^{-kax})^2}$;

$$f''(x) = \frac{-a^3bk^2e^{-kax}(1+be^{-kax})^2 - a^2bke^{-kax}[2(1+be^{-kax})(-abke^{-kax})]}{(1+be^{-kax})^4}$$

$$= \frac{a^3bk^2e^{-kax}(-1+be^{-kax})}{(1+be^{-kax})^3}$$

b. Since a, b, and k are positive constants, $f''(c) = 0$ if $-1 + be^{-kac} = 0$, that is, $b = e^{kac}$, so $kac = \ln b$. Thus $c = (\ln b)/ka$.

c. $f((\ln b)/ka) = \dfrac{a}{1 + be^{-ka((\ln b)/ka)}} = \dfrac{a}{1 + be^{-\ln b}} = \dfrac{a}{1+1} = \dfrac{a}{2}$

3.6 Implicit Differentiation

1. $6y\dfrac{dy}{dx} = 8x^3$; $\dfrac{dy}{dx} = \dfrac{4x^3}{3y}$

3. $2y\dfrac{dy}{dx} + \dfrac{dy}{dx} = \dfrac{(1-x)-(1+x)(-1)}{(1-x)^2} = \dfrac{2}{(1-x)^2}$; $\dfrac{dy}{dx} = \dfrac{2}{(2y+1)(1-x)^2}$

5. $\sec y \tan y \dfrac{dy}{dx} - \sec^2 x = 0$; $\dfrac{dy}{dx} = \dfrac{\sec^2 x}{\sec y \tan y}$

7. $\dfrac{(\cos y)(dy/dx)(y^2+1) - (\sin y)2y(dy/dx)}{(y^2+1)^2} = 3$; $\dfrac{dy}{dx} = \dfrac{3(y^2+1)^2}{(y^2+1)\cos y - 2\sin y}$

9. $2x + \left(2xy^2 + 2x^2y\dfrac{dy}{dx}\right) + 3y^2\dfrac{dy}{dx} = 0$; $\dfrac{dy}{dx} = \dfrac{-2x - 2xy^2}{2x^2y + 3y^2}$

11. $2x + 2y\dfrac{dy}{dx} = \dfrac{(2y(dy/dx))x^2 - y^2(2x)}{x^4} = \dfrac{2y}{x^2}\dfrac{dy}{dx} - \dfrac{2y^2}{x^3}$; $\dfrac{dy}{dx} = \dfrac{x + (y^2/x^3)}{(y/x^2) - y} = \dfrac{x^4 + y^2}{xy - x^3y}$

13. $\dfrac{1}{2}(xy)^{1/2}\left(y + x\dfrac{dy}{dx}\right) + \dfrac{1}{2}(x+2y)^{-1/2}\left(1 + 2\dfrac{dy}{dx}\right) = 0$; $\dfrac{dy}{dx} = \dfrac{-y(xy)^{-1/2} - (x+2y)^{-1/2}}{x(xy)^{-1/2} + 2(x+2y)^{-1/2}}$

15. $e^y + xe^y\dfrac{dy}{dx} = \dfrac{dy}{dx} + 2x$; $\dfrac{dy}{dx} = \dfrac{2x - e^y}{xe^y - 1}$

17. $\dfrac{dy}{dx} + \dfrac{1}{2\sqrt{y}}\dfrac{dy}{dx}(\ln x) + \dfrac{\sqrt{y}}{x} = 2x + 2y\dfrac{dy}{dx}$; $\dfrac{dy}{dx} = \dfrac{2x - \sqrt{y}/x}{1 + (\ln x)/(2\sqrt{y}) - 2y}$

19. $2x + 2y\dfrac{dy}{dx} = \dfrac{dy}{dx}$; $\dfrac{dy}{dx} = \dfrac{2x}{1-2y}$. At $(0,1)$, $\dfrac{dy}{dx} = \dfrac{0}{1-2} = 0$.

21. $y + x\dfrac{dy}{dx} = 0$; $\dfrac{dy}{dx} = -\dfrac{y}{x}$. At $(-2,-1)$, $\dfrac{dy}{dx} = \dfrac{-(-1)}{-2} = -\dfrac{1}{2}$.

23. $3x^2 + 2\left(y + x\dfrac{dy}{dx}\right) = 0$; $\dfrac{dy}{dx} = -\dfrac{3x^2+2y}{2x}$. At $(1,2)$, $\dfrac{dy}{dx} = \dfrac{3+2\cdot 2}{-2} = -\dfrac{7}{2}$.

25. $2x + \dfrac{y - x(dy/dx)}{y^2} = 0$; $\dfrac{dy}{dx} = \dfrac{2x + (1/y)}{x/y^2} = \dfrac{2xy^2 + y}{x}$. At $(1, -\frac{1}{3})$, $\dfrac{dy}{dx} = \dfrac{2(1)(-\frac{1}{3})^2 + (-\frac{1}{3})}{1} = -\dfrac{1}{9}$.

27. $\dfrac{1}{2\sqrt{x}}(\sqrt{y}+2) + (\sqrt{x}+1)\left(\dfrac{1}{2\sqrt{y}}\dfrac{dy}{dx}\right) = 0$; $\dfrac{dy}{dx} = \dfrac{\sqrt{y}+2}{2\sqrt{x}} \cdot \dfrac{-2\sqrt{y}}{\sqrt{x}+1} = -\dfrac{y + 2\sqrt{y}}{x + \sqrt{x}}$.

At $(1,4)$, $\dfrac{dy}{dx} = -\dfrac{4 + 2\cdot 2}{1+1} = -4$.

29. $2e^{x^2y}\left(2xy + x^2\dfrac{dy}{dx}\right) = 1$, so $4xye^{x^2y} + 2x^2e^{x^2y}\dfrac{dy}{dx} = 1$; $\dfrac{dy}{dx} = \dfrac{1 - 4xye^{x^2y}}{2x^2e^{x^2y}}$

At $(2,0)$, $\dfrac{dy}{dx} = \dfrac{1 - 4(2)(0)e^0}{2(2^2)e^0} = \dfrac{1}{8}$.

31. $y^2 + 2xy\dfrac{dy}{dx} = 0$; $\dfrac{dy}{dx} = -\dfrac{y}{2x}$. At $(2, -3)$, $\dfrac{dy}{dx} = -\dfrac{(-3)}{2(2)} = \dfrac{3}{4}$.
Thus l: $y - (-3) = \frac{3}{4}(x-2)$, or $y = \frac{3}{4}x - \frac{9}{2}$.

33. $\cos(x+y)\left(1+\dfrac{dy}{dx}\right) = 2$; $\dfrac{dy}{dx} = \dfrac{2}{\cos(x+y)} - 1$. At $(0, \pi)$, $\dfrac{dy}{dx} = \dfrac{2}{\cos(0+\pi)} - 1 = -2 - 1 = -3$.
Thus l: $y - \pi = -3(x-0)$, or $y = -3x + \pi$.

35. $2x - 4y^3\dfrac{dy}{dx} = 0$; $\dfrac{dy}{dx} = \dfrac{x}{2y^3}$;

$$\frac{d^2y}{dx^2} = \frac{2y^3 - x(6y^2(dy/dx))}{(2y^3)^2} = \frac{2y^3 - 6xy^2[x/(2y^3)]}{4y^6} = \frac{2y^3 - (3x^2/y)}{4y^6} = \frac{2y^4 - 3x^2}{4y^7}$$

37. $2x\sin 2y + 2x^2(\cos 2y)\dfrac{dy}{dx} = 0$; $\dfrac{dy}{dx} = -\dfrac{2x\sin 2y}{2x^2\cos 2y} = -\dfrac{\tan 2y}{x}$

$$\frac{d^2y}{dx^2} = -\frac{(\sec^2 2y)(2(dy/dx))x - \tan 2y}{x^2} = -\frac{(\sec^2 2y)2(-(\tan 2y)/x)x - \tan 2y}{x^2} = \frac{\tan 2y\,(2\sec^2 2y + 1)}{x^2}$$

39. $2y\dfrac{dy}{dt} - 2x\dfrac{dx}{dt} = 0$, so $\dfrac{dy}{dt} = \dfrac{2x(dx/dt)}{2y} = \dfrac{x}{y}\dfrac{dx}{dt}$

41. $\dfrac{dx}{dt}\sin y + x\cos y\dfrac{dy}{dt} = 0$, so $\dfrac{dy}{dt} = -\dfrac{\sin y}{x\cos y}\dfrac{dx}{dt} = -\dfrac{\tan y}{x}\dfrac{dx}{dt}$

43. $\dfrac{dy}{dt} = -\sin(xy^2)\left[\dfrac{dx}{dt}y^2 + 2xy\dfrac{dy}{dt}\right]$, so $\dfrac{dy}{dt} = -\dfrac{y^2\sin(xy^2)(dx/dt)}{1 + 2xy\sin(xy^2)}$.

45. a. $3y^2\dfrac{dy}{dx} = 2x$; $\dfrac{dy}{dx} = \dfrac{2x}{3y^2}$. Also $y = x^{2/3}$; $\dfrac{dy}{dx} = \dfrac{2}{3}x^{-1/3} = \dfrac{2x}{3x^{4/3}} = \dfrac{2x}{3y^2}$.

b. $\dfrac{-8}{y^2}\dfrac{dy}{dx} = 2x$; $\dfrac{dy}{dx} = \dfrac{-xy^2}{4}$. Also $y = \dfrac{8}{x^2+4}$; $\dfrac{dy}{dx} = \dfrac{-16x}{(x^2+4)^2} = \dfrac{-x}{4}\left(\dfrac{8}{x^2+4}\right)^2 = \dfrac{-xy^2}{4}$.

c. $3y^2\dfrac{dy}{dx} = \dfrac{2x(x^2-1) - x^2(2x)}{(x^2-1)^2} = \dfrac{-2x}{(x^2-1)^2}$; $\dfrac{dy}{dx} = \dfrac{-2x}{3y^2(x^2-1)^2}$.

Also $y = \left(\dfrac{x^2}{x^2-1}\right)^{1/3}$;

$$\frac{dy}{dx} = \frac{1}{3}\left(\frac{x^2}{x^2-1}\right)^{-2/3}\left[\frac{2x(x^2-1) - x^2(2x)}{(x^2-1)^2}\right] = \frac{1}{3}y^{-2}\left[\frac{-2x}{(x^2-1)^2}\right] = \frac{-2x}{3y^2(x^2-1)^2}.$$

47. Differentiating the equation $x^3 + y^3 = 2xy$ in (3) implicitly, we find that

$$3x^2 + 3y^2\frac{dy}{dx} = 2y + 2x\frac{dy}{dx}$$

Now $dy/dx = -1$, provided that $3x^2 - 3y^2 = 2y - 2x$, so that $3(x-y)(x+y) = -2(x-y)$. Thus either $x - y = 0$ or $3(x+y) = -2$. Since Q is to lie in the first quadrant, and hence $x > 0$ and $y > 0$, it follows that $3(x+y)$ cannot equal -2. Therefore $x - y = 0$, so $x = y$. Substituting this into $x^3 + y^3 = 2xy$ yields $2x^3 = 2x^2$, so $x = 1$ since $x > 0$. Consequently Q is the point $(1, 1)$.

49. $\frac{2}{3}x^{-1/3}+\frac{2}{3}y^{-1/3}\frac{dy}{dx}=0$; $\frac{dy}{dx}=-\left(\frac{y}{x}\right)^{1/3}$, and at $(2\sqrt{2},2\sqrt{2})$, $\frac{dy}{dx}=-1$.
Thus l: $y-2\sqrt{2}=-1(x-2\sqrt{2})$, or $y=-x+4\sqrt{2}$. The x intercept is $4\sqrt{2}$, and the y intercept is $4\sqrt{2}$, so the area of the triangle is $\frac{1}{2}(4\sqrt{2})(4\sqrt{2})=16$.

51. Differentiating implicitly, we obtain $2(x^2+y^2)(2x+2y(dy/dx))=2x-2y(dy/dx)$. If $dy/dx=0$, then $(x^2+y^2)(2x)=x$, that is, $(2x^2+2y^2-1)x=0$. Since $x\neq 0$ by hypothesis, it follows that $2x^2+2y^2-1=0$, or $y^2=\frac{1}{2}-x^2$. Substituting for y^2 in the original equation, we obtain $(x^2+\frac{1}{2}-x^2)^2=x^2-(\frac{1}{2}-x^2)$, so that $\frac{1}{4}=2x^2-\frac{1}{2}$, or $x=\pm\sqrt{\frac{3}{8}}=\pm\sqrt{6}/4$. Then $y^2=\frac{1}{2}-x^2=\frac{1}{2}-\frac{3}{8}=\frac{1}{8}$. Thus the points at which the tangent lines are horizontal are $(\sqrt{6}/4,\sqrt{2}/4)$, $(\sqrt{6}/4,-\sqrt{2}/4)$, $(-\sqrt{6}/4,\sqrt{2}/4)$, and $(-\sqrt{6}/4,-\sqrt{2}/4)$.

3.7 Related Rates

1. As in (2), $dV/dt=4\pi r^2(dr/dt)$, and we are to determine dV/dt at the instant t_0 when $r=4$. Since $dr/dt=-\frac{1}{2}$ by hypothesis, we have

$$\left.\frac{dV}{dt}\right|_{t=t_0}=4\pi r^2\left.\frac{dr}{dt}\right|_{t=t_0}=4\pi(4^2)(-\tfrac{1}{2})=-32\pi.$$

Thus the volume decreases at the rate of 32π cubic centimeters per minute when $t=t_0$.

3. Here

$$4\pi r^2\frac{dr}{dt}=\frac{dV}{dt}=-\frac{2}{V}=-\frac{2}{4\pi r^3/3}=-\frac{3}{2\pi r^3}$$

and we are to determine dr/dt at the instant t_0 when $r=\frac{1}{2}$. We have

$$\left.\frac{dr}{dt}\right|_{t=t_0}=-\frac{3}{(2\pi r^3)4\pi r^2}=-\frac{3}{8\pi^2r^5}=-\frac{3}{8\pi^2(1/32)}=-\frac{12}{\pi^2}.$$

Thus the radius decreases at the rate of $12/\pi^2$ inches per hour.

5. Here $4\pi r^2(dr/dt)=dV/dt=(d/dt)(4\sqrt{t})=2/\sqrt{t}$ for $t>0$, and we are to determine dr/dt when $t=64$. Since $V=4\sqrt{64}=32$ for $t=64$, we have $32=V=\frac{4}{3}\pi r^3$ at that instant, so $r^3=96/(4\pi)=24/\pi$ and thus $r=(24/\pi)^{1/3}$ when $t=64$. Then

$$\left.\frac{dr}{dt}\right|_{t=64}=\frac{2}{\sqrt{t}}\frac{1}{4\pi r^2}=\frac{2}{\sqrt{64}}\frac{1}{4\pi(24/\pi)^{2/3}}=\frac{1}{64\pi^{1/3}3^{2/3}}.$$

Thus after 64 seconds the radius increases at the rate of $1/(64\pi^{1/3}3^{2/3})$ centimeters per second.

7. Let A denote the area and r the radius of the circular pool. Then $A=\pi r^2$ and $dA/dt=2\pi r(drdt)$, and we are to determine dr/dt at the instant t_0 when $r=10$. Since $dA/dt=3$ by hypothesis, we have

$$3=\left.\frac{dA}{dt}\right|_{t=t_0}=2\pi r\left.\frac{dr}{dt}\right|_{t=t_0}=2\pi(10)\left.\frac{dr}{dt}\right|_{t=t_0},\quad\text{so}\quad\left.\frac{dr}{dt}\right|_{t=t_0}=\frac{3}{2\pi(10)}=\frac{3}{20\pi}.$$

Thus the radius increases at the rate of $3/(20\pi)$ centimeters per minute when $t=t_0$.

9. Let x be the distance from the bottom of the ladder to the wall and y the distance from the ground to the top of the ladder. Then $x^2+y^2 = 13^2$, so $2x(dx/dt)+2y(dy/dt) = 0$. We are to find dx/dt at the instant t_0 when $x = 3$. Because $x^2+y^2 = 13^2$, it follows that if $x = 3$, then $y^2 = 13^2 - x^2 = 13^2 - 3^2 = 160$, so $y = \sqrt{160} = 4\sqrt{10}$. Since $dy/dt = 1$ by hypothesis, for $t = t_0$ the equation $2x(dx/dt) + 2y(dy/dt) = 0$ becomes

$$2(3)\left.\frac{dx}{dt}\right|_{t=t_0} + 2(4\sqrt{10}) = 0, \quad \text{so that} \quad \left.\frac{dx}{dt}\right|_{t=t_0} = -\frac{8\sqrt{10}}{6} = -\frac{4}{3}\sqrt{10}.$$

Thus the base of the ladder approaches the wall at the rate of $\frac{4}{3}\sqrt{10}$ feet per second when $t = t_0$.

11. Let x be the distance from the bottom of the board to the wall and y the distance from the ground to the top of the board, so $x^2 + y^2 = 5^2$, and thus $2x(dx/dt) + 2y(dy/dt) = 0$.

a. We are to find dx/dt at the instant t_0 when $x = 4$ and $dy/dt = -2$. If $x = 4$, then $y^2 = 5^2 - x^2 = 5^2 - 4^2 = 9$, so $y = 3$. We obtain

$$2(4)\left.\frac{dx}{dt}\right|_{t=t_0} + 2(3)(-2) = 0, \quad \text{so} \quad \left.\frac{dx}{dt}\right|_{t=t_0} = \frac{-2(3)(-2)}{2(4)} = \frac{3}{2}.$$

Thus the bottom end slides at the rate of 3/2 feet per second when $t = t_0$.

b. Let A denote the area of the region, so $A = \frac{1}{2}xy$, and $dA/dt = \frac{1}{2}(dx/dt)y + \frac{1}{2}x(dy/dt)$. We are to find dA/dt at the instant t_0 when $x = 4$, $y = 3$, $dx/dt = \frac{3}{2}$, and $dy/dt = -2$. We obtain

$$\left.\frac{dA}{dt}\right|_{t=t_0} = \frac{1}{2}\left(\frac{3}{2}\right)3 + \frac{1}{2}(4)(-2) = -\frac{7}{4}.$$

Thus the area shrinks at the rate of $\frac{7}{4}$ square feet per second when $t = t_0$.

13. Let x be the depth of the water, y the width of the top of the water, and V the volume at any time. Then $V = \frac{1}{2}xy(12) = 6xy$, and we are to find dx/dt at the instant t_0 when $x = \frac{1}{2}$. From the figure, $y/2 = x\tan(\pi/6) = (\sqrt{3}/3)x$, so that $y = (2\sqrt{3}/3)x$ and thus $V = 6x(2\sqrt{3}/3)x = 4\sqrt{3}x^2$. Then $dV/dt = 8\sqrt{3}x(dx/dt)$. Since $dV/dt = 3$ by hypothesis, we obtain

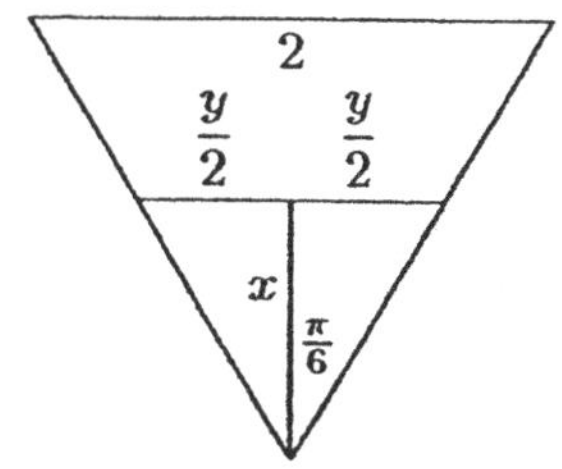

$$3 = 8\sqrt{3}\left(\frac{1}{2}\right)\left.\frac{dx}{dt}\right|_{t=t_0}, \quad \text{so that} \quad \left.\frac{dx}{dt}\right|_{t=t_0} = \frac{3}{8\sqrt{3}\left(\frac{1}{2}\right)} = \frac{\sqrt{3}}{4}.$$

Thus the water level rises at the rate of $\sqrt{3}/4$ feet per minute when $t = t_0$.

15. Since $pV = c$, we have $(dp/dt)V + p(dV/dt) = 0$. We are to find dp/dt at the instant t_0 when $p = 100$ and $V = 20$. Since $dV/dt = -10$ by hypothesis, we obtain

$$\left(\left.\frac{dp}{dt}\right|_{t=t_0}\right)20 + 100(-10) = 0, \quad \text{so that} \quad \left.\frac{dp}{dt}\right|_{t=t_0} = 50.$$

Thus the pressure increases at the rate of 50 pounds per square centimeter per second when $t = t_0$.

17. We are to find dx/dt at the moment t_0 when the nucleus is at $(2c, c\sqrt{3})$, that is, $x = 2c$ and $y = c\sqrt{3}$. Now $2x(dx/dt) - 2y(dy/dt) = 0$, and by hypothesis, $dy/dt = 2$. Therefore

$$\left.\frac{dx}{dt}\right|_{t=t_0} = \frac{y}{x}\frac{dy}{dt} = \frac{c\sqrt{3}}{2c}(2) = \sqrt{3}.$$

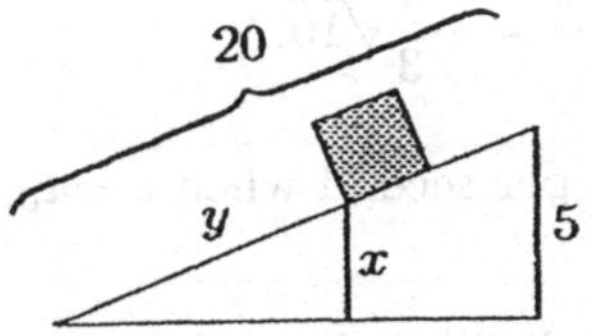

Exercise 19

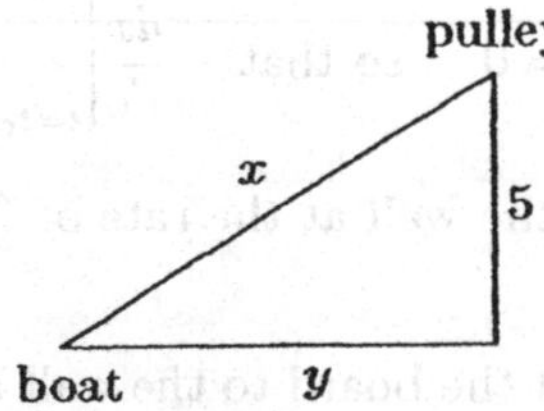

Exercise 21

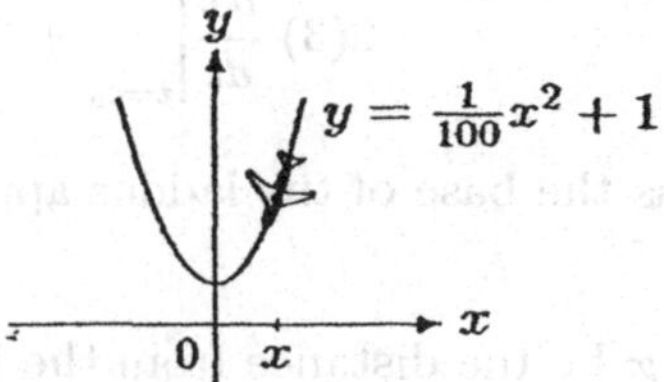

Exercise 23

19. Let x be the height of the bottom end of the box, and y the distance the bottom end has traveled at any time (see the figure on the next page). Then by similar triangles, $x/5 = y/20$, so that $x = \frac{1}{4}y$. We are to find dx/dt. Since $dy/dt = 3$ by hypothesis, we obtain $dx/dt = \frac{1}{4}(dy/dt) = \frac{1}{4}(3) = \frac{3}{4}$. Thus the bottom of the box rises at the rate of $\frac{3}{4}$ foot per second.

21. Let x be the distance along the pulley and y the distance from the bow to the dock at any given time. Then from the figure, $x^2 = y^2 + 5^2$, so that $2x(dx/dt) = 2y(dy/dt)$. We are to find dx/dt at the instant t_0 when $y = 12$. If $y = 12$, then $x^2 = y^2 + 5^2 = 12^2 + 5^2 = 169$, so $x = 13$. Since $dy/dt = -2$ by hypothesis, we obtain

$$2(13)\left.\frac{dx}{dt}\right|_{t=t_0} = 2(12)(-2), \quad \text{so} \quad \left.\frac{dx}{dt}\right|_{t=t_0} = -\frac{24}{13}.$$

Thus the rope is pulled in at the rate of 24/13 feet per second when $t = t_0$.

23. Assume that $x > 0$, as in the figure. By hypothesis, $y = \frac{1}{100}x^2 + 1$, so that $dy/dt = \frac{1}{50}x(dx/dt)$. We are to find dx/dt at the instant t_0 when $y = 2501$. If $y = 2501$, then $\frac{1}{100}x^2 = y - 1 = 2501 - 1 = 2500$, so $x = 500$. Since $dy/dt = -100$ by hypothesis, we obtain

$$-100 = \frac{1}{50}(500)\left.\frac{dx}{dt}\right|_{t=t_0}, \quad \text{so that} \quad \left.\frac{dx}{dt}\right|_{t=t_0} = -10.$$

Thus the shadow moves at the rate of 10 feet per second when $t = t_0$.

25. Let x be the distance between the runner and third base. Then $\tan\theta = x/90$, so that $(\sec^2\theta)(d\theta/dt) = \frac{1}{90}(dx/dt)$. We are to find $d\theta/dt$ at the instant t_0 when $x = 30$. Since the runner is approaching third base at the rate of 24 feet per second, we have $dx/dt = -24$. Moreover, if $x = 30$, then $\tan\theta = \frac{30}{90} = \frac{1}{3}$, and thus $\sec^2\theta = 1 + \tan^2\theta = 1 + (\frac{1}{3})^2 = \frac{10}{9}$. Therefore

$$\left.\frac{d\theta}{dt}\right|_{t=t_0} = \frac{1}{90\sec^2\theta}\left.\frac{dx}{dt}\right|_{t=t_0} = \frac{1}{90(\frac{10}{9})}(-24) = -\frac{6}{25}.$$

Consequently the angle θ is changing at the rate of $\frac{6}{25}$ radians per second when the runner is 30 feet from third base.

27. Let x be the distance between the sports car and the intersection, and y the distance between the police car and sports car (see the figure in Exercise 26). Then $y^2 = x^2 + (\frac{1}{4})^2$, so $2y(dy/dt) = 2x(dx/dt)$. We are to find x under the condition that $dx/dt = -50$ and $dy/dt = -30$. This yields $2y(-30) = 2x(-50)$, so that $y = \frac{5}{3}x$. Since $x^2 = y^2 - (\frac{1}{4})^2$, we substitute $y = \frac{5}{3}x$ to obtain $x^2 = (\frac{5}{3}x)^2 - (\frac{1}{4})^2$. Thus $x^2 = \frac{25}{9}x^2 - \frac{1}{16}$, so that $\frac{16}{9}x^2 = \frac{1}{16}$, or $x^2 = \frac{9}{16^2}$. Therefore $x = \frac{3}{16}$. Consequently the sports car would be $\frac{3}{16}$ mile from the intersection.

29. Let x be the length of string let out, and y the ground distance from the holder to a position directly below the kite. Then $x^2 = y^2 + 100^2$, so that $2x(dx/dt) = 2y(dy/dt)$. We are to find dx/dt at the instant t_0 when $x = 200$. If $x = 200$, then $y^2 = x^2 - 100^2 = 200^2 - 100^2 = 30{,}000$, so $y = 100\sqrt{3}$. Since $dy/dt = 10$ by hypothesis, we have

$$2(200)\left.\frac{dx}{dt}\right|_{t=t_0} = 2(100\sqrt{3})(10), \quad \text{so that} \quad \left.\frac{dx}{dt}\right|_{t=t_0} = 5\sqrt{3}.$$

Thus the string must be let out at the rate of $5\sqrt{3}$ feet per second when $t = t_0$.

Exercise 31

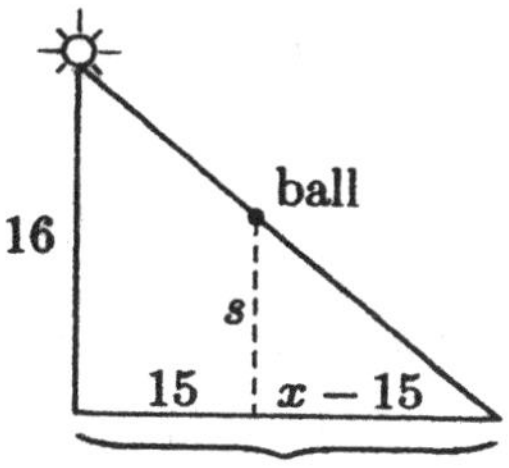

Exercise 33

31. Let x be the altitude of the rocket, and θ the angle of elevation (see the figure). Then $\tan\theta = x/5$, and we are to find $d\theta/dt$ at the instant t_0 when $x = 2$. Now $(\sec^2\theta)(d\theta/dt) = \frac{1}{5}(dx/dt)$, and if $x = 2$, then $\sec\theta = \sqrt{x^2+5^2}/5 = \sqrt{2^2+5^2}/5 = \sqrt{29}/5$, so that $\sec^2\theta = \frac{29}{25}$. Since at that instant $dx/dt = 300$, we obtain

$$\frac{29}{25}\left.\frac{d\theta}{dt}\right|_{t=t_0} = \frac{1}{5}(300), \quad \text{so that} \quad \left.\frac{d\theta}{dt}\right|_{t=t_0} = \frac{1500}{29}.$$

Thus the angle of elevation increases at the rate of of $\frac{1500}{29}$ radians per hour when $t = t_0$.

33. Let x be the distance from the shadow to the base of the street light, and s the height of the ball. We are to find dx/dt when $s = 6$. From the figure, we find that $s/(x-15) = 16/x$, so that $sx = 16x - 240$. Therefore $x = 240/(16-s)$. By the Chain Rule,

$$\frac{dx}{dt} = \frac{240}{(16-s)^2}\frac{ds}{dt}$$

By hypothesis, $ds/dt = -10$ when $s = 6$. Therefore

$$\left.\frac{dx}{dt}\right|_{x=6} = \frac{240}{10^2}(-10) = -24$$

We conclude that when the ball is 6 feet high, its shadow is moving along the ground toward the street light at the rate of 24 feet per second.

35. Using the notation in the figure, we have $x = 3000 \tan\theta$, so that $dx/dt = 3000(\sec^2\theta)(d\theta/dt)$. When the distance between the helicopter and the searchlight is 5000 feet, we have $\sec\theta = \frac{5000}{3000} = \frac{5}{3}$. Since $dx/dt = 100$, it follows that $100 = 3000(\frac{5}{3})^2(d\theta/dt)$, so that $d\theta/dt = \frac{3}{250}$ (radians per second) when the distance between the helicopter and the searchlight is 5000 feet.

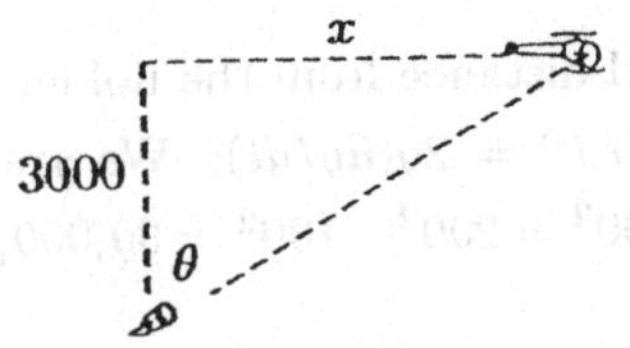

Exercise 35

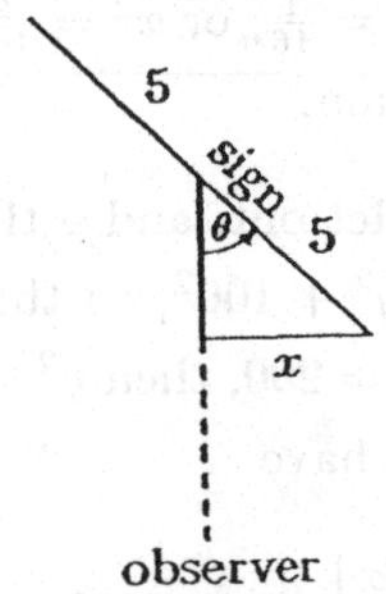

Exercise 37

37. Let x be one half the width of the sign as seen by the observer, and θ the angle shown in the figure. We are to find dx/dt at the instant t_0 when $x = 3$ and $dx/dt > 0$. Since the sign makes 10 revolutions per minute and appears to grow when $t = t_0$, we have $d\theta/dt = 20\pi$. Notice that $x = 5\sin\theta$, so that $dx/dt = (5\cos\theta)(d\theta/dt)$. If $x = 3$, then $\cos\theta = \frac{4}{5}$. Therefore

$$\left.\frac{dx}{dt}\right|_{t=t_0} = (5\cos\theta)\left.\frac{d\theta}{dt}\right|_{t=t_0} = 5\left(\frac{4}{5}\right)(20\pi) = 80\pi.$$

Thus half the width changes at the rate of 80π feet per minute, and the total width changes at the rate of 160π feet per minute when $t = t_0$.

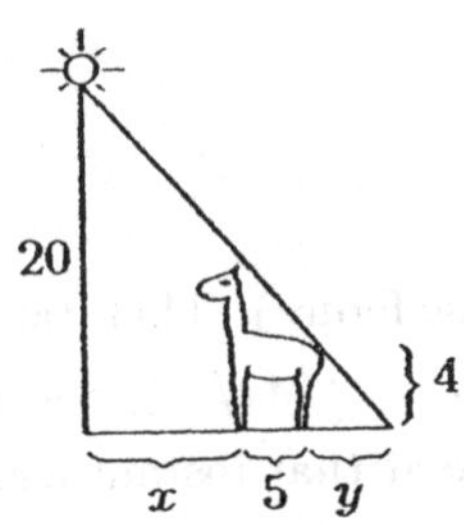

Light ray over deer's rump

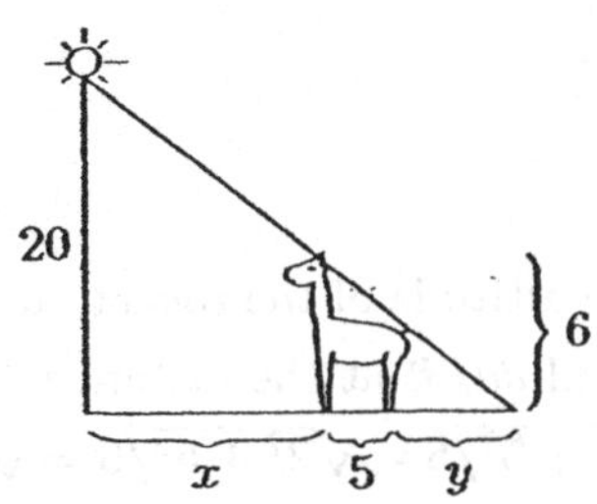

Light ray over deer's head

39. Let x be the distance from the base of the street light to forelegs of the deer, and let y be the distance from the rear legs of the deer to the tip of the shadow. We want y as a function of x.

Notice that if the deer is close to the base of the light, the tip is formed by a ray just passing over the deer's rump. Since the rump is 4 feet high and the deer is 5 feet long, by using similar triangles we have

$$\frac{y}{4} = \frac{x+y+5}{20}, \quad \text{or} \quad y = \frac{x+5}{4}.$$

Thus for "small" x, $dy/dt = \frac{1}{4}(dx/dt) = -\frac{3}{4}$ (feet per second).

Now if the deer is far from the base of the light, the tip of the shadow is formed by light just passing over the deer's head. Since the head is 6 feet high and the deer is 5 feet long, by using similar triangles we have

$$\frac{y+5}{6} = \frac{x+y+5}{20}, \quad \text{or} \quad y = \frac{3}{7}x - 5.$$

Thus for "large" x, $dy/dt = \frac{3}{7}(dx/dt) = -9/7$ (feet per second).

We now determine the boundary between "x small" and "x large". That boundary occurs when the same ray of light just passes over the head and just passes over the rump. At that point $(x+5)/4 = y = \frac{3}{7}x - 5$. Solving for x, we find the boundary at $x = 35$ (feet). For (a) we conclude that when the deer is 48 feet from the street light, x is "large", so the shadow is decreasing at $\frac{9}{7}$ feet per second. For (b) we conclude that when the deer is 24 feet from the street light, x is "small", so the shadow is decreasing at $\frac{3}{4}$ feet per second.

3.8 Polynomial Approximation and the Newton Raphson Method

1. Let $f(x) = \sqrt{x}$, $a = 100$, and $h = 1$. Then $f'(x) = \frac{1}{2}x^{-1/2}$, $f'(100) = \frac{1}{20}$, and $\sqrt{101} \approx f(100) + f'(100)h = 10 + \frac{1}{20}(1) = 10.05$.

3. Let $f(x) = \sqrt[3]{x}$, $a = 27$, and $h = 2$. Then $f'(x) = \frac{1}{3}x^{-2/3}$, $f'(27) = \frac{1}{27}$, and $\sqrt[3]{29} \approx f(27) + f'(27)h = 3 + \frac{1}{27}(2) \approx 3.074074074$.

5. Let $f(x) = x^{4/3}$, $a = 27$, and $h = 1$. Then $f'(x) = \frac{4}{3}x^{1/3}$, $f'(27) = 4$, and $(28)^{4/3} \approx f(27) + f'(27)h = 81 + 4(1) = 85$.

7. Let $f(x) = \cos x$, $a = \pi/6$, and $h = -\pi/78$. Then

$$f'(x) = -\sin x, \quad f'(x)\left(\frac{\pi}{6}\right) = \frac{-1}{2}$$

and

$$\cos\left(\frac{2\pi}{13}\right) \approx f\left(\frac{\pi}{6}\right) + f'\left(\frac{\pi}{6}\right)h = \frac{\sqrt{3}}{2} + \left(\frac{-1}{2}\right)\left(\frac{-\pi}{78}\right) = \frac{\sqrt{3}}{2} + \frac{\pi}{156} \approx .8861638182.$$

9. Let $f(x) = \sec x$, $a = \pi/4$, and $h = -\pi/68$. Then $f'(x) = \sec x \tan x$, $f'(\pi/4) = \sqrt{2}$, and $\sec(4\pi/17) \approx f(\pi/4) + f'(\pi/4)h = \sqrt{2} + \sqrt{2}(-\pi/68) \approx 1.348877049$.

11. Let $f(x) = e^x$, $a = 0$, and $h = 0.1$. Then $f'(x) = e^x$, $f'(0) = e^0 = 1$, and $e^{0.1} \approx f(0) + f'(0)h = 1 + 1(0.1) = 1.1$.

13. Let $f(x) = \ln x$, $a = 1$, and $h = 0.1$. Then $f'(x) = 1/x$, $f'(1) = 1$, and $\ln 1.1 \approx f(1) + f'(1)h = 0 + 1(0.1) = 0.1$.

15. $df = f'(4)h = \left(\frac{1}{2\sqrt{4}}\right)(.2) = .05$

17. $df = f'(2)h = \frac{1}{2}(1 + 2^3)^{-1/2}(12)(.01) = .02$

19. $df = 15x^2\,dx$

21. $df = -\sin x \cos(\cos x)\,dx$

23. $du = 2x^3(1 + x^4)^{-1/2}\,dx$

25. Let $f(x) = \cos x$, $a = \pi/6$, and $h = 2\pi/13 - \pi/6 = -\pi/78$. Then $f'(x) = -\sin x$ and $f''(x) = -\cos x$. Next, $|f''(x)| = |-\cos x| \le 1$ for $-\pi/78 \le x \le 0$, so we can let $M = 1$. Therefore by (3),

$$\text{error} \le \frac{1}{2}(1)\left(-\frac{\pi}{78}\right)^2 = \frac{\pi^2}{12{,}168}.$$

Consequently $\pi^2/12{,}168$ is an upper bound for the error.

27. Let $f(x) = \sqrt[3]{x}$, $a = 27$, and $h = 28 - 27 = 1$. Then $f'(x) = \frac{1}{3}x^{-2/3}$ and $f''(x) = -\frac{2}{9}x^{-5/3}$. Next, $|f''(x)| = \left|-\frac{2}{9}x^{-5/3}\right| \leq \frac{2}{9}(27)^{-5/3} = \frac{2}{2187}$ for $27 \leq x \leq 28$, so we can let $M = \frac{2}{2187}$. Therefore by (3),

$$\text{error} \leq \frac{1}{2}\left(\frac{2}{2187}\right)1^2 = \frac{1}{2187}.$$

Consequently 1/2187 is an upper bound for the error.

29. Let $f(x) = \sqrt[3]{x} = x^{1/3}$, and let $a = 27$, which is close to 29 and has a convenient cube root. Then $f'(x) = x^{-2/3}/3$ and $f''(x) = -2x^{-5/3}/9$. Therefore $f(a) = 3$, $f'(a) = 1/27$, and $f''(a) = -2/2187$. By (8), $p_2(x) = 3 + (x - 27)/27 - (x - 27)^2/2187$. Now we use $p_2(29)$ to approximate $f(29) = \sqrt[3]{29}$:

$$\sqrt[3]{29} \approx p_2(29) = 3 + 2/27 - 4/2187 \approx 3.072245085$$

31. Let $f(x) = \cos x$, and let $a = \pi/6$, which is close to $2\pi/13$ and has a convenient cosine value. Then $f'(x) = -\sin x$ and $f''(x) = -\cos x$. Therefore $f(a) = \cos(\pi/6) = \sqrt{3}/2$, $f'(a) = -\sin(\pi/6) = -1/2$, and $f''(a) = -\cos(\pi/6) = -\sqrt{3}/2$. By (8),

$$p_2(x) = \frac{\sqrt{3}}{2} - \frac{1}{2}(x - \frac{\pi}{6}) - \frac{\sqrt{3}}{2}(x - \frac{\pi}{6})^2$$

We use $p_2(2\pi/13)$ to approximate $f(2\pi/13) = \cos(2\pi/13)$:

$$\cos\frac{2\pi}{13} \approx p_2(\frac{2\pi}{13}) = \frac{\sqrt{3}}{2} - \frac{1}{2}(\frac{2\pi}{13} - \frac{\pi}{6}) - \frac{\sqrt{3}}{2}(\frac{2\pi}{13} - \frac{\pi}{6})^2 \approx 0.8847589319$$

33. Let $f(x) = \ln x$ and $a = 1$. Then $f'(x) = 1/x$ and $f''(x) = -1/x^2$. Therefore $f(a) = \ln 1 = 0$, $f'(a) = 1$, and $f''(a) = -1$. By (8), $p_2(x) = (x-1) - (x-1)^2/2$. We use $p_2(1.1)$ to approximate $f(1.1) = \ln 1.1$:

$$\ln 1.1 \approx p_2(1.1) = 0.1 - \frac{1}{2}(0.1)^2 = 0.095$$

35. Let $f(x) = x^3 - 3x - 1$, so that $f'(x) = 3x^2 - 3$. Letting the initial value of c be 2, we obtain 1.879385242 for the desired approximate solution.

37. Let $f(x) = x^3 - 2x - 5$, so that $f'(x) = 3x^2 - 2$. Letting the initial value of c be 2, we obtain 2.094551482 for the desired approximate solution.

39. Let $f(x) = 2x^3 - 5x - 3$, so that $f'(x) = 6x^2 - 5$. Letting the initial value of c be -0.5, we obtain -0.8228756555 for the desired approximate solution.

41. Let $f(x) = \tan x - x$, so that $f'(x) = \sec^2 x - 1$. Letting the initial value of c be 4.5, we obtain 4.493409458 for the desired approximate solution.

43. Let $f(x) = e^{-x} - x$, so that $f'(x) = -e^{-x} - 1$. Letting the initial value of c be 0, we obtain 0.5671432904 for the desired approximate solution.

45. Let $f(x) = x^2 - 15$, so that $f'(x) = 2x$. Using the Newton-Raphson method with initial value 4 for c, we find that $\sqrt{15} \approx 3.872983346$.

47. Let $f(x) = x^3 - 9$, so that $f'(x) = 3x^2$. Using the Newton-Raphson method with initial value 2 for c, we find that $\sqrt[3]{9} \approx 2.080083823$.

49. Notice that $f'(x) = 4x^3 + 4x - 1$. Letting the initial value of c be first -0.5 and then 0.5, we obtain $-.1823735451$ and $.6001766211$, respectively, for the approximate zeros of f.

51. Notice that $f'(x) = 4x^3 + 2x + 8$. Letting the initial value of c be -1, we obtain 0.1230777986 for an approximate zero of f. However, 0.1230777986 lies outside the interval $[-2, 0]$, and hence is not the desired zero.

53. If the initial value of c is $1/\sqrt{5}$, then successive values of c oscillate, $-1/\sqrt{5}$, $1/\sqrt{5}$, $-1/\sqrt{5}$, $1/\sqrt{5}$, and so on. (Actually, f has no zeros in $(0, 1)$.)

55. Since $p_2(x) = f(a) + f'(a)(x-a) + \frac{1}{2}f''(a)(x-a)^2$ and since $f(a)$, $f'(a)$, and $f''(a)$ are constants, we find that $p_2'(x) = f'(a) + f''(a)(x-a)$ and hence that $p_2''(x) = f''(a)$. Substituting a for x, we conclude that $p_2(a) = f(a)$, $p_2'(a) = f'(a)$, and $p_2''(a) = f''(a)$.

57. The volume of a hemispherical dome of radius r is given by $V(r) = \frac{2}{3}\pi r^3$. Since $V'(r) = 2\pi r^2$, we use (2) to determine that the volume of paint necessary is

$$V\left(20 + \frac{1}{1200}\right) - V(20) \approx V'(20)\frac{1}{1200} = 2\pi(20)^2\frac{1}{1200} = \frac{2}{3}\pi \approx 2.094395102 \text{ (cubic feet).}$$

59. a. Let $f(v) = 1/\sqrt{1-(v^2/c^2)}$. Then $f(c/2) = 1/\sqrt{1-(1/4)} = (2\sqrt{3})/3 \approx 1.154700538$.

b. Let $f(x) = 1/\sqrt{1-x}$, $a = 0$, and $h = v^2/c^2 = 1/(3600)^2$. Then $f'(x) = \frac{1}{2}(1-x)^{-3/2}$, $f'(0) = \frac{1}{2}$, and

$$\frac{1}{\sqrt{1-(v^2/c^2)}} = f\left(\frac{v^2}{c^2}\right) = f\left(\frac{1}{(3600)^2}\right) \approx f(0) + f'(0)h = 1 + \frac{1}{2}\frac{1}{(3600)^2} \approx 1.00000004.$$

61. a. By hypothesis, $P(0) = 500$; by Figure 3.58, $P'(0) = 5$. Letting $a = 0$ and $h = 1$ in (2), we have $P(1) \approx P(0) + P'(0)h = 500 + 5(1) = 505$.

b. From (a) we find that $P(1) = 505$, and from Figure 3.58, $P'(1) = 5$. Letting $a = 1$ and $h = 1$ in (2), we have $P(2) \approx P(1) + P'(1)h = 505 + 5(1) = 510$.

c. By using our data and Figure 3.58, we obtain in succession

$$P(3) \approx P(2) + P'(2)h = 510 + 10(1) = 520$$
$$P(4) \approx P(3) + P'(3)h = 520 + 20(1) = 540$$
$$P(5) \approx P(4) + P'(4)h = 540 + 30(1) = 570$$
$$P(6) \approx P(5) + P'(5)h = 570 + 0(1) = 570$$
$$P(7) \approx P(6) + P'(6)h = 570 - 10(1) = 560$$

Chapter 3 Review

1. $f'(x) = -12x^2 - \dfrac{4}{x^3}$

3. $g'(x) = \dfrac{(2x-1)^2 - x[2(2x-1)(2)]}{(2x-1)^4} = -\dfrac{2x+1}{(2x-1)^3}$

5. $f'(t) = (-\sin t)\sin 2t + (\cos t)(2\cos 2t) = -\sin t \sin 2t + 2\cos t \cos 2t$

7. $f'(t) = 5\tan t + 5t\sec^2 t + 9\sec 3t \tan 3t$

9. $f'(t) = 2e^{2t}\ln(3+e^t) + e^{2t}\dfrac{1}{3+e^t}e^t = 2e^{2t}\ln(3+e^t) + \dfrac{e^{3t}}{3+e^t}$

11. $\dfrac{dy}{dx} = 12x^2 - \sqrt{3} - \dfrac{2}{5x^2}$

13. $\dfrac{dy}{dx} = \dfrac{\cos x(1-\sec x) - \sin x(-\sec x\tan x)}{(1-\sec x)^2} = \dfrac{\cos x - 1 + \tan^2 x}{(1-\sec x)^2}$

15. $\dfrac{dy}{dx} = 3x^2\sqrt{x^2-4} + x^3[\frac{1}{2}(x^2-4)^{-1/2}(2x)] = 3x^2\sqrt{x^2-4} + x^4/\sqrt{x^2-4}$

17. $\dfrac{dy}{dx} = e^x + xe^x + 5e^{-x}$

19. $f'(x) = 9x^2 - 4x$, so $f'(1) = 5$. Thus l: $y - 5 = 5(x-1)$, or $y = 5x$.

21. $f'(x) = \cos\sqrt{2}x - x(\sqrt{2}\sin\sqrt{2}x)$, so $f'(0) = 1$. Thus l: $y - 0 = 1(x-0)$, or $y = x$.

23. $f'(x) = \sqrt{x-1} + x\dfrac{1}{2\sqrt{x-1}}$, so $f'(5) = 2 + \frac{5}{4} = \frac{13}{4}$. Thus l: $y - 10 = \frac{13}{4}(x-5)$, or $y = \frac{13}{4}x - \frac{25}{4}$.

25. $\displaystyle\lim_{x\to 0^-}\frac{f(x)-f(0)}{x-0} = \lim_{x\to 0^-}\frac{2\sin x - 0}{x} = 2\lim_{x\to 0^-}\frac{\sin x}{x} = 2;$

$\displaystyle\lim_{x\to 0^+}\frac{f(x)-f(0)}{x-0} = \lim_{x\to 0^+}\frac{3x^2+2x-0}{x} = \lim_{x\to 0^+}(3x+2) = 2.$

Therefore $f'(0) = 2$. Thus l: $y - 0 = 2(x-0)$, or $y = 2x$.

27. $f'(x) = 3x^{11} - 36x^5$; $f''(x) = 33x^{10} - 180x^4$

29. $f'(t) = 3t(t^2+9)^{1/2}$; $f''(t) = 3(t^2+9)^{1/2} + 3t^2(t^2+9)^{-1/2}$

31. $f'(x) = 2x + \ln x + 1$; $f''(x) = 2 + 1/x$

33. $9y^2\dfrac{dy}{dx} - \left(8xy + 4x^2\dfrac{dy}{dx}\right) + \left(y + x\dfrac{dy}{dx}\right) = 0$; $\dfrac{dy}{dx} = \dfrac{8xy - y}{9y^2 - 4x^2 + x}$

35. $\dfrac{dy}{dx}(\sqrt{x}+1) + y\left(\dfrac{1}{2\sqrt{x}}\right) = 1$; $\dfrac{dy}{dx} = \dfrac{2\sqrt{x} - y}{2\sqrt{x}(\sqrt{x}+1)}$

37. $3y^2\dfrac{dy}{dx} + \cos(xy^2)\left[y^2 + 2xy\dfrac{dy}{dx}\right] = 0$; $\dfrac{dy}{dx} = \dfrac{-y^2\cos xy^2}{3y^2 + 2xy\cos xy^2} = -\dfrac{y\cos xy^2}{3y + 2x\cos xy^2}$

39. $6x^2 - (4\cos 4y)\dfrac{dy}{dx} = 2xy + x^2\dfrac{dy}{dx}$; $\dfrac{dy}{dx} = \dfrac{6x^2 - 2xy}{x^2 + 4\cos 4y}$. At $(1,0)$, $\dfrac{dy}{dx} = \dfrac{6}{1+4} = \dfrac{6}{5}$.

41. $\dfrac{dx}{dt}y + x\dfrac{dy}{dt} = 0$, so $\dfrac{dy}{dt} = -\dfrac{y}{x}\dfrac{dx}{dt}$

43. $df = (2x\cos x - x^2\sin x)\,dx$

45. $df = 5(x - e^x)^4(1 - e^x)\,dx$

47. Let $f(x) = 1 + \sqrt{x}$. Then $f'(x) = x^{-1/2}/2$ and $f''(x) = -x^{-3/2}/4$. For the linear approximation we use (2) in Section 3.8 with $a = 9$ and $h = 1$. We find the $1 + \sqrt{10} = f(10) \approx f(9) + f'(9)(1) = 4 + 1/6 \approx 4.166666667$. For the quadratic approximation we use $p_2(10)$, where p_2 is given by (8) in Section 3.8 with $a = 9$. Since

$$p_2(x) = f(9) + f'(9)(x-9) + \frac{1}{2}f''(9)(x-9)^2 = 4 + \frac{1}{6}(x-9) - \frac{1}{216}(x-9)^2$$

we find that

$$1 + \sqrt{10} \approx p_2(10) = 4 + \frac{1}{6} - \frac{1}{216} \approx 4.162037037$$

49. Notice that $f'(x) = 2x - 4\sin x$ and $f''(x) = 2 - 4\cos x$. Letting the initial value of c be 2, we obtain 1.895494267 as an approximate solution of $f'(x) = 0$. Since f' is an odd function, -1.895494267 is also an approximate solution. Finally, 0 is a solution.

51. Notice that $f'(x) = \ln x - 1/x$ and $f''(x) = 1/x + 1/x^2$. Letting the initial value of c be 1, we obtain 1.763222834 as an approximate solution of $f'(x) = 0$.

53. To find the point of intersection, we solve the equation

$$x^2 + 1 = x^2 - \cos\left(\frac{\pi}{x^2+1}\right), \quad \text{or} \quad \cos\left(\frac{\pi}{x^2+1}\right) = -1$$

so that $\pi/(x^2+1) = \pi + 2n\pi$ for some integer n. Since $0 < \pi/(x^2+1) \le \pi$, the only solution is $x = 0$. Since

$$f'(x) = 2x \quad \text{and} \quad g'(x) = 2x - \frac{2\pi x}{(x^2+1)^2}\sin\left(\frac{\pi}{x^2+1}\right)$$

we have $f'(0) = 0 = g'(0)$. Thus the tangent lines at the point of intersection are identical.

55. $\displaystyle\lim_{x\to a}\frac{f(x)-f(a)}{x^{1/2}-a^{1/2}} = \lim_{x\to a}\left[\frac{f(x)-f(a)}{x^{1/2}-a^{1/2}}\,\frac{x^{1/2}+a^{1/2}}{x^{1/2}+a^{1/2}}\right] = \lim_{x\to a}\frac{f(x)-f(a)}{x-a}\cdot\lim_{x\to a}(x^{1/2}+a^{1/2}) = 2a^{1/2}f'(a)$

57. $g'(x) = f'(x^2)(2x)$, so that $g'(0) = [f'(0)](0) = 0$.

59. Since $dy/dx = -1/x^2$, the tangent line at $(a, 1/a)$ is

$$y - \frac{1}{a} = \frac{-1}{a^2}(x - a), \quad \text{or} \quad y = \frac{-1}{a^2}x + \frac{2}{a}.$$

To find the tangent line that passes through $(4, 0)$, we substitute $x = 4$, $y = 0$ in the equation of the tangent line: $0 = (-1/a^2)(4) + (2/a) = (-4 + 2a)/a^2$. Thus $a = 2$, so the particular tangent line is $y = (-1/4)x + 1$. The x and y intercepts of this tangent line are 4 and 1, respectively. The area of the triangle is $\frac{1}{2}(4)(1) = 2$.

61. Let h denote the height (or depth) of the water. If water is flowing in at a constant rate, then h is increasing but at an ever slower rate. Thus the rate of change of h, which is h', is decreasing, and therefore the graph in Figure 3.61(c) is reasonable and the graph in Figure 3.61(a) is unreasonable. Since the rate of change never becomes 0, neither of the graphs in Figures 3.61 (b) and (d) could be the graph of the rate of change.

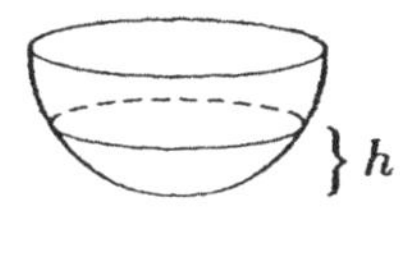

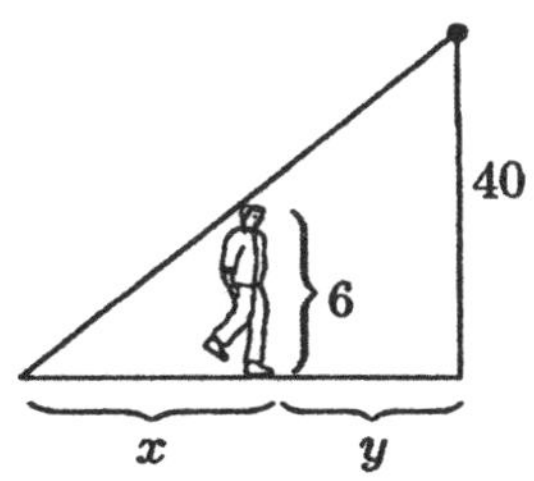

Exercise 63

63. Let x be the length of the shadow and y the distance from the person to the tower. By the figure, $(x + y)/40 = x/6$, so that $6x + 6y = 40x$ and thus $x = \frac{3}{17}y$. We are to show that dx/dt is constant and negative. Now $dx/dt = \frac{3}{17}(dy/dt)$, and since $dy/dt = -150$ by hypothesis, we obtain $dx/dt = \frac{3}{17}(-150) = -\frac{450}{17}$, a negative constant.

65. $m_C(x) = C'(x) = 3 + 1/\sqrt{x}$ and $m_R(x) = R'(x) = 5/\sqrt{x}$, so we need to determine the value of x for which $0 \le x \le 2$ and $3 + 1/\sqrt{x} = 5/\sqrt{x}$, or equivalently, $3 = 4\sqrt{x}$. Therefore $\sqrt{x} = \frac{4}{3}$, so $x = \frac{16}{9}$. Consequently if $a = \frac{16}{9}$, then $m_C(a) = m_R(a)$.

Cumulative Review(Chapters 1–2)

1. The given inequality is equivalent to

$$\frac{x(x+\sqrt{3})(x-\sqrt{3})}{(1-x)^3} > 0.$$

From the diagram we see that the solution is the union of $(-\sqrt{3}, 0)$ and $(1, \sqrt{3})$.

x	− − − − − − − 0 + + + + + +
$x+\sqrt{3}$	− 0 + + + + + + + + + + +
$x-\sqrt{3}$	− − − − − − − − − − − − 0 +
$(1-x)^3$	+ + + + + + + + + 0 − − − −
$\frac{x(x+\sqrt{3})(x-\sqrt{3})}{(1-x)^3}$	− 0 + + + + 0 − + + 0 −

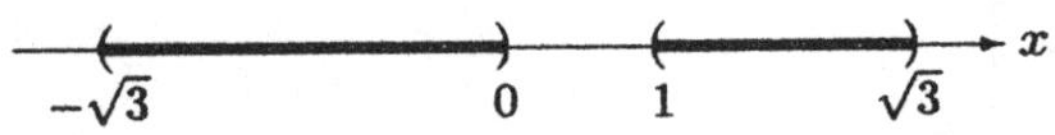

3. The given inequality is equivalent to $3-2 < 1/|x| < 3+2$, or $\frac{1}{5} < |x| < 1$. Thus the solution is the union of $(-1, -\frac{1}{5})$ and $(\frac{1}{5}, 1)$.

5. Since $f(x) = 6x^2 - x - 2 = (2x+1)(3x-2)$, the inequality $f(x) > 0$ is equivalent to $(2x+1)(3x-2) > 0$. From the diagram we see that the solution is the union of $(-\infty, -\frac{1}{2})$ and $(\frac{2}{3}, \infty)$.

$2x+1$	− − − 0 + + + + + + + +
$3x-2$	− − − − − − − − 0 + + +
$(2x+1)(3x-2)$	+ + + 0 − − − − 0 + + +

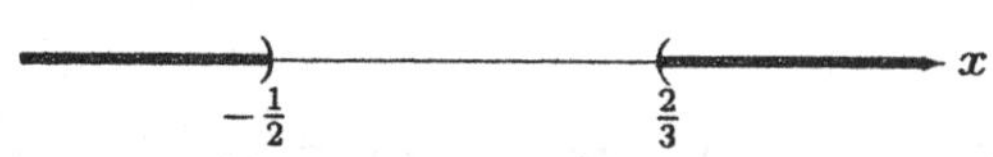

7. The domain consists of all x for which $x^2 - 1 \geq 0$ (hence $x \leq -1$ or $x \geq 1$) and $\sqrt{x^2-1} - x \geq 0$. But if $x \geq 1$, then the fact that $x^2 - 1 < x^2$ implies that $\sqrt{x^2-1} < x$ (so $\sqrt{x^2-1} - x < 0$); if $x \leq -1$, then $-x \geq 0$, so $\sqrt{x^2-1} - x \geq 0$. Thus the domain is $(-\infty, -1]$.

9. $\lim_{x\to 2} \frac{x^2-3x+2}{x^2-5x+6} = \lim_{x\to 2} \frac{(x-1)(x-2)}{(x-2)(x-3)} = \lim_{x\to 2} \frac{x-1}{x-3} = -1.$

11. Since $|x^2| = x^2$, we have

$$\lim_{x\to 0}\frac{|x^3|-x^2}{x^3+x^2} = \lim_{x\to 0}\frac{|x|x^2-x^2}{x^3+x^2} = \lim_{x\to 0}\frac{|x|-1}{x+1} = \frac{0-1}{0+1} = -1.$$

13. $$\lim_{x\to 2}\frac{f(x)-f(2)}{x-2} = \lim_{x\to 2}\frac{\sqrt{2x^2-4}-2}{x-2} = \lim_{x\to 2}\frac{\sqrt{2x^2-4}-2}{x-2}\,\frac{\sqrt{2x^2-4}+2}{\sqrt{2x^2-4}+2}$$
$$= \lim_{x\to 2}\frac{2x^2-4-4}{(x-2)(\sqrt{2x^2-4}+2)} = \lim_{x\to 2}\frac{2(x-2)(x+2)}{(x-2)(\sqrt{2x^2-4}+2)}$$
$$= \lim_{x\to 2}\frac{2(x+2)}{\sqrt{2x^2-4}+2} = 2$$

15.

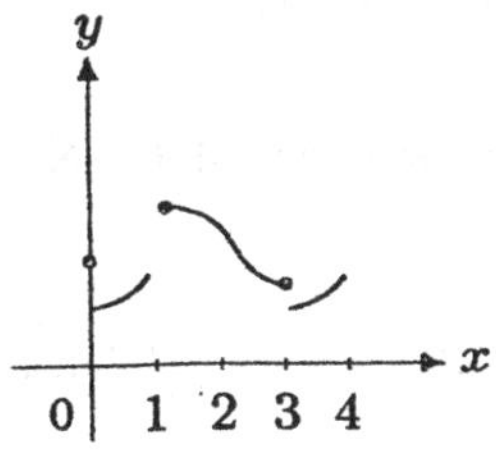

17. $f(\sqrt{a}) = \frac{1}{2}\left(\sqrt{a}+\frac{a}{\sqrt{a}}\right) = \frac{1}{2}(\sqrt{a}+\sqrt{a}) = \sqrt{a}.$

Chapter 4

Applications of the Derivative

4.1 Maximum and Minimum Values

1. $f'(x) = 2x + 4 = 2(x+2)$, so $f'(x) = 0$ for $x = -2$. Critical number: -2.

3. $f'(x) = 12x^3 + 12x^2 - 24x = 12x(x+2)(x-1)$, so $f'(x) = 0$ for $x = -2$, 0, or 1. Critical numbers: -2, 0, 1.

5. $g'(x) = 1 - (1/x^2) = (x^2 - 1)/x^2$, so $g'(x) = 0$ for $x = \pm 1$. Critical numbers: -1, 1.

7. $k'(t) = \dfrac{\sqrt{t^2+1} - (t^2/\sqrt{t^2+1})}{t^2+1} = \dfrac{1}{(t^2+1)^{3/2}}$, so $k'(t)$ is never 0 and exists for all t. No critical numbers.

9. $f'(x) = \cos x$, so $f'(x) = 0$ for $x = (\pi/2) + n\pi$ for any integer n. Critical numbers: $(\pi/2) + n\pi$ for any integer n.

11. $f'(x) = 1 + \cos x$, so $f'(x) = 0$ for $x = \pi + 2n\pi = (2n+1)\pi$ for any integer n. Critical numbers: $(2n+1)\pi$ for any integer n.

13. $f'(z) = 1$ for $z > 2$, $f'(z) = -1$ for $z < 2$, and $f'(2)$ does not exist. Critical number: 2.

15. $f'(x) = 2xe^x + x^2e^x = x(2+x)e^x$, so $f'(x) = 0$ for $x = -2$ or $x = 0$. Critical numbers: -2, 0.

17. $f'(x) = \ln x + x(1/x) = 1 + \ln x$, so $f'(x) = 0$ for $x = 1/e$. Critical number: $1/e$.

19. Since $f'(x) = 2x - 1$, the only critical number in $(0, 2)$ is $\frac{1}{2}$. Thus the extreme values of f on $[0, 2]$ can occur only at 0, $\frac{1}{2}$, or 2. Since $f(0) = 0$, $f(\frac{1}{2}) = -\frac{1}{4}$ and $f(2) = 2$, the minimum value of f on $[0, 2]$ is $f(\frac{1}{2})$, which equals $-\frac{1}{4}$, and the maximum value is $f(2)$, which equals 2.

21. Since $f'(t) = 1/(2t^2)$ is never zero and is defined for every positive number t, there are no critical numbers. Because f can have an extreme value on $(0, \infty)$ only at a critical number or at an endpoint in the interval, and because neither exists, f has neither maximum nor minimum values.

23. Since $k'(z) = z/\sqrt{1+z^2}$, the only critical number of k in $(-2,3)$ is 0. Thus the extreme values of k on $[-2,3]$ can occur only at -2, 0, or 3. Since $k(-2) = \sqrt{5}$, $k(0) = 1$ and $k(3) = \sqrt{10}$, the minimum value of k on $[-2,3]$ is $k(0)$, which equals 1, and the maximum value is $k(3)$, which equals $\sqrt{10}$.

25. Since $f'(x) = 1/(2\sqrt{x})$ for $x > 0$ and $f'(x) = -1/(2\sqrt{-x})$ for $x < 0$ and $f'(0)$ does not exist, the only critical number of f in $(-1,2)$ is 0. Thus the only possible extreme value of f on $(-1,2)$ can occur at 0. Observe that for $-1 < x < 2$ we have $f(x) = \sqrt{|x|} \geq 0 = f(0)$, so that the minimum value of f on $(-1,2)$ is $f(0)$, which equals 0, and there is no maximum value.

27. Since $f'(x) = -\frac{1}{3}x^{-2/3}\cos\sqrt[3]{x}$ for $x \neq 0$ and $f'(0)$ does not exist, the only critical number of f in $(-\pi^3/27, \pi^3/8)$ is 0. Thus the extreme values of f on $[-\pi^3/27, \pi^3/8]$ can occur at $-\pi^3/27$, 0, or $\pi^3/8$. Since $f(-\pi^3/27) = -\sin(-\pi/3) = \frac{1}{2}\sqrt{3}$, $f(0) = 0$, and $f(\pi^3/8) = -\sin(\pi/2) = -1$, the minimum value of f on $[-\pi^3/27, \pi^3/8]$ is $f(\pi^3/8)$, which equals -1, and the maximum value is $f(-\pi^3/27)$, which equals $\frac{1}{2}\sqrt{3}$.

29. Since $f'(x) = \frac{1}{2}\sec^2(x/2) > 0$ for x in $(-\pi/2, \pi/6)$, f has no critical numbers in $(-\pi/2, \pi/6)$. Along with the fact that $(-\pi/2, \pi/6)$ has no endpoints, this means that f has no extreme value on $(-\pi/2, \pi/6)$.

31. Since $f'(x) = e^x + xe^x = (1+x)e^x$, the only critical number of f in $(-2,0)$ is -1. Thus the extreme values of f on $[-2,0]$ can occur only at -2, -1, or 0. Since $f(-2) = -2e^{-2}$, $f(-1) = -e^{-1}$, and $f(0) = 0$, the minimum value of f on $[-2,0]$ is $f(-1)$, which equals $-e^{-1}$, and the maximum value of f on $[-2,0]$ is $f(0)$, which equals 0.

33. Since $f'(x) = 1 - 2/x$, f has no critical numbers in $(\frac{1}{2}, 2)$. Thus the extreme values of f on $[\frac{1}{2}, 2]$ occur at the endpoints $\frac{1}{2}$ and 2. Since $f'(x) < 0$ for $\frac{1}{2} < x < 2$, f is decreasing on $[\frac{1}{2}, 2]$. Thus the minimum value of f on $[\frac{1}{2}, 2]$ is $f(2)$, which equals $2 - 2\ln 2$, and the maximum value of f on $[\frac{1}{2}, 2]$ is $f(\frac{1}{2})$, which equals $\frac{1}{2} - 2\ln\frac{1}{2} = \frac{1}{2} + 2\ln 2$.

35. $f'(x) = x^3 + 3x^2 - 1$ and $f''(x) = 3x^2 + 6x$. The Newton-Raphson method yields (approximate) critical numbers -2.879385242 and $.5320888862$.

37. $f'(x) = 2x + e^x$ and $f''(x) = 2 + e^x$. The Newton-Raphson method yields approximately $-.3517337112$ as the only critical number of f in $[-1,1]$. Since $f(-1) = 1 + e^{-1}$, $f(-.3517337112) \approx .8271840261$, and $f(1) = 1 + e$, we conclude that $f(-.3517337112)$, which is approximately $.8271840261$, is the (approximate) minimum value of f, and $f(1)$, which is $1 + e$, is the maximum value of f on $[-1,1]$.

39. $f'(x) = \dfrac{a(cx+d) - (ax+b)c}{(cx+d)^2} = \dfrac{ad-bc}{(cx+d)^2}$

If $ad - bc \neq 0$, then $f'(x) \neq 0$ for all x in the domain of f. Thus if $ad - bc \neq 0$, then f has no critical numbers. Now assume that $ad - bc = 0$. If $d \neq 0$, then $a = bc/d$, so that

$$f(x) = \frac{(bc/d)x + b}{cx+d} = \frac{b}{d}\left(\frac{cx+d}{cx+d}\right) = \frac{b}{d}$$

and thus f is a constant function. If $d = 0$, then $c \neq 0$ (since not both c and d are 0 by hypothesis), so $b = 0$ (because $ad - bc = 0$) and thus $f(x) = ax/(cx) = a/c$. Therefore if $ad - bc = 0$, then f is a constant function.

41. Assume that $f'(c) = \lim_{x\to c}[f(x) - f(c)]/(x - c) < 0$. Then for x in some open interval I about c the inequality $[f(x) - f(c)]/(x - c) < 0$ holds. If x is in I and $x > c$, then $x - c > 0$, so that

$$f(x) - f(c) = (x - c)\left(\frac{f(x) - f(c)}{x - c}\right) < 0.$$

Therefore $f(x) < f(c)$, so $f(c)$ is not a minimum value. If x is in I and $x < c$, then $x - c < 0$, so that

$$f(x) - f(c) = (x - c)\left(\frac{f(x) - f(c)}{x - c}\right) > 0.$$

Therefore $f(x) > f(c)$, so $f(c)$ is not a maximum value. Thus if $f'(c) < 0$, then f does not have an extreme value at c.

43. Let $g(x) = f(x) - mx$ for $a \le x \le b$. Since $g'(x) = f'(x) - m$ and $f'(a) < m < f'(b)$, we have $g'(a) = f'(a) - m < 0$ and $g'(b) = f'(b) - m > 0$. Since $g'(a) < 0$, we must have $[g(x) - g(a)]/(x - a) < 0$ for x in an interval $(a, a + \delta)$. But then

$$(x - a)\left(\frac{g(x) - g(a)}{x - a}\right) < 0$$

so $g(x) < g(a)$ for x in $(a, a + \delta)$. Similarly, since $g'(b) > 0$, we must have

$$\frac{[g(x) - g(b)]}{x - b} > 0$$

for x in an interval $(b - \delta, b)$. But then

$$g(x) - g(b) = (x - b)\left(\frac{g(x) - g(b)}{x - b}\right) < 0 \quad \text{for } b - \delta < x < b$$

and hence $g(x) < g(b)$ for x in $(b - \delta, b)$. Thus g does not assume its minimum value at a or b. But since g is continuous on $[a, b]$, the Maximum-Minimum Theorem says that g has a minimum value on $[a, b]$, which must occur on (a, b). Since g is differentiable on (a, b), Theorem 4.3 says that there is a number c in (a, b) such that $g'(c) = 0$. But $g'(c) = f'(c) - m$. Thus $f'(c) = m$.

45. a. Since the length (20 feet) of fencing is shorter than the width (30 feet) of the house, the solution for this part is identical to the solution of Example 2, with "2000 meters" replaced by "20 feet." The solution of Example 2 therefore implies that the dimensions that maximize the area of the garden are a length of 10 feet parallel to the house and a width of 5 feet perpendicular to the house.

b. There are two cases to consider. In the first case, the house forms one boundary of the garden, as in the first diagram below, so fencing is used only on three sides of the garden. In this case we let x and y be the dimensions of the garden, as in the diagram. Then the area is given by $A = xy$, and x and y must satisfy $2x + y = 80$, or equivalently, $y = 80 - 2x$. Note that $0 \le y \le 30$, so that $25 \le x \le 40$. Therefore the area of the garden is given by $A(x) = x(80 - 2x)$, so that $A'(x) = 80 - 4x$. Since $A'(x) < 0$ for $25 \le x \le 40$, $A(x)$ decreases with x, so the maximum area is $A(25) = 750$ square feet, with $x = 25$ feet and $y = 80 - 2(25) = 30$ feet.

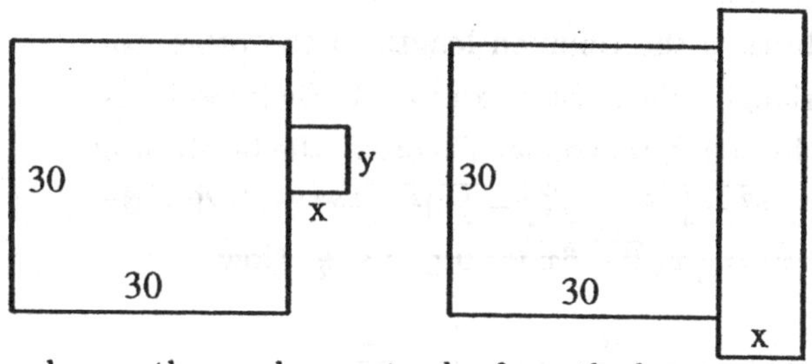

In the second case, the garden protrudes from the house, as in the second diagram above. We let x and y be the dimensions shown in the diagram. Then the area A_1 is given by $A_1 = xy$, where x and y must satisfy $2x + y + (y - 30) = 80$, so $x + y = 55$ and $A_1(x) = x(55 - x)$. Since $y > 30$, it follows that $0 \le x \le 25$. We find that $A_1'(x) = 55 - 2x$, so that $A_1'(x) > 0$ for $0 \le x \le 25$. Therefore $A_1(x)$ increases with x, and hence the maximum area of the garden for this case is $A_1(25) = 750$ square feet.

Comparing the two cases, we see that the maximum area of the garden is 750 square feet in each case!

47. Since $x(t + \pi) = x(t)$, the maximum value of x equals its maximum value on $[0, \pi]$. Now $x'(t) = 2\cos 2t - 2\sqrt{3}\sin 2t$, so that $x'(t) = 0$ only if $2\cos 2t - 2\sqrt{3}\sin 2t = 0$, or $\tan 2t = 1/\sqrt{13}$. Thus the only critical number of x in $(0, \pi)$ is $\pi/12$. Thus the maximum value of x on $[0, \pi]$ can occur only at 0, $\pi/12$, or π. Since $x(0) = x(\pi) = \sqrt{3}$, and $x(\pi/12) = \frac{1}{2} + \frac{3}{2} = 2$, the maximum value of x is 2. Consequently the maximum distance from the origin is 2.

49. $dM/dx = \frac{1}{2}w(L - x) - \frac{1}{2}wx = \frac{1}{2}w(L - 2x)$, so $dM/dx = 0$ only for $x = L/2$. Thus the extreme values of M on $[0, L]$ can occur only at 0, $L/2$, or L. Since $M = 0$ if $x = 0$ or $x = L$, M is largest for $x = L/2$, that is, at the midpoint of the beam.

51. We must determine the maximum value of P on $[0, 4]$. Now $P'(t) = -6(20 - t)^2 t + 2(20 - t)^3 = (20 - t)^2(-6t + 40 - 2t) = (20 - t)^2(40 - 8t)$, so $P'(t) > 0$ for $0 \le t \le 4$. Thus the net profit increases the longer the stocks are kept, so the company should retain them the full four-year period.

53. Let x be the length of the sides of the base of the crate and y the height. We are given that the volume $V = x^2 y$ of the crate is 6 cubic feet, and we would like to minimize the cost $C = 5x^2 + 2(4xy) + 1(x^2)$. Since $x^2 y = V = 6$, we have $y = 6/x^2$, so that $C = 6x^2 + 8x(6/x^2) = 6x^2 + 48/x$. Since the length of the base is to be between 1 and 2, the domain of C is $[1, 2]$. We have $C'(x) = 12x - 48/x^2$, so that $C'(x) = 0$ if $12x^3 = 48$, or $x = \sqrt[3]{4}$. Thus the extreme values of C on $[1, 2]$ can occur only at 1, $\sqrt[3]{4}$, or 2. Since $C(1) = 54$, $C(\sqrt[3]{4}) = 6\sqrt[3]{16} + 48/\sqrt[3]{4} \approx 45.4$, and $C(2) = 48$, the minimum value of C on $[1, 2]$ occurs for $x = \sqrt[3]{4}$. The corresponding value of y is $6/x^2 = 6/\sqrt[3]{16}$. Thus for minimum cost the base should be $\sqrt[3]{4}$ feet square and the height should be $6/\sqrt[3]{16}$ feet.

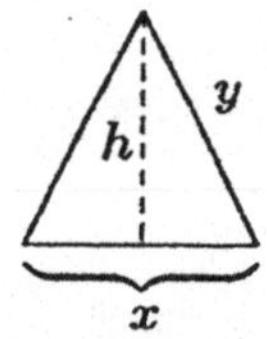

55. Let x be the length of the base, y the common length of the other two sides, and A the area of the triangle. We must maximize A. By hypothesis, $x+2y=3$, so $y=\frac{3}{2}-\frac{1}{2}x$, so by the Pythagorean Theorem, the height h of the triangle is given by $h=\sqrt{y^2-\frac{1}{4}x^2}=\sqrt{(\frac{3}{2}-\frac{1}{2}x)^2-\frac{1}{4}x^2}=\frac{1}{2}\sqrt{9-6x}$. Thus $A=\frac{1}{2}xh=\frac{1}{2}x(\frac{1}{2}\sqrt{9-6x})=\frac{1}{4}x\sqrt{9-6x}$ for $0\le x\le\frac{3}{2}$. Now

$$A'(x)=\frac{1}{4}\sqrt{9-6x}-\frac{3x}{4\sqrt{9-6x}}=\frac{9-9x}{4\sqrt{9-6x}}$$

so $A'(x)=0$ if $9-9x=0$, that is, $x=1$. Thus the extreme values of A on $[0,\frac{3}{2}]$ can occur only at 0, 1, or $\frac{3}{2}$. Since $A(0)=0$, $A(1)=\frac{1}{4}\sqrt{3}$, and $A(\frac{3}{2})=0$, the area A is maximum if $x=1$. Then $y=\frac{3}{2}-\frac{1}{2}\cdot 1=1$, making the triangle equilateral.

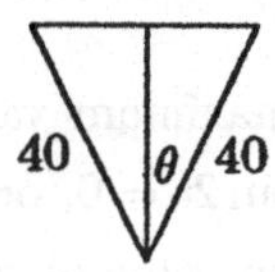

Exercise 57

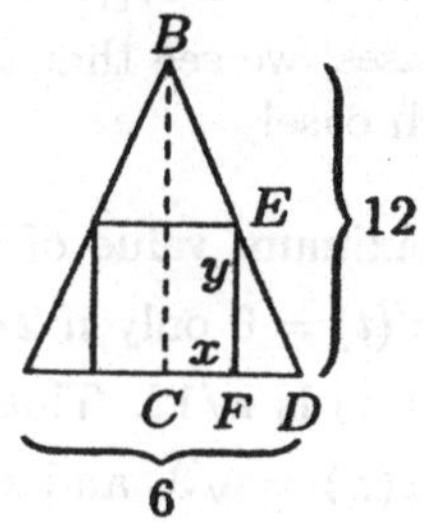

Exercise 59

57. a. The cross-section of the trough is the triangle in the figure above, with height equal to $40\cos\theta$ and width equal to $80\sin\theta$. Its area equals $\frac{1}{2}(80\sin\theta)(40\cos\theta)=1600\sin\theta\cos\theta$. Thus $V=1000(1600\sin\theta\cos\theta)=1.6\times10^6\sin\theta\cos\theta=.8\times10^6\sin 2\theta$ for $0\le\theta\le\pi/2$. Thus $V'(\theta)=1.6\times10^5\cos 2\theta$, so that $V'(\theta)=0$ if $\cos 2\theta=0$, or $2\theta=\pi/2$, or $\theta=\pi/4$. Thus the extreme value of V on $[0,\pi/2]$ can occur only at 0, $\pi/4$, or $\pi/2$. Since $V(0)=0$, $V(\pi/4)=800$, and $V(\pi/2)=0$ the maximum value of V occurs for $\theta=\pi/4$. The corresponding depth is $40\cos(\pi/4)=20\sqrt{2}$ (centimeters).

b. By (a), the maximum volume is $V(\pi/4)=800$ (cubic centimeters).

59. Let x be $\frac{1}{2}$ the length of the base of the rectangle, y the height, and A the area. We are to maximize A. Since triangles BCD and EFD are similar, it follows that $\frac{12}{3}=y/(3-x)$, so that $y=4(3-x)=12-4x$. Thus $A=2xy=2x(12-4x)=24x-8x^2$ for $0\le x\le 3$. Now $A'(x)=24-16x$, so $A'(x)=0$ if $x=\frac{3}{2}$. Thus the extreme values of A on $[0,3]$ can occur only at 0, $\frac{3}{2}$, or 3. Since $A(0)=0$, $A(\frac{3}{2})=24(\frac{3}{2})-8(\frac{3}{2})^2=18$, and $A(3)=0$, the maximum area of an inscribed rectangle is 18.

61. Let V be the volume of the cylinder, h its height, and r its radius. We are to maximize V. From similar triangles ABC and ADE, we deduce that $H/R = h/(R-r)$, so that $h = (H/R)(R-r)$ and thus $V = \pi r^2 h = \pi r^2[(H/R)(R-r)] = \pi H r^2 - (\pi H/R)r^3$ for $0 \le r \le R$. Now $V'(r) = 2\pi Hr - (3\pi H/R)r^2$, so $V'(r) = 0$ if $2\pi Hr - (3\pi H/R)r^2 = 0$, that is, $r = 0$ or $r = \frac{2}{3}R$. Thus the extreme values of V on $[0, R]$ can occur only at 0, $\frac{2}{3}R$, or R. Since $V(0) = 0 = V(R)$ and $V(\frac{2}{3}R) = \pi H(\frac{2}{3}R)^2 - (\pi H/R)(\frac{2}{3}R)^3 = \frac{4}{27}\pi R^2 H$, it follows that the maximum possible volume is $\frac{4}{27}\pi R^2 H$.

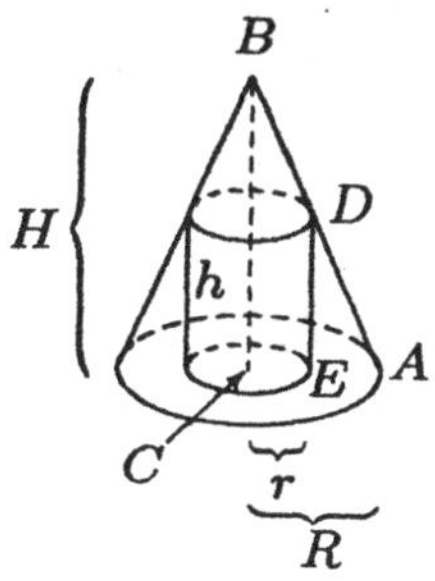

Exercise 61

4.2 The Mean Value Theorem

1. $\dfrac{f(4)-f(0)}{4-0} = \dfrac{-8-0}{4} = -2$; $f'(x) = 2x - 6$

   We must find c in $(0,4)$ such that $f'(c) = -2$, which means that $2c - 6 = -2$, or $c = 2$.

3. $\dfrac{f(0)-f(-2)}{0-(-2)} = \dfrac{0-4}{2} = -2$; $f'(x) = 3x^2 - 6$

   We must find c in $(-2,0)$ such that $f'(c) = -2$, which means that $3c^2 - 6 = -2$, or $c^2 = \frac{4}{3}$. Since $-2 < c < 0$, we have $c = -\frac{2}{3}\sqrt{3}$.

5. $\dfrac{f(1)-f(-2)}{1-(-2)} = \dfrac{5-(-4)}{3} = 3$; $f'(x) = 3x^2$

   We must find c in $(-2,1)$ such that $f'(c) = 3$, which means that $3c^2 = 3$, or $c^2 = 1$. Since $-2 < c < 1$, we have $c = -1$.

7. $\dfrac{f(2)-f(-2)}{2-(-2)} = \dfrac{3-(-25)}{4} = 7$; $f'(x) = 3x^2 - 6x + 3$

   We must find c in $(-2,2)$ such that $f'(c) = 7$, which means that $3c^2 - 6c + 3 = 7$, or $3c^2 - 6c - 4 = 0$. The roots are $1 + \sqrt{21}/3$ and $1 - \sqrt{21}/3$. Since $-2 < c < 2$, we have $c = 1 - \sqrt{21}/3$.

9. $\dfrac{f(8)-f(1)}{8-1} = \dfrac{3-2}{7} = \dfrac{1}{7}$; $f'(x) = \frac{1}{3}x^{-2/3}$

   We must find c in $(1,8)$ such that $f'(c) = \frac{1}{7}$, which means that $\frac{1}{3}c^{-2/3} = \frac{1}{7}$, so that $c = (\frac{7}{3})^{3/2}$.

11. $m = \dfrac{f(2) - f(1)}{2 - 1} = \dfrac{3 - (-1)}{1} = 4$; $f'(x) = 2x + 2/x^2$

Thus we seek a solution in $(1, 2)$ of $2c + 2/c^2 = 4$, or $-4 + 2c + 2/c^2 = 0$. By the Newton-Raphson method we find that $c \approx 1.618033989$.

13. a. Since $f'(x) = 2Ax + B$, c must satisfy the equation

$$2Ac + B = \frac{(Ab^2 + Bb + C) - (Aa^2 + Ba + C)}{b - a} = \frac{A(b^2 - a^2) + B(b - a)}{b - a} = A(b + a) + B.$$

Therefore $c = (b + a)/2$, the midpoint of $[a, b]$.

b. It means geometrically that the line that connects any two distinct points P and Q on a parabola is parallel to the line that is tangent to the point on the parabola whose x coordinate is the average of the x coordinates of P and Q.

15. By the Mean Value Theorem, there exists a number c in (a, b) such that

$$\frac{f(b) - f(a)}{b - a} = f'(c), \quad \text{so that} \quad \left|\frac{f(b) - f(a)}{b - a}\right| = |f'(c)| \le M.$$

Thus $|f(b) - f(a)| \le M|b - a|$. Equivalently, $f(a) - M(b - a) \le f(b) \le f(a) + M(b - a)$.

17. If $f(x) = x^{2/3}$, then $f'(x) = \frac{2}{3}x^{-1/3}$, and for $27 \le x \le 28$, we have $|f'(x)| \le \frac{2}{9}$. From Exercise 15, $|28^{2/3} - 27^{2/3}| \le \frac{2}{9}(28 - 27) = \frac{2}{9}$. Thus $9 - \frac{2}{9} \le 28^{2/3} \le 9 + \frac{2}{9} = \frac{83}{9}$. But $28^{2/3} > 9$, so $9 < 28^{2/3} \le \frac{83}{9}$.

19. Let $f(x) = \sqrt{x}$. If x is in $[2.89, 3]$, then

$$|f'(x)| = \left|\frac{1}{2\sqrt{x}}\right| \le \frac{1}{2\sqrt{2.89}} = \frac{1}{3.4}.$$

With $a = 2.89$ and $b = 3$, Exercise 15 implies that $|\sqrt{3} - 1.7| = |f(3) - f(2.89)| \le (1/3.4)(3 - 2.89) = \frac{11}{340} \approx 0.0323529412$.

21. Let $f(x) = (x-1)\sin x$. Then f is continuous on $[0, 1]$ and differentiable on $(0, 1)$. Since $f(0) = f(1) = 0$ and $f'(x) = \sin x + (x - 1)\cos x$, Rolle's Theorem implies the existence of a number c in $(0, 1)$ such that $\sin c + (c - 1)\cos c = 0$, or $\tan c = 1 - c$.

23. Let $f(x) = x + \sin x$, so that $f(\pi) = \pi$. Then $f'(x) = 1 + \cos x$, so that $0 < f'(c) < 1$ for $\pi/2 < c < \pi$ and for $\pi < c < 3\pi/2$. For any x with $\pi/2 < x < \pi$ or $\pi < x < 3\pi/2$, the Mean Value Theorem implies that there is c between x and π such that $f(x) - f(\pi) = f'(c)(x - \pi)$. Therefore $|x + \sin x - \pi| = f'(c)|x - \pi| < |x - \pi|$.

25. a. Suppose b and d are fixed points, with $b < d$. Then $f(b) = b$ and $f(d) = d$. Now apply the Mean Value Theorem to f on the interval $[b, d]$ to find that there is a number c such that

$$f'(c) = \frac{f(d) - f(b)}{d - b} = \frac{d - b}{d - b} = 1$$

which contradicts the assumption that $f'(x) \neq 1$ for every number x. Thus f has at most one fixed point.

b. Since $f'(x) = \frac{1}{2}\cos 2x \leq \frac{1}{2} < 1$, we conclude from part (a) that f has at most one fixed point. Since $f(0) = \sin 0 = 0$, 0 is the only fixed point.

27. a. Since $f(-1) = 0$ and $f'(-1) = 1$, $f(x)$ must be positive for all x in some interval $(-1, c)$. Since $f(1) = 0$ and $f'(1) = 1$, $f(x)$ must be negative for all x in some interval $(d, 1)$. Thus f assumes both positive and negative values on $(-1, 1)$. The Intermediate Value Theorem implies that f must also assume the value 0 on $(-1, 1)$.

b. We know $f(1) = 0 = f(-1)$, and by part (a), there is a number r in $(-1, 1)$ such that $f(r) = 0$. Applying the Mean Value Theorem to f on $[-1, r]$ and on $[r, 1]$, we find that there are numbers s and t in $(-1, r)$ and $(r, 1)$, respectively, such that $f'(s) = 0 = f'(t)$. Both s and t are in $(-1, 1)$, so f' has at least two zeros in $(-1, 1)$.

29. f satisfies the hypothesis of Rolle's Theorem. Thus there is c in $(0, 1)$ such that $f'(c) = 0$. Since $f'(x) = mx^{m-1}(x-1)^n + nx^m(x-1)^{n-1} = x^{m-1}(x-1)^{n-1}[m(x-1) + nx]$, the only value of c in $(0, 1)$ for which $f'(c) = 0$ satisfies $m(c-1) + nc = 0$, that is, $c = m/(m+n)$. The point $m/(m+n)$ divides the interval $[0, 1]$ into the intervals $[0, m/(m+n)]$ and $[m/(m+n), 1]$. The lengths of these intervals are $m/(m+n)$ and $1 - m/(m+n) = n/(m+n)$. Thus the ratio of the lengths is m/n.

31. Since 2400 feet equals $\frac{2400}{5280}$ miles, and 12 seconds equals $\frac{12}{3600}$ hours, the average velocity of the racing car was $\frac{2400}{5280} \cdot \frac{3600}{12}$, or rather, $\frac{1500}{11}$, miles per hour. Notice that $\frac{1500}{11} > 130$. Assuming that the function giving the position of the car at any instant is differentiable, we use the Mean Value Theorem to conclude that at some instant the car must have been traveling at least 130 miles per hour.

33. Let the parabola have equation $y = ax^2 + bx + c$. Since $(0, 2)$, $(1, 0)$, and $(2, 1)$ are on the parabola, we have $2 = a \cdot 0^2 + b \cdot 0 + c$, $0 = a \cdot 1^2 + b \cdot 1 + c$, and $1 = a \cdot 2^2 + b \cdot 2 + c$. The first of these equations implies that $c = 2$, so that the other two become $a + b = -2$ and $4a + 2b = -1$. These imply that $a = \frac{3}{2}$ and $b = -\frac{7}{2}$. Therefore an equation of the parabola is $y = \frac{3}{2}x^2 - \frac{7}{2}x + 2$. Since the slope of l is $(2-1)/(0-2) = -\frac{1}{2}$, an equation of l is $y = 2 - \frac{1}{2}x$. Thus for $0 \leq x \leq 2$, the vertical distance V between the point on l and the point on the parabola corresponding to x is given by

$$V = \left(2 - \frac{1}{2}x\right) - \left(\frac{3}{2}x^2 - \frac{7}{2}x + 2\right) = 3x - \frac{3}{2}x^2.$$

Since $dV/dx = 3 - 3x$, we have $dV/dx = 0$ for $x = 1$. Consequently V can assume its extreme values on $[0, 2]$ only at 0, 1, or 2. Since $V(0) = 0 = V(2)$ and $V(1) = \frac{3}{2}$, the vertical distance is maximum for $x = 1$. Thus the point $(1, 0)$ on the parabola has maximum vertical distance from the line l.

4.3 Consequences of the Mean Value Theorem

1. C

3. $\frac{3}{2}x^2 + C$

5. $-\frac{1}{3}x^3 + C$

7. $-\cos x + C$

9. $\frac{1}{2}\sin^2 x + C$

11. $e^x + C$

13. Let $g(x) = -2x$. Then $g'(x) = -2 = f'(x)$, so $f(x) = g(x) + C = -2x + C$ for the appropriate constant C. Since $f(0) = 0$, we have $0 = f(0) = -2(0) + C = C$. Thus $f(x) = -2x$.

15. Let $g(x) = \frac{1}{3}x^3$. Then $g'(x) = x^2 = f'(x)$, so $f(x) = g(x) + C = \frac{1}{3}x^3 + C$ for the appropriate constant C. Since $f(0) = -5$, we have $-5 = f(0) = \frac{1}{3}(0)^3 + C = C$, so that $C = -5$. Thus $f(x) = \frac{1}{3}x^3 - 5$.

17. Let $g(x) = \sin x$. Then $g'(x) = \cos x = f'(x)$, so $f(x) = g(x) + C = \sin x + C$ for the appropriate constant C. Since $f(\pi/3) = 1$, we have $1 = f(\pi/3) = \sin(\pi/3) + C = (\sqrt{3}/2) + C$, so $C = 1 - (\sqrt{3}/2)$. Thus $f(x) = \sin x + 1 - \sqrt{3}/2$.

19. Let $g(x) = e^x$. Then $g'(x) = e^x = f'(x)$, so $f(x) = g(x) + C = e^x + C$ for the appropriate constant C. Since $f(0) = 10$, we have $10 = f(0) = e^0 + C = 1 + C$, so that $C = 9$. Thus $f(x) = e^x + 9$.

21. Let $g(x) = 0$. Then $g'(x) = 0 = f''(x) = (f')'(x)$, so by Theorem 4.6, $f'(x) = g(x) + C_1 = C_1$ for some constant C_1. Now let $h(x) = C_1 x$. Then $h'(x) = C_1 = f'(x)$, so by Theorem 4.6, $f(x) = h(x) + C_2 = C_1 x + C_2$ for some constant C_2.

23. By Exercise 21, $f(x) = C_1 x + C_2$, where C_1 and C_2 are constants. Thus $f'(x) = C_1$. Since $f'(0) = -1$ and $f(0) = 2$ by hypothesis, it follows that $C_1 = -1$ and $2 = f(0) = C_1 \cdot 0 + C_2 = C_2$. Thus $f(x) = -x + 2$.

25. Let $g(x) = -\cos x$. Then $g'(x) = \sin x = f''(x) = (f')'(x)$, so by Theorem 4.6, $f'(x) = g(x) + C_1 = -\cos x + C_1$ for some constant C_1. Now let $h(x) = -\sin x + C_1 x$. Then $h'(x) = -\cos x + C_1 = f'(x)$, so by Theorem 4.6, $f(x) = h(x) + C_2 = -\sin x + C_1 x + C_2$ for some constant C_2. Since $f'(\pi) = -2$, we have $-2 = f'(\pi) = -\cos\pi + C_1 = 1 + C_1$, so $C_1 = -3$. Since $f(0) = 4$, we have $4 = f(0) = -\sin 0 + C_1 \cdot 0 + C_2$, so $C_2 = 4$. Thus $f(x) = -\sin x - 3x + 4$.

27. Since $f^{(4)}(x) = (f')^{(3)}(x)$, it follows from the solution of Exercise 26 that $f'(x) = D_1x^2 + D_2x + D_3$, where D_1, D_2, and D_3 are constants. Now let $g(x) = \frac{1}{3}D_1x^3 + \frac{1}{2}D_2x^2 + D_3x$. Then $g'(x) = D_1x^2 + D_2x + D_3 = f'(x)$, so by Theorem 4.6, $f(x) = g(x) + D_4 = \frac{1}{3}D_1x^3 + \frac{1}{2}D_2x^2 + D_3x + D_4$, where D_4 is some constant. Letting $C_1 = \frac{1}{3}D_1$, $C_2 = \frac{1}{2}D_2$, $C_3 = D_3$, and $C_4 = D_4$, we have $f(x) = C_1x^3 + C_2x^2 + C_3x + C_4$.

29. $f'(x) = 2x + 1$, so $f'(x) < 0$ for $x < -\frac{1}{2}$ and $f'(x) > 0$ for $x > -\frac{1}{2}$. Moreover, $f'(x) = 0$ only for $x = -\frac{1}{2}$. By Theorem 4.7, f is decreasing on $(-\infty, -\frac{1}{2}]$ and increasing on $[-\frac{1}{2}, \infty)$.

31. $f'(x) = 3x^2 - 2x + 1 = 3(x - \frac{1}{3})^2 + \frac{2}{3} > 0$ for all x. By Theorem 4.7, f is increasing on $(-\infty, \infty)$.

33. $f'(x) = 4x^3 - 6x^2 = 2x^2(2x - 3)$, so $f'(x) < 0$ for $x < \frac{3}{2}$, and $f'(x) > 0$ for $x > \frac{3}{2}$. Moreover, $f'(x) = 0$ only for $x = 0$ or $\frac{3}{2}$. By Theorem 4.7, f is decreasing on $(-\infty, \frac{3}{2}]$ and is increasing on $[\frac{3}{2}, \infty)$.

35. $f'(x) = 5x^4 + 3x^2 - 2 = (x^2 + 1)(5x^2 - 2)$, so $f'(x) < 0$ for $-\sqrt{2/5} < x < \sqrt{2/5}$, and $f'(x) > 0$ for $x < -\sqrt{2/5}$ and for $x > \sqrt{2/5}$. Moreover, $f'(x) = 0$ only for $x = -\sqrt{2/5}$ or $\sqrt{2/5}$. By Theorem 4.7, f is increasing on $(-\infty, -\sqrt{2/5}]$ and on $[\sqrt{2/5}, \infty)$, and is decreasing on $[-\sqrt{2/5}, \sqrt{2/5}]$.

37. Notice that the domain of g is $[-4, 4]$. Also $g'(x) = -x/\sqrt{16 - x^2}$, so $g'(x) > 0$ for $-4 < x < 0$ and $g'(x) < 0$ for $0 < x < 4$. Moreover, $g'(x) = 0$ only for $x = 0$. By Theorem 4.7, g is increasing on $[-4, 0]$ and is decreasing on $[0, 4]$.

39. Notice that -3 is not in the domain of g. Also $g'(x) = -1/(x + 3)^2 < 0$ for $x \neq -3$. We conclude from Theorem 4.7 that g is decreasing on $(-\infty, -3)$ and on $(-3, \infty)$.

41. $k'(x) = -2x/(x^2 + 1)^2$, so $k'(x) > 0$ for $x < 0$ and $k'(x) < 0$ for $x > 0$. Moreover, $k'(x) = 0$ only for $x = 0$. By Theorem 4.7, k is increasing on $(-\infty, 0]$ and is decreasing on $[0, \infty)$

43. Notice that $f(t)$ is not defined for $t = \pi/2 + n\pi$ for any integer n. Also $f'(t) = \sec^2 t > 0$ for all t in the domain of f. By Theorem 4.7, f is increasing on each interval of the form $(\pi/2 + n\pi, \pi/2 + (n+1)\pi)$, where n is any integer.

45. $f'(t) = -2\sin t - 1$, so $f'(t) < 0$ if $\sin t > -\frac{1}{2}$ and $f'(t) > 0$ if $\sin t < -\frac{1}{2}$. Thus $f'(t) < 0$ for $-\pi/6 + 2n\pi < t < 7\pi/6 + 2n\pi$ for any integer n, and $f'(t) > 0$ for $7\pi/6 + 2n\pi < t < 11\pi/6 + 2n\pi$ for any integer n. By Theorem 4.7, f is decreasing on $[-\pi/6 + 2n\pi, 7\pi/6 + 2n\pi]$ for any integer n and is increasing on $[7\pi/6 + 2n\pi, 11\pi/6 + 2n\pi]$ for any integer n.

47. $f'(x) = e^x + xe^x = (1 + x)e^x$. Since $e^x > 0$ for all x, we have $f'(x) < 0$ for $x < -1$, and $f'(x) > 0$ for $x > -1$. By Theorem 4.7, f is decreasing on $(-\infty, -1]$ and is increasing on $[-1, \infty)$.

49. Notice that the domain of f is $(0, \infty)$. We have $f'(x) = 1 - 1/x$, so $f'(x) > 0$ for $x > 1$, and $f'(x) < 0$ for $0 < x < 1$. By Theorem 4.7, f is decreasing on $(0, 1]$ and is increasing on $[1, \infty)$.

51. f is decreasing on $[-1, 1]$.

53. f is increasing on $[-1, 0]$, and f is decreasing on $[0, 1]$.

55. Let $f(x) = x^4 - 4x$. Then $f'(x) = 4x^3 - 4 = 4(x^3 - 1) > 0$ for $x > 1$. By Theorem 4.7, f is increasing on $[1, \infty)$. Therefore $x^4 - 4x = f(x) > f(1) = -3$ for $x > 1$.

57. Let $f(x) = \frac{1}{4}x + 1/x$. Then $f'(x) = \frac{1}{4} - 1/x^2 = (x^2 - 4)/4x^2 > 0$ for $x > 2$. By Theorem 4.7, f is increasing on $[2, \infty)$. Therefore $\frac{1}{4}x + 1/x = f(x) > f(2) = 1$ for $x > 2$.

59. a. Let $f(x) = e^x - (1 + x + \frac{1}{2}x^2)$. Then $f'(x) = e^x - (1 + x) > 0$ by Example 7. By Theorem 4.7, f is increasing on $[0, \infty)$. Therefore $f(x) > f(0) = 0$ for $x > 0$. Thus $e^x - (1 + x + \frac{1}{2}x^2) > 0$, or equivalently, $e^x > 1 + x + \frac{1}{2}x^2$ for $x > 0$.

b. Let $g(x) = e^x - (1 + x + \frac{1}{2}x^2 + \frac{1}{6}x^3)$. Then $g'(x) = e^x - (1 + x + \frac{1}{2}x^2) > 0$ for $x > 0$ by part (a). By Theorem 4.7, g is increasing on $[0, \infty)$. Therefore $g(x) > g(0) = 0$ for $x > 0$. Thus $e^x - (1 + x + \frac{1}{2}x^2 + \frac{1}{6}x^3) > 0$, or equivalently, $e^x > 1 + x + \frac{1}{2}x^2 + \frac{1}{6}x^3$ for $x > 0$.

61. Let $h(x) = f(x) - g(x)$, so that by hypothesis, $h(a) = f(a) - g(a) \geq 0$. Fix $z > a$. Then $h'(x) = f'(x) - g'(x) > 0$ for all x in (a, z), so by Theorem 4.7, h is increasing on $[a, z]$. Thus $f(z) - g(z) = h(z) > h(a) = f(a) - g(a) \geq 0$, or $f(z) > g(z)$. Since z was an arbitrary number greater than a, $f(x) > g(x)$ for all $x > a$.

63. Let $f(x) = (1 + x)^n - (1 + nx)$. Then $f'(x) = n(1 + x)^{n-1} - n$, so that if $x > 0$, then $f'(x) > 0$. By Theorem 4.7, f is increasing on $[0, \infty)$, so $f(x) > f(0) = 0$ for x in $(0, \infty)$. Hence $(1 + x)^n > 1 + nx$ for x in $(0, \infty)$.

65. Let $f(x) = \sin x$ and $g(x) = x - x^3/6$. Then $f(0) = 0 = g(0)$, and by Exercise 64, $f'(x) = \cos x > 1 - x^2/2 = g'(x)$ for $x > 0$. By Exercise 61, $\sin x = f(x) > g(x) = x - x^3/6$ for $x > 0$.

67. Since $f'(x) = 3x^2$ for $-1 < x < 0$, we must have $f(x) = x^3 + C_1$, for $-1 < x < 0$ and for an appropriate constant C_1. Since $f(0) = 0$ and f is continuous at 0, we have $0 = f(0) = \lim_{x \to 0^-} (x^3 + C_1) = C_1$. Thus $C_1 = 0$, so $f(x) = x^3$ for $-1 < x < 0$. Similarly, since $f'(x) = 2e^x$ for $0 < x < 1$, we must have $f(x) = 2e^x + C_2$ for $0 < x < 1$ and for an appropriate constant C_2. Since $f(0) = 0$ and f is continuous at 0, it follows that $0 = f(0) = \lim_{x \to 0^+} (2e^x + C_2) = 2 + C_2$. Thus $C_2 = -2$, so $f(x) = 2e^x - 2$ for $0 < x < 1$. Therefore

$$f(x) = \begin{cases} x^3 & \text{for } -1 < x < 0 \\ 0 & \text{for } x = 0 \\ 2e^x - 2 & \text{for } 0 < x < 1 \end{cases}$$

69.

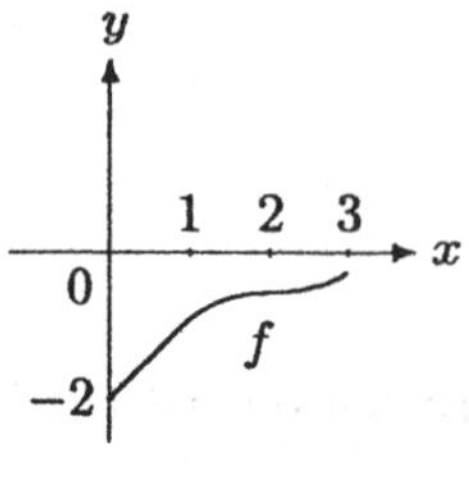

71. a. Since $F' = f$ and $G' = g$, we have $(F+G)' = F' + G' = f + g$. Thus $F + G$ is an antiderivative of $f + g$.

b. Since $F' = f$ we have $(cF)' = cF' = cf$. Thus cF is an antiderivative of cf.

c. No. If $f(x) = g(x) = x$, then $(fg)(x) = x^2$. If $F(x) = G(x) = \frac{1}{2}x^2$, then $F' = f$ and $G' = g$, $(FG)(x) = \frac{1}{4}x^4$, and $(FG)'(x) = x^3$. Thus $(FG)' \neq fg$.

d. No. If $f(x) = g(x) = x$, then $(f/g)(x) = 1$ for $x \neq 0$. If $F(x) = G(x) = \frac{1}{2}x^2$, then $F' = f$ and $G' = g$, $(F/G)(x) = 1$ and $(F/G)'(x) = 0$ for $x \neq 0$. Thus $(F/G)' \neq f/g$.

73. Taking $x = y = 0$ in the equation $f(x+y) = f(x) + f(y)$, we have $f(0) = f(0) + f(0)$, so that $f(0) = 0$. Taking $y = h$ in the equation $f(x+y) = f(x) + f(y)$, we find that $f(x+h) = f(x) + f(h)$, so that $f(x+h) - f(x) = f(h)$. It now follows from (2) in Section 3.2 that for any x,

$$f'(x) = \lim_{h\to 0} \frac{f(x+h) - f(x)}{h} = \lim_{h\to 0} \frac{f(h)}{h} = \lim_{h\to 0} \frac{f(h) - f(0)}{h} = f'(0).$$

Letting $f'(0) = c$, we have $f'(x) = c$, so that by Theorem 4.6, $f(x) = cx + C$ for some constant C. Since $f(0) = 0$, it follows that $C = 0$, and thus that $f(x) = cx$ for all x.

75. $v'(x) = k(a-x) - kx = ka - 2kx$, so that $v'(x) > 0$ for $0 < x < a/2$ and $v'(x) < 0$ for $a/2 < x < a$. Thus v is increasing on $[0, a/2]$ and decreasing on $[a/2, a]$.

77. $$\frac{d}{d\omega} i_{\max} = \left(\frac{-V_0}{2}\right)\left[\left(\frac{1}{\omega C} - \omega L\right)^2 + R^2\right]^{-3/2}\left[2\left(\frac{1}{\omega C} - \omega L\right)\left(\frac{-1}{\omega^2 C} - L\right)\right]$$

Thus $\dfrac{d}{d\omega} i_{\max} > 0$ if and only if $\left(\dfrac{1}{\omega C} - \omega L\right)\left(\dfrac{1}{\omega^2 C} + L\right) > 0$, that is, $1/(\omega C) - \omega L > 0$, since all constants are positive. Consequently

$$\frac{d}{d\omega} i_{\max} > 0 \quad \text{if and only if} \quad 0 < \omega < \sqrt{\frac{1}{CL}}$$

and $i_{\max}$ is therefore increasing on $(0, \sqrt{1/(CL)}]$ by Theorem 4.7.

79. Let $U = -\frac{1}{2}m\omega^2 r^2/l$. Then $dU/dr = -m\omega^2 r/l$, so that $dU/dr = dT/dr$. By Theorem 4.6, $T = U + C = -\frac{1}{2}m\omega^2 r^2/l + C$ for the appropriate constant C. Since $T = 0$ for $r = l$, we have $0 = 0 + C$, so that $C = 0$. Therefore $T = -\frac{1}{2}m\omega^2 r^2/l$.

4.4 Exponential Growth and Decay

1. a. By (4) we have $f(t) = 4f(0)$ if $f(0)e^{kt} = 4f(0)$, or $e^{kt} = 4$. Since $k = \frac{1}{2}\ln 2$ by the solution of Example 1, this means that $e^{(t/2)\ln 2} = 4$, or $2^{t/2} = 4$. Thus $t/2 = 2$, or $t = 4$. Therefore it takes 4 days for the algae to quadruple in number.

   b. As in (a), if $f(t) = 3f(0)$, then $2^{t/2} = 3$, so $(t/2)\ln 2 = \ln 3$, or $t = (2\ln 3)/\ln 2 \approx 3.17$. Therefore it takes approximately 3.17 days for the algae to triple in number.

3. Let $f(t)$ be the number of beetles t days after there are 1200 beetles. Then $f(0) = 1200$, and we seek the value of t for which $f(t) = 1500$. Since the doubling time is 6 days and 20 hours, which is the same as $\frac{41}{6}$ days, we have $f(\frac{41}{6}) = 2f(0)$, so by (4), $2f(0) = f(\frac{41}{6}) = f(0)e^{41k/6}$. Thus $e^{41k/6} = 2$, so that $k = (6\ln 2)/41$. If $f(t) = 1500$, then by (4), $1500 = f(0)e^{kt} = 1200e^{kt}$, or $e^{kt} = \frac{5}{4}$, so that $t = (1/k)\ln 1.25 = (41\ln 1.25)/(6\ln 2) \approx 2.2$. Thus there were 1200 beetles approximately 2.2 days ago.

5. Let $f(t)$ be the population in millions of the country with a doubling time of 20 years t years after its population is 50,000,000, and let $g(t)$ be the population of the country with a doubling time of 10 years. From the hypotheses we notice that $f(0) = 50$ and $g(0) = 20$. We seek the value of t for which $f(t) = g(t)$. By (4), there are constants k_1 and k_2 such that $f(t) = f(0)e^{k_1 t} = 50e^{k_1 t}$ and $g(t) = 20e^{k_2 t}$. Since $f(20) = 2f(0)$, we have $2f(0) = f(20) = f(0)e^{20k_1}$, or $e^{20k_1} = 2$, so that $k_1 = \frac{1}{20}\ln 2$. Similarly, since $g(10) = 2g(0)$, we have $k_2 = \frac{1}{10}\ln 2$. Now if $f(t) = g(t)$, then $50e^{k_1 t} = 20e^{k_2 t}$, or $e^{(k_2 - k_1)t} = 2.5$, so that

$$t = \frac{\ln 2.5}{k_2 - k_1} = \frac{\ln 2.5}{\frac{1}{10}\ln 2 - \frac{1}{20}\ln 2} = \frac{20\ln 2.5}{\ln 2} \approx 26.44.$$

Thus it will be approximately 26.44 years until the two countries have the same population.

7. a. The population halves in any time interval of duration $-(\ln 2)/k$ if for any time t, $f(t + (-\ln 2)/k) = \frac{1}{2}f(t)$. By hypothesis, $f(t)$ decays exponentially, so that

$$f\left(t - \frac{\ln 2}{k}\right) = f(0)e^{k[t-(\ln 2)/k]} = f(0)e^{kt - \ln 2} = f(0)e^{kt}e^{-\ln 2} = f(t)\frac{1}{2} = \frac{1}{2}f(t).$$

   b. Since $h = -(\ln 2)/k$, we have $k = -(\ln 2)/h$, so that $f(t) = f(0)e^{kt} = f(0)e^{-(\ln 2)t/h} = f(0)(e^{-\ln 2})^{t/h} = f(0)(\frac{1}{2})^{t/h}$.

9. Let $f(t)$ be the amount of radium remaining after t years. Then $f(t) \approx f(0)e^{kt}$, where $1590 \approx (-\ln 2)/k$, or $k \approx (-\ln 2)/1590$, by Exercise 7. We seek t such that $f(t)/f(0) = \frac{9}{10}$. We have $\frac{9}{10} = f(t)/f(0) \approx e^{(-\ln 2/1590)t}$, so that $t \approx [-1590/(\ln 2)]\ln \frac{9}{10} \approx 241.7$ (years).

11. Let $f(t)$ be the amount of C^{14} present t years after 13,000 B.C. and let $g(t)$ be the amount of C^{14} present t years after 12,300 B.C. Then $f(t) = f(0)e^{kt}$ and $g(t) = g(0)e^{kt}$ for $t \geq 0$, where $k = -(\ln 2)/5730$ (by Example 2) and $f(0) = g(0)$. The amounts present in 2000 A.D. are $f(15{,}000)$ and $g(14{,}300)$. Furthermore,
$$\frac{g(14{,}300)}{g(0)} - \frac{f(15{,}000)}{f(0)} = e^{14{,}300k} - e^{15{,}000k} \approx 0.0143962542.$$
Thus the difference is approximately 1.44%.

13. Let $f(t)$ be the amount (in milligrams) of iodine 131 t days after delivery. We are to determine $f(0)$, and are given that $f(2) = 100$ and $f(8.14) = \frac{1}{2}f(0)$. By (4), $\frac{1}{2}f(0) = f(8.14) = f(0)e^{8.14k}$, or $e^{8.14k} = \frac{1}{2}$, so that $k = (1/8.14)\ln\frac{1}{2}$. Also by (4) and the hypothesis, $100 = f(2) = f(0)e^{2k}$, so that $f(0) = 100e^{-2k} = 100e^{-(2/8.14)\ln(1/2)} = 100e^{(\ln 2)/4.07} \approx 119$. Thus approximately 119 milligrams of iodine 131 should be purchased.

15. By (8), $f(t) \approx e^{(-1.25\times 10^{-4})t}$, so that $f(1600) \approx e^{(-1.25\times 10^{-4})1600} = e^{-0.2} \approx 0.819$.

17. a. $p(0) \approx 29.92$ (inches of mercury)

b. $p(5) \approx (29.92)e^{(-.2)5} \approx 11.01$ (inches of mercury)

c. $p(10) \approx (29.92)e^{(-.2)10} \approx 4.049$ (inches of mercury)

19. Let $f(t)$ be the amount of sodium pentobarbitol in the blood stream after t hours. Then $f(t) = f(0)e^{kt}$ for $t \geq 0$. By hypothesis, $f(5) = \frac{1}{2}f(0)$, so that $k = (-\ln 2)/5$. To anesthetize a 10 kilogram dog for one-half hour we need $f(\frac{1}{2}) = 20(10) = 200$ milligrams, so that $200 = f(\frac{1}{2}) = f(0)e^{-[(\ln 2)/5](1/2)} = f(0)e^{-(\ln 2)/10}$ and thus $f(0) = 200e^{(\ln 2)/10} \approx 214$ (milligrams).

21. For $A(10) = 2S$, we need p to satisfy $2 = A(10)/S = e^{10p/100} = e^{p/10}$, and thus $p = 10\ln 2 \approx 6.93$ (%).

23. Let $f(t)$ be the amount of sugar after t minutes. Then $f(t) = f(0)e^{kt} = e^{kt}$ for $t \geq 0$. By hypothesis $f(1) = \frac{3}{4}$, so that $\frac{3}{4} = e^{1k} = e^{k}$, and thus $k = \ln\frac{3}{4}$. Consequently $f(t) = \frac{1}{2} = e^{(\ln 3/4)t}$ if $t = (\ln\frac{1}{2})/(\ln\frac{3}{4}) \approx 2.41$ (minutes).

25. Since $D(t) = P(0) - P(t)$ and $P(0) = e^{\lambda t}P(t)$, we have $D(t) = P(t)(e^{\lambda t} - 1)$, or $D(t)/P(t) = e^{\lambda t} - 1$. Thus
$$\lambda t = \ln\left(\frac{D(t)}{P(t)} + 1\right), \quad \text{or} \quad t = \frac{1}{\lambda}\ln\left(\frac{D(t)}{P(t)} + 1\right).$$

27. a. $t = (1.885)10^9 \ln\left[9.068\left(\frac{1.95\times 10^{-12}}{2.885\times 10^{-8}}\right) + 1\right] \approx 1{,}150{,}000$ (years)

b. $\dfrac{D(t)}{P(t)} = \dfrac{1}{9.068}(e^{4.19/1.885} - 1) \approx 0.908$

4.5 The First and Second Derivative Tests

1. $f'(x) = 2x + 6 = 2(x+3)$, so f' changes from negative to positive at -3.

3. $f'(x) = 8x^3 - 8x = 8x(x-1)(x+1)$, so f' changes from negative to positive at -1 and 1, and from positive to negative at 0.

5. $f'(t) = \dfrac{(2t-1)(t^2+t+1) - (t^2-t+1)(2t+1)}{(t^2+t+1)^2} = \dfrac{2(t+1)(t-1)}{(t^2+t+1)^2}$
so that f' changes from positive to negative at -1 and from negative to positive at 1.

7. $f'(t) = \cos t + \frac{1}{2}$, so $f'(t) > 0$ if $\cos t > -\frac{1}{2}$, and $f'(t) < 0$ if $\cos t < -\frac{1}{2}$. Thus f' changes from positive to negative at $2\pi/3 + 2n\pi$ for any integer n, and changes from negative to positive at $4\pi/3 + 2n\pi$ for any integer n.

9. $f'(x) = -6x + 3 = 3(1-2x)$, so f' changes from positive to negative at $\frac{1}{2}$. By the First Derivative Test, $f(\frac{1}{2}) = \frac{31}{4}$ is a relative maximum value of f.

11. $f'(x) = 3x^2 + 6x = 3x(x+2)$, so f' changes from positive to negative at -2 and from negative to positive at 0. By the First Derivative Test, $f(-2) = 8$ is a relative maximum value and $f(0) = 4$ is a relative minimum value of f.

13. $g'(x) = 8x + \dfrac{1}{x^2} = \dfrac{8x^3+1}{x^2} = \dfrac{(2x+1)(4x^2-2x+1)}{x^2}$
Since $4x^2 - 2x + 1 > 0$ for all x, g' changes from negative to positive at $-\frac{1}{2}$. By the First Derivative Test, $g(-\frac{1}{2}) = 3$ is a relative minimum value of g.

15. $f'(x) = \dfrac{(16+x^3) - x(3x^2)}{(16+x^3)^2} = \dfrac{16-2x^3}{(16+x^3)^2} = \dfrac{-2(x-2)(x^2+2x+4)}{(16+x^3)^2}$
Since $x^2 + 2x + 4 > 0$ for all x, f' changes from positive to negative at 2. By the First Derivative Test, $f(2) = 1/12$ is a relative maximum value of f.

17. $f'(x) = \sqrt{1-x^2} - \dfrac{x^2}{\sqrt{1-x^2}} = \dfrac{1-2x^2}{\sqrt{1-x^2}} = \dfrac{(1-\sqrt{2}\,x)(1+\sqrt{2}\,x)}{\sqrt{1-x^2}}$
so f' changes from negative to positive at $-\frac{1}{2}\sqrt{2}$ and from positive to negative at $\frac{1}{2}\sqrt{2}$. By the First Derivative Test, $f(-\frac{1}{2}\sqrt{2}) = -\frac{1}{2}$ is a relative minimum value and $f(\frac{1}{2}\sqrt{2}) = \frac{1}{2}$ is a relative maximum value of f.

19. $k'(x) = -\sin x + \frac{1}{2}$, so $k'(x) > 0$ if $\sin x < \frac{1}{2}$, and $k'(x) < 0$ if $\sin x > \frac{1}{2}$. Thus k' changes from positive to negative at $\pi/6 + 2n\pi$ for any integer n, and from negative to positive at $5\pi/6 + 2n\pi$ for any integer n. By the First Derivative Test, $k(\pi/6 + 2n\pi) = \frac{1}{2}\sqrt{3} + \pi/12 + n\pi$ is a relative maximum value of k for any integer n, and $k(5\pi/6 + 2n\pi) = -\frac{1}{2}\sqrt{3} + 5\pi/12 + n\pi$ is a relative minimum value of k for any integer n.

21. $k'(x) = \cos\left(\dfrac{x^2}{1+x^2}\right)\dfrac{2x(1+x^2)-x^2(2x)}{(1+x^2)^2} = \dfrac{2x}{(1+x^2)^2}\cos\left(\dfrac{x^2}{1+x^2}\right)$

Since $0 \le x^2/(1+x^2) \le 1 < \pi/2$, we have $\cos(x^2/(1+x^2)) > 0$ for all x. Thus k' changes from negative to positive at 0. By the First Derivative Test, $k(0) = 0$ is a relative minimum value of k.

23. $f'(x) = 2xe^{-x} - x^2e^{-x} = xe^{-x}(2-x)$. Thus f' changes from negative to positive at 0, and from positive to negative at 2. By the First Derivative Test, $f(0) = 0$ is a relative minimum value and $f(2) = 4e^{-2}$ is a relative maximum value.

25. $f'(x) = -8x+3$, so $f'(x) = 0$ if $x = \frac{3}{8}$. Since $f''(x) = -8 < 0$ for all x, the Second Derivative Test implies that $f(\frac{3}{8}) = -\frac{7}{16}$ is a relative maximum value of f.

27. $f'(x) = 3x^2 - 6x - 24 = 3(x-4)(x+2)$, so $f'(x) = 0$ if $x = 4$ or $x = -2$. Since $f''(x) = 6x - 6$, so that $f''(4) = 18 > 0$ and $f''(-2) = -18 < 0$, the Second Derivative Test implies that $f(4) = -79$ is a relative minimum value of f, and $f(-2) = 29$ is a relative maximum value of f.

29. $f'(x) = 12x^3 - 12x^2 - 9x = 3x(2x+1)(2x-3)$, so $f'(x) = 0$ if $x = 0$, $x = -\frac{1}{2}$, or $x = \frac{3}{2}$. Since $f''(x) = 36x^2 - 24x - 9$, so that $f''(0) = -9 < 0$, $f''(-\frac{1}{2}) = 12 > 0$ and $f''(\frac{3}{2}) = 36 > 0$, the Second Derivative Test implies that $f(0) = \frac{1}{2}$ is a relative maximum value of f, $f(-\frac{1}{2}) = \frac{1}{16}$ is a relative minimum value of f, and $f(\frac{3}{2}) = -\frac{127}{16}$ is a relative minimum value of f.

31. $f'(t) = 2t - 1/t^2$, so $f'(t) = 0$ if $2t = 1/t^2$, or $t^3 = \frac{1}{2}$, or $t = 1/\sqrt[3]{2}$. Since $f''(t) = 2 + 2/t^3$, so that $f''(1/\sqrt[3]{2}) = 2 + 2/\frac{1}{2} = 6 > 0$, the Second Derivative Test implies that $f(1/\sqrt[3]{2}) = 1/(2^{2/3}) + \sqrt[3]{2} + 1$ is a relative minimum value of f.

33. $f'(t) = \cos t - \sin t$, so $f'(t) = 0$ if $\sin t = \cos t$, that is, if $x = \pi/4 + n\pi$ for any integer n. Next, $f''(t) = \sin t - \cos t$. If n is even, then $f''(\pi/4 + n\pi) = f''(\pi/4) = -\sqrt{2} < 0$, so the Second Derivative Test implies that $f(\pi/4 + n\pi) = \sqrt{2}$ is a relative maximum value of f. Analogously, if n is odd, then $f''(\pi/4 + n\pi) = f''(\pi/4 + \pi) = f''(5\pi/4) = \sqrt{2} > 0$, so the Second Derivative Test implies that $f(\pi/4 + n\pi) = -\sqrt{2}$ is a relative minimum value of f.

35. $f'(t) = e^t - (e^{-t})(-1) = e^t + e^{-t}$, so that $f'(t) > 0$ for all t. Therefore f is increasing on $(-\infty, \infty)$ and hence has no relative extreme values.

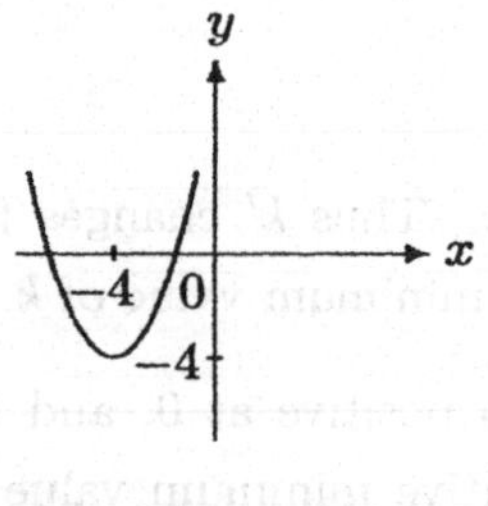

Exercise 37

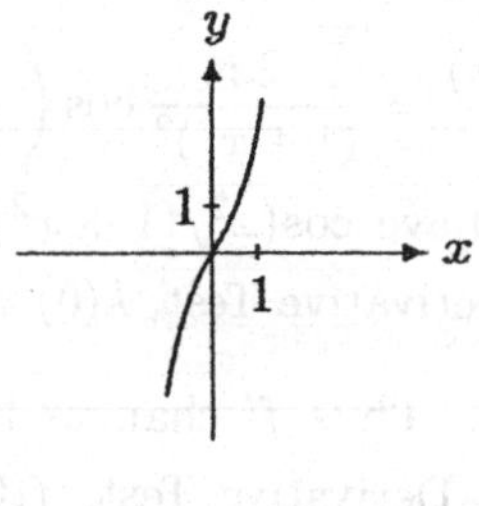

Exercise 39

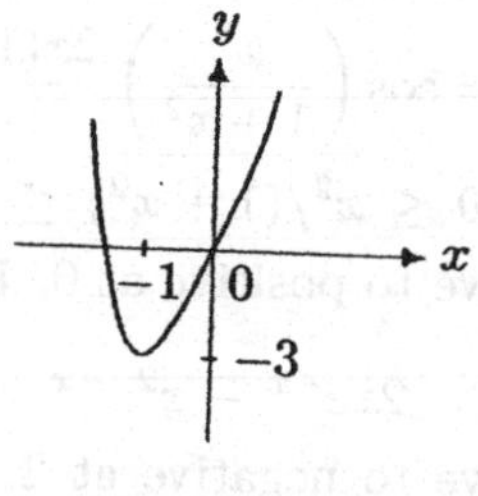

Exercise 41

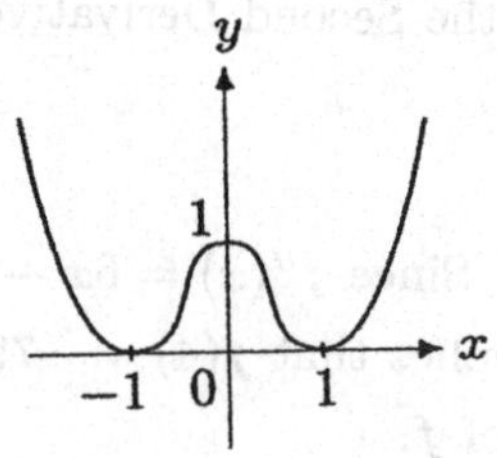

Exercise 43

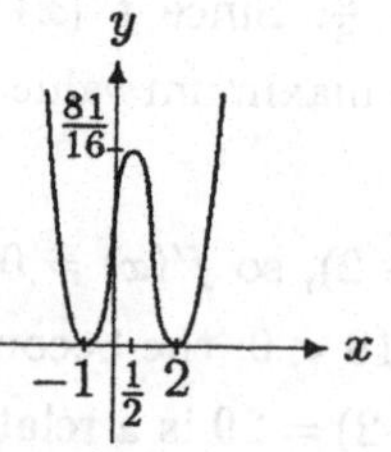

Exercise 45

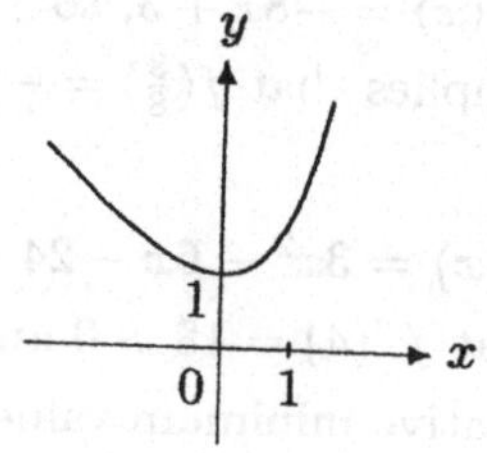

Exercise 47

37. $f'(x) = 2x + 8 = 2(x + 4)$, so that $f'(x) = 0$ for $x = -4$. Also $f''(x) = 2 > 0$ for all x. By the Second Derivative Test, $f(-4) = -4$ is a relative minimum value of f.

39. $f'(x) = 3x^2 + 3 = 3(x^2 + 1) > 0$ for all x, so that f has no critical numbers, and therefore no relative extreme values.

41. $f'(x) = 4x^3 + 4 = 4(x + 1)(x^2 - x + 1)$, so that since $x^2 - x + 1 > 0$ for all x, $f'(x) = 0$ for $x = -1$. Next, $f''(x) = 12x^2$, so that $f''(-1) = 12 > 0$. By the Second Derivative Test, $f(-1) = -3$ is a relative minimum value of f.

43. $f'(x) = 2(x^2 - 1)(2x) = 4x(x + 1)(x - 1)$, so that $f'(x) = 0$ for $x = 0$, $x = 1$, or $x = -1$. Next, $f''(x) = 12x^2 - 4$, so that $f''(0) = -4 < 0$, $f''(1) = 8 > 0$, and $f''(-1) = 8 > 0$. By the Second Derivative Test, $f(0) = 1$ is a relative maximum value of f, and $f(1) = f(-1) = 0$ is a relative minimum value of f.

45. $f'(x) = 2(x - 2)(x + 1)^2 + (x - 2)^2 2(x + 1) = 2(x - 2)(x + 1)(2x - 1)$, so that $f'(x) = 0$ for $x = 2$, $x = -1$, or $x = \frac{1}{2}$. Since f' changes from negative to positive at -1 and 2, and positive to negative at $\frac{1}{2}$, the First Derivative Test implies that $f(-1) = f(2) = 0$ is a relative minimum value of f, and $f(\frac{1}{2}) = \frac{81}{16}$ is a relative maximum value of f. (The graph in this exercise is the same as that in Exercise 44 shifted right 2 units.)

47. $f'(x) = e^x - 1$, so that $f'(0) = 0$. Since f' changes from negative to positive at 0, the First Derivative Test implies that $f(0) = 1$ is a relative minimum value of f.

49. $f'(x) = 5x^4$ and $f''(x) = 20x^3$, so that $f'(0) = f''(0) = 0$. But $f'(x) > 0$ if $x \neq 0$, so f is increasing, and thus has no relative extreme values.

51. $f'(x) = 5x^4 - 3x^2 = x^2(5x^2 - 3)$ and $f''(x) = 20x^3 - 6x = 2x(10x^2 - 3)$, so that $f'(0) = f''(0) = 0$. But $f'(x) < 0$ for x in $(-\sqrt{\frac{3}{5}}, \sqrt{\frac{3}{5}})$ and $x \neq 0$, so that f is decreasing on $[-\sqrt{\frac{3}{5}}, \sqrt{\frac{3}{5}}]$, and thus f has no relative extreme value at 0.

53. $f'(x) = 3x^2e^x + x^3e^x = x^2e^x(3+x)$ and $f''(x) = 6xe^x + 3x^2e^x + 3x^2e^x + x^3e^x = xe^x(6 + 6x + x^2)$, so that $f'(0) = f''(0) = 0$. But $f'(x) > 0$ for $x > -3$, so that f is increasing on $[-3, \infty)$, and thus f has no relative extreme value at 0.

55. Relative maximum value at -0.5; relative minimum value at 0.

57. Relative maximum value at -5.0 and at 1.2; relative minimum value at -1.2 and at 5.0.

59. Assume that $f''(c) > 0$. Since $f'(c) = 0$, $f'(x)/(x-c) = [f'(x) - f'(c)]/(x-c) > 0$ for all x in some interval $(c-\delta, c+\delta)$. Therefore if $c - \delta < x < c$, then $f'(x) < 0$ since $x - c < 0$, while if $c < x < c+\delta$, then $f'(x) > 0$ since $x - c > 0$. This means that f' changes from negative to positive at c, and hence by the First Derivative Test, f has a relative minimum value at c.

61. a. Let $f(x) = x^{4/3}$. Then $f'(x) = (\frac{4}{3})x^{1/3}$, so $f'(0) = 0$. Since

$$\lim_{x\to 0} \frac{f'(x) - f'(0)}{x - 0} = \lim_{x\to 0} \left(\frac{4}{3}\right) x^{-2/3} = \infty$$

$f''(0)$ does not exist. Now $f'(x) < 0$ for $x < 0$ and $f'(x) > 0$ for $x > 0$, so $f(0) = 0$ is a relative minimum value for f.

b. Let $f(x) = x^{5/3}$. Then $f'(x) = (\frac{5}{3})x^{2/3}$, so $f'(0) = 0$. But $[f'(x) - f'(0)]/(x - 0) = (\frac{5}{3})x^{-1/3}$, and since $\lim_{x\to 0} x^{-1/3}$ does not exist, likewise $\lim_{x\to 0}[f'(x) - f'(0)]/(x-0)$ and hence $f''(0)$ do not exist. Since $f(x) < 0$ if $x < 0$ and $f(x) > 0$ if $x > 0$, $f(0)$ is not a relative extreme value.

4.6 Extreme Values on an Arbitrary Interval

1. Let $r(x) = kx(a-x)$. Then $r'(x) = ka - 2kx$, so $r'(x) = 0$ if $ka - 2kx = 0$, that is, $x = a/2$. Since $r''(x) = -2k < 0$, (1) and the Second Derivative Test imply that $r(x)$ is maximized for $x = a/2$. Thus the value of x for which dx/dt is maximum is $a/2$.

3. $\dfrac{dP}{dR} = \dfrac{E^2(R+r)^2 - 2E^2R(R+r)}{(R+r)^4} = \dfrac{E^2(r-R)}{(R+r)^3}$
Therefore $dP/dR = 0$ only for $R = r$, and dP/dR changes from positive to negative at r. By (1) and the First Derivative Test, the maximum value of P occurs for $R = r$.

5. $P'(r) = \dfrac{\pi}{6a^5}\left[4r^3e^{-r/a} + r^4e^{-r/a}\left(-\dfrac{1}{a}\right)\right] = \dfrac{\pi r^3}{6a^5}e^{-r/a}\left(4 - \dfrac{r}{a}\right)$
Therefore $P'(r) = 0$ only for $r = 4a$, and P' changes from positive to negative at $4a$. By (1) and the First Derivative Test, the maximum value of P occurs at $4a$. Thus the most probable distance of the electron from the center of the atom is $4a$ (exactly 4 times the most probable state, as in Example 3).

7. $ds/dx = -2kx\ln x - kx^2(1/x) = -kx(2\ln x + 1)$, so that $ds/dx = 0$ only if $\ln x = -\frac{1}{2}$, or $x = e^{-1/2}$. Since ds/dx changes from positive to negative at $e^{-1/2}$, it follows from (1) and the First Derivative Test that the maximum value of s occurs for $x = e^{-1/2}$.

9. Let x and y be the positive numbers, so $0 < x < 18$ and $0 < y < 18$. If P denotes their product, we seek to maximize P. Since $x + y = 18$, we have $P = xy = x(18-x) = 18x - x^2$. Now $P'(x) = 18 - 2x$, so $P'(x) = 0$ for $x = 9$. Since $P''(x) = -2 < 0$ for all x, by (1) and the Second Derivative Test we know that the maximum value of P occurs for $x = 9$. The corresponding value of y is $y = 18 - 9 = 9$. The numbers are 9 and 9.

11. Let x be the length of a side of the base, h the height, V the volume, and S the surface area. We must minimize S. By hypothesis $S = 4xh + x^2$ and $4 = V = x^2h$, so $h = 4/x^2$. Thus $S = 4x(4/x^2) + x^2 = 16/x + x^2$. Next, $S'(x) = -16/x^2 + 2x$, so $S'(x) = 0$ if $-16/x^2 + 2x = 0$, that is, $2x^3 = 16$, or $x = 2$. Since $S''(x) = 32/x^3 + 2 > 0$ for all $x > 0$, it follows from (1) and the Second Derivative Test that the surface area is minimum for $x = 2$. Then $h = 4/2^2 = 1$, so the dimensions are 2 meters on a side of the base, and 1 meter in height.

13. Let r be the radius of the two semicircles, and x the length of the rectangular portion of the field. Our goal is to maximize the area $A = 2rx$ of the rectangular portion. Since the perimeter of the entire field is to be 440 yards, we have $2x + 2\pi r = 440$, so that $x = 220 - \pi r$. Therefore $A = 2rx = 2r(220 - \pi r) = 440r - 2\pi r^2$ for $0 \le r \le 220/\pi$. Thus $dA/dr = 440 - 4\pi r$, so that $dA/dr = 0$ only for $r = 110/\pi$. Since $d^2A/dr^2 = -4\pi < 0$, it follows from (1) and the Second Derivative Test that the maximum value of A occurs for $r = 110/\pi$. The corresponding value of x is $220 - \pi(110/\pi) = 110$.

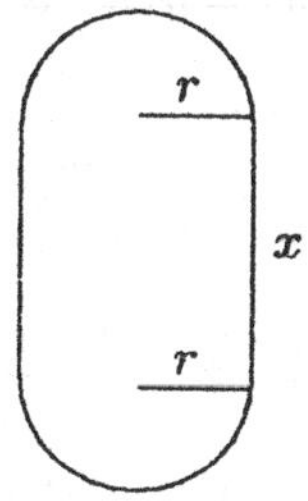

Thus the area of the rectangular portion is maximum if the length of the rectangle is 110 yards and the radius of the semicircles is $110/\pi$ yards.

15. Let x and h be as in the figure, and let P be the perimeter and A the area. We are to maximize A. By hypothesis the triangle is equilateral, so its altitude is $\frac{1}{2}\sqrt{3}\,x$. Thus $A = xh + \frac{1}{2}x(\frac{1}{2}\sqrt{3}\,x) = xh + \frac{1}{4}\sqrt{3}\,x^2$ and $12 = P = 2h + 3x$. Then $h = 6 - \frac{3}{2}x$, so $A = x(6 - \frac{3}{2}x) + \frac{1}{4}\sqrt{3}\,x^2 = 6x + (\frac{1}{4}\sqrt{3} - \frac{3}{2})x^2$ for $0 < x \le 4$. Now $A'(x) = 6 + (\frac{1}{2}\sqrt{3} - 3)x$, so $A'(x) = 0$ if $x = -6/(\sqrt{3}/2 - 3) = 12/(6 - \sqrt{3})$. Since $A''(x) = \frac{1}{2}\sqrt{3} - 3 < 0$ for all x, it follows from (1) and the Second Derivative Test that A is maximum if $x = 12/(6 - \sqrt{3})$. Since

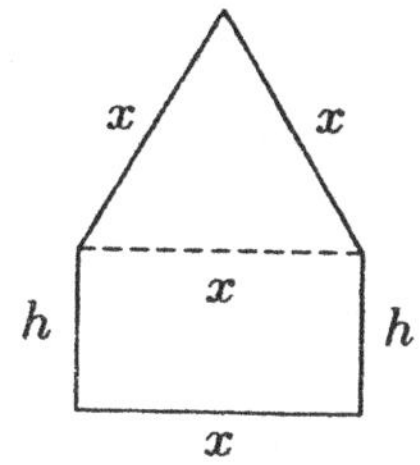

$$h = 6 - \frac{3}{2}\,\frac{12}{6 - \sqrt{3}} = \frac{18 - 6\sqrt{3}}{6 - \sqrt{3}}$$

the maximum amount of light will enter if

$$x = \frac{12}{6 - \sqrt{3}} \approx 2.8\,(\text{feet}) \quad \text{and} \quad h = \frac{18 - 6\sqrt{3}}{6 - \sqrt{3}} \approx 1.8\,(\text{feet}).$$

17. Let x be the length in kilometers of the side along the highway, and y the length in kilometers of the sides perpendicular to the highway. Let k be the cost per kilometer of fence for the cheaper wood. Then the cost of the two sides perpendicular to the highway is $ky + ky = 2ky$, and the cost of the other two sides is $kx + 3kx = 4kx$. By hypothesis, $xy = 1$, so that $y = 1/x$. Thus the total cost of the fence is given by $C = 4kx + 2ky = 4kx + 2k/x$ for $x > 0$. Then $C'(x) = 4k - 2k/x^2$, so that $C'(x) = 0$ if $x^2 = \frac{1}{2}$, or $x = \frac{1}{2}\sqrt{2}$. Since $C''(x) = 4k/x^3 > 0$ for $x > 0$, it follows from (1) and the Second Derivative Test that the cost is minimized if $x = \frac{1}{2}\sqrt{2}$. The corresponding value of y is $1/(\frac{1}{2}\sqrt{2}) = \sqrt{2}$. Thus the cost is minimized if the side of the fence along the highway is $\frac{1}{2}\sqrt{2}$ kilometers long and the sides perpendicular to the highway are $\sqrt{2}$ kilometers long.

19. $f'(\theta) = q\csc^2\theta - \dfrac{r\cos\theta}{\sin^2\theta} = \dfrac{q - r\cos\theta}{\sin^2\theta}$ so that $f'(\theta) = 0$ if $\cos\theta = q/r$. Also

$$f''(\theta) = \frac{r\sin\theta(\sin^2\theta) - (q - r\cos\theta)2\sin\theta\cos\theta}{\sin^4\theta} = \frac{r(\sin^2\theta + 2\cos^2\theta) - 2q\cos\theta}{\sin^3\theta}$$

$$= \frac{r(1+\cos^2\theta) - 2q\cos\theta}{\sin^3\theta} > \frac{r(1+\cos^2\theta - 2\cos\theta)}{\sin^3\theta} = \frac{r(1-\cos\theta)^2}{\sin^3\theta} > 0 \quad \text{for } \theta \text{ in } (0, \pi/2).$$

By (1) and the Second Derivative Test, $f(\theta)$ is minimum for the value of θ in $(0, \pi/2)$ satisfying $\cos\theta = q/r$.

21. If we let $p = 6ab$, $q = 3b^2/2$, and $r = 3\sqrt{3}b^2/2$ in Exercise 19, then $q < r$, so Exercise 19 implies that the minimum value of S is attained for θ_0 such that $\cos\theta_0 = q/r = 1/\sqrt{3}$.

23. a. E is continuous on $[0, \infty)$, and since $E'(r) = c > 0$ for $0 < r < a$ and $E'(r) = -ca^2/r^2 < 0$ for $r > a$, E is increasing on $[0, a]$ and is decreasing on $[a, \infty)$. Therefore $E(a)$ is the maximum value of E.

b. Since

$$\lim_{r\to a^-} \frac{E(r) - E(a)}{r - a} = \lim_{r\to a^-} \frac{c(r-a)}{r-a} = c$$

and

$$\lim_{r\to a^+} \frac{E(r) - E(a)}{r - a} = \lim_{r\to a^+} \frac{(ca^2/r) - ca}{r - a} = \lim_{r\to a^+} \frac{-ca}{r} = -c$$

E is not differentiable at a.

25. The distance between a point (x, y) on the parabola and the point $(-1, 0)$ is given by

$$D = \sqrt{(x+1)^2 + y^2} = \sqrt{x^2 + 2x + 1 + y^2} = \sqrt{y + 1 + y^2}.$$

This distance is minimized for the same value of y that minimizes the square E of the distance, so we let $E = D^2 = y + 1 + y^2$. Then $E'(y) = 1 + 2y$, so $E'(y) = 0$ if $y = -\frac{1}{2}$. Since $E'(y) < 0$ for $y < -\frac{1}{2}$ and $E(y) > 0$ for $y > -\frac{1}{2}$, it follows from (1) and the First Derivative Test that E has its minimum value at $-\frac{1}{2}$. If $y = -\frac{1}{2}$, then $-\frac{1}{2} + 1 = x^2 + 2x + 1 = (x+1)^2$, so that $x + 1 = \sqrt{\frac{1}{2}} = \sqrt{2}/2$ or $x + 1 = -\sqrt{\frac{1}{2}} = -\sqrt{2}/2$. Thus $x = -1 + \sqrt{2}/2$ or $x = -1 - \sqrt{2}/2$. Therefore $(-1 + \sqrt{2}/2, -\frac{1}{2})$ and $(-1 - \sqrt{2}/2, -\frac{1}{2})$ are the points on the parabola $y = x^2 + 2x$ closest to the point $(-1, 0)$.

27. a. For any point (x, x^2) on the parabola $y = x^2$, let $f(x)$ denote the square of the distance between $(0, p)$ and (x, x^2). Thus $f(x) = (x-0)^2 + (x^2-p)^2 = x^2 + (x^2-p)^2$, so that $f'(x) = 2x + 2(x^2-p)(2x) = 4x^3 + 2x(1-2p) = 2x[2x^2 + (1-2p)]$. This implies that $f'(0) = 0$ and if $1 - 2p \ge 0$, then $f'(x) = 0$ only for $x = 0$, and f' changes from negative to positive at 0. Thus if $1 - 2p \ge 0$, then (1) and the First Derivative Test imply that the maximum value of f occurs for $x = 0$. For any $x \ne 0$, we have $f'(x) \ne 0$, so f assumes its maximum only at 0. Thus if $1 - 2p \ge 0$, or equivalently, $p \le \frac{1}{2}$, then the origin is the only point on the parabola that is closest to $(0, p)$.

b. Recall from part (a) that $f'(x) = 2x[2x^2 + (1-2p)]$. Thus if $1-2p < 0$, then $f'(x) = 0$ for $x = 0$, for $x = -\sqrt{p-\frac{1}{2}}$, and for $x = \sqrt{p-\frac{1}{2}}$. Since $f''(x) = 12x^2 + 2(1-2p)$, we have $f''(0) = 2(1-2p) < 0$, so by the Second Derivative Test, f has a relative maximum value at 0. But $f''(\sqrt{p-\frac{1}{2}}) = 12(p-\frac{1}{2}) + 2(1-2p) > 0$. Since $\sqrt{p-\frac{1}{2}}$ is the only critical number of f in $(0,\infty)$, it follows from (1) and the Second Derivative Test that f assumes its maximum value on $[0,\infty)$ at $\sqrt{p-\frac{1}{2}}$. Since $f(-x) = f(x)$, it follows that f assumes its maximum value on $(-\infty,\infty)$ at $-\sqrt{p-\frac{1}{2}}$ and $\sqrt{p-\frac{1}{2}}$. Thus if $1-2p < 0$, or equivalently, $p > \frac{1}{2}$, then there are two points on the parabola that are closest to $(0,p)$.

29. a. Following the solution of Example 4 but with 10 replacing 5, we find that T is given by

$$T = \frac{\sqrt{x^2+4}}{3} + \frac{10-x}{4} \quad \text{for } 0 < x < 10.$$

Thus $T'(x)$ is the same as in Example 4, so $T'(x) = 0$ for $x = \frac{6}{7}\sqrt{7}$. As in Example 4, it follows from (1) and the Second Derivative Test that T is minimized for $x = \frac{6}{7}\sqrt{7}$. Thus the ranger should walk toward the point $\frac{6}{7}\sqrt{7}$ miles down the road.

b. This time T is given by

$$T = \frac{\sqrt{x^2+4}}{3} + \frac{\frac{1}{2}-x}{4} \quad \text{for } 0 \le x \le \tfrac{1}{2}.$$

Thus $T'(x)$ is the same as in Example 4. However, since $\frac{6}{7}\sqrt{7} > \frac{1}{2}$, there are no critical numbers of T in $(0,\frac{1}{2})$, so the minimum value of T must occur at 0 or $\frac{1}{2}$. Since

$$T(0) = \frac{\sqrt{0^2+4}}{3} + \frac{\frac{1}{2}-0}{4} = \frac{19}{24} \quad \text{and} \quad T(\tfrac{1}{2}) = \frac{\sqrt{(\frac{1}{2})^2+4}}{3} + 0 = \frac{\sqrt{17}}{6}$$

it follows that T is minimized for $x = \frac{1}{2}$. Thus the ranger should walk directly toward the car.

c. T is given by

$$T = \frac{\sqrt{x^2+4}}{3} + \frac{c-x}{4} \quad \text{for } 0 \le x \le c$$

so $T'(x)$ is as given in Example 4. If $c > \frac{6}{7}\sqrt{7}$, it follows as in Example 4 that the ranger should walk toward the point $\frac{6}{7}\sqrt{7}$ miles down the road. However, if $c \le \frac{6}{7}\sqrt{7}$, it follows as in part (b) that the ranger should walk directly toward the car.

31. Let x and y denote, respectively, the width and length of the printing on the page. We are to minimize the area A of the page. By hypothesis $xy = 35$, so $y = 35/x$, and thus $A = (x+2)(y+4) = (x+2)(35/x+4) = 43+70/x+4x$ for $x > 0$. Now $A'(x) = -70/x^2 + 4$, so $A'(x) = 0$ if $x^2 = \frac{70}{4}$, or $x = \frac{1}{2}\sqrt{70}$. Since $A''(x) = 140/x^3 > 0$ for $x > 0$, it follows from (1) and the Second Derivative Test that the minimum value of A is

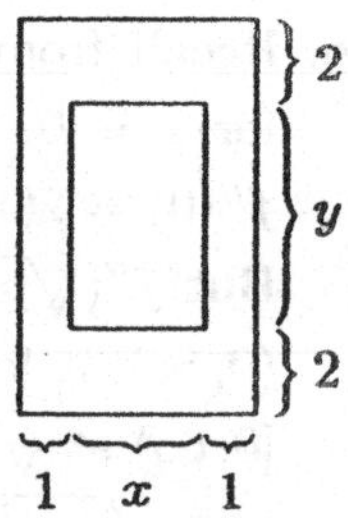

$$A(\tfrac{1}{2}\sqrt{70}) = 43 + 70/(\tfrac{1}{2}\sqrt{70}) + 4(\tfrac{1}{2}\sqrt{70}) = 43 + 4\sqrt{70}.$$

Thus the minimum area is $43 + 4\sqrt{70}$ square inches.

33. Let h be the height, r the radius of the base, and V the volume of the cylinder. We are to maximize V. By the figure, $r^2 = R^2 - (h/2)^2$, so that $V = \pi r^2 h = \pi(R^2 - h^2/4)h = \pi R^2 h - (\pi/4)h^3$. Now $V'(h) = \pi R^2 - (3\pi/4)h^2$, so $V'(h) = 0$ if $\pi R^2 - (3\pi/4)h^2 = 0$, that is, $h = \frac{2}{3}\sqrt{3}\,R$. Since $V''(h) = -(3\pi/2)h < 0$ for $h > 0$, it follows from (1) and the Second Derivative Test that the maximum volume is attained for $h = \frac{2}{3}\sqrt{3}\,R$.

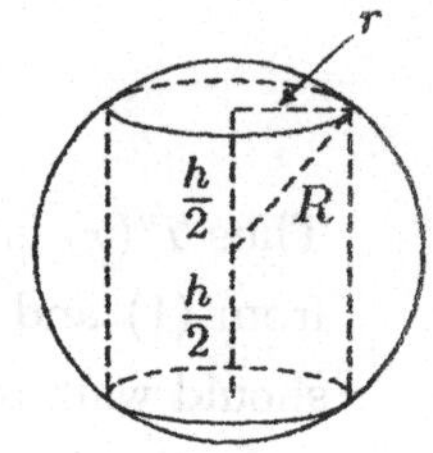

35. In order for the pipe to fit around the corner, its length must be less than or equal to the length of any pipe that touches the two outside walls (as in the figure), no matter what angle θ it makes with respect to the inside wall. Thus we must find the minimum length of a pipe that touches the walls, as in the figure, with $0 < \theta < \pi/2$. Then the length L of the pipe is given by $L = L_1 + L_2 = 10\csc\theta + 10\sec\theta$. Therefore

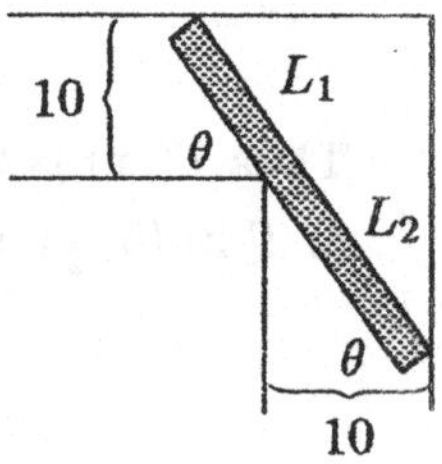

$$\frac{dL}{d\theta} = -10\csc\theta\,\cot\theta + 10\sec\theta\,\tan\theta = 10\left(-\frac{\cos\theta}{\sin^2\theta} + \frac{\sin\theta}{\cos^2\theta}\right) = \frac{10}{\sin^2\theta\,\cos^2\theta}(\sin^3\theta - \cos^3\theta)$$

so that $dL/d\theta = 0$ if and only if $\sin^3\theta = \cos^3\theta$, or $\tan^3\theta = 1$, or $\tan\theta = 1$, or $\theta = \pi/4$. Since $dL/d\theta < 0$ for $0 < \theta < \pi/4$ and $dL/d\theta > 0$ for $\pi/4 < \theta < \pi/2$, it follows from (1) and the First Derivative Test that L assumes its minimum value for $\theta = \pi/4$. Thus the longest pipe that can be carried around the corner has length $10\csc(\pi/4) + 10\sec(\pi/4) = 20\sqrt{2}$ (feet).

37. Let x be the loan interest rate. Then the amount the firm can lend is k/x^2, where $k > 0$, and the profit is given by $P(x) = (k/x^2)(x - .05) = (k/x) - [k(.05)/x^2]$ for $.05 \le x$. Thus

$$P'(x) = \frac{-k}{x^2} + \frac{k(.1)}{x^3} = \frac{k}{x^3}(-x + .1)$$

and $P'(x) = 0$ if $x = .1$. Since $P'(x) > 0$ if $.05 < x < .1$, and $P'(x) < 0$ if $x > .1$, it follows from (1) and the First Derivative Test that the profit is maximized if $x = .1$, that is, if the loan interest rate is 10%.

39. The annual stocking cost is $bx/2$ and the annual reordering cost is $1000c/x$. The inventory cost is given by $f(x) = bx/2 + 1000c/x$. Now $f'(x) = b/2 - 1000c/x^2$, so that $f'(x) = 0$ only if $x^2 = 2000c/b$, or $x = 20\sqrt{5c/b}$. Since $f'(x) < 0$ if $x < 20\sqrt{5c/b}$ and $f'(x) > 0$ if $x > 20\sqrt{5c/b}$, it follows from (1) and the First Derivative Test that f has a minimum value at $x = 20\sqrt{5c/b}$.

41. Let a and b be the distances from A and B, respectively, to the line separating the two media, and let l be the distance shown in the diagram. Then a, b, and l are constants, and $a\tan\theta + b\tan\varphi = l$. Differentiating implicitly with respect to θ, we find that

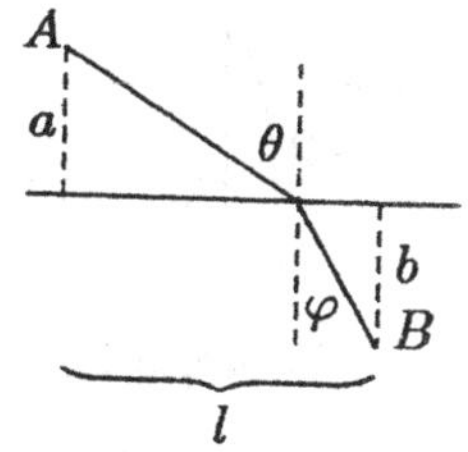

$$a\sec^2\theta + b\sec^2\varphi\frac{d\varphi}{d\theta} = 0 \quad \text{so that} \quad \frac{d\varphi}{d\theta} = -\frac{a\sec^2\theta}{b\sec^2\varphi}.$$

Our goal is to minimize the time T it takes the light to travel from A to B. In the first medium the distance traveled is $a\sec\theta$, so the time the light travels in the first medium is distance/velocity = $(a\sec\theta)/u$. Similarly the time of travel in the second medium is $(b\sec\varphi)/v$. Thus

$$T = \frac{a\sec\theta}{u} + \frac{b\sec\varphi}{v}$$

so that

$$\frac{dT}{d\theta} = \frac{a\sec\theta\tan\theta}{u} + \frac{b\sec\varphi\tan\varphi}{v}\frac{d\varphi}{d\theta} = \frac{a\sec\theta\tan\theta}{u} + \frac{b\sec\varphi\tan\varphi}{v}\left(-\frac{a\sec^2\theta}{b\sec^2\varphi}\right)$$

$$= a\sec^2\theta\left(\frac{\tan\theta}{u\sec\theta} - \frac{\tan\varphi}{v\sec\varphi}\right) = a\sec^2\theta\left(\frac{\sin\theta}{u} - \frac{\sin\varphi}{v}\right).$$

Thus

$$\frac{dT}{d\theta} = 0 \quad \text{if and only if} \quad \frac{\sin\theta}{u} = \frac{\sin\varphi}{v}.$$

Because $d\varphi/d\theta = -(a\sec^2\theta)/(b\sec^2\varphi) < 0$, φ is a decreasing function of θ. It follows that $(\sin\theta)/u - (\sin\varphi)/v$ is an increasing function of θ. Therefore $dT/d\theta$ is an increasing function of θ and hence $d^2T/d\theta^2 > 0$. By (1) and the Second Derivative Test, T is minimized if θ and φ satisfy $(\sin\theta)/u = (\sin\varphi)/v$.

4.7 Concavity and Inflection Points

1. $f'(x) = -3x + 1$; $f''(x) = -3$. Thus the graph is concave downward on $(-\infty, \infty)$.

3. $g'(x) = 3x^2 - 12x + 12 = 3(x-2)^2$; $g''(x) = 6x - 12 = 6(x-2)$. Thus the graph is concave upward on $(2, \infty)$ and concave downward on $(-\infty, 2)$.

5. $f'(x) = \dfrac{(x^2+1) - 2x^2}{(x^2+1)^2} = \dfrac{1-x^2}{(x^2+1)^2} = \dfrac{(1+x)(1-x)}{(x^2+1)^2}$

$$f''(x) = \frac{(x^2+1)^2(-2x) - 4x(x^2+1)(1-x^2)}{(x^2+1)^4} = \frac{2x(x^2-3)}{(x^2+1)^3}$$

Thus the graph is concave upward on $(-\sqrt{3}, 0)$ and $(\sqrt{3}, \infty)$ and concave downward on $(-\infty, -\sqrt{3})$ and $(0, \sqrt{3})$.

7. $g'(x) = e^x + xe^x = (1+x)e^x$; $g''(x) = e^x + (1+x)e^x = (2+x)e^x$. Since $e^x > 0$ for all x, the graph is concave upward on $(-2, \infty)$ and concave downward on $(-\infty, -2)$.

9. The domain of f consists of all positive numbers, and $f'(x) = \ln x + x(1/x) = \ln x + 1$ and $f''(x) = 1/x$. Thus the graph is concave upward on $(0, \infty)$.

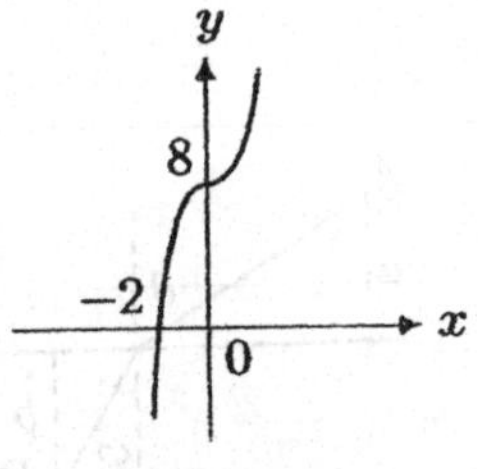

Exercise 11

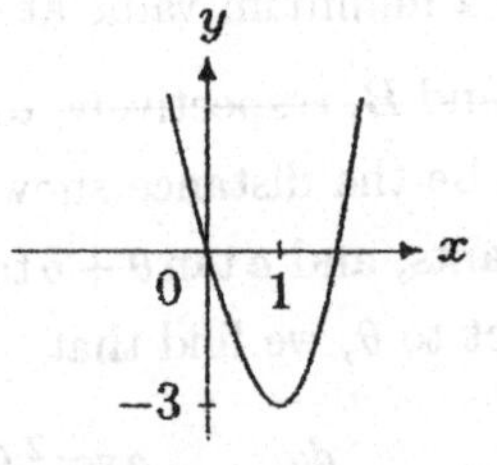

Exercise 13

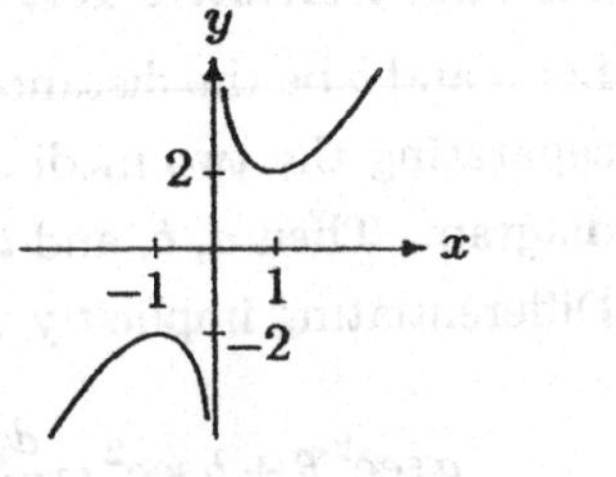

Exercise 15

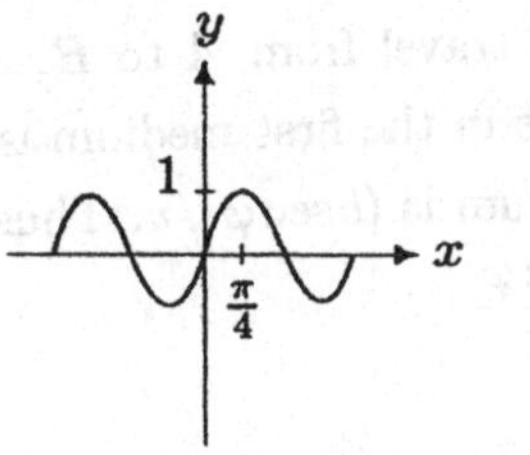

Exercise 17

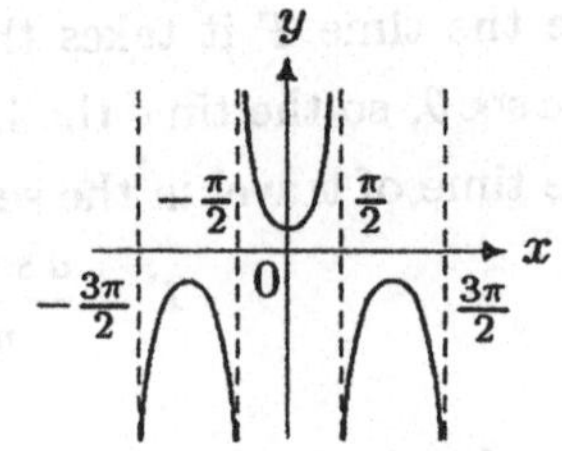

Exercise 19

11. $f'(x) = 3x^2$; $f''(x) = 6x$. Thus the graph is concave upward on $(0, \infty)$ and concave downward on $(-\infty, 0)$.

13. $g'(x) = 4x^3 - 4 = 4(x-1)(x^2+x+1)$; $g''(x) = 12x^2$. Thus the graph is concave upward on $(-\infty, \infty)$. Also, $g(1) = -3$ is the minimum value of g.

15. $f'(x) = 1 - (1/x)^2$; $f''(x) = 2/x^3$. Thus the graph is concave upward on $(0, \infty)$ and concave downward on $(-\infty, 0)$. Also, $f(-1) = -2$ is a relative maximum value, and $f(1) = 2$ is a relative minimum value.

17. Refer to Example 4 and the paragraph following it. We have $f'(x) = 2\cos 2x$ and $f''(x) = -4\sin 2x = -4f(x)$. Thus the graph is concave upward on $(n\pi + \pi/2, (n+1)\pi)$ for any integer n and concave downward on $(n\pi, n\pi + \pi/2)$ for any integer n. Furthermore, for any integer n, $f(n\pi + \pi/4) = 1$ is the maximum value, and $f(n\pi - \pi/4) = -1$ the minimum value of f.

19. $f'(x) = \sec x \tan x$; $f''(x) = \sec x \tan^2 x + \sec^3 x = \sec x\left((\sec^2 x - 1) + \sec^2 x\right) = \sec x\,(2\sec^2 x - 1)$. Since $|\sec x| \geq 1$ for all x in the domain, $2\sec^2 x - 1 > 0$ for all x in the domain, so f and f'' have the same sign. Thus the graph of f is concave upward on $(-\pi/2 + 2n\pi, \pi/2 + 2n\pi)$ and concave downward on $(\pi/2 + 2n\pi, 3\pi/2 + 2n\pi)$ for any integer n. Also $f(2n\pi) = 1$ is a relative minimum value and $f((2n+1)\pi) = -1$ is a relative maximum value of f.

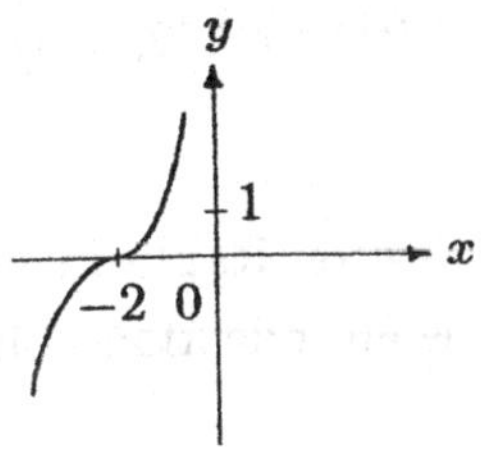

Exercise 21

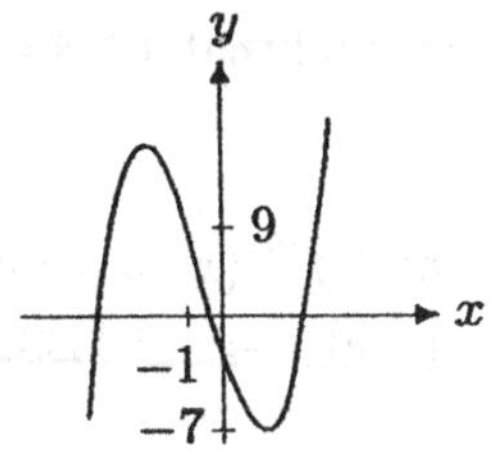

Exercise 23

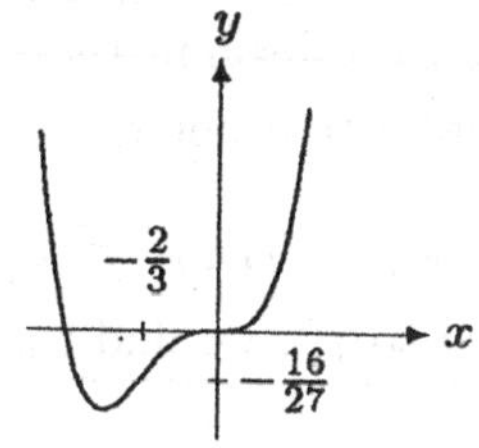

Exercise 25

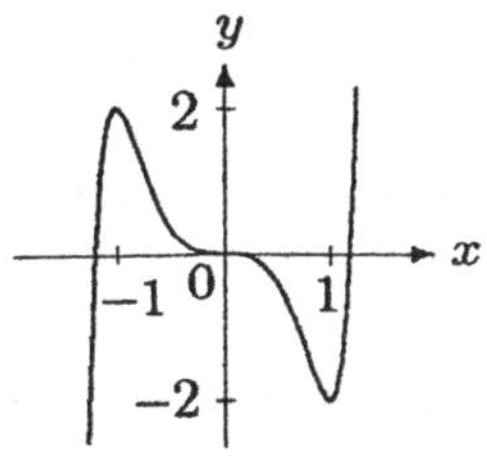

Exercise 27

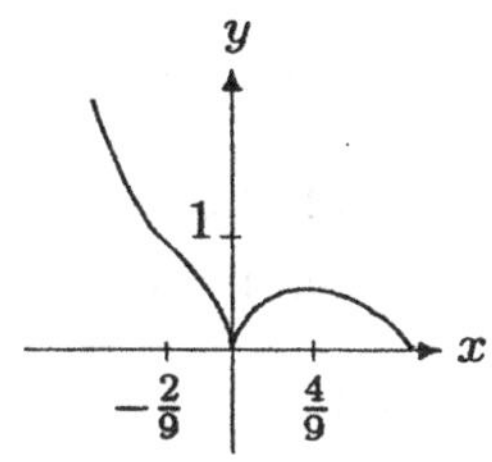

Exercise 29

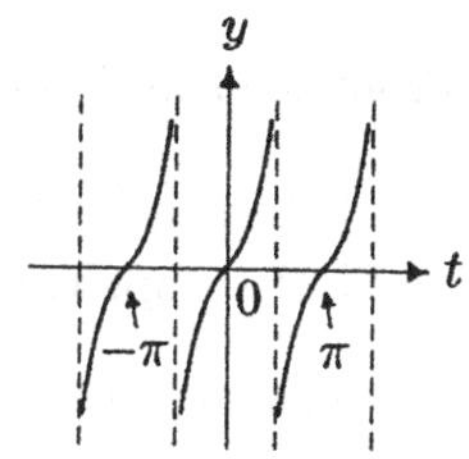

Exercise 31

21. $f'(x) = 3(x+2)^2$ and $f''(x) = 6(x+2)$. Thus f'' changes sign at -2, so $(-2, 0)$ is an inflection point.

23. $f'(x) = 3x^2 + 6x - 9 = 3(x+3)(x-1)$ and $f''(x) = 6x + 6 = 6(x+1)$. Thus f'' changes sign at -1, so $(-1, 9)$ is an inflection point. Also $f(-3) = 25$ is a relative maximum value and $f(1) = -7$ is a relative minimum value of f.

25. $g'(x) = 12x^3 + 12x^2 = 12x^2(x+1)$ and $g''(x) = 36x^2 + 24x = 12x(3x+2)$. Thus g'' changes sign at 0 and $-\frac{2}{3}$, so $(0,0)$ and $(-\frac{2}{3}, -\frac{16}{27})$ are inflection points. Also $g(-1) = -1$ is the minimum value of g.

27. $g'(x) = 9x^8 - 9x^2 = 9x^2(x^6 - 1)$ and $g''(x) = 72x^7 - 18x = 18x(4x^6 - 1)$. Thus g'' changes sign at $-1/\sqrt[3]{2}$, 0, and $1/\sqrt[3]{2}$, so that $(-1/\sqrt[3]{2}, \frac{11}{8})$, $(0,0)$, and $(1/\sqrt[3]{2}, -\frac{11}{8})$ are inflection points. Also $g(-1) = 2$ is a relative maximum value and $g(1) = -2$ is a relative minimum value of g.

29. $g'(x) = \frac{4}{9}x^{-1/3} - x^{2/3} = x^{-1/3}(\frac{4}{9} - x)$ and $g''(x) = -\frac{4}{27}x^{-4/3} - \frac{2}{3}x^{-1/3} = -\frac{2}{3}x^{-4/3}(\frac{2}{9} + x)$ for $x \neq 0$. Since g'' changes from positive to negative at $-\frac{2}{9}$, the point $(-\frac{2}{9}, g(-\frac{2}{9})) = (-\frac{2}{9}, \frac{4}{5}(\frac{2}{9})^{2/3})$ is a point of inflection. Finally $g(\frac{4}{9}) = \frac{2}{5}(\frac{4}{9})^{2/3}$ is a relative maximum value of g.

31. $f'(t) = \sec^2 t$ and $f''(t) = 2\sec^2 t \tan t$. Thus f'' changes sign at $n\pi$, for any integer n, so $(n\pi, 0)$ is an inflection point for any integer n.

33. $f'(x) = 7x^6 - 5x^4 + 3x^2 + 6x$, $f''(x) = 42x^5 - 20x^3 + 6x + 6$, and $f^{(3)}(x) = 210x^4 - 60x^2 + 6$. By the Newton-Raphson method applied to f'', we find that f'' changes sign at $c \approx -.7492409172$, so $(c, f(c))$ is an inflection point.

35. $f'(x) = e^x - \cos x$, $f''(x) = e^x + \sin x$, and $f^{(3)}(x) = e^x + \cos x$. By the Newton-Raphson method applied to f'', we find that f'' changes sign at $c \approx -.588532744$, so $(c, f(c))$ is an inflection point.

37. a. Since f' is positive on (a, c), it follows that f is increasing on $[a, c]$. Therefore $f(c) > f(b) > f(a)$. Since f' is negative on $[c, d]$, it follows that f is decreasing on $[c, d]$. Therefore $f(c) \geq f(d)$. Thus f has the largest value at c.

b. From the solution of part (a), only $f(a)$ or $f(d)$ could be the smallest.

c. Since $f'' = (f')'$, we consider the slope of f' at a, b, c, d. At c and d the slope is negative. At a and b the slope is positive and larger at b. Thus f'' is largest at b.

d. Since the graph of f' is steeper at d and both $f''(c)$ and $f''(d)$ are negative, it follows that $f''(d)$ is the smallest value of f'' at a, b, c, and d.

e. The graph of f has an inflection point at a number r if f'' changes from positive to negative at r. This is the case if f' has a relative maximum at r. This is true if $r = e$. Thus f has an inflection point at e.

39. a. Since the graph of f appears to be concave downward on $(-3, -1)$ and on $(1, 3)$, with f' decreasing on those two intervals, we deduce that $f'' < 0$ on $(-3, -1)$ and on $(1, 3)$. Similarly, since the graph of f appears to be concave upward on $(-1, 1)$, with f' increasing on $(-1, 1)$, we conclude that $f'' > 0$ on $(-1, 1)$.

b. From part (a) the graph changes concavity at $(-1, f(-1))$ and at $(1, f(1))$. Thus $(-1, f(-1))$ and $(1, f(1))$ would be inflection points of the graph of f.

c. Since $g'' = f'$, inflection points of the graph of g would occur for x at which f' changes sign, which by the First Derivative Test means those x for which f has a relative extreme value. From Figure 4.78 we conclude that $x = -2, 0$, and 2 have these properties.

41. $f'(x) = \dfrac{-2x}{(1+x^2)^2}$; $f''(x) = \dfrac{-2(1+x^2)^2 - (-2x)2(1+x^2)(2x)}{(1+x^2)^4} = \dfrac{6x^2-2}{(1+x^2)^3}$; $f''(0) = -2$;
$g'(x) = -xe^{-x^2/2}$; $g''(x) = -e^{-x^2/2} + (-x)e^{-x^2/2}(-x) = (x^2-1)e^{-x^2/2}$; $g''(0) = -1$.
Thus $|f''(0)| > |g''(0)|$, which corresponds to the greater concavity of the graph of f at $(0, 1)$ than that of g.

43. Let $g = -f$. Then $g' = -f'$ and $g'' = -f''$. Thus the concavity of the graph of $-f$ is the opposite of the concavity of the graph of f.

45. For $x \neq 0$, $f'(x) = (\frac{1}{3})x^{-2/3}$ and $f''(x) = (-\frac{2}{9})x^{-5/3}$. Since

$$\lim_{x \to 0} \frac{f(x) - f(0)}{x - 0} = \lim_{x \to 0} x^{-2/3} = \infty,$$

$f'(0)$ does not exist, although the line $x = 0$ is tangent to the graph of f at $(0, 0)$. Since $f'(0)$ does not exist, $f''(0)$ cannot exist. But f'' changes sign at 0, so $(0, f(0)) = (0, 0)$ is an inflection point.

47. The graph of f has either one or three inflection points.

49. If f is a polynomial of degree n, $n \geq 2$, then f'' is a polynomial of degree $n - 2$. Since f'' is defined for all x and has at most $n - 2$ real zeros, the graph of f can have at most $n - 2$ inflection points.

51. Suppose that $f''(c) > 0$. Since f'' is continuous on an open interval containing c, there is an open interval I centered at c such that $f'' > 0$ on I. Thus f'' does not change sign at c, and hence f does not have an inflection point at $(c, f(c))$, as assumed. Similarly, the assumption that $f''(c) < 0$ leads to a contradiction. Therefore $f''(c) = 0$.

4.8 Limits at Infinity

1. $\lim_{x \to \infty} \frac{2}{x - 3} = \lim_{x \to \infty} \frac{2/x}{1 - 3x} = \frac{0}{1 - 0} = 0$

3. $\lim_{x \to \infty} \frac{x}{3x + 2} = \lim_{x \to \infty} \frac{1}{3 + 2/x} = \frac{1}{3 + 0} = \frac{1}{3}$

5. $\lim_{x \to \infty} \frac{2x^2 + x - 1}{x^2 - x + 4} = \lim_{x \to \infty} \frac{2 + 1/x - 1/x^2}{1 - 1/x + 4/x^2} = \frac{2 + 0 - 0}{1 - 0 + 0} = 2$

7. $\lim_{t \to \infty} \frac{t}{t^{1/2} + 2t^{-1/2}} = \lim_{t \to \infty} \left(\frac{t}{t + 2} \cdot t^{1/2}\right) = \lim_{t \to \infty} \left(\frac{1}{1 + 2/t} \cdot t^{1/2}\right) = \infty$

9. Since

$$-\frac{1}{\sqrt{x^2 - 1}} \leq \frac{\cos x}{\sqrt{x^2 - 1}} \leq \frac{1}{\sqrt{x^2 - 1}} \quad \text{for } x < -1$$

and since

$$\lim_{x \to -\infty} -\frac{1}{\sqrt{x^2 - 1}} = \lim_{x \to -\infty} \frac{1}{\sqrt{x^2 - 1}} = 0$$

the Squeezing Theorem for limits at $-\infty$ implies that $\lim_{x \to -\infty} (\cos x)/\sqrt{x^2 - 1} = 0$.

11. $\lim_{x\to-\infty} \frac{x^2}{4x^3-9} = \lim_{x\to-\infty} \frac{1/x}{4-9/x^3} = \frac{0}{4-0} = 0$

13. For $x > \frac{1}{2}$,

$$x - \sqrt{4x^2-1} = (x - \sqrt{4x^2-1})\frac{x+\sqrt{4x^2-1}}{x+\sqrt{4x^2-1}} = \frac{1-3x^2}{x+\sqrt{4x^2-1}} = \frac{1/x-3x}{1+\sqrt{4-1/x^2}}$$

so

$$\lim_{x\to\infty}(x-\sqrt{4x^2-1}) = \lim_{x\to\infty}\frac{1/x-3x}{1+\sqrt{4-1/x^2}} = -\infty.$$

The exercise can also be solved as follows:

$$\lim_{x\to\infty}(x-\sqrt{4x^2-1}) = \lim_{x\to\infty}\left(x - x\sqrt{4-\frac{1}{x^2}}\right) = \lim_{x\to\infty} x\left(1-\sqrt{4-\frac{1}{x^2}}\right) = -\infty.$$

15. Since $\lim_{x\to\infty} 1/x = 0$, the Substitution Theorem for limits at ∞ (with $y = 1/x$) implies that $\lim_{x\to\infty} \tan 1/x = \lim_{y\to 0^+} \tan y = \lim_{y\to 0} \tan y = \tan 0 = 0$.

17. Since $\lim_{x\to\infty} e^x = \infty$, we have $\lim_{x\to\infty} e^{-x} = \lim_{x\to\infty} 1/e^x = 1/\lim_{x\to\infty} e^x = 0$.

19. Since $-1/x > 0$ if $x < 0$, and since $\lim_{x\to-\infty}(-1/x) = 0$, the Substitution Theorem for limits at ∞ (with $y = -1/x$) implies that $\lim_{x\to-\infty} e^{-1/x} = \lim_{y\to 0^+} e^y = \lim_{y\to 0} e^y = e^0 = 1$.

21. Since $\lim_{x\to\infty} \ln x = \infty$, the Substitution Theorem for limits at ∞ (with $y = \ln x$) implies that $\lim_{x\to\infty} 1/\ln x = \lim_{y\to\infty} 1/y = 0$.

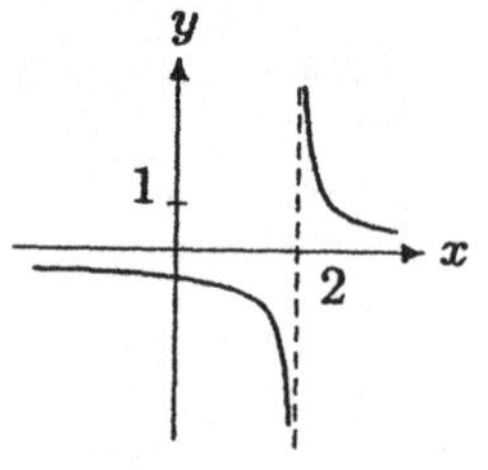

Exercise 23

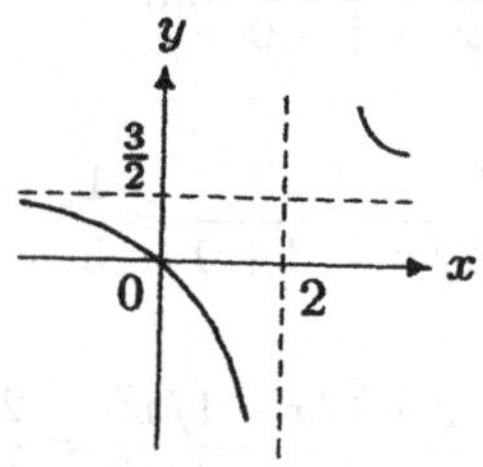

Exercise 25

23. $\lim_{x\to\infty} \frac{1}{x-2} = \lim_{x\to-\infty} \frac{1}{x-2} = 0$

so $y = 0$ is a horizontal asymptote; $f'(x) = -1/(x-2)^2 < 0$ for $x \neq 2$, so f is decreasing on $(-\infty, 2)$ and $(2, \infty)$; $f''(x) = 2/(x-2)^3$, so the graph of f is concave upward on $(2, \infty)$ and concave downward on $(-\infty, 2)$;

$$\lim_{x\to 2^+} \frac{1}{x-2} = \infty \quad \text{and} \quad \lim_{x\to 2^-} \frac{1}{x-2} = -\infty$$

so $x = 2$ is a vertical asymptote.

25. $\lim_{x\to\infty} \frac{3x}{2x-4} = \lim_{x\to-\infty} \frac{3x}{2x-4} = \frac{3}{2}$

so $y = \frac{3}{2}$ is a horizontal asymptote;

$$f'(x) = \frac{3(2x-4) - 3x(2)}{(2x-4)^2} = \frac{-12}{(2x-4)^2} = \frac{-3}{(x-2)^2} \quad \text{for } x \neq 2$$

so f is decreasing on $(-\infty, 2)$ and $(2, \infty)$; $f''(x) = 6/(x-2)^3$, so the graph of f is concave upward on $(2, \infty)$ and concave downward on $(-\infty, 2)$;

$$\lim_{x\to2^+} \frac{3x}{2x-4} = \infty \quad \text{and} \quad \lim_{x\to2^-} \frac{3x}{2x-4} = -\infty$$

so $x = 2$ is a vertical asymptote.

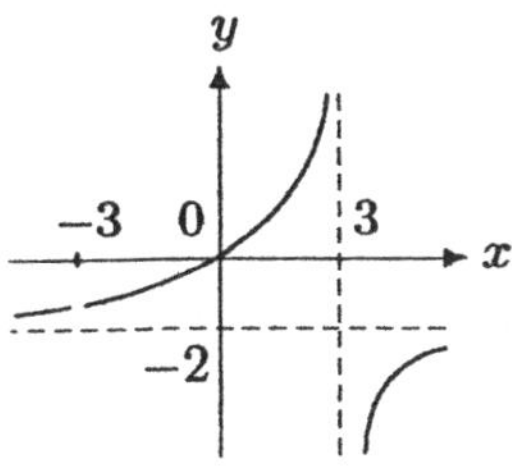

Exercise 27

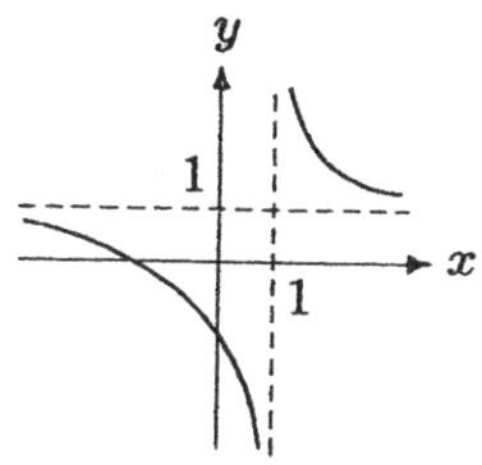

Exercise 29

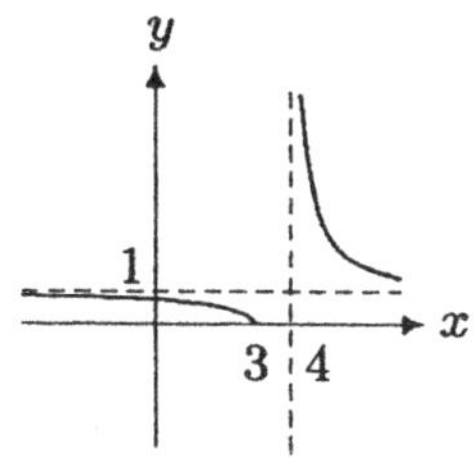

Exercise 31

27. For $x \neq -3$ and 3, $f(x) = \frac{2x(x+3)}{(3-x)(3+x)} = \frac{2x}{3-x}$,

so $\lim_{x\to\infty} f(x) = \lim_{x\to-\infty} f(x) = -2$ and thus $y = -2$ is a horizontal asymptote;

$$f'(x) = \frac{2(3-x) - 2x(-1)}{(3-x)^2} = \frac{6}{(3-x)^2} > 0 \quad \text{for } x \neq -3 \text{ and } 3$$

so f is increasing on $(-\infty, -3)$, $(-3, 3)$, and $(3, \infty)$; $f''(x) = 12/(3-x)^3$ for $x \neq -3$ and 3, so the graph of f is concave upward on $(-\infty, -3)$ and $(-3, 3)$ and concave downward on $(3, \infty)$; $\lim_{x\to3^+} 2x/(3-x) = -\infty$ and $\lim_{x\to3^-} 2x/(3-x) = \infty$, so $x = 3$ is a vertical asymptote.

29. $\lim_{x\to\infty} \frac{x+2}{x-1} = \lim_{x\to-\infty} \frac{x+2}{x-1} = 1$

so $y = 1$ is a horizontal asymptote;

$$f'(x) = \frac{1(x-1) - (x+2)(1)}{(x-1)^2} = \frac{-3}{(x-1)^2} < 0 \quad \text{for } x \neq 1$$

so f is decreasing on $(-\infty, 1)$ and $(1, \infty)$; $f''(x) = 6/(x-1)^3$, so the graph of f is concave upward on $(1, \infty)$ and concave downward on $(-\infty, 1)$;

$$\lim_{x\to1^+} \frac{x+2}{x-1} = \infty \quad \text{and} \quad \lim_{x\to1^-} \frac{x+2}{x-1} = -\infty$$

so $x = 1$ is a vertical asymptote.

31. The domain of f consists of all x for which $(3-x)/(4-x) \geq 0$, that is, all x for which $3-x=0$, or $3-x$ and $4-x$ have the same sign. Thus the domain of f consists of $(-\infty, 3]$ and $(4, \infty)$.

$$\lim_{x\to\infty}\sqrt{\frac{3-x}{4-x}} = \lim_{x\to-\infty}\sqrt{\frac{3-x}{4-x}} = 1$$

so $y=1$ is a horizontal asymptote;

$$f'(x) = \frac{1}{2}\left(\frac{3-x}{4-x}\right)^{-1/2}\left[\frac{-(4-x)-(3-x)(-1)}{(4-x)^2}\right] = \frac{-1}{2(4-x)^2}\left(\frac{3-x}{4-x}\right)^{-1/2}$$

so f is decreasing on $(-\infty, 3]$ and $(4, \infty)$;

$$f''(x) = \frac{-1}{(4-x)^3}\left(\frac{3-x}{4-x}\right)^{-1/2} + \frac{1}{4(4-x)^2}\left(\frac{3-x}{4-x}\right)^{-3/2}\left[\frac{-(4-x)-(3-x)(-1)}{(4-x)^2}\right]$$

$$= \frac{-1}{(4-x)^3}\left(\frac{3-x}{4-x}\right)^{-1/2}\left[1+\frac{1}{4(4-x)}\left(\frac{3-x}{4-x}\right)^{-1}\right] = \frac{-1}{(4-x)^3}\left(\frac{3-x}{4-x}\right)^{-1/2}\left[1+\frac{1}{4(3-x)}\right]$$

$$= \frac{4x-13}{4(3-x)(4-x)^3}\left(\frac{3-x}{4-x}\right)^{-1/2}$$

so the graph of f is concave upward on $(4, \infty)$ and concave downward on $(-\infty, 3)$;

$\lim_{x\to 4^+}\sqrt{(3-x)/(4-x)} = \infty$, so $x=4$ is a vertical asymptote.

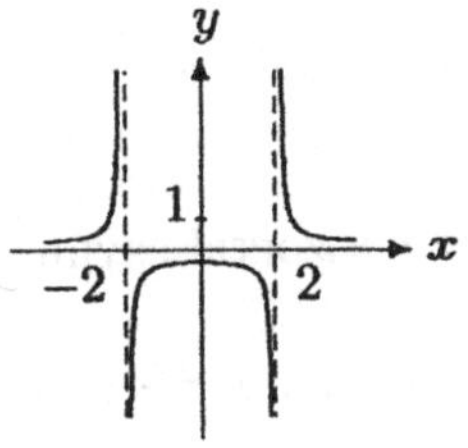

Exercise 33

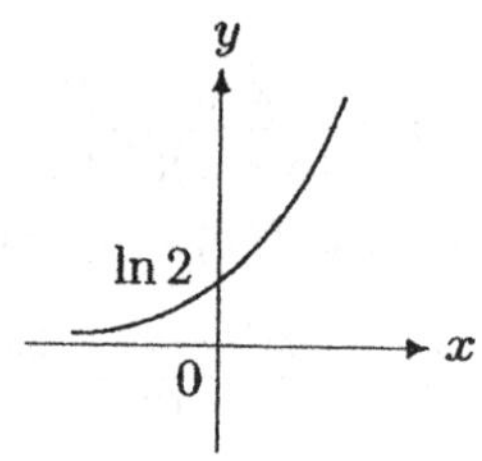

Exercise 35

33. $\lim_{x\to\infty}\frac{1}{x^2-4} = \lim_{x\to-\infty}\frac{1}{x^2-4} = 0$

so $y=0$ is a horizontal asymptote; $f'(x) = -2x/(x^2-4)^2$, so f is increasing on $(-\infty,-2)$ and $(-2,0]$, and decreasing on $[0,2)$ and $(2,\infty)$;

$$f''(x) = \frac{-2(x^2-4)^2-(-2x)2(x^2-4)(2x)}{(x^2-4)^4} = \frac{2(3x^2+4)}{(x^2-4)^3}$$

so the graph of f is concave upward on $(-\infty,-2)$ and $(2,\infty)$ and concave downward on $(-2,2)$;

$$\lim_{x\to2^+}\frac{1}{x^2-4} = \lim_{x\to-2^-}\frac{1}{x^2-4} = \infty \quad\text{and}\quad \lim_{x\to2^-}\frac{1}{x^2-4} = \lim_{x\to-2^+}\frac{1}{x^2-4} = -\infty$$

so $x=2$ and $x=-2$ are vertical asymptotes.

35. $\lim_{x\to\infty} \ln(1+e^x) = \infty$ and $\lim_{x\to-\infty} \ln(1+e^x) = \ln 1 = 0$

so $y = 0$ is a horizontal asymptote; $f'(x) = e^x/(1+e^x) > 0$ for all x, so f is increasing on $(-\infty, \infty)$;

$$f''(x) = \frac{e^x(1+e^x) - e^x e^x}{(1+e^x)^2} = \frac{e^x}{(1+e^x)^2} > 0 \quad \text{for all } x$$

so the graph of f is concave upward on $(-\infty, \infty)$.

37. Horizontal asymptote: $y = 1/\sqrt{2}$; vertical asymptotes: $x = -4$ and $x = 4$.

39. No horizontal asymptotes; vertical asymptotes: $x = -1$ and $x = 1$.

41. No horizontal asymptotes; vertical asymptote: $x = 0$.

43. $\lim_{x\to\infty} \frac{2x^2+1}{3x^2-5} = \lim_{x\to 0+} \frac{2(1/x)^2+1}{3(1/x)^2+5} = \lim_{x\to 0} \frac{2+x^2}{3+5x^2} = \frac{2}{3}$

45. $\lim_{x\to\infty} \frac{\sqrt{1+x^2}}{x} = \lim_{x\to 0+} \frac{\sqrt{1+(1/x)2}}{1/x} = \lim_{x\to 0+} \sqrt{x^2}\sqrt{1+1/x^2} = \lim_{x\to 0} \sqrt{x^2+1} = 1$

47. a. undefined b. 0 c. undefined d. $-\infty$ e. undefined
f. $-\infty$ g. $-\infty$ h. $-\infty$ i. $x = -2$ and $x = 0$ j. $y = 0$

49. Since $\lim_{x\to\infty} 2/x^2 = 0$ and $\lim_{x\to-\infty} 2/x^2 = 0$, it follows that

$$\lim_{x\to\infty} \frac{\sqrt{2+4x^2}}{x} = \lim_{x\to\infty} \sqrt{\frac{2}{x^2}+4} = 2 \quad \text{and} \quad \lim_{x\to-\infty} \frac{\sqrt{2+4x^2}}{x} = \lim_{x\to-\infty} -\sqrt{\frac{2}{x^2}+4} = -2.$$

Therefore the lines $y = 2$ and $y = -2$ are horizontal asymptotes of the graph of f.

51. Let $\varepsilon > 0$. If $L = 0$, we choose $M_1 \geq M$ so that if $x \geq M_1$, then $[f(x)]^2 < \varepsilon^2$. This implies that if $x \geq M_1$, then $0 \leq f(x) \leq \varepsilon$, so that $\lim_{x\to\infty} f(x) = 0 = \sqrt{L}$. If $L > 0$, we choose $M_1 \geq M$ so that if $x \geq M_1$, then $|[f(x)]^2 - L| < \varepsilon\sqrt{L}$. This implies that if $x \geq M_1$, then

$$|f(x) - \sqrt{L}| = \frac{|[f(x)]^2 - L|}{f(x)+\sqrt{L}} \leq \frac{|[f(x)]^2 - L|}{\sqrt{L}} < \frac{\varepsilon\sqrt{L}}{\sqrt{L}} = \varepsilon$$

so that $\lim_{x\to\infty} f(x) = \sqrt{L}$.

53. a. On earth, $a = g$, and hence Newton's Second Law of Motion becomes $F = ma = mg$. For the earth's force of attraction on an object on the earth's surface, we have $F = GMm/R^2$. Equating the two expressions for the force, we have $mg = GMm/R^2$.

b. By (4), $v_{esc} = \sqrt{2GM/R}$, and by part (a), $GM/R = gR$. Therefore $v_{esc} = \sqrt{2gR}$.

c. Using $g = 9.8$ (meters per second squared) and $R = 6.37 \times 10^6$ (meters), we find from (b) that $v_{esc} \approx 11,433$ (meters per second), or equivalently, 11.4 (kilometers per second), which is close to the value of 11.2 obtained form (4).

55. a. Since $a > 0$, $\lim_{t\to\infty} e^{-at} = 0$. Therefore $\lim_{t\to\infty} v(t) = \lim_{t\to\infty} v^*(1 - e^{-at}) = v^*(1-0) = v^*$.

b. As time progresses, the velocity approaches the number v^*. Thus v^* is the "terminal," or "limiting," velocity of the falling parachutist.

57. $\lim_{t\to\infty} S(t) = \lim_{t\to\infty} \left(\frac{a}{t} + b\right) = b$

59. $\lim_{t\to\infty} \left(I(t) - 3\sin\frac{30t}{\pi}\right) = \lim_{t\to\infty} \frac{100}{1+t^2} = 0$

4.9 Graphing

1. $f'(x) = 3(x^2+1)$; $f''(x) = 6x$; increasing on $(-\infty, \infty)$; concave upward on $(0, \infty)$ and concave downward on $(-\infty, 0)$; inflection point is $(0, 2)$.

3. $f'(x) = 4x^3 + 24x^2 + 72x$; $f''(x) = 12x^2 + 48x + 72$; relative minimum value is $f(0) = -3$; increasing on $[0, \infty)$ and decreasing on $(-\infty, 0]$; concave upward on $(-\infty, \infty)$.

5. $g'(x) = 1 - 4/x^2$; $g''(x) = 8/x^3$; relative maximum value is $g(-2) = -4$; relative minimum value is $g(2) = 4$; increasing on $(-\infty, -2]$ and $[2, \infty)$, and decreasing on $[-2, 0)$ and $(0, 2]$; concave upward on $(0, \infty)$ and concave downward on $(-\infty, 0)$; vertical asymptote is $x = 0$; symmetry with respect to the origin.

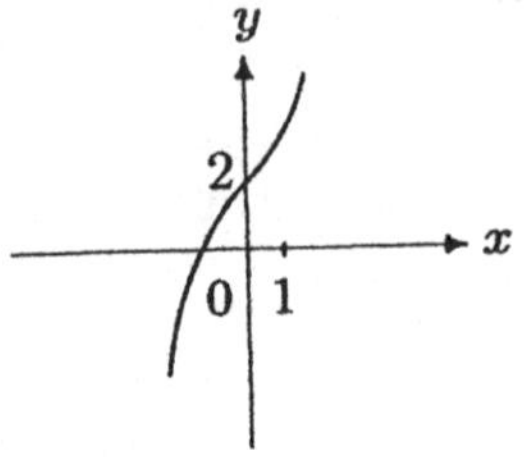

Exercise 1

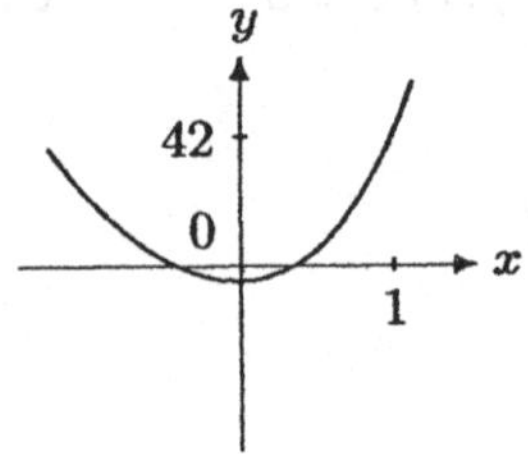

Exercise 3

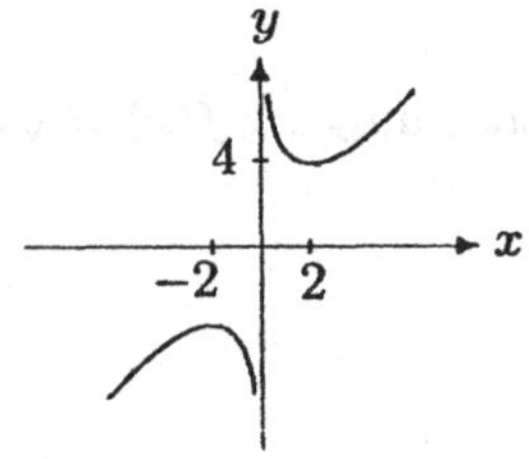

Exercise 5

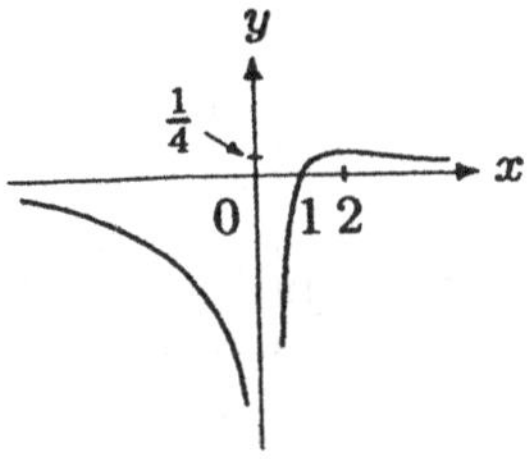

Exercise 7

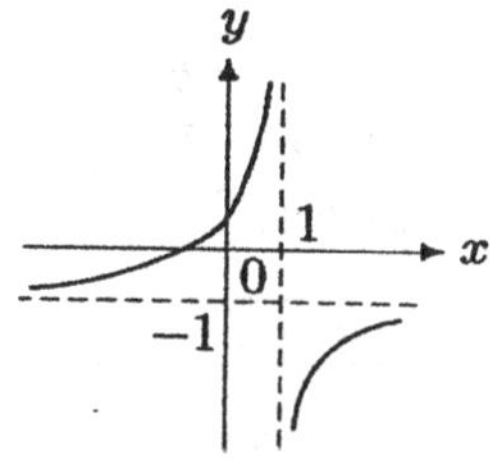

Exercise 9

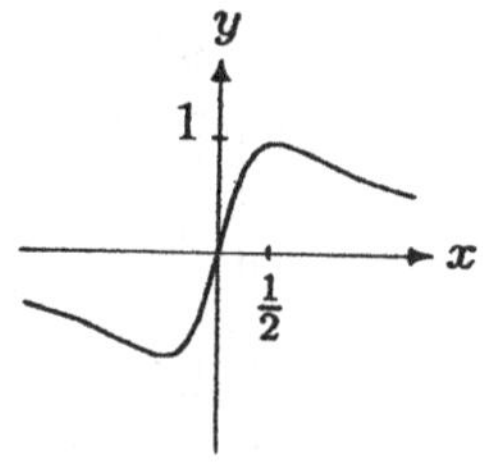

Exercise 11

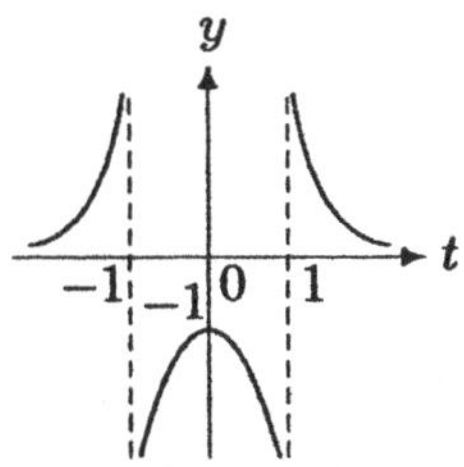

Exercise 13

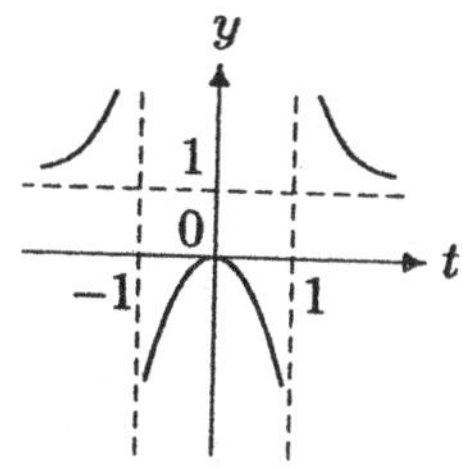

Exercise 15

7. $g'(x) = \dfrac{2-x}{x^3}$; $g''(x) = \dfrac{2(x-3)}{x^4}$

relative maximum value is $g(2) = \frac{1}{4}$; increasing on $(0, 2]$, and decreasing on $(-\infty, 0)$ and on $[2, \infty)$; concave upward on $(3, \infty)$ and concave downward on $(-\infty, 0)$ and $(0, 3)$; inflection point is $(3, \frac{2}{9})$; vertical asymptote is $x = 0$; horizontal asymptote is $y = 0$.

9. $k'(x) = \dfrac{2}{(1-x)^2}$; $k''(x) = \dfrac{4}{(1-x)^3}$

increasing on $(-\infty, 1)$ and $(1, \infty)$; concave upward on $(-\infty, 1)$ and concave downward on $(1, \infty)$; vertical asymptote is $x = 1$; horizontal asymptote is $y = -1$.

11. $k'(x) = \dfrac{4(1-4x^2)}{(1+4x^2)^2}$; $k''(x) = \dfrac{32x(-3+4x^2)}{(1+4x^2)^3}$

relative maximum value is $k(\frac{1}{2}) = 1$; relative minimum value is $k(-\frac{1}{2}) = -1$; increasing on $[-\frac{1}{2}, \frac{1}{2}]$, and decreasing on $(-\infty, -\frac{1}{2}]$ and $[\frac{1}{2}, \infty)$; concave upward on $(-\sqrt{3}/2, 0)$ and $(\sqrt{3}/2, \infty)$, and concave downward on $(-\infty, -\sqrt{3}/2)$ and $(0, \sqrt{3}/2)$; inflection points are $(-\sqrt{3}/2, -\sqrt{3}/2)$, $(0, 0)$, and $(\sqrt{3}/2, \sqrt{3}/2)$; horizontal asymptote is $y = 0$; symmetry with respect to the origin.

13. $f'(t) = \dfrac{-2t}{(t^2-1)^2}$; $f''(t) = \dfrac{6t^2+2}{(t^2-1)^3}$

relative maximum value is $f(0) = -1$; increasing on $(-\infty, -1)$ and $(-1, 0]$, and decreasing on $[0, 1)$ and $(1, \infty)$; concave upward on $(-\infty, -1)$ and $(1, \infty)$, and concave downward on $(-1, 1)$; vertical asymptotes are $t = -1$ and $t = 1$; horizontal asymptote is $y = 0$; symmetry with respect to the y axis.

15. Since $t^2/(t^2-1) = 1 + 1/(t^2-1)$, the graph is the same as the graph in Exercise 13, but translated upward one unit.

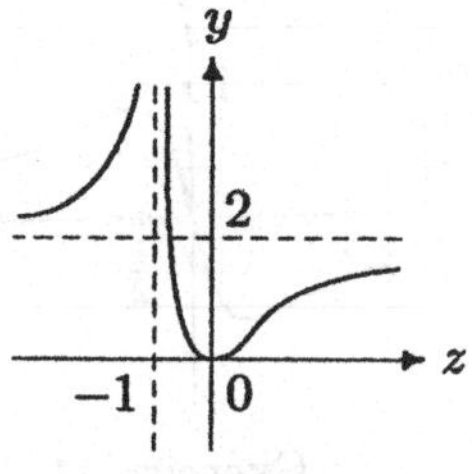

Exercise 17

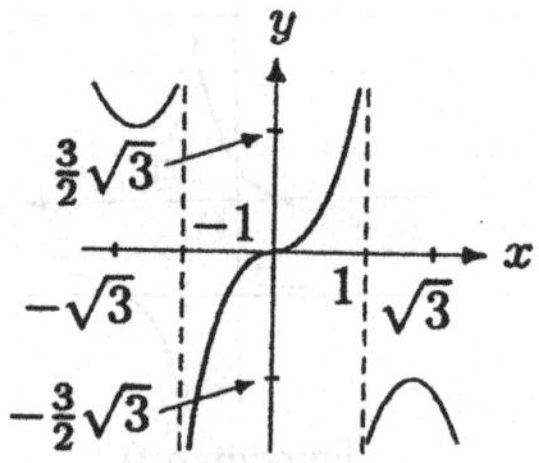

Exercise 19

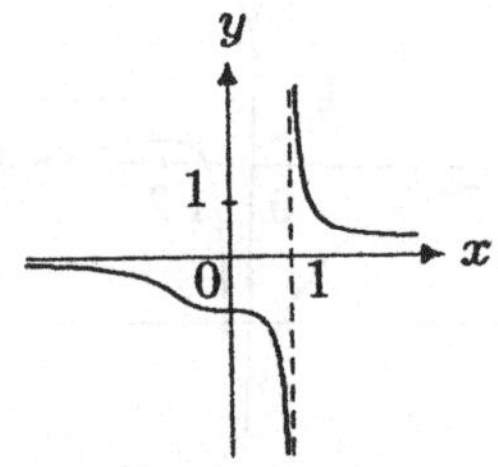

Exercise 21

17. $f'(z) = \dfrac{4z}{(z+1)^3}$; $f''(z) = \dfrac{4(1-2z)}{(z+1)^4}$

relative minimum value is $f(0) = 0$; increasing on $(-\infty, -1)$ and $[0, \infty)$, and decreasing on $(-1, 0]$; concave upward on $(-\infty, -1)$ and $(-1, \frac{1}{2})$, and concave downward on $(\frac{1}{2}, \infty)$; inflection point is $(\frac{1}{2}, \frac{2}{9})$; vertical asymptote is $z = -1$; horizontal asymptote is $y = 2$.

19. $f'(x) = \dfrac{x^2(3-x^2)}{(1-x^2)^2}$; $f''(x) = \dfrac{2x(x^2+3)}{(1-x^2)^3}$

relative maximum value is $f(\sqrt{3}) = -\frac{3}{2}\sqrt{3}$; relative minimum value is $f(-\sqrt{3}) = \frac{3}{2}\sqrt{3}$; increasing on $[-\sqrt{3}, -1)$, $(-1, 1)$, and $(1, \sqrt{3}]$, and decreasing on $(-\infty, -\sqrt{3}]$ and on $(\sqrt{3}, \infty)$; concave upward on $(-\infty, -1)$ and on $(0, 1)$,and concave downward on $(-1, 0)$ and $(1, \infty)$; inflection point is $(0, 0)$; vertical asymptotes are $x = -1$ and $x = 1$; symmetry with respect to the origin.

21. $f'(x) = \dfrac{-3x^2}{(x^3-1)^2}$; $f''(x) = \dfrac{6x(2x^3+1)}{(x^3-1)^3}$

decreasing on $(-\infty, 1)$ and $(1, \infty)$; concave upward on $(-\sqrt[3]{\frac{1}{2}}, 0)$ and $(1, \infty)$, and concave downward on $(-\infty, -\sqrt[3]{\frac{1}{2}})$ and $(0, 1)$; inflection points are $(-\sqrt[3]{\frac{1}{2}}, -\frac{2}{3})$ and $(0, -1)$; vertical asymptote is $x = 1$; horizontal asymptote is $y = 0$.

23. $f'(x) = \frac{1}{2}(x-1)^{-1/2} + \frac{1}{2}(x+1)^{-1/2}$; $f''(x) = -\frac{1}{4}(x-1)^{-3/2} - \frac{1}{4}(x+1)^{-3/2}$;
domain is $[1, \infty)$; increasing on $[1, \infty)$; concave downward on $(1, \infty)$.

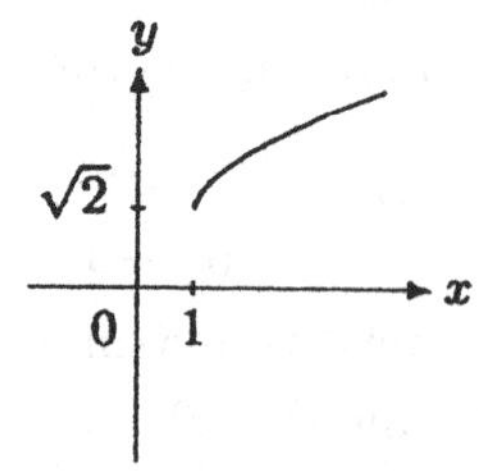

Exercise 23

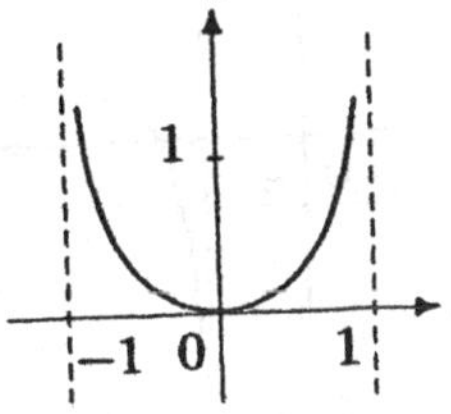

Exercise 25

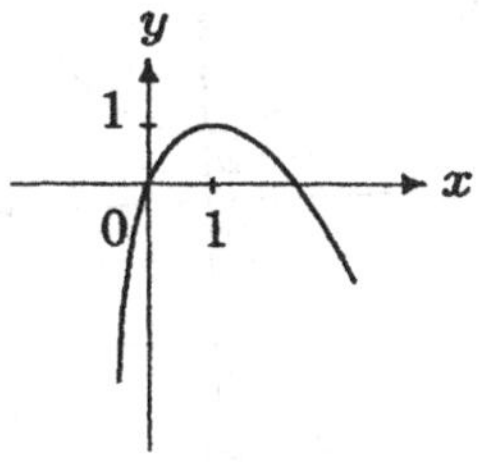

Exercise 27

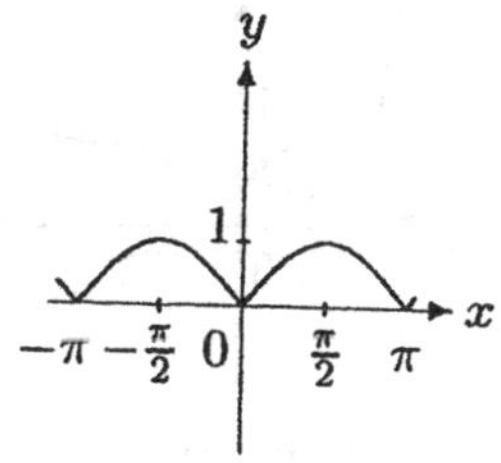

Exercise 29

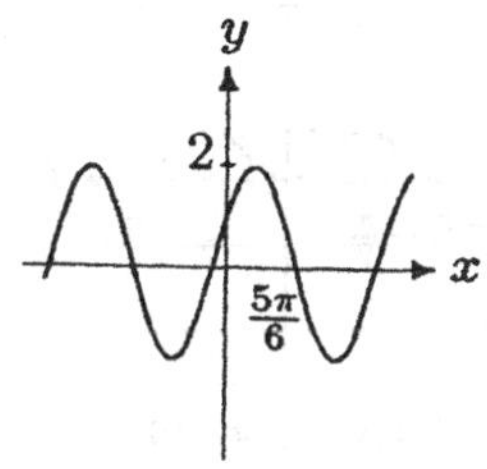

Exercise 31

25. $f'(x) = \dfrac{x(2-x^2)}{(1-x^2)^{3/2}}$; $f''(x) = \dfrac{2+x^2}{(1-x^2)^{5/2}}$

domain is $(-1,1)$; relative minimum value is $f(0) = 0$; increasing on $[0,1)$ and decreasing on $(-1,0]$; concave upward on $(-1,1)$; vertical asymptotes are $x=-1$ and $x=1$; symmetry with respect to the y axis.

27. $f'(x) = 4(1-x^{1/3})$; $f''(x) = -\frac{4}{3}x^{-2/3}$

relative maximum value is $f(1) = 1$; increasing on $(-\infty, 1]$ and decreasing on $[1,\infty)$; concave downward on $(-\infty,\infty)$.

29. $g'(x) = \left\{\begin{array}{ll} \cos x & \text{for } 2n\pi < x < (2n+1)\pi \\ -\cos x & \text{for } (2n+1)\pi < x < (2n+2)\pi \end{array}\right\}$, for any integer n

$g''(x) = -|\sin x|$ for $x \neq n\pi$, for any integer n.

relative maximum value is $g(\pi/2 + n\pi) = 1$; relative minimum value is $g(n\pi) = 0$; increasing on $[n\pi, \pi/2+n\pi]$ and decreasing on $[\pi/2+n\pi, (n+1)\pi]$; concave downward on $(n\pi, (n+1)\pi)$; symmetry with respect to the y axis.

31. $g'(x) = \sqrt{3}\cos x - \sin x$; $g''(x) = -\sqrt{3}\sin x - \cos x = -g(x)$

relative maximum value is $g(\pi/3 + 2n\pi) = 2$ for any integer n; relative minimum value is $g(-2\pi/3 + 2n\pi) = -2$ for any integer n; increasing on $[-2\pi/3 + 2n\pi, \pi/3 + 2n\pi]$ and decreasing on $[\pi/3 + 2n\pi, 4\pi/3 + 2n\pi]$; concave upward on $(-7\pi/6 + 2n\pi, -\pi/6 + 2n\pi)$ and concave downward on $(-\pi/6 + 2n\pi, 5\pi/6 + 2n\pi)$; inflection points are $(-\pi/6 + n\pi, 0)$.

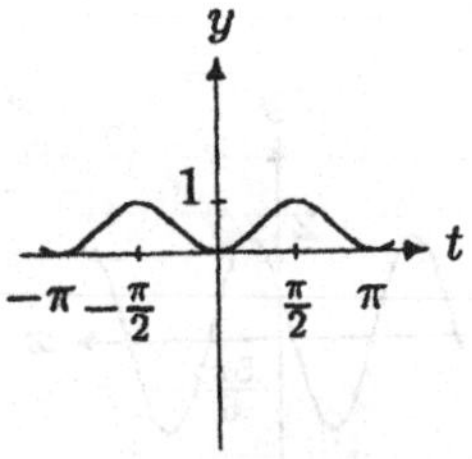

Exercise 33

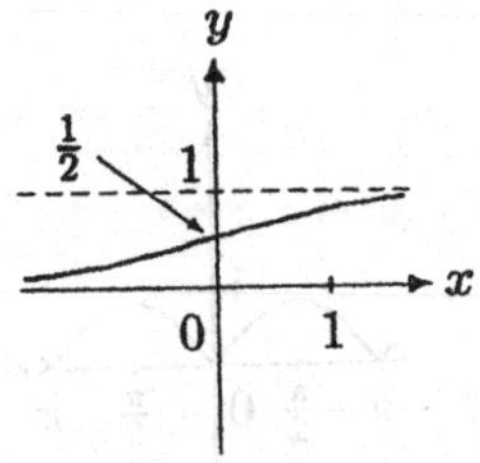

Exercise 35

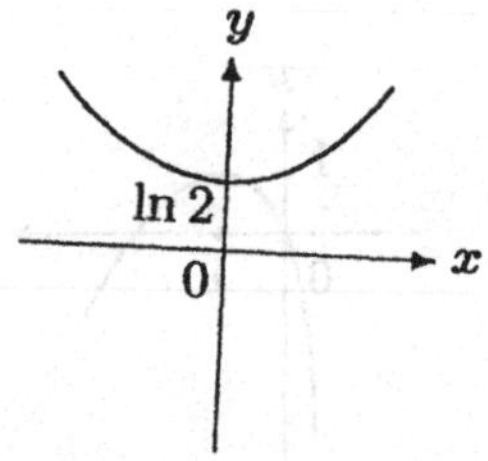

Exercise 37

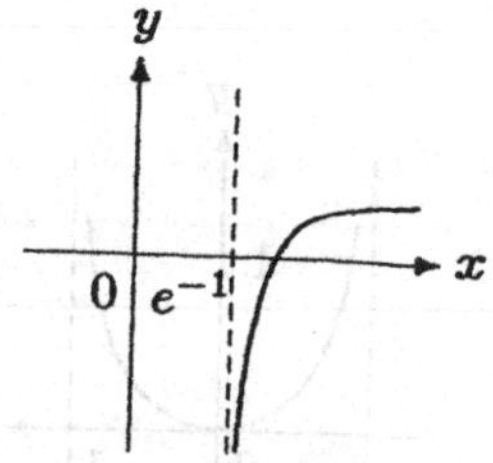

Exercise 39

33. $g'(t) = 2\sin t \cos t = \sin 2t$; $g''(t) = 2\cos 2t$

relative maximum value is $g(n\pi + \pi/2) = 1$ for any integer n; relative minimum value is $g(n\pi) = 0$ for any integer n; increasing on $[n\pi, n\pi + \pi/2]$ and decreasing on $[n\pi + \pi/2, (n+1)\pi]$; concave upward on $(n\pi - \pi/4, n\pi + \pi/4)$ and concave downward on $(n\pi + \pi/4, n\pi + 3\pi/4)$; inflection points are $(\pi/4 + n\pi/2, \frac{1}{2})$; symmetry with respect to the y axis.

35. $f'(x) = \dfrac{e^x}{(1+e^x)^2}$; $f''(x) = \dfrac{e^x - e^{2x}}{(1+e^x)^3}$

increasing on $(-\infty, \infty)$; $f''(x) = 0$ if $e^x(1 - e^x) = 0$, or $x = 0$; concave upward on $(-\infty, 0)$ and concave downward on $(0, \infty)$; inflection point is $(0, \frac{1}{2})$;

$$\lim_{x\to\infty} \frac{e^x}{1+e^x} = \lim_{x\to\infty} \frac{1}{e^{-x}+1} = 1 \quad \text{and} \quad \lim_{x\to-\infty} \frac{e^x}{1+e^x} = 0$$

so that $y = 1$ and $y = 0$ are horizontal asymptotes.

37. $f'(x) = \dfrac{e^x - e^{-x}}{e^x + e^{-x}} = \dfrac{e^{-x}(e^{2x} - 1)}{e^x + e^{-x}}$; $f''(x) = \dfrac{4}{(e^x + e^{-x})^2}$

increasing on $[0, \infty)$ and decreasing on $(-\infty, 0]$; minimum value is $f(0) = \ln 2$; concave upward on $(-\infty, \infty)$; symmetry with respect to the y axis.

39. $f'(x) = \dfrac{1}{x(1 + \ln x)}$; $f''(x) = \dfrac{-(2 + \ln x)}{x^2(1 + \ln x)^2}$

domain is (e^{-1}, ∞); increasing on (e^{-1}, ∞); concave downward on (e^{-1}, ∞); vertical asymptote is $x = e^{-1}$.

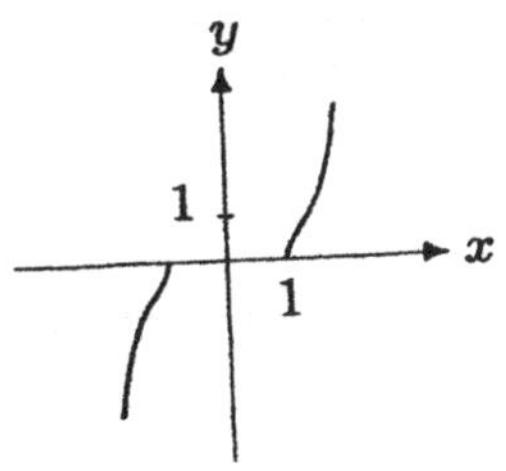

Exercise 41

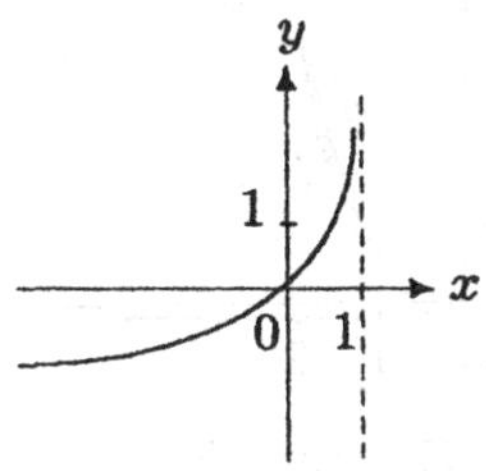

Exercise 43

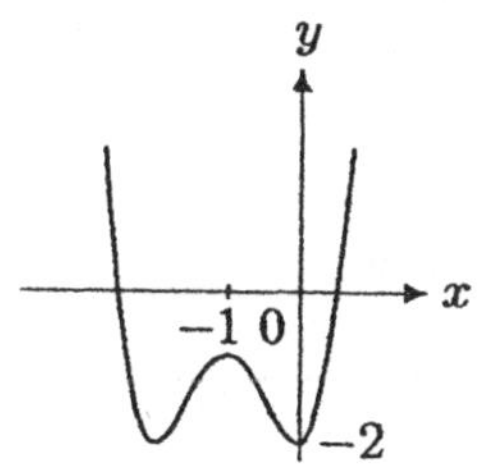

Exercise 45

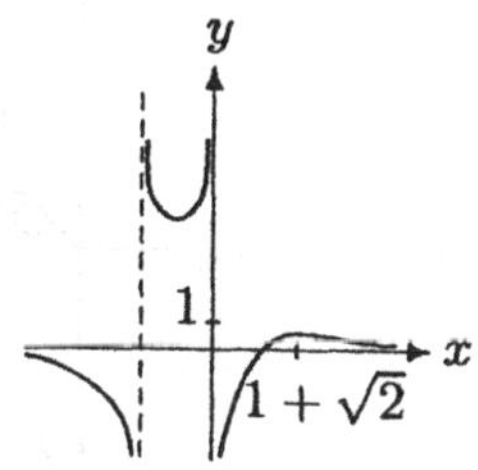

Exercise 47

41. $f'(x) = \dfrac{2x^2-1}{\sqrt{x^2-1}}$; $f''(x) = \dfrac{x(2x^2-3)}{(x^2-1)^{3/2}}$
increasing on $(-\infty,-1]$ and $[1,\infty)$; concave upward on $(-\sqrt{\frac{3}{2}},-1)$ and $(\sqrt{\frac{3}{2}},\infty)$, and concave downward on $(-\infty,-\sqrt{\frac{3}{2}})$ and $(1,\sqrt{\frac{3}{2}})$; inflection points are $(\sqrt{\frac{3}{2}},\sqrt{\frac{3}{4}})$ and $(-\sqrt{\frac{3}{2}},-\sqrt{\frac{3}{4}})$; symmetry with respect to the origin.

43. $f'(x) = \dfrac{2-x}{2(1-x)^{3/2}}$; $f''(x) = \dfrac{4-x}{4(1-x)^{5/2}}$
domain is $(-\infty,1)$; increasing on $(-\infty,1)$; concave upward on $(-\infty,1)$; vertical asymptote is $x=1$.

45. $f'(x) = 4x^3+12x^2+8x = 4x(x+1)(x+2)$; $f''(x) = 12x^2+24x+8 = 12(x+1-\frac{1}{3}\sqrt{3})(x+1+\frac{1}{3}\sqrt{3})$
by the Newton-Raphson method, the zeros of f are approximately -2.554 and 0.554; relative maximum value is $f(-1)=-1$; relative minimum value is $f(-2)=f(0)=-2$; increasing on $[-2,-1]$ and $[0,\infty)$, and decreasing on $(-\infty,-2]$ and $[-1,0]$; concave upward on $(-\infty,-1-\frac{1}{3}\sqrt{3})$ and $(-1+\frac{1}{3}\sqrt{3},\infty)$, and concave downward on $(-1-\frac{1}{3}\sqrt{3},-1+\frac{1}{3}\sqrt{3})$; inflection points are $(-1-\frac{1}{3}\sqrt{3},-\frac{14}{9})$ and $(-1+\frac{1}{3}\sqrt{3},-\frac{14}{9})$.

47. $f'(x) = \dfrac{-x^2+2x+1}{x^2(x+1)^2}$; $f''(x) = \dfrac{2x^3-6x^2-6x-2}{x^3(x+1)^3}$
relative minimum value is $f(1-\sqrt{2}) = \sqrt{2}/(3\sqrt{2}-4)$; relative maximum value is $f(1+\sqrt{2}) = \sqrt{2}/(4+3\sqrt{2})$; increasing on $[1-\sqrt{2},0)$ and $(0,1+\sqrt{2})$, and decreasing on $(-\infty,-1)$, $(-1,1-\sqrt{2}]$ and $(1+\sqrt{2},\infty)$; by the Newton-Raphson method the zero c of f'' is approximately 3.847; concave upward on $(-1,0)$ and (c,∞), and concave downward on $(-\infty,-1)$ and $(0,c)$; inflection point is $(c,f(c))$, which is approximately $(3.847, 0.153)$; vertical asymptotes are $x=-1$ and $x=0$; horizontal asymptote is $y=0$.

49. Vertical asymptotes: $x=0$, $x=1$, and $x=2$; horizontal asymptote: $y=0$; no relative extreme values.

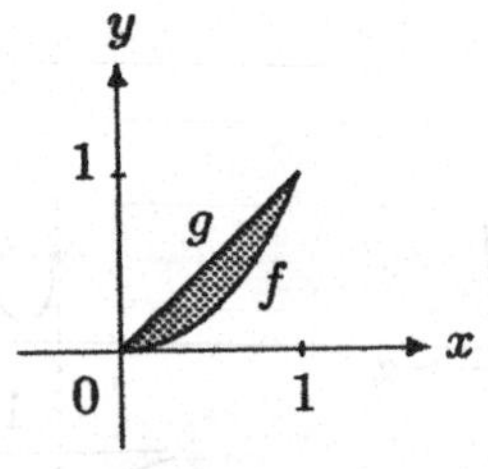

Exercise 51

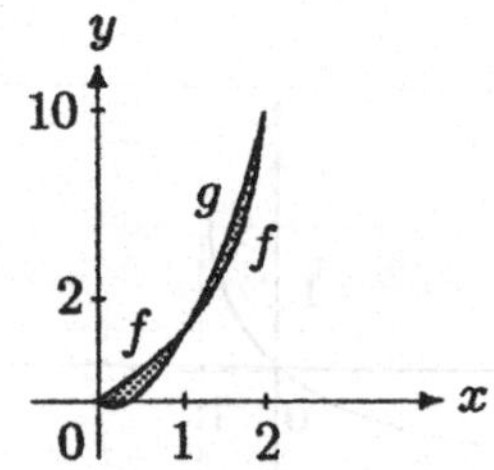

Exercise 53

51. The graphs intersect at (x, y) such that $x^2 = y = x$, or $x^2 - x = 0$, or $x = 0$ and $x = 1$. For $0 \le x \le 1$, $f(x) \le g(x)$. Since $g'(x) = 1$ and $f'(x) = 2x$, both functions are increasing on $[0, 1]$. The graph of f is concave upward on $(0, 1)$.

53. The graphs intersect at (x, y) such that $x^3 + x = y = 3x^2 - x$, or $x^3 - 3x^2 + 2x = 0$, or $x = 0$, 1, and 2. For $0 < x < 1$, $f(x) \ge g(x)$, and for $1 < x < 2$, $f(x) \le g(x)$. Since $f'(x) = 3x^2 + 1$ and $f''(x) = 6x$, f is increasing on $[0, 2]$ and its graph is concave upward on $(0, 2)$. Since $g'(x) = 6x - 1$ and $g''(x) = 6$, g is increasing on $[\frac{1}{6}, 2]$ and decreasing on $[0, \frac{1}{6}]$; relative minimum value is $g(\frac{1}{6}) = -\frac{1}{12}$; its graph is concave upward on $(0, 2)$.

55. The graphs intersect at (x, y) such that

$$\frac{2x}{\sqrt{1+x^2}} = y = \frac{x}{\sqrt{1-x^2}}, \quad \text{or} \quad \frac{4x^2}{1+x^2} = \frac{x^2}{1-x^2},$$

or $3x^2 = 5x^4$ so that $x = 0$ and $x = \pm\sqrt{\frac{3}{5}}$. Since

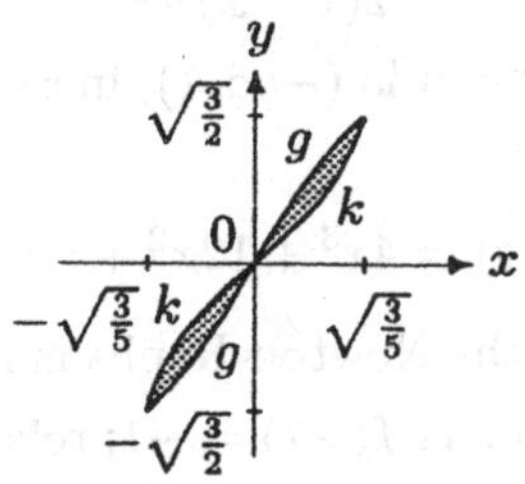

$$g'(x) = 2/(1+x^2)^{3/2} \quad \text{and} \quad g''(x) = -6x/(1+x^2)^{5/2}$$

g is increasing on $[-\sqrt{\frac{3}{5}}, \sqrt{\frac{3}{5}}]$, and the graph of g is concave upward on $[-\sqrt{\frac{3}{5}}, 0)$ and concave downward on $(0, \sqrt{\frac{3}{5}})$. Since

$$k'(x) = \frac{1}{(1-x^2)^{3/2}} \quad \text{and} \quad k''(x) = \frac{3x}{(1-x^2)^{5/2}}$$

k is increasing on $[-\sqrt{\frac{3}{5}}, \sqrt{\frac{3}{5}}]$, and the graph of k is concave downward on $(-\sqrt{\frac{3}{5}}, 0)$ and concave upward on $(0, \sqrt{\frac{3}{5}})$.

Chapter 4 Review

1. The domain of f is $(-\infty, 2]$;

$$f'(x) = 2x\sqrt{2-x} - \frac{x^2}{2\sqrt{2-x}} = \frac{x(8-5x)}{2\sqrt{2-x}}$$

so $f'(x) = 0$ for $x = 0$ and $\frac{8}{5}$, and $f'(x)$ is undefined for $x = 2$. Critical numbers: 0, $\frac{8}{5}$, 2.

3. Since $f'(x) = 2x+1$, the only critical number in $(-2,2)$ is $-\frac{1}{2}$. Thus the extreme values of f on $[-2,2]$ can occur only at -2, $-\frac{1}{2}$ and 2. Since $f(-2) = 3$, $f(-\frac{1}{2}) = \frac{3}{4}$, and $f(2) = 7$, the minimum value is $f(-\frac{1}{2})$, which is $\frac{3}{4}$, and the maximum value is $f(2)$, which is 7.

5. Since $f'(x) = 1 + x/\sqrt{1-x^2}$, the only critical number in $(-1,1)$ is $-\frac{1}{2}\sqrt{2}$. Thus the extreme values of f on $[-1,1]$ can occur only at -1, $-\frac{1}{2}\sqrt{2}$, and 1. Since $f(-1) = -1$, $f(-\frac{1}{2}\sqrt{2}) = -\sqrt{2}$, and $f(1) = 1$, the minimum value is $f(-\frac{1}{2}\sqrt{2})$, which is $-\sqrt{2}$, and the maximum value is $f(1)$, which is 1.

7. a. $f(-1) = (-1)+1 = 0$ and $f(1) = 1-1 = 0$, and $f'(x) = 1$ for x in $(-1,1)$ with $x \neq 0$. Also $f'(0)$ does not exist.

   b. This does not contradict Rolle's Theorem because f is not continuous at 0.

9. Let $g(x) = \frac{1}{3}x^3 + \cos x$. Then $g'(x) = x^2 - \sin x = f'(x)$, so by Theorem 4.6, $f(x) = g(x) + C = \frac{1}{3}x^3 + \cos x + C$.

11. Let $g(x) = \frac{1}{3}x^3 - 4x$. Then $g'(x) = x^2 - 4 = f''(x) = (f')'(x)$ so by Theorem 4.6, $f'(x) = g(x) + C_1 = \frac{1}{3}x^3 - 4x + C_1$ for some constant C_1. Now let $h(x) = \frac{1}{12}x^4 - 2x^2 + C_1x$. Then $h'(x) = \frac{1}{3}x^3 - 4x + C_1 = f'(x)$, so by Theorem 4.6, $f(x) = h(x) + C_2 = \frac{1}{12}x^4 - 2x^2 + C_1x + C_2$ for some constant C_2.

13. $f'(x) = x^2 - 2x + 1 = (x-1)^2$, so $f'(x) \geq 0$ for all x, and $f'(x) > 0$ except for $x = 1$. By Theorem 4.7, f is increasing on $(-\infty, \infty)$.

15. $f'(x) = \cos x - \frac{1}{8}\sec^2 x$, so $f'(x) > 0$ if $\cos x > \frac{1}{8}\sec^2 x$, or $\cos^3 x > \frac{1}{8}$, or $\cos x > \frac{1}{2}$. By Theorem 4.7, f is increasing on $[2n\pi - \pi/3, 2n\pi + \pi/3]$ and decreasing on $[2n\pi + \pi/3, 2n\pi + \pi/2)$, on $(2n\pi + \pi/2, 2n\pi + 3\pi/2)$, and on $(2n\pi + 3\pi/2, 2n\pi + 5\pi/3]$, for any integer n.

17. Let $f(x) = \sqrt{x+3} - \sqrt{3} - x/4$ for $0 \leq x \leq 1$. Then $f'(x) = 1/(2\sqrt{x+3}) - \frac{1}{4}$. Thus f' is decreasing, and $f'(x) > 1/(2\sqrt{1+3}) - \frac{1}{4} = 0$ for $0 < x < 1$. We conclude from Theorem 4.7 that f is increasing on $[0,1]$. Since $f(0) = 0$, it follows that $\sqrt{x+3} \geq \sqrt{3} + x/4$ for $0 \leq x \leq 1$.

19. $f'(x) = 12x^3 - 30x^2 + 12x = 6x(2x-1)(x-2)$; $f''(x) = 36x^2 - 60x + 12$. The relative extreme values (if any) must occur at the critical numbers 0, $\frac{1}{2}$, and 2. Since $f''(0) = 12 > 0$, $f''(\frac{1}{2}) = -9 < 0$, and $f''(2) = 36 > 0$, the Second Derivative Test implies that $f(0) = 3$ is a relative minimum value, $f(\frac{1}{2}) = \frac{55}{16}$ is a relative maximum value, and $f(2) = -5$ is a relative minimum value.

21. $f'(x) = 2(x+1)(x-2)^4 + 4(x+1)^2(x-2)^3 = 6x(x+1)(x-2)^3$. The relative extreme values (if any) must occur at the critical numbers -1, 0, and 2. Since $f'(x) < 0$ for x in $(-\infty, -1)$ or $(0,2)$, and since $f'(x) > 0$ for x in $(-1,0)$ or $(2,\infty)$, the First Derivative Test implies that $f(-1) = f(2) = 0$ is a relative minimum value, and $f(0) = 16$ is a relative maximum value.

23. $f'(x) = 2x^3 + 3x^2 - 12x$; $f''(x) = 6x^2 + 6x - 12 = 6(x+2)(x-1)$. Thus the graph is concave upward on $(-\infty, -2)$ and $(1, \infty)$, and concave downward on $(-2, 1)$.

25. $f'(x) = \dfrac{-4x^3}{(1+x^4)^2}$ and $f''(x) = \dfrac{4x^2(5x^4-3)}{(1+x^4)^3}$

Thus the graph is concave upward on $(-\infty, -\sqrt[4]{\frac{3}{5}})$ and on $(\sqrt[4]{\frac{3}{5}}, \infty)$, and concave downward on $(-\sqrt[4]{\frac{3}{5}}, \sqrt[4]{\frac{3}{5}})$.

27. Since $f'(x) = (x-1)^3(x+1)$ is positive on $(-\infty, -1)$ and $(1, \infty)$, f is increasing on $(-\infty, -1]$ and $[1, \infty)$. Since $f'(x) < 0$ on $(-1, 1)$, f is decreasing on $[-1, 1]$. Also $f''(x) = 3(x-1)^2(x+1) + (x-1)^3 = (x-1)^2(4x+2)$. Thus the graph of f is concave upward on $(-\frac{1}{2}, \infty)$ and concave downward on $(-\infty, -\frac{1}{2})$.

29. $f'(x) = 3x^2 - 6$; $f''(x) = 6x$; relative maximum value is $f(-\sqrt{2}) = 4\sqrt{2} - 1$; relative minimum value is $f(\sqrt{2}) = -4\sqrt{2} - 1$; increasing on $(-\infty, -\sqrt{2}]$ and $[\sqrt{2}, \infty)$, and decreasing on $[-\sqrt{2}, \sqrt{2}]$; concave upward on $(0, \infty)$ and concave downward on $(-\infty, 0)$; inflection point is $(0, -1)$.

31. $f'(x) = \dfrac{x^2-3}{x^4}$; $f''(x) = \dfrac{12-2x^2}{x^5}$

relative maximum value is $f(-\sqrt{3}) = \frac{2}{9}\sqrt{3}$; relative minimum value is $f(\sqrt{3}) = -\frac{2}{9}\sqrt{3}$; increasing on $(-\infty, -\sqrt{3}]$ and $[\sqrt{3}, \infty)$, and decreasing on $[-\sqrt{3}, 0)$ and $(0, \sqrt{3}]$; concave upward on $(-\infty, -\sqrt{6})$ and $(0, \sqrt{6})$, and concave downward on $(-\sqrt{6}, 0)$ and $(\sqrt{6}, \infty)$; inflection points are $(-\sqrt{6}, \frac{5}{36}\sqrt{6})$ and $(\sqrt{6}, -\frac{5}{36}\sqrt{6})$; vertical asymptote is $x = 0$; horizontal asymptote is $y = 0$; symmetry with respect to the origin.

33. $k'(x) = \dfrac{-10x}{(x^2-4)^2}$; $k''(x) = \dfrac{10(3x^2+4)}{(x^2-4)^3}$

relative maximum value is $k(0) = -\frac{1}{4}$; increasing on $(-\infty, -2)$ and $(-2, 0]$, and decreasing on $[0, 2)$ and $(2, \infty)$; concave upward on $(-\infty, -2)$ and $(2, \infty)$, and concave downward on $(-2, 2)$; vertical asymptotes are $x = -2$ and $x = 2$; horizontal asymptote is $y = 1$; symmetry with respect to the y axis.

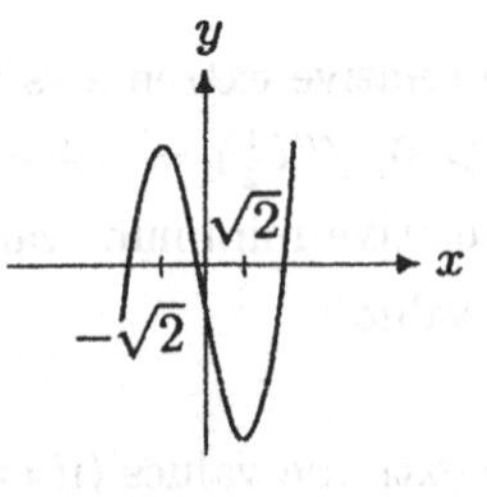

Exercise 29

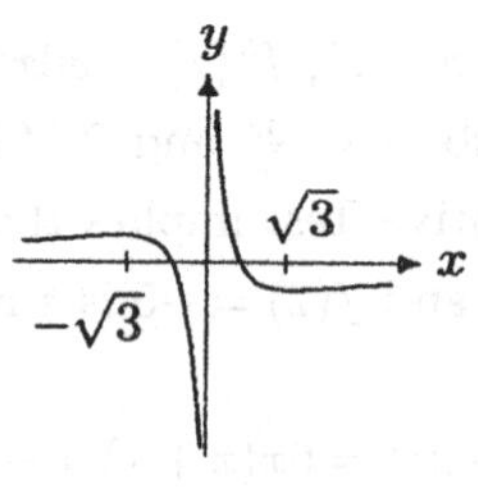

Exercise 31

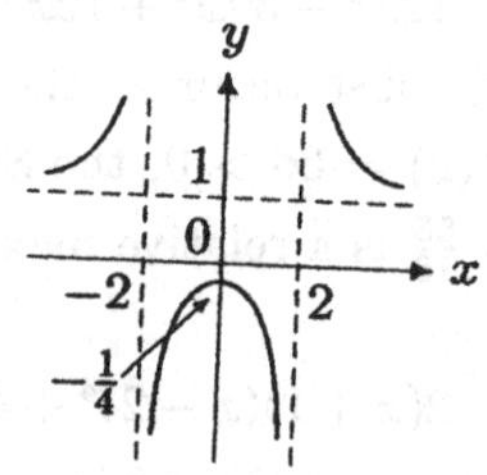

Exercise 33

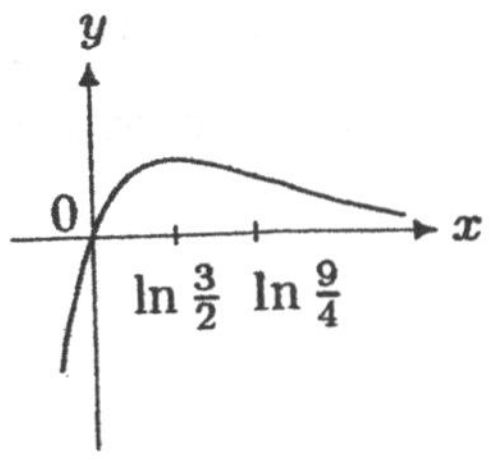

Exercise 35

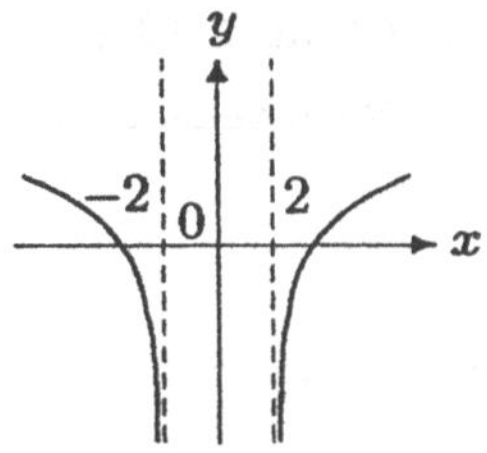

Exercise 37

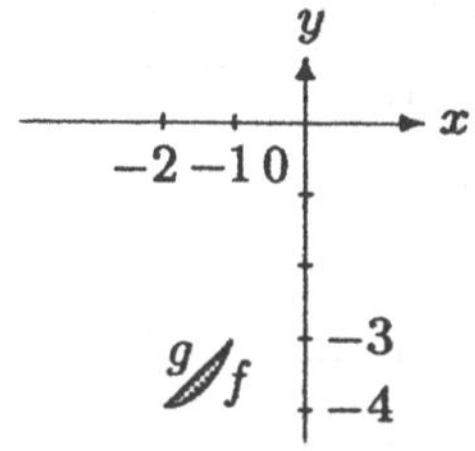

Exercise 39

35. $f'(x) = -2e^{-2x} + 3e^{-3x} = e^{-3x}(3 - 2e^x)$; $f''(x) = 4e^{-2x} - 9e^{-3x} = e^{-3x}(4e^x - 9)$
relative maximum value is $f(\ln\frac{3}{2}) = -\frac{9}{8}$; increasing on $(-\infty, \ln\frac{3}{2}]$, and decreasing on $[\ln\frac{3}{2}, \infty)$; concave upward on $(\ln\frac{9}{4}, \infty)$ and concave downward on $(-\infty, \ln\frac{9}{4})$; inflection point is $(\ln\frac{9}{4}, \frac{80}{729})$.

37. $f'(x) = \dfrac{2x}{x^2-4}$; $f''(x) = \dfrac{-2(x^2+4)}{(x^2-4)^2}$

domain is the union of $(-\infty, -2)$ and $(2, \infty)$; increasing on $(2, \infty)$ and decreasing on $(-\infty, -2)$; concave downward on $(-\infty, -2)$ and $(2, \infty)$; vertical asymptotes are $x = -2$ and $x = 2$; symmetry with respect to the y axis.

39. The graphs intersect at (x, y) such that $x^2 + 4x = y = x - 2$, or $x^2 + 3x + 2 = 0$, or $x = -2$ or $x = -1$. For $-2 \le x \le -1$, $f(x) \le g(x)$. Since $f'(x) = 2x + 4$ and $f''(x) = 2$, f has a relative minimum value at -2, and the graph of f is concave upward on $(-2, -1)$; g is increasing on $[-2, -1]$.

41. $f'(x) = mx^{m-1}(x-1)^n + nx^m(x-1)^{n-1} = x^{m-1}(x-1)^{n-1}[m(x-1) + nx]$, so $f'(x) = 0$ for $x = 0$, 1, and $m/(m+n)$. If $f(\frac{1}{4})$ is to be a relative extreme value, then $m/(m+n) = \frac{1}{4}$, so that $4m = m + n$, or $n = 3m$. Furthermore, if $f(0)$ and $f(1)$ are relative extreme values, then f' must change sign at 0 and 1, which requires $m - 1$ and $n - 1$ to be odd, that is, m and n even. Thus if $f(0)$, $f(\frac{1}{4})$, and $f(1)$ are relative extreme values, then $n = 3m$ with m even.

43. The area A of the rectangle in Figure 4.88 is given by $A = xy = xe^{-x}$ for $x > 0$. Thus $dA/dx = e^{-x} - xe^{-x} = (1 - x)e^{-x}$, so that $dA/dx = 0$ only for $x = 1$. Since $dA/dx > 0$ for $0 < x < 1$ and $dA/dx < 0$ for $x > 1$, (1) of Section 4.6 and the First Derivative Test imply that A is maximum for $x = 1$. The maximum area is $(1)e^{-1} = e^{-1}$.

45. a. By Theorem 4.8, the amount of Uranium 238 is given by $f(t) = f(0)e^{kt}$ for some constant k. Since $f(10^9) = .85719f(0)$, we have $f(0)e^{k\cdot 10^9} = .85719f(0)$, so that $e^{k\cdot 10^9} = .85719$, or $k = 10^{-9}\ln .85719$. We seek the value of t for which $f(t) = \frac{1}{2}f(0)$, or $f(0)e^{kt} = \frac{1}{2}f(0)$, so that

$$t = \frac{\ln(1/2)}{k} = \frac{-\ln 2}{10^{-9}\ln .85719} \approx 4.4982 \times 10^9.$$

Thus the half-life of Uranium 238 is approximately 4.5 billion years.

b. Taking $t = 10^7$ in $f(t) = f(0)e^{kt}$, we find that $f(10^7) = f(0)e^{k\cdot 10^7} = f(0)e^{10^{-2}\ln .85719}$. Thus $f(10^7)/f(0) = e^{10^{-2}\ln .85719} \approx .99846$. This means that after 10 million years, approximately 99.8% of the Uranium 238 would remain.

47. $D'(v) = \dfrac{\sqrt{3}}{48}\left(120v^{1/2} - \dfrac{5}{2}v^{3/2}\right) = \dfrac{5\sqrt{3}}{96}v^{1/2}(48 - v)$

so $D'(v) = 0$ only if $v = 48$. Since $D'(v) > 0$ for $0 < v < 48$ and $D'(v) < 0$ for $48 < v < 75$, (1) of Section 4.6 and the First Derivative Test imply that the maximum value of D occurs for $v = 48$ (miles per hour).

49. Let x be the length in feet of one side of the base, y the height in feet of the toolshed, and C the cost of the material. We are to minimize C. Since the volume is 800 cubic feet, we have $x^2y = 800$, so that $y = 800/x^2$. Since the floor costs $6x^2$ dollars, the roof costs $2x^2$ dollars, and each side $5xy$ dollars, C is given by $C(x) = 6x^2 + 2x^2 + 4(5xy) = 8x^2 + 20x(800/x^2) = 8(x^2 + 2000/x)$ for $x > 0$. Thus $C'(x) = 8(2x - 2000/x^2)$, so that $C'(x) = 0$ only if $2x = 2000/x^2$, that is, for $x^3 = 1000$, or $x = 10$. Since $C''(x) = 16 + 32{,}000/x^3$ for $x > 0$, (1) of Section 4.6 and the Second Derivative Test imply that the minimum value of C is $C(10)$. For $x = 10$ we have $y = 800/10^2 = 8$, so the base should be 10 feet on a side, and the shed should be 8 feet tall.

51. The revenue the company receives from x passengers is given by $R(x) = x(1000 - 2x) = 1000x - 2x^2$ We are to maximize R. Since $R'(x) = 1000 - 4x$, we find that $R'(x) = 0$ only for $x = 250$. Since $R''(x) = -4 < 0$ for $x > 0$, (1) of Section 4.6 and the Second Derivative Test imply that the maximum revenue is $R(250) = 125{,}000$ dollars.

53. $f'(v) = \dfrac{v}{c^2}\left(1 - \dfrac{v^2}{c^2}\right)^{-3/2}$

and $f''(v) = \dfrac{1}{c^2}\left(1 - \dfrac{v^2}{c^2}\right)^{-3/2} + \dfrac{3v^2}{c^4}\left(1 - \dfrac{v^2}{c^2}\right)^{-5/2}$

$= \dfrac{1}{c^2}\left(1 + \dfrac{2v^2}{c^2}\right)\left(1 - \dfrac{v^2}{c^2}\right)^{-5/2}$

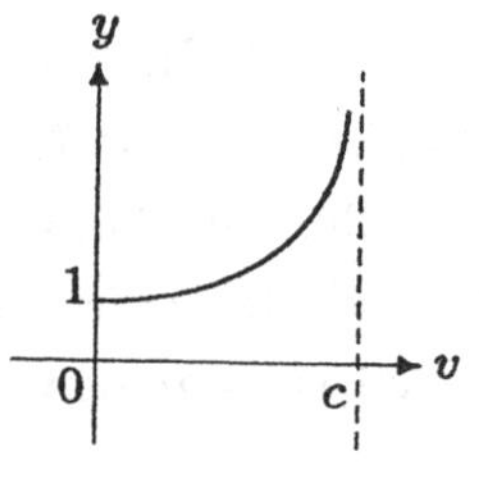

The function is increasing and concave upward on $(0, c)$. The line $v = c$ is a vertical asymptote.

55. Let x be the length of the fence parallel to the river, y the length of each fence perpendicular to the river, and P the total length of fence. We are to minimize P. Then $xy = 6400$, so that $y = 6400/x$, and thus $P(x) = x + 4y = x + 4(6400/x) = x + 25{,}600/x$ for $x > 0$. Since $P'(x) = 1 - 25{,}600/x^2$, we have $P'(x) = 0$ only if $x = 160$. Since $P''(x) = 51{,}200/x^3$, we have $P''(x) > 0$ for all $x > 0$. Thus (1) of Section 4.6 and the Second Derivative Test imply that the total length of fence is minimized if $x = 160$ feet.

57. a. Since $C(x) = xC_a(x)$, we have $C'(x) = C_a(x) + xC_a'(x)$. Thus if $C_a(x_0) = C'(x_0)$, then $x_0 C_a'(x_0) = 0$. Since $x_0 \neq 0$, this means that $C_a'(x_0) = 0$ if and only if $C_a(x_0) = C'(x_0)$.

b. Since $C_a'(x_0) = 0$, and since $C_a''(x_0) > 0$ by hypothesis, the Second Derivative Test implies that C_a has a relative minimum value at x_0. Assuming that there is only one number x such that $C_a(x) = C'(x_0)$, we conclude from part (a) that C_a has only one critical number, so by (1) in Section 4.6, $C_a(x_0)$ is the minimum value of C_a.

59. a. If the pigeon flies across the river perpendicular to the shore, then the angle θ in Figure 4.91 is 0. If the pigeon flies directly toward the nest, then θ is the angle θ_{max} with $\theta_{\text{max}} = k/a$. Suppose the pigeon lies at an angle θ with $0 \le \theta \le \theta_{\text{max}}$. Then the distance flown over water is $a \sec\theta$, so that the time of flight over water is $(a\sec\theta)/v$. The distance flown over land is $k - a\tan\theta$, so that the time of flight over land is $(k - a\tan\theta)/w$. Thus

$$T = \frac{a\sec\theta}{v} + \frac{k - a\tan\theta}{w} \quad \text{for } 0 \le \theta \le \theta_{\text{max}}.$$

b. Since

$$T'(\theta) = \frac{a\sec\theta\tan\theta}{v} - \frac{a\sec^2\theta}{w} = a\sec\theta\left(\frac{\tan\theta}{v} - \frac{\sec\theta}{w}\right)$$

$$= \frac{a}{\cos\theta}\left(\frac{\sin\theta}{v\cos\theta} - \frac{1}{w\cos\theta}\right) = \frac{a}{\cos^2\theta}\left(\frac{\sin\theta}{v} - \frac{1}{w}\right)$$

it follows that $T'(\theta) = 0$ if $\sin\theta = v/w$.

c. For Jo, $a = 1000$, $k = 3000$, $v = 20$, and $w = 24$. Thus $\tan\theta_{\text{max}} = 3000/1000 = 3$. By part (b), $T'(\theta_0) = 0$ if $\sin\theta_0 = \frac{20}{24} = \frac{5}{6}$. Then $\tan\theta_0 = \sin\theta_0/\cos\theta_0 = (5/6)/\sqrt{1-(5/6)^2} = 5/\sqrt{11}$ and $\sec\theta_0 = 6/\sqrt{11}$. Since $\tan\theta_0 = 5/\sqrt{11} < 3 = \tan\theta_{\text{max}}$, we have $0 < \theta_0 < \theta_{\text{max}}$. By the formula for $T'(\theta)$ in part (b), it follows that T' changes from negative to positive at θ_0. Thus (1) in Section 4.6 and the First Derivative Test imply that, for Jo, the time is minimized for $\theta = \theta_0$. Since

$$T(\theta_0) = \frac{1000(6/\sqrt{11})}{20} + \frac{3000 - 1000(5/\sqrt{11})}{24} \approx 153 \text{ (seconds)}$$

it takes Jo about 153 seconds to fly to the nest.

For Mo, $a = 1000$, $k = 3000$, $v = 18$, and $w = 26$. Thus $\tan\theta_{\text{max}} = 3$. By part (b), $T'(\theta_1) = 0$ if $\sin\theta_1 = \frac{18}{26} = \frac{9}{13}$. Then $\tan\theta_1 = \sin\theta_1/\cos\theta_1 = (9/13)/\sqrt{1-(9/13)^2} = 9/\sqrt{88}$ and $\sec\theta_1 = 13/\sqrt{88}$. Since $\tan\theta_1 = 9/\sqrt{88} < 3 = \tan\theta_{\text{max}}$, we have $0 < \theta_1 < \theta_{\text{max}}$. By the formula for $T'(\theta)$ in part (b) it follows that T' changes from negative to positive at θ_1. Thus (1) in Section 4.6 and the First Derivative Test imply that, for Mo, the time is minimized for $\theta = \theta_1$. Since

$$T(\theta_1) = \frac{1000(13/\sqrt{88})}{18} + \frac{3000 - 1000(9/\sqrt{88})}{26} \approx 155 \text{ (seconds)}$$

it takes Mo about 155 seconds to fly to the nest. Therefore Jo arrives at the nest first.

61. Let the radius of the barrel be r, and V the volume. We are to maximize V. The length of the barrel is $2\sqrt{l^2-4r^2}$, and the volume is given by $V(r) = \pi r^2(2\sqrt{l^2-4r^2}) = 2\pi r^2\sqrt{l^2-4r^2}$. Now

$$V'(r) = 4\pi r\sqrt{l^2-4r^2} - \frac{8\pi r^3}{\sqrt{l^2-4r^2}} = \frac{4\pi r(l^2-6r^2)}{\sqrt{l^2-4r^2}}$$

so that $V'(r) = 0$ for $r > 0$ only if $r = (l/6)\sqrt{6}$. Since $V'(r) > 0$ if $0 < r < (l/6)\sqrt{6}$, and $V'(r) < 0$ if $(l/6)\sqrt{6} < r < l/2$, (1) of Section 4.6 and the First Derivative Test imply that the barrel has maximum volume if the radius is $(l/6)\sqrt{6}$; the length is then $2\sqrt{l^2 - \frac{4}{6}l^2} = \frac{2}{3}\sqrt{3}\,l$.

63. Let x be the distance shown in the figure to the right, and L the total length of wire required. We are to minimize L. Now L is given by $L(x) = \sqrt{x^2+50^2} + \sqrt{(150-x)^2+75^2}$ for $0 \le x \le 150$, so that

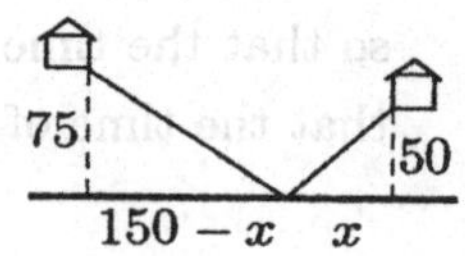

$$L'(x) = \frac{x}{\sqrt{x^2+50^2}} + \frac{x-150}{\sqrt{(150-x)^2+75^2}}.$$

Thus $L'(x) = 0$ if

$$\frac{x}{\sqrt{x^2+50^2}} = \frac{150-x}{\sqrt{(150-x)^2+75^2}}$$

or $x\sqrt{(150-x)^2+75^2} = (150-x)\sqrt{x^2+50^2}$, or $x^2(150-x)^2+75^2x^2 = (150-x)^2x^2+(150-x)^2\cdot 50^2$, or $75^2x^2 = (150-x)^2\cdot 50^2$, or $75x = (150-x)50$, or $x = 60$. If we replace "=" by ">" in the preceding argument, the resulting inequalities are valid. Thus $L'(x) > 0$ if $x > 60$. Similarly, $L'(x) < 0$ if $x < 60$. Thus (1) of Section 4.6 and the First Derivative Test imply that the minimum value of L occurs for $x = 60$. Consequently the telephone pole should be located 60 feet down the street from the house that is 50 feet from the street.

65. Let $t = 0$ correspond to the time of the accident. Since the amount of alcohol in the person's system decreases exponentially with time, it is given as a function of time t (in hours) by $f(t) = f(0)e^{kt}$ for some constant k. We are given that $f(2) = .0007$ and $f(4) = .0005$. Thus

$$f(0)e^{2k} = .0007 \quad\text{and}\quad f(0)e^{4k} = .0005.$$

Dividing and canceling $f(0)$, we have

$$\frac{e^{4k}}{e^{2k}} = \frac{.0005}{.0007}, \quad\text{or}\quad e^{2k} = \frac{5}{7}.$$

Thus $.0007 = f(0)e^{2k} = \frac{5}{7}f(0)$, so that $f(0) = \frac{7}{5}(.0007) = .00098 < .0001$. We conclude that the driver was not intoxicated at the time of the accident.

Cumulative Review(Chapters 1–3)

1. Since $4x^6 - 1 = 0$ for $x = -1/\sqrt[3]{2}$ and $x = 1/\sqrt[3]{2}$, it follows that $4x^6 - 1 > 0$ for $x < -1/\sqrt[3]{2}$ and for $x > 1/\sqrt[3]{2}$, and $4x^6 - 1 < 0$ for $-1/\sqrt[3]{2} < x < 1/\sqrt[3]{2}$. From the diagram we see that the solution of the given inequality is the union of $(-1/\sqrt[3]{2}, 0)$ and $(1/\sqrt[3]{2}, \infty)$.

x	$- - - - - 0 + + + + +$
$4x^6 - 1$	$+ + 0 - - - - - - 0 + +$
$18x(4x^6 - 1)$	$- - 0 + + 0 - - 0 + +$

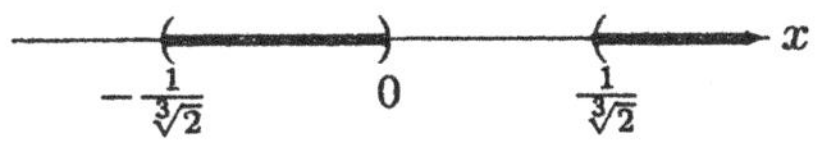

3. The given inequality is equivalent to $2 \le 1 + x^4 < 16$, or $1 \le x^4 < 15$. Thus the solution is the union of $(-\sqrt[4]{15}, -1]$ and $[1, \sqrt[4]{15})$.

5. a. The domain of f is $(-3, \infty)$ and the domain of g consists of all numbers except $\frac{1}{2}$. Thus the domain of $f \circ g$ consists of all x except $\frac{1}{2}$ such that

$$\frac{1}{2x-1} > -3, \quad \text{or} \quad \frac{1}{2x-1} + 3 > 0, \quad \text{or} \quad \frac{2(3x-1)}{2x-1} > 0.$$

From the diagram we see that the solution of this inequality, and hence the domain of $f \circ g$, is the union of $(-\infty, \frac{1}{3})$ and $(\frac{1}{2}, \infty)$.

$3x - 1$	$- - 0 + + + + + + + +$
$2x - 1$	$- - - - - - - - - 0 + +$
$\dfrac{2(3x-1)}{2x-1}$	$+ + 0 - - - - - -\quad + +$

b. $$(f \circ g)(x) = f(g(x)) = f\left(\frac{1}{2x-1}\right) = \frac{1}{\sqrt{1/(2x-1) + 3}} = \sqrt{\frac{2x-1}{6x-2}}$$

7. $$\lim_{x \to 4^+} \frac{x(x+4)}{16 - x^2} = \lim_{x \to 4^+} \frac{x(x+4)}{(4-x)(4+x)} = \lim_{x \to 4^+} \frac{x}{4-x} = -\infty$$

9. $$\lim_{x \to 0^+} \frac{\sin 2x - 2\sqrt{x}\,\sin x + 4x^2}{x} = \lim_{x \to 0^+} \left(\frac{\sin 2x}{x} - 2\sqrt{x}\,\frac{\sin x}{x} + 4x\right)$$

$$= \lim_{x \to 0^+} 2\,\frac{\sin 2x}{2x} - 2 \lim_{x \to 0^+} \sqrt{x} \lim_{x \to 0^+} \frac{\sin x}{x} + 4 \lim_{x \to 0^+} x = 2(1) - 2(0)(1) + 4(0) = 2$$

11. Since the slope of the line $y - 2x = 8$ is 2, we seek (a, b) on the graph of f at which the slope is the negative reciprocal of 2, that is, $-\frac{1}{2}$. But $f'(x) = 2(x-4)$, so a must satisfy $2(a-4) = -\frac{1}{2}$. Thus $a = 4 - \frac{1}{4} = \frac{15}{4}$, and $b = f(a) = (\frac{15}{4} - 4)^2 + 1 = \frac{17}{16}$. Therefore the point is $(\frac{15}{4}, \frac{17}{16})$.

13. $$f'(4) = \lim_{x\to 4} \frac{f(x) - f(4)}{x - 4} = \lim_{x\to 4} \frac{2/x^{1/2} - 1}{x - 4} = \lim_{x\to 4} \frac{2 - x^{1/2}}{x^{1/2}(x^{1/2} - 2)(x^{1/2} + 2)}$$
$$= \lim_{x\to 4} \frac{-1}{x^{1/2}(x^{1/2} + 2)} = \frac{-1}{2(2+2)} = -\frac{1}{8}$$

15. $$f'(x) = \frac{-(1 - e^{\tan x}\sec^2 x)}{(x - e^{\tan x})^2}$$

17. $$\frac{d}{dx}[3(2x+1)^{5/2} - 5(2x+1)^{3/2}] = 3\left(\frac{5}{2}\right)(2x+1)^{3/2}(2) - 5\left(\frac{3}{2}\right)(2x+1)^{1/2}(2)$$
$$= 15(2x+1)^{1/2}[(2x+1) - 1] = 30x(2x+1)^{1/2}$$

19. Differentiating the given equation implicitly, we have
$$y^2 + 2xy\frac{dy}{dx} + 3\frac{dy}{dx} - 4 = 0, \quad \text{so} \quad \frac{dy}{dx} = \frac{4 - y^2}{2xy + 3}.$$

21. The area A is given by $A = \pi r^2$. Differentiating implicitly, we have $dA/dt = 2\pi r(dr/dt)$. Since $dA/dt = 2\pi\sqrt{r}$ by hypothesis, it follows that $2\pi\sqrt{r} = dA/dt = 2\pi r(dr/dt)$, so $dr/dt = 1/\sqrt{r}$. When the radius is increasing at the rate of 2 feet per minute, we have $2 = dr/dt = 1/\sqrt{r}$, so $r = (\frac{1}{2})^2 = \frac{1}{4}$. At that time, $A = \pi(\frac{1}{4})^2 = \frac{1}{16}\pi$ (square feet).

23. By (1) of Section 1.3, the height of the baton in meters is given by $h(t) = -4.9t^2 + v_0 t$. Thus $h'(t) = -9.8t + v_0$, so $h'(t) = 0$ for $t = v_0/9.8$. Thus $h(v_0/9.8)$ is the maximum height. Since there are $7 \times 24 \times 60 \times 60 = 604{,}800$ seconds in a week, we have $0 = h(604{,}800) = -4.9(604{,}800)^2 + v_0(604{,}800)$, so that $v_0 = 4.9(604{,}800)$. Thus $h(v_0/9.8) = h(302{,}400) = -4.9(302{,}400)^2 + 4.9(604{,}800)(302{,}400) = 4.9(302{,}400)^2 \approx 4.5 \times 10^{11}$ (meters). Thus the baton would have gone about 4.5×10^8 kilometers high!

Chapter 5

The Integral

5.1 Preparation for the Definite Integral

1. From the figures, $\Delta x_1 = \Delta x_2 = \Delta x_3 = 1$. Moreover, $m_1 = -1/2, m_2 = 1/2$, and $m_3 = 3/2$; $M_1 = 1/2, M_2 = 3/2$, and $M_3 = 5/2$. We obtain

$$L_f(P) = -\frac{1}{2}(1) + \frac{1}{2}(1) + \frac{3}{2}(1) = \frac{3}{2} \quad \text{and} \quad U_f(P) = \frac{1}{2}(1) + \frac{3}{2}(1) + \frac{5}{2}(1) = \frac{9}{2}$$

3. From the figures, $\Delta x_1 = \Delta x_2 = \cdots = \Delta x_8 = \pi/4$. Moreover,

$$L_f(P) = \frac{\pi}{4}\left(\cos\frac{\pi}{4} + \cos\frac{\pi}{2} + \cos\frac{3\pi}{4} + \cos\pi + \cos\pi + \cos\frac{5\pi}{4} + \cos\frac{3\pi}{2} + \cos\frac{7\pi}{4}\right)$$

$$= \frac{\pi}{4}\left(\frac{\sqrt{2}}{2} + 0 - \frac{\sqrt{2}}{2} - 1 - 1 - \frac{\sqrt{2}}{2} + 0 + \frac{\sqrt{2}}{2}\right) = -\frac{\pi}{2}$$

and

$$U_f(P) = \frac{\pi}{4}\left(\cos 0 + \cos\frac{\pi}{4} + \cos\frac{\pi}{2} + \cos\frac{3\pi}{4} + \cos\frac{5\pi}{4} + \cos\frac{3\pi}{2} + \cos\frac{7\pi}{4} + \cos 2\pi\right)$$

$$= \frac{\pi}{4}\left(1 + \frac{\sqrt{2}}{2} + 0 - \frac{\sqrt{2}}{2} - \frac{\sqrt{2}}{2} + 0 + \frac{\sqrt{2}}{2} + 1\right) = \frac{\pi}{2}$$

5. left sum: $f(1)(1) + f(2)(1) + f(3)(1) = 1 + \frac{1}{2} + \frac{1}{3} = \frac{11}{6}$
right sum: $f(2)(1) + f(3)(1) + f(4)(1) = \frac{1}{2} + \frac{1}{3} + \frac{1}{4} = \frac{13}{12}$

7. Since $x + 2$ is increasing on $[-1, 2]$,

$$L_f(P) = 1\left(\frac{-1}{2} - (-1)\right) + \frac{3}{2}\left(0 - \left(\frac{-1}{2}\right)\right) + 2\left(\frac{1}{2} - 0\right) + \frac{5}{2}\left(1 - \frac{1}{2}\right) + 3\left(\frac{3}{2} - 1\right) + \frac{7}{2}\left(2 - \frac{3}{2}\right) = \frac{27}{4},$$

$$U_f(P) = \frac{3}{2}\left(\frac{-1}{2} - (-1)\right) + 2\left(0 - \left(\frac{-1}{2}\right)\right) + \frac{5}{2}\left(\frac{1}{2} - 0\right) + 3\left(1 - \frac{1}{2}\right) + \frac{7}{2}\left(\frac{3}{2} - 1\right) + 4\left(2 - \frac{3}{2}\right) = \frac{33}{4}.$$

9. Since $-1/x$ is increasing on $[-4,-1]$,

$$L_f(P) = \frac{1}{4}(-3-(-4)) + \frac{1}{3}(-2-(-3)) + \frac{1}{2}(-1-(-2)) = \frac{13}{12},$$

$$U_f(P) = \frac{1}{3}(-3-(-4)) + \frac{1}{2}(-2-(-3)) + 1(-1-(-2)) = \frac{11}{6}.$$

11. Since $f'(x) = \cos x - \sin x$, so that $f'(x) > 0$ on $(0,\pi/4)$ and $f'(x) < 0$ on $(\pi/4,\pi/2)$, we know that f is increasing on $[0,\pi/4]$ and decreasing on $[\pi/4,\pi/2]$. Thus

$$L_f(P) = 1\left(\frac{\pi}{4}-0\right) + 1\left(\frac{\pi}{2}-\frac{\pi}{4}\right) = \frac{\pi}{2} \quad \text{and} \quad U_f(P) = \sqrt{2}\left(\frac{\pi}{4}-0\right) + \sqrt{2}\left(\frac{\pi}{2}-\frac{\pi}{4}\right) = \frac{\sqrt{2}\,\pi}{2}.$$

13. Since $\Delta x_k = \frac{1}{9}$ for $1 \le k \le 9$ and f is increasing on $[0,1]$,

$$\begin{aligned} L_f(P) &= 0\left(\frac{1}{9}\right) + \frac{1}{3}\left(\frac{1}{9}\right) + \frac{\sqrt{2}}{3}\left(\frac{1}{9}\right) + \frac{\sqrt{3}}{3}\left(\frac{1}{9}\right) + \frac{\sqrt{4}}{3}\left(\frac{1}{9}\right) + \frac{\sqrt{5}}{3}\left(\frac{1}{9}\right) \\ &\quad + \frac{\sqrt{6}}{3}\left(\frac{1}{9}\right) + \frac{\sqrt{7}}{3}\left(\frac{1}{9}\right) + \frac{\sqrt{8}}{3}\left(\frac{1}{9}\right) \approx 0.6039259454 \\ U_f(P) &= \frac{1}{3}\left(\frac{1}{9}\right) + \frac{\sqrt{2}}{3}\left(\frac{1}{9}\right) + \frac{\sqrt{3}}{3}\left(\frac{1}{9}\right) + \frac{\sqrt{4}}{3}\left(\frac{1}{9}\right) + \frac{\sqrt{5}}{3}\left(\frac{1}{9}\right) \\ &\quad + \frac{\sqrt{6}}{3}\left(\frac{1}{9}\right) + \frac{\sqrt{7}}{3}\left(\frac{1}{9}\right) + \frac{\sqrt{8}}{3}\left(\frac{1}{9}\right) + \frac{\sqrt{9}}{3}\left(\frac{1}{9}\right) \approx 0.7150370565. \end{aligned}$$

15. Since $\ln x$ is increasing on $[0.5, 2]$,

$$L_f(P) = (\ln 0.5)(0.25) + (\ln 0.75)(0.25) + \cdots + (\ln 1.5)(0.25) + (\ln 1.75)(0.25) \approx 0.0518487986;$$

$$U_f(P) = (\ln 0.75)(0.25) + (\ln 1)(0.25) + \cdots + (\ln 1.75)(0.25) + (\ln 2)(0.25) \approx 0.3984223889.$$

17. left sum: $f(-2)(1) + f(-1)(1) + f(0)(1) = -7 - 5 - 3 - 15$

right sum: $f(-1)(1) + f(0)(1) + f(1)(1) = -5 - 3 - 1 = -9$

midpoint sum: $f(-\frac{3}{2})(1) + f(-\frac{1}{2})(1) + f(\frac{1}{2})(1) = -6 - 4 - 2 = -12$

19. left sum: $f(1)(\frac{1}{2}) + f(\frac{3}{2})(\frac{1}{2}) + f(2)(\frac{1}{2}) + f(\frac{5}{2})(\frac{1}{2}) = \frac{1}{2}(1 + \frac{4}{9} + \frac{1}{4} + \frac{4}{25}) \approx 0.927222222$

right sum: $f(\frac{3}{2})(\frac{1}{2}) + f(2)(\frac{1}{2}) + f(\frac{5}{2})(\frac{1}{2}) + f(3)(\frac{1}{2}) = \frac{1}{2}(\frac{4}{9} + \frac{1}{4} + \frac{4}{25} + \frac{1}{9}) = \frac{869}{1800} \approx$ 0.482777778

midpoint sum: $f(\frac{5}{4})(\frac{1}{2}) + f(\frac{7}{4})(\frac{1}{2}) + f(\frac{9}{4})(\frac{1}{2}) + f(\frac{11}{4})(\frac{1}{2}) = \frac{1}{2}(\frac{16}{25} + \frac{16}{49} + \frac{16}{81} + \frac{16}{121}) \approx$ 0.6481464407

21. left sum: approximately 4.163897657
right sum: approximately 4.163897657
midpoint sum: approximately 4.15942373

23. left sum: approximately 2.124017421
right sum: approximately 2.193332139
midpoint sum: approximately 2.158987223

25. Using the left sum with $n = 4$, we obtain $A \approx f(1)(\frac{1}{2}) + f(\frac{3}{2})(\frac{1}{2}) +$ $f(2)(\frac{1}{2}) + f(\frac{5}{2})(\frac{1}{2}) = \frac{1}{2}(1 + \frac{4}{9} + \frac{1}{4} + \frac{4}{25}) = \frac{1669}{1800} \approx 0.927222222$

27. Using a calculator for the upper sum with $n = 50$, we obtain $A \approx 0.7113948156$.

29. Let $P = \{x_0, x_1, \ldots, x_n\}$. In order to find $U_f(P)$ we must be able to find the maximum value of f on $[x_0, x_1] = [0, x_1]$, and this is not possible since $\lim_{x\to 0^+} 1/x = \infty$. We can find the minimum value of f on each subinterval determined by P. Thus we can find $L_f(P)$.

31. Let $P = \{x_0, x_1, \ldots, x_n\}$ be a partition of $[a, b]$ such that $\Delta x_k = (b - a)/n$ for $1 \le k \le n$. Since f is increasing on $[a, b]$,

$$U_f(P) = f(x_1)\frac{b-a}{n} + f(x_2)\frac{b-a}{n} + \cdots + f(x_n)\frac{b-a}{n}$$

$$L_f(P) = f(x_0)\frac{b-a}{n} + f(x_1)\frac{b-a}{n} + \cdots + f(x_{n-1})\frac{b-a}{n}.$$

Using the fact that $x_0 = a$ and $x_n = b$, we conclude that

$$U_f(P) - L_f(P) = f(x_n)\frac{b-a}{n} - f(x_0)\frac{b-a}{n} = [f(b) - f(a)]\frac{b-a}{n}.$$

33. Notice that

$$f'(x) = \frac{10^4(1+x^2) - 10^4(2x^2)}{(1+x^2)^2} = \frac{10^4(1+x)(1-x)}{(1+x^2)^2},$$

so that f is increasing on $[0, 1]$ and decreasing on $[1, 4]$.

a. Let $P = \{0, 1, 2, 3, 4\}$. Then

$$L_f(P) = 0(1) + \frac{10^4(2)}{5}(1) + \frac{10^4(3)}{10}(1) + \frac{10^4(4)}{17}(1) = \frac{159}{170} \times 10^4 \approx 0.9352941176 \times 10^4.$$

b. Let $P' = \{0, \frac{1}{2}, 1, \frac{3}{2}, 2, \frac{5}{2}, 3, \frac{7}{2}, 4\}$. Then

$$L_f(P') = 0\left(\frac{1}{2}\right) + \frac{10^4(\frac{1}{2})}{\frac{5}{4}}\left(\frac{1}{2}\right) + \frac{10^4(\frac{3}{2})}{\frac{13}{4}}\left(\frac{1}{2}\right) + \frac{10^4(2)}{5}\left(\frac{1}{2}\right) + \frac{10^4(\frac{5}{2})}{\frac{29}{4}}\left(\frac{1}{2}\right)$$

$$+\frac{10^4(3)}{10}\left(\frac{1}{2}\right) + \frac{10^4(\frac{7}{2})}{\frac{53}{4}}\left(\frac{1}{2}\right) + \frac{10^4(4)}{17}\left(\frac{1}{2}\right) \approx 1.202905554 \times 10^4.$$

5.2 The Definite Integral

1. Since $2x$ is increasing on $[-1,3]$, $L_f(P) = -2(1)+0(1)+2(1)+4(1) = 4$, $U_f(P) = 0(1)+2(1)+4(1)+6(1) = 12$.

3. Since $3\sin x$ is increasing on $[-\pi/4, \pi/4]$,

$$L_f(P) = \left(\frac{-3\sqrt{2}}{2}\right)\left(\frac{\pi}{4}\right) + 0\left(\frac{\pi}{4}\right) = \frac{-3\sqrt{2}\,\pi}{8} \quad \text{and} \quad U_f(P) = 0\left(\frac{\pi}{4}\right) + \frac{3\sqrt{2}}{2}\left(\frac{\pi}{4}\right) = \frac{3\sqrt{2}\,\pi}{8}.$$

5. Let $f(x) = x^2 - x$ for $1 \le x \le 3$.
left sum: $f(1)(1) + f(2)(1) = 0(1) + 2(1) = 2$
right sum: $f(2)(1) + f(3)(1) = 2(1) + 6(1) = 8$
midpoint sum: $f(\frac{3}{2})(1) + f(\frac{5}{2})(1) = \frac{3}{4}(1) + \frac{15}{4}(1) = \frac{9}{2}$

7. Let $f(x) = \cos x$ for $-\pi \le x \le 0$.

left sum: $f(-\pi)\left(\frac{\pi}{3}\right) + f\left(-\frac{2\pi}{3}\right)\left(\frac{\pi}{3}\right) + f\left(-\frac{\pi}{3}\right)\left(\frac{\pi}{3}\right) = (\cos(-\pi))\left(\frac{\pi}{3}\right) + \left(\cos\left(-\frac{2\pi}{3}\right)\right)\left(\frac{\pi}{3}\right) + \left(\cos\left(-\frac{\pi}{3}\right)\right)\left(\frac{\pi}{3}\right) = -\frac{\pi}{3} - \frac{\pi}{6} + \frac{\pi}{6} = -\frac{\pi}{3}$

right sum: $f\left(-\frac{2\pi}{3}\right)\left(\frac{\pi}{3}\right) + f\left(-\frac{\pi}{3}\right)\left(\frac{\pi}{3}\right) + f(0)\left(\frac{\pi}{3}\right) = \left(\cos\left(-\frac{2\pi}{3}\right)\right)\left(\frac{\pi}{3}\right) + \left(\cos\left(-\frac{\pi}{3}\right)\right)\left(\frac{\pi}{3}\right) + (\cos 0)\left(\frac{\pi}{3}\right) = -\frac{\pi}{6} + \frac{\pi}{6} + \frac{\pi}{3} = \frac{\pi}{3}$

midpoint sum: $f\left(-\frac{5\pi}{6}\right)\left(\frac{\pi}{3}\right) + f\left(-\frac{\pi}{2}\right)\left(\frac{\pi}{3}\right) + f\left(-\frac{\pi}{6}\right)\left(\frac{\pi}{3}\right) = \left(\cos\left(-\frac{5\pi}{6}\right)\right)\left(\frac{\pi}{3}\right) + \left(\cos\left(-\frac{\pi}{2}\right)\right)\left(\frac{\pi}{3}\right) + \left(\cos\left(-\frac{\pi}{6}\right)\right)\left(\frac{\pi}{3}\right) = -\frac{\sqrt{3}}{2}\left(\frac{\pi}{3}\right) + 0\left(\frac{\pi}{3}\right) + \frac{\sqrt{3}}{2}\left(\frac{\pi}{3}\right) = 0$

9. The midpoint sum yields

$$\int_0^{\pi} \sin x\, dx \approx \frac{\pi - 0}{10}\left[\sin\frac{\pi}{20} + \sin\frac{3\pi}{20} + \sin\frac{5\pi}{20} + \cdots + \sin\frac{19\pi}{20}\right] \approx 2.008248408.$$

11. The right sum yields

$$\int_{1.1}^{1.2} \ln(1+e^x)\, dx \approx \frac{1.2 - 1.1}{20}\left[\ln(1+e^{1.105}) + \ln(1+e^{1.11}) + \cdots + \ln(1+e^{1.2})\right]$$
$$\approx 0.1427055747$$

13. First we notice that e^{x^2} is symmetric about the y axis, so we can obtain the required answer by approximating the integral $\int_0^1 e^{x^2}\,dx$ by the lower sum with $n = 10$, and multiplying the answer by 2. Next, we observe that e^{x^2} is increasing on $[0, 1]$, so on $[0, 1]$ the lower sum is the left sum. It follows that

$$\int_{-1}^{1} e^{x^2}\,dx = 2\int_0^1 e^{x^2}\,dx \approx (2)\frac{1-0}{10}\left[e^{0^2} + e^{(0.1)^2} + e^{(0.2)^2} \cdots + e^{(0.9)^2}\right]$$

$$\approx 2.762521203$$

15. By Example 2, $\int_{-3}^{3} x\,dx = \frac{1}{2}[3^2 - (-3)^2] = 0$.

17. By (4), $\int_{-5}^{0} -3x^2\,dx = -\frac{3}{3}[0^3 - (-5)^3] = -125$.

19. $A \approx f(0)(\frac{1}{2}) + f(\frac{1}{2})(\frac{1}{2}) + f(1)(1) = 0(\frac{1}{2}) + \frac{1}{3}(\frac{1}{2}) + \frac{1}{2}(1) = \frac{2}{3}$

21. $A = \int_{-2}^{3} \frac{5}{2}\,dx = \frac{5}{2}(3 - (-2)) = \frac{25}{2}$

23. $A = \int_1^4 x\,dx = \frac{1}{2}(4^2 - 1^2) = \frac{15}{2}$

25. For any partition P of $[a, b]$ and any choice of $t_1, t_2, \ldots, t_n$ with t_k in $[x_{k-1}, x_k]$, we have

$$\sum_{k=1}^{n} f(t_k)\Delta x_k \le \sum_{k=1}^{n} g(t_k)\Delta x_k$$

It follows from Definition 5.4 that

$$\int_a^b f(x)\,dx = \lim_{\|P\|\to 0}\sum_{k=1}^{n} f(t_k)\Delta x_k \le \lim_{\|P\|\to 0}\sum_{k=1}^{n} g(t_k)\Delta x_k = \int_a^b g(x)\,dx$$

27. For $x \ge 1$ we have $x \le x^6$. Thus $\int_1^2 x\,dx \le \int_1^2 x^6\,dx$.

29. For $x \ge 0$ we have $\sin x \le x$. Thus $\int_0^{\pi/2} \sin x\,dx \le \int_0^{\pi/2} x\,dx$.

31. Let $I = \frac{1}{2}(b^2 - a^2) + 4(b - a)$, and let $f(x) = x + 4$. We will show that the midpoint sum of f for any partition $P = \{x_0, x_1, \ldots, x_n\}$ of $[a, b]$ actually equals I. For $1 \le k \le n$, we choose

$$t_k = \frac{1}{2}(x_k + x_{k-1})$$

which is the midpoint of $[x_{k-1}, x_k]$. Then, as in Examples 1 and 2, the resulting midpoint sum is

$$\begin{aligned}
&\sum_{k=1}^{n} f(t_k)\Delta x_k \\
&= [\tfrac{1}{2}(x_1 + x_0) + 4]\,\Delta x_1 + [\tfrac{1}{2}(x_2 + x_1) + 4]\,\Delta x_2 + \cdots + [\tfrac{1}{2}(x_n + x_{n-1}) + 4]\,\Delta x_n \\
&= \frac{1}{2}(x_1 + x_0)(x_1 - x_0) + \cdots + \frac{1}{2}(x_n + x_{n-1})(x_n - x_{n-1}) + 4(\Delta x_1 + \cdots + \Delta x_n) \\
&= \frac{1}{2}(x_1^2 - x_0^2) + \cdots + \frac{1}{2}(x_n^2 - x_{n-1}^2) + 4(b - a) \\
&= \frac{1}{2}(x_n^2 - x_0^2) + 4(b - a) = \frac{1}{2}(b^2 - a^2) + 4(b - a)
\end{aligned}$$

Therefore $\displaystyle\int_a^b (x + 4)\,dx = \frac{1}{2}(b^2 - a^2) + 4(b - a)$.

33. a. $\int_1^3 4x\,dx = \frac{4}{2}(3^2 - 1^2) = 16$

b. $\int_2^6 -\frac{1}{2}x\,dx = \frac{-1/2}{2}(6^2 - 2^2) = -8$

c. $\int_{-5}^5 \sqrt{3}x\,dx = (\sqrt{3}/2)[5^2 - (-5)^2] = 0$

d. $\int_{-1/\pi}^0 \pi x\,dx = (\pi/2)[0^2 - (-1/\pi)^2] = -1/(2\pi)$

35. Let $f(x) = x^2$, and let $I = \frac{1}{3}(b^3 - a^3)$. Next, let $P = \{x_0, x_1, \ldots, x_n\}$ be any partition of $[a, b]$, and let $t_k = \frac{1}{3}(x_k^2 + x_k x_{k-1} + x_{k-1}^2)$ for $1 \le k \le n$. We will prove that the corresponding Riemann sum $\sum_{k=1}^n f(t_k)\Delta x_k$ equals I.

First observe that since $0 \le x_{k-1} \le x_k$, we have

$$x_{k-1}^2 = \tfrac{1}{3}(x_{k-1}^2 + x_{k-1}^2 + x_{k-1}^2) \le \tfrac{1}{3}(x_{k-1}^2 + x_k x_{k-1} + x_k^2) \le \tfrac{1}{3}(x_k^2 + x_k^2 + x_k^2) = x_k^2$$

Thus $x_{k-1}^2 \le t_k^2 \le x_k^2$, and since $0 \le x_0$, it follows that $x_{k-1} \le t_k \le x_k$. For the Riemann sum $\sum_{k=1}^n f(t_k)\Delta x_k$ we calculate that

$$\begin{aligned}
\sum_{k=1}^{n} f(t_k)\Delta x_k &= t_1^2\Delta x_1 + t_2^2\Delta x_2 + \cdots + t_n^2\Delta x_n \\
&= \frac{1}{3}(x_1^2 + x_1x_0 + x_0^2)(x_1 - x_0) + \cdots + \frac{1}{3}(x_n^2 + x_n x_{n-1} + x_{n-1}^2)(x_n - x_{n-1}) \\
&= \frac{1}{3}(x_1^3 - x_0^3) + \cdots + \frac{1}{3}(x_n^3 - x_{n-1}^3) = \frac{1}{3}(b^3 - a^3)
\end{aligned}$$

We conclude, as in Example 2, that $\int_a^b x^2\,dx = \frac{1}{3}(b^3 - a^3)$.

37. $\int_a^b x^n\,dx = \frac{1}{n+1}(b^{n+1} - a^{n+1})$

39. a. Let $f(x) = 1/x$, so $f'(x) = -1/x^2$ and $f''(x) = 2/x^3$. Since $|f''|$ is decreasing on $[1,2]$, $K = |f''(1)| = 2$. Thus

$$E_n^M \le \frac{2}{24n^2}(2-1)^3 = \frac{1}{12n^2}.$$

b. Let $f(x) = \sin x$, so $f'(x) = \cos x$ and $f''(x) = -\sin x$. Since $|f''| \le 1$ and $|f''(\pi/2)| = 1$, let $K = 1$. Then

$$E_n^M \le \frac{1}{24n^2}(3-0)^3 = \frac{9}{8n^2}.$$

c. Let $f(x) = e^{-x^2}$, so $f'(x) = -2xe^{-x^2}$ and $f''(x) = -2e^{-x^2} - 2x(-2xe^{-x^2}) = 2e^{-x^2}(-1+2x^2)$. Since $|e^{-x^2}| \le 1$ for $0 \le x \le 1$ and $|-1+2x^2| \le 1$ for $0 \le x \le 1$, we have

$$|f''(x)| = 2|e^{-x^2}|\,|-1+2x^2| \le 2\cdot 1\cdot 1 = 2 \quad \text{for } 0 \le x \le 1.$$

Thus we let $K = 2$. Then

$$E_n^M \le \frac{2}{24n^2}(1-0)^3 = \frac{1}{12n^2}.$$

41. Let $f(x) = e^x$, and let $P = \{x_0, x_1, \ldots, x_n\}$ be a partition of $[a,b]$. For any $k = 1, 2, \ldots, n$, apply the Mean Value Theorem to e^x on $[x_{k-1}, x_k]$ to obtain a t_k in $[x_{k-1}, x_k]$ such that

$$\frac{e^{x_k} - e^{x_{k-1}}}{x_k - x_{k-1}} = e^{t_k}$$

This means that $e^{t_k}(x_k - x_{k-1}) = e^{x_k} - e^{x_{k-1}}$. Since e^x is an increasing function, $e^{x_{k-1}} \le e^{t_k} \le e^{x_k}$. Therefore

$$e^{x_{k-1}}(x_k - x_{k-1}) \le e^{t_k}(x_k - x_{k-1}) = e^{x_k} - e^{x_{k-1}} \le e^{x_k}(x_k - x_{k-1})$$

Thus

$$L_f(P) \le \sum_{k=1}^n e^{x_{k-1}}(x_k - x_{k-1}) \le \sum_{k=1}^n (e^{x_k} - e^{x_{k-1}}) \le \sum_{k=1}^n e^{x_k}(x_k - x_{k-1}) \le U_f(P)$$

Since

$$\sum_{k=1}^n (e^{x_k} - e^{x_{k-1}}) = (e^{x_1} - e^{x_0}) + (e^{x_2} - e^{x_1}) + \cdots + (e^{x_n} - e^{x_{n-1}}) = e^{x_n} - e^{x_0} = e^b - e^a$$

it follows that $L_f(P) \le e^b - e^a \le U_f(P)$ for all partitions P. Consequently $\int_a^b e^x\,dx = e^b - e^a$.

43. a. midpoint sum ≈ 18.85539331 b. $A = \pi ab$

45. The area is approximately $(100)(200) + (125)(200) + (124)(200) = 69,800$ (square feet).

5.3 Special Properties of the Definite Integral

1. $\int_3^5 7\,dx = 7(5-3) = 14$

3. $\int_2^{-1} -10\,du = -10(-1-2) = 30$

5. $\int_0^1 x\,dx = \frac{1}{2}(1^2-0^2) = \frac{1}{2}$, $\int_1^2 x\,dx = \frac{1}{2}(2^2-1^2) = \frac{3}{2}$, so $\int_0^1 x\,dx + \int_1^2 x\,dx = \frac{1}{2}+\frac{3}{2} = 2$;
$\int_0^2 x\,dx = \frac{1}{2}(2^2-0^2) = 2$

7. $\int_1^0 y^2\,dy = -\int_0^1 y^2\,dy = -\frac{1}{3}(1^3-0^3) = -\frac{1}{3}$, $\int_0^2 y^2\,dy = \frac{1}{3}(2^3-0) = \frac{8}{3}$,
so $\int_1^0 y^2\,dy + \int_0^2 y^2\,dy = -\frac{1}{3}+\frac{8}{3} = \frac{7}{3}$; $\int_1^2 y^2\,dy = \frac{1}{3}(2^3-1^3) = \frac{7}{3}$

9. $\int_0^2 f(x)\,dx + \int_3^0 f(x)\,dx = \int_3^0 f(x)\,dx + \int_0^2 f(x)\,dx = \int_3^2 f(x)\,dx$, so $a=3$ and $b=2$.

11. $\int_a^b f(t)\,dt = \int_5^3 f(t)\,dt + \int_3^1 f(t)\,dt = \int_5^1 f(t)\,dt$, so $a=5$ and $b=1$.

13. $m = \frac{1}{3}$, $M = \frac{1}{2}$; $\frac{1}{3} = \frac{1}{3}(3-2) \le \int_2^3 \frac{1}{x}\,dx \le \frac{1}{2}(3-2) = \frac{1}{2}$

15. $m = \frac{1}{2}$, $M = \frac{\sqrt{2}}{2}$; $\frac{\pi}{24} = \frac{1}{2}\left(\frac{\pi}{3}-\frac{\pi}{4}\right) \le \int_{\pi/4}^{\pi/3} \cos x\,dx \le \frac{\sqrt{2}}{2}\left(\frac{\pi}{3}-\frac{\pi}{4}\right) = \frac{\sqrt{2}\,\pi}{24}$

17. By (1), $f_{av} = \frac{1}{1-0}\int_0^1 x\,dx = \frac{1}{2}(1^2-0^2) = \frac{1}{2}$

19. By (1), $f_{av} = \frac{1}{1-(-1)}\int_{-1}^1 x^2\,dx = \frac{1}{2}\left\{\frac{1}{3}[1^3-(-1)^3]\right\} = \frac{1}{3}$

21. The mean value of f on $[a,b]$ is

$$\frac{1}{b-a}\int_a^b f(x)\,dx = \frac{1}{b-a}\int_a^b x\,dx = \frac{1}{b-a}\left[\frac{1}{2}(b^2-a^2)\right] = \frac{1}{2}(a+b).$$

23. $A = \int_{-1}^1 f(x)\,dx = \int_{-1}^0 -x\,dx + \int_0^1 x^2\,dx = -\frac{1}{2}[0^2-(-1)^2] + \frac{1}{3}[1^3-0^3] = \frac{5}{6}$

25. b. By Figure 5.30,

$$\text{left sum} = f(-1)(0.2) + f(-0.8)(0.2) + \cdots + f(0.6)(0.2) + f(0.8)(0.2)$$
$$\approx (0.2)(1+1.6+2.02+2.18+2.15+2+1.75+1.5+1.25+1.07) \approx 3.30.$$

Thus the mean value of f on $[0,1] = \frac{1}{2}\int_{-1}^1 f(x)\,dx \approx \frac{3.30}{2} = 1.65$.

27. If $c < a < b$, then by the Addition Property we have

$$\int_c^b f(x)\,dx = \int_c^a f(x)\,dx + \int_a^b f(x)\,dx$$

so that

$$\int_a^b f(x)\,dx = -\int_c^a f(x)\,dx + \int_c^b f(x)\,dx = \int_a^c f(x)\,dx + \int_c^b f(x)\,dx.$$

If $b < a < c$, then by the Addition Property we have

$$\int_b^c f(x)\,dx = \int_b^a f(x)\,dx + \int_a^c f(x)\,dx$$

so that

$$\int_a^b f(x)\,dx = -\int_b^a f(x)\,dx = \int_a^c f(x)\,dx - \int_b^c f(x)\,dx = \int_a^c f(x)\,dx + \int_c^b f(x)\,dx.$$

29. Let T_{mv} denote the mean temperature during the year. Then

$$T_{mv} = \frac{1}{12}\int_0^{12} T(t)\,dt = \frac{1}{12}\left(\int_0^1 T(t)\,dt + \int_1^2 T(t)\,dt + \cdots + \int_{11}^{12} T(t)\,dt\right)$$

Since $\int_{k-1}^{k} T(t)\,dt$ is the mean temperature during the kth month, it follows that T_{mv} is the average of the mean temperatures during the 12 months of the year. Using this and the values for the monthly mean temperatures given in the table, we conclude that

$$T_{mv} \approx \frac{32+33+41+53+62+71+77+75+68+58+47+36}{12}$$

$$\approx 54.4 \text{ (degrees Fahrenheit)}$$

5.4 The Fundamental Theorem of Calculus

1. $F'(x) = x(1+x^3)^{29}$

3. $F(y) = \int_y^2 \frac{1}{t^3}\,dt = -\int_2^y \frac{1}{t^3}\,dt$, so $F'(y) = -\frac{1}{y^3}$.

5. Let $G(x) = \int_0^x t\sin t\,dt$, so $F(x) = G(x^2)$. Since $G'(x) = x\sin x$, the Chain Rule implies that $F'(x) = [G'(x^2)](2x) = (x^2\sin x^2)(2x) = 2x^3\sin x^2$.

7. Notice that

$$G(y) = \int_y^{y^2} (1+t^2)^{1/2}\,dt = \int_y^0 (1+t^2)^{1/2}\,dt + \int_0^{y^2} (1+t^2)^{1/2}\,dt$$

$$= -\int_0^y (1+t^2)^{1/2}\,dt + \int_0^{y^2} (1+t^2)^{1/2}\,dt.$$

Let

$$H(y) = \int_0^y (1+t^2)^{1/2}\,dt \quad\text{and}\quad K(y) = \int_0^{y^2} (1+t^2)^{1/2}\,dt.$$

Then $K(y) = H(y^2)$ and $G(y) = -H(y) + K(y)$, so that $G'(y) = -H'(y) + K'(y)$. Now $H'(y) = (1+y^2)^{1/2}$ and by the Chain Rule, $K'(y) = [H'(y^2)](2y) = 2y(1+y^4)^{1/2}$. Therefore

$$G'(y) = -(1+y^2)^{1/2} + 2y(1+y^4)^{1/2}.$$

9. Let $G(x) = \int_0^x (1+t^2)^{4/5}\,dt$. Then $G'(x) = (1+x^2)^{4/5}$ and

$$F(x) = \frac{d}{dx}G(4x) = [G'(4x)](4) = 4(1+16x^2)^{4/5}$$

so that $F'(x) = \frac{16}{5}(1+16x^2)^{-1/5}(32x) = \frac{512}{5}x(1+16x^2)^{-1/5}$.

11. $\displaystyle\int_0^1 4\,dx = 4x\Big|_0^1 = 4$

13. $\displaystyle\int_1^3 -y\,dy = -\frac{1}{2}y^2\Big|_1^3 = -\frac{9}{2} - \left(-\frac{1}{2}\right) = -4$

15. $\displaystyle\int_1^{-3} 3u\,du = \frac{3}{2}u^2\Big|_1^{-3} = \frac{27}{2} - \frac{3}{2} = 12$

17. $\displaystyle\int_0^1 x^{100}\,dx = \frac{1}{101}x^{101}\Big|_0^1 = \frac{1}{101}$

19. $\displaystyle\int_{-1}^1 u^{1/3}\,du = \frac{3}{4}u^{4/3}\Big|_{-1}^1 = \frac{3}{4} - \frac{3}{4}(-1)^{4/3} = 0$

21. $\displaystyle\int_1^4 x^{-7/9}\,dx = \frac{9}{2}x^{2/9}\Big|_1^4 = \frac{9}{2}(4^{2/9} - 1)$

23. $\displaystyle\int_{-1.5}^{2\pi} (5-x)\,dx = \left(5x - \frac{1}{2}x^2\right)\Big|_{-1.5}^{2\pi} = (10\pi - 2\pi^2) - (-7.5 - 2.25/2) = 10\pi - 2\pi^2 + 8.625$

25. $\displaystyle\int_{-4}^{-1} (5x+14)\,dx = \left(\frac{5}{2}x^2 + 14x\right)\Big|_{-4}^{-1} = \left(\frac{5}{2} - 14\right) - (40 - 56) = \frac{9}{2}$

27. $\displaystyle\int_{-\pi}^{\pi/3} \cos x\,dx = \sin x\Big|_{-\pi}^{\pi/3} = \frac{1}{2}\sqrt{3} - 0 = \frac{1}{2}\sqrt{3}$

29. $\displaystyle\int_{\pi/3}^{-\pi/4} \sin t\,dt = -\cos t\Big|_{\pi/3}^{-\pi/4} = -\frac{1}{2}\sqrt{2} - \left(-\frac{1}{2}\right) = \frac{1}{2} - \frac{1}{2}\sqrt{2}$

31. $\displaystyle\int_1^2 \frac{1}{y^4}\,dy = \frac{-1}{3y^3}\bigg|_1^2 = \frac{-1}{24} - \left(\frac{-1}{3}\right) = \frac{7}{24}$

33. $\displaystyle\int_2^4 \frac{1}{x}\,dx = \ln x|_2^4 = \ln 4 - \ln 2 = 2\ln 2 - \ln 2 = \ln 2$

35. $\displaystyle\int_0^2 e^x\,dx = e^x|_0^2 = e^2 - e^0 = e^2 - 1$

37. $\displaystyle\int_{\pi/6}^{\pi/2} \csc^2 t\,dt = -\cot t|_{\pi/6}^{\pi/2} = 0 - (-\sqrt{3}) = \sqrt{3}$

39. $\displaystyle\int_0^{\pi/2} \left(\frac{d}{dx}\sin^5 x\right) dx = \sin^5 x|_0^{\pi/2} = 1 - 0 = 1$

41. $\displaystyle A = \int_{-1}^1 x^4\,dx = \frac{1}{5}x^5\bigg|_{-1}^1 = \frac{1}{5} - \left(-\frac{1}{5}\right) = \frac{2}{5}$

43. $\displaystyle A = \int_0^{2\pi/3} \sin x\,dx = -\cos x|_0^{2\pi/3} = \frac{1}{2} - (-1) = \frac{3}{2}$

45. $\displaystyle A = \int_1^4 x^{1/2}\,dx = \frac{2}{3}x^{3/2}\bigg|_1^4 = \frac{16}{3} - \frac{2}{3} = \frac{14}{3}$

47. $\displaystyle A = \int_0^{\pi/4} \sec^2 x\,dx = \tan x|_0^{\pi/4} = 1 - 0 = 1$

49. $\displaystyle A = \int_{1/e}^1 \frac{1}{x}\,dx = \ln x|_{1/e}^1 = \ln 1 - \ln(1/e) = 0 - (-1) = 1$

51. $\displaystyle A = \int_0^{\pi/2} \cos^2 x\,dx = \left(\frac{x}{2} + \frac{\sin 2x}{4}\right)\bigg|_0^{\pi/2} = \frac{\pi}{4} - 0 = \frac{\pi}{4}$

53. $\displaystyle\int_{-1}^1 x^n\,dx = \frac{1}{n+1}x^{n+1}\bigg|_{-1}^1 = \frac{1}{n+1}[1^{n+1} - (-1)^{n+1}] = \frac{1}{n+1}[1 - (-1)^{n+1}]$

a. If n is odd, then $(-1)^{n+1} = (-1)^{\text{even}} = 1$, so that $\int_{-1}^1 x^n\,dx = [1/(n+1)](1-1) = 0$.

b. If n is even, then $(-1)^{n+1} = (-1)^{\text{odd}} = -1$, so that $\int_{-1}^1 x^n\,dx = [1/(n+1)][1-(-1)] = 2/(n+1)$.

55. a. $\int_0^x f(t)\,dt = \int_0^x t\,dt = \frac{1}{2}x^2$; $\frac{d}{dx}\int_0^x f(t)\,dt = \frac{d}{dx}\left(\frac{1}{2}x^2\right) = x = f(x)$

b. $\int_0^x f(t)\,dt = \int_0^x -2t^2\,dt = -\frac{2}{3}t^3\Big|_0^x = -\frac{2}{3}x^3$; $\frac{d}{dx}\int_0^x f(t)\,dt = \frac{d}{dx}\left(-\frac{2}{3}x^3\right) = -2x^2 = f(x)$

c. $\int_0^x f(t)\,dt = \int_0^x -\sin t\,dt = \cos t|_0^x = \cos x - 1$; $\frac{d}{dx}\int_0^x f(t)\,dt = \frac{d}{dx}(\cos x - 1) = -\sin x = f(x)$

d. $\int_0^x f(t)\,dt = \int_0^x 10t^4\,dt = 2t^5|_0^x = 2x^5$; $\frac{d}{dx}\int_0^x f(t)\,dt = \frac{d}{dx}(2x^5) = 10x^4 = f(x)$

57. Since $x^2 + 4x$ is an increasing function on $[1, 2]$, the first sum is the lower sum of $\int_1^2 (x^2 + 4x)\,dx$ for the partition P, and the last sum is the upper sum of $\int_1^2 (x^2 + 4x)\,dx$ for P. Since $L_f(P) \le I \le U_f(P)$ for every partition P, it follows that

$$I = \int_1^2 (x^2 + 4x)\,dx = \left(\frac{1}{3}x^3 + 2x^2\right)\Big|_1^2 = \left(\frac{8}{3} + 8\right) - \left(\frac{1}{3} + 2\right) = \frac{25}{3}.$$

59. a. For $0 \le t \le 10$, we have $f(t) - f(0) = \int_0^t v(s)\,ds = \int_0^t (10s - s^2)\,ds = (5s^2 - \frac{1}{3}s^3)|_0^t = 5t^2 - \frac{1}{3}t^3$.

b. $a(t) = v'(t) = 10 - 2t$, so that $a(t) = 0$ if $t = 5$. Since the car is at the origin at $t = 0$, we have $f(0) = 0$, so it follows from (a) that $f(5) = f(5) - f(0) = 5(5^2) - \frac{1}{3}(5^3) = \frac{250}{3}$.

61. The amount of flow per day is $\int_0^{24} F'(t)\,dt$, and by (5) and the hint,

$$\int_0^{24} F'(t)\,dt = -\frac{336{,}000}{\pi}\cos\frac{\pi t}{24}\Big|_0^{24} = -\frac{336{,}000}{\pi}(\cos\pi - \cos 0) = \frac{672{,}000}{\pi}\ \text{(tons)}.$$

63. $R(x) - R(1) = \int_1^x m_R(t)\,dt = \int_1^x (4 - 0.02t)\,dt = (4t - 0.01t^2)|_1^x = (4x - 0.01x^2) - 3.99$

a. Since $R(1) = 4$, it follows that $R(30) = 4 + [(120 - 9) - 3.99] = 111.01$ (dollars).

b. $m_R(x) = 0$ if $x = 200$; $R(200) = 4 + [(800 - 400) - 3.99] = 400.01$ (dollars).

65. a. $v(t) = v(0) + \int_0^t a(s)\,ds = v(0) + \int_0^t -32\,ds = v(0) - 32t$

$f(t) = f(0) + \int_0^t v(s)\,ds = f(0) + \int_0^t (v(0) - 32s)\,ds = 528 + (v(0)s - 16s^2)|_0^t = 528 + v(0)t - 16t^2$

Since $f(6) = 0$, we have $0 = 528 + v(0)6 - 576$, so that $v(0) = 8$ (feet per second).

b. As in part (a), $f(t) = f(0) + \int_0^t (v(0) - 32s)\,ds$. Since $f(0) = 992$ and $v(0) = 4$, we have $f(t) = 992 + \int_0^t (4 - 32s)\,ds = 992 + (4s - 16s^2)|_0^t = 992 + 4t - 16t^2 = 4(4t + 31)(-t + 8)$, so $f(t) = 0$ if $t = 8$ (seconds).

67. The amount lost between time t_1 and t_2 is $\int_{t_1}^{t_2} R(t)\,dt$.

69. a. By the Fundamental Theorem (or by (7)), $C(b) - C(a) = \int_a^b m_C(x)\,dx$. Then

$$\frac{\int_a^b m_C(x)\,dx}{b-a} = \frac{C(b)-C(a)}{b-a} = \text{average cost between the } a\text{th and } b\text{th units produced}$$

as defined in Section 3.1.

b. $\dfrac{\int_1^4 m_C(x)\,dx}{4-1} = \dfrac{\int_1^4 (1/x^{1/2})\,dx}{4-1} = \dfrac{2x^{1/2}\big|_1^4}{3} = \dfrac{4-2}{3} = \dfrac{2}{3}$ (thousand dollars per thousand umbrellas)

71. $$W = (2500\pi)(62.5)\int_0^{100}(100-y)\,dy = 156{,}250\pi\left(100y - \frac{y^2}{2}\right)\Bigg|_0^{100}$$
$$= 156{,}250\pi(10{,}000 - 5{,}000)) = 781{,}250{,}000\pi \text{ (foot-pounds)}$$

73. Since $a(t) = -10$ by hypothesis, (10) implies that

$$v(t) - v(0) = \int_0^t -10\,du = -10u\big|_0^t = -10t$$

for all t until the plane stops. Since $v(0) = 150$ by assumption, $v(t) = v(0) - 10t = 150 - 10t$. Notice that $v(t) = 0$ if $150 - 10t = 0$, which occurs if $t = 15$. Next, let $f(t)$ denote the distance the plane travels during t seconds. By (9),

$$f(15) - f(0) = \int_0^{15} v(u)\,du = \int_0^{15}(150 - 10u)\,du = (150u - 5u^2)\big|_0^{15} = 150(15) - 5(15)^2 = 1125.$$

Since $f(0) = 0$, we have $f(15) = 1125$, which is the stopping distance. Since the airstrip must be 60% longer than the stopping distance, the airstrip needs to be $(1.6)(1125) = 1800$ feet long.

75. a. $p(t_2) - p(t_1) = \displaystyle\int_{t_1}^{t_2} \frac{dp}{dt}\,dt = \int_{t_1}^{t_2} F\,dt$

b. The impulse of the force during the 10^{-3} seconds while the ball is in contact with the floor is given by $p(10^{-3}) - p(0) = (0.2)(4.5) - (0.1)(-5) = 0.95$. Thus the impulse is 0.95 kilogram meter per second. The mean force is $0.95/10^{-3} = 950$ Newtons.

5.5 Indefinite Integrals and Integration Rules

1. $\displaystyle\int (2x - 7)\,dx = x^2 - 7x + C$

3. $\displaystyle\int (2x^{1/3} - 3x^{3/4} + x^{2/5})\,dx = \frac{3}{2}x^{4/3} - \frac{12}{7}x^{7/4} + \frac{5}{7}x^{7/5} + C$

5. $\int\left(t^5-\frac{1}{t^4}\right)dt=\frac{t^6}{6}+\frac{1}{3t^3}+C$

7. $\int(2\cos x-5x)\,dx=2\sin x-\frac{5}{2}x^2+C$

9. $\int(3\csc^2 x-x)\,dx=-3\cot x-\frac{1}{2}x^2+C$

11. $\int(2t+1)^2\,dt=\int(4t^2+4t+1)\,dt=\frac{4}{3}t^3+2t^2+t+C$

13. $\int\left(1+\frac{1}{x}\right)^2dx=\int\left(1+\frac{2}{x}+\frac{1}{x^2}\right)dx=x+2\ln x-\frac{1}{x}+C$

15. $\int_{-1}^{2}(3x-4)\,dx=\left(\frac{3}{2}x^2-4x\right)\Big|_{-1}^{2}=-2-\frac{11}{2}=-\frac{15}{2}$

17. $\int_{\pi/4}^{\pi/2}(-7\sin x+3\cos x)\,dx=(7\cos x+3\sin x)|_{\pi/4}^{\pi/2}=3-5\sqrt{2}$

19. $\int_{-\pi/4}^{-\pi/2}\left(3x-\frac{1}{x^2}+\sin x\right)dx=\left(\frac{3}{2}x^2+\frac{1}{x}-\cos x\right)\Big|_{-\pi/4}^{-\pi/2}$

$=\left(\frac{3\pi^2}{8}-\frac{2}{\pi}\right)-\left(\frac{3\pi^2}{32}-\frac{4}{\pi}-\frac{\sqrt{2}}{2}\right)=\frac{9\pi^2}{32}+\frac{2}{\pi}+\frac{\sqrt{2}}{2}$

21. $\int_{\pi/3}^{\pi/4}(3\sec^2\theta+4\csc^2\theta)\,d\theta=(3\tan\theta-4\cot\theta)|_{\pi/3}^{\pi/4}=-1-\left(3\sqrt{3}-\frac{4\sqrt{3}}{3}\right)=-1-\frac{5}{3}\sqrt{3}$

23. $\int_{1}^{1/3}(3t+2)^3\,dt=\int_{1}^{1/3}(27t^3+54t^2+36t+8)\,dt=\left(\frac{27}{4}t^4+18t^3+18t^2+8t\right)\Big|_{1}^{1/3}=\frac{-136}{3}$

25. $\int_{\pi/2}^{\pi}\left(\pi\sin x-2x+\frac{5}{x^2}+2\pi\right)dx=\left(-\pi\cos x-x^2-\frac{5}{x}+2\pi x\right)\Big|_{\pi/2}^{\pi}=\pi+\frac{5}{\pi}+\frac{\pi^2}{4}$

27. $\int_{-1}^{1}(2x+5)(2x-5)\,dx=\int_{-1}^{1}(4x^2-25)\,dx=\left(\frac{4}{3}x^3-25x\right)\Big|_{-1}^{1}=-\frac{142}{3}$

29. $\int_{4}^{7}|x-5|\,dx=\int_{4}^{5}(5-x)\,dx+\int_{5}^{7}(x-5)\,dx=\left(5x-\frac{1}{2}x^2\right)\Big|_{4}^{5}+\left(\frac{1}{2}x^2-5x\right)\Big|_{5}^{7}=\frac{1}{2}+2=\frac{5}{2}$

31. $\int_{0}^{\pi}(\sin x-2e^x)\,dx=(-\cos x-2e^x)|_0^{\pi}=(-\cos\pi-2e^{\pi})-(-\cos 0-2e^0)$

$=(1-2e^{\pi})-(-1-2)=4-2e^{\pi}$

33. $\displaystyle\int_4^6 f(x)\,dx = \int_4^5 2x\,dx + \int_5^6 (20-2x)\,dx = x^2\Big|_4^5 + (20x - x^2)\Big|_5^6 = 9 + 9 = 18$

35. Since $F'(x) = 20x(1+x^2)^9$, we have $\int 20x(1+x^2)^9\,dx = (1+x^2)^{10} + C$.

37. Since $F'(x) = x\cos x + 2\sin x$, we have $\int (x\cos x + 2\sin x)\,dx = x\sin x - \cos x + C$.

39. Since $F'(x) = 21\sin^6 x\,\cos x$, we have $\int 21\sin^6 x\,\cos x\,dx = 3\sin^7 x + C$.

41. Since $F'(x) = 2xe^{(x^2)} + e^{-x}$, we have $\int (2xe^{(x^2)} + e^{-x})\,dx = e^{(x^2)} - e^{-x} + C$.

43. $\displaystyle A = \int_{-1}^1 (3x^2+4)\,dx = (x^3+4x)\Big|_{-1}^1 = 5 - (-5) = 10$

45. $\displaystyle A = \int_1^4 \left(3\sqrt{x} - \frac{1}{\sqrt{x}}\right)dx = (2x^{3/2} - 2x^{1/2})\Big|_1^4 = (16-4) - (2-2) = 12$

47. $$A = \int_{\pi/4}^{\pi/2} (2\sin x + 3\cos x)\,dx = (-2\cos x + 3\sin x)\Big|_{\pi/4}^{\pi/2}$$
$$= (0+3) - \left[-2\left(\frac{\sqrt{2}}{2}\right) + 3\left(\frac{\sqrt{2}}{2}\right)\right] = 3 - \frac{\sqrt{2}}{2}$$

49. $$A = \int_2^4 \left(2x - \frac{4}{x}\right)dx = (x^2 - 4\ln x)\Big|_2^4 = (16 - 4\ln 4) - (4 - 4\ln 2)$$
$$= 12 - 4\ln 4 + 4\ln 2 = 12 - 8\ln 2 + 4\ln 2 = 12 - 4\ln 2$$

51. a. Since $0 \le \sin x \le x$ for $0 \le x \le 1$, we have $0 \le \sin x^2 \le x^2$ for $0 \le x \le 1$. Then by Corollary 5.19, $0 \le \int_0^1 \sin(x^2)\,dx \le \int_0^1 x^2\,dx = \frac{1}{3}x^3\Big|_0^1 = \frac{1}{3}$.

b. Since $0 \le \sin x \le x$ for $0 \le x \le 1$, we have $0 \le \sin^{3/2} x \le x^{3/2}$ for $0 \le x \le 1$. Then by Corollary 5.19, $0 \le \int_0^{\pi/6} \sin^{3/2} x\,dx \le \int_0^{\pi/6} x^{3/2}\,dx = \frac{2}{5}x^{5/2}\Big|_0^{\pi/6} = \frac{2}{5}(\pi/6)^{5/2}$.

53. $\int_0^2 f(x)\,dx = \int_0^2 (1-x)\,dx = (x - \frac{1}{2}x^2)\Big|_0^2 = (2-2) - 0 = 0$, so $\left|\int_0^2 f(x)\,dx\right| = 0$. However, $\int_0^2 |f(x)|\,dx = \int_0^1 (1-x)\,dx + \int_1^2 -(1-x)\,dx = (x - \frac{1}{2}x^2)\Big|_0^1 - (x - \frac{1}{2}x^2)\Big|_1^2 = (\frac{1}{2} - 0) - (0 - \frac{1}{2}) = 1$. Thus
$$\left|\int_0^2 f(x)\,dx\right| < \int_0^2 |f(x)|\,dx.$$

55. Since $\tan x$ is an increasing function on $(-\pi/2, \pi/2)$, it follows that $-\sqrt{3} = \tan(-\pi/3) \le \tan x \le \tan(-\pi/4) = -1$ for x in $[-\pi/3, -\pi/4]$. Thus for x in $[-\pi/3, -\pi/4]$, we have $|\tan x| \le \sqrt{3}$. By Exercise 54,
$$\left|\int_{-\pi/3}^{-\pi/4} \tan x\,dx\right| \le \sqrt{3}\left(-\frac{\pi}{4} + \frac{\pi}{3}\right) = \frac{1}{12}\sqrt{3}\,\pi.$$

57. a. Fix any x in $[a, b]$. By the Mean Value Theorem there is a number c in (a, x) such that $[f(x) - f(a)]/(x-a) = f'(c)$. Since $|f'(c)| \le M$ and $f(a) = 0$ by hypothesis, we have

$$\frac{|f(x)|}{|x-a|} = \frac{|f(x) - f(a)|}{|x-a|} = |f'(c)| \le M, \quad \text{or} \quad |f(x)| \le M|x-a| = M(x-a).$$

b. By (a) and (5) we have

$$\left|\int_a^b f(x)\,dx\right| \le \int_a^b |f(x)|\,dx \le \int_a^b M(x-a)\,dx = M\int_a^b (x-a)\,dx = M\left.\frac{(x-a)^2}{2}\right|_a^b = M\frac{(b-a)^2}{2}.$$

59. Let $f(x) = ax^3 + bx^2 + cx + d$. Then $f(-1/\sqrt{3}) + f(1/\sqrt{3}) = a(-1/\sqrt{3})^3 + b(-1/\sqrt{3})^2 + c(-1/\sqrt{3}) + d + a(1/\sqrt{3})^3 + b(1/\sqrt{3})^2 + c(1/\sqrt{3}) + d = \frac{2}{3}b + 2d$. Since $\int_{-1}^1 x^n\,dx = 0$ whenever n is odd (see Exercise 53(a) of Section 5.4), we have $\int_{-1}^1 ax^3\,dx = 0 = \int_{-1}^1 cx\,dx$, so that

$$\int_{-1}^1 f(x)\,dx = \int_{-1}^1 (ax^3 + bx^2 + cx + d)\,dx = \int_{-1}^1 (bx^2 + d)\,dx$$

$$= \left(\frac{b}{3}x^3 + dx\right)\Big|_{-1}^1 = \left(\frac{b}{3} + d\right) - \left(-\frac{b}{3} - d\right) = \frac{2}{3}b + 2d.$$

Thus (6) is valid.

61. If $g(x) \le f(x)$ for $b \le x \le a$, then $-f(x) \le -g(x)$, so that Corollary 5.19 implies that

$$\int_a^b g(x)\,dx = \int_b^a -g(x)\,dx \ge \int_b^a -f(x)\,dx = \int_a^b f(x)\,dx.$$

63. We will use induction. The result is valid for $n = 2$ by Theorem 5.16. Assume the result is valid for $n-1$ functions, and let $f_1, f_2, \ldots, f_n$ be continuous on an interval I. Then

$$\int [f_1(x) + f_2(x) + \cdots + f_{n-1}(x) + f_n(x)]\,dx = \int [f_1(x) + f_2(x) + \cdots + f_{n-1}(x)]\,dx + \int f_n(x)\,dx$$

$$= \left[\int f_1(x)\,dx + \int f_2(x)\,dx + \cdots + \int f_{n-1}(x)\,dx\right] + \int f_n(x)\,dx$$

which is equivalent to the desired equation.

65. a. Since $(x/2) + 2y(dy/dx) = 0$, so that $dy/dx = -x/4y$, the slope of the tangent line at $(\sqrt{2}, 1/\sqrt{2})$ is

$$\left.\frac{dy}{dx}\right|_{x=\sqrt{2}} = \frac{-\sqrt{2}}{4/\sqrt{2}} = \frac{-1}{2}.$$

The equation of the tangent line is $y - 1/\sqrt{2} = (-1/2)(x - \sqrt{2})$, or $y = \sqrt{2} - \frac{1}{2}x$. Since $\sqrt{1 - (x^2/4)} \le \sqrt{2} - \frac{1}{2}x$ for $0 \le x \le 2$, we have

$$\int_0^2 \sqrt{1 - \frac{x^2}{4}}\,dx \le \int_0^2 \left(\sqrt{2} - \frac{1}{2}x\right)dx = \left(\sqrt{2}\,x - \frac{1}{4}x^2\right)\Big|_0^2 = 2\sqrt{2} - 1.$$

b. The triangle inscribed in the quarter ellipse has vertices $(0, 0)$, $(2, 0)$, $(0, 1)$, so the area of the triangle is 1.

67. Let $p(r) = \int_a^b [f(x) + rg(x)]^2\,dx$. Then by the Addition Property and (2),

$$p(r) = \int_a^b [(f(x))^2 + 2rf(x)g(x) + r^2(g(x))^2]\,dx$$

$$= \int_a^b (f(x))^2\,dx + 2r\int_a^b f(x)g(x)\,dx + r^2\int_a^b (g(x))^2\,dx = Ar^2 + Br + C$$

where $A = \int_a^b (g(x))^2\,dx$, $B = 2\int_a^b f(x)g(x)\,dx$ and $C = \int_a^b (f(x))^2\,dx$. For any given real number r we have $[f(x) + rg(x)]^2 \geq 0$ for all x in $[a, b]$, so by Corollary 5.10, $p(r) \geq 0$ for any r. But then p has at most one zero, so by the quadratic formula the discriminant of p, which is $B^2 - 4AC$, is nonpositive. Thus $B^2 \leq 4AC$. This is equivalent to the desired inequality.

5.6 Integration by Substitution

1. Let $u = 4x - 5$, so that $du = 4\,dx$. Then

$$\int \sqrt{4x-5}\,dx = \int \sqrt{u}\,\frac{1}{4}\,du = \frac{1}{4}\int \sqrt{u}\,du = \frac{1}{4}\left(\frac{2}{3}u^{3/2}\right) + C = \frac{1}{6}(4x-5)^{3/2} + C.$$

3. Let $u = \pi x$, so that $du = \pi\,dx$. Then

$$\int \cos \pi x\,dx = \int (\cos u)\frac{1}{\pi}\,du = \frac{1}{\pi}\int \cos u\,du = \frac{1}{\pi}\sin u + C = \frac{1}{\pi}\sin \pi x + C.$$

5. Let $u = x^2$, so that $du = 2x\,dx$. Then

$$\int x\cos x^2\,dx = \int (\cos u)\frac{1}{2}\,du = \frac{1}{2}\int \cos u\,du = \frac{1}{2}\sin u + C = \frac{1}{2}\sin x^2 + C.$$

7. Let $u = \cos t$, so that $du = -\sin t\,dt$. Then

$$\int \cos^{-4} t\,\sin t\,dt = \int u^{-4}(-1)\,du = \frac{1}{3}u^{-3} + C = \frac{1}{3}\cos^{-3} t + C.$$

9. Let $u = t^2 - 3t + 1$, so that $du = (2t-3)\,dt$. Then

$$\int \frac{2t-3}{(t^2-3t+1)^{7/2}}\,dt = \int \frac{1}{u^{7/2}}\,du = \int u^{-7/2}\,du = -\frac{2}{5}u^{-5/2} + C = -\frac{2}{5}\frac{1}{(t^2-3t+1)^{5/2}} + C.$$

11. Let $v = x + 1$, so that $dv = dx$ and $x - 1 = v - 2$. Then

$$\int (x-1)\sqrt{x+1}\,dx = \int (v-2)\sqrt{v}\,dv = \int (v^{3/2} - 2v^{1/2})\,dv$$

$$= \frac{2}{5}v^{5/2} - 2\left(\frac{2}{3}v^{3/2}\right) + C = \frac{2}{5}(x+1)^{5/2} - \frac{4}{3}(x+1)^{3/2} + C.$$

13. Let $u = 3 + \sec x$, so that $du = \sec x\,\tan x\,dx$. Then

$$\int \sec x\,\tan x\sqrt{3+\sec x}\,dx = \int \sqrt{u}\,du = \frac{2}{3}u^{3/2} + C = \frac{2}{3}(3+\sec x)^{3/2} + C.$$

15. Let $u = x^2$, so that $du = 2x\,dx$. Then

$$\int xe^{(x^2)}\,dx = \int e^u \frac{1}{2}\,du = \frac{1}{2}e^u + C = \frac{1}{2}e^{(x^2)} + C.$$

17. Let $u = x^2 + 1$, so that $du = 2x\,dx$. Then

$$\int \frac{x}{x^2+1}\,dx = \int \frac{1}{u}\frac{1}{2}\,du = \frac{1}{2}\ln u + C = \frac{1}{2}\ln(x^2+1) + C.$$

19. Let $u = x^3 + 1$, so that $du = 3x^2\,dx$. Then

$$\int 3x^2(x^3+1)^{12}\,dx = \int u^{12}\,du = \frac{1}{13}u^{13} + C = \frac{1}{13}(x^3+1)^{13} + C.$$

21. Let $u = 4 + x^{3/2}$, so that $du = \frac{3}{2}x^{1/2}\,dx$. Then

$$\int \sqrt{x}(4+x^{3/2})\,dx = \int u\frac{2}{3}\,du = \frac{1}{3}u^2 + C = \frac{1}{3}(4+x^{3/2})^2 + C.$$

23. Let $u = 3x + 7$, so that $du = 3\,dx$. Then

$$\int \sqrt{3x+7}\,dx = \int u^{1/2}\frac{1}{3}\,du = \frac{1}{3}\int u^{1/2}\,du = \frac{1}{3}\left(\frac{2}{3}u^{3/2}\right) + C = \frac{2}{9}(3x+7)^{3/2} + C.$$

25. Let $u = 1 + 2x + 4x^2$, so that $du = (2+8x)\,dx$. Then

$$\int (1+4x)\sqrt{1+2x+4x^2}\,dx = \int (\sqrt{u})\frac{1}{2}\,du = \frac{1}{2}\int u^{1/2}\,du = \frac{1}{2}\left(\frac{2}{3}u^{3/2}\right) + C = \frac{1}{3}(1+2x+4x^2)^{3/2} + C.$$

27. Let $u = \pi x$, so that $du = \pi\,dx$. If $x = -1$, then $u = -\pi$; if $x = 3$, then $u = 3\pi$. Thus

$$\int_{-1}^{3} \sin \pi x\,dx = \int_{-\pi}^{3\pi} (\sin u)\frac{1}{\pi}\,du = \frac{1}{\pi}\int_{-\pi}^{3\pi} \sin u\,du = \frac{1}{\pi}(-\cos u)\Big|_{-\pi}^{3\pi} = \frac{1}{\pi}[-(-1) + (-1)] = 0.$$

29. Let $u = \sin t$, so that $du = \cos t\,dt$. Then

$$\int \sin^6 t \cos t\,dt = \int u^6\,du = \frac{1}{7}u^7 + C = \frac{1}{7}\sin^7 t + C.$$

31. Let $u = \sin 2z$, so that $du = 2\cos 2z\,dz$. Then

$$\int \sqrt{\sin 2z}\,\cos 2z\,dz = \int \sqrt{u}\,\frac{1}{2}\,du = \frac{1}{2}\int \sqrt{u}\,du = \frac{1}{2}\left(\frac{2}{3}u^{3/2}\right) + C = \frac{1}{3}(\sin 2z)^{3/2} + C.$$

33. Let $u = \cos z$, so that $du = -\sin z\,dz$. If $z = 0$, then $u = 1$; if $z = \pi/4$, then $u = \sqrt{2}/2$. Thus

$$\int_0^{\pi/4} \frac{\sin z}{\cos^2 z}\,dz = \int_1^{\sqrt{2}/2} \frac{1}{u^2}(-1)\,du = \frac{1}{u}\Big|_1^{\sqrt{2}/2} = \frac{1}{\sqrt{2}/2} - 1 = \sqrt{2} - 1.$$

35. Let $u = \sqrt{z}$, so that $du = (1/2\sqrt{z})\,dz$. Then

$$\int \frac{1}{\sqrt{z}} \sec^2 \sqrt{z}\,dz = \int (\sec^2 u)(2)\,du = 2\int \sec^2 u\,du = 2\tan u + C = 2\tan\sqrt{z} + C.$$

37. Let $u = w^2 + 1$, so that $du = 2w\,dw$. Then

$$\int w\left(\sqrt{w^2+1} + \frac{1}{\sqrt{w^2+1}}\right)dw = \int\left(\sqrt{u} + \frac{1}{\sqrt{u}}\right)\frac{1}{2}\,du = \frac{1}{2}\int\left(\sqrt{u} + \frac{1}{\sqrt{u}}\right)du$$

$$= \frac{1}{2}\left(\frac{2}{3}u^{3/2} + 2u^{1/2}\right) + C = \frac{1}{3}(w^2+1)^{3/2} + (w^2+1)^{1/2} + C.$$

39. Let $u = 1 + 4x^{1/3}$, so that $du = \frac{4}{3}x^{-2/3}$. If $x = 1$, then $u = 5$; if $x = 8$, then $u = 9$. Thus

$$\int_1^8 x^{-2/3}\sqrt{1+4x^{1/3}}\,dx = \int_5^9 \sqrt{u}\left(\frac{3}{4}\right)du = \frac{3}{4}\int_5^9 \sqrt{u}\,du = \frac{3}{4}\left(\frac{2}{3}u^{3/2}\right)\Big|_5^9 = \frac{1}{2}(27 - 5\sqrt{5}).$$

41. Let $u = 1 + e^{2x}$, so that $du = 2e^{2x}\,dx$. Then

$$\int e^{2x}\sin(1+e^{2x})\,dx = \int(\sin u)\frac{1}{2}\,du = -\frac{1}{2}\cos u + C = -\frac{1}{2}\cos(1+e^{2x}) + C.$$

43. Let $u = 1 + x^4$, so that $du = 4x^3\,dx$. Then

$$\int \frac{x^3}{1+x^4}\,dx = \int \frac{1}{u}\frac{1}{4}\,du = \frac{1}{4}\ln u + C = \frac{1}{4}\ln(1+x^4) + C.$$

45. Let $u = x + 2$, so that $du = dx$ and $x = u - 2$. Then

$$\int x\sqrt{x+2}\,dx = \int (u-2)\sqrt{u}\,du = \int (u^{3/2} - 2u^{1/2})\,du$$

$$= \frac{2}{5}u^{5/2} - \frac{4}{3}u^{3/2} + C = \frac{2}{5}(x+2)^{5/2} - \frac{4}{3}(x+2)^{3/2} + C.$$

47. Let $u = 6 - 2x$, so that $du = -2\,dx$ and $4x = 2(6-u)$. If $x = 1$, then $u = 4$; if $x = 3$, then $u = 0$. Thus

$$\int_1^3 4x\sqrt{6-2x}\,dx = \int_4^0 2(6-u)\sqrt{u}\left(-\frac{1}{2}\right)du = -\int_4^0 (6u^{1/2} - u^{3/2})\,du$$

$$= \int_0^4 (6u^{1/2} - u^{3/2})\,du = \left(4u^{3/2} - \frac{2}{5}u^{5/2}\right)\Big|_0^4 = \left[4(8) - \frac{2}{5}(32)\right] - 0 = \frac{96}{5}.$$

49. Let $u = 1 - 8t$, so that $du = -8\,dt$ and $t^2 = \frac{1}{64}(1-u)^2$. Then

$$\int t^2\sqrt{1-8t}\,dt = \int \frac{1}{64}(1-u)^2\sqrt{u}\left(-\frac{1}{8}\right)du = -\frac{1}{512}\int (u^{1/2} - 2u^{3/2} + u^{5/2})\,du$$

$$= -\frac{1}{512}\left(\frac{2}{3}u^{3/2} - \frac{4}{5}u^{5/2} + \frac{2}{7}u^{7/2}\right) + C = -\frac{1}{256}\left[\frac{1}{3}(1-8t)^{3/2} - \frac{2}{5}(1-8t)^{5/2} + \frac{1}{7}(1-8t)^{7/2}\right] + C.$$

51. Let $u = t + 2$, so that $du = dt$ and $t^2 = (u-2)^2$. If $t = -1$, then $u = 1$; if $t = 2$, then $u = 4$. Thus

$$\int_{-1}^{2} \frac{t^2}{\sqrt{t+2}}\,dt = \int_1^4 \frac{(u-2)^2}{\sqrt{u}}\,du = \int_1^4 (u^{3/2} - 4u^{1/2} + 4u^{-1/2})\,du$$

$$= \left(\frac{2}{5}u^{5/2} - \frac{8}{3}u^{3/2} + 8u^{1/2}\right)\Big|_1^4 = \left[\frac{2}{5}(32) - \frac{8}{3}(8) + 8(2)\right] - \left[\frac{2}{5} - \frac{8}{3} + 8\right] = \frac{26}{15}.$$

53. $A = \int_0^3 \sqrt{x+1}\,dx$. Let $u = x + 1$, so that $du = dx$. If $x = 0$, then $u = 1$; if $x = 3$, then $u = 4$. Thus

$$A = \int_0^3 \sqrt{x+1}\,dx = \int_1^4 \sqrt{u}\,du = \frac{2}{3}\left(4^{3/2} - 1^{3/2}\right) = \frac{14}{3}.$$

55. $A = \int_1^2 [x/(x^2+1)^2]\,dx$. Let $u = x^2 + 1$, so that $du = 2x\,dx$. If $x = 1$, then $u = 2$; if $x = 2$, then $u = 5$. Thus

$$A = \int_1^2 \frac{x}{(x^2+1)^2}\,dx = \int_2^5 \frac{1}{u^2}\left(\frac{1}{2}\right)du = \frac{1}{2}\int_2^5 \frac{1}{u^2}\,du = -\frac{1}{2}\frac{1}{u}\Big|_2^5 = -\frac{1}{2}\left(\frac{1}{5} - \frac{1}{2}\right) = \frac{3}{20}.$$

57. $A = \int_{1/8}^{1/3} (1/x^2)(1 + 1/x)^{1/2}\,dx$. Let $u = 1 + 1/x$, so that $du = -(1/x^2)\,dx$. If $x = \frac{1}{8}$, then $u = 9$; if $x = \frac{1}{3}$, then $u = 4$. Thus

$$A = \int_{1/8}^{1/3} \frac{1}{x^2}\left(1 + \frac{1}{x}\right)^{1/2} dx = \int_9^4 u^{1/2}(-1)\,du = -\int_9^4 u^{1/2}\,du = -\frac{2}{3}u^{3/2}\Big|_9^4 = \frac{38}{3}.$$

59. By the trigonometric identity $\sin^2 x = \frac{1}{2} - \frac{1}{2}\cos 2x$ we have

$$\int \sin^2 x\,dx = \int \left(\frac{1}{2} - \frac{1}{2}\cos 2x\right) dx = \int \frac{1}{2}\,dx - \frac{1}{2}\int \cos 2x\,dx.$$

Let $u = 2x$, so that $du = 2\,dx$. Then

$$\int \sin^2 x\,dx = \frac{1}{2}x - \frac{1}{2}\int (\cos u)\frac{1}{2}\,du = \frac{1}{2}x - \frac{1}{4}\sin u + C = \frac{1}{2}x - \frac{1}{4}\sin 2x + C.$$

61. a. Let $u = 1 - x$, so that $du = -dx$. If $x = 0$, then $u = 1$; if $x = 1$, then $u = 0$. Thus $\int_0^1 x^n(1-x)^m\,dx = \int_1^0 (1-u)^n u^m(-1)\,du = \int_0^1 x^m(1-x)^n\,dx$.

b. By part (a),

$$\int_0^1 x^2(1-x)^{10}\,dx = \int_0^1 x^{10}(1-x)^2\,dx = \int_0^1 (x^{10} - 2x^{11} + x^{12})\,dx$$

$$= \left(\frac{1}{11}x^{11} - \frac{2}{12}x^{12} + \frac{1}{13}x^{13}\right)\Big|_0^1 = \frac{1}{11} - \frac{1}{6} + \frac{1}{13} = \frac{1}{858} \approx 0.0011655012.$$

63. a. Let $u = ax + b$, so that $du = a\,dx$. Then

$$\int f(ax+b)\,dx = \int f(u)\frac{1}{a}\,du = \frac{1}{a}\int f(u)\,du = \frac{1}{a}F(u) + C = \frac{1}{a}F(ax+b) + C.$$

b. If $f(x) = \sin x$, then since $\int \sin x\,dx = -\cos x + C$, part (a) implies that

$$\int \sin(ax+b)\,dx = -\frac{1}{a}\cos(ax+b) + C.$$

c. If $f(x) = x^n$ with $n \neq 0, -1$, then since $\int x^n\,dx = [1/(n+1)]x^{n+1} + C$, part (a) implies that

$$\int (ax+b)^n\,dx = \frac{1}{a}\left[\frac{1}{n+1}(ax+b)^{n+1}\right] + C = \frac{1}{a(n+1)}(ax+b)^{n+1} + C.$$

65. The problem with the argument is that if $u = 1/x$, then u is not continuous (much less differentiable) on $[-1, 1]$, as required by Theorem 5.20 and also (4).

67. a. Let $u = 2\pi x$, so that $du = 2\pi\,dx$. If $x = 0$, then $u = 0$; if $x = \frac{1}{4}$, then $u = \pi/2$. Then by (3),

$$\int_0^{1/4} 2\sin^2(2\pi x)\,dx = \int_0^{\pi/2} (2\sin^2 u)\frac{1}{2\pi}\,du = \frac{1}{\pi}\int_0^{\pi/2} \sin^2 u\,du$$

$$= \frac{1}{\pi}\left(\frac{1}{2}u - \frac{1}{4}\sin 2u\right)\Big|_0^{\pi/2} = \frac{1}{\pi}\left(\frac{\pi}{4} - \frac{1}{4}\sin\pi\right) = \frac{1}{4}.$$

b. Using the result of part (a) with $u = 2\pi$ when $x = 1$, we have

$$\int_0^1 |\psi(x)|^2\,dx = \int_0^1 2\sin^2(2\pi x)\,dx = \frac{1}{\pi}\left(\frac{1}{2}u - \frac{1}{4}\sin 2u\right)\Big|_0^{2\pi} = 1.$$

The physical significance of this result is that the electron is certain to be in the interval $[0, 1]$.

69. Let $u = x - a$, so that $du = dx$. If $x = x_1$, then $u = x_1 - a$; if $x = x_2$, then $u = x_2 - a$. Thus

$$W = \int_{x_1}^{x_2} \frac{Gm_1m_2}{(x-a)^2}\,dx = \int_{x_1-a}^{x_2-a} \frac{Gm_1m_2}{u^2}\,du = Gm_1m_2\int_{x_1-a}^{x_2-a} \frac{1}{u^2}\,du = Gm_1m_2\left(-\frac{1}{u}\right)\Big|_{x_1-a}^{x_2-a}$$

$$= Gm_1m_2\left(-\frac{1}{x_2-a} + \frac{1}{x_1-a}\right) = \frac{Gm_1m_2(x_2-x_1)}{(x_1-a)(x_2-a)}.$$

5.7 The Logarithm

1. $\displaystyle\int_2^8 \frac{1}{x}\,dx = \ln x|_2^8 = \ln 8 - \ln 2 = \ln\frac{8}{2} = \ln 4$

3. $\displaystyle\int_{-4}^{-12} \frac{2}{t}\,dt = 2\int_{-4}^{-12} \frac{1}{t}\,dt = 2\ln|t|\Big|_{-4}^{-12} = 2(\ln 12 - \ln 4) = 2\ln 3$

5. The domain is $(-1, \infty)$; $f'(x) = 1/(x+1)$.

7. The domain consists of all x such that $(x-3)/(x-2) > 0$, that is, the union of $(-\infty, 2)$ and $(3, \infty)$; since $f(x) = \frac{1}{2}\ln[(x-3)/(x-2)]$, it follows that

$$f'(x) = \frac{1}{2}\frac{1}{(x-3)/(x-2)}\frac{1(x-2)-1(x-3)}{(x-2)^2} = \frac{1}{2(x-3)(x-2)}.$$

9. The domain is $(0, \infty)$; $f'(t) = [\cos(\ln t)](1/t)$.

11. The domain consists of all x such that $\ln x > 0$, that is, $(1, \infty)$; $f'(x) = [1/(\ln x)](1/x) = 1/(x \ln x)$.

13. By implicit differentiation,

$$\ln(y^2+x) + x\left[\frac{1}{y^2+x}\left(2y\frac{dy}{dx}+1\right)\right] = 5\frac{dy}{dx}$$

so that

$$\left(5 - \frac{2xy}{y^2+x}\right)\frac{dy}{dx} = \ln(y^2+x) + \frac{x}{y^2+x}, \quad \text{and thus} \quad \frac{dy}{dx} = \frac{(y^2+x)\ln(y^2+x)+x}{5(y^2+x)-2xy}.$$

15. The domain consists of $(-\infty, 0)$ and $(0, \infty)$.

$$f(x) = \begin{cases} \ln x & \text{for } x > 0 \\ \ln(-x) & \text{for } x < 0 \end{cases}$$

$$f'(x) = \frac{1}{x}; \quad f''(x) = \frac{-1}{x^2}$$

No critical numbers or inflection points; concave downward on $(-\infty, 0)$ and on $(0, \infty)$; vertical asymptote is $x = 0$; symmetric with respect to the y axis.

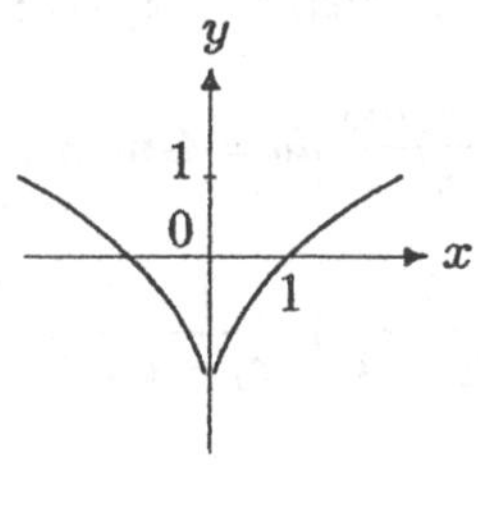

17. The domain is $(-\infty, \infty)$.

$$f'(x) = \frac{2}{x^2+1}; \quad f''(x) = \frac{2-2x^2}{(x^2+1)^2}$$

Relative minimum value is $f(0) = 0$; inflection points are $(-1, \ln 2)$ and $(1, \ln 2)$; concave downward on $(-\infty, -1)$ and $(1, \infty)$, and concave upward on $(-1, 1)$; symmetric with respect to the y axis.

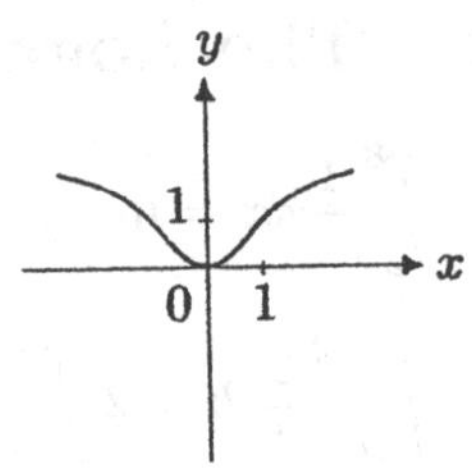

19. Let

$$g(x) = x - \frac{f(x)}{f'(x)} = x - \frac{x^{-2} - \ln x}{-2x^{-3} - 1/x} = \frac{3x + x^3 - x^3 \ln x}{2 + x^2},$$

and begin the Newton-Raphson method by letting $c_1 = 1$. A zero of f is approximately 1.531584394.

21. Let $u = x - 1$, so that $du = dx$. Then

$$\int \frac{1}{x-1}\,dx = \int \frac{1}{u}\,du = \ln|u| + C = \ln|x-1| + C.$$

23. Let $u = x^2 + 4$, so that $du = 2x\,dx$. Then

$$\int \frac{x}{x^2+4}\,dx = \int \frac{1}{u}\frac{1}{2}\,du = \frac{1}{2}\int \frac{1}{u}\,du = \frac{1}{2}\ln|u| + C = \frac{1}{2}\ln(x^2+4) + C.$$

25. Let $u = 1 - 3\cos x$, so that $du = 3\sin x\,dx$. If $x = 0$, then $u = -2$; if $x = \pi/3$, then $u = -\frac{1}{2}$. Thus

$$\int_0^{\pi/3} \frac{\sin x}{1 - 3\cos x}\,dx = \int_{-2}^{-1/2} \frac{1}{u}\left(\frac{1}{3}\right) du = \frac{1}{3}\int_{-2}^{-1/2} \frac{1}{u}\,du = \frac{1}{3}\ln|u|\Big|_{-2}^{-1/2} = \frac{1}{3}\left(\ln\frac{1}{2} - \ln 2\right) = -\frac{2}{3}\ln 2.$$

27. Let $u = x^2 + 4x - 1$, so that $du = (2x+4)\,dx = 2(x+2)\,dx$. If $x = -1$, then $u = -4$; if $x = 0$, then $u = -1$. Thus

$$\int_{-1}^{0} \frac{x+2}{x^2+4x-1}\,dx = \int_{-4}^{-1} \frac{1}{u}\left(\frac{1}{2}\right) du = \frac{1}{2}\int_{-4}^{-1} \frac{1}{u}\,du = \frac{1}{2}\ln|u|\Big|_{-4}^{-1} = -\frac{1}{2}\ln 4 = -\ln 2.$$

29. Let $u = \ln z$, so that $du = (1/z)\,dz$. Then

$$\int \frac{\ln z}{z}\,dz = \int u\,du = \frac{1}{2}u^2 + C = \frac{1}{2}(\ln z)^2 + C.$$

31. Let $u = \ln(\ln t)$, so that

$$du = \frac{1}{\ln t}\cdot\frac{1}{t}\,dt = \frac{1}{t\ln t}\,dt.$$

Then

$$\int \frac{\ln(\ln t)}{t\ln t}\,dt = \int u\,du = \frac{1}{2}u^2 + C = \frac{1}{2}\left(\ln(\ln t)\right)^2 + C.$$

33. Note that $\int \cot t\,dt = \int(\cos t/\sin t)\,dt$. Let $u = \sin t$, so that $du = \cos t\,dt$. Then

$$\int \cot\,dt = \int \frac{\cos t}{\sin t}\,dt = \int \frac{1}{u}\,du = \ln|u| + C = \ln|\sin t| + C.$$

35. Note that

$$\int \frac{x}{1 + x\tan x}\,dx = \int \frac{x}{1 + x\dfrac{\sin x}{\cos x}}\,dx = \int \frac{x\cos x}{\cos x + x\sin x}\,dx.$$

Let $u = \cos x + x\sin x$, so that $du = (-\sin x + \sin x + x\cos x)\,dx = x\cos x\,dx$. Then

$$\int \frac{x}{1 + x\tan x}\,dx = \int \frac{x\cos x}{\cos x + x\sin x}\,dx = \int \frac{1}{u}\,du = \ln|u| + C = \ln|\cos x + x\sin x| + C.$$

37. $A = \int_e^{e^2} (1/x)\,dx = \ln|x|\Big|_e^{e^2} = \ln e^2 - \ln e = 2 - 1 = 1$

39. $A = \displaystyle\int_{\pi/4}^{\pi/3} \frac{\sin^3 x}{\cos x}\,dx = \int_{\pi/4}^{\pi/3} \frac{(1-\cos^2 x)\sin x}{\cos x}\,dx = \int_{\pi/4}^{\pi/3} (\tan x - \sin x \cos x)\,dx$

$$= \left(-\ln|\cos x| - \frac{1}{2}\sin^2 x\right)\Bigg|_{\pi/4}^{\pi/3} = \left[-\ln\frac{1}{2} - \frac{1}{2}\left(\frac{1}{2}\sqrt{3}\right)^2\right] - \left[-\ln\frac{1}{2}\sqrt{2} - \frac{1}{2}\left(\frac{1}{2}\sqrt{2}\right)^2\right] = \frac{1}{2}\ln 2 - \frac{1}{8}$$

41. Since $\ln|f(x)| = \ln|x+1|^{1/5} + \ln|2x+3|^2 + \ln|7-4x|^{-1/2} = \frac{1}{5}\ln|x+1| + 2\ln|2x+3| - \frac{1}{2}\ln|7-4x|$, we have

$$\frac{d}{dx}\ln|f(x)| = \frac{1}{5}\,\frac{1}{x+1} + \frac{4}{2x+3} + \frac{2}{7-4x}.$$

Then (18) yields

$$f'(x) = \left[(x+1)^{1/5}(2x+3)^2(7-4x)^{-1/2}\right]\left[\frac{1}{5(x+1)} + \frac{4}{2x+3} + \frac{2}{7-4x}\right].$$

43. Since $\ln|y| = \ln|x+3|^{2/3} + \ln|2x-1|^{1/3} - \ln|4x+5|^{4/3} = \frac{2}{3}\ln|x+3| + \frac{1}{3}\ln|2x-1| - \frac{4}{3}\ln|4x+5|$, we have

$$\frac{d}{dx}\ln|y| = \frac{2}{3}\,\frac{1}{x+3} + \frac{1}{3}\,\frac{2}{2x-1} - \frac{4}{3}\,\frac{4}{4x+5}.$$

Then (18) yields

$$\frac{dy}{dx} = \sqrt[3]{\frac{(x+3)^2(2x-1)}{(4x+5)^4}}\left(\frac{2}{3x+9} + \frac{2}{6x-3} - \frac{16}{12+15}\right).$$

45. Since $\ln|y| = \ln x^{3/2} + \ln e^{-x^2} - \ln|1-e^x| = \frac{3}{2}\ln|x| - x^2 - \ln|1-e^x|$, we have

$$\frac{d}{dx}\ln|y| = \frac{3}{2x} - 2x - \frac{-e^x}{1-e^x}.$$

Then (18) yields

$$\frac{dy}{dx} = \frac{x^{3/2}e^{-x^2}}{1-e^x}\left(\frac{3}{2x} - 2x + \frac{e^x}{1-e^x}\right).$$

47. Let $f(x) = \ln x^r$ and $g(x) = r\ln x$. Then $f'(x) = (1/x^r)rx^{r-1} = r/x$ and $g'(x) = r/x$, so $f'(x) = g'(x)$. Moreover, $f(1) = \ln 1^r = \ln 1 = 0$ and $g(1) = r\ln 1 = 0$. By Theorem 4.6(b), $f = g$, so $\ln b^r = r\ln b$.

49. a. By the Law of Logarithms, $\ln(b/c) = \ln b(1/c) = \ln b + \ln(1/c)$, and by (12), $\ln(1/c) = -\ln c$. Therefore $\ln(b/c) = \ln b - \ln c$.

b. Let $f(x) = \ln(x/c)$ and $g(x) = \ln x - \ln c$. Then

$$f'(x) = \frac{1}{x/c}\left(\frac{1}{c}\right) = \frac{1}{x} \quad\text{and}\quad g'(x) = \frac{1}{x}$$

so $f'(x) = g'(x)$. Moreover, $f(c) = \ln(c/c) = \ln 1 = 0$ and $g(c) = \ln c - \ln c = 0$. By Theorem 4.6(b), $f = g$, so $\ln(b/c) = \ln b - \ln c$.

51. We have $\ln n = \int_1^n (1/t)\, dt$. Let P be the partition of $[1, n]$ with subintervals of length 1. Then

$$\text{lower sum} = \frac{1}{2} + \frac{1}{3} + \cdots + \frac{1}{n} \quad \text{and} \quad \text{upper sum} = 1 + \frac{1}{2} + \frac{1}{3} + \cdots + \frac{1}{n-1}.$$

Thus

$$\frac{1}{2} + \frac{1}{3} + \frac{1}{4} + \cdots + \frac{1}{n} < \ln n < 1 + \frac{1}{2} + \frac{1}{3} + \cdots + \frac{1}{n-1}.$$

53. a. Since $1/t \le 1/\sqrt{t}$ for $t \ge 1$, we have

$$\ln x = \int_1^x \frac{1}{t}\, dt \le \int_1^x \frac{1}{\sqrt{t}}\, dt = 2\sqrt{t}\Big|_1^x = 2\left(\sqrt{x} - 1\right) \quad \text{for } x \ge 1.$$

b. Since $0 \le \ln x \le 2(\sqrt{x} - 1)$ for $x \ge 1$ by (a), we have $0 \le (\ln x)/x \le \left(2\sqrt{x} - 2\right)/x$. But $\lim_{x\to\infty} \left(2\sqrt{x} - 2\right)/x = \lim_{x\to\infty} \left(2/\sqrt{x} - 2/x\right) = 0$. Thus by the Squeezing Theorem,

$$\lim_{x\to\infty} \frac{\ln x}{x} = 0.$$

c. By (b) and (12),

$$\lim_{x\to 0^+} x \ln x = \lim_{y\to\infty} \frac{1}{y} \ln \frac{1}{y} = \lim_{y\to\infty} \frac{-\ln y}{y} = 0.$$

55. a. $f'(x) = \dfrac{(1/x)x - \ln x}{x^2} = \dfrac{1 - \ln x}{x^2}$

Since $\ln e = 1$ and $\ln x$ is increasing, we have $f'(x) > 0$ for $0 < x < e$ and $f'(x) < 0$ for $x > e$. Thus f is increasing on $(0, e]$ and decreasing on $[e, \infty)$.

b. By (a), $f(e) = 1/e$ is the maximum value of f.

c. $f''(x) = \dfrac{-(1/x)x^2 - 2x(1 - \ln x)}{x^4} = \dfrac{-3 + 2\ln x}{x^3}$

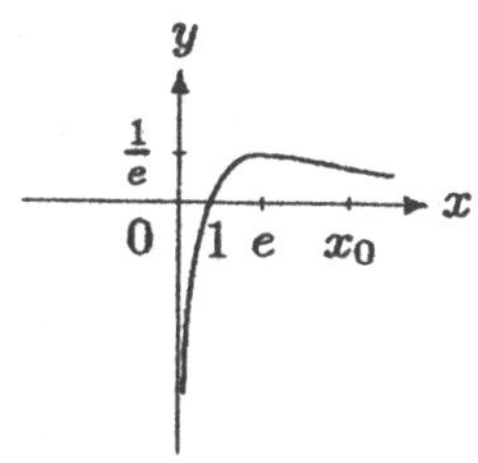

If $\ln x_0 = \frac{3}{2}$, then $x_0 \approx 4.5$, and the graph is concave downward on $(0, x_0)$ and concave upward on (x_0, ∞). By Exercise 53(b), $y = 0$ is a horizontal asymptote of f. Since $\lim_{x\to 0^+} \ln x = -\infty$ and $\lim_{x\to 0^+} 1/x = \infty$, we have $\lim_{x\to 0^+} (\ln x)/x = -\infty$, so that the line $x = 0$ is a vertical asymptote.

57. $W = \displaystyle\int_{V_1}^{V_2} P\, dV = \int_{V_1}^{V_2} \frac{c}{V}\, dV = (c \ln V)\Big|_{V_1}^{V_2} = c(\ln V_2 - \ln V_1) = c \ln \frac{V_2}{V_1}$

59. a. $\ln w = 4.4974 + 3.135 \ln 1 = 4.4974$, so $w \approx 89.7834$ (kilograms)

b. $\ln w = 4.4974 + 3.135 \ln \frac{1}{2}$, so $w \approx 10.2204$ (kilograms)

5.8 Another Look at Area

1. $x^2 - 2x \le 0$ on $[0, 2]$. Thus

$$A = \int_0^2 -(x^2 - 2x)\,dx = -\left(\frac{1}{3}x^3 - x^2\right)\Big|_0^2 = -\left(\frac{8}{3} - 4\right) = \frac{4}{3}.$$

3. The function $y = x\sqrt{1-x^2}$ is an odd function on $[-1, 1]$, and $x\sqrt{1-x^2} \ge 0$ on $[0, 1]$. Thus $A = 2\int_0^1 x\sqrt{1-x^2}\,dx$. Let $u = 1 - x^2$, so that $du = -2x\,dx$. If $x = 0$, then $u = 1$; if $x = 1$, then $u = 0$. Thus

$$A = 2\int_0^1 x\sqrt{1-x^2}\,dx = 2\int_1^0 \sqrt{u}\left(-\frac{1}{2}\right)du = -\frac{2}{3}u^{3/2}\Big|_1^0 = \frac{2}{3}.$$

5. Let $f(x) = |x|$ and $g(x) = x^2$. Notice that f and g are even functions, and $[-1, 1]$ is symmetric with respect to the origin. Thus $A = 2\int_0^1 |f(x) - g(x)|\,dx$. Since $f(x) \ge g(x)$ for $0 \le x \le 1$, we find that

$$A = 2\int_0^1 (f(x) - g(x))\,dx = 2\int_0^1 (x - x^2)\,dx = 2\left(\frac{1}{2}x^2 - \frac{1}{3}x^3\right)\Big|_0^1 = 2\left(\frac{1}{6}\right) = \frac{1}{3}.$$

7. $x^2 + 2x = x(x+2) \le 0$ on $[-1, 0]$, whereas $x^2 + 2x \ge 0$ on $[0, 3]$. Thus

$$A = \int_{-1}^0 -(x^2 + 2x)\,dx + \int_0^3 (x^2 + 2x)\,dx = -\left(\frac{1}{3}x^3 + x^2\right)\Big|_{-1}^0 + \left(\frac{1}{3}x^3 + x^2\right)\Big|_0^3 = \frac{2}{3} + 18 = \frac{56}{3}.$$

9. $x/\sqrt{1+x^2} \ge 0$ on $[0, \sqrt{7}]$, whereas $x/\sqrt{1+x^2} \le 0$ on $[-1, 0]$. Thus

$$A = \int_{-1}^0 \frac{-x}{\sqrt{1+x^2}}\,dx + \int_0^{\sqrt{7}} \frac{x}{\sqrt{1+x^2}}\,dx \overset{u=1+x^2}{=} \int_2^1 -\frac{1}{2\sqrt{u}}\,du + \int_1^8 \frac{1}{2\sqrt{u}}\,du$$

$$= -\sqrt{u}\Big|_2^1 + \sqrt{u}\Big|_1^8 = (\sqrt{2} - 1) + (\sqrt{8} - 1) = 3\sqrt{2} - 2.$$

11. $(\ln x)/x \le 0$ on $[\frac{1}{2}, 1]$, whereas $(\ln x)/x \ge 0$ on $[1, 2]$. Thus

$$A = \int_{1/2}^1 -\frac{\ln x}{x}\,dx + \int_1^2 \frac{\ln x}{x}\,dx \overset{u=\ln x}{=} \int_{-\ln 2}^0 -u\,du + \int_0^{\ln 2} u\,du$$

$$= -\frac{1}{2}u^2\Big|_{-\ln 2}^0 + \frac{1}{2}u^2\Big|_0^{\ln 2} = \frac{1}{2}(\ln 2)^2 + \frac{1}{2}(\ln 2)^2 = (\ln 2)^2.$$

13. $f(x) \ge g(x)$ for $-2 \le x \le 1$, so

$$A = \int_{-2}^1 (x^2 - x^3)\,dx = \left(\frac{1}{3}x^3 - \frac{1}{4}x^4\right)\Big|_{-2}^1 = \frac{1}{12} + \frac{20}{3} = \frac{27}{4}.$$

15. $g(x) - k(x) = x^2 + 3x + 2 = (x+1)(x+2)$, so $g(x) - k(x) \geq 0$ for $-3 \leq x \leq -2$ and for $-1 \leq x \leq 0$, whereas $g(x) - k(x) \leq 0$ for $-2 \leq x \leq -1$. Thus

$$A = \int_{-3}^{-2} (x^2 + 3x + 2)\, dx + \int_{-2}^{-1} (-x^2 - 3x - 2)\, dx + \int_{-1}^{0} (x^2 + 3x + 2)\, dx$$

$$= \left(\frac{1}{3}x^3 + \frac{3}{2}x^2 + 2x\right)\Bigg|_{-3}^{-2} + \left(\frac{-1}{3}x^3 - \frac{3}{2}x^2 - 2x\right)\Bigg|_{-2}^{-1} + \left(\frac{1}{3}x^3 + \frac{3}{2}x^2 + 2x\right)\Bigg|_{-1}^{0} = \frac{5}{6} + \frac{1}{6} + \frac{5}{6} = \frac{11}{6}.$$

17. Since $\sec x \geq \tan x$ and $\sec x \geq 0$ for $-\pi/3 \leq x \leq \pi/6$, we have $f(x) \geq g(x)$ for $-\pi/3 \leq x \leq \pi/6$. Thus

$$A = \int_{-\pi/3}^{\pi/6} (\sec^2 x - \sec x \tan x)\, dx = (\tan x - \sec x)\Big|_{-\pi/3}^{\pi/6}$$

$$= \left(\frac{1}{3}\sqrt{3} - \frac{2}{3}\sqrt{3}\right) - \left(-\sqrt{3} - 2\right) = \frac{2}{3}\sqrt{3} + 2.$$

19. $g(x) - k(x) = \sin^2 x - (\sin x)/(\cos x) = (\sin x)(\sin x - \sec x)$. Since $\sin x - \sec x < 0$ for all x in $[-\pi/4, \pi/4]$, it follows that $g(x) \geq k(x)$ for $-\pi/4 \leq x \leq 0$ and $g(x) \leq k(x)$ for $0 \leq x \leq \pi/4$. Thus

$$A = \int_{-\pi/4}^{0} \left(\frac{1}{2} - \frac{1}{2}\cos 2x - \tan x\right) dx + \int_{0}^{\pi/4} -\left(\frac{1}{2} - \frac{1}{2}\cos 2x - \tan x\right) dx$$

$$= \left(\frac{x}{2} - \frac{\sin 2x}{4} + \ln|\cos x|\right)\Bigg|_{-\pi/4}^{0} + \left(-\frac{x}{2} + \frac{\sin 2x}{4} - \ln|\cos x|\right)\Bigg|_{0}^{\pi/4}$$

$$= \left(\frac{\pi}{8} - \frac{1}{4} - \ln\frac{\sqrt{2}}{2}\right) + \left(-\frac{\pi}{8} + \frac{1}{4} - \ln\frac{\sqrt{2}}{2}\right) = -2\ln\frac{\sqrt{2}}{2} = \ln 2.$$

21. $g(x) - f(x) = x^2 - x\sqrt{2x+3} = x(x - \sqrt{2x+3})$. Now if $x \geq 0$, then $x \geq \sqrt{2x+3}$ if $x^2 \geq 2x + 3$, or $x^2 - 2x - 3 \geq 0$, or $(x-3)(x+1) \geq 0$, which happens if $x \geq 3$. Thus $f(x) \geq g(x)$ for $0 \leq x \leq 3$. If $x < 0$, then $x - \sqrt{2x+3} \leq 0$, so $x(x - \sqrt{2x+3}) \geq 0$. Hence $g(x) \geq f(x)$ for $-1 \leq x \leq 0$. Thus

$$A = \int_{-1}^{0} (x^2 - x\sqrt{2x+3})\, dx + \int_{0}^{3} -(x^2 - x\sqrt{2x+3})\, dx$$

$$= \frac{1}{3}x^3\Big|_{-1}^{0} - \int_{-1}^{0} x\sqrt{2x+3}\, dx - \frac{1}{3}x^3\Big|_{0}^{3} + \int_{0}^{3} x\sqrt{2x+3}\, dx$$

$$\overset{u=2x+3}{=} \frac{1}{3} - \int_{1}^{3} \frac{1}{2}(u-3)\sqrt{u}\,\frac{1}{2}\, du - 9 + \int_{3}^{9} \frac{1}{2}(u-3)\sqrt{u}\,\frac{1}{2}\, du$$

$$= \frac{1}{3} - \frac{1}{4}\left(\frac{2}{5}u^{5/2} - 2u^{3/2}\right)\Bigg|_{1}^{3} - 9 + \frac{1}{4}\left(\frac{2}{5}u^{5/2} - 2u^{3/2}\right)\Bigg|_{3}^{9}$$

$$= \frac{1}{3} - \frac{1}{4}\left[\left(\frac{18\sqrt{3}}{5} - 6\sqrt{3}\right) - \left(\frac{2}{5} - 2\right)\right] - 9 + \frac{1}{4}\left[\left(\frac{2}{5}\cdot 243 - 54\right) - \left(\frac{18\sqrt{3}}{3} - 6\sqrt{3}\right)\right] = \frac{6}{5}\sqrt{3} + \frac{26}{15}.$$

23. The graphs intersect if $e^{2x} = e^x$, which occurs for $x = 0$. Since $2x \le x$ for $-1 \le x \le 0$ and $2x \ge x$ for $0 \le x \le 1$, we have $f(x) \le g(x)$ for $-1 \le x \le 0$, whereas $f(x) \ge g(x)$ for $0 \le x \le 1$. Thus

$$A = \int_{-1}^{0} (e^x - e^{2x})\,dx + \int_0^1 (e^{2x} - e^x)\,dx = \left(e^x - \frac{1}{2}e^{2x}\right)\Big|_{-1}^{0} + \left(\frac{1}{2}e^{2x} - e^x\right)\Big|_0^1$$

$$= \left[\left(1 - \frac{1}{2}\right) - \left(e^{-1} - \frac{1}{2}e^{-2}\right)\right] + \left[\left(\frac{1}{2}e^2 - e\right) - \left(\frac{1}{2} - 1\right)\right] = 1 - e^{-1} + \frac{1}{2}e^{-2} + \frac{1}{2}e^2 - e.$$

25. The graphs intersect at (x, y) if $x^3 = y = x^{1/3}$, or $x^9 = x$, or $x = -1$, 0, or 1. Also $f(x) \ge g(x)$ on $[-1, 0]$ and $g(x) \ge f(x)$ on $[0, 1]$. Thus

$$A = \int_{-1}^{0} (x^3 - x^{1/3})\,dx + \int_0^1 (x^{1/3} - x^3)\,dx = \left(\frac{1}{4}x^4 - \frac{3}{4}x^{4/3}\right)\Big|_{-1}^{0} + \left(\frac{3}{4}x^{4/3} - \frac{1}{4}x^4\right)\Big|_0^1 = \frac{1}{2} + \frac{1}{2} = 1.$$

27. The graphs intersect at (x, y) if $x^2 + 1 = y = 2x + 9$, or $x^2 - 2x - 8 = 0$, or $x = -2$ or $x = 4$. Also $g(x) \ge f(x)$ on $[-2, 4]$. Thus

$$A = \int_{-2}^{4} [(2x + 9) - (x^2 + 1)]\,dx = \int_{-2}^{4} (2x - x^2 + 8)\,dx = \left(x^2 - \frac{1}{3}x^3 + 8x\right)\Big|_{-2}^{4}$$

$$= \left(16 - \frac{64}{3} + 32\right) - \left(4 + \frac{8}{3} - 16\right) = 36.$$

29. The graphs intersect at (x, y) if $x^3 + 1 = y = (x + 1)^2$, or $x^3 - x^2 - 2x = 0$, or $x(x + 1)(x - 2) = 0$, or $x = -1$, 0, or 2. Also$f(x) \ge g(x)$ on $[-1, 0]$ and $g(x) \ge f(x)$ on $[0, 2]$. Thus

$$A = \int_{-1}^{0} [(x^3+1) - (x+1)^2]\,dx + \int_0^2 [(x+1)^2 - (x^3+1)]\,dx = \int_{-1}^{0} (x^3 - x^2 - 2x)\,dx + \int_0^2 (-x^3 + x^2 + 2x)\,dx$$

$$= \left(\frac{1}{4}x^4 - \frac{1}{3}x^3 - x^2\right)\Big|_{-1}^{0} + \left(-\frac{1}{4}x^4 + \frac{1}{3}x^3 + x^2\right)\Big|_0^2 = -\left(\frac{1}{4} + \frac{1}{3} - 1\right) + \left(-4 + \frac{8}{3} + 4\right) = \frac{37}{12}.$$

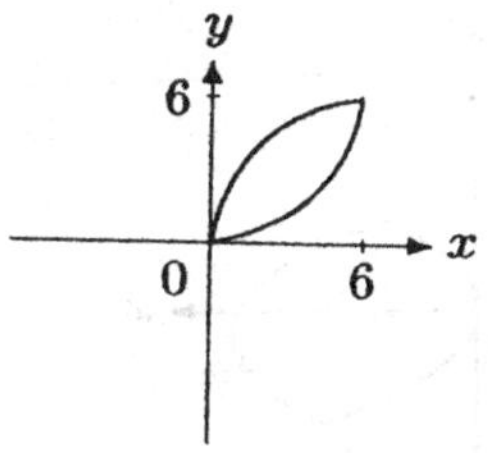

Exercise 31

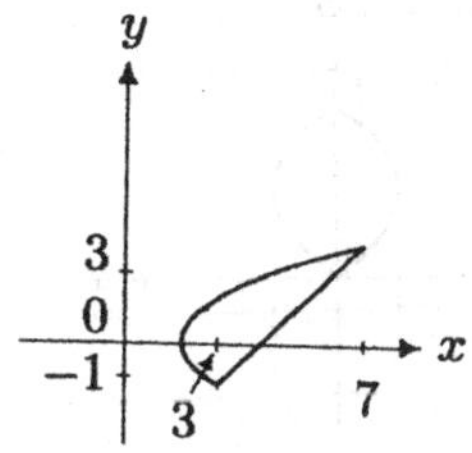

Exercise 33

31. The graphs intersect at (x, y) if $6x = y^2 = (x^2/6)^2$, or $x^4 = 216x$, or $x = 0$ or 6. Thus the graphs intersect at $(0,0)$ and $(6,6)$, so

$$A = \int_0^6 \left(\sqrt{6x} - \frac{1}{6}x^2\right) dx = \left(\frac{2\sqrt{6}}{3}x^{3/2} - \frac{1}{18}x^3\right)\Bigg|_0^6 = \left(\frac{2\sqrt{6}}{3}\right)(6\sqrt{6}) - 12 = 12.$$

33. The graphs intersect at (x, y) if $\frac{1}{2}(y^2 + 5) = x = y + 4$, or $y^2 - 2y - 3 = 0$, or $y = -1$ or 3. Thus the graphs intersect at $(3, -1)$ and $(7, 3)$. Thus

$$A = \int_{-1}^3 \left[(y+4) - \frac{1}{2}(y^2+5)\right] dy = \int_{-1}^3 \left(y - \frac{1}{2}y^2 + \frac{3}{2}\right) dy$$

$$= \left(\frac{1}{2}y^2 - \frac{1}{6}y^3 + \frac{3}{2}y\right)\Bigg|_{-1}^3 = \left(\frac{9}{2} - \frac{9}{2} + \frac{9}{2}\right) - \left(\frac{1}{2} + \frac{1}{6} - \frac{3}{2}\right) = \frac{16}{3}.$$

35. The graphs of $y = x + 2$ and $y = \frac{1}{3}(2 - x)$ intersect at $(-1, 1)$; the graphs of $y = x + 2$ and $y = -3x + 6$ intersect at $(1, 3)$; the graphs of $y = \frac{1}{3}(2 - x)$ and $y = -3x + 6$ intersect at $(2, 0)$. We obtain

$$A = \int_{-1}^1 \left[(x+2) - \frac{1}{3}(2-x)\right] dx + \int_1^2 \left[(-3x+6) - \frac{1}{3}(2-x)\right] dx$$

$$= \int_{-1}^1 \left(\frac{4}{3}x + \frac{4}{3}\right) dx + \int_1^2 \left(\frac{16}{3} - \frac{8}{3}x\right) dx$$

$$= \frac{1}{3}(2x^2 + 4x)\Bigg|_{-1}^1 + \frac{1}{3}(16x - 4x^2)\Bigg|_1^2 = \frac{8}{3} + \frac{4}{3} = 4.$$

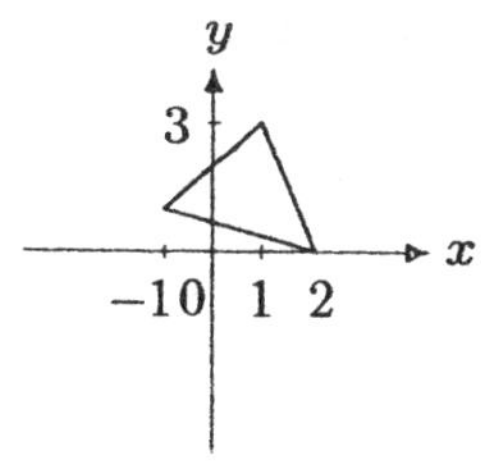

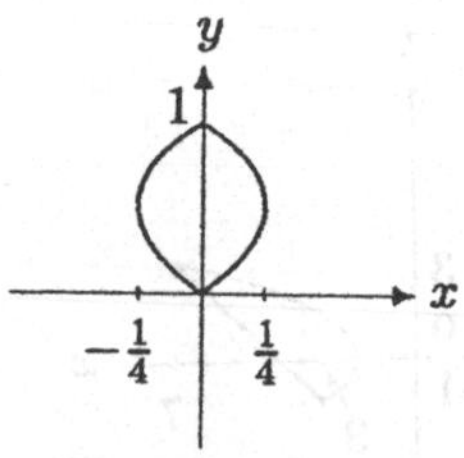

Exercise 37

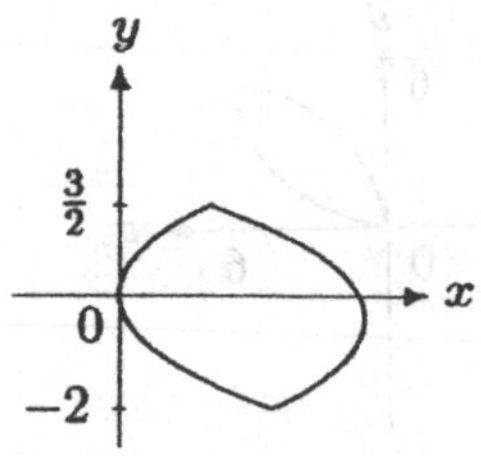

Exercise 39

37. The graphs intersect at (x, y) if $y^2 - y = x = y - y^2$, or $2(y^2 - y) = 0$, or $y = 0$ or 1. Thus

$$A = \int_0^1 [(y - y^2) - (y^2 - y)]\, dy = \int_0^1 2(y - y^2)\, dy = \left(y^2 - \frac{2}{3}y^3\right)\Big|_0^1 = \frac{1}{3}.$$

39. The graphs intersect at (x, y) if $y^2 = x = 6 - y - y^2$, or $2y^2 + y - 6 = 0$, or $(2y - 3)(y + 2) = 0$, or $y = -2$, or $\frac{3}{2}$. Thus

$$A = \int_{-2}^{3/2} [(6 - y - y^2) - y^2]\, dy = \int_{-2}^{3/2} (6 - y - 2y^2)\, dy = \left(6y - \frac{1}{2}y^2 - \frac{2}{3}y^3\right)\Big|_{-2}^{3/2}$$

$$= \left(9 - \frac{9}{8} - \frac{9}{4}\right) - \left(-12 - 2 + \frac{16}{3}\right) = \frac{343}{24}.$$

41. Let A_R denote the area of R_a, so that $A_R = \frac{1}{2}(\text{base})(\text{height}) = \frac{1}{2}(2a)(a^2) = a^3$. Next, let A_S denote the area of S_a. Since the line joining $(0, 0)$ and (a, a^2) has equation $y = ax$, we find that

$$A_S = 2\int_0^a (ax - x^2)\, dx = 2\left(\frac{1}{2}ax^2 - \frac{1}{3}x^3\right)\Big|_0^a = 2\left(\frac{1}{2}a^3 - \frac{1}{3}a^3\right) = \frac{1}{3}a^3.$$

Thus $r_a = A_R/A_S = a^3/(a^3/3) = 3$ so that r_a is independent of a.

43. The profit is given by

$$\int_0^4 [(\sqrt{t} + 3) - (t^{1/3} + 2)]\, dt = \int_0^4 (\sqrt{t} - t^{1/3} + 1)\, dt = \left(\frac{2}{3}t^{3/2} - \frac{3}{4}t^{4/3} + t\right)\Big|_0^4$$

$$= \frac{2}{3}(8) - \frac{3}{4}(4)^{4/3} + 4 = \frac{28}{3} - 3(4)^{1/3} \approx 4.57113.$$

Thus it is not possible to make a profit of \$5000 during the four-month period.

Chapter 5 Review

1. left sum:

$$f(0)\left(\frac{\pi}{3}\right)+f\left(\frac{\pi}{3}\right)\left(\frac{\pi}{3}\right)+f\left(\frac{2\pi}{3}\right)\left(\frac{\pi}{3}\right)+f(\pi)\left(\frac{\pi}{2}\right)$$

$$=0\left(\frac{\pi}{3}\right)+\left(\frac{\pi}{3}\frac{\sqrt{3}}{2}\right)\left(\frac{\pi}{3}\right)+\left(\frac{2\pi}{3}\frac{\sqrt{3}}{2}\right)\left(\frac{\pi}{3}\right)+0\left(\frac{\pi}{2}\right)=\frac{\pi^2\sqrt{3}}{6}$$

right sum:

$$f\left(\frac{\pi}{3}\right)\left(\frac{\pi}{3}\right)+f\left(\frac{2\pi}{3}\right)\left(\frac{\pi}{3}\right)+f(\pi)\left(\frac{\pi}{3}\right)+f\left(\frac{3\pi}{2}\right)\left(\frac{\pi}{2}\right)$$

$$=\left(\frac{\pi}{3}\frac{\sqrt{3}}{2}\right)\left(\frac{\pi}{3}\right)+\left(\frac{2\pi}{3}\frac{\sqrt{3}}{2}\right)\left(\frac{\pi}{3}\right)+0\left(\frac{\pi}{3}\right)+\left(-\frac{3\pi}{2}\right)\left(\frac{\pi}{2}\right)=\frac{\pi^2\sqrt{3}}{6}-\frac{3\pi^2}{4}$$

midpoint sum:

$$f\left(\frac{\pi}{6}\right)\left(\frac{\pi}{3}\right)+f\left(\frac{\pi}{2}\right)\left(\frac{\pi}{3}\right)+f\left(\frac{5\pi}{6}\right)\left(\frac{\pi}{3}\right)+f\left(\frac{5\pi}{4}\right)\left(\frac{\pi}{2}\right)$$

$$=\left(\frac{\pi}{6}\frac{1}{2}\right)\left(\frac{\pi}{3}\right)+\left(\frac{\pi}{2}\right)\left(\frac{\pi}{3}\right)+\left(\frac{5\pi}{6}\frac{1}{2}\right)\left(\frac{\pi}{3}\right)+\left(-\frac{5\pi}{4}\frac{\sqrt{2}}{2}\right)\left(\frac{\pi}{2}\right)=\frac{\pi^2}{3}-\frac{5\sqrt{2}\pi^2}{16}$$

3. $\displaystyle\int\left(x^{3/5}-8x^{5/3}\right)dx=\frac{5}{8}x^{8/5}-8\left(\frac{3}{8}x^{8/3}\right)+C=\frac{5}{8}x^{8/5}-3x^{8/3}+C$

5. $\displaystyle\int\left(x^3-3x+2-\frac{2}{x}\right)dx=\frac{1}{4}x^4-\frac{3}{2}x^2+2x-2\ln|x|+C$

7. Let $u=1+\sqrt{x+1}$, so that $du=1/(2\sqrt{x+1})\,dx$. Then

$$\int\frac{1+\sqrt{x+1}}{\sqrt{x+1}}\,dx=\int u(2)\,du=2\int u\,du=2\left(\frac{1}{2}u^2\right)+C=(1+\sqrt{x+1})^2+C.$$

9. Let $u=\cos 3t$, so that $du=-3\sin 3t\,dt$. Then

$$\int\cos^3 3t\,\sin 3t\,dt=\int u^3\left(-\frac{1}{3}\right)du=-\frac{1}{3}\int u^3\,du=-\frac{1}{3}\left(\frac{1}{4}u^4\right)+C=-\frac{1}{12}\cos^4 3t+C.$$

11. Let $u=1+\sqrt{x}$, so that $du=1/(2\sqrt{x})\,dx$ and $2\sqrt{x}=2(u-1)$. Then

$$\int\sqrt{1+\sqrt{x}}\,dx=\int\sqrt{1+\sqrt{x}}\,2\sqrt{x}\left(\frac{1}{2\sqrt{x}}\right)dx=\int\sqrt{u}\,[2(u-1)]\,du=\int 2(u^{3/2}-u^{1/2})\,du$$

$$=2\left(\frac{2}{5}u^{5/2}-\frac{2}{3}u^{3/2}\right)+C=\frac{4}{5}(1+\sqrt{x})^{5/2}-\frac{4}{3}(1+\sqrt{x})^{3/2}+C.$$

13. $\displaystyle\int_{-1}^{-2}\left(\frac{5}{3}x^{2/3}-\frac{4}{x^3}\right)dx=\left(x^{5/3}+\frac{2}{x^2}\right)\Big|_{-1}^{-2}=\left[(-2)^{5/3}+\frac{1}{2}\right]-(-1+2)$

$\displaystyle=-\frac{1}{2}-2^{5/3}$

15. Let $u=x^3+9x+1$, so that $du=(3x^2+9)\,dx=3(x^2+3)\,dx$. If $x=0$, then $u=1$; if $x=2$, then $u=27$. Thus

$$\int_0^2(x^2+3)(x^3+9x+1)^{1/3}\,dx=\int_1^{27}u^{1/3}\frac{1}{3}\,du=\frac{1}{3}\int_1^{27}u^{1/3}\,du=\frac{1}{3}\left(\frac{3}{4}u^{4/3}\right)\Big|_1^{27}=\frac{1}{4}(81-1)=20.$$

17. $\displaystyle\int_{-8}^{-2}\frac{-1}{5u}\,du=-\frac{1}{5}\ln|u|\Big|_{-8}^{-2}=-\frac{1}{5}\ln 2+\frac{1}{5}\ln 8=-\frac{1}{5}\ln 2+\frac{1}{5}\ln 2^3=-\frac{1}{5}\ln 2+\frac{3}{5}\ln 2=\frac{2}{5}\ln 2$

19. Let $u=x+e^x$, so that $du=(1+e^x)\,dx$. If $x=0$, then $u=1$; if $x=1$, then $u=1+e$. Thus

$$\int_0^1\frac{1+e^x}{x+e^x}\,dx=\int_1^{1+e}\frac{1}{u}\,du=\ln u\Big|_1^{1+e}=\ln(1+e)-\ln 1=\ln(1+e).$$

21. Let $u=1+\sin t$, so that $du=\cos t\,dt$. If $t=-\pi/4$ then $u=1-\sqrt{2}/2$, and if $t=\pi/2$ then $u=2$. Thus

$$\int_{-\pi/4}^{\pi/2}\frac{\cos t}{1+\sin t}\,dt=\int_{1-\sqrt{2}/2}^{2}\frac{1}{u}\,du=\ln u\Big|_{1-\sqrt{2}/2}^{2}=\ln 2-\ln\left(\frac{2-\sqrt{2}}{2}\right)=2\ln 2-\ln(2-\sqrt{2}).$$

23. Let $u=x/(x+1)$, so that

$$du=\frac{(x+1)-x}{(x+1)^2}\,dx=\frac{1}{(x+1)^2}\,dx.$$

If $x=\frac{1}{26}$ then $u=\frac{1}{27}$, and if $x=\frac{1}{7}$ then $u=\frac{1}{8}$. Thus

$$\int_{1/26}^{1/7}\frac{1}{x^2}\left(\frac{x+1}{x}\right)^{1/3}dx=\int_{1/26}^{1/7}\left(\frac{x+1}{x}\right)^2\left(\frac{x+1}{x}\right)^{1/3}\frac{1}{(x+1)^2}\,dx=\int_{1/27}^{1/8}u^{-7/3}\,du$$

$$=-\frac{3}{4}u^{-4/3}\Big|_{1/27}^{1/8}=-\frac{3}{4}(16-81)=\frac{195}{4}.$$

25. $\displaystyle A=\int_2^4\left(\frac{7}{4}x^2\sqrt{x}+\frac{1}{\sqrt{x}}\right)dx=\int_2^4\left(\frac{7}{4}x^{5/2}+x^{-1/2}\right)dx=\left(\frac{1}{2}x^{7/2}+2x^{1/2}\right)\Big|_2^4$

$=(64+4)-(4\sqrt{2}+2\sqrt{2})=68-6\sqrt{2}$

27. Notice that

$$\frac{e^x - e^{2x}}{1+e^x} = \frac{e^x(1-e^x)}{1+e^x}.$$

Let $u = 1 + e^x$, so that $du = e^x\,dx$ and $1 - e^x = 2 - (1+e^x) = 2 - u$. If $x = 1$, then $u = 1 + e$; if $x = \ln 3$, then $u = 1 + e^{\ln 3} = 1 + 3 = 4$. Finally, $(e^x - e^{2x})/(1+e^x) < 0$ on $[1, \ln 3]$. Thus

$$A = \int_1^{\ln 3} -\frac{e^x - e^{2x}}{1+e^x}\,dx = \int_1^{\ln 3} -\frac{e^x(1-e^x)}{1+e^x}\,dx = \int_{1+e}^{4} -\frac{2-u}{u}\,du = \int_{1+e}^{4}\left(1 - \frac{2}{u}\right)du$$

$$= (u - 2\ln u)\Big|_{1+e}^{4} = (4 - 2\ln 4) - \big(1 + e - 2\ln(1+e)\big) = 3 - e + \ln\frac{(1+e)^2}{16} \approx 0.1356528243.$$

29. The graphs intersect at (x, y) if $2x^5 + 5x^4 = y = 2x^5 + 20x^2$, or $5x^4 - 20x^2 = 0$, or $x = -2$, 0, or 2. Also $g(x) \geq f(x)$ on $[-2, 2]$. Thus

$$A = \int_{-2}^{2} [(2x^5 + 20x^2) - (2x^5 + 5x^4)]\,dx = \int_{-2}^{2} (20x^2 - 5x^4)\,dx = \left(\frac{20}{3}x^3 - x^5\right)\Big|_{-2}^{2}$$

$$= \left(\frac{160}{3} - 32\right) - \left(-\frac{160}{3} + 32\right) = \frac{128}{3}.$$

31. The graphs intersect at (x, y) if $2y^3 + y^2 + 5y - 7 = x = y^3 + 4y^2 + 3y - 7$, or $y^3 - 3y^2 + 2y = 0$, or $y = 0$, 1, or 2. Also $2y^3 + y^2 + 5y - 7 \geq y^3 + 4y^2 + 3y - 7$ on $[0, 1]$, and $y^3 + 4y^2 + 3y - 7 \geq 2y^3 + y^2 + 5y - 7$ on $[1, 2]$. Thus

$$A = \int_0^1 [(2y^3 + y^2 + 5y - 7) - (y^3 + 4y^2 + 3y - 7)]\,dy + \int_1^2 [(y^3 + 4y^2 + 3y - 7) - (2y^3 + y^2 + 5y - 7)]\,dy$$

$$= \int_0^1 (y^3 - 3y^2 + 2y)\,dy + \int_1^2 (-y^3 + 3y^2 - 2y)\,dy$$

$$= \left(\frac{1}{4}y^4 - y^3 + y^2\right)\Big|_0^1 + \left(-\frac{1}{4}y^4 + y^3 - y^2\right)\Big|_1^2$$

$$= \left(\frac{1}{4} - 1 + 1\right) + \left[(-4 + 8 - 4) - \left(-\frac{1}{4} + 1 + 1\right)\right] = \frac{1}{2}.$$

33. The domain is the union of $(-\infty, -2)$ and $(2, \infty)$.

$$f'(x) = \frac{2x}{x^2-4}; \quad f''(x) = \frac{-8-2x^2}{(x^2-4)^2}$$

decreasing on $(-\infty, -2)$ and increasing on $(2, \infty)$; concave downward on $(-\infty, -2)$ and $(2, \infty)$; vertical asymptotes are $x = -2$ and $x = 2$; symmetric with respect to the y axis.

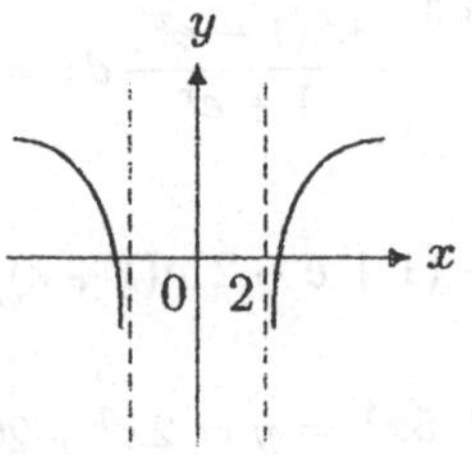

35. $F'(x) = x\sqrt{1+x^5}$

37. Let $G(x) = \int_1^x (1/t)\,dt$, so that $F(x) = G(\ln x)$. Since $G'(x) = 1/x$, the Chain Rule implies that

$$F'(x) = [G'(\ln x)]\frac{1}{x} = \frac{1}{\ln x} \cdot \frac{1}{x} = \frac{1}{x \ln x}.$$

39. $f'(x) = \dfrac{1}{x} + \dfrac{1}{1/x}\left(\dfrac{-1}{x^2}\right) = \dfrac{1}{x} - \dfrac{1}{x} = 0.$
Alternatively, $f(x) = \ln x - \ln x = 0$, so $f'(x) = 0$.

41. Since

$$\begin{aligned}\ln|f(x)| &= \ln(4-\cos x)^{1/3} + \ln|2x-5|^{1/2} - \ln|x+5|^{1/3} \\ &= \frac{1}{3}\ln(4-\cos x) + \frac{1}{2}\ln|2x-5| - \frac{1}{3}\ln|x+5|\end{aligned}$$

we have

$$\frac{d}{dx}\ln|f(x)| = \frac{1}{3} \cdot \frac{1}{4-\cos x} \cdot \sin x + \frac{1}{2} \cdot \frac{1}{2x-5} \cdot 2 - \frac{1}{3} \cdot \frac{1}{x+5}.$$

Thus (18) of Section 5.7 yields

$$f'(x) = \frac{(4-\cos x)^{1/3}\sqrt{2x-5}}{\sqrt[3]{x+5}}\left(\frac{\sin x}{12-3\cos x} + \frac{1}{2x-5} - \frac{1}{3x+15}\right).$$

43. The integrals in (a) and (c) cannot be easily evaluated by substitution. For the integral in (b) let $u = x^2 + 6$, so that $du = 2x\,dx$. Then

$$\int x\sqrt{x^2+6}\,dx = \int \sqrt{u}\,\frac{1}{2}\,du = \frac{1}{3}u^{3/2} + C = \frac{1}{3}(x^2+6)^{3/2} + C.$$

45. The integrals in (a) and (b) cannot be easily evaluated by substitution. For the integral in (c) let $u = \ln(x+1)$, so that $du = [1/(x+1)]\,dx$. Then

$$\int \frac{\ln(x+1)}{x+1}\,dx = \int u\,du = \frac{1}{2}u^2 + C = \frac{1}{2}[\ln(x+1)]^2 + C.$$

47. By the General Comparison Property,

$$\frac{26}{3} = \frac{1}{3}x^3\Big|_1^3 = \int_1^3 \sqrt{x^4}\,dx \le \int_1^3 \sqrt{1+x^4}\,dx \le \int_1^3 \sqrt{2x^4}\,dx = \frac{\sqrt{2}}{3}x^3\Big|_1^3 = \frac{26}{3}\sqrt{2}.$$

49. a. If $1 \le t \le x$, then $1/t \le 1$, so the General Comparison Property implies that $\int_1^x (1/t)\,dt \le \int_1^x 1\,dt$.

b. Since $\int_1^x (1/t)\,dt = \ln x$ and $\ln x \ge 0$ for $x \ge 1$, and since $\int_1^x 1\,dt = x - 1$, part (a) implies that $0 \le \ln x \le x - 1$.

c. From (b) we find that if $x \ge 1$, then $0 \le x \ln x \le x(x-1) = x^2 - x$. Thus $0 \le \int_1^2 x \ln x\,dx \le \int_1^2 (x^2 - x)\,dx = \left(\frac{1}{3}x^3 - \frac{1}{2}x^2\right)\Big|_1^2 = \frac{5}{6}$.

51. Let $f(x) = rx + s$ for $a \le x \le b$, and let $P = \{x_0, x_1, \ldots, x_n\}$ be a partition of $[a, b]$. Then

$$\begin{aligned}
\frac{1}{2}(\text{left sum} + \text{right sum}) &= \frac{1}{2}\{[f(x_0)\Delta x_1 + f(x_1)\Delta x_2 + \cdots + f(x_{n-1})\Delta x_n] \\
&\quad + [f(x_1)\Delta x_1 + f(x_2)\Delta x_2 + \cdots + f(x_n)\Delta x_n]\} \\
&= \frac{f(x_0) + f(x_1)}{2}\Delta x_1 + \cdots + \frac{f(x_{n-1}) + f(x_n)}{2}\Delta x_n \\
&= \frac{(rx_0 + s) + (rx_1 + s)}{2}\Delta x_1 + \cdots + \frac{(rx_{n-1} + s) + (rx_n + s)}{2}\Delta x_n \\
&= \left[r\left(\frac{x_0 + x_1}{2}\right) + s\right]\Delta x_1 + \cdots + \left[r\left(\frac{x_{n-1} + x_n}{2}\right) + s\right]\Delta x_n \\
&= f\left(\frac{x_0 + x_1}{2}\right)\Delta x_1 + \cdots + f\left(\frac{x_{n-1} + x_n}{2}\right)\Delta x_n \\
&= \text{midpoint sum.}
\end{aligned}$$

53. a. Since $G'(x) = f(x)$ by Theorem 5.12, positivity of f on I would imply that $G'(x) > 0$ for x in I and hence that f is increasing on I.

b. b is a critical number of G if $G'(b) = 0$. Now $G'(b) = f(b)$ by Theorem 5.12. Therefore b is a critical number of G provided that $f(b) = 0$.

c. By Theorem 4.13, if $G'' > 0$ on I, then the graph of G is concave upward. If f' exists on I, then $G'' = f'$, so that the graph of G is concave upward if $f' > 0$, that is, if f is increasing on I.

55. $a(t) = -4$, so that $v(t) = v(0) + \int_0^t a(s)\,ds = 44 + \int_0^t -4\,ds = 44 - 4t$, so that $v(t) = 0$ if $t = 11$. Let $f(t)$ denote the position of the car at time t. Then the distance traveled by the car before coming to a stop is $f(11) - f(0) = \int_0^{11} v(t)\,dt = \int_0^{11}(44 - 4t)\,dt = (44t - 2t^2)\Big|_0^{11} = 484 - 242 = 242$ (feet).

57. Let H denote the average rate of heat production, so

$$H = \frac{\int_0^{1/60}(110\sin 120\pi t)^2 R\,dt}{1/60} = 60(110)^2 R \int_0^{1/60} \sin^2 120\pi t\,dt.$$

Let $u = 120\pi t$, so that $du = 120\pi\,dt$. Then

$$H = \frac{60(110)^2 R}{120\pi}\int_0^{2\pi} \sin^2 u\,du = \frac{6050}{\pi} R\left(\frac{1}{2}u - \frac{1}{4}\sin 2u\right)\Bigg|_0^{2\pi} = 6050R.$$

59. a. Since $dv/dm = -u_0/m$, integration yields $v = -u_0 \ln m + C$. By hypothesis, at $t = 0$ we have $v = v_0$ and $m = m_0$. Therefore $v_0 = -u_0 \ln m_0 + C$, so $C = v_0 + u_0 \ln m_0$. Consequently $v = -u_0 \ln m + (v_0 + u_0 \ln m_0) = v_0 + u_0 \ln(m_0/m)$.

b. By hypothesis, $u_0 = 3 \times 10^3$ and $v - v_0 = 3 \times 10^7$, so the equation $v = v_0 + u_0 \ln(m_0/m)$ becomes $3 \times 10^7 = (3 \times 10^3)\ln(m_0/m)$, or $10^4 = \ln(m_0/m)$. Thus $m_0/m = e^{(10^4)}$, so that $m/m_0 = e^{-(10^4)} \approx 0$.

61. $$\int_{20}^{50} H\,dt = \int_{20}^{50} 1945\left(\frac{T}{474}\right)^3 dt = \frac{1945}{(474)^3}\int_{20}^{50} T^3\,dt = \frac{1945}{4(474)^3}T^4\Bigg|_{20}^{50}$$
$$= \frac{1945}{4(474)^3}(50^4 - 20^4) \approx 27.80621535$$

Thus it takes approximately 28 joules per mole.

Cumulative Review(Chapters 1–4)

1. Since $2 + \sin x > 0$ for all x, the given inequality is equivalent to

$$-\frac{1}{(x-3)^2}\left(\frac{2-x}{4-x}\right)^{-1} < 0, \quad \text{or} \quad \frac{x-4}{(x-3)^2(2-x)} < 0.$$

From the diagram we see that the solution is the union of $(-\infty, 2)$ and $(4, \infty)$.

$x-4$	− − − − − − − − − 0 + +
$2-x$	+ + 0 − − − − − − − − −
$(x-3)^2$	+ + + + + 0 + + + + +
$\frac{x-4}{(x-3)^2(2-x)}$	− − + + + + 0 − −

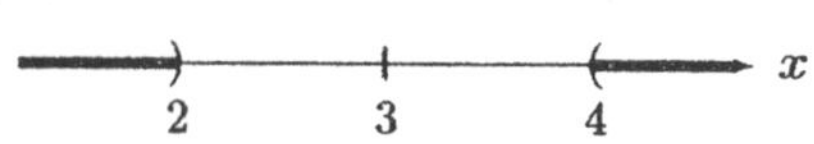

3. $$\lim_{x\to\pi/2^-}\frac{\cos x}{\sin x - 1} = \lim_{x\to\pi/2^-}\left(\frac{\cos x}{\sin x - 1}\,\frac{\sin x + 1}{\sin x + 1}\right) = \lim_{x\to\pi/2^-}\frac{\cos x\,(\sin x + 1)}{\sin^2 x - 1}$$
$$= \lim_{x\to\pi/2^-}\frac{\cos x\,(\sin x + 1)}{-\cos^2 x} = \lim_{x\to\pi/2^-}\frac{\sin x + 1}{-\cos x} = -\infty$$

5. Let $f(x) = \tan x$, so that $f'(x) = \sec^2 x$. Then
$$\lim_{x\to\pi/4}\frac{\tan x - 1}{x - \pi/4} = \lim_{x\to\pi/4}\frac{f(x) - f(\pi/4)}{x - \pi/4} = f'\left(\frac{\pi}{4}\right) = \sec^2\left(\frac{\pi}{4}\right) = 2.$$

7. $f'(x) = e^{1/(x^2+1)}\left[\dfrac{-2x}{(x^2+1)^2}\right]$

9. $f'(x) = 6x^2 - 18x + 14 = 6\left(x - \frac{3}{2}\right)^2 + \frac{1}{2} > 0$ for all x, so f is increasing.

11. Since $f'(x) = c + 3/x^4$, the equations $f(x) = 0$ and $f'(x) = 0$ become $cx - 1/x^3 - \frac{1}{2} = 0$ and $c + 3/x^4 = 0$. From the second of these we conclude that $cx = -3/x^3$, so the first becomes $-3/x^3 - 1/x^3 - \frac{1}{2} = 0$, or $x^3 = -8$, or $x = -2$. Since $cx = -3/x^3$, this implies that $c = -3/(-2)^4 = -\frac{3}{16}$. Thus if $c = -\frac{3}{16}$, then the equations $f(x) = 0$ and $f'(x) = 0$ have the same root, namely -2.

13.

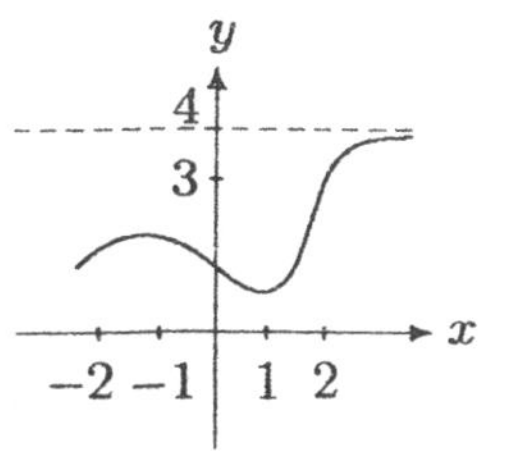

15. $f'(x) = \dfrac{1}{x^2} - \dfrac{2}{x^3}$; $f''(x) = -\dfrac{2}{x^3} + \dfrac{6}{x^4}$

relative minimum value is $f(2) = \frac{3}{4}$; increasing on $(-\infty, 0)$ and $[2, \infty)$, and decreasing on $(0, 2]$; concave upward on $(-\infty, 0)$ and $(0, 3)$, and concave downward on $(3, \infty)$; inflection point is $(3, \frac{7}{9})$; vertical asymptote is $x = 0$; horizontal asymptote is $y = 1$.

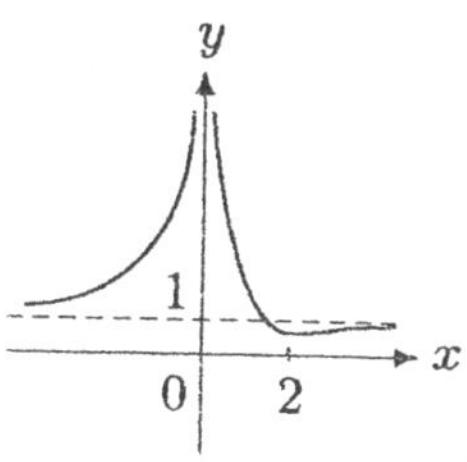

17. The height of the pot t seconds after it is dropped is given by $h(t) = -16t^2 + h_0$. Let t^* be such that $h(t^*) = 128$. Then the velocity at $t = t^*$ is -32, so $-32t^* = -32$, or $t^* = 1$. Therefore $128 = h(1) = -16(1)^2 + h_0$, so $h_0 = 128 + 16 = 144$. Consequently $h(t) = -16t^2 + 144$.

a. One second later $t = 2$, and $v(2) = -32(2) = -64$ (feet per second).

b. The pot hits the ground at t such that $h(t) = 0$. Now $0 = h(t) = -16t^2 + 144$, so that $t = 3$. Since the pot is 128 feet above the ground when $t = 1$, it takes 2 additional seconds for the pot to fall the last 128 feet to the ground.

19. From the given information it follows that if the toll is set at $3+x$ dollars, then the number of cars using the toll road per day would be $24{,}000-(20x)300 = 24{,}000-6000x$. Thus the total revenue R would be given by $R = (24{,}000-6000x)(x+3) = -6000x^2+6000x+72{,}000$. Then $R'(x) = -12{,}000x+6000 = 0$ for $x = \frac{1}{2}$. Since $R''(x) = -12{,}000 < 0$, the revenue is maximized for $x = \frac{1}{2}$, which means that the toll would be \$3.50.

21. We need to find $d\theta/dt$ at the moment that $\varphi = \pi/4$. By Snell's Law, $\sin\theta = \mu\sin\varphi$, so that $(\cos\theta)(d\theta/dt) = (\mu\cos\varphi)(d\varphi/dt)$. Now when $\varphi = \pi/4$, Snell's Law with $\mu = 1.33$ yields $\sin\theta = 1.33\sin(\pi/4) = 1.33\sqrt{2}/2$. Letting $\varphi = \pi/4$, $\mu = 1.33$ and $d\varphi/dt = -\frac{1}{10}$, we obtain

$$\frac{d\theta}{dt} = \frac{\mu\cos\varphi}{\cos\theta}\frac{d\varphi}{dt} = \frac{1.33\cos(\pi/4)}{\sqrt{1-\sin^2\theta}}\left(-\frac{1}{10}\right) = \frac{1.33\sqrt{2}/2}{\sqrt{1-(1.33\sqrt{2}/2)^2}}\left(-\frac{1}{10}\right) \approx -0.2766633735.$$

Since 0.2766633735 radians is approximately 15.9 degrees, the angle of incidence is changing at approximately 15.9 degrees per second.

Chapter 6

Applications of the Integral

6.1 Volume

1. $V = \int_0^1 \pi (x^{3/2})^2\,dx = \pi \int_0^1 x^3\,dx = \frac{\pi}{4}x^4\Big|_0^1 = \frac{\pi}{4}$

3. $V = \int_0^{\pi/6} \pi(\sqrt{\cos x})^2\,dx = \pi \int_0^{\pi/6} \cos x\,dx = \pi \sin x\,\big|_0^{\pi/6} = \frac{\pi}{2}$

5. $V = \int_{-\pi/4}^0 \pi \sec^2 x\,dx = \pi \int_{-\pi/4}^0 \sec^2 x\,dx = \pi \tan x\,\big|_{-\pi/4}^0 = \pi$

7. $V = \int_1^2 \pi[x(x^3+1)^{1/4}]^2\,dx = \pi \int_1^2 x^2(x^3+1)^{1/2}\,dx \overset{u=x^3+1}{=} \pi \int_2^9 u^{1/2}\cdot\frac{1}{3}\,du$

$$= \frac{\pi}{3}\int_2^9 u^{1/2}\,du = \frac{\pi}{3}\frac{2}{3}u^{3/2}\Big|_2^9 = \frac{2\pi}{9}(27 - 2\sqrt{2})$$

9. $V = \int_1^2 \pi(\sqrt{1+y^3})^2\,dy = \pi \int_1^2 (1+y^3)\,dy = \pi(y + \frac{1}{4}y^4)\Big|_1^2 = \frac{19}{4}\pi$

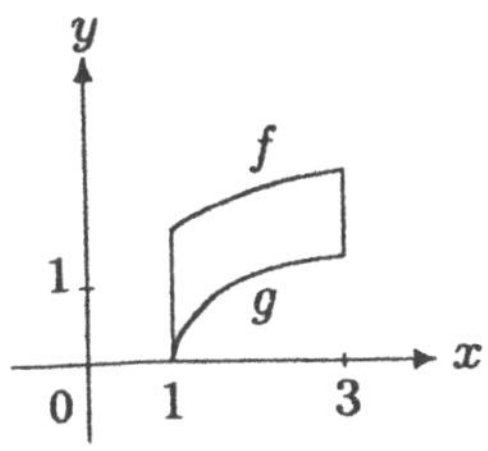

Exercise 11

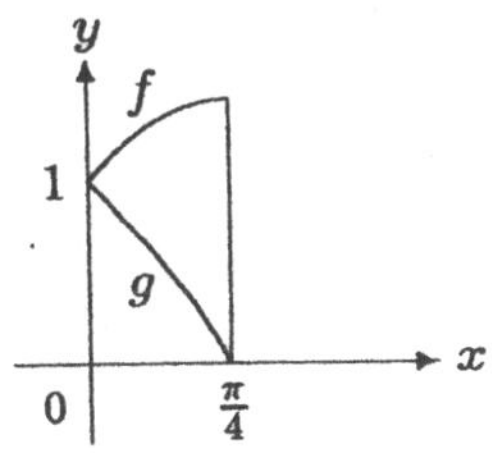

Exercise 13

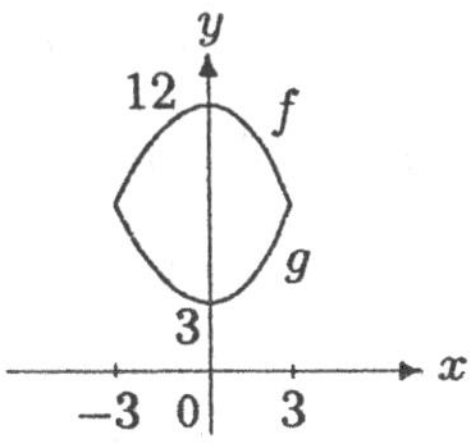

Exercise 15

11. $V = \int_1^3 \pi\left[(\sqrt{x+1})^2 - (\sqrt{x-1})^2\right]\,dx = \pi \int_1^3 2\,dx = 2\pi x\big|_1^3 = 4\pi$

13. $V = \int_0^{\pi/4} \pi[(\cos x + \sin x)^2 - (\cos x - \sin x)^2]\,dx$

$$= \pi \int_0^{\pi/4} (4\cos x\,\sin x)\,dx \overset{u=\sin x}{=} \pi \int_0^{\sqrt{2}/2} 4u\,du = 2\pi u^2\big|_0^{\sqrt{2}/2} = \pi$$

15. The graphs of $y = x^2/2 + 3$ and $y = 12 - x^2/2$ intersect for (x, y) such that $x^2/2 + 3 = y = 12 - x^2/2$, or $x^2 = 9$, so that $x = -3$ or $x = 3$. Since $x^2/2 + 3 \le 12 - x^2/2$ for $-3 \le x \le 3$, it follows that

$$V = \int_{-3}^3 \pi\left[\left(12 - \frac{x^2}{2}\right)^2 - \left(\frac{x^2}{2} + 3\right)^2\right]\,dx = \pi \int_{-3}^3 (135 - 15x^2)\,dx = \pi(135x - 5x^3)\big|_{-3}^3 = 540\pi.$$

17. The graphs of $y = 5x$ and $y = x^2 + 2x + 2$ intersect for (x, y) such that $5x = y = x^2 + 2x + 2$, or $x^2 - 3x + 2 = 0$, so that $x = 1$ or $x = 2$. Since $5x \geq x^2 + 2x + 2$ for $1 \leq x \leq 2$, it follows that

$$V = \int_1^2 \pi[(5x)^2 - (x^2+2x+2)^2]\,dx = \pi \int_1^2 (-x^4 - 4x^3 + 17x^2 - 8x - 4)\,dx$$

$$= \pi \left(-\frac{1}{5}x^5 - x^4 + \frac{17}{3}x^3 - 4x^2 - 4x\right)\Big|_1^2$$

$$= \pi \left[\left(-\frac{32}{5} - 16 + \frac{136}{3} - 16 - 8\right) - \left(-\frac{1}{5} - 1 + \frac{17}{3} - 4 - 4\right)\right] = \frac{37}{15}\pi.$$

19. $V = \int_0^2 2\pi x(x-1)^2\,dx = 2\pi \int_0^2 (x^3 - 2x^2 + x)\,dx = 2\pi \left(\frac{1}{4}x^4 - \frac{2}{3}x^3 + \frac{1}{2}x^2\right)\Big|_0^2$
$= 2\pi(4 - \frac{16}{3} + 2) = \frac{4}{3}\pi$

21. $V = \int_0^{\sqrt{3}/2} 2\pi x \frac{1}{\sqrt{1-x^2}}\,dx \overset{u=1-x^2}{=} \pi \int_1^{1/4} \frac{1}{u^{1/2}}(-1)\,du = -2\pi u^{1/2}\Big|_1^{1/4}$
$= -2\pi(\frac{1}{2} - 1) = \pi$

23. $V = \int_1^2 2\pi y \frac{\ln y}{y^2}\,dy = 2\pi \int_1^2 \frac{\ln y}{y}\,dy \overset{u=\ln y}{=} 2\pi \int_0^{\ln 2} u\,du = \pi u^2\Big|_0^{\ln 2} = \pi(\ln 2)^2$

25. Since $f(x) = 1 \geq x - 2 = g(x)$ for $1 \leq x \leq 3$,
$V = \int_1^3 2\pi x[1 - (x-2)]\,dx = 2\pi \int_1^3 (3x - x^2)\,dx = 2\pi \left(\frac{3}{2}x^2 - \frac{1}{3}x^3\right)\Big|_1^3$
$= 2\pi \left[\left(\frac{27}{9} - 9\right) - \left(\frac{3}{2} - \frac{1}{3}\right)\right] = \frac{20}{3}\pi$

27. Let $0 \leq y \leq 1$. Notice that $y^2 + 1 \geq y\sqrt{1+y^3}$ if $y^4 + 2y^2 + 1 = (y^2+1)^2 \geq y^2(1+y^3) = y^2 + y^5$, or equivalently, $y^4 - y^5 + y^2 + 1 \geq 0$, which is valid for $0 \leq y \leq 1$. Thus $f(y) \geq g(y)$ for $0 \leq y \leq 1$, so that

$$V = \int_0^1 2\pi y \left[(y^2+1) - y\sqrt{1+y^3}\right] dy$$

$$= 2\pi \left[\left(\frac{1}{4}y^4 + \frac{1}{2}y^2\right)\Big|_0^1 - \int_0^1 y^2\sqrt{1+y^3}\,dy\right]$$

$$\overset{u=1+y^3}{=} 2\pi \left(\frac{3}{4} - \int_1^2 \sqrt{u}\cdot\frac{1}{3}\,du\right) = 2\pi \left(\frac{3}{4} - \frac{1}{3}\cdot\frac{2}{3}u^{3/2}\Big|_1^2\right)$$

$$= 2\pi \left(\frac{3}{4} - \frac{4}{9}\sqrt{2} + \frac{2}{9}\right) = \frac{\pi}{18}\left(35 - 16\sqrt{2}\right).$$

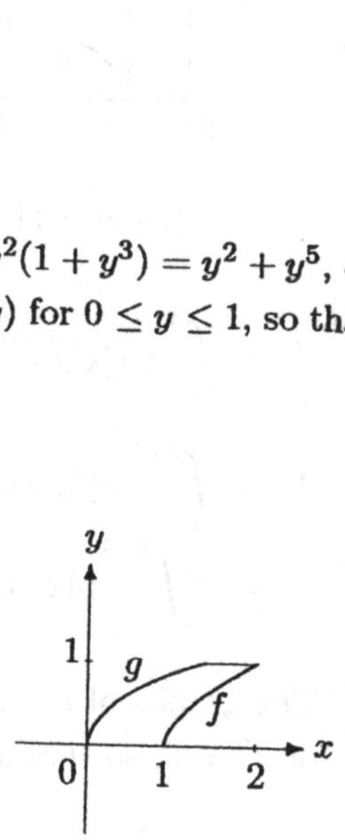

29. The graphs intersect for (x, y) such that $2x = y = x^2$, so that $x = 0$ or $x = 2$. Since $2x \geq x^2$ for $0 \leq x \leq 2$, we have $V = \int_0^2 2\pi x(2x - x^2)\,dx = 2\pi \int_0^2 (2x^2 - x^3)\,dx = 2\pi \left(\frac{2}{3}x^3 - \frac{1}{4}x^4\right)\Big|_0^2 = 2\pi\left(\frac{16}{3} - 4\right) = \frac{8}{3}\pi$.

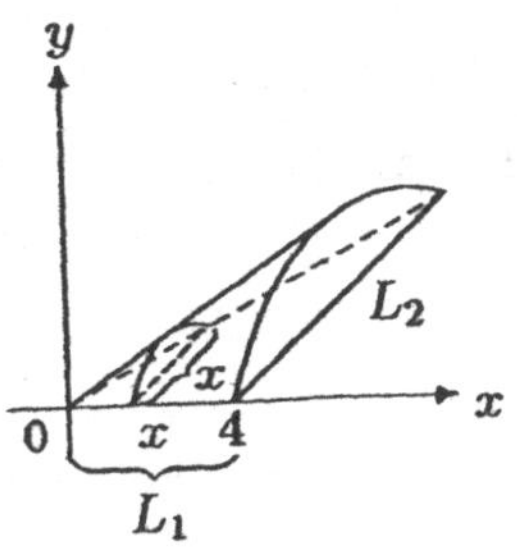

Exercise 31

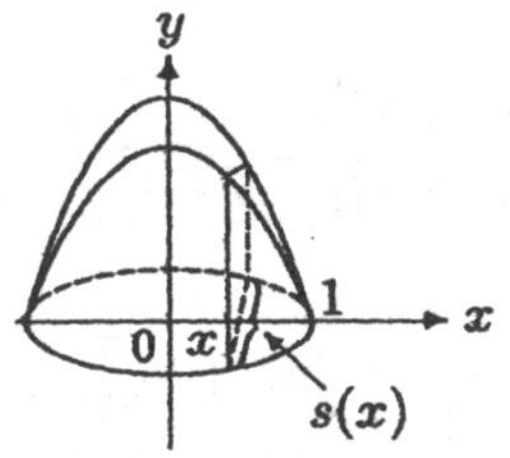

Exercise 33

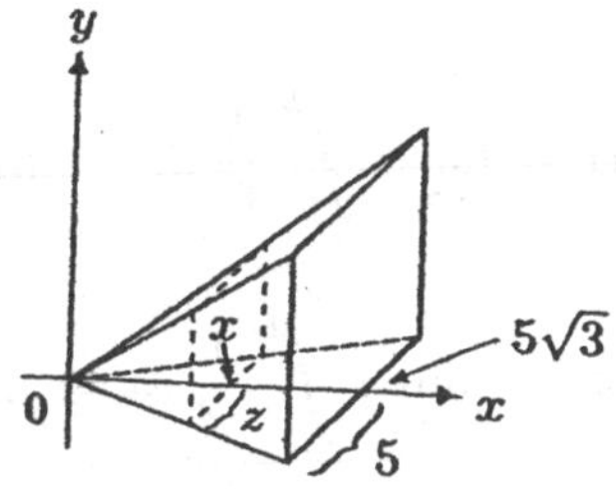

Exercise 35

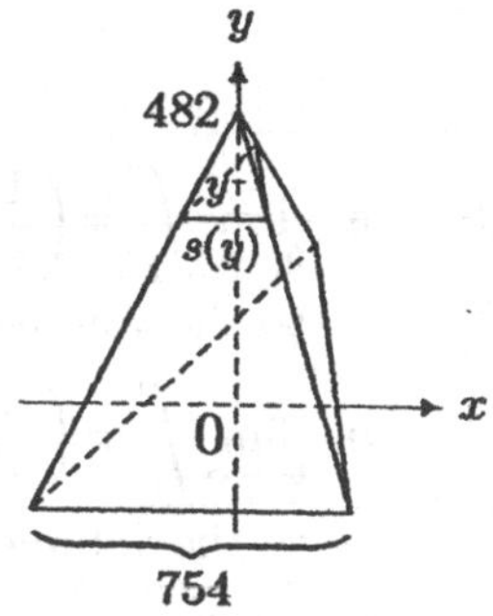

Exercise 43

31. Place the base so that L_1 lies along the positive x axis with the vertex opposite L_2 at the origin. Then the diameter of the semicircle x units from the origin is x and the cross-sectional area is given by $A(x) = \frac{1}{2}\pi(x/2)^2 = \frac{1}{8}\pi x^2$, so $V = \int_0^4 \frac{1}{8}\pi x^2\,dx = \frac{1}{24}\pi x^3\Big|_0^4 = \frac{8}{3}\pi$.

33. Let a square cross-section x units from the center have a side $s(x)$ units long. The points $(x, -\sqrt{1-x^2})$ and $(x, \sqrt{1-x^2})$ are on the circular base, so $s(x) = 2\sqrt{1-x^2}$. Thus the cross-sectional area of the corresponding square is given by $A(x) = \left(s(x)\right)^2 = 4(1-x^2)$, so $V = \int_{-1}^{1} 4(1-x^2)\,dx = (4x - \frac{4}{3}x^3)\Big|_{-1}^{1} = \frac{16}{3}$.

35. Place the base so that the given altitude lies along the positive x axis with the vertex at the origin. Then the length of a side of the square x units from the origin is $2z$, where by similar triangles $z/x = 5/(5\sqrt{3})$, or $z = x/\sqrt{3}$. The area of that square is given by $A(x) = (2z)^2 = \frac{4}{3}x^2$. Thus $V = \int_0^{5\sqrt{3}} \frac{4}{3}x^2\,dx = \frac{4}{9}x^3\Big|_0^{5\sqrt{3}} = \frac{4}{9}(125)(3\sqrt{3}) = \frac{500}{3}\sqrt{3}$.

37. $V = \displaystyle\int_a^b \pi[f(x) - c]^2\,dx$

39. $V = \displaystyle\int_0^1 \pi(1-e^{-2x})^2\,dx = \pi\int_0^1 (1-2e^{-2x}+e^{-4x})\,dx = \pi\left(x + e^{-2x} - \frac{1}{4}e^{-4x}\right)\Bigg|_0^1$

$$= \pi\left[\left(1 + e^{-2} - \frac{1}{4}e^{-4}\right) - \left(1 - \frac{1}{4}\right)\right] = \pi\left(\frac{1}{4} + e^{-2} - \frac{1}{4}e^{-4}\right)$$

41. The graphs of $y = x^2-x+1$ and $y = 2x^2-4x+3$ intersect for (x, y) such that $x^2-x+1 = y = 2x^2-4x+3$, or $x^2-3x+2 = 0$, so that $x = 1$ or $x = 2$. Since $x^2-x+1 \ge 2x^2-4x+3$ for $1 \le x \le 2$, it follows from Exercise 40 that $V = \int_1^2 \pi[(x^2-x+1)-1]^2\,dx - \int_1^2 \pi[(2x^2-4x+3)-1]^2\,dx = \pi\int_1^2 (x^4-2x^3+x^2)\,dx - \pi\int_1^2 4(x-1)^4\,dx = \pi\left(\frac{1}{5}x^5 - \frac{1}{2}x^4 + \frac{1}{3}x^3\right)\Big|_1^2 - \frac{4}{5}\pi(x-1)^5\Big|_1^2 = \pi\left[\left(\frac{32}{5} - 8 + \frac{8}{3}\right) - \left(\frac{1}{5} - \frac{1}{2} + \frac{1}{3}\right)\right] - \frac{4}{5}\pi = \frac{7}{30}\pi$.

43. Let a triangular cross-section y feet above the base have a side $s(y)$ feet long. By similar triangles, we have $s(y)/(482-y) = \frac{754}{482}$, so $s(y) = \frac{754}{482}(482-y)$. Thus the cross-sectional area is given by $A(y) = \frac{1}{2}s(y)\left((\sqrt{3}/2)s(y)\right) = (\sqrt{3}/4)\left(s(y)\right)^2$, which is $\sqrt{3}/4$ of the cross-sectional area of the Cheops pyramid. Since the height is still 482 feet, the volume V is $\sqrt{3}/4$, or approximately 43.3%, of the original volume.

45. a. $\lim_{b\to\infty}\int_1^b \pi\left(\frac{1}{x}\right)^2 dx = \lim_{b\to\infty}\pi\int_1^b \frac{1}{x^2}\,dx = \lim_{b\to\infty}\pi\left(-\frac{1}{x}\right)\Big|_1^b = \lim_{b\to\infty}\pi\left(-\frac{1}{b}+1\right) = \pi,$
so the volume is π.

b. $\lim_{b\to\infty}\int_1^b \pi\left(\frac{1}{x^2}\right)^2 dx = \lim_{b\to\infty}\pi\int_1^b \frac{1}{x^4}\,dx = \lim_{b\to\infty}\frac{-\pi}{3x^3}\Big|_1^b = \lim_{b\to\infty}\left(-\frac{\pi}{3b^3}+\frac{\pi}{3}\right) = \frac{\pi}{3},$
so the volume is $\pi/3$.

47. $\boldsymbol{V = \int_a^b 2\pi(x-c)f(x)\,dx}$

49. Using the formula derived in Exercise 47, we find that $V = \int_2^3 2\pi(x-2)\,x^2\,dx$

$$= 2\pi\int_2^3 (x^3 - 2x^2)\,dx = 2\pi\left(\frac{1}{4}x^4 - \frac{2}{3}x^3\right)\Big|_2^3 = 2\pi\left[\left(\frac{81}{4} - 18\right) - \left(4 - \frac{16}{3}\right)\right]$$
$$= \frac{43}{6}\pi$$

51. The graphs intersect for (x,y) such that $2x = y = -2x^2 + 4x$, or $2x^2 - 2x = 0$, so that $x = 0$ or $x = 1$. Since $-2x^2 + 4x \ge 2x$ for $0 \le x \le 1$, we have $V = \int_0^1 2\pi(x+1)[(-2x^2+4x) - 2x]\,dx = 2\pi\int_0^1 (2x - 2x^3)\,dx = 2\pi\left(x^2 - \frac{1}{2}x^4\right)\Big|_0^1 = \pi$.

53. Let a square cross-section x feet from the center have a side $s(x)$ feet long. Then $s(x) = 20\sqrt{1 - x^2/400} = \sqrt{400 - x^2}$. Thus the cross-sectional area is given by $A(x) = (s(x))^2 = 400 - x^2$, so that $V = \int_{-20}^{20}(400 - x^2)\,dx = \left(400x - \frac{1}{3}x^3\right)\Big|_{-20}^{20} = \frac{32{,}000}{3}$ (cubic feet).

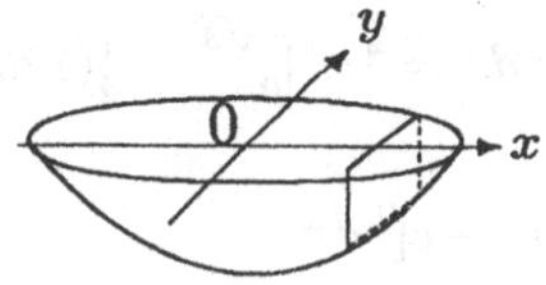

55. We can use the result of Exercise 54 with $a = \frac{1}{2}$ and $b = 1$. Then $V = \frac{4}{3}\pi(\frac{1}{2})(1)^2 = 2\pi/3$ (cubic centimeters).

57. In order to integrate along the y axis, we notice that $x = \sqrt{y/6}$ for $0 \le y \le 6$. For any depth h of soda, the volume of the soda is given by

$$V = \int_0^h \pi\left(\sqrt{\frac{y}{6}}\right)^2 dy = \pi\int_0^h \frac{y}{6}\,dy = \pi\frac{h^2}{12}.$$

Therefore $dV/dt = \pi(h/6)(dh/dt)$. Since $dV/dt = -\frac{1}{2}$, it follows that when $h = \frac{3}{2}$ we have $dh/dt = (-\frac{1}{2})[6/\pi(\frac{3}{2})] = -2/\pi$. Thus the depth is decreasing at the rate of $2/\pi$ inches per second.

59. $V = \int_0^5 \pi\left[(f(x))^2 - (g(x))^2\right] dx$

$$= \pi\int_0^{1/2}\frac{x^2}{4}\,dx + \pi\int_{1/2}^{7/2}\frac{1}{16}\,dx + \pi\int_{7/2}^{9/2}\frac{1}{16}\left[1 + \left(x - \frac{7}{2}\right)^2\right]^2 dx + \pi\int_{9/2}^{5}\left[\frac{1}{4} - \left(x - \frac{9}{2}\right)^2\right] dx$$

$$= \frac{\pi}{12}x^3\Big|_0^{1/2} + \frac{\pi}{16}x\Big|_{1/2}^{7/2} + \frac{\pi}{16}\int_{7/2}^{9/2}\left[1+2\left(x-\frac{7}{2}\right)^2+\left(x-\frac{7}{2}\right)^4\right]dx + \pi\left(\frac{1}{4}x - \frac{1}{3}\left(x-\frac{9}{2}\right)^3\right)\Big|_{9/2}^{5}$$

$$= \frac{\pi}{96} + \frac{3\pi}{16} + \frac{\pi}{16}\left[x + \frac{2}{3}\left(x-\frac{7}{2}\right)^3 + \frac{1}{5}\left(x-\frac{7}{2}\right)^5\right]\Big|_{7/2}^{9/2} + \frac{\pi}{12}$$

$$= \frac{\pi}{96} + \frac{3\pi}{16} + \frac{\pi}{16}\left(\frac{28}{15}\right) + \frac{\pi}{12} = \frac{191}{480}\pi \text{ (cubic centimeters)}$$

61. If we substitute $x = 20$ into $y = \frac{1}{40}x^2$, we obtain $y = 10$. Thus the wok is 10 centimeters high. By (5), its volume V is given by $V = \int_0^{20} 2\pi x\left(10 - \frac{1}{40}x^2\right)dx = 2\pi\int_0^{20}\left(10x - \frac{1}{40}x^3\right)dx = 2\pi\left(5x^2 - \frac{1}{160}x^4\right)\Big|_0^{20} = 2000\pi$ (cubic centimeters).

6.2 Length of a Curve

1. $L = \int_1^5 \sqrt{1+(2)^2}\,dx = \int_1^5 \sqrt{5}\,dx = 4\sqrt{5}$

3. $$L = \int_2^3 \sqrt{1+\left(2x - \frac{1}{8x}\right)^2}\,dx = \int_2^3 \sqrt{1 + 4x^2 - \frac{1}{2} + \frac{1}{64x^2}}\,dx$$

$$= \int_2^3 \sqrt{\left(2x + \frac{1}{8x}\right)^2}\,dx = \int_2^3\left(2x + \frac{1}{8x}\right)dx = \left(x^2 + \frac{1}{8}\ln|x|\right)\Big|_2^3 = 5 + \frac{1}{8}\ln\frac{3}{2}$$

5. $$L = \int_1^2 \sqrt{1+\left(4x^3 - \frac{1}{16x^3}\right)^2}\,dx = \int_1^2 \sqrt{1 + 16x^6 - \frac{1}{2} + \frac{1}{256x^6}}\,dx$$

$$= \int_1^2 \sqrt{\left(4x^3 + \frac{1}{16x^3}\right)^2}\,dx = \int_1^2\left(4x^3 + \frac{1}{16x^3}\right)dx = \left(x^4 - \frac{1}{32x^2}\right)\Big|_1^2$$

$$= \left(16 - \frac{1}{128}\right) - \left(1 - \frac{1}{32}\right) = \frac{1923}{128}$$

7. $$L = \int_1^2 \sqrt{1+\left[\frac{2x}{1+x^2} - \frac{1}{8}\left(x + \frac{1}{x}\right)\right]^2}\,dx = \int_1^2 \sqrt{1 + \frac{4x^2}{(1+x^2)^2} - \frac{1}{2} + \frac{1}{64}\frac{(1+x^2)^2}{x^2}}\,dx$$

$$= \int_1^2 \sqrt{\left[\frac{2x}{1+x^2} + \frac{1}{8}\left(\frac{1+x^2}{x}\right)\right]^2}\,dx = \int_1^2\left[\frac{2x}{1+x^2} + \frac{1}{8}\left(\frac{1}{x} + x\right)\right]dx$$

$$= \left[\ln(1+x^2) + \frac{1}{8}\left(\ln|x| + \frac{1}{2}x^2\right)\right]\Big|_1^2 = \left(\ln 5 + \frac{1}{8}\ln 2 + \frac{1}{4}\right) - \left(\ln 2 + \frac{1}{16}\right) = \ln 5 - \frac{7}{8}\ln 2 + \frac{3}{16}$$

9. $L = \int_{\pi/4}^{\pi/3} \sqrt{1 + \left(\frac{-1}{4}\cos x + \sec x\right)^2}\, dx = \int_{\pi/4}^{\pi/3} \sqrt{1 + \frac{1}{16}\cos^2 x - \frac{1}{2} + \sec^2 x\, dx}$

$= \int_{\pi/4}^{\pi/3} \sqrt{\left(\frac{1}{4}\cos x + \sec x\right)^2}\, dx = \int_{\pi/4}^{\pi/3} \left(\frac{1}{4}\cos x + \sec x\right) dx$

$= \left(\frac{1}{4}\sin x + \ln|\sec x + \tan x|\right)\Big|_{\pi/4}^{\pi/3} = \left[\frac{1}{8}\sqrt{3} + \ln(2+\sqrt{3})\right] - \left[\frac{1}{8}\sqrt{2} + \ln(\sqrt{2}+1)\right]$

$= \frac{1}{8}\left(\sqrt{3} - \sqrt{2}\right) + \ln\frac{2+\sqrt{3}}{\sqrt{2}+1}$

11. $L = \int_0^1 \sqrt{1 + \left[x^2 + 1 - \frac{1}{4(1+x^2)}\right]^2}\, dx = \int_0^1 \sqrt{1 + (x^2+1)^2 - \frac{1}{2} + \left[\frac{1}{4(1+x^2)}\right]^2}\, dx$

$= \int_0^1 \sqrt{(x^2+1)^2 + \frac{1}{2} + \left[\frac{1}{4(1+x^2)}\right]^2}\, dx = \int_0^1 \sqrt{\left[x^2 + 1 + \frac{1}{4(1+x^2)}\right]^2}\, dx$

$= \int_0^1 \left[x^2 + 1 + \frac{1}{4(1+x^2)}\right] dx = \left(\frac{1}{3}x^3 + x + \frac{1}{4}\tan^{-1} x\right)\Big|_0^1 = \frac{4}{3} + \frac{\pi}{16}$

13. $L = \int_0^{\pi/4} \sqrt{1 + \tan^2 x}\, dx = \int_0^{\pi/4} \sec x\, dx = \ln|\sec x + \tan x|\big|_0^{\pi/4} = \ln(\sqrt{2}+1)$

15. $L = \int_2^4 \sqrt{1 + (\sqrt{x^n - 1})^2}\, dx = \int_2^4 x^{n/2}\, dx = \frac{1}{(n/2)+1} x^{(n+2)/2}\Big|_2^4 = \frac{2}{n+2}\left(4^{(n+2)/2} - 2^{(n+2)/2}\right)$

17. $L = \int_1^{2^n} \sqrt{1 + (\sqrt{x^{1/n} - 1})^2}\, dx = \int_1^{2^n} x^{1/(2n)}\, dx = \frac{1}{(1/2n)+1} x^{(1+2n)/(2n)}\Big|_1^{2^n}$

$= \frac{2n}{2n+1}\left(2^{n[(1+2n)/(2n)]} - 1\right) = \frac{2n}{2n+1}\left(\sqrt{2}\, 2^n - 1\right)$

19. a. $y = (r^{2/3} - x^{2/3})^{3/2}$ for $0 \le x \le r$.

b. From (a), $y = (r^{2/3} - x^{2/3})^{3/2}$. Now if $f(x) = (r^{2/3} - x^{2/3})^{3/2}$ then

$$f'(x) = \frac{3}{2}(r^{2/3} - x^{2/3})^{1/2}(-\frac{2}{3}x^{-1/3}) = -x^{-1/3}(r^{2/3} - x^{2/3})^{1/2} \text{ for } x > 0$$

so that if L is the total length of the astroid, then

$\frac{1}{4}L = \lim_{\epsilon\to 0^+} \int_\epsilon^r \sqrt{1 + [f'(x)]^2}\, dx = \lim_{\epsilon\to 0^+} \int_\epsilon^r \sqrt{1 + \left[-x^{-1/3}(r^{2/3} - x^{2/3})^{1/2}\right]^2}\, dx$

$= \lim_{\epsilon\to 0^+} \int_\epsilon^r \sqrt{1 + x^{-2/3}(r^{2/3} - x^{2/3})}\, dx = \lim_{\epsilon\to 0^+} \int_\epsilon^r \sqrt{x^{-2/3} r^{2/3}}\, dx$

$= \lim_{\epsilon\to 0^+} \int_\epsilon^r x^{-1/3} r^{1/3}\, dx = \lim_{\epsilon\to 0^+} \frac{3}{2} x^{2/3} r^{1/3}\Big|_\epsilon^r = \lim_{\epsilon\to 0^+} \left(\frac{3}{2}r - \frac{3}{2}\epsilon^{2/3} r^{1/3}\right)$

$= \frac{3}{2}r$

Thus $L = 6r$.

6.3 Area of a Surface

1. $$S = \int_{-1/2}^{3/2} 2\pi\sqrt{4-x^2}\sqrt{1+\left(\frac{-x}{\sqrt{4-x^2}}\right)^2}\,dx$$
$$= 2\pi\int_{-1/2}^{3/2}\sqrt{4-x^2}\sqrt{1+\frac{x^2}{4-x^2}}\,dx = 2\pi\int_{-1/2}^{3/2} 2\,dx = 4\pi x\Big|_{-1/2}^{3/2} = 8\pi$$

3. $$S = \int_2^6 2\pi\sqrt{x}\sqrt{1+\left(\frac{1}{2\sqrt{x}}\right)^2}\,dx = 2\pi\int_2^6 \sqrt{x}\sqrt{1+\frac{1}{4x}}\,dx = \pi\int_2^6 \sqrt{4x+1}\,dx$$
$$= \frac{\pi}{6}(4x+1)^{3/2}\Big|_2^6 = \frac{\pi}{6}(125-27) = \frac{49}{3}\pi$$

5. $$S = \int_0^1 2\pi\left[\frac{1}{2}(e^x+e^{-x})\right]\sqrt{1+\left[\frac{1}{2}(e^x-e^{-x})\right]^2}\,dx$$
$$= \int_0^1 \pi(e^x+e^{-x})\sqrt{1+\frac{1}{4}(e^{2x}-2+e^{-2x})^2}\,dx$$
$$= \int_0^1 \pi(e^x+e^{-x})\sqrt{\frac{1}{4}(e^{2x}+2+e^{-2x})^2}\,dx$$
$$= \int_0^1 \pi(e^x+e^{-x})\sqrt{\frac{1}{4}(e^x+e^{-x})^2}\,dx = \int_0^1 \frac{\pi}{2}(e^x+e^{-x})(e^x+e^{-x})\,dx$$
$$= \int_0^1 \frac{\pi}{2}(e^{2x}+2+e^{-2x})\,dx = \frac{\pi}{2}\left(\frac{1}{2}e^{2x}+2-\frac{1}{2}e^{-2x}\right)\Big|_0^1$$
$$= \frac{\pi}{2}\left(\frac{1}{2}e^2+2x-\frac{1}{2}e^{-2}\right)-\frac{\pi}{2}\left(\frac{1}{2}+0-\frac{1}{2}\right) = \frac{\pi}{4}(e^2+4e^{-2})$$

7. $$S = \int_1^{\sqrt{2}} 2\pi\left(\frac{1}{4}x^4+\frac{1}{8x^2}\right)\sqrt{1+\left(x^3-\frac{1}{4x^3}\right)^2}\,dx$$
$$= 2\pi\int_1^{\sqrt{2}}\left(\frac{1}{4}x^4+\frac{1}{8x^2}\right)\sqrt{x^6+\frac{1}{2}+\frac{1}{16x^6}}\,dx$$
$$= \frac{\pi}{2}\int_1^{\sqrt{2}}\left(x^4+\frac{1}{2x^2}\right)\left(x^3+\frac{1}{4x^3}\right)\,dx$$
$$= \frac{\pi}{2}\int_1^{\sqrt{2}}\left(x^7+\frac{3}{4}x+\frac{1}{8x^5}\right)\,dx = \frac{\pi}{2}\left(\frac{1}{8}x^8+\frac{3}{8}x^2-\frac{1}{32x^4}\right)\Big|_1^{\sqrt{2}}$$
$$= \frac{\pi}{16}\left[\left(16+6-\frac{1}{16}\right)-\left(1+3-\frac{1}{4}\right)\right] = \frac{291}{256}\pi$$

9. a. $$S = \int_{-a}^{a} 2\pi\sqrt{r^2-x^2}\sqrt{1+\left(\frac{-x}{\sqrt{r^2-x^2}}\right)^2}\,dx$$
$$= 2\pi\int_{-a}^{a}\sqrt{r^2-x^2}\sqrt{1+\frac{x^2}{r^2-x^2}}\,dx = 2\pi\int_{-a}^{a} r\,dx = 4\pi ra$$

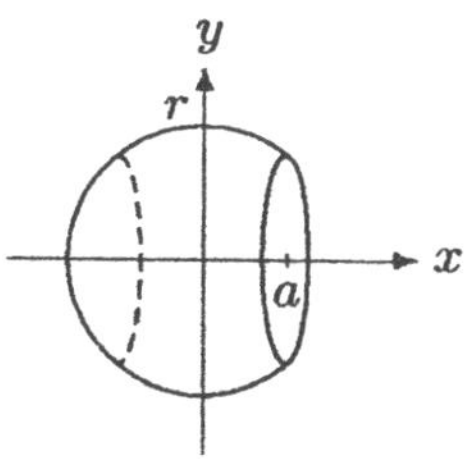

b. Yes.

11. For any ϵ, we have

$$S_\epsilon = \int_\epsilon^1 2\pi(1-x^{2/3})^{3/2}\sqrt{1+\left[\frac{3}{2}(1-x^{2/3})^{1/2}\left(-\frac{2}{3}x^{-1/3}\right)\right]^2}\,dx$$
$$= 2\pi\int_\epsilon^1 \frac{1}{x^{1/3}}(1-x^{2/3})^{3/2}\,dx \overset{u=1-x^{2/3}}{=} 2\pi\int_{1-\epsilon^{2/3}}^0 \left(-\frac{3}{2}u^{3/2}\right)du$$
$$= -2\pi\left(\frac{3}{5}u^{5/2}\right)\Big|_{1-\epsilon^{2/3}}^0 = \frac{6\pi}{5}(1-\epsilon^{2/3})^{5/2}$$

It follows that

$$S = \lim_{\epsilon\to 0^+} S_\epsilon = \lim_{\epsilon\to 0^+}\frac{6\pi}{5}(1-\epsilon^{2/3})^{5/2} = \frac{6\pi}{5}$$

13. The wok is obtained by revolving about the y axis the graph of $y = \frac{1}{40}x^2$ for $0 \le x \le 20$. The graph is equivalent to the graph of $x = \sqrt{40y}$ for $0 \le y \le 10$. Thus

$$S = \int_0^{10} 2\pi\sqrt{40y}\sqrt{1+\left(\frac{20}{\sqrt{40y}}\right)^2}\,dy = 2\pi\int_0^{10}\sqrt{40y}\sqrt{1+\frac{400}{40y}}\,dy = 2\pi\sqrt{40}\int_0^{10}\sqrt{y+10}\,dy$$
$$= 2\pi\sqrt{40}\left[\frac{2}{3}(y+10)^{3/2}\right]\Big|_0^{10} = \frac{4}{3}\pi\sqrt{40}\left(20^{3/2}-10^{3/2}\right) = \frac{800}{3}\pi(2\sqrt{2}-1) \text{ (square centimeters).}$$

6.4 Work

1. $W = \int_0^{60} 60(1 - x^2/20{,}000)\,dx = 60(x - x^3/60{,}000)\big|_0^{60} = 3384$ (joules)

3. $f(x) = 10$; $W = \int_0^8 10\,dx = 10x\big|_0^8 = 80$ (foot-pounds)

5. $W = \int_0^\pi 10^4 \sin x\,dx = -10^4\cos x\big|_0^\pi = 2\times 10^4$ (newton-kilometers). Thus $W = 2\times 10^7$ joules.

7. $W = \int_0^5 2(10^6)(5-x)\,dx = 2(10^6)\left(5x - \frac{1}{2}x^2\right)\big|_0^5 = 2(10^6)(25-12.5) = 2.5\times 10^7$ (ergs)

9. Using the information of Exercise 8, we have $W = \int_{20}^{30} 4(10^5)x\,dx = 2(10^5)x^2\big|_{20}^{30} = 1000(10^5) = 10^8$ (ergs).

11. By hypothesis, $6\times 10^7 = W = \int_0^{-5} kx\,dx$. Since $\int_0^{-5} kx\,dx = \frac{1}{2}kx^2\big|_0^{-5} = \frac{25}{2}k$, it follows that $k = \frac{12}{25}\times 10^7$. Thus the work necessary to stretch the spring 2 centimeters is given by $W = \int_0^2 \left(\frac{12}{25}\times 10^7\right)x\,dx = \left(\frac{12}{25}\times 10^7\right)\left(\frac{1}{2}x^2\right)\big|_0^2 = \frac{24}{25}\times 10^7 = 9.6\times 10^6$ (ergs).

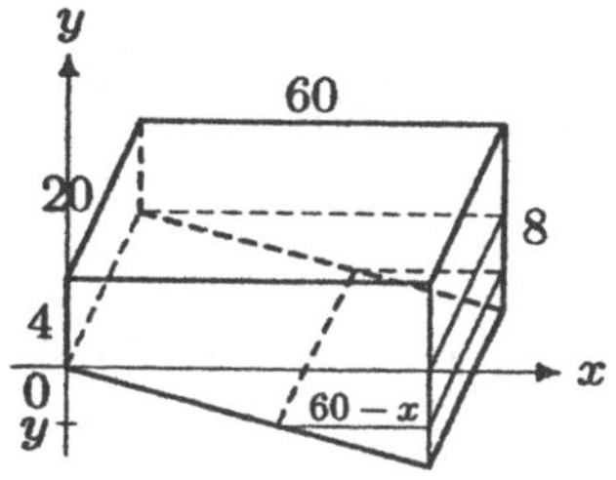

Exercise 13

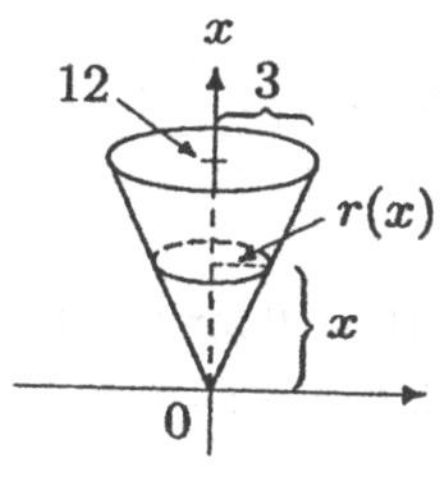

Exercise 15

13. We place the upper edge of the pool floor at the origin, as in the accompanying figure. Then $l = 5$, and a particle of water y feet from the bottom is to be raised $5 - y$ feet. For $0 \le y \le 4$, the cross-sectional area $A(y)$ is given by $A(y) = (60)(20) = 1200$. For $-4 \le y \le 0$ the floor meets the side wall in the line $y = \frac{-4}{60}x = \frac{-1}{15}x$, so that $A(y) = (60 - x)20 = (60 + 15y)20 = 1200 + 300y$. Thus $W = \int_{-4}^{0} 62.5(5-y)(1200+300y)\,dy + \int_0^4 62.5(5-y)(1200)\,dy = 18{,}750 \int_{-4}^{0}(-y^2 + y + 20)\,dy + 75{,}000 \int_0^4 (5-y)\,dy = 18{,}750\left(-\frac{1}{3}y^3 + \frac{1}{2}y^2 + 20y\right)\Big|_{-4}^{0} + 75{,}000\left(5y - \frac{1}{2}y^2\right)\Big|_0^4 = 18{,}750\left(-\frac{64}{3} - 8 + 80\right) + 75{,}000(20 - 8) = 1{,}850{,}000$ (foot-pounds).

15. We position the x axis as in the figure, with the origin at the vertex of the cone. Then $l = 12$, and a particle of water x feet from the vertex is to be raised $12 - x$ feet. Moreover, the water to be pumped extends from 0 to 12 on the x axis. By similar triangles, we have $r(x)/x = \frac{3}{12} = \frac{1}{4}$, so that $r(x) = x/4$ and hence $A(x) = \pi[r(x)]^2 = \pi x^2/16$. Thus by (4),

$$W = \int_0^{12} 62.5(12 - x)\frac{\pi}{16}x^2\,dx = \frac{62.5\pi}{16}\int_0^{12}(12x^2 - x^3)\,dx$$
$$= \frac{62.5\pi}{16}\left(4x^3 - \frac{1}{4}x^4\right)\Big|_0^{12} = 6750\pi \text{ (foot-pounds)}.$$

17. If the tank is positioned with respect to the x axis as in the figure, then $l = 0$ and the gasoline to be pumped extends from -4 to 0 on the x axis. Moreover, the width $w(x)$ of the cross-section at x is given by $w(x) = 2\sqrt{16 - x^2}$, so $A(x) = 10w(x) = 20\sqrt{16 - x^2}$. Thus by (4) with 62.5 replaced by 42, we have

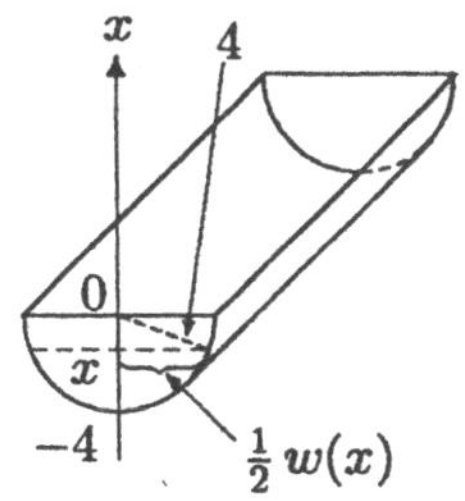

$$W = \int_{-4}^{0} 42(0 - x)20\sqrt{16 - x^2}\,dx \overset{u = 16 - x^2}{=} 840\int_0^{16}\sqrt{u}\left(\frac{1}{2}\right)du$$
$$= 420\left(\frac{2}{3}u^{3/2}\right)\Big|_0^{16} = 17{,}920 \text{ (foot-pounds)}.$$

19. Let W_1 denote the work required to extend the spring 6 centimeters from its natural length, and W_2 the work required to compress it to its natural length. Then $W_1 = \int_0^6 kx\,dx$ and $W_2 = \int_6^0 kx\,dx$, where k is the spring constant. Then the total work W done in extending and then compressing the spring is given by $W = W_1 + W_2 = \int_0^6 kx\,dx + \int_6^0 kx\,dx = \int_0^0 kx\,dx = 0$.

21. $W \approx \frac{20}{10}(2.90 + 3.01 + 3.52 + 3.41 + 3.19 + 2.78 + 2.76 + 2.83 + 2.90 + 2.84)$
$= 60.28$ (ergs).

23. a. $h'(t) = 8 - 32t$, so $h'(t) = 0$ for $t = \frac{1}{4}$. The maximum height is $h(\frac{1}{4}) = 7$ (feet). Since $f(h) = -0.2$, we have $W = \int_7^0 -0.2\,dh = (-0.2)h\big|_7^0 = 1.4$ (foot-pounds).

b. We change to 6 the upper limit on the integral in the solution of part (a). Thus $W = \int_7^6 -0.2\,dh = (-0.2)h\big|_7^6 = 0.2$ (foot-pounds).

25. a. $f(x) = 200$; $W = \int_0^{80} 200\,dx = 200(80) = 16{,}000$ (foot-pounds).

b. When the bucket is x feet above the ground, the chain is $80 - x$ feet long, so $f(x) = 200 + (80 - x) = 280 - x$. Thus $W = \int_0^{80}(280 - x)\,dx = (280x - \frac{1}{2}x^2)\big|_0^{80} = 19{,}200$ (foot-pounds).

27. When the bucket has been raised x feet, $x/2$ seconds have elapsed, so that $\frac{1}{2}(x/2)$ pounds of water have leaked out. Therefore the total weight of water and bucket is given by $f(x) = 21 - \frac{1}{4}x$. Since the bucket is empty when $\frac{1}{4}x = 20$, or $x = 80$, we have $W = \int_0^{80}(21 - \frac{1}{4}x)\,dx = (21x - \frac{1}{8}x^2)\big|_0^{80} = 880$ (foot-pounds).

29. Let $s(x)$ denote the length of the side of a cross-section x feet above the ground. Then $s(x)/754 = (482 - x)/482$, so $s(x) = \frac{754}{482}(482 - x)$. The cross-sectional area is given by $A(x) = (s(x))^2 = (\frac{754}{482})^2(482 - x)^2$. Let $P = \{x_0, x_1, \ldots, x_n\}$ be any partition of $[0, 482]$, and for $1 \le k \le n$ let t_k be any point in $[x_{k-1}, x_k]$. Then the amount of work ΔW_k required to lift the portion of the pyramid that will reside between the heights x_{k-1} and x_k is approximately $(150A(t_k)\Delta x_k)(t_k)$ (weight × distance). Thus the total work W, which is the sum of $\Delta W_1, \Delta W_2, \ldots, \Delta W_n$, is approximately $\sum_{k=1}^n 150 t_k A(t_k)\Delta x_k$, so that

$$W = \int_0^{482} 150xA(x)\,dx = 150\int_0^{482} x\left(\frac{754}{482}\right)^2(482 - x)^2\,dx = \frac{150(754)^2}{(482)^2}\int_0^{482}\left((482)^2x - 964x^2 + x^3\right)dx$$

$$= \frac{150(754)^2}{(482)^2}\left[\frac{(482)^2}{2}x^2 - \frac{964}{3}x^3 + \frac{1}{4}x^4\right]\Bigg|_0^{482} = \frac{150(754)^2(482)^2}{12} \approx 1.651 \times 10^{12} \text{ (foot-pounds)}.$$

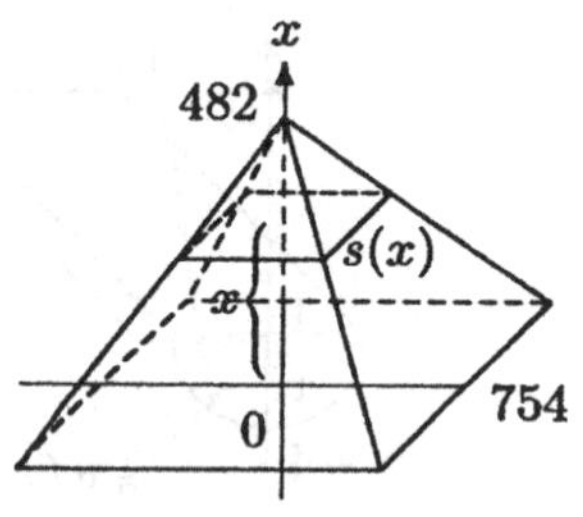

31. Set up a coordinate system as in the figure, and let $P = \{x_0, x_1, \ldots, x_n\}$ be a partition of $[0, 4]$. The portion of the triangle in the interval $[x_{k-1}, x_k]$ has approximate weight $2y_k\Delta x_k$ units, and must rise approximately x_k feet. By similar triangles, $y_k/(4 - x_k) = \frac{3}{4}$, so that $y_k = \frac{3}{4}(4 - x_k)$. Thus the work Δx_k required to raise the portion in the interval $[x_{k-1}, x_k]$ is approximately $2\left[\frac{3}{4}(4 - x_k)\Delta x_k\right](x_k)$. Then the total work S, which is the sum of ΔW_1, ΔW_2, $\ldots$, ΔW_n, is approximately $\sum_{k=1}^{n} \frac{3}{2}(4 - x_k)x_k\Delta x_k$, which is a Riemann sum for $\int_0^4 \frac{3}{2}(4 - x)x\,dx$. Therefore

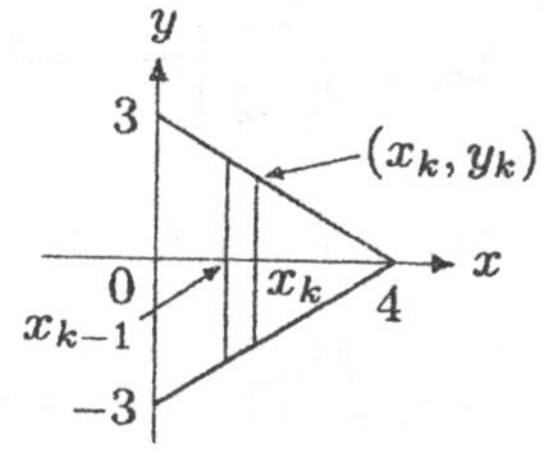

$$W = \int_0^4 \frac{3}{2}(4 - x)x\,dx = \frac{3}{2}\int_0^4 (4x - x^2)\,dx = \frac{3}{2}\left(2x^2 - \frac{1}{3}x^3\right)\Bigg|_0^4 = 16 \text{ (foot-pounds)}.$$

6.5 Moments and Center of Gravity

1. Take the origin of the x axis at the axis of revolution of the seesaw, with the positive x axis pointing toward the 20-kilogram child. Then the moment of the two children on the seesaw is $15(-2) + 20(2) = 10$. For equilibrium the moment of the 10-kilogram child must be -10. Thus that child should sit on the same side as the 15-kilogram child, one meter from the axis of revolution.

3. Let the origin be the center of mass of the system consisting of the earth and moon, with both earth and moon on the x axis. Let x denote the distance between the center of mass and the earth. Then

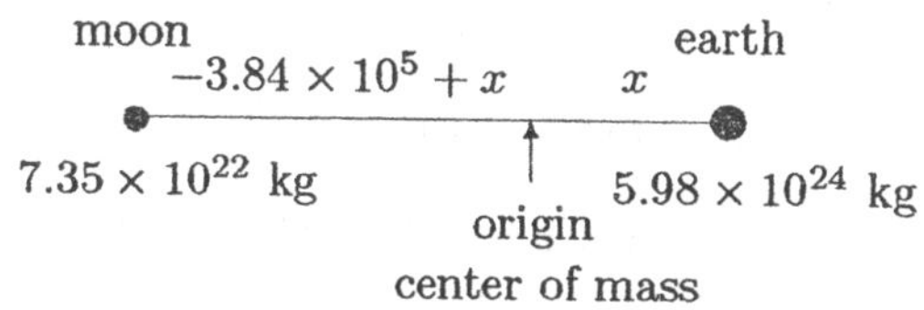

$$\underbrace{(7.35 \times 10^{22})}_{\text{moon mass}}\underbrace{(-3.84 \times 10^5 + x)}_{\text{moon coordinate}} + \underbrace{(5.98 \times 10^{24})}_{\text{earth mass}}\underbrace{x}_{\text{earth coordinate}} = 0.$$

Solving for x, we find that

$$x = \frac{(7.35 \times 10^{22})(3.84 \times 10^5)}{(7.35 \times 10^{22}) + (5.98 \times 10^{24})} \approx 4662.426695.$$

Thus the center of the earth is approximately 4662 kilometers from the center of mass. Since the radius of the earth is (approximately) 6.37×10^3 kilometers, the center of mass is inside the earth.

5. $M_x = \int_0^2 \frac{1}{2}\left[x^2 - (-2)^2\right]\,dx = \frac{1}{2}\left(\frac{1}{3}x^3 - 4x\right)\Bigg|_0^2 = -\frac{8}{3};$

$$M_y = \int_0^2 x[x - (-2)]\,dx = \int_0^2 \left(x^2 + 2x\right)\,dx = \left(\frac{1}{3}x^3 + x^2\right)\Bigg|_0^2 = \frac{20}{3}$$

$$A = \int_0^2 ((x - (-2))\,dx = \left(\frac{1}{2}x^2 + 2x\right)\Bigg|_0^2 = 6$$

$$\bar{x} = \frac{M_y}{A} = \frac{20/3}{6} = \frac{10}{9};\ \bar{y} = \frac{M_x}{A} = \frac{-8/3}{6} = \frac{-4}{9};\ (\bar{x}, \bar{y}) = \left(\frac{10}{9}, \frac{-4}{9}\right)$$

7. $M_x = \int_0^2 \frac{1}{2}\left[(2-x)^2 - (-(2-x))^2\right] dx = 0$

$M_y = \int_0^2 x[(2-x) - (-(2-x))]\, dx = \int_0^2 (4x - 2x^2)\, dx = \left(2x^2 - \frac{2}{3}x^3\right)\Big|_0^2 = \frac{8}{3}$

$A = \int_0^2 [(2-x) - (-(2-x))]\, dx = \int_0^2 (4-2x)\, dx = (4x - x^2)\big|_0^2 = 4$

$\bar{x} = \frac{M_y}{A} = \frac{8/3}{4} = \frac{2}{3};\ \bar{y} = \frac{M_x}{A} = \frac{0}{4} = 0;\ (\bar{x}, \bar{y}) = \left(\frac{2}{3}, 0\right)$

9. $M_x = \int_1^2 \frac{1}{2}\left[(x+1)^4 - (x-1)^4\right] dx = \frac{1}{2}\left[\frac{1}{5}(x+1)^5 - \frac{1}{5}(x-1)^5\right]\Big|_1^2 = 21$

$M_y = \int_1^2 x\left[(x+1)^2 - (x-1)^2\right] dx = \int_1^2 4x^2\, dx = \frac{4}{3}x^3\Big|_1^2 = \frac{28}{3}$

$A = \int_1^2 \left[(x+1)^2 - (x-1)^2\right] dx = \int_1^2 4x\, dx = 2x^2\big|_1^2 = 6$

$\bar{x} = \frac{M_y}{A} = \frac{28/3}{6} = \frac{14}{9};\ \bar{y} = \frac{M_x}{A} = \frac{21}{6} = \frac{7}{2};\ (\bar{x}, \bar{y}) = \left(\frac{14}{9}, \frac{7}{2}\right)$

11. $M_x = \int_2^5 \frac{1}{2}\left(\frac{1}{x-1} - \frac{1}{x+1}\right) dx = \frac{1}{2}[\ln(x-1) - \ln(x+1)]\Big|_2^5$

$= \frac{1}{2}[(\ln 4 - \ln 6) - (\ln 1 - \ln 3)] = \frac{1}{2}\ln 2$

In the evaluation of the integrals for M_y we will make two substitutions: $u = x - 1$ and $v = x + 1$.

$M_y = \int_2^5 x\left(\frac{1}{\sqrt{x-1}} - \frac{1}{\sqrt{x+1}}\right) dx = \int_2^5 \frac{x}{\sqrt{x-1}}\, dx - \int_2^5 \frac{x}{\sqrt{x+1}}\, dx$

$= \int_1^4 \frac{u+1}{\sqrt{u}}\, du - \int_3^6 \frac{v-1}{\sqrt{v}}\, dv = \int_1^4 \left(u^{1/2} + \frac{1}{u^{1/2}}\right) du - \int_3^6 \left(v^{1/2} - \frac{1}{v^{1/2}}\right) du$

$= \left(\frac{2}{3}u^{3/2} + 2u^{1/2}\right)\Big|_1^4 - \left(\frac{2}{3}v^{3/2} - 2v^{1/2}\right)\Big|_3^6$

$= (\frac{16}{3} + 4 - \frac{2}{3} - 2) - (4\sqrt{6} - 2\sqrt{6} - 2\sqrt{3} + 2\sqrt{3}) = \frac{20}{3} - 2\sqrt{6}$

$A = \int_2^5 \left(\frac{1}{\sqrt{x-1}} - \frac{1}{\sqrt{x+1}}\right) dx = (2\sqrt{x-1} - 2\sqrt{x+1}\big|_2^5$

$= (4 - 2\sqrt{6}) - (2 - 2\sqrt{3}) = 2 - 2\sqrt{6} + 2\sqrt{3}$

$\bar{x} = \frac{\frac{20}{3} - 2\sqrt{6}}{2 - 2\sqrt{6} + 2\sqrt{3}} = \frac{10 - 3\sqrt{6}}{3(1 - \sqrt{6} + \sqrt{3})};\ \bar{y} = \frac{\frac{1}{2}\ln 2}{2 - 2\sqrt{6} + 2\sqrt{3}} = \frac{\ln 2}{4(1 - \sqrt{6} + \sqrt{3})}$

$(\bar{x}, \bar{y}) = \left(\frac{10 - 3\sqrt{6}}{3(1 - \sqrt{6} + \sqrt{3})}, \frac{\ln 2}{4(1 - \sqrt{6} + \sqrt{3})}\right)$

13. The graphs of f and g intersect for (x, y) such that $2 - x^2 = y = |x|$, which means $2 - x^2 = x$ for $x \geq 0$ and $2 - x^2 = -x$ for $x < 0$. Thus $x = -1$ or $x = 1$.

$$M_x = \int_{-1}^{1} \frac{1}{2}\left[(2 - x^2)^2 - |x|^2\right] dx = \frac{1}{2}\int_{-1}^{1} (4 - 5x^2 + x^4)\, dx = \frac{1}{2}\left(4x - \frac{5}{3}x^3 + \frac{1}{5}x^5\right)\Big|_{-1}^{1} = \frac{38}{15}$$

$$M_y = \int_{-1}^{1} x\left[(2 - x^2) - |x|\right] dx = \int_{-1}^{0} (2x + x^2 - x^3)\, dx + \int_{0}^{1} (2x - x^2 - x^3)\, dx$$

$$= \left(x^2 + \frac{x^3}{3} - \frac{x^4}{4}\right)\Big|_{-1}^{0} + \left(x^2 - \frac{x^3}{3} - \frac{x^4}{4}\right)\Big|_{0}^{1} = \frac{-5}{12} + \frac{5}{12} = 0$$

$$A = \int_{-1}^{1} (2 - x^2 - |x|)\, dx = \int_{-1}^{0} (2 + x - x^2)\, dx + \int_{0}^{1} (2 - x - x^2)\, dx$$

$$= \left(2x + \frac{x^2}{2} - \frac{x^3}{3}\right)\Big|_{-1}^{0} + \left(2x - \frac{x^2}{2} - \frac{x^3}{3}\right)\Big|_{0}^{1} = \frac{7}{6} + \frac{7}{6} = \frac{7}{3}$$

$$\bar{x} = \frac{M_y}{A} = 0; \quad \bar{y} = \frac{M_x}{A} = \frac{38/15}{7/3} = \frac{38}{35}; \quad (\bar{x}, \bar{y}) = \left(0, \frac{38}{35}\right)$$

15. The graphs of the lines $y = x + 2$, $y = -3x + 6$, and $y = (2 - x)/3$ intersect in pairs for (x, y) such that $x + 2 = -3x + 6$, $x + 2 = (2 - x)/3$ or $-3x + 6 = (2 - x)/3$. From these equations we obtain $x = 1$, $x = -1$, and $x = 2$, respectively. To use Definition 8.5 we let

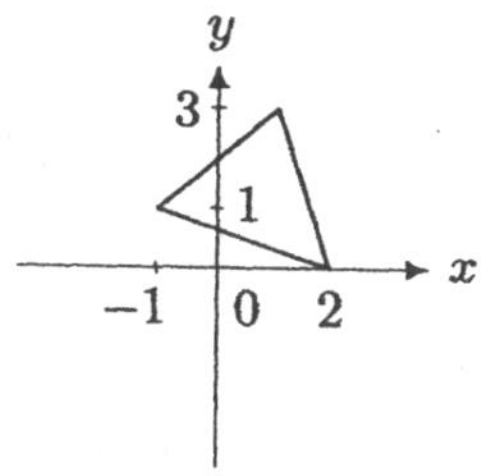

$$f(x) = \begin{cases} x + 2 & \text{for } -1 \leq x \leq 1 \\ -3x + 6 & \text{for } 1 \leq x \leq 2 \end{cases}$$

and $g(x) = (2 - x)/3$ for $-1 \leq x \leq 2$.

$$M_x = \int_{-1}^{1} \frac{1}{2}\left[(x + 2)^2 - \left(\frac{2 - x}{3}\right)^2\right] dx + \frac{1}{2}\int_{1}^{2} \left[(-3x + 6)^2 - \left(\frac{2 - x}{3}\right)^2\right] dx$$

$$= \frac{4}{9}\int_{-1}^{1} (x^2 + 5x + 4)\, dx + \frac{49}{9}\int_{1}^{2} (x^2 - 4x + 4)\, dx$$

$$= \frac{4}{9}\left(\frac{1}{3}x^3 + \frac{5}{2}x^2 + 4x\right)\Big|_{-1}^{1} + \frac{40}{9}\left(\frac{1}{3}x^3 - 2x^2 + 4x\right)\Big|_{1}^{2} = \frac{104}{27} + \frac{40}{27} = \frac{16}{3}$$

$$M_y = \int_{-1}^{1} x\left[(x + 2) - \left(\frac{2 - x}{3}\right)\right] dx + \int_{1}^{2} x\left[(-3x + 6) - \left(\frac{2 - x}{3}\right)\right] dx$$

$$= \frac{4}{3}\int_{-1}^{1} (x^2 + x)\, dx + \frac{8}{3}\int_{1}^{2} (-x^2 + 2x)\, dx = \frac{4}{3}\left(\frac{1}{3}x^3 + \frac{1}{2}x^2\right)\Big|_{-1}^{1} + \frac{8}{3}\left(\frac{-1}{3}x^3 + x^2\right)\Big|_{1}^{2} = \frac{8}{9} + \frac{16}{9} = \frac{8}{3}$$

$$A = \int_{-1}^{1} \left[(x + 2) - \left(\frac{2 - x}{3}\right)\right] dx + \int_{1}^{2} \left[(-3x + 6) - \left(\frac{2 - x}{3}\right)\right] dx$$

$$= \frac{4}{3}\int_{-1}^{1} (x + 1)\, dx + \frac{8}{3}\int_{1}^{2} (-x + 2)\, dx = \frac{4}{3}\left(\frac{1}{2}x^2 + x\right)\Big|_{-1}^{1} + \frac{8}{3}\left(\frac{-1}{2}x^2 + 2x\right)\Big|_{1}^{2} = \frac{8}{3} + \frac{4}{3} = 4$$

$$\bar{x} = \frac{M_y}{A} = \frac{8/3}{4} = \frac{2}{3}; \quad \bar{y} = \frac{M_x}{A} = \frac{16/3}{4} = \frac{4}{3}; \quad (\bar{x}, \bar{y}) = \left(\frac{2}{3}, \frac{4}{3}\right)$$

17. $(\bar{x}, \bar{y}) = (0, 3)$

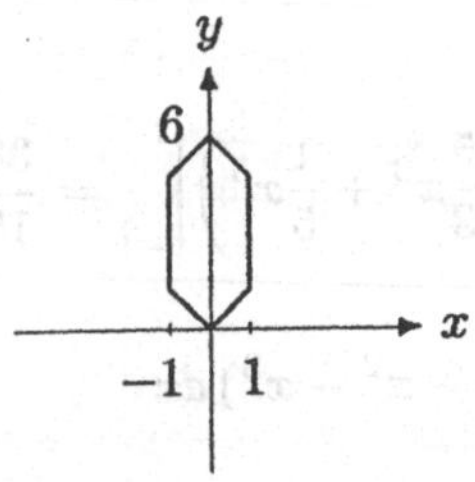

19. $(\bar{x}, \bar{y}) = (0, 0)$

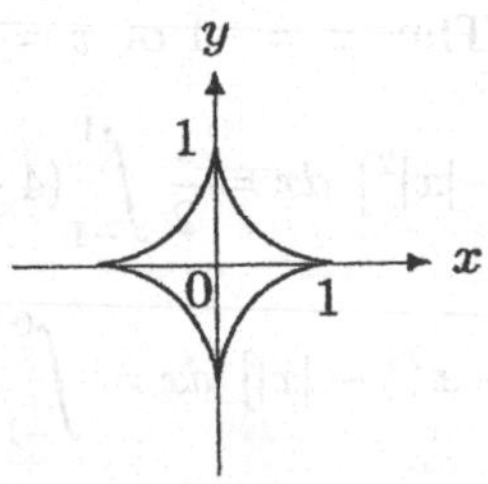

21. The triangle is symmetric with respect to the y axis, so $\bar{x} = 0$. Let

$$f(x) = \begin{cases} h + \dfrac{2h}{b}x & \text{for } \dfrac{-b}{2} \le x \le 0 \\ h - \dfrac{2h}{b}x & \text{for } 0 \le x \le \dfrac{b}{2} \end{cases}$$

Then

$$M_x = \int_{-b/2}^{0} \frac{1}{2}\left(h + \frac{2h}{b}x\right)^2 dx + \int_{0}^{b/2} \frac{1}{2}\left(h - \frac{2h}{b}x\right)^2 dx$$

$$= \frac{1}{2}\left(\frac{b}{2h}\right)\left(\frac{1}{3}\right)\left(h + \frac{2h}{b}x\right)^3\Big|_{-b/2}^{0} + \frac{1}{2}\left(\frac{-b}{2h}\right)\left(\frac{1}{3}\right)\left(h - \frac{2h}{b}x\right)^3\Big|_{0}^{b/2} = \frac{bh^2}{6}$$

$$A = \frac{bh}{2}; \quad \bar{y} = \frac{bh^2/6}{bh/2} = \frac{h}{3};$$

$(\bar{x}, \bar{y}) = (0, h/3)$, which is the "centroid" of the triangle.

23. Let $f(x) = (h/a^2)x^2$ and $g(x) = 0$.

$$M_x = \int_0^a \frac{1}{2}\left(\frac{h}{a^2}x^2\right)^2 dx = \frac{h^2}{2a^4}\left(\frac{1}{5}x^5\right)\Big|_0^a = \frac{ah^2}{10}; \quad M_y = \int_0^a x\left(\frac{h}{a^2}x^2\right) dx = \frac{hx^4}{4a^2}\Big|_0^a = \frac{a^2h}{4}$$

$$A = \int_0^a \frac{h}{a^2}x^2\,dx = \frac{hx^3}{3a^2}\Big|_0^a = \frac{ah}{3}; \quad \bar{x} = \frac{a^2h/4}{ah/3} = \frac{3a}{4}; \quad \bar{y} = \frac{ah^2/10}{ah/3} = \frac{3h}{10};$$

thus $(\bar{x}, \bar{y}) = \left(\dfrac{3a}{4}, \dfrac{3h}{10}\right)$.

25. The area A of the semicircular region and the volume V of the sphere generated by revolving the semicircular region R about its diameter are given, respectively, by $A = \frac{1}{2}\pi r^2$ and $V = \frac{4}{3}\pi r^3$. The Theorem of Pappus and Guldin then implies that $\frac{4}{3}\pi r^3 = 2\pi\bar{x}(\frac{1}{2}\pi r^2)$, so that $\bar{x} = 4r/3\pi$. Since $\bar{y} = 0$ by symmetry, the center of gravity of the semicircular region is $(4r/3\pi, 0)$.

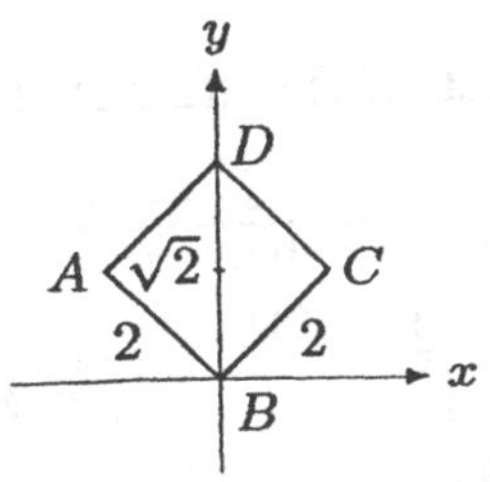

Exercise 27

27. Set up a coordinate system as in the figure, so that the center of gravity of the square is $(0, \sqrt{2})$ and the square is revolved about the x axis. Since $\bar{y} = \sqrt{2}$ and since the area of the square is 4, the Theorem of Pappus and Guldin says that the volume V of the solid is $V = 2\pi\bar{y}(\text{area}) = 2\pi\sqrt{2}\,(4) = 8\pi\sqrt{2}$.

29. Substituting $u = -x$, we obtain

$$\int_{-a}^{0} xf(x)\,dx = -\int_{a}^{0}(-u)f(-u)\,du = -\int_{0}^{a} uf(u)\,du = -\int_{0}^{a} xf(x)\,dx.$$

Thus

$$M_y = \int_{-a}^{a} xf(x)\,dx = \int_{-a}^{0} xf(x)\,dx + \int_{0}^{a} xf(x)\,dx = 0.$$

31. If l is any line tangent to the circle which has radius r, and if the area of R is A, then the Theorem of Pappus and Guldin says that the volume V of the solid of revolution is given by $V = (2\pi r)A$.

6.6 Hydrostatic Force

1. Following the solution of Example 1, but with $c=3$ instead of 4, we find that $F = \int_0^3 62.5(3-x)(\frac{1}{2}x)\,dx = 31.25\int_0^3 (3x - x^2)\,dx = 31.25\left(\frac{3}{2}x^2 - \frac{1}{3}x^3\right)\Big|_0^3 = (31.25)\frac{9}{2} = 140.625$ (pounds).

3. We take the origin at water level. Then $c=0$ and $w(x)/2 = [x-(-1)]/\sqrt{3}$, so that $w(x) = [(2\sqrt{3})/3](x+1)$;

$$\begin{aligned} F &= \int_{-1}^{0} 62.5(0-x)\frac{2\sqrt{3}}{3}(x+1)\,dx \\ &= \frac{125\sqrt{3}}{3}\left(-\frac{1}{3}x^3 - \frac{1}{2}x^2\right)\Big|_{-1}^{0} \\ &= \frac{125\sqrt{3}}{18} \approx 12.0281 \text{ (pounds).} \end{aligned}$$

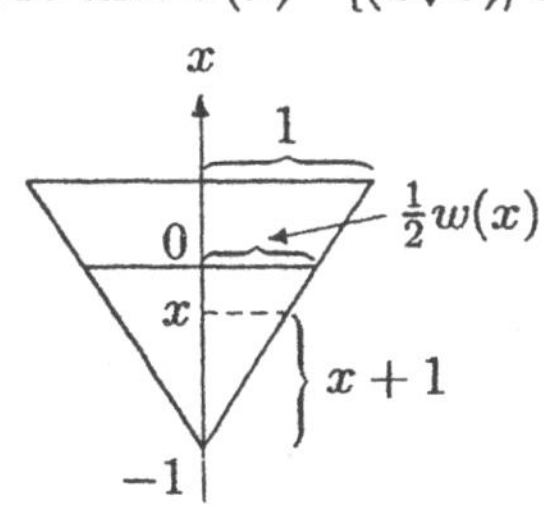

5. For the triangles pointing downward, $c = 0$ and $w(x)/2 = (x+3)/\sqrt{3}$, so that $w(x) = [(2\sqrt{3})/3](x+3)$;

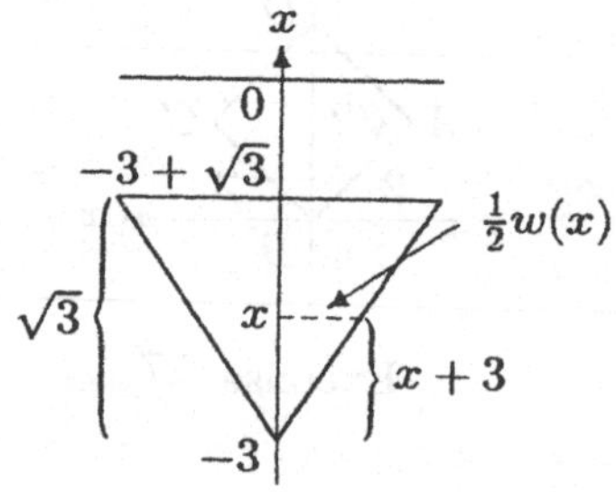

$$F = \int_{-3}^{-3+\sqrt{3}} 62.5(0-x)\frac{2\sqrt{3}}{3}(x+3)\,dx$$
$$= \frac{125\sqrt{3}}{3}\left(-\frac{1}{3}x^3 - \frac{3}{2}x^2\right)\Big|_{-3}^{-3+\sqrt{3}}$$
$$= \frac{125\sqrt{3}}{3}\left(\frac{9}{2} - \sqrt{3}\right) \approx 199.760 \text{ (pounds).}$$

For the triangles pointing upward, $c = 0$ and

$$\frac{w(x)}{2} = \frac{-3+\sqrt{3}-x}{\sqrt{3}}, \quad \text{so that} \quad w(x) = \frac{2\sqrt{3}}{3}\left(-3+\sqrt{3}-x\right);$$

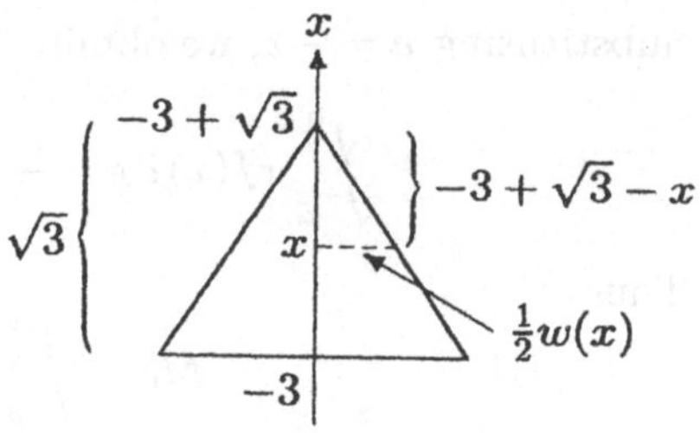

$$F = \int_{-3}^{-3+\sqrt{3}} 62.5(0-x)\frac{2\sqrt{3}}{3}\left(-3+\sqrt{3}-x\right)dx$$
$$= \frac{125\sqrt{3}}{3}\left[\frac{(3-\sqrt{3})}{2}x^2 + \frac{1}{3}x^3\right]\Big|_{-3}^{-3+\sqrt{3}}$$
$$= \frac{125\sqrt{3}}{3}\left(\frac{9}{2} - \frac{\sqrt{3}}{2}\right) \approx 262.260 \text{ (pounds).}$$

7. If the top of the block is x feet from the surface of the water, the hydrostatic force on the top is $F_t = 62.5(x)(1) = 62.5x$. The hydrostatic force on the bottom is $F_b = 62.5(x+1)(1) = 62.5(x+1)$. The difference between these forces is $F_b - F_t = 62.5(x+1) - 62.5x = 62.5$ (pounds).

9. The portion of the plate between x_{k-1} and x_k has area approximately equal to $w(t_k)(\Delta x_k \sec\theta)$. Thus the force on that portion of the plate is approximately equal to $(62.5)(c-t_k)(w(t_k)\sec\theta\,\Delta x_k)$. The hydrostatic force F on the plate is approximately equal to $\sum_{k=1}^{n}(62.5)(c-t_k)(\sec\theta\, w(t_k)\,\Delta x_k)$, which is a Riemann sum for $62.5(c-x)(\sec\theta)w$. It follows that the hydrostatic force is given by $\int_a^b (62.5\sec\theta)(c-x)w(x)\,dx$.

11. a. We place the origin at the base of the smaller end of the pool. Then by (2) the hydrostatic force F is given by $F = \int_0^4 62.5(4-x)20\,dx = 1250\left(-\frac{1}{2}\right)(4-x)^2\big|_0^4 = 10{,}000$ (pounds).

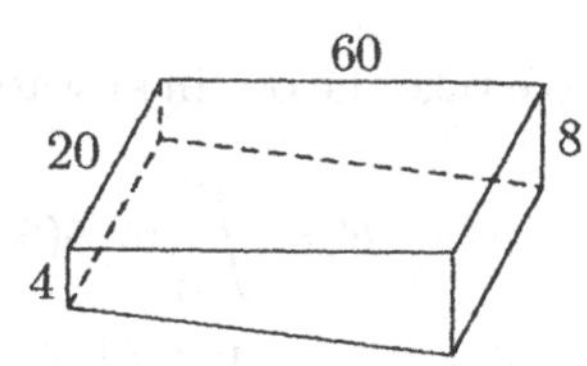

b. We place the origin at the base of the larger end of the pool. Then by (2) the hydrostatic force F is given by $F = \int_0^8 62.5(8-x)20\,dx = 1250\left(-\frac{1}{2}\right)(8-x)^2\big|_0^8 = 40{,}000$ (pounds).

c. We place the origin at the base of the larger end of the pool. For $4 \le x \le 8$, the width $w(x)$ of each side at x is 60 and for $0 \le x \le 4$, the width $w(x)$ of each side at x is $15x$. Thus by (2) the hydrostatic force F on each side of the pool is given by $F = \int_0^8 62.5(8-x)w(x)\,dx$ $= \int_0^4 62.5(8-x)15x\,dx + \int_4^8 62.5(8-x)60\,dx = 937.5\left(4x^2 - \frac{1}{3}x^3\right)\big|_0^4 + 3750\left(-\frac{1}{2}\right)(8-x)^2\big|_4^8 = 937.5\left(64 - \frac{64}{3}\right) + 1875(4^2 - 0^2) = 70{,}000$ (pounds).

d. We place the origin at the base of the larger end of the pool. By Exercise 9 with $\sec\theta = \frac{1}{4}\sqrt{60^2+4^2} = \sqrt{226}$, the hydrostatic force F is given by $F = \int_0^4 62.5\sqrt{226}\,(8-x)20\,dx = 1250\sqrt{226}\,(-\frac{1}{2})\,(8-x)^2\Big|_0^4 = 625\sqrt{226}\,(8^2-4^2) = 30{,}000\sqrt{226} \approx 451{,}000$ (pounds).

13. Since the diameter is 3 (rather than 40 as in Exercise 12) and the height is 100 (rather than 30 as in Exercise 11), we have $F = 62.5\int_0^{100} 3\pi x\,dx = (62.5)\frac{3}{2}\pi x^2\Big|_0^{100} = 937{,}500\pi \approx 2{,}945{,}000$ (pounds).

6.7 Parametrized Curves

1. $x^2+y^2=4$ for $0 \le x \le 2$, $0 \le y \le 2$

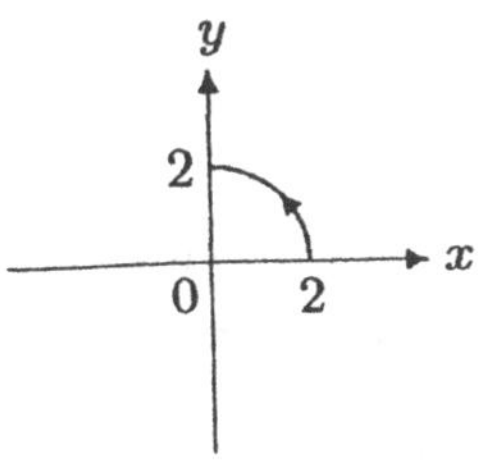

3. $(x-2)^2+(y+1)^2=1$

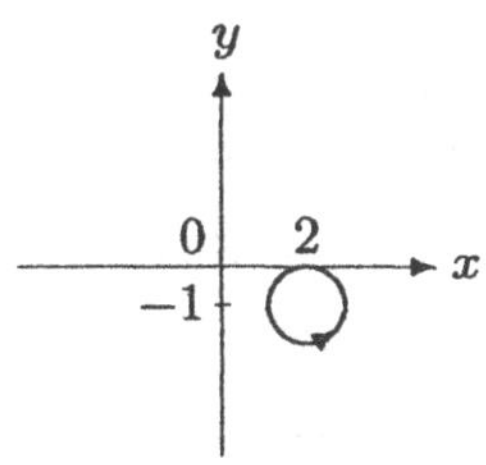

5. $x=-y$

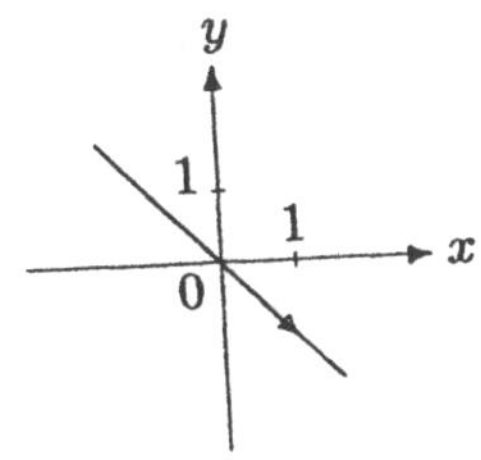

7. $x=3$ for $-2 \le y \le -1$

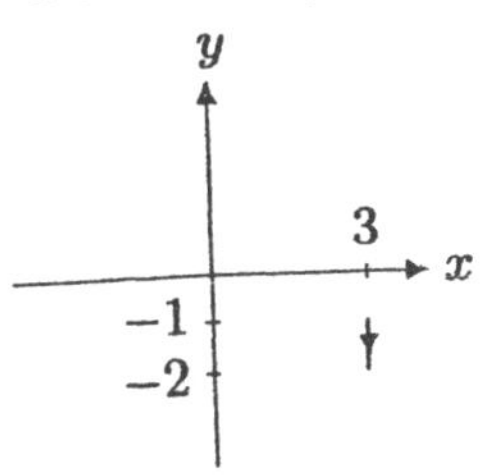

9. $y = x^{2/3}$

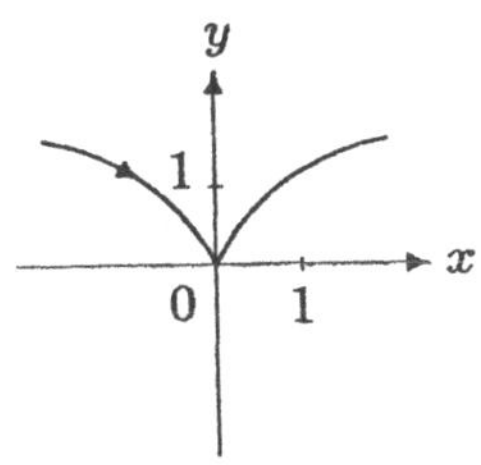

11. $y = 1/x^3$ for $x > 0$

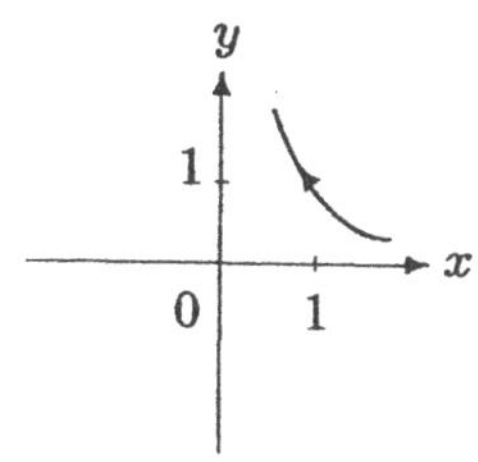

13. a. Notice that for $0 \le t \le 2\pi$, $\sin t$ and $\cos t$ have the same sign only if $0 < t < \pi/2$ or $\pi < t < 3\pi/2$. The points on the two curves coincide if $t = \pi/4$ or $t = 5\pi/4$.

b. For any value of t for which the corresponding points on the two curves coincide, we have $x = \cos t = -\sin t$ and $y = \sin t = \cos t$. These imply that $\sin t = -\sin t$ and $\cos t = -\cos t$. Thus $\sin t = 0$ and $\cos t = 0$, which is impossible. Thus there is no value of t for which the corresponding points on the two curves coincide.

c. For any value of t for which the corresponding points on the two curves coincide, we have $x = 1+\cos t = \sin t$ and $y = \sin t = \cos t$. These imply that $1+\cos t = \sin t = \cos t$, or $1 = 0$. Thus there is no value of t for which the corresponding points on the two curves coincide.

15. a. We have

$$x^3 = \frac{27t^3}{(1+t^3)^3}, \quad y^3 = \frac{27t^6}{(1+t^3)^3} \quad \text{and} \quad xy = \frac{9t^3}{(1+t^3)^2}.$$

Thus

$$x^3 + y^3 = \frac{27t^3}{(1+t^3)^3} + \frac{27t^6}{(1+t^3)^3} = \frac{27t^3(1+t^3)}{(1+t^3)^3} = \frac{27t^3}{(1+t^3)^2} = 3xy,$$

so an equation of the Folium of Descartes in rectangular coordinates is $x^3 + y^3 = 3xy$.

b. If $t > -1$, then $y > 0$, and if $t < -1$, then $y < 0$. Therefore the part of the folium above the x axis is traced out as t increases without bound, and the part of the folium below the x axis is traced out as t increases toward -1.

17. a. As n increases, the number of loops increases.

b. By the Chain Rule, $dy/dt = (dy/dx)(dx/dt)$, so that

$$\frac{dy}{dx} = \frac{dy/dt}{dx/dt} = \frac{n\cos nt}{\cos t}.$$

The outer loops correspond to x near 1 or -1. Since $\sin t = x$ is near 1 or -1, it follows that $\cos t$ is near 0, and because n is even, $\cos nt$ is near -1 or 1. It follows that $dy/dx = (n\cos nt)/(\cos t)$ is large. Thus a small interval of x values is sufficient for y to cycle through its range of values. Therefore the outer loops are thin.

19. b. Since

$$x^2 = \frac{4t^2}{(1+t^2)^2} \quad \text{and} \quad y^2 = \frac{(1-t^2)^2}{(1+t^2)^2} = \frac{1-2t^2+t^4}{(1+t^2)^2},$$

we have

$$x^2 + y^2 = \frac{4t^2}{(1+t^2)^2} + \frac{1-2t^2+t^4}{(1+t^2)^2} = \frac{1+2t^2+t^4}{(1+t^2)^2} = 1.$$

Thus (x, y) lies on the circle $x^2 + y^2 = 1$. The graph consists of the entire circle except the point $(0, -1)$, since $y = (1-t^2)/(1+t^2)$ cannot equal -1.

21. $$x = \begin{cases} 0 & \text{for } -\frac{\pi}{2} \le t \le \frac{\pi}{2} \\ -\frac{\pi}{4}\cos t & \text{for } \frac{\pi}{2} < t \le \frac{3\pi}{2} \\ \frac{\pi}{4}\cos t & \text{for } \frac{3\pi}{2} < t \le \frac{5\pi}{2} \end{cases} \qquad y = \begin{cases} t & \text{for } -\frac{\pi}{2} \le t \le \frac{\pi}{2} \\ \frac{\pi}{4}(1+\sin t) & \text{for } \frac{\pi}{2} < t \le \frac{3\pi}{2} \\ -\frac{\pi}{4}(1+\sin t) & \text{for } \frac{3\pi}{2} < t \le \frac{5\pi}{2} \end{cases}$$

6.8 Length and Surface Area for Parametrized Curves

1. $L = \int_0^1 \sqrt{(-2t)^2 + (3t^2)^2}\, dt = \int_0^1 t\sqrt{4+9t^2}\, dt \overset{u=4+9t^2}{=} \int_4^{13} \sqrt{u} \cdot \frac{1}{18}\, du$
$= \frac{1}{18}\left(\frac{2}{3}u^{3/2}\right)\Big|_4^{13} = \frac{1}{27}(13\sqrt{13} - 8)$

3. $L = \int_0^{\pi/2} \sqrt{(\cos t - \cos t + t\sin t)^2 + (\sin t + t\cos t - \sin t)^2}\, dt$
$= \int_0^{\pi/2} \sqrt{t^2(\sin^2 t + \cos^2 t)}\, dt = \int_0^{\pi/2} t\, dt = \frac{1}{2}t^2\Big|_0^{\pi/2} = \frac{1}{8}\pi^2$

5. $S = \int_{-3/4}^0 (2\pi)\frac{2}{3}(1+t)^{3/2}\sqrt{[-(1-t)^{1/2}]^2 + [(1+t)^{1/2}]^2}\, dt = \frac{4\sqrt{2}\,\pi}{3}\int_{-3/4}^0 (1+t)^{3/2}\, dt$
$= \frac{8\sqrt{2}\,\pi}{15}(1+t)^{5/2}\Big|_{-3/4}^0 = \frac{8\sqrt{2}\,\pi}{15}\left(1 - \frac{1}{32}\right) = \frac{31}{60}\sqrt{2}\,\pi$

7. a. By (12),

$$S = \int_\alpha^\beta 2\pi r \sin t \sqrt{(-r\sin t)^2 + (r\cos t)^2}\, dt = \int_\alpha^\beta 2\pi r \sin t \sqrt{r^2(\sin^2 t + \cos^2 t)}\, dt$$

$$= \int_\alpha^\beta 2\pi r^2 \sin t\, dt = -2\pi r^2 \cos t\Big|_\alpha^\beta = 2\pi r^2(\cos\alpha - \cos\beta).$$

If $\alpha = 0$ and $\beta = \pi/4$, then $\beta - \alpha = \pi/4$ and $S = 2\pi r^2(1 - \sqrt{2}/2)$. If $\alpha = \pi/4$ and $\beta = \pi/2$, then $\beta - \alpha = \pi/4$ and $S = 2\pi r^2(\sqrt{2}/2 - 0)$. Since $1 - \sqrt{2}/2 \neq \sqrt{2}/2$, we see that S does not depend only on $\beta - \alpha$.

b. The results of parts (a) and (b) do not contradict the remark following Example 2 of Section **6.3**. There $b - a$ referred to a difference of x values. Here $\beta - \alpha$ refers to a difference of t values, not x values.

9. a. By (2),

$$C_{ab} = \int_0^{2\pi} \sqrt{(-a\sin t)^2 + (b\cos t)^2}\, dt = \int_0^{2\pi} \sqrt{a^2\sin^2 t + b^2\cos^2 t}\, dt.$$

b. If $a = b > 0$, then by (a),

$$C_{ab} = \int_0^{2\pi} \sqrt{a^2\sin^2 t + a^2\cos^2 t}\, dt = \int_0^{2\pi} a\, dt = 2\pi a.$$

11. $S = \int_0^{\pi/2} 2\pi \sin^n t \sqrt{(-n\cos^{n-1} t\, \sin t)^2 + (n\sin^{n-1} t\, \cos t)^2}\, dt$
$= 2n\pi \int_0^{\pi/2} \sin^n t\, \sin t\, \cos t \sqrt{\cos^{2n-4} t + \sin^{2n-4} t}\, dt$

We can carry out the integration if $2n - 4 = 0$ or 2, which means that $n = 2$ or 3. (Exercise 10 (c) uses $n = 3$.)

Chapter 6 Review

1. Since $1+\sqrt{x} \ge 1 \ge e^{-x}$ for $0 \le x \le 1$,

$$V = \int_0^1 \pi\left[(1+\sqrt{x})^2 - (e^{-x})^2\right] dx = \pi \int_0^1 \left(1 + 2\sqrt{x} + x - e^{-2x}\right) dx$$

$$= \pi\left(x + \frac{4}{3}x^{3/2} + \frac{1}{2}x^2 + \frac{1}{2}e^{-2x}\right)\Big|_0^1 = \pi\left(\frac{7}{3} + \frac{1}{2}e^{-2}\right).$$

3. Since $e^{x^2} \ge 1 \ge e^{-x^2}$ for $0 \le x \le 1$,

$$V = \int_0^1 2\pi x(e^{x^2} - e^{-x^2})\, dx \overset{u=x^2}{=} 2\pi \int_0^1 (e^u - e^{-u})\frac{1}{2}\, du = \pi(e^u - e^{-u})\Big|_0^1 = \pi(e + e^{-1} - 2).$$

5. a. $V = \int_0^2 2\pi x(x^3)\, dx = \frac{2}{5}\pi x^5\Big|_0^2 = \frac{64}{5}\pi$

b. $M_x = \int_0^2 \frac{1}{2}(x^3)^2\, dx = \frac{1}{14}x^7\Big|_0^2 = \frac{64}{7}$; $M_y = \int_0^2 x(x^3)\, dx = \frac{1}{5}x^5\Big|_0^2 = \frac{32}{5}$

$A = \int_0^2 x^3\, dx = \frac{1}{4}x^4\Big|_0^2 = 4$

$\bar{x} = \frac{32/5}{4} = \frac{8}{5}$; $\bar{y} = \frac{64/7}{4} = \frac{16}{7}$; $(\bar{x}, \bar{y}) = \left(\frac{8}{5}, \frac{16}{7}\right)$

c. $A = 4$, $b = \frac{8}{5}$; $V = (2\pi)\left(\frac{8}{5}\right)(4) = \frac{64}{5}\pi$

7. a. $V_1 = \int_1^3 2\pi x\left[\left(x + \frac{c}{x}\right) - x\right] dx = \int_1^3 2\pi c\, dx = 4\pi c$

b. $V_2 = \int_1^3 \pi\left[\left(x + \frac{c}{x}\right)^2 - x^2\right] dx = \pi \int_1^3 \left(2c + \frac{c^2}{x^2}\right) dx = \pi\left(2cx - \frac{c^2}{x}\right)\Big|_1^3$

$= \pi\left[\left(6c - \frac{c^2}{3}\right) - (2c - c^2)\right] = 4\pi c + \frac{2}{3}\pi c^2$

c. $4\pi c + \frac{2}{3}\pi c^2 = V_2 = V_1 = 4\pi c$ only if $c = 0$.

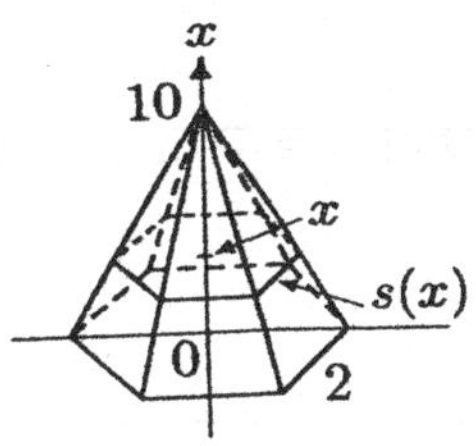

Exercise 9

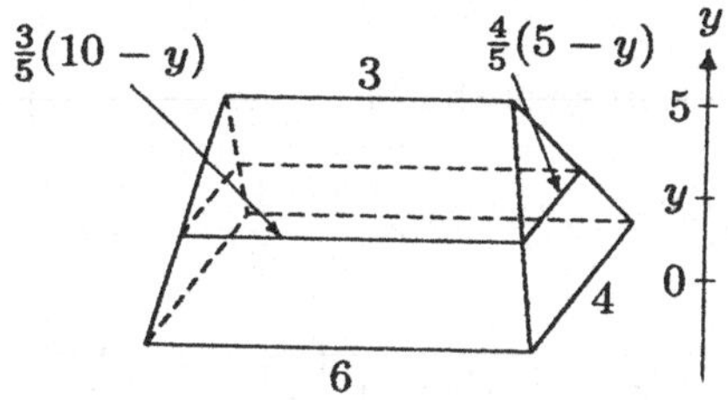

Exercise 11

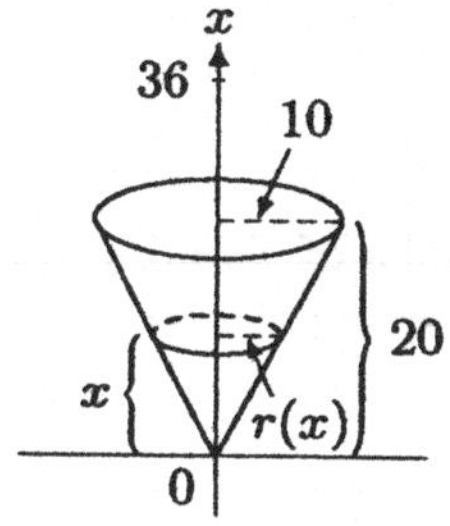

Exercise 19

9. Let a hexagonal cross-section x feet from the base have a side $s(x)$ feet long. Then $s(x)/2 = (10-x)/10$, so that $s(x) = \frac{1}{5}(10-x)$. The area is $A(x) = 6(\sqrt{3}/4)(s(x))^2 = [(3\sqrt{3})/50](10-x)^2$. Thus the volume is given by

$$V = \int_0^{10} \frac{3\sqrt{3}}{50}(10-x)^2\,dx = \frac{3\sqrt{3}}{50}\left(\frac{-1}{3}\right)(10-x)^3\Big|_0^{10} = 20\sqrt{3}\,\text{(cubic feet)}.$$

11. We place the origin of the y axis at the base of the solid. As shown in the figure, the cross section at y has length $\frac{3}{5}(10-y)$ and width $\frac{4}{5}(5-y)$. Thus the cross-sectional area at y is given by $A(y) = \frac{3}{5}(10-y)\frac{4}{5}(5-y) = \frac{12}{25}(50-15y+y^2)$. By (2) of Section 6.1, the volume is given by $V = \int_0^5 A(y)\,dy = \int_0^5 \frac{12}{25}(50-15y+y^2)\,dy = \frac{12}{25}\left(50y - \frac{15}{2}y^2 + \frac{1}{3}y^3\right)\Big|_0^5 = \frac{12}{25}\left(250 - \frac{375}{2} + \frac{125}{3}\right) = 50$.

13. $$L = \int_0^1 \sqrt{1+\left(e^x - \frac{1}{4}e^{-x}\right)^2}\,dx = \int_0^1 \sqrt{e^{2x} + \frac{1}{2} + \frac{1}{16}e^{-2x}}\,dx = \int_0^1 \sqrt{\left(e^x + \frac{1}{4}e^{-x}\right)^2}\,dx$$
$$= \int_0^1 \left(e^x + \frac{1}{4}e^{-x}\right)dx = \left(e^x - \frac{1}{4}e^{-x}\right)\Big|_0^1 = \left(e - \frac{1}{4}e^{-1}\right) - \left(1 - \frac{1}{4}\right) = e - \frac{1}{4}e^{-1} - \frac{3}{4}$$

15. $$S = \int_0^1 2\pi\left(e^x + \frac{1}{4}e^{-x}\right)\sqrt{1+\left(e^x - \frac{1}{4}e^{-x}\right)^2}\,dx = 2\pi\int_0^1 \left(e^x + \frac{1}{4}e^{-x}\right)\sqrt{1+\left(e^{2x} - \frac{1}{2} + \frac{1}{16}e^{-2x}\right)}\,dx$$
$$= 2\pi\int_0^1 \left(e^x + \frac{1}{4}e^{-x}\right)\sqrt{e^{2x} + \frac{1}{2} + \frac{1}{16}e^{-2x}}\,dx = 2\pi\int_0^1 \left(e^x + \frac{1}{4}e^{-x}\right)\left(e^x + \frac{1}{4}e^{-x}\right)dx$$
$$= 2\pi\int_0^1 \left(e^{2x} + \frac{1}{2} + \frac{1}{16}e^{-2x}\right)dx = 2\pi\left(\frac{1}{2}e^{2x} + \frac{1}{2}x - \frac{1}{32}e^{-2x}\right)\Big|_0^1 = \pi\left(e^2 - \frac{1}{16}e^{-2} + \frac{1}{16}\right)$$

17. We have $4 = \int_0^{1/2} kx\,dx = (k/2)x^2\big|_0^{1/2} = k/8$, so that $k = 32$. Thus $W = \int_1^{3/2} 32x\,dx = 16x^2\big|_0^{3/2} = 20$ (foot-pounds).

19. If the tank is placed with respect to the axes as in the figure, then $l = 36$, and the radius $r(x)$ of a cross-section x feet from the bottom satisfies $r(x)/10 = x/20$, or $r(x) = \frac{1}{2}x$, so that the cross-sectional area is given by $A(x) = \pi(\frac{1}{2}x)^2 = (\pi/4)x^2$. Next we find the original height of water in the tank. If

21. The graphs of f and g intersect for (x, y) such that $x = y = x^2$, which means that $x = 0$ or $x = 1$.

$$M_x = \int_0^1 \frac{1}{2}[x^2 - (x^2)^2]\,dx = \frac{1}{2}\left(\frac{1}{3}x^3 - \frac{1}{5}x^5\right)\Big|_0^1 = \frac{1}{15}$$

$$M_y = \int_0^1 x(x - x^2)\,dx = \left(\frac{1}{3}x^3 - \frac{1}{4}x^4\right)\Big|_0^1 = \frac{1}{12}$$

$$A = \int_0^1 (x - x^2)\,dx = \left(\frac{1}{2}x^2 - \frac{1}{3}x^3\right)\Big|_0^1 = \frac{1}{6}$$

$$\bar{x} = \frac{1/12}{1/6} = \frac{1}{2}; \; \bar{y} = \frac{1/15}{1/6} = \frac{2}{5}; \; (\bar{x}, \bar{y}) = \left(\frac{1}{2}, \frac{2}{5}\right)$$

23. a. $$M_x = \int_1^3 \frac{1}{2}\left[\left(x + \frac{c}{x}\right)^2 - x^2\right] dx = \frac{1}{2}\int_1^3 \left(2c + \frac{c^2}{x^2}\right) dx = \frac{1}{2}\left(2cx - \frac{c^2}{x}\right)\Big|_1^3$$

$$= \frac{1}{2}\left[\left(6c - \frac{c^2}{3}\right) - (2c - c^2)\right] = 2c + \frac{1}{3}c^2$$

$$M_y = \int_1^3 x\left[\left(x + \frac{c}{x}\right) - x\right] dx = \int_1^3 c\,dx = 2c$$

$$A = \int_1^3 \left[\left(x + \frac{c}{x}\right) - x\right] dx = \int_1^3 \frac{c}{x}\,dx = c\ln x\Big|_1^3 = c\ln 3$$

$$\bar{x} = \frac{M_y}{A} = \frac{2c}{c\ln 3} = \frac{2}{\ln 3}; \; \bar{y} = \frac{M_x}{A} = \frac{2c + c^2/3}{c\ln 3} = \frac{6 + c}{3\ln 3}$$

Thus the center of gravity is $\left(\frac{2}{\ln 3}, \frac{6+c}{3\ln 3}\right)$.

b. Using the Theorem of Pappus and Guldin, we find that the volume V_1 of the solid obtained by revolving R about the y axis is given by

$$V_1 = 2\pi\bar{x}A = 2\pi\left(\frac{2}{\ln 3}\right)(c\ln 3) = 4\pi c.$$

Similarly, the volume V_2 of the solid obtained by revolving R about the x axis is given by

$$V_2 = 2\pi\bar{y}A = 2\pi\left(\frac{6+c}{3\ln 3}\right)(c\ln 3) = 4\pi c + \frac{2}{3}\pi c^2.$$

25.

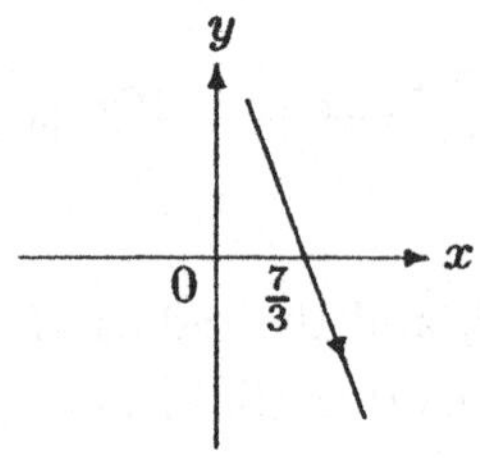

27. $L = \int_0^1 \sqrt{\left[\frac{1}{2}(e^{2t} + e^{-2t}) - 1\right]^2 + (e^t - e^{-t})^2}\, dt$

$$= \int_0^1 \sqrt{\frac{1}{4}(e^{4t} + 2 + e^{-4t}) - (e^{2t} + e^{-2t}) + 1 + (e^{2t} - 2 + e^{-2t})}\, dt$$

$$= \int_0^1 \sqrt{\frac{1}{4}e^{4t} - \frac{1}{2} + \frac{1}{4}e^{-4t}}\, dt = \int_0^1 \sqrt{\left(\frac{1}{2}e^{2t} - \frac{1}{2}e^{-2t}\right)^2}\, dt$$

$$= \int_0^1 \left(\frac{1}{2}e^{2t} - \frac{1}{2}e^{-2t}\right) dt = \frac{1}{4}(e^{2t} + e^{-2t})\Big|_0^1$$

$$= \frac{1}{4}(e^2 + e^{-2}) - \frac{1}{4}(1+1) = \frac{1}{4}(e^2 - 2 + e^{-2}) = \frac{1}{4}(e + e^{-1})^2$$

29. $S = \int_1^{\sqrt{3}} (2\pi)\frac{1}{2}\sqrt{(t^2)^2 + (t)^2}\, dt = 2\pi \int_1^{\sqrt{3}} \frac{1}{2}t^2\sqrt{t^4 + t^2}\, dt = 2\pi \int_1^{\sqrt{3}} \frac{1}{2}t^3\sqrt{t^2+1}\, dt$

$$\overset{u = t^2+1}{=} 2\pi \int_2^4 \frac{1}{4}(u-1)\sqrt{u}\, du = \frac{1}{2}\pi \int_2^4 (u^{3/2} - u^{1/2})\, du = \frac{1}{2}\pi \left(\frac{2}{5}u^{5/2} - \frac{2}{3}u^{3/2}\right)\Big|_2^4$$

$$= \frac{1}{2}\pi\left(\frac{64}{5} - \frac{16}{3}\right) - \frac{1}{2}\pi\left(\frac{8}{5}\sqrt{2} - \frac{4}{3}\sqrt{2}\right) = \frac{1}{15}\pi(56 - 2\sqrt{2})$$

31. a. Since $V = xA$, we have $F = pA = p(V/x) = c/x$.

b. $W = \int_1^2 \frac{c}{x}\, dx = c\ln|x|\Big|_1^2 = c\ln 2$

33. a. The force F on the bottom that is due to the olive oil is given by

$$F = (57.3)(\text{depth of barrel})(\text{area of bottom}) = (57.3)(3)[\pi(1)^2] = 171.9\pi \text{ (pounds)}.$$

b. We take the origin to be at the center of the end of the barrel, so $c = 1$. Then $w(x) = 2\sqrt{1-x^2}$, so that

$$F = \int_{-1}^1 (57.3)(1-x)[2\sqrt{1-x^2}]\, dx$$

$$= 114.6\int_{-1}^1 \sqrt{1-x^2}\, dx - 114.6\int_{-1}^1 x\sqrt{1-x^2}\, dx.$$

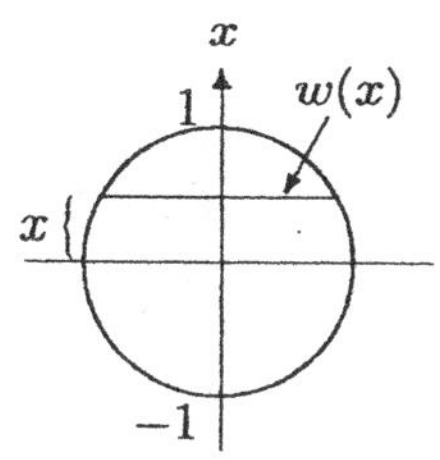

Now $\int_{-1}^1 \sqrt{1-x^2}\, dx$ = area of semicircle of radius $1 = \pi/2$. Since $x\sqrt{1-x^2}$ is an odd function on $[-1, 1]$, $\int_{-1}^1 x\sqrt{1-x^2}\, dx = 0$. Thus $F = 114.6(\pi/2) + 0 = 57.3\pi$ (pounds).

35. a. By (3) in Section 6.1 with y replacing x, we have $V = \int_0^h \pi[f(y)]^2\, dy$, so

$$\frac{dV}{dt} = \frac{dV}{dh}\frac{dh}{dt} = \pi[f(h)]^2\frac{dh}{dt}.$$

b. Using the result of part (a), the equation $dV/dt = cA\sqrt{h}$, and the assumption that $dh/dt = k$, we find that $cA\sqrt{h} = dV/dt = \pi[f(h)]^2(dh/dt) = \pi[f(h)]^2k$. Solving for $f(h)$, we find that

$$f(h) = \left(\frac{cA\sqrt{h}}{\pi k}\right)^{1/2} = \sqrt{\frac{cA}{\pi k}}\, h^{1/4} \quad \text{for } 0 \le h \le b.$$

37. Consider the partition $P = \{r_0, \ldots, r_n\}$ of the interval $[0, R]$. For any k with $1 \le k \le n$, the volume ΔV_k of the spherical shell each of whose points lies at a distance r from the center of the star, where $r_{k-1} \le r \le r_k$, is approximately $\frac{4}{3}\pi r_k^3 - \frac{4}{3}\pi r_{k-1}^3 = \frac{4}{3}\pi(r_k^2 + r_k r_{k-1} + r_{k-1}^2)\Delta r_k$. Thus the mass M_k of the shell is approximately $D(r_k)\Delta V_k$. Therefore the mass M of the star is approximately

$$\sum_{k=1}^{n} D(r_k)\Delta V_k = \sum_{k=1}^{n} D(r_k)\left[\frac{4}{3}\pi(r_k^2 + r_k r_{k-1} + r_{k-1}^2)\right]\Delta r_k$$

which is not a Riemann sum, but which approximates the Riemann sum $\sum_{k=1}^{n} D(r_k)\left[\frac{4}{3}\pi(3r_k^2)\right]\Delta r_k$ of $\int_0^R 4\pi r^2 D(r)\,dr$. Consequently $M = \int_0^R 4\pi r^2 D(r)\,dr$.

Cumulative Review(Chapters 1–5)

1. $$\lim_{x\to 3^+} \frac{\sqrt{x^2-9}}{x-3} = \lim_{x\to 3^+} \frac{\sqrt{x-3}\,\sqrt{x+3}}{(\sqrt{x-3})^2} = \lim_{x\to 3^+} \frac{\sqrt{x+3}}{\sqrt{x-3}} = \infty$$

3. Since

$$-|\sin x| \le \sin x \sin\frac{1}{x} \le |\sin x| \quad \text{and} \quad \lim_{x\to 0} |\sin x| = 0 = \lim_{x\to 0}(-|\sin x|)$$

the Squeezing Theorem implies that $\lim_{x\to 0} \sin x \sin(1/x) = 0$.

5. a. Since division by 0 is undefined and 0 is not in the domain of the natural logarithm, the domain consists of all x such that $1 + x \ne 0$ and $1 - x \ne 0$, that is, all numbers except -1 and 1.

b. Since $\ln(1/b) = -\ln b$, we have

$$f(-x) = \ln\left|\frac{1-(-x)}{1-x}\right| = \ln\left|\frac{1+x}{1-x}\right| = -\ln\left|\frac{1-x}{1+x}\right| = -f(x).$$

Therefore f is an odd function.

c. Since f is continuous at every number in its domain, the only possible vertical asymptotes are $x = -1$ and $x = 1$. By the version of the Substitution Rule for infinite limits (with $y = 1 - x$),

$$\lim_{x\to 1} \ln\left|\frac{1-x}{1+x}\right| = \lim_{y\to 0} \ln\left|\frac{y}{2-y}\right| = -\infty.$$

Therefore the line $x = 1$ is a vertical asymptote. Since the graph of f is symmetric with respect to the origin (because f is odd), the line $x = -1$ is also a vertical asymptote.

7. $f'(x) = \cos x - (x-1)\sin x$

$f''(x) = -\sin x - \sin x - (x-1)\cos x = -2\sin x - (x-1)\cos x$

$f^{(3)}(x) = -2\cos x - \cos x + (x-1)\sin x = -3\cos x + (x-1)\sin x$

$f^{(4)}(x) = 3\sin x + \sin x + (x-1)\cos x = 4\sin x + (x-1)\cos x$

In general, $f^{(2n)}(x) = (-1)^n[2n\sin x + (x-1)\cos x]$. Thus $f^{(24)}(x) = 24\sin x + (x-1)\cos x$.

9. If h denotes the height of the equilateral triangle, then the area A is given by $A = (\frac{1}{2}h)(2h/\sqrt{3}) = h^2/\sqrt{3}$. Since the area is by hypothesis growing at the rate of 9 square inches per minute, $9 = dA/dt = (2h/\sqrt{3})(dh/dt)$, so $dh/dt = (9\sqrt{3})/2h$. When $A = \sqrt{3}$, we have $\sqrt{3} = h^2/\sqrt{3}$, so $h = \sqrt{3}$, and thus $dh/dt = (9\sqrt{3})/(2\sqrt{3}) = \frac{9}{2}$ (inches per minute).

11. $f'(x) = 5x^4 - 10x^2 + 5 = 5(x^2-1)^2$; $f''(x) = 20x^3 - 20x = 20x(x^2-1)$;

increasing on $(-\infty, \infty)$; concave upward on $(-1, 0)$ and $(1, \infty)$, and concave downward on $(-\infty, -1)$ and $(0, 1)$; inflection points are $(-1, -\frac{8}{3})$, $(0, 0)$, and $(1, \frac{8}{3})$; symmetry with respect to origin.

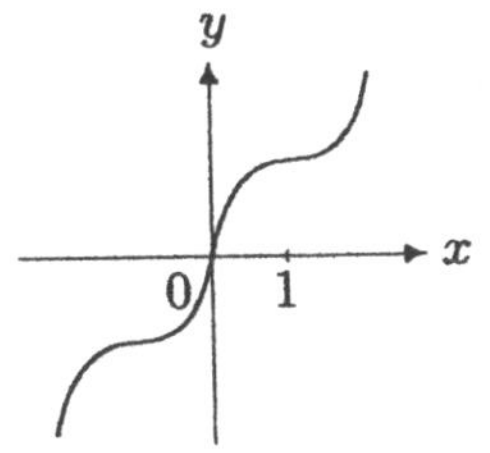

13. $a(t) = \frac{1}{2}t + \cos t$, and we are to find the maximum value of a on $[0, \pi/2]$. Now $a'(t) = \frac{1}{2} - \sin t$, so that $a'(t) = 0$ if $\sin t = \frac{1}{2}$. Since t must lie in $[0, \pi/2]$, it follows that $t = \pi/6$. Since $a''(t) = -\cos t < 0$ for t in $(0, \pi/2)$, we conclude from (1) in Section 4.6 and the Second Derivative Test that $a(\pi/6)$ is the maximum value, so that the acceleration is maximum in $[0, \pi/2]$ for $t = \pi/6$.

15. It suffices to find the points on the graph of $y = 2/(1+x^2)$ for which the square S of the distance to the origin is minimized. For (x, y) on the graph, S is given by

$$S = (x-0)^2 + (y-0)^2 = x^2 + \left(\frac{2}{1+x^2}\right)^2 = x^2 + \frac{4}{(1+x^2)^2}.$$

Thus

$$S'(x) = 2x - \frac{16x}{(1+x^2)^3} = 2x\left(1 - \frac{8}{(1+x^2)^3}\right)$$

so that $S'(x) = 0$ if $x = -1$, $x = 0$, or $x = 1$. Since S' changes from negative to positive at -1 and 1, and from positive to negative at 0, it follows that S assumes its minimum value at -1 and 1 (at which points S takes on the same value, namely 2). Thus the points $(-1, 1)$ and $(1, 1)$ are the points on the graph that are closest to the origin.

17. Since $g(x) = \int_x^{x+\pi} \sin^{2/3} t\, dt = \int_0^{x+\pi} \sin^{2/3} t\, dt - \int_0^x \sin^{2/3} t\, dt$, we have

$$g'(x) = \sin^{2/3}(x+\pi) - \sin^{2/3} x = (-\sin x)^{2/3} - \sin^{2/3} x = 0.$$

Thus g is a constant function, by Theorem 4.6(a).

19. Let $u = 2 + \cos x$, so that $du = -\sin x\, dx$; if $x = 0$, then $u = 3$, and if $x = \pi$, then $u = 1$;

$$\int_0^\pi \frac{\sin x}{2+\cos x}\, dx = \int_3^1 \frac{1}{u}(-1)\, du = -\ln u\Big|_3^1 = \ln 3.$$

21. Let $u = \ln t$, so that $du = (1/t)\, dt$;

$$\int \frac{1}{t}(\ln t)^{5/3}\, dt = \int u^{5/3}\, du = \frac{3}{8}u^{8/3} + C = \frac{3}{8}(\ln t)^{8/3} + C.$$

Chapter 7

Inverse Functions, l' Hôpital's Rule, and Differential Equations

7.1 Inverse Functions

1. $f'(x) = 5x^4 \geq 0$, and $f'(x) > 0$ for $x \neq 0$, so f has an inverse. Domain of f^{-1}: $(-\infty, \infty)$; range of f^{-1}: $(-\infty, \infty)$.

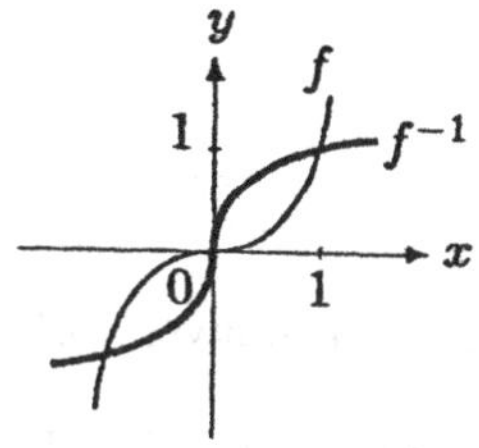

3. $f(-x) = f(x)$ for all x, so f does not have an inverse.

5. $f'(t) = -\frac{1}{2}(4-t)^{-1/2} < 0$ for $t < 4$, so f has an inverse. Domain of f^{-1}: $[0, \infty)$; range of f^{-1}: $(-\infty, 4]$.

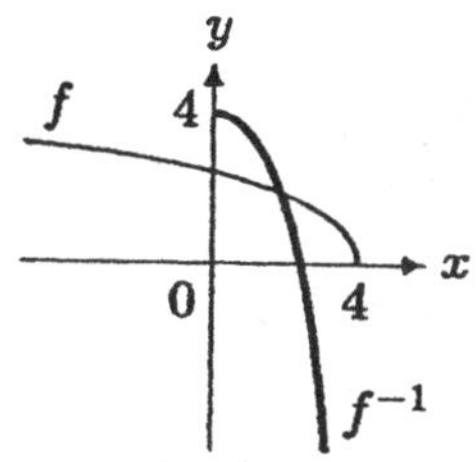

7. $f(x) = 0$ for $x \leq 0$, so f does not have an inverse.

9. $f'(x) = 1 - \cos x \geq 0$, and $f'(x) > 0$ for $x \neq 2n\pi$ for any integer n. Thus f is increasing on any bounded interval and hence on $(-\infty, \infty)$. Therefore f has an inverse. Domain of f^{-1}: $(-\infty, \infty)$; range of f^{-1}: $(-\infty, \infty)$.

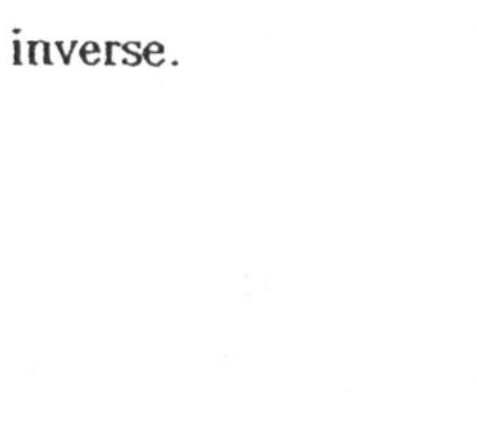

11. $f(n\pi) = 0$ for every integer n, so f does not have an inverse.

13. $g'(t) = -1/(3-t) < 0$, so g has an inverse. Domain of g^{-1}: $(-\infty, \infty)$; range of g^{-1}: $(-\infty, 3)$.

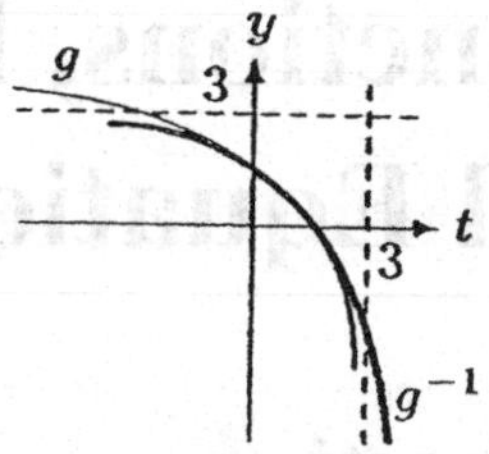

15. f has no inverse.

17. f has an inverse.

19. k is the inverse of f; h is the inverse of g.

21. Let x_0 be any number in the domain of f. Then $f(a+x_0) = f(x_0)$, but $a + x_0 \neq x_0$ since $a \neq 0$. Thus f does not satisfy (3) and hence does not have an inverse.

23. $f(\pi + x) = f(x)$ for all x in the domain of f, so f does not have an inverse by Exercise 21.

25. $y = -4x^3 - 1$; $x^3 = \dfrac{y+1}{-4}$; $x = -\sqrt[3]{\dfrac{y+1}{4}}$, so $f^{-1}(x) = -\sqrt[3]{\dfrac{x+1}{4}}$.

27. $y = \sqrt{1+x}$; $y^2 = 1 + x$; $x = y^2 - 1$, so $g^{-1}(x) = x^2 - 1$ for $x \geq 0$.

29. $y = \dfrac{t-1}{t+1}$; $y(t+1) = t - 1$; $t(y-1) = -1 - y$; $t = \dfrac{y+1}{1-y}$, so $k^{-1}(t) = \dfrac{t+1}{1-t}$.

31. $f'(x) = 2x$. Since $f'(x) > 0$ for $x > 0$, f has an inverse on $[0, \infty)$. Since $f'(x) < 0$ for $x < 0$, f has an inverse on $(-\infty, 0]$.

33. $f'(x) = 3x^2 - 5$. Since $f'(x) > 0$ for $x > \sqrt{\frac{5}{3}}$, f has an inverse on $\left[\sqrt{\frac{5}{3}}, \infty\right)$. Since $f'(x) > 0$ for $x < -\sqrt{\frac{5}{3}}$, f has an inverse on $\left(-\infty, -\sqrt{\frac{5}{3}}\right]$. Since $f'(x) < 0$ for $-\sqrt{\frac{5}{3}} < x < \sqrt{\frac{5}{3}}$, f has an inverse on $\left[-\sqrt{\frac{5}{3}}, \sqrt{\frac{5}{3}}\right]$.

35. $f'(x) = -2x/(1+x^2)^2$. Since $f'(x) > 0$ for $x < 0$, f has an inverse on $(-\infty, 0]$. Since $f'(x) < 0$ for $x > 0$, f has an inverse on $[0, \infty)$.

37. $f'(x) = -\sin x$. If n is any integer, then $f'(x) < 0$ for $2n\pi < x < (2n+1)\pi$, and $f'(x) > 0$ for $(2n+1)\pi < x < (2n+2)\pi$. Thus f has an inverse on any interval of the form $[n\pi, (n+1)\pi]$, where n is an integer.

39. $f'(x) = 2\sin x \cos x = \sin 2x$. If n is any integer, then $f'(x) > 0$ for $n\pi < x < n\pi + \pi/2$, and $f'(x) < 0$ for $n\pi + \pi/2 < x < (n+1)\pi$. Thus f has an inverse on any interval of the form $[n\pi/2, (n+1)\pi/2]$, where n is an integer.

41. $f(-1) = 6$. Since $f'(x) = 3x^2$, we have $f'(-1) = 3$. Thus $(f^{-1})'(6) = 1/[f'(-1)] = \frac{1}{3}$.

43. $f(0) = 0$. Since $f'(x) = 1 + \cos x$, we have $f'(0) = 2$. Thus $(f^{-1})'(0) = 1/[f'(0)] = \frac{1}{2}$.

45. $f(1) = 0$. Since $f'(x) = 4/x$, we have $f'(1) = 4$. Thus $(f^{-1})'(0) = 1/[f'(1)] = \frac{1}{4}$.

47. $f(-1) = -2$. Since $f'(t) = 3 + 3/t^4$, we have $f'(-1) = 6$. Thus $(f^{-1})'(-2) = 1/[f'(-1)] = \frac{1}{6}$.

49. $\dfrac{dx}{dy} = \dfrac{1}{dy/dx} = \dfrac{1}{9x^8 + 7}$.

51. $\dfrac{dx}{dy} = \dfrac{1}{dy/dx} = \dfrac{1}{[1/(x^3+1)]3x^2} = \dfrac{x^3+1}{3x^2}$

53. $\dfrac{dx}{dy} = \dfrac{1}{dy/dx} = \dfrac{1}{\cos x}$ for $-\dfrac{\pi}{2} < x < \dfrac{\pi}{2}$

55. We apply the criterion that follows Theorem 7.3. Since $f'(x) > 0$ for all x in $(4, 6)$ or in $(8, 9)$, f has an inverse on $[4, 6]$ and on $[8, 9]$. Since $f'(x) < 0$ for all x in $(1, 4)$ or in $(6, 8)$, f has an inverse on $[1, 4]$ and on $[6, 8]$.

57. a. By Theorem 5.12, $f'(x) = \sqrt{1 + x^4} > 0$ for all x, so f has an inverse.

 b. Since $f'(x) = \sqrt{1 + x^4}$, we have $f'(1) = \sqrt{2}$. Thus $(f^{-1})'(c) = 1/[f'(1)] = 1/\sqrt{2} = \sqrt{2}/2$.

59. a. If f is a polynomial of even degree, then $f(-x) = f(x)$ for all x, so f cannot have an inverse.

 b. If f is a polynomial of odd degree, then f can have an inverse. An example is given by $f(x) = x^3 + 5x$.

61. Let $y = (g \circ f)(x)$. Then

$$(f^{-1} \circ g^{-1})(y) = (f^{-1} \circ g^{-1})((g \circ f)(x)) = f^{-1}\Big(g^{-1}\big(g(f(x))\big)\Big) = f^{-1}(f(x)) = x.$$

Let $x = (f^{-1} \circ g^{-1})(y)$. Then

$$(g \circ f)(x) = (g \circ f)(f^{-1} \circ g^{-1})(y) = g\Big(f\big(f^{-1}(g^{-1}(y))\big)\Big) = g(g^{-1}(y)) = y.$$

Thus $g \circ f$ has an inverse, and $(g \circ f)^{-1} = f^{-1} \circ g^{-1}$ by Definition 6.1.

63. Since f^{-1} exists, Exercise 63 says that $g \circ f$ has an inverse, and since $f^{-1}(x) = -x$, we find that

$$(g \circ f)^{-1}(x) = (f^{-1} \circ g^{-1})(x) = f^{-1}(g^{-1}(x)) = -g^{-1}(x).$$

65. Let $k(x) = f^{-1}(x) - a$. Then $g\big(k(x)\big) = f\big((f^{-1}(x) - a) + a\big) = f\big(f^{-1}(x)\big) = x$, and $k\big(g(x)\big) = f^{-1}\big(f(x) + a)\big) - a = x + a - a = x$. Thus g has an inverse, and $g^{-1}(x) = k(x) = f^{-1}(x) - a$.

67. $\left.\dfrac{dy}{dx}\right|_{x=2} = 12$ and $\left.\dfrac{dx}{dy}\right|_{y=8} = \dfrac{1}{12} = \dfrac{1}{(dy/dx)|_{x=2}}$ but $\left.\dfrac{dx}{dy}\right|_{y=2} = \dfrac{1}{3(2)^{2/3}} \neq \dfrac{1}{12} = \dfrac{1}{(dy/dx)|_{x=2}}$.

69. a. $f(\pi) = \pi$, $f'(x) = 1 + \cos x$, and $f'(\pi) = 1 - 1 = 0$. By Exercise 70, $(f^{-1})(\pi)$ does not exist.

b. $f(0) = -4$, $f'(x) = 5x^4 + 3x^2$, and $f'(0) = 5 \cdot 0 + 3 \cdot 0 = 0$. By Exercise 70, $(f^{-1})'(-4)$ does not exist.

71. Let $f(x) = \left[(x-1)^{1/3} + 1\right]^{1/2}$. If $y = f(x)$, then $y^2 = (x-1)^{1/3} + 1$, so $x = (y^2 - 1)^3 + 1$. Thus f^{-1} exists, and $f^{-1}(y) = (y^2 - 1)^3 + 1$. Now we use Exercise 72(b), with $a = 0$ and $b = 1$. Since $f(0) = 0$ and $f(1) = 1$, we obtain

$$\int_0^1 \left[(x-1)^{1/3} + 1\right]^{1/2} dx = \int_0^1 f(x)\,dx = 1f(1) - 0f(0) - \int_0^1 f^{-1}(y)\,dy$$

$$= 1 - \int_0^1 [(y^2-1)^3 + 1]\,dy = 1 - \int_0^1 (y^6 - 3y^4 + 3y^2)\,dy = 1 - \left(\frac{1}{7}y^7 - \frac{3}{5}y^5 + y^3\right)\Big|_0^1 = \frac{16}{35}.$$

73. From (8) with $a = f^{-1}(x)$ and $c = x$, we have

$$(f^{-1})'(x) = \frac{1}{f'(f^{-1}(x))}.$$

Differentiating both sides of this equation, we obtain

$$(f^{-1})''(x) = \frac{-f''(f^{-1}(x)) \cdot (f^{-1})'(x)}{[f'(f^{-1}(x))]^2} = \frac{-f''(f^{-1}(x))}{[f'(f^{-1}(x))]^2} \cdot \frac{1}{f'(f^{-1}(x))} = \frac{-f''(f^{-1}(x))}{[f'(f^{-1}(x))]^3}.$$

75. The length of the graph of f is the same as the length of the graph of f^{-1}. Now $f^{-1}(x) = x^{3/2}$ for $0 \le x \le 4$, and $(f^{-1})'(x) = \frac{3}{2}x^{1/2}$. Then $L = \int_0^4 \sqrt{1 + [(f^{-1})'(x)]^2}\,dx = \int_0^4 \sqrt{1 + \frac{9}{4}x}\,dx = \frac{2}{3}\left(1 + \frac{9}{4}x\right)^{3/2}\left(\frac{4}{9}\right)\Big|_0^4 = \frac{8}{27}(10)^{3/2} - \frac{8}{27} = \frac{8}{27}(10\sqrt{10} - 1)$.

77. For $x \ge 0$ and $y \ge 0$ we have

$$y = 2.54x \quad \text{if and only if} \quad x = \frac{1}{2.54}y.$$

Thus $y = f(x)$ if and only if $x = g(y)$. Therefore f and g are inverses of one another.

7.2 The Natural Exponential Function

1. $f'(x) = \dfrac{1}{e^x + e^{-x}}(e^x - e^{-x}) = \dfrac{e^x - e^{-x}}{e^x + e^{-x}}$

3. $\dfrac{dy}{dx} = \dfrac{e^x(e^x - 1) - (e^x + 1)(e^x)}{(e^x - 1)^2} = \dfrac{-2e^x}{(e^x - 1)^2}$

5. $2xe^y + x^2e^y\dfrac{dy}{dx} = \dfrac{1}{xy}\left(y + x\dfrac{dy}{dx}\right)$, so $\left(x^2e^y - \dfrac{1}{y}\right)\dfrac{dy}{dx} = \dfrac{1}{x} - 2xe^y$ and thus

$$\frac{dy}{dx} = \frac{1/x - 2xe^y}{x^2e^y - 1/y} = \frac{y - 2x^2ye^y}{x^3ye^y - x}.$$

7. a. $f(x) = e^x \sin x$; $f'(x) = e^x(\sin x + \cos x)$; $f''(x) = e^x(\sin x + \cos x) + e^x(\cos x - \sin x) = 2e^x \cos x$; $f^{(3)}(x) = 2e^x \cos x - 2e^x \sin x = 2e^x(\cos x - \sin x)$; $f^{(4)}(x) = 2e^x(\cos x - \sin x) + 2e^x(-\sin x - \cos x) = -4e^x \sin x = -4f(x)$. Thus $f^{(8)}(x) = -4f^{(4)}(x) = (-4)(-4f(x)) = 16f(x) = 16e^x \sin x$.

   b. Every 8th derivative of f multiplies f by 16. Thus $f^{(80)}(x) = 16^{10}f(x) = 16^{10}e^x \sin x$.

9. $\dfrac{dy}{dx} = e^{2x} + 2xe^{2x} = (1 + 2x)e^{2x}$, so $x\dfrac{dy}{dx} = (xe^{2x})(1 + 2x) = y(1 + 2x)$.

11. $f'(x) = e^x - e^{-x}$; $f''(x) = e^x + e^{-x}$; $f'(x) = 0$ if $e^x = e^{-x}$, or $e^{2x} = 1$, or $x = 0$; $f''(0) = 2 > 0$; relative minimum value is $f(0) = 2$; concave upward on $(-\infty, \infty)$; symmetric with respect to the y axis.

Exercise 11

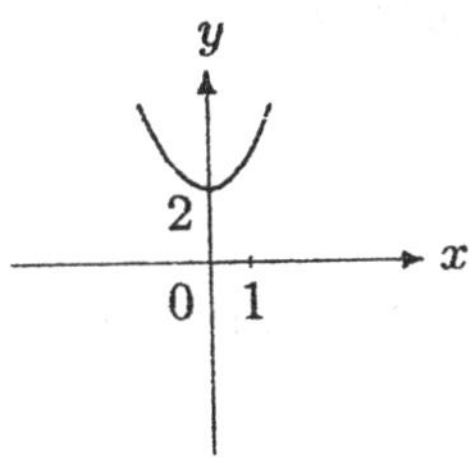

13. $\displaystyle\int e^{ex}\,dx \overset{u=ex}{=} \int e^u \cdot \frac{1}{e}\,du = \frac{1}{e}e^u + C = \frac{1}{e}e^{ex} + C = e^{ex-1} + C$

15. $\displaystyle\int_0^{\pi/3} \frac{e^{\tan y}}{\cos^2 y}\,dy = \int_0^{\pi/3} e^{\tan y}\sec^2 y\,dy \overset{u=\tan y}{=} \int_0^{\sqrt{3}} e^u\,du = e^u\Big|_0^{\sqrt{3}} = e^{\sqrt{3}} - e^0 = e^{\sqrt{3}} - 1$

17. $\displaystyle\int \frac{e^{2t}}{\sqrt{e^{2t} - 4}}\,dt \overset{u=e^{2t}-4}{=} \int \frac{1}{2\sqrt{u}}\,du = \sqrt{u} + C = \sqrt{e^{2t} - 4} + C$

19. $\displaystyle\int \frac{e^{-t}\ln(1 + e^{-t})}{1 + e^{-t}}\,dt \overset{u=\ln(1+e^{-t})}{=} \int -u\,du = -\frac{1}{2}u^2 + C = -\frac{1}{2}[\ln(1 + e^{-t})]^2 + C$

21. $\displaystyle\int \frac{1}{1 + e^{-x}}\,dx = \int \frac{e^x}{e^x + 1}\,dx \overset{u=e^x+1}{=} \int \frac{1}{u}\,du = \ln|u| + C = \ln(e^x + 1) + C$

23. $\displaystyle\int e^{(x - e^x)}\,dx = \int e^x e^{-(e^x)}\,dx \overset{u=e^x}{=} \int e^{-u}\,du = -e^{-u} + C = -e^{-(e^x)} + C$

25. The slope of the line $y = -4x - 7$ is -4, so we seek the value of x for which $(d/dx)(e^{x^2-4}) = -4$. This is equivalent to $2xe^{x^2-4} = -4$, or $xe^{x^2-4} = -2$. Now this equation is satisfied if $x = -2$. The corresponding point on the graph of the equation $y = e^{x^2-4}$ is $(-2, 1)$.

27. $y = \dfrac{e^x - 1}{e^x + 1}$; $(e^x + 1)y = e^x - 1$; $e^x = \dfrac{1+y}{1-y}$; $x = \ln\left(\dfrac{1+y}{1-y}\right)$, so $f^{-1}(x) = \ln\left(\dfrac{1+x}{1-x}\right)$.

29. The equation $e^{-x} = 2 - x$ is equivalent to $e^{-x} + x - 2 = 0$. Using the Newton-Raphson method with initial value 1, we obtain 1.84140566 as an approximate solution of $e^{-x} = 2 - x$.

33. The graphs intersect at (x, y) if $e^{2x} = y = e^{-2x}$, or $e^{4x} = 1$, or $x = 0$. Since $e^{2x} \geq e^{-2x}$ on $[0, \frac{1}{2}]$, we have

$$A = \int_0^{1/2} (e^{2x} - e^{-2x})\,dx = \left(\frac{1}{2}e^{2x} + \frac{1}{2}e^{-2x}\right)\Big|_0^{1/2} = \frac{1}{2}(e + e^{-1}) - 1.$$

35. $V = \displaystyle\int_0^1 2\pi x \left(e^{(x^2)} - e^{-x^2}\right) dx \overset{u=x^2}{=} \int_0^1 \pi(e^u - e^{-u})\,du$

$= \pi(e^u + e^{-u})\big|_0^1 = \pi[(e + e^{-1}) - (1+1)] = \pi(e + e^{-1} - 2)$

37. $L = \displaystyle\int_0^\pi \sqrt{(e^t \sin t + e^t \cos t)^2 + (e^t \cos t - e^t \sin t)^2}\,dt$

$= \displaystyle\int_0^\pi \sqrt{e^{2t}(\sin^2 t + 2\sin t \cos t + 2\cos^2 t - 2\cos t \sin t + \sin^2 t)}\,dt$

$= \displaystyle\int_0^\pi e^t\sqrt{2}\,dt = \sqrt{2}\,e^t\Big|_0^\pi = \sqrt{2}\,(e^\pi - 1)$

39. Let $f(x) = \ln x$, so that $f^{-1}(y) = e^y$. Also let $a = 1$ and $b = e$, so that $f(a) = \ln 1 = 0$ and $f(b) = \ln e = 1$. Therefore by the formula,

$$\int_1^e \ln x\,dx = e \ln e - 1 \ln 1 - \int_0^1 f^{-1}(y)\,dy = e - \int_0^1 e^y\,dy = e - e^y\big|_0^1 = e - (e - 1) = 1.$$

41. a. Since $f(x + h) = f(x)f(h)$, $f(0) = 1$, and $f'(0) = 1$, we find that for any x,

$$f'(x) = \lim_{h\to 0} \frac{f(x+h) - f(x)}{h} = \lim_{h\to 0} \frac{f(x)f(h) - f(x)}{h} = f(x) \lim_{h\to 0} \frac{f(h) - 1}{h}$$

$$= f(x) \lim_{h\to 0} \frac{f(h) - f(0)}{h - 0} = f(x)f'(0) = f(x) \cdot 1 = f(x).$$

b. By (a) and the Product Rule,

$$\frac{d}{dx}(e^{-x} f(x)) = -e^{-x} f(x) + e^{-x} f'(x) = -e^{-x} f(x) + e^{-x} f(x) = 0.$$

c. By (b) and Theorem 4.6, $e^{-x} f(x) = C$ for some constant C. Therefore $f(x) = e^x e^{-x} f(x) = e^x C = Ce^x$. Since $f(0) = 1$, we have $C \cdot e^0 = 1$, so that $C = 1$ and hence $f(x) = Ce^x = e^x$.

43. Let $f(x) = \dfrac{1}{\sigma\sqrt{2\pi}}\,e^{-(x-\mu)^2/2\sigma^2}$. Then

$$f(\mu) = \frac{1}{\sigma\sqrt{2\pi}}\,e^0 = \frac{1}{\sigma\sqrt{2\pi}}.$$

For $x \neq \mu$, we have $-(x-\mu)^2/2\sigma^2 < 0$, so that $e^{-(x-\mu)^2/2\sigma^2} < e^0 = 1$, and hence

$$f(x) = \frac{1}{\sigma\sqrt{2\pi}} e^{-(x-\mu)^2/2\sigma^2} < \frac{1}{\sigma\sqrt{2\pi}}.$$

Therefore the maximum value of f is $f(\mu) = 1/\sigma\sqrt{2\pi}$.

45. a. Since $f'(t) = 2f(t)$ and $g'(t) = 6g(t)$, it follows that $x = f(t) = C_1e^{2t}$ and $y = g(t) = C_2e^{6t}$, where C_1 and C_2 are constants. Then $e^{2t} = x/C_1$, so that $2t = \ln(x/C_1)$, or equivalently, $6t = 3\ln(x/C_1)$. Therefore

$$y = C_2e^{6t} = C_2e^{3\ln(x/C_1)} = C_2[e^{\ln(x/C_1)}]^3 = C_2\left(\frac{x}{C_1}\right)^3 = \left(\frac{C_2}{C_1^3}\right)x^3$$

b. By (a), $f(t) = C_1e^{2t}$, so that $2 = f(0) = C_1e^0 = C_1$. Similarly, $g(t) = C_2e^{6t}$, so that $3 = g(0) = C_2e^0 = C_2$. By part (a), $y = (C_2/C_1^3)x^3 = \frac{3}{8}x^3$. Since $x = f(t) = 2e^{2t} > 0$ and $y = 3e^{6t} > 0$, the path C of the particle is as in the figure.

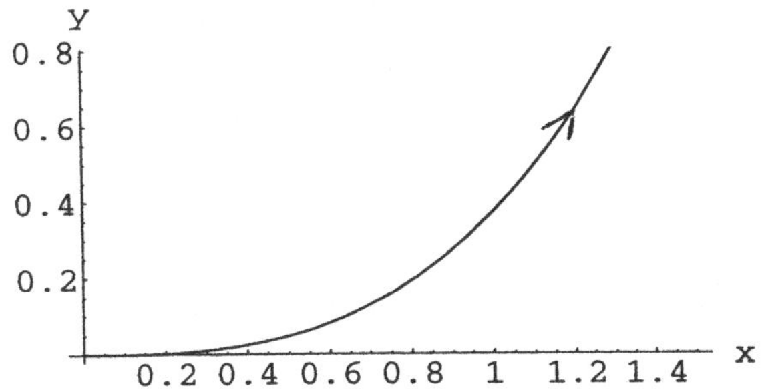

47. a. $0 = Q(0) = VC + c$, so $c = -VC$.

b. $Q(t) = VC(1 - e^{-t/RC})$; $Q'(t) = \dfrac{V}{R}e^{-t/RC}$; $Q''(t) = \dfrac{-V}{R^2C}e^{-t/RC}$; no relative extreme values; increasing on $[0,\infty)$; concave downward on $(0,\infty)$; $\lim_{t\to\infty} Q(t) = \lim_{t\to\infty} VC(1 - e^{-t/RC}) = VC$, so that $y = VC$ is a horizontal asymptote.

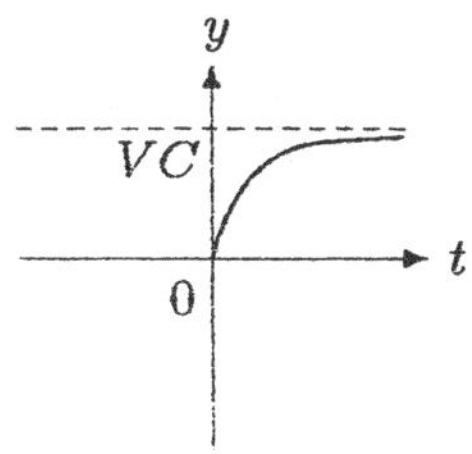

c. If $0.9VC = Q(t) = VC(1 - e^{-t/RC})$, then $.9 = 1 - e^{-t/RC}$, or $e^{-t/RC} = .1$, or $-t/RC = \ln\frac{1}{10} = -\ln 10$, or $t = RC\ln 10$.

7.3 General Exponential and Logarithmic Functions

1. By (6) and the Chain Rule, $g'(x) = (\ln 2)2^{-x}(-1) = (-\ln 2)2^{-x}$.

3. By (9), $y = t^t = e^{t\ln t}$, so $dy/dt = e^{t\ln t}(\ln t + t \cdot 1/t) = t^t(\ln t + 1)$.

5. By (9), $y = t^{2/t} = e^{(2\ln t)/t}$, so

$$\frac{dy}{dt} = e^{(2\ln t)/t}\left(\frac{(2/t)t - 2\ln t}{t^2}\right) = t^{2/t}\left(\frac{2 - 2\ln t}{t^2}\right).$$

7. By (9), $f(x) = (\cos x)^{\cos x} = e^{(\cos x)\ln(\cos x)}$, so

$$f'(x) = e^{(\cos x)\ln(\cos x)}\left[(-\sin x)\ln(\cos x) + (\cos x)\left(\frac{-\sin x}{\cos x}\right)\right] = (-\sin x)(\cos x)^{\cos x}[\ln(\cos x) + 1].$$

9. Since $(2x)^{\sqrt{2}} = 2^{\sqrt{2}}x^{\sqrt{2}}$, it follows from (8) that $f'(x) = 2^{\sqrt{2}}\sqrt{2}\,x^{\sqrt{2}-1} = 2\sqrt{2}\,(2x)^{\sqrt{2}-1}$.

11. $f'(x) = (\ln 2)(2^x + 2^{-x}) > 0$; increasing on $(-\infty, \infty)$;
$f''(x) = (\ln 2)^2(2^x - 2^{-x})$; $f''(x) = 0$ if $2^x = 2^{-x}$, or $2^{2x} = 1$, or $x = 0$; $f''(x) < 0$ if $x < 0$ and $f''(x) > 0$ if $x > 0$; concave downward on $(-\infty, 0)$ and concave upward on $(0, \infty)$; inflection point: $(0, 0)$; symmetric with respect to the origin.

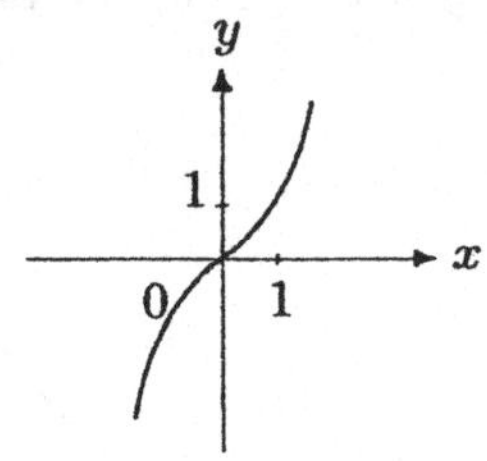

Exercise 11

13. By (10), $\int 2^x\,dx = \frac{1}{\ln 2}2^x + C$.

15. By (10), $\int_{-2}^{0} 3^{-x}\,dx \overset{u=-x}{=} \int_{2}^{0} 3^u(-1)\,du = \frac{-1}{\ln 3}3^u\Big|_2^0 = -\frac{1}{\ln 3}(1 - 3^2) = \frac{8}{\ln 3}$.

17. By (10), $\int x \cdot 5^{-x^2}\,dx \overset{u=-x^2}{=} \int -\frac{1}{2}5^u\,du = -\frac{1}{2}\cdot\frac{1}{\ln 5}5^u + C = \frac{-1}{2\ln 5}5^{-x^2} + C$.

19. By (11), $\int x^{2\pi}\,dx = \frac{1}{2\pi + 1}x^{2\pi+1} + C$.

21. By (10), $A = \int_1^2 x \cdot 2^{(x^2)}\,dx \overset{u=x^2}{=} \int_1^4 2^u\left(\frac{1}{2}\right)du = \frac{1}{2\ln 2}2^u\Big|_1^4 = \frac{1}{2\ln 2}(2^4 - 2^1) = \frac{7}{\ln 2}$.

23. $3^x \geq 2^x$ on $[0, 2]$, so by (10),

$$A = \int_0^2 (3^x - 2^x)\,dx = \left(\frac{1}{\ln 3}3^x - \frac{1}{\ln 2}2^x\right)\Big|_0^2 = \left(\frac{3^2}{\ln 3} - \frac{2^2}{\ln 2}\right) - \left(\frac{1}{\ln 3}3^0 - \frac{1}{\ln 2}2^0\right) = \frac{8}{\ln 3} - \frac{3}{\ln 2}.$$

25. By (14), $\log_3 5 = \frac{\ln 5}{\ln 3} \approx 1.464973521$.

27. By (14), $\log_\pi e = \dfrac{\ln e}{\ln \pi} = \dfrac{1}{\ln \pi} \approx 0.8735685268.$

29. $V = \displaystyle\int_1^3 \pi \left(\frac{\log_3 x}{\sqrt{x}}\right)^2 dx = \int_1^3 \frac{\pi}{x} (\log_3 x)^2\, dx$

We make the substitution $u = \log_3 x$, so that $du = \frac{1}{x \ln 3}\, dx$. Thus

$$V = \int_0^1 \pi u^2 \ln 3\, du = \frac{\pi}{3} u^3 \ln 3\Big|_0^1 = \frac{\pi}{3} \ln 3$$

31. Suppose $0 < b < a \le e$. Since $(\ln x)/x$ is increasing on $(0, e]$, we have $(\ln a)/a > (\ln b)/b$, so $b \ln a > a \ln b$ and thus $\ln a^b > \ln b^a$. Since $\ln x$ is increasing, it follows that $a^b > b^a$. Now suppose $e \le a < b$. Since $(\ln x)/x$ is decreasing on $[e, \infty)$, it follows that $(\ln a)/a > (\ln b)/b$, so that $b \ln a > a \ln b$ and thus $\ln a^b > \ln b^a$. Because $\ln x$ is increasing, we conclude that $a^b > b^a$.

33. By (14), $f(x) = (x \ln x)/\ln a$, so $f'(x) = (1/\ln a)(\ln x + x \cdot 1/x) = (1/\ln a)(\ln x + 1)$ and $f''(x) = 1/(x \ln a)$. Thus the graph of f is concave upward on $(0, \infty)$ if $\ln a > 0$, that is, $a > 1$, and the graph of f is concave downward on $(0, \infty)$ if $\ln a < 0$, that is, $0 < a < 1$.

35. We seek b such that $1.5 + \log_2 x = \log_2(bx) = \log_2 b + \log_2 x$. This implies that $\log_2 b = 1.5$, so that $b = 2^{1.5} = 2\sqrt{2}$.

37. a. By (14), $\log_b a \log_a x = \dfrac{\ln a}{\ln b}\dfrac{\ln x}{\ln a} = \dfrac{\ln x}{\ln b} = \log_b x$

b. Taking $b = 4$ and $a = 7$ in part (a), we find that $\log_4 x = \log_4 7 \log_7 x$.

39. a. Consider any rectangle with one vertex at the origin and the opposite vertex, denoted (b, c), on the graph of $y = a/x^a$, as in the figure. Then $c = a/b^a$, and the rectangle is a square if and only if $b = c = a/b^a$, or equivalently, $b^{a+1} = a$, or $b = a^{1/(a+1)}$. Thus there is only one such square.

b. Using the notation in part (a), we find that the area $A = b^2 = a^{2/(a+1)}$.

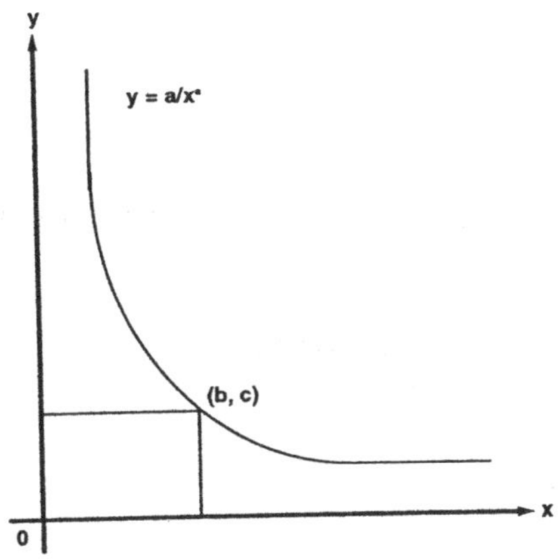

41. For the Sierpinski carpet, $m = 8$ and $r = \frac{1}{3}$, so the fractal dimension is $\log_{1/r} m = \log_3 8 = 3\log_3 2 \approx 1.89$.

43. a. By (6) and the Chain Rule,

$$\frac{dy}{dt} = k(\ln a)a^{(b^t)}(\ln b)b^t = k(\ln a)(\ln b)b^t a^{(b^t)}.$$

Since $0 < a < 1$ and $0 < b < 1$, both $\ln a$ and $\ln b$ are negative. Since $k > 0$, $b^t > 0$, and $a^{(b^t)} > 0$, we conclude that $dy/dt > 0$ for all t.

b. The level of diffusion is increasing.

c. Since $0 < b < 1$, so that $\ln b < 0$, we have $\lim_{t\to\infty} b^t = \lim_{t\to\infty} e^{t\ln b} = 0$. By the Substitution Rule with $y = b^t$, $\lim_{t\to\infty} ka^{(b^t)} = \lim_{y\to 0} ka^y = ka^0 = k$.

7.4 Hyperbolic Functions

1. $\sinh 0 = \dfrac{e^0 - e^{-0}}{2} = 0$

3. $\tanh 0 = \dfrac{\sinh 0}{\cosh 0} = \dfrac{e^0 - e^{-0}}{e^0 + e^{-0}} = \dfrac{0}{1} = 0$

5. $\coth(-1) = \dfrac{\cosh(-1)}{\sinh(-1)} = \dfrac{e^{-1} + e^1}{e^{-1} - e^1} = \dfrac{1 + e^2}{1 - e^2}$

7. $\sinh(\ln 3) = \dfrac{e^{\ln 3} - e^{-\ln 3}}{2} = \dfrac{3 - \frac{1}{3}}{2} = \dfrac{4}{3}$

9. $\coth(\ln 4) = \dfrac{\cosh(\ln 4)}{\sinh(\ln 4)} = \dfrac{e^{\ln 4} + e^{-\ln 4}}{e^{\ln 4} - e^{-\ln 4}} = \dfrac{4 + \frac{1}{4}}{4 - \frac{1}{4}} = \dfrac{17}{15}$

11. $\operatorname{sech}(\ln \sqrt{2}) = \dfrac{1}{\cosh(\ln \sqrt{2})} = \dfrac{2}{e^{\ln \sqrt{2}} + e^{-\ln \sqrt{2}}} = \dfrac{2}{\sqrt{2} + 1/\sqrt{2}} = \dfrac{2\sqrt{2}}{3}$

13. $\sinh(\ln x) = \dfrac{e^{\ln x} - e^{-\ln x}}{2} = \dfrac{x - 1/x}{2} = \dfrac{x^2 - 1}{2x}$

15. $\tanh(\ln x) = \dfrac{\sinh(\ln x)}{\cosh(\ln x)} = \left(\dfrac{x^2 - 1}{2x}\right)\Big/\left(\dfrac{x^2 + 1}{2x}\right) = \dfrac{x^2 - 1}{x^2 + 1}$ from Exercises 13 and 14

17. $\dfrac{d}{dx}\tanh x = \dfrac{d}{dx}\dfrac{\sinh x}{\cosh x} = \dfrac{\cosh^2 x - \sinh^2 x}{\cosh^2 x} = \dfrac{1}{\cosh^2 x} = \operatorname{sech}^2 x$

19. $\dfrac{d}{dx}\operatorname{sech} x = \dfrac{d}{dx}\dfrac{1}{\cosh x} = \dfrac{-\sinh x}{\cosh^2 x} = -\operatorname{sech} x \tanh x$

21. $f'(x) = (-\operatorname{sech}\sqrt{x}\tanh\sqrt{x})\left(\dfrac{1}{2\sqrt{x}}\right) = \dfrac{-1}{2\sqrt{x}}\operatorname{sech}\sqrt{x}\tanh\sqrt{x}$

23. $f'(x) = \left(2\sinh\sqrt{1-x^2}\cosh\sqrt{1-x^2}\right)\left(\dfrac{-x}{\sqrt{1-x^2}}\right) = \dfrac{-2x}{\sqrt{1-x^2}}\sinh\sqrt{1-x^2}\cosh\sqrt{1-x^2}$

25. $f'(x) = \left[\sinh(\tan e^{2x})\right]\left(\sec^2 e^{2x}\right)\left(2e^{2x}\right) = 2e^{2x}\left(\sec^2 e^{2x}\right)\sinh(\tan e^{2x})$

27. Since

$$\frac{dy}{dx} = \frac{\dfrac{1}{\sqrt{1+x^2}}\sqrt{1+x^2} - (\sinh^{-1}x)\left(\dfrac{x}{\sqrt{1+x^2}}\right)}{1+x^2} = \frac{1}{1+x^2} - \frac{xy}{1+x^2},$$

it follows that $(1+x^2)(dy/dx) = 1 - xy$, or $(1+x^2)(dy/dx) + xy = 1$.

29. $\displaystyle\int \operatorname{sech}^2 x\,dx = \tanh x + C$

31. $\displaystyle\int 2^x \sinh 2^x\,dx \overset{u=2^x}{=} \int \sinh u \cdot \frac{1}{\ln 2}\,du = \frac{\cosh u}{\ln 2} + C = \frac{\cosh 2^x}{\ln 2} + C$

33. $\displaystyle\int_5^{10} \frac{1}{\sqrt{x^2+1}}\,dx = \sinh^{-1}10 - \sinh^{-1}5 = \ln(10+\sqrt{101}) - \ln(5+\sqrt{26}) = \ln\left(\frac{10+\sqrt{101}}{5+\sqrt{26}}\right)$

35. $\displaystyle A = \int_{-4}^{4} 4\cosh\frac{x}{4}\,dx \overset{u=x/4}{=} 4\int_{-1}^{1}(\cosh u)\cdot 4\,du = 16\sinh u\Big|_{-1}^{1}$

$= 16\left(\sinh 1 - \sinh(-1)\right) = 32\sinh 1 = 16(e - e^{-1})$

37. The length would be given by

$$L = \int_0^{\ln 2}\sqrt{1+\cosh^2 x}\,dx = \int_0^{\ln 2}\sqrt{1+\frac{1}{4}\left(e^{2x}+2+e^{-2x}\right)}\,dx$$

Neither integrand can be simplified as in Exercise 36, so the integration appears to be impossible.

39. $\displaystyle M_x = \int_0^1 \frac{1}{2}\frac{1}{1+x^2}\,dx = \frac{1}{2}\tan^{-1}x\Big|_0^1 = \frac{\pi}{8}$

$\displaystyle M_y = \int_0^1 x\frac{1}{\sqrt{1+x^2}}\,dx \overset{u=1+x^2}{=} \int_1^2 \frac{1}{u^{1/2}}\cdot\frac{1}{2}\,du = u^{1/2}\Big|_1^2 = \sqrt{2}-1$

$\displaystyle A = \int_0^1 \frac{1}{\sqrt{1+x^2}}\,dx = \sinh^{-1}\Big|_0^1 = \sinh^{-1}1 - \sinh^{-1}0$

$= \ln\left(1+\sqrt{1^2+1}\right) - 0 = \ln\left(1+\sqrt{2}\right)$

$$\bar{x} = \frac{M_y}{A} = \frac{\sqrt{2}-1}{\ln(1+\sqrt{2})}; \quad \bar{y} = \frac{M_x}{A} = \frac{\pi/8}{\ln(1+\sqrt{2})} = \frac{\pi}{8\ln(1+\sqrt{2})};$$

$$(\bar{x}, \bar{y}) = \left(\frac{\sqrt{2}-1}{\ln(1+\sqrt{2})}, \frac{\pi}{8\ln(1+\sqrt{2})}\right)$$

41. If $x \geq 0$, then $\cosh x = \frac{1}{2}(e^x + e^{-x}) > \frac{1}{2}e^x = \frac{1}{2}e^{|x|}$.
If $x < 0$, then $\cosh x = \frac{1}{2}(e^x + e^{-x}) > \frac{1}{2}e^{-x} = \frac{1}{2}e^{|x|}$.
Thus $\cosh x > \frac{1}{2}e^{|x|}$ for all x.

43. a. From Exercise 42(a), $(\cosh x + \sinh x)^n = (e^x)^n = e^{nx} = \cosh nx + \sinh nx$ for all x.

b. From Exercise 42(b), $(\cosh x - \sinh x)^n = (e^{-x})^n = e^{-(nx)} = \cosh nx - \sinh nx$ for all x.

45. $$f(x) = \frac{1}{2}(1+\tanh x) = \frac{1}{2}\left(1+\frac{\sinh x}{\cosh x}\right) = \frac{1}{2}\left(1+\frac{(e^x-e^{-x})/2}{(e^x+e^{-x})/2}\right)$$
$$= \frac{1}{2}\left(1+\frac{e^x-e^{-x}}{e^x+e^{-x}}\right) = \frac{1}{2}\left(\frac{e^x+e^{-x}+e^x-e^{-x}}{e^x+e^{-x}}\right) = \frac{e^x}{e^x+e^{-x}} = \frac{1}{1+e^{-2x}}$$

47. Let the shaded region have area A. Since the area of the triangle with vertices $(0,0)$, $(\cosh t, 0)$, and $(\cosh t, \sinh t)$ is $\frac{1}{2}(\sinh t)(\cosh t)$, it follows that

$$A = \frac{1}{2}(\sinh t)(\cosh t) - \int_1^{\cosh t} \sqrt{x^2-1}\,dx.$$

From Exercise 44(a) we know that $2\sinh t\, \cosh t = \sinh 2t$, and from Exercise 42 we have

$$\int_1^{\cosh t} \sqrt{x^2-1}\,dx = \frac{\sinh 2t}{4} - \frac{t}{2}.$$

Thus

$$A = \frac{1}{4}\sinh 2t - \left(\frac{\sinh 2t}{4} - \frac{t}{2}\right) = \frac{t}{2}.$$

49. a. Since the minimum value of $\cosh x$ is 1, the maximum value of $694 - 69\cosh(x/100)$ is $694 - 69 = 625$.

b. $y = 0$ if $694 - 69\cosh(x/100) = 0$, or $\cosh(x/100) = 694/69$. This is equivalent to

$$\frac{e^{x/100} + e^{-x/100}}{2} = \frac{694}{69},$$

or $(e^{x/100})^2 + 1 = (1388/69)e^{x/100}$, or $(e^{x/100})^2 - (1388/69)e^{x/100} + 1 = 0$.

Thus $e^{x/100}$ must be one of the two solutions, say z_1 and z_2, of the quadratic equation $z^2 - (1388/69)z + 1 = 0$. Therefore x must be $100\ln z_1$ or $100\ln z_2$. Using the quadratic formula and a calculator, we find that $x \approx -300$ or $x \approx 300$. Thus the distance between the two intercepts is approximately 600.

7.5 The Inverse Trigonometric Functions

1. Since $\sin\frac{\pi}{3} = \frac{\sqrt{3}}{2}$ and $-\frac{\pi}{2} \le \frac{\pi}{3} \le \frac{\pi}{2}$, it follows that $\sin^{-1}\frac{\sqrt{3}}{2} = \frac{\pi}{3}$.

3. Since $\cos\frac{\pi}{4} = \frac{\sqrt{2}}{2}$ and $0 \le \frac{\pi}{4} \le \pi$, it follows that $\cos^{-1}\frac{\sqrt{2}}{2} = \frac{\pi}{4}$.

5. Since $\tan\left(-\frac{\pi}{6}\right) = -\frac{1}{\sqrt{3}}$ and $-\frac{\pi}{2} < -\frac{\pi}{6} < \frac{\pi}{2}$, it follows that $\tan^{-1}\left(-\frac{1}{\sqrt{3}}\right) = -\frac{\pi}{6}$.

7. Since $\cot\frac{\pi}{6} = \sqrt{3}$ and $0 < \frac{\pi}{6} < \pi$, it follows that $\cot^{-1}\sqrt{3} = \frac{\pi}{6}$.

9. Since $\sec\frac{5\pi}{4} = -\sqrt{2}$ and $\pi \le \frac{5\pi}{4} < \frac{3\pi}{2}$, it follows that $\sec^{-1}(-\sqrt{2}) = \frac{5\pi}{4}$.

11. $\sin\left(\sin^{-1}\left(-\frac{1}{2}\right)\right) = \sin\left(-\frac{\pi}{6}\right) = -\frac{1}{2}$

13. $\tan(\sec^{-1}\sqrt{2}) = \tan\frac{\pi}{4} = 1$

15. $\csc(\cot^{-1}(-\sqrt{3})) = \csc\frac{5\pi}{6} = 2$

17. $\sin^{-1}\left(\cos\frac{\pi}{6}\right) = \sin^{-1}\frac{1}{2}\sqrt{3} = \frac{\pi}{3}$

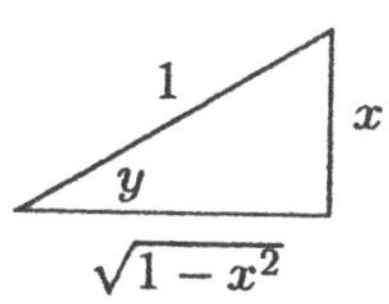

Exercise 19

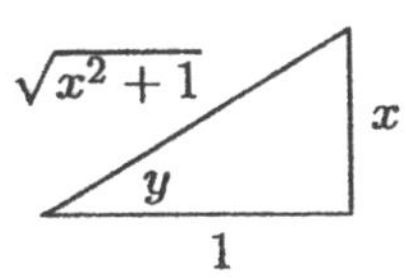

Exercise 21

19. We will evaluate $\cos(\sin^{-1}x)$ by evaluating $\cos y$ for the value of y in $[-\pi/2, \pi/2]$ such that $\sin^{-1}x = y$, that is, $\sin y = x$. From the diagram we have $\cos(\sin^{-1}x) = \cos y = \sqrt{1-x^2}$.

21. We will evaluate $\sec(\tan^{-1}x)$ by evaluating $\sec y$ for the value of y in $(-\pi/2, \pi/2)$ such that $\tan^{-1}x = y$, that is, $\tan y = x$. From the diagram we have $\sec(\tan^{-1}x) = \sec y = \sqrt{x^2+1}$.

23. We will evaluate $\cos(\cot^{-1}x^2)$ by evaluating $\cos y$ for the value of y in $(0, \pi)$ such that $\cot^{-1}x^2 = y$, that is, $\cot y = x^2$. Since $x^2 \ge 0$, we have $0 < y \le \pi/2$. We find from the diagram that $\cos(\cot^{-1}x^2) = \cos y = x^2/\sqrt{x^4+1}$.

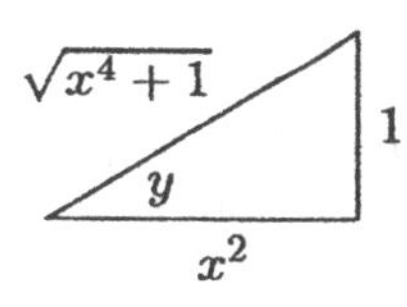

25. Using a trigonometric identity for $\cos 2a$ and the fact that $\cos(\sin^{-1}x) = \sqrt{1-x^2}$ (see Exercise 19), we find that $\cos(2\sin^{-1}x) = \cos^2(\sin^{-1}x) - \sin^2(\sin^{-1}x) = (\sqrt{1-x^2})^2 - x^2 = (1-x^2) - x^2 = 1 - 2x^2$.

27. Using (7), we find that $\frac{d}{dx}\cos^{-1}x = \frac{-1}{\sqrt{1-x^2}}$, so that by the Chain Rule,

$$f'(x) = \frac{-1}{\sqrt{1-(-3x)^2}}(-3) = \frac{3}{\sqrt{1-9x^2}}.$$

29. Using (4), we find that $f'(t) = \frac{1}{(\sqrt{t})^2+1}\left(\frac{1}{2\sqrt{t}}\right) = \frac{1}{2(t+1)\sqrt{t}}$.

31. By (8), $\frac{d}{dx}\cot^{-1}x = \frac{-1}{x^2+1}$, so that if $f(x) = \cot^{-1}\sqrt{1-x^2}$, then

$$f'(x) = \frac{-1}{(\sqrt{1-x^2})^2+1}\left(\frac{-x}{\sqrt{1-x^2}}\right) = \frac{x}{(2-x^2)\sqrt{1-x^2}}.$$

33. $\int \frac{1}{x^2+16}\,dx = \frac{1}{4}\tan^{-1}\frac{x}{4} + C$

35. $\int \frac{1}{9x^2+16}\,dx = \frac{1}{9}\int \frac{1}{x^2+\frac{16}{9}}\,dx = \frac{1}{9}\cdot\frac{3}{4}\tan^{-1}\frac{3x}{4} + C = \frac{1}{12}\tan^{-1}\frac{3x}{4} + C$

37. $\int \frac{1}{2x^2+4x+6}\,dx = \frac{1}{2}\int \frac{1}{(x+1)^2+2}\,dx \overset{u=x+1}{=} \frac{1}{2}\int \frac{1}{u^2+2}\,du$

$$= \frac{1}{2}\left(\frac{1}{\sqrt{2}}\tan^{-1}\frac{u}{\sqrt{2}}\right) + C = \frac{\sqrt{2}}{4}\tan^{-1}\frac{x+1}{\sqrt{2}} + C$$

39. $\int \frac{1}{\sqrt{9-4x^2}}\,dx = \frac{1}{2}\int \frac{1}{\sqrt{\frac{9}{4}-x^2}}\,dx = \frac{1}{2}\sin^{-1}\frac{2x}{3} + C$

41. $\int \frac{1}{x\sqrt{x^2-25}}\,dx = \frac{1}{5}\sec^{-1}\frac{x}{5} + C$

43. $\int \frac{x^3}{\sqrt{1-x^8}}\,dx \overset{u=x^4}{=} \frac{1}{4}\int \frac{1}{\sqrt{1-u^2}}\,du = \frac{1}{4}\sin^{-1}u + C = \frac{1}{4}\sin^{-1}x^4 + C$

45. $\int \frac{e^{-x}}{1+e^{-2x}}\,dx \overset{u=e^{-x}}{=} \int \frac{-1}{1+u^2}\,du = -\tan^{-1}u + C = -\tan^{-1}e^{-x} + C$

47. $\int \frac{\tan^{-1}2x}{1+4x^2}\,dx \overset{u=\tan^{-1}2x}{=} \frac{1}{2}\int u\,du = \frac{1}{4}u^2 + C = \frac{1}{4}(\tan^{-1}2x)^2 + C$

49. $\int \frac{\cos t}{9+\sin^2 t}\,dt \overset{u=\sin t}{=} \int \frac{1}{9+u^2}\,du = \frac{1}{3}\tan^{-1}\frac{u}{3} + C = \frac{1}{3}\tan^{-1}\left(\frac{1}{3}\sin t\right) + C$

51. $$\int \frac{\cos 4x}{(\sin 4x)\sqrt{16\sin^2 4x - 4}}\,dx \overset{u=4\sin 4x}{=} \int \frac{1}{\frac{1}{4}u\sqrt{u^2-4}}\left(\frac{1}{16}\right)du = \frac{1}{4}\int \frac{1}{u\sqrt{u^2-4}}\,du$$
$$= \frac{1}{8}\sec^{-1}\frac{u}{2} + C = \frac{1}{8}\sec^{-1}(2\sin 4x) + C$$

53. $$\int_0^2 \frac{1}{\sqrt{16-x^2}}\,dx = \sin^{-1}\frac{x}{4}\bigg|_0^2 = \sin^{-1}\frac{1}{2} - \sin^{-1}0 = \frac{1}{6}\pi$$

55. $$\int_{4\sqrt{3}/3}^4 \frac{1}{x\sqrt{x^2-4}}\,dx = \frac{1}{2}\sec^{-1}\frac{x}{2}\bigg|_{4\sqrt{3}/3}^4 = \frac{1}{2}\left(\sec^{-1}2 - \sec^{-1}\frac{2\sqrt{3}}{3}\right) = \frac{1}{2}\left(\frac{\pi}{3}-\frac{\pi}{6}\right) = \frac{1}{12}\pi$$

57. The graphs intersect at (x, y) if $1/(x^2-2x+4) = y = \frac{1}{3}$, or $3 = x^2 - 2x + 4$, or $x^2 - 2x + 1 = 0$, or $x = 1$. Since $\frac{1}{3} \geq 1/(x^2-2x+4)$ on $[0, 1]$, we have
$$A = \int_0^1 \left(\frac{1}{3} - \frac{1}{x^2-2x+4}\right)dx = \int_0^1 \left(\frac{1}{3} - \frac{1}{(x-1)^2+3}\right)dx = \left(\frac{x}{3} - \frac{1}{\sqrt{3}}\tan^{-1}\frac{x-1}{\sqrt{3}}\right)\bigg|_0^1$$
$$= \left(\frac{1}{3} - 0\right) - \left(0 - \frac{1}{\sqrt{3}}\tan^{-1}\frac{-1}{\sqrt{3}}\right) = \frac{1}{3} - \frac{1}{18}\pi\sqrt{3}.$$

59. $$V = \int_0^{\sqrt{2}/2} 2\pi \frac{y}{\sqrt{1-y^4}}\,dy \overset{u=y^2}{=} 2\pi\int_0^{1/2} \frac{1}{\sqrt{1-u^2}}\cdot\frac{1}{2}\,du = \pi\sin^{-1}u\big|_0^{1/2} = \frac{\pi^2}{6}$$

61. Let $f(x) = \sin^{-1}x$, so that $f^{-1}(y) = \sin y$. Also let $a = 0$ and $b = 1$, so that $f(a) = \sin^{-1}0 = 0$ and $f(b) = \sin^{-1}1 = \pi/2$. Therefore by the formula,
$$\int_0^1 \sin^{-1}x\,dx = 1\sin^{-1}1 - 0\sin^{-1}0 - \int_0^{\pi/2}\sin y\,dy = \frac{\pi}{2} + \cos y\big|_0^{\pi/2} = \frac{\pi}{2} - 1.$$

63. Let $f(x) = \tan^{-1}x$. Then $f'(x) = 1/(1+x^2)$ and $f''(x) = -2x/(1+x^2)^2$. Since $f'(0)$ exists, $f''(x) > 0$ for $x < 0$, and $f''(x) < 0$ for $x > 0$, there is an inflection point at $x = 0$.

65. $$\frac{d}{dx}\left(\tan^{-1}\frac{x}{\sqrt{1-x^2}}\right) = \frac{1}{1+[x/\sqrt{1-x^2}]^2}\cdot\frac{\sqrt{1-x^2}+(x^2/\sqrt{1-x^2})}{1-x^2}$$
$$= (1+x^2)\cdot\frac{1}{(1-x^2)^{3/2}} = \frac{1}{\sqrt{1-x^2}} = \frac{d}{dx}(\sin^{-1}x)$$

Thus by Theorem 4.6 there is a constant C such that $\tan^{-1}\left(x/\sqrt{1-x^2}\right) = \sin^{-1}x + C$. For $x = 0$ we obtain $0 = \tan^{-1}0 = \sin^{-1}0 + C = C$, so that $C = 0$ and thus
$$\tan^{-1}\frac{x}{\sqrt{1-x^2}} = \sin^{-1}x.$$

67. a. $\tan^{-1}\frac{1}{2}+\tan^{-1}\frac{1}{3}=\tan^{-1}\frac{\frac{1}{2}+\frac{1}{3}}{1-(\frac{1}{2})(\frac{1}{3})}=\tan^{-1}1=\frac{\pi}{4}$

b. $2\tan^{-1}\frac{1}{3}+\tan^{-1}\frac{1}{7}=\tan^{-1}\frac{\frac{1}{3}+\frac{1}{3}}{1-(\frac{1}{3})(\frac{1}{3})}+\tan^{-1}\frac{1}{7}=\tan^{-1}\frac{3}{4}+\tan^{-1}\frac{1}{7}$

$$=\tan^{-1}\frac{\frac{3}{4}+\frac{1}{7}}{1-(\frac{3}{4})(\frac{1}{7})}=\tan^{-1}1=\frac{\pi}{4}$$

c. $\tan^{-1}\frac{120}{119}-\tan^{-1}\frac{1}{239}=\tan^{-1}\frac{\frac{120}{119}-\frac{1}{239}}{1+(\frac{120}{119})(\frac{1}{239})}=\tan^{-1}1=\frac{\pi}{4}$

d. Using Exercise 66 twice, we find that

$$4\tan^{-1}\frac{1}{5}=2\tan^{-1}\frac{\frac{1}{5}+\frac{1}{5}}{1-(\frac{1}{5})(\frac{1}{5})}=2\tan^{-1}\frac{5}{12}=\tan^{-1}\frac{\frac{5}{12}+\frac{5}{12}}{1-(\frac{5}{12})(\frac{5}{12})}=\tan^{-1}\frac{120}{119}.$$

Now by (c) we obtain $4\tan^{-1}\frac{1}{5}-\tan^{-1}\frac{1}{239}=\tan^{-1}\frac{120}{119}-\tan^{-1}\frac{1}{239}=\pi/4$.

69. $f'(x)=\dfrac{1}{1+x^2}+\dfrac{1}{1+1/x^2}\left(-\dfrac{1}{x^2}\right)=\dfrac{1}{1+x^2}-\dfrac{1}{1+x^2}=0$
on $(-\infty,0)$ and on $(0,\infty)$. Thus on $(-\infty,0)$ there is a constant c_1 such that $f(x)=c_1$; since $f(-1)=\tan^{-1}(-1)+\tan^{-1}(-1)=-\pi/4-\pi/4=-\pi/2$, it follows that on $(-\infty,0)$, $f(x)=-\pi/2$. Similarly, on $(0,\infty)$ there is a constant c_2 such that $f(x)=c_2$; since $f(1)=\tan^{-1}1+\tan^{-1}1=\pi/4+\pi/4=\pi/2$, it follows that on $(0,\infty)$, $f(x)=\pi/2$.

71. We will use trigonometric identities for $\cos 2a$ and $\sin(\pi/2-a)$, along with the fact that $\cos(\sin^{-1}x)=\sqrt{1-x^2}$ (see Exercise 19).

a. Notice that

$$\sin\left(\frac{\pi}{2}-2\sin^{-1}\sqrt{1-\frac{x}{6}}\right)=\cos\left(2\sin^{-1}\sqrt{1-\frac{x}{6}}\right)$$
$$=\cos^2\left(\sin^{-1}\sqrt{1-\frac{x}{6}}\right)-\sin^2\left(\sin^{-1}\sqrt{1-\frac{x}{6}}\right)$$
$$=\left[\sqrt{1-\left(\sqrt{1-\frac{x}{6}}\right)^2}\right]^2-\left(\sqrt{1-\frac{x}{6}}\right)^2=\left[1-\left(1-\frac{x}{6}\right)\right]-\left(1-\frac{x}{6}\right)=\frac{x}{3}-1.$$

Since $-\pi/2\le\pi/2-2\sin^{-1}\sqrt{1-x/6}\le\pi/2$, it follows that

$$\sin^{-1}\left(\frac{x}{3}-1\right)=\frac{\pi}{2}-2\sin^{-1}\sqrt{1-\frac{x}{6}}.$$

b. Notice that

$$\sin\left(2\sin^{-1}\frac{\sqrt{x}}{\sqrt{6}}-\frac{\pi}{2}\right)=-\sin\left(\frac{\pi}{2}-2\sin^{-1}\frac{\sqrt{x}}{\sqrt{6}}\right)=-\cos\left(2\sin^{-1}\frac{\sqrt{x}}{\sqrt{6}}\right)$$

$$= \sin^2\left(\sin^{-1}\frac{\sqrt{x}}{\sqrt{6}}\right) - \cos^2\left(\sin^{-1}\frac{\sqrt{x}}{\sqrt{6}}\right)$$

$$= \left(\frac{\sqrt{x}}{\sqrt{6}}\right)^2 - \left[\sqrt{1-\left(\frac{\sqrt{x}}{\sqrt{6}}\right)^2}\,\right]^2 = \frac{x}{6} - \left(1-\frac{x}{6}\right) = \frac{x}{3} - 1.$$

Since $-\pi/2 \le 2\sin^{-1}(\sqrt{x}/\sqrt{6}) - \pi/2 \le \pi/2$, it follows that $\sin^{-1}(x/3-1) = 2\sin^{-1}(\sqrt{x}/\sqrt{6}) - \pi/2$.

73. a. Let $f(x) = \sin^{-1}x + \cos^{-1}x$. Then $f'(x) = 1/\sqrt{1-x^2} - 1/\sqrt{1-x^2} = 0$, so by part (a) of Theorem 4.6, there is a constant c such that $f(x) = c$ for $-1 \le x \le 1$.

b. If $x = 0$ in part (a), we have $c = f(0) = \sin^{-1}0 + \cos^{-1}0 = 0 + \pi/2 = \pi/2$.

75. Let x denote the distance in feet from the floor to the bottom of the painting, and let V denote the viewing angle. We are to maximize V. If the bottom of the painting is below eye level, as in the first figure, then

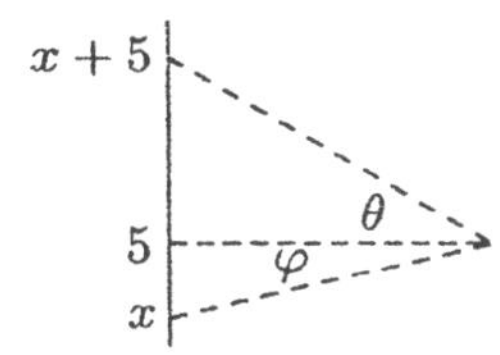

$$V = \theta + \varphi = \tan^{-1}\frac{x}{10} + \tan^{-1}\frac{5-x}{10}$$

$$= \tan^{-1}\frac{x}{10} - \tan^{-1}\frac{x-5}{10} \quad \text{for } x > 5.$$

If the bottom of the painting is above eye level, as in the second figure, then

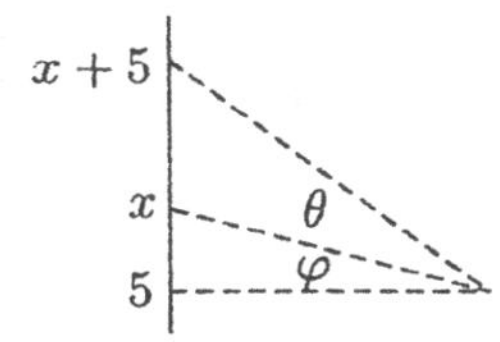

$$V = \theta = (\theta+\varphi) - \varphi = \tan^{-1}\frac{x}{10} - \tan^{-1}\frac{x-5}{10} \quad \text{for } 0 \le x \le 5.$$

Therefore we are to maximize V with respect to x, where

$$V(x) = \tan^{-1}\frac{x}{10} - \tan^{-1}\frac{x-5}{10} \quad \text{with } 0 \le x.$$

Now

$$V'(x) = \frac{1}{1+(x/10)^2}\cdot\frac{1}{10} - \frac{1}{1+[(x-5)/10]^2}\cdot\frac{1}{10}$$

so that $V'(x) = 0$ if

$$1 + \frac{x^2}{10^2} = 1 + \frac{(x-5)^2}{10^2}$$

or $x^2 = (x-5)^2$, or $x^2 = x^2 - 10x + 25$, or $x = 2.5$. Since $V'(x) > 0$ for $0 \le x < 2.5$ and $V'(x) < 0$ for $x > 2.5$, (1) of Section 4.6 and the First Derivative Test implies that V is maximum for $x = 2.5$. Consequently the bottom of the painting should be 2.5 feet above the floor. If 10 is replaced by 15 (or any positive value), a similar analysis would lead again to the equation $x^2 = (x-5)^2$, so the answer would be the same.

77. a. Using the notation in the diagram, we wish to maximize the angle θ and we have

$$\alpha + \theta + \beta = \pi, \quad \tan\alpha = \frac{15}{30-x}, \quad \text{and} \quad \tan\beta = \frac{20}{x}.$$

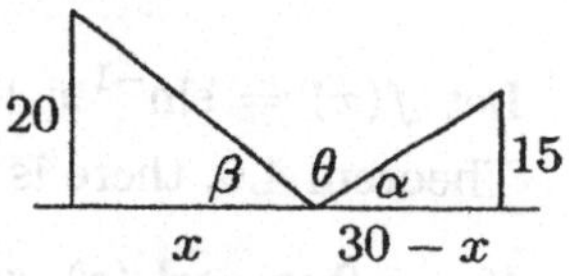

Thus $\theta = \pi - \tan^{-1}(20/x) - \tan^{-1}[15/(30-x)]$, so that

$$\frac{d\theta}{dx} = \frac{-1}{1+(20/x)^2}\left(-\frac{20}{x^2}\right) - \frac{1}{1+\left(15/(30-x)\right)^2}\cdot\frac{15}{(30-x)^2} = \frac{20}{x^2+400} - \frac{15}{(30-x)^2+225}.$$

Thus $d\theta/dx = 0$ if $20[(30-x)^2 + 225] = 15(x^2+400)$, or $4(30-x)^2 + 900 = 3x^2 + 1200$, or $x^2 - 240x + 3300 = 0$. Since $0 \le x \le 30$, it follows from the quadratic formula that $x = \frac{1}{2}\left(240 - \sqrt{(240)^2 - 4(1)(3300)}\right) \approx 14.64346247$. We have

$$\frac{d^2\theta}{dx^2} = \frac{-20}{(x^2+400)^2}(2x) - \frac{15\cdot 2(30-x)}{\left((30-x)^2+225\right)^2}$$

so that $d^2\theta/dx^2 < 0$ for $0 < x < 30$. Therefore the value of x for which $d\theta/dx = 0$ yields the maximum value of θ. Thus the tomatoes should be planted approximately $30 - 14.6 = 15.4$ meters from the smaller building.

b. Using the formula in (a) for θ along with the value of x that maximizes θ, we find by calculator that the maximum value of θ is approximately 1.43 radians. Since an angle of 2π radians corresponds to 24 hours, an angle of 1.43 radians corresponds to $(1.43/2\pi)24 \approx 5.46$ hours. Thus the plants will be in sunlight for approximately 5 hours and 28 minutes.

7.6 L'Hôpital's Rule

1. $\lim_{x\to a}(x^{16} - a^{16}) = 0 = \lim_{x\to a}(x-a)$; $\lim_{x\to a}\frac{x^{16}-a^{16}}{x-a} = \lim_{x\to a}\frac{16x^{15}}{1} = 16a^{15}$

3. $\lim_{x\to 0}(\cos x - 1) = 0 = \lim_{x\to 0} x$; $\lim_{x\to 0}\frac{\cos x - 1}{x} = \lim_{x\to 0}\frac{-\sin x}{1} = 0$

5. $\lim_{x\to 0^+}\sin 8\sqrt{x} = 0 = \lim_{x\to 0^+}\sin 5\sqrt{x}$; $\lim_{x\to 0^+}\frac{\sin 8\sqrt{x}}{\sin 5\sqrt{x}} = \lim_{x\to 0^+}\frac{(\cos 8\sqrt{x})(4/\sqrt{x})}{(\cos 5\sqrt{x})[5/(2\sqrt{x})]} = \lim_{x\to 0^+}\frac{8\cos 8\sqrt{x}}{5\cos 5\sqrt{x}} = \frac{8}{5}$

7. $\lim_{x\to\infty}(e^{1/x}-1)=0=\lim_{x\to\infty}\dfrac{1}{x}$; $\lim_{x\to\infty}\dfrac{e^{1/x}-1}{1/x}=\lim_{x\to\infty}\dfrac{e^{1/x}(-1/x^2)}{-1/x^2}=\lim_{x\to\infty}e^{1/x}=1$

9. $\lim_{x\to\pi/2^-}\tan x=\infty=\lim_{x\to\pi/2^-}(\sec x+1)$;

$$\lim_{x\to\pi/2^-}\frac{\tan x}{\sec x+1}=\lim_{x\to\pi/2^-}\frac{\sec^2 x}{\sec x\,\tan x}=\lim_{x\to\pi/2^-}\frac{\sec x}{\tan x}=\lim_{x\to\pi/2^-}\frac{1}{\sin x}=1$$

11. $\lim_{x\to\infty}x=\infty=\lim_{x\to\infty}\ln x$; $\lim_{x\to\infty}\dfrac{x}{\ln x}=\lim_{x\to\infty}\dfrac{1}{1/x}=\lim_{x\to\infty}x=\infty$

13. $\lim_{x\to\infty}\ln(1+x)=\infty=\lim_{x\to\infty}\ln x$;

$$\lim_{x\to\infty}\frac{\ln(1+x)}{\ln x}=\lim_{x\to\infty}\frac{1/(1+x)}{1/x}=\lim_{x\to\infty}\frac{x}{1+x}=\lim_{x\to\infty}\frac{1}{(1/x)+1}=1$$

15. $\lim_{x\to0}(1-\cos x)=0=\lim_{x\to0}\sin x$; $\lim_{x\to0}\dfrac{1-\cos x}{\sin x}=\lim_{x\to0}\dfrac{\sin x}{\cos x}=0$

17. $\lim_{x\to\pi/2^-}\tan 4x=0=\lim_{x\to\pi/2^-}\tan 2x$; $\lim_{x\to\pi/2^-}\dfrac{\tan 4x}{\tan 2x}=\lim_{x\to\pi/2^-}\dfrac{4\sec^2 4x}{2\sec^2 2x}=2$

19. The conditions for applying l'Hôpital's Rule twice are met;

$$\lim_{x\to0}\frac{1-\cos 2x}{1-\cos 3x}=\lim_{x\to0}\frac{2\sin 2x}{3\sin 3x}=\lim_{x\to0}\frac{4\cos 2x}{9\cos 3x}=\frac{4}{9}.$$

21. $\lim_{x\to0}\left(\sqrt{1+x}-\sqrt{1-x}\right)=0=\lim_{x\to0}x$;

$$\lim_{x\to0}\frac{\sqrt{1+x}-\sqrt{1-x}}{x}=\lim_{x\to0}\frac{1/(2\sqrt{1+x})+1/(2\sqrt{1-x})}{1}=1$$

23. $\lim_{x\to0^+}\sin x=0=\lim_{x\to0^+}(e^{\sqrt{x}}-1)$; $\lim_{x\to0^+}\dfrac{\sin x}{e^{\sqrt{x}}-1}=\lim_{x\to0^+}\dfrac{\cos x}{(1/2\sqrt{x})e^{\sqrt{x}}}=\lim_{x\to0^+}\dfrac{2\sqrt{x}\cos x}{e^{\sqrt{x}}}=0$

25. $\lim_{x\to0}(5^x-3^x)=1-1=0=\lim_{x\to0}x$; $\lim_{x\to0}\dfrac{5^x-3^x}{x}=\lim_{x\to0}\dfrac{(\ln 5)5^x-(\ln 3)3^x}{1}=\ln 5-\ln 3$

27. $\lim_{x\to0}\sin^{-1}x=0=\lim_{x\to0}x$; $\lim_{x\to0}\dfrac{\sin^{-1}x}{x}=\lim_{x\to0}\dfrac{1/\sqrt{1-x^2}}{1}=1$

29. The conditions for applying l'Hôpital's Rule three times are met;

$$\lim_{x\to0}\frac{x^3}{x-\sin x}=\lim_{x\to0}\frac{3x^2}{1-\cos x}=\lim_{x\to0}\frac{6x}{\sin x}=\lim_{x\to0}\frac{6}{\cos x}=6.$$

31. The conditions for applying l'Hôpital's Rule twice are met;

$$\lim_{x\to0}\frac{\tanh x-\sinh x}{x^2}=\lim_{x\to0}\frac{\operatorname{sech}^2 x-\cosh x}{2x}=\lim_{x\to0}\frac{-2\operatorname{sech}^2 x\,\tanh x-\sinh x}{2}=0.$$

33. $\lim_{x\to 0}(\csc x - \cot x) = \lim_{x\to 0}\dfrac{1-\cos x}{\sin x} = 0$ by Exercise 15

35. $\lim_{x\to 0^+}\ln(\sin x) = -\infty$; $\lim_{x\to 0^+}\csc x = \infty$

$$\lim_{x\to 0^+}\sin x\ln(\sin x) = \lim_{x\to 0^+}\frac{\ln(\sin x)}{\csc x} = \lim_{x\to 0^+}\frac{\cos x/\sin x}{-\csc x\cot x} = \lim_{x\to 0^+}(-\sin x) = 0$$

37. $\lim_{x\to 0^+}\left(\ln\dfrac{1}{x}\right)^x = \lim_{x\to 0^+} e^{x\ln(\ln(1/x))} = e^{\lim_{x\to 0^+} x\ln(\ln(1/x))} = e^{\lim_{x\to 0^+}[\ln(\ln(1/x))/(1/x)]}$;

$$\lim_{x\to 0^+}\ln\left(\ln\frac{1}{x}\right) = \infty = \lim_{x\to 0^+}\frac{1}{x};\ \lim_{x\to 0^+}\frac{\ln(\ln(1/x))}{1/x} = \lim_{x\to 0^+}\frac{\dfrac{1}{\ln(1/x)}\dfrac{1}{1/x}\left(\dfrac{-1}{x^2}\right)}{-1/x^2} = \lim_{x\to 0^+}\frac{x}{\ln(1/x)} = 0$$

Thus $\lim_{x\to 0^+}(\ln(1/x))^x = e^0 = 1$.

39. $\lim_{x\to \pi/2^-}\ln(\cos x) = -\infty$, $\lim_{x\to \pi/2^-}\tan x = \infty$;

$$\lim_{x\to \pi/2^-}\frac{\ln\cos x}{\tan x} = \lim_{x\to \pi/2^-}\frac{(-\sin x)/(\cos x)}{\sec^2 x} = \lim_{x\to \pi/2^-}(-\sin x\cos x) = 0$$

41. $\lim_{x\to\infty} x\sin\dfrac{1}{x} = \lim_{x\to\infty}\dfrac{\sin(1/x)}{1/x}$; $\lim_{x\to\infty}\sin\dfrac{1}{x} = 0 = \lim_{x\to\infty}\dfrac{1}{x}$;

$$\lim_{x\to\infty} x\sin\frac{1}{x} = \lim_{x\to\infty}\frac{\sin(1/x)}{1/x} = \lim_{x\to\infty}\frac{(-1/x^2)\cos(1/x)}{-1/x^2} = \lim_{x\to\infty}\cos\frac{1}{x} = 1$$

43. $\lim_{x\to\infty}\ln(x^2+1) = \infty = \lim_{x\to\infty} x$; $\lim_{x\to\infty}\dfrac{\ln(x^2+1)}{\ln x} = \lim_{x\to\infty}\dfrac{2x/(x^2+1)}{1/x} = \lim_{x\to\infty}\dfrac{2x^2}{x^2+1} = \lim_{x\to\infty}\dfrac{2}{1+1/x^2} = 2$

45. $\lim_{x\to\infty} e^{(e^x)} = \infty = \lim_{x\to\infty} e^x$; $\lim_{x\to\infty}\dfrac{e^{(e^x)}}{e^x} = \lim_{x\to\infty}\dfrac{e^x e^{(e^x)}}{e^x} = \lim_{x\to\infty} e^{(e^x)} = \infty$

47. $\lim_{x\to\infty}\left(1+\dfrac{1}{x^2}\right)^x = \lim_{x\to\infty} e^{x\ln(1+1/x^2)} = e^{\lim_{x\to\infty} x\ln(1+1/x^2)} = e^{\lim_{x\to\infty}[\ln(1+1/x^2)]/(1/x)}$;

$$\lim_{x\to\infty}\frac{\ln(1+1/x^2)}{1/x} = \lim_{x\to\infty}\frac{[x^2/(x^2+1)](-2/x^3)}{-1/x^2} = \lim_{x\to\infty}\frac{2x}{x^2+1} = \lim_{x\to\infty}\frac{2/x}{1+1/x^2} = 0$$

Thus $\lim_{x\to\infty}(1+1/x^2)^x = e^0 = 1$.

49. $\lim_{x\to\infty}\ln(\ln x) = \infty = \lim_{x\to\infty}\sqrt{x}$;

$$\lim_{x\to\infty} x^{-1/2}\ln(\ln x) = \lim_{x\to\infty}\frac{\ln(\ln x)}{\sqrt{x}} = \lim_{x\to\infty}\frac{[1/(\ln x)](1/x)}{1/(2\sqrt{x})} = \lim_{x\to\infty}\frac{2}{\sqrt{x}\ln x} = 0$$

51. $\lim_{x\to a}(a^2 - ax) = 0 = \lim_{x\to a}(a-\sqrt{ax})$; $\lim_{x\to a}\dfrac{a^2-ax}{a-\sqrt{ax}} = \lim_{x\to a}\dfrac{-a}{-a/(2\sqrt{ax})} = \lim_{x\to a} 2\sqrt{ax} = 2a$

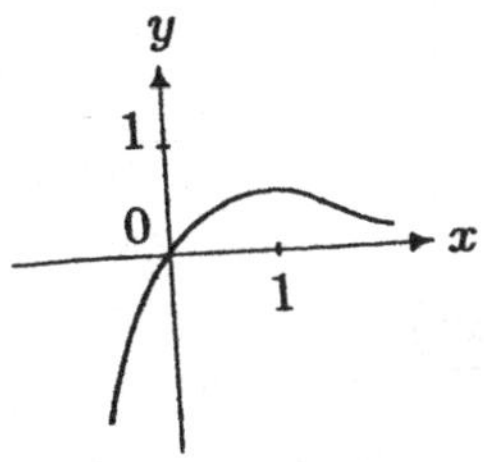

Exercise 53

53. $f'(x) = e^{-x} - xe^{-x}$; $f''(x) = -2e^{-x} + xe^{-x}$; relative maximum value is $f(1) = e^{-1}$; concave downward on $(-\infty, 2)$; concave upward on $(2, \infty)$; inflection point: $(2, 2e^{-2})$; $\lim_{x\to\infty} xe^{-x} = \lim_{x\to\infty}(x/e^x) = \lim_{x\to\infty}(1/e^x) = 0$, so $y = 0$ is a horizontal asymptote. Also $\lim_{x\to-\infty} xe^{-x} = -\infty$.

55. The conditions for applying l'Hôpital's Rule twice are met;

$$\lim_{x\to 0^+} f(x) = \lim_{x\to 0^+} x(\ln x)^2 = \lim_{x\to 0^+} \frac{(\ln x)^2}{1/x} = \lim_{x\to 0^+} \frac{(2\ln x)(1/x)}{-1/x^2}$$
$$= \lim_{x\to 0^+} \frac{2\ln x}{-1/x} = \lim_{x\to 0^+} \frac{2/x}{1/x^2} = \lim_{x\to 0^+} 2x = 0 = f(0).$$

57. a. $\lim_{x\to\pi/2} \sin x \neq 0$ and $\lim_{x\to\pi/2} x \neq 0$. Thus l'Hôpital's Rule does not apply.

b. By the Quotient Rule for Limits,

$$\lim_{x\to\pi/2} \frac{\sin x}{x} = \frac{\lim_{x\to\pi/2} \sin x}{\lim_{x\to\pi/2} x} = \frac{\sin \pi/2}{\pi/2} = \frac{2}{\pi}.$$

59. Applying l'Hôpital's Rule several times, we find that

$$\lim_{x\to 0} \frac{e^{-1/x^2}}{x} = \lim_{x\to 0} \frac{e^{-1/x^2}(2/x^3)}{1} = \lim_{x\to 0} \frac{2e^{-1/x^2}}{x^3} = \lim_{x\to 0} \frac{2e^{-1/x^2}(2/x^3)}{3x^2}$$
$$= \lim_{x\to 0} \frac{4e^{-1/x^2}}{3x^5} = \lim_{x\to 0} \frac{4e^{-1/x^2}(2/x^3)}{15x^4} = \lim_{x\to 0} \frac{8e^{-1/x^2}}{15x^7}.$$

Thus each time we apply l'Hôpital's Rule we obtain a fraction with a higher power of x in the denominator, so that we are unable to evaluate the limit in this manner.

61. $\lim_{x\to 0^+} (x^x)^x = \lim_{x\to 0^+} e^{x\ln(x^x)} = \lim_{x\to 0^+} e^{x^2 \ln x} = e^{\lim_{x\to 0^+} x^2 \ln x}$

Since $\lim_{x\to 0^+} x^2 \ln x = \lim_{x\to 0^+} (\ln x)/(1/x^2)$, and since $\lim_{x\to 0^+} \ln x = -\infty$ and $\lim_{x\to 0^+} 1/x^2 = \infty$, l'Hôpital's Rule implies that

$$\lim_{x\to 0^+} \frac{\ln x}{1/x^2} = \lim_{x\to 0^+} \frac{1/x}{-2/x^3} = \lim_{x\to 0^+} \left(-\frac{x^2}{2}\right) = 0.$$

Thus $\lim_{x\to 0^+} (x^x)^x = e^0 = 1$.

63. $$\lim_{x\to 0}\frac{\int_0^x (x-t)\sin(t^2)\,dt}{\ln(1+x^4)} = \lim_{x\to 0}\frac{x\int_0^x \sin(t^2)\,dt - \int_0^x t\sin(t^2)\,dt}{\ln(1+x^4)}$$

$$\overset{\text{l'Hôpital's Rule}}{=} \lim_{x\to 0}\frac{\int_0^x \sin(t^2)\,dt + x\cdot\sin(x^2) - x\sin(x^2)}{[1/(1+x^4)]4x^3}$$

$$= \lim_{x\to 0}\frac{(1+x^4)\int_0^x \sin(t^2)\,dt}{4x^3}$$

$$\overset{\text{l'Hôpital's Rule}}{=} \lim_{x\to 0}\frac{4x^3\int_0^x \sin(t^2)\,dt + (1+x^4)\sin(x^2)}{12x^2}$$

$$= \lim_{x\to 0}\frac{x}{3}\int_0^x \sin(t^2)\,dt + \lim_{x\to 0}(1+x^4)\frac{\sin x^2}{12x^2}$$

$$= 0 + \lim_{x\to 0}(1+x^4)\lim_{x\to 0}\frac{\sin x^2}{12x^2} = \frac{1}{12}$$

65. a. Since the hypotenuse l of T is tangent to the graph of e^{-x} at (z, e^{-z}), a point-slope equation of l is $y - e^{-z} = -e^{-z}(x-z)$. If $x = 0$ then $y = e^{-z} - e^{-z}(-z) = (1+z)e^{-z}$, so the y intercept of l is $(1+z)e^{-z}$. If $y = 0$ then $-e^{-z} = -e^{-z}(x-z)$, so $1 = x - z$ and hence $x = 1+z$; thus the x intercept of l is $1+z$. Therefore the area A of T is given by $A = \frac{1}{2}[(1+z)e^{-z}](1+z) = \frac{1}{2}(1+z)^2e^{-z}$. By l'Hôpital's Rule,

$$\lim_{z\to\infty}\frac{1}{2}(1+z)^2e^{-z} = \lim_{z\to\infty}\frac{1}{2}\frac{(1+z)^2}{e^z} = \lim_{z\to\infty}\frac{1+z}{e^z} = \lim_{z\to\infty}\frac{1}{e^z} = 0.$$

Consequently for any given $\varepsilon > 0$ the area A of T is less than ε if z is large enough.

b. By part (a) the area A of T is given by $A = \frac{1}{2}(1+z)^2e^{-z}$. Thus $dA/dz = (1+z)e^{-z} - \frac{1}{2}(1+z)^2e^{-z} = (1+z)[1 - \frac{1}{2}(1+z)]e^{-z} = \frac{1}{2}(1+z)(1-z)e^{-z}$. Since $dA/dz > 0$ if $0 < z < 1$ and $dA/dz < 0$ if $z > 1$, it follows that A is maximum if $z = 1$. Since by part (a) the x intercept of l is $1+z$, the area of T is maximal if the base of T has length $1 + 1 = 2$.

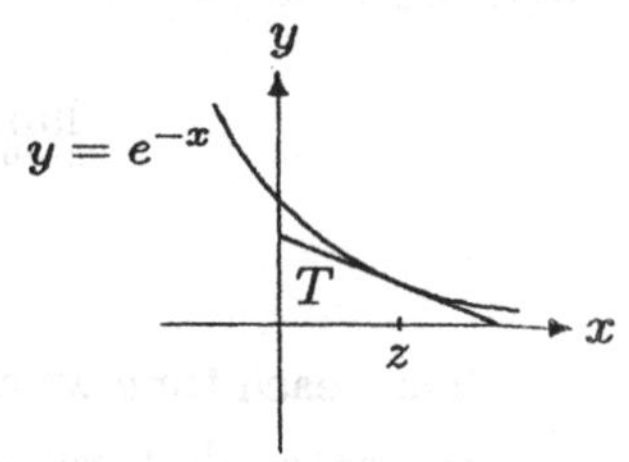

7.7 Introduction to Differential Equations

1. If $y = \frac{1}{3}e^{3t}$, then $dy/dt = 3(\frac{1}{3}e^{3t}) = e^{3t}$, so that $dy/dt = e^{3t}$.

3. If $y = \tan t + \sec t$ for $0 < t < \pi/2$, then $dy/dt = \sec^2 t + \sec t\tan t$, so that $2(dy/dt) - y^2 = 2(\sec^2 t + \sec t\,\tan t) - (\tan t + \sec t)^2 = 2\sec^2 t + 2\sec t\,\tan t - (\tan^2 t + 2\sec t\,\tan t + \sec^2 t) = \sec^2 t - \tan^2 t = 1$. Thus $2(dy/dt) - y^2 = 1$.

5. If $y = \sin 2x - \cos 2x$, then $dy/dx = 2\cos 2x + 2\sin 2x$, and $d^2y/dx^2 = -4\sin 2x + 4\cos 2x$. Thus $d^2y/dx^2 + 4y = -4\sin 2x + 4\cos 2x + 4(\sin 2x - \cos 2x) = 0$, so that $d^2y/dx^2 + 4y = 0$.

7. If $y = xe^{-2x}$, then $dy/dx = e^{-2x} - 2xe^{-2x}$ and $d^2y/dx^2 = -2e^{-2x} - 2e^{-2x} + 4xe^{-2x}$. Thus $d^2y/dx^2 + 4(dy/dx) + 4y = (-2e^{-2x} - 2e^{-2x} + 4xe^{-2x}) + 4(e^{-2x} - 2xe^{-2x}) + 4xe^{-2x} = 0$, so that $d^2y/dx^2 + 4(dy/dx) + 4y = 0$.

9. If $y = -2e^{-3x}$, then $dy/dx = 6e^{-3x}$, so that $dy/dx + 5y = 6e^{-3x} + 5(-2e^{-3x}) = -4e^{-3x}$. Thus y satisfies the differential equation. Since $y(0) = -2e^{-3(0)} = -2$, y also satisfies the initial condition.

11. If $y = x\int_0^x \sqrt{1+t^4}\,dt$, then $dy/dx = \int_0^x \sqrt{1+t^4}\,dt + x\sqrt{1+x^4}$. Thus

$$x\frac{dy}{dx} - y = x\left(\int_0^x \sqrt{1+t^4}\,dt + x\sqrt{1+x^4}\right) - x\int_0^x \sqrt{1+t^4}\,dt = x^2\sqrt{1+x^4}$$

so that y satisfies the differential equation. Since $y(0) = 0\int_0^0 \sqrt{1+t^4}\,dt = 0$ and $y'(0) = \int_0^0 \sqrt{1+t^4}\,dt + 0\sqrt{1+0^4} = 0$, y also satisfies the initial conditions.

13. Part of the slope field of the differential equation $dy/dx = 2x$.

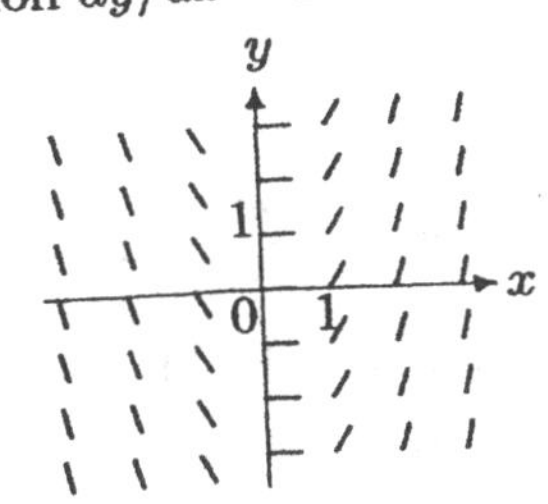

15. Part of the slope field of the differential equation $dy/dx = -x/y$.

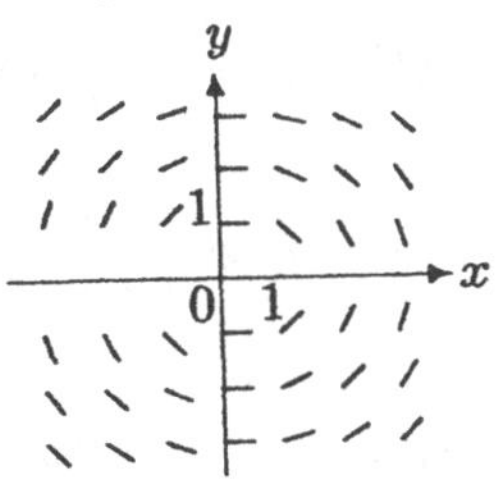

17. Part of the slope field of the differential equation $dy/dx = x/y$.

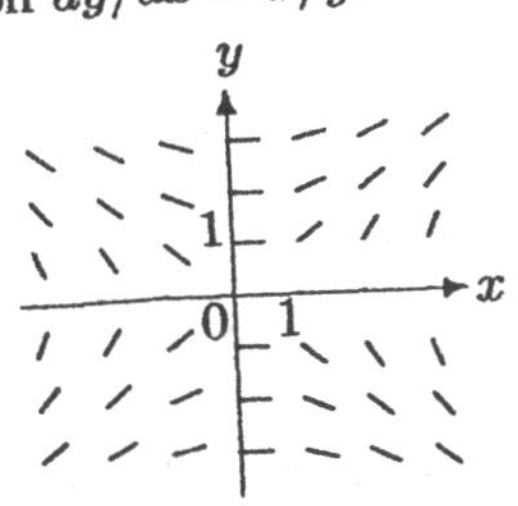

19. If $y = \sqrt{r^2 - x^2}$, then $dy/dx = -x/\sqrt{r^2 - x^2} = -x/y$, so that $y(dy/dx) + x = y(-x/y) + x = 0$. Thus y satisfies the differential equation.

21. Let $y = x^2$. Then $dy/dx = 2x$. Substituting into the differential equation, we obtain $c(2x)^2 - x(2x) + x^2 = 0$, or $(4c - 1)x^2 = 0$. Thus $c = \frac{1}{4}$.

23. a. We have $dy/dx = g(y) > 0$ for all y, so that y is an increasing function. Thus the graph of any solution will rise from left to right.

b. By the Chain Rule, we have

$$\frac{d^2y}{dx^2} = \frac{d}{dx}(g(y)) = \frac{d}{dy}(g(y))\frac{dy}{dx} = g'(y)g(y) > 0.$$

Therefore the graph of any solution will be concave upward (in addition to the property obtained in (a)).

25. Since $dy/dx)^2 \geq 0$, c must be nonnegative. If $c \geq 0$ and $y = \sqrt{c}\,x$, then $dy/dx = \sqrt{c}$, so y is a solution of $(dy/dx)^2 = c$.

27. If $C = c(1 - e^{-at})$, then $dC/dt = ace^{-at}$. Thus C is a solution of $dC/dt + aC = b$ if and only if $ace^{-at} + ac(1 - e^{-at}) = b$, or $ac = b$. Thus $c(1 - e^{-at})$ is a solution if and only if $c = b/a$.

7.8 Methods of Solving Differential Equations

1. $\dfrac{dy}{dx} = \dfrac{x}{y}$, so $y\,dy = x\,dx$; thus $\frac{1}{2}y^2 = \frac{1}{2}x^2 + C_1$, or $y^2 - x^2 = C$.

3. $(1 + x^2)\,dy = (1 + y^2)\,dx$, so $\dfrac{1}{1 + y^2}\,dy = \dfrac{1}{1 + x^2}\,dx$; thus $\tan^{-1} y = \tan^{-1} x + C$.

5. $\dfrac{1 + e^x}{1 - e^{-y}}\,dy + e^{x+y}\,dx = 0$, so $\dfrac{e^{-y}}{1 + e^x}\left(\dfrac{1 + e^x}{1 - e^{-y}}\,dy + e^{x+y}\,dx\right) = 0$, or $\dfrac{e^{-y}}{1 - e^{-y}}\,dy + \dfrac{e^x}{1 + e^x}\,dx = 0$; thus $\ln|1 - e^{-y}| + \ln(1 + e^x) = C_1$, so that if $C_1 = \ln C$, then $\ln|1 - e^{-y}| + \ln(1 + e^x) = \ln C$, or $\ln[|1 - e^{-y}|(1 + e^x)] = \ln C$. Therefore $|1 - e^{-y}|(1 + e^x) = C$, or $|1 - e^{-y}| = C/(1 + e^x)$.

7. $e^{-2y}\,dy = (x - 2)\,dx$, and thus $-\frac{1}{2}e^{-2y} = \frac{1}{2}x^2 - 2x + C$. If $y(0) = 0$, then $-\frac{1}{2}e^0 = 0 - 0 + C$, so that $C = -\frac{1}{2}$. Therefore the particular solution with $y(0) = 0$ is $-\frac{1}{2}e^{-2y} = \frac{1}{2}x^2 - 2x - \frac{1}{2}$, or $y = -\frac{1}{2}\ln(-x^2 + 4x + 1)$.

9. $y^2x\dfrac{dy}{dx} - x + 1 = 0$, so $y^2\,dy = \dfrac{1}{x}(x - 1)\,dx = \left(1 - \dfrac{1}{x}\right)dx$; thus $\frac{1}{3}y^3 = x - \ln|x| + C$. If $y(1) = 3$, then $\frac{1}{3}(3)^3 = 1 - \ln 1 + C$, so that $C = 8$. Therefore the particular solution with $y(1) = 3$ is $\frac{1}{3}y^3 = x - \ln|x| + 8$.

11. $P(x) = 1/x^2$ and $Q(x) = 0$. Since $-1/x$ is an antiderivative of P, $S(x) = -1/x$, so (12) implies that $y = e^{1/x}\int e^{-1/x}0\,dx = e^{1/x}(C) = Ce^{1/x}$.

13. $P(x) = 2$ and $Q(x) = 4$. Since $2x$ is an antiderivative of P, $S(x) = 2x$, so (12) implies that $y = e^{-2x}\int e^{2x}4\,dx = 4e^{-2x}\int e^{2x}\,dx = 4e^{-2x}(\frac{1}{2}e^{2x} + C_1) = 2 + Ce^{-2x}$.

15. $P(x) = -1$ and $Q(x) = 1/(1 - e^{-x})$. Since $-x$ is an antiderivative of P, $S(x) = -x$, so (12) implies that
$$y = e^x \int e^{-x} \frac{1}{1 - e^{-x}}\, dx = e^x(\ln|1 - e^{-x}| + C).$$

17. $P(x) = \tan x$ and $Q(x) = \tan x$. Since $-\ln\cos x$ is an antiderivative of P for $-\pi/2 < x < \pi/2$ (since $\cos x > 0$ for such x), we have that $S(x) = -\ln\cos x$. Then (12) implies that
$$y = e^{\ln\cos x} \int e^{-\ln\cos x} \tan x\, dx = \cos x \int \sec x \tan x\, dx = \cos x\,(\sec x + C) = 1 + C\cos x.$$

19. $P(x) = 5$ and $Q(x) = -4e^{-3x}$. Since $5x$ is an antiderivative of P, $S(x) = 5x$, so (12) implies that $y = e^{-5x}\int e^{5x}(-4e^{-3x})\, dx = -4e^{-5x}\int e^{2x}\, dx = -4e^{-5x}(\frac{1}{2}e^{2x} + C)$. If $y(0) = -4$, then $-4 = -4e^{-5(0)}(\frac{1}{2}e^{2(0)} + C)$, so that $C = \frac{1}{2}$. Therefore the particular solution with $y(0) = -4$ is $y = -4e^{-5x}(\frac{1}{2}e^{2x} + \frac{1}{2}) = -2e^{-3x} - 2e^{-5x}$.

21. The equation is equivalent to $dy/dx + (1/\cos x)y = 1/\cos x$, so that $P(x) = 1/\cos x = \sec x = Q(x)$. Since $\ln(\sec x + \tan x)$ is an antiderivative of P for $0 < x < \pi/2$, $S(x) = \ln(\sec x + \tan x)$, so (12) implies that
$$y = e^{-\ln(\sec x + \tan x)} \int e^{\ln(\sec x + \tan x)} \sec x\, dx = \frac{1}{\sec x + \tan x}\int (\sec x + \tan x)\sec x\, dx$$
$$= \frac{1}{\sec x + \tan x}\int (\sec^2 x + \tan x \sec x)\, dx = \frac{1}{\sec x + \tan x}(\tan x + \sec x + C) = 1 + \frac{C}{\sec x + \tan x}.$$
If $y(\pi/4) = 2$, then $2 = 1 + C/(\sqrt{2} + 1)$, so that $C = \sqrt{2} + 1$. Therefore the particular solution with $y(\pi/4) = 2$ is $y = 1 + (\sqrt{2} + 1)/(\sec x + \tan x)$.

23. a. Since $v = y/x$, we have $y = xv$, so that $dy/dx = x(dv/dx) + v$.

b. By (15), $f(v) = f(y/x) = dy/dx \overset{(15)}{=} x(dv/dx) + v$, so that $f(v) - v = x(dv/dx)$ and thus $[1/(f(v) - v)]\, dv = (1/x)\, dx$.

25. If we multiply both sides of the differential equation by $(x^2 + y^2)/y$, we obtain $y + 2x(dy/dx) = 0$, and can be rewritten as $(2/y)\, dy = -(1/x)\, dx$. The solution is given by $2\ln y = -\ln x + C_1$, or $\ln y^2 + \ln x = C_1$, or $\ln(xy^2) = C_1$, or $xy^2 = C$, where $C = e^{C_1}$.

27. a. Letting $v = mg/b$ in (13), we obtain $-mg + b(mg/b) = m(dv/dt)$, so $dv/dt = 0$.

b. When the velocity reaches the limiting velocity, it no longer changes, so $dv/dt = 0$.

29. a. Separating the variables in (18), we obtain $(1/\sqrt{s})\, ds = \sqrt{2g}\, dt$, so that $2\sqrt{s} = \sqrt{2g}\,t + C$. Since $s(0) = 0$, we have $C = 0$. Thus $2\sqrt{s} = \sqrt{2g}\,t$, or $s = \frac{1}{2}gt^2$.

b. Since $s(0) = 0$, the initial height h_0 of the object is 0, and since the object is dropped, the initial velocity v_0 is 0. Thus (1) of Section 1.3, which applies when height is measured in meters and time in seconds, becomes $h(t) = -4.9t^2 = \frac{1}{2}(-9.8)t^2 = \frac{1}{2}gt^2$. (Similarly, (2) of Section 1.3, which applies when height is measured in feet and time in seconds, becomes $h(t) = -16t^2 = \frac{1}{2}(-32)t^2 = \frac{1}{2}gt^2$. This also agrees with the result in part (a).)

31. Suppose $y = Cx^k$, as in (7). Substituting first $x = 25$ and $y = .44$, and then $x = 110$ and $y = 3$, we obtain $.44 = C\,25^k$ and $3 = C\,110^k$. Dividing these two equations to eliminate C, we find that $.44/3 = (25/110)^k = (5/22)^k$. Taking natural logarithms, we find that $\ln(.44/3) = k\ln(5/22)$. Using a calculator, we obtain $k \approx 1.2956$. Then $C = 3/110^k \approx .0068$. If we substitute the values 460, 1400, and 6000 of x in the table into the equation $y = Cx^k$, we obtain approximately 19, 81, and 534, respectively, for the corresponding values of y. These values are reasonably close to the values of y given in the table, so the data seems to (nearly) satisfy an allometric relation.

33. a. Separating the variables in $dy/dx = c(1+y^2)^{1/2}$, we obtain $[1/(1+y^2)^{1/2}]\,dy = c\,dx$. By (9) in Section 7.4, this implies $\sinh^{-1} y = cx + C$, where C is a constant. Therefore $y = \sinh(cx + C)$.

b. By (a), $f'(x) = y = \sinh(cx + C)$. Therefore since

$$\frac{d}{dx}\left(\frac{1}{c}\cosh(cx+C)\right) = \sinh(cx+C),$$

Theorem 4.6 implies that

$$f(x) = \frac{1}{c}\cosh(cx+C) + D, \quad \text{where } D \text{ is a constant.}$$

Chapter 7 Review

1. $f'(x) = 9x^2 + 25x^4 \geq 0$ for all x, and $f'(x) = 0$ only for $x = 0$, so f has an inverse.

3. $f'(x) = 1/x^2 > 0$ for x in $(-\infty, 0)$ and $(0, \infty)$, and $f(x) > 1$ for $x < 0$, whereas $f(x) < 1$ for $x > 0$. Thus f has an inverse.

5. $g(x + 2\pi) = g(x)$ for all x, so g does not have an inverse.

7. $y = \dfrac{3x-2}{-x+1}$; $(-x+1)y = 3x - 2$; $x(y+3) = y + 2$; $x = \dfrac{y+2}{y+3}$; $f^{-1}(x) = \dfrac{x+2}{x+3}$

9. Since $f(-x) = f(x)$ for every x in $(-a, a)$, f cannot have an inverse by (3) of Section 7.1.

11. $\dfrac{dy}{dx} = \dfrac{1}{1+2^x}(\ln 2)\,2^x = \dfrac{(\ln 2)\,2^x}{1+2^x}$

13. $\dfrac{dy}{dx} = \dfrac{1}{1+\sinh^2 x}\cosh x = \dfrac{\cosh x}{\cosh^2 x} = \dfrac{1}{\cosh x}$

15. $\dfrac{dy}{dx} = \dfrac{1}{\sqrt{1-(1-x^2)^{2/3}}}\,\dfrac{1}{3}(1-x^2)^{-2/3}(-2x) = \dfrac{-2x}{3(1-x^2)^{2/3}\sqrt{1-(1-x^2)^{2/3}}}$

17. Since $y = \dfrac{\ln(\tan^{-1} x^2)}{\ln 4}$, it follows that $\dfrac{dy}{dx} = \dfrac{1}{(\ln 4)(\tan^{-1} x^2)} \cdot \dfrac{2x}{1+x^4} = \dfrac{2x}{(\ln 4)(1+x^4)\tan^{-1} x^2}$.

19. $\displaystyle\int \frac{e^x}{\sqrt{1+e^x}}\,dx \overset{u=1+e^x}{=} \int \frac{1}{\sqrt{u}}\,du = 2\sqrt{u}+C = 2\sqrt{1+e^x}+C$

21. $\displaystyle\int \frac{e^x}{\sqrt{1+e^{2x}}}\,dx \overset{u=e^x}{=} \int \frac{1}{\sqrt{1+u^2}}\,du = \sinh^{-1}u + C = \sinh^{-1}(e^x)+C$

23. $$\int \frac{e^x}{e^x+e^{-x}}\,dx \overset{u=e^x}{=} \int \frac{1}{u+1/u}\,du = \int \frac{u}{u^2+1}\,du \overset{v=u^2}{=} \int \frac{1}{v+1}\frac{1}{2}\,dv = \frac{1}{2}\int \frac{1}{v+1}\,dv$$
$$= \frac{1}{2}\ln|v+1|+C = \frac{1}{2}\ln(u^2+1)+C = \frac{1}{2}\ln(e^{2x}+1)+C$$

25. $\displaystyle\int_0^1 x^2 5^{-x^3}\,dx \overset{u=x^3}{=} \int_0^1 5^{-u}\frac{1}{3}\,du = \left.\frac{-5^{-u}}{3\ln 5}\right|_0^1 = \frac{1}{3\ln 5}\left(1-\frac{1}{5}\right) = \frac{4}{15\ln 5}$

27. $\displaystyle\int \frac{3}{1+4t^2}\,dt \overset{u=2t}{=} \int \frac{3}{1+u^2}\frac{1}{2}\,du = \frac{3}{2}\tan^{-1}u + C = \frac{3}{2}\tan^{-1}2t + C$

29. $\displaystyle\int_{-5/4}^{5/4} \frac{1}{\sqrt{25-4t^2}}\,dt = \frac{1}{2}\int_{-5/4}^{5/4} \frac{1}{\sqrt{25/4-t^2}}\,dt = \left.\frac{1}{2}\sin^{-1}\frac{2t}{5}\right|_{-5/4}^{5/4} = \frac{1}{2}\left[\sin^{-1}\frac{1}{2} - \sin^{-1}\left(\frac{-1}{2}\right)\right] = \frac{\pi}{6}$

31. $$\int \operatorname{sech} x\,dx = \int \frac{1}{\cosh x}\,dx = \int \frac{2}{e^x+e^{-x}}\,dx = \int \frac{2e^x}{e^{2x}+1}\,dx$$
$$\overset{u=e^x}{=} \int \frac{2}{u^2+1}\,du = 2\tan^{-1}u + C = 2\tan^{-1}e^x + C$$

33. $$\int \frac{x}{x^4+4x^2+10}\,dx = \int \frac{x}{(x^2+2)^2+6}\,dx \overset{u=x^2+2}{=} \int \frac{1}{u^2+6}\frac{1}{2}\,du$$
$$= \frac{1}{2}\frac{1}{\sqrt{6}}\tan^{-1}\frac{u}{\sqrt{6}} + C = \frac{\sqrt{6}}{12}\tan^{-1}\left(\frac{x^2+2}{\sqrt{6}}\right)+C$$

35. $\displaystyle\lim_{t\to 0}\sinh at = 0 = \lim_{t\to 0} t;\ \lim_{t\to 0}\frac{\sinh at}{t} = \lim_{t\to 0}\frac{a\cosh at}{1} = a$

37. The conditions for applying l'Hôpital's Rule twice are met;
$$\lim_{x\to 0+}\frac{\ln x}{\ln(\sin x)} = \lim_{x\to 0+}\frac{1/x}{(\cos x)/(\sin x)} = \lim_{x\to 0+}\frac{\tan x}{x} = \lim_{x\to 0+}\frac{\sec^2 x}{1} = 1.$$

39. $\displaystyle\lim_{x\to 0+}\left(\frac{1}{x}-\frac{1}{\tan^{-1}x}\right) = \lim_{x\to 0+}\frac{\tan^{-1}x - x}{x\tan^{-1}x}$;
the conditions for applying l'Hôpital's Rule twice are met;
$$\lim_{x\to 0+}\left(\frac{1}{x}-\frac{1}{\tan^{-1}x}\right) = \lim_{x\to 0+}\frac{\tan^{-1}x - x}{x\tan^{-1}x} = \lim_{x\to 0+}\frac{\dfrac{1}{x^2+1}-1}{\tan^{-1}x + \dfrac{x}{x^2+1}}$$

$$= \lim_{x\to 0^+} \frac{\dfrac{-2x}{(x^2+1)^2}}{\dfrac{1}{x^2+1} + \dfrac{1-x^2}{(x^2+1)^2}} = \lim_{x\to 0^+}(-x) = 0.$$

41. $\displaystyle\lim_{x\to\infty}\left(\frac{x+1}{x-1}\right)^x = \lim_{x\to\infty} e^{x\ln[(x+1)/(x-1)]} = e^{\lim_{x\to\infty} x\ln[(x+1)/(x-1)]}$;

$\displaystyle\lim_{x\to\infty}\left[x\ln\frac{x+1}{x-1}\right] = \lim_{x\to\infty}\frac{\ln[(x+1)/(x-1)]}{1/x}$, and $\displaystyle\lim_{x\to\infty}\ln\frac{x+1}{x-1} = 0 = \lim_{x\to\infty}\frac{1}{x}$;

thus

$$\lim_{x\to\infty} x\ln\frac{x+1}{x-1} = \lim_{x\to\infty}\frac{\ln\dfrac{x+1}{x-1}}{1/x} = \lim_{x\to\infty}\frac{\dfrac{1}{x+1}-\dfrac{1}{x-1}}{-1/x^2} = \lim_{x\to\infty}\frac{2x^2}{x^2-1} = 2$$

so that $\displaystyle\lim_{x\to\infty}\left(\frac{x+1}{x-1}\right)^x = e^2$.

43. $e^{(y^2)}\,dx + x^2y\,dy = 0$, so $(1/x^2)\,dx + ye^{-(y^2)}\,dy = 0$, which is a separable differential equation. By integration we obtain the general solution $-1/x - (1/2)e^{-(y^2)} = C$.

45. $P(x) = 2/x$ and $Q(x) = x^2 + 6$. Since $2\ln x$ is an antiderivative of P for $x > 0$, $S(x) = 2\ln x$, so by (12) in Section 7.8, the general solution is

$$y = e^{-2\ln x}\int e^{2\ln x}(x^2+6)\,dx = \frac{1}{x^2}\int x^2(x^2+6)\,dx = \frac{1}{x^2}\int (x^4+6x^2)\,dx$$
$$= \frac{1}{x^2}\left(\frac{1}{5}x^5 + 2x^3 + C\right) = \frac{1}{5}x^3 + 2x + \frac{C}{x^2}.$$

47. $2xy\,dx = (y+1)\,dy$, so $2x\,dx = (1+1/y)\,dy$, which is a separable differential equation. By integration we obtain the general solution $x^2 = y + \ln y + C$. If $y(0) = 1$, then $0^2 = 1 + \ln 1 + C$, so $C = -1$. Consequently the particular solution is $x^2 = y + \ln y - 1$.

49. $P(x) = -2$ and $Q(x) = 3$. Since $-2x$ is an antiderivative of P, $S(x) = -2x$, so by (12) of Section 7.8, the general solution is $y = e^{2x}\int e^{-2x}\,3\,dx = e^{2x}\left(-\frac{3}{2}e^{-2x} + C\right) = -\frac{3}{2} + Ce^{2x}$. If $y(0) = 2$, then $2 = -\frac{3}{2} + C$, so $C = \frac{7}{2}$. Consequently the particular solution is $y = -\frac{3}{2} + \frac{7}{2}e^{2x} = \frac{1}{2}(7e^{2x} - 3)$.

51. $f'(x) = (2x)^x(\ln 2x + 1)$, $f''(x) = (2x)^x[(\ln 2x+1)^2 + 1/x]$; relative minimum value is $f(\frac{1}{2}e^{-1}) = 1/e^{1/(2e)} \approx 0.83$; concave upward on $(0,\infty)$. Finally

$$\lim_{x\to 0^+} f(x) = \lim_{x\to 0^+}(2x)^x = \lim_{x\to 0^+} 2^x \lim_{x\to 0^+} x^x.$$

Since $\lim_{x\to 0^+} 2^x = 2^0 = 1$, and since $\lim_{x\to 0^+} x^x = 1$ by Example 8 of Section 7.6, it follows that $\lim_{x\to 0^+}(2x)^x = 1\cdot 1 = 1$. Also $\lim_{x\to\infty}(2x)^x = \infty$.

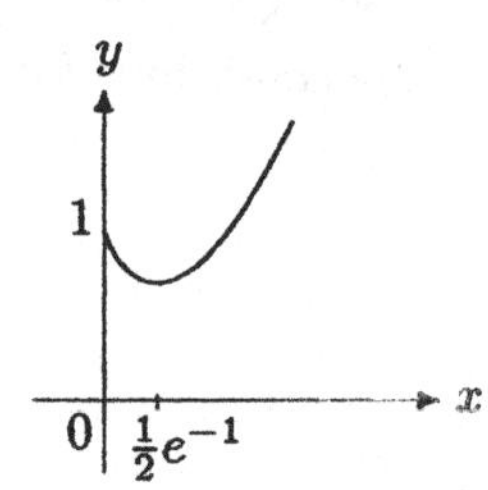

53. Let $y = \sinh^{-1} x$, so $x = \sinh y$. Note that $\cosh y = \sqrt{1+\sinh^2 y} = \sqrt{1+x^2}$. Then

$$\frac{d}{dx}\sinh^{-1} x = \frac{1}{\frac{d}{dy}\sinh y} = \frac{1}{\cosh y} = \frac{1}{\sqrt{1+x^2}}.$$

55. a. Replacing x by $-x$ in the inequality $e^x > 1+x$ for $x \neq 0$, we see that $e^{-x} > 1-x$ for $x \neq 0$. If $x < 1$, then $1-x > 0$, so it follows that

$$\frac{1}{e^{-x}} < \frac{1}{1-x}, \quad \text{or} \quad e^x < \frac{1}{1-x} \quad \text{for } x < 1.$$

b. Replacing x by $1/x$ in the inequalities $e^x > 1+x$ and $e^x < 1/(1-x)$ in (a) yields

$$e^{1/x} > 1 + \frac{1}{x} = \frac{x+1}{x} \quad \text{and} \quad e^{1/x} < \frac{1}{1-1/x} = \frac{x}{x-1}.$$

These hold in particular for $x > 1$. Taking natural logarithms, we obtain

$$\frac{1}{x} > \ln\frac{x+1}{x} \quad \text{and} \quad \frac{1}{x} < \ln\frac{x}{x-1}, \quad \text{so that} \quad \ln\frac{x+1}{x} < \frac{1}{x} < \ln\frac{x}{x-1} \quad \text{for } x > 1.$$

57. a. From the graphs we conjecture that $\lim_{x\to 0^+}[f(x)/g(x)] = 1$.

b. $$\lim_{x\to 0^+}\frac{f(x)}{g(x)} = \lim_{x\to 0^+}\frac{(\sin x)^x}{(\sin x)^{\sin x}} = \lim_{x\to 0^+}(\sin x)^{x-\sin x}$$
$$= \lim_{x\to 0^+} e^{(x-\sin x)\ln\sin x} = e^{\lim_{x\to 0^+}(x-\sin x)\ln\sin x};$$

$\lim_{x\to 0^+}(x-\sin x)\ln\sin x = \lim_{x\to 0^+}(\ln\sin x)/[1/(x-\sin x)]$.
The conditions for applying l'Hôpital's Rule are met;

$$\lim_{x\to 0^+}\frac{\ln\sin x}{\frac{1}{x-\sin x}} = \lim_{x\to 0^+}\frac{\frac{\cos x}{\sin x}}{-\frac{1-\cos x}{(x-\sin x)^2}} = \lim_{x\to 0^+}\frac{(x-\sin x)^2\cos x}{-(\sin x)(1-\cos x)}$$

$$= \lim_{x\to 0^+}\frac{(x-\sin x)^2(\cos x)(1+\cos x)}{-(\sin x)(1-\cos x)(1+\cos x)} = \lim_{x\to 0^+}\frac{(x-\sin x)^2}{-\sin^3 x}\lim_{x\to 0^+}(\cos x)(1+\cos x)$$

provided both of the latter two limits exist. Now $\lim_{x\to 0^+}(\cos x)(1+\cos x) = 1(1+1) = 2$ and by l'Hôpital's Rule,

$$\lim_{x\to 0^+}\frac{(x-\sin x)^2}{-\sin^3 x} = \lim_{x\to 0^+}\frac{2(x-\sin x)(1-\cos x)}{-3(\sin^2 x)(\cos x)} \lim_{x\to 0^+}\frac{2(x-\sin x)(1-\cos x)}{-3(1+\cos x)(1-\cos x)(\cos x)}$$

$$= \lim_{x\to 0^+}\frac{2(x-\sin x)}{-3(1+\cos x)(\cos x)} = \frac{2(0-0)}{-3(1+1)(1)} = 0.$$

Therefore $\lim_{x\to 0^+}(x-\sin x)\ln\sin x = \lim_{x\to 0^+}(\ln\sin x)/[1/(x-\sin x)] = 0\cdot 1 = 0$. Consequentl $\lim_{x\to 0^+} f(x)/g(x) = e^0 = 1$.

59. $\lim_{x\to\infty} \dfrac{f(x)}{g(x)} = \lim_{x\to\infty} \dfrac{x+\sin x}{x} = \lim_{x\to\infty}\left(1+\dfrac{\sin x}{x}\right) = 1+0 = 1$; $\dfrac{f'(x)}{g'(x)} = \dfrac{1+\cos x}{1} = 1+\cos x$, but since $\lim_{x\to\infty} 1 = 1$ and $\lim_{x\to\infty}\cos x$ does not exist, it follows that $\lim_{x\to\infty}[f'(x)/g'(x)]$ does not exist.

61. Since $\sqrt{x} \ge 1/\sqrt{x}$ on $[4,9]$, it follows that $f(x) \ge g(x)$ on $[4,9]$. Thus

$$A = \int_4^9 \left(\frac{\sqrt{x}}{x-1}e^{\sqrt{x}} - \frac{1}{\sqrt{x}\,(x-1)}e^{\sqrt{x}}\right) dx = \int_4^9 \frac{x-1}{\sqrt{x}\,(x-1)}e^{\sqrt{x}}\,dx$$

$$= \int_4^9 \frac{e^{\sqrt{x}}}{\sqrt{x}}\,dx \overset{u=\sqrt{x}}{=} \int_2^3 e^u(2)\,du = 2e^u\Big|_2^3 = 2(e^3-e^2).$$

63. Let R be the rectangle with base $[-x,x]$ on the x axis, with $x>0$. Also let A be the area of R. Then $A = 2xe^{-x^2}$. Now $dA/dx = 2e^{-x^2} - 4x^2e^{-x^2}$, so $dA/dx = 0$ if $4x^2 = 2$. Therefore $x = 1/\sqrt{2}$. Since $dA/dx > 0$ for $0 < x < 1/\sqrt{2}$ and $dA/dx < 0$ for $1/\sqrt{2} < x$, it follows from (1) of Section 4.6 and the First Derivative Test that A is maximum for $x = 1/\sqrt{2}$. Since $(d/dx)e^{-x^2} = -2xe^{-x^2}$ and $(d^2/dx^2)e^{-x^2} = -2e^{-x^2} + 4x^2e^{-x^2}$, it follows that $(1/\sqrt{2}, e^{-1/2})$ is an inflection point of the graph of e^{-x^2}.

65. a. By (6) in Section 7.5,

$$\int_{-L/2}^{L/2} \frac{1}{x^2+R^2}\,dx = \frac{1}{R}\tan^{-1}\frac{x}{R}\Big|_{-L/2}^{L/2} = \frac{1}{R}\tan^{-1}\frac{L}{2R} - \frac{1}{R}\tan^{-1}\left(-\frac{L}{2R}\right) = \frac{2}{R}\tan^{-1}\frac{L}{2R}.$$

Therefore

$$F = \frac{GmM}{L}\int_{-L/2}^{L/2}\frac{1}{x^2+R^2}\,dx = \frac{2GmM}{RL}\tan^{-1}\frac{L}{2R}.$$

b. By l'Hôpital's Rule,

$$\lim_{L\to 0^+} F = \lim_{L\to 0^+}\frac{2GmM}{RL}\tan^{-1}\frac{L}{2R} = \frac{2GmM}{R}\lim_{L\to 0^+}\frac{\tan^{-1}(L/2R)}{L}$$

$$\overset{\text{l'Hôpital's Rule}}{=} \frac{2GmM}{R}\lim_{L\to 0^+}\frac{\frac{1}{1+(L/2R)^2}\frac{1}{2R}}{1} = \frac{2GmM}{R}\frac{1}{2R} = \frac{GmM}{R^2}.$$

Cumulative Review(Chapters 1–6)

1. $\lim_{x\to\sqrt{2}} \dfrac{2\sqrt{2}-2x}{8-4x^2} = \lim_{x\to\sqrt{2}} \dfrac{2(\sqrt{2}-x)}{4(\sqrt{2}-x)(\sqrt{2}+x)} = \lim_{x\to\sqrt{2}} \dfrac{1}{2(\sqrt{2}+x)} = \dfrac{1}{4\sqrt{2}} = \dfrac{1}{8}\sqrt{2}$

3. $f'(x) = \dfrac{-e^{-x}(1+e^x)-(1+e^{-x})e^x}{(1+e^x)^2} = \dfrac{-e^{-x}-2-e^x}{(1+e^x)^2}$

5. $f'(x) = 1/x - \frac{1}{4}x$, so

$$\sqrt{1+(f'(x))^2} = \sqrt{1+\left(\frac{1}{x}-\frac{1}{4}x\right)^2} = \sqrt{1+\frac{1}{x^2}-\frac{1}{2}+\frac{1}{16}x^2}$$

$$= \sqrt{\frac{1}{x^2}+\frac{1}{2}+\frac{1}{16}x^2} = \sqrt{\left(\frac{1}{x}+\frac{1}{4}x\right)^2} = \frac{1}{x}+\frac{1}{4}x.$$

7. a. The inequality $(x-4)/(2x+6) > \frac{3}{20}$ is equivalent to $(x-4)/(2x+6) - \frac{3}{20} > 0$, or $[14(x-7)]/[40(x+3)] > 0$. From the diagram we see that the solution is the union of $(-\infty,-3)$ and $(7,\infty)$. Thus $f(x) > \frac{3}{20}$ for $x > 7$.

$x-7$	− − − − − − − − − 0 + +
$x+3$	− − 0 + + + + + + + +
$\dfrac{14(x-7)}{40(x+3)}$	+ + − − − − − − 0 + +

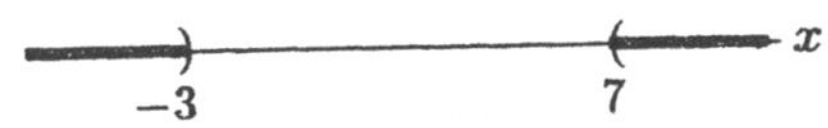

b. $f'(x) = \dfrac{2x+6-(x-4)(2)}{(2x+6)^2} = \dfrac{14}{(2x+6)^2} > 0$ for $x > 7$.
Thus f is increasing on $[7,\infty)$. Since $f(7) = \frac{3}{20}$, it follows that $f(x) > \frac{3}{20}$ for $x > 7$.

9. $f'(x) = -e^x e^{(1-e^x)}$; $f''(x) = -e^x e^{(1-e^x)} + e^{2x} e^{(1-e^x)} = e^x e^{(1-e^x)}(-1+e^x)$; $f'(x) < 0$ for all x, so f is a decreasing function; $f''(x) > 0$ if $-1+e^x > 0$, or equivalently, $x > 0$; $f''(x) < 0$ if $x < 0$; concave downward on $(-\infty,0)$ and concave upward on $(0,\infty)$; inflection point is $(0,1)$. Let $y = 1-e^x$. Then $\lim_{x\to-\infty}(1-e^x) = 1$, so that $\lim_{x\to-\infty} e^{(1-e^x)} = \lim_{y\to1^-} e^y = e$. Similarly, $\lim_{x\to\infty}(1-e^x) = -\infty$, so that $\lim_{x\to\infty} e^{(1-e^x)} = \lim_{y\to-\infty} e^y = 0$. Thus the horizontal asymptotes are $y = 0$ and $y = e$.

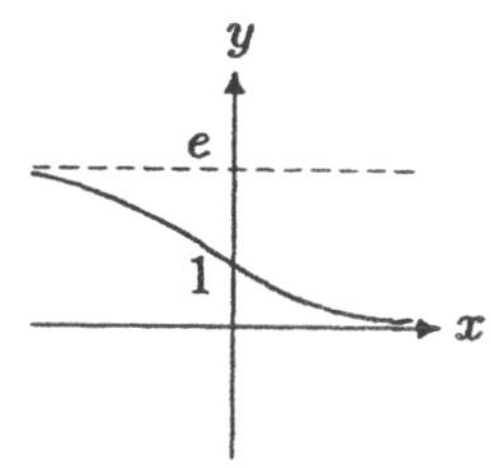

11. Using the notation in the figure and the Law of Cosines, we have

$$D^2 = (200)^2 + (200)^2 - 2(200)(200)\cos\theta = 80{,}000(1-\cos\theta).$$

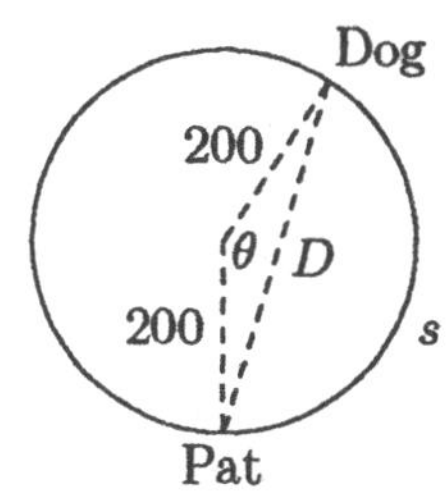

We are to find dD/dt at the moment that $D = 200$. Differentiating $D^2 = 80{,}000(1-\cos\theta)$ implicitly, we find that

$$2D\frac{dD}{dt} = 80{,}000(\sin\theta)\frac{d\theta}{dt}.$$

To determine $d\theta/dt$, we use the fact that if s is the length of the circular arc corresponding to D, then $s = 200\theta$. Since the dog moves at the rate of 50 feet per second, it follows that $50 = ds/dt = 200(d\theta/dt)$, so that $d\theta/dt = \frac{1}{4}$. At the instant $D = 200$, the triangle in the figure is equilateral, so $\theta = \pi/3$. Therefore at that instant

$$\frac{dD}{dt} = \frac{80{,}000(\sin(\pi/3))\frac{1}{4}}{2(200)} = 25\sqrt{3}\,(\text{feet per second}).$$

13. Let $u = 4 + x^2$, so that $du = 2x\,dx$;

$$\int \frac{x}{4+x^2}\,dx = \int \frac{1}{u}\left(\frac{1}{2}\right)du = \frac{1}{2}\ln|u| + C = \frac{1}{2}\ln(4+x^2) + C.$$

15. Let $f(t)$ be the position of the car at time t. Then

$$f(2)-f(0) = \int_0^2 f'(t)\,dt = \int_0^2 v(t)\,dt = \int_0^2 \left(40 + \frac{40}{4+t^2}\right)dt = \left(40t + 20\tan^{-1}\frac{t}{2}\right)\Big|_0^2$$
$$= (80+20\tan^{-1}1) - (0+20\tan^{-1}0) = 80+20(\frac{\pi}{4}) = 80+5\pi \text{ (miles)}$$

17. For $x > 0$ and $y \geq 0$, we have $y^2 - x^7 = 8x^4$, so $y = \sqrt{8x^4 + x^7} = x^2\sqrt{8+x^3}$. Thus $A = \int_1^2 x^2\sqrt{8+x^3}\,dx$. Let $u = 8 + x^3$, so that $du = 3x^2\,dx$. If $x = 1$, then $u = 9$, and if $x = 2$, then $u = 16$. Thus

$$A = \int_1^2 x^2\sqrt{8+x^3}\,dx = \int_9^{16} \sqrt{u}\,(\frac{1}{3})\,du = \frac{2}{9}u^{3/2}\Big|_9^{16} = \frac{2}{9}(64-27) = \frac{74}{9}$$

19. a. Since the figure is symmetric with respect to the y axis, and since the graphs of the two functions meet when $x = -2$ and $x = 2$, we find that

$$V = 2\int_0^2 \pi\{(4-x^2)^2 - [1-(|x|-1)^2]^2\}\,dx = 2\pi\int_0^2 [(4-x^2)^2 - (2x-x^2)^2]\,dx$$
$$= 2\pi\int_0^2 (16 - 12x^2 + 4x^3)\,dx = 2\pi(16x - 4x^3 + x^4)\big|_0^2$$
$$= 2\pi(32 - 32 + 16) = 32\pi$$

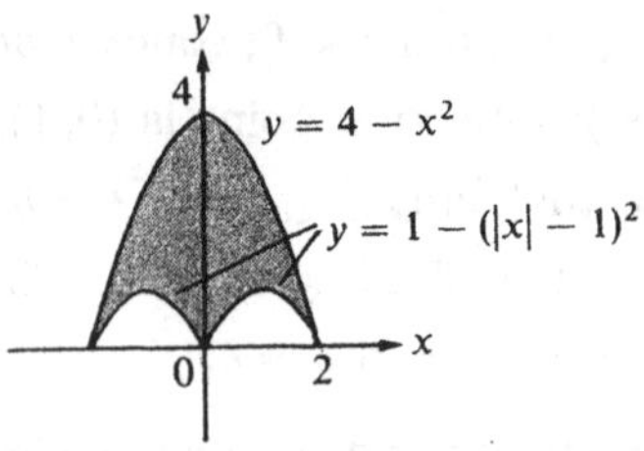

b. Because the figure is symmetric with respect to the y axis, $\bar{x} = 0$. To find $\bar{y}$ we will use the Theorem of Pappus and Guldin, which says that $V = 2\pi\bar{y}A$, where $V = 32\pi$ by part (a), and where A is the area of R. Now

$$A = 2\int_0^2 \{(4-x^2) - [1-(|x|-1)^2]\}\,dx = 2\int_0^2 (4-2x)\,dx = (8x - 2x^2)\big|_0^2 = 8$$

Thus $\bar{y} = V/(2\pi A) = 32\pi/[(2\pi)8] = 2$. Therefore $(\bar{x}, \bar{y}) = (0, 2)$.

Chapter 8

Techniques of Integration

8.1 Integration by Parts

1. $u = x$, $dv = \sin x\, dx$; $du = dx$, $v = -\cos x$; $\int x \sin x\, dx = x(-\cos x) - \int(-\cos x)\, dx = -x\cos x + \sin x + C$.

3. $u = \ln x$, $dv = x\, dx$; $du = (1/x)\, dx$, $v = \frac{1}{2}x^2$; $\int x \ln x\, dx = \frac{1}{2}x^2 \ln x - \int(\frac{1}{2}x^2)(1/x)\, dx = \frac{1}{2}x^2 \ln x - \frac{1}{2}\int x\, dx = \frac{1}{2}x^2 \ln x - \frac{1}{4}x^2 + C$.

5. $u = (\ln x)^2$, $dv = dx$; $du = [(2\ln x)/x]\, dx$, $v = x$; $\int(\ln x)^2\, dx = (\ln x)^2 x - \int x[(2\ln x)/x]\, dx = x(\ln x)^2 - 2\int \ln x\, dx$. By Example 3, $-2\int \ln x\, dx = -2x\ln x + 2x + C$. Thus $\int(\ln x)^2\, dx = x(\ln x)^2 - 2x\ln x + 2x + C$.

7. $u = \ln x$, $dv = x^3\, dx$; $du = (1/x)\, dx$, $v = \frac{1}{4}x^4$; $\int x^3 \ln x\, dx = (\ln x)(\frac{1}{4}x^4) - \int(\frac{1}{4}x^4)(1/x)\, dx = \frac{1}{4}x^4 \ln x - \frac{1}{4}\int x^3\, dx = \frac{1}{4}x^4 \ln x - \frac{1}{16}x^4 + C$.

9. $u = x^2$, $dv = e^{4x}\, dx$; $du = 2x\, dx$, $v = \frac{1}{4}e^{4x}$; $\int x^2 e^{4x}\, dx = \frac{1}{4}x^2 e^{4x} - \int \frac{1}{2}xe^{4x}\, dx = \frac{1}{4}x^2 e^{4x} - \frac{1}{2}\int xe^{4x}\, dx$. For $\int xe^{4x}\, dx$, let $u = x$, $dv = e^{4x}\, dx$; $du = dx$, $v = \frac{1}{4}e^{4x}$; $\int xe^{4x}\, dx = \frac{1}{4}xe^{4x} - \int \frac{1}{4}e^{4x}\, dx = \frac{1}{4}xe^{4x} - \frac{1}{16}e^{4x} + C_1$. Thus $\int x^2 e^{4x}\, dx = \frac{1}{4}x^2 e^{4x} - \frac{1}{8}xe^{4x} + \frac{1}{32}e^{4x} + C$.

11. $u = x^3$, $dv = \cos x\, dx$; $du = 3x^2\, dx$, $v = \sin x$; $\int x^3 \cos x\, dx = x^3 \sin x - \int 3x^2 \sin x\, dx = x^3 \sin x - 3\int x^2 \sin x\, dx$. Since $\int x^2 \sin x\, dx = -x^2 \cos x + 2x\sin x + 2\cos x + C_1$ by the solution of Exercise 10, it follows that $\int x^3 \cos x\, dx = x^3 \sin x + 3x^2 \cos x - 6x\sin x - 6\cos x + C$.

13. $u = \cos 3x$, $dv = e^{3x}\, dx$; $du = -3\sin 3x\, dx$, $v = \frac{1}{3}e^{3x}$; $\int e^{3x}\cos 3x\, dx = \frac{1}{3}e^{3x}\cos 3x + \int e^{3x}\sin 3x\, dx$. For $\int e^{3x}\sin 3x\, dx$, let $u = \sin 3x$, $dv = e^{3x}\, dx$; $du = 3\cos 3x\, dx$, $v = \frac{1}{3}e^{3x}$; $\int e^{3x}\sin 3x\, dx = \frac{1}{3}e^{3x}\sin 3x - \int e^{3x}\cos 3x\, dx$. Thus $\int e^{3x}\cos 3x\, dx = \frac{1}{3}e^{3x}\cos 3x + \frac{1}{3}e^{3x}\sin 3x - \int e^{3x}\cos 3x\, dx$, so $2\int e^{3x}\cos 3x\, dx = \frac{1}{3}e^{3x}\cos 3x + \frac{1}{3}e^{3x}\sin 3x + C_1$ and therefore $\int e^{3x}\cos 3x\, dx = \frac{1}{6}e^{3x}\cos 3x + \frac{1}{6}e^{3x}\sin 3x + C$.

15. $u = t$, $dv = 2^t\, dt$; $du = dt$, $v = (1/\ln 2)2^t$;

$$\int t \cdot 2^t\, dt = \frac{t}{\ln 2}2^t - \int \frac{1}{\ln 2}2^t\, dt = \frac{t}{\ln 2}2^t - \frac{1}{(\ln 2)^2}2^t + C.$$

17. $u = t^2$, $dv = 4^t\,dt$; $du = 2t\,dt$, $v = (1/\ln 4)4^t$;

$$\int t^2 4^t\,dt = \frac{t^2}{\ln 4}4^t - \int \frac{2t}{\ln 4}4^t\,dt = \frac{t^2}{\ln 4}4^t - \frac{2}{\ln 4}\int t\cdot 4^t\,dt.$$

For $\int t\cdot 4^t\,dt$, let $u = t$, $dv = 4^t\,dt$; $du = dt$, $v = (1/\ln 4)4^t$;

$$\int t\cdot 4^t\,dt = \frac{t}{\ln 4}4^t - \int \frac{1}{\ln 4}4^t\,dt = \frac{t}{\ln 4}4^t - \frac{1}{(\ln 4)^2}4^t + C_1.$$

Thus

$$\int t^2\,4^t\,dt = \left(\frac{t^2}{\ln 4} - \frac{2t}{(\ln 4)^2} + \frac{2}{(\ln 4)^3}\right)4^t + C.$$

19. $u = t$, $dv = \sinh t\,dt$; $du = dt$, $v = \cosh t$; $\int t\sinh t\,dt = t\cosh t - \int \cosh t\,dt = t\cosh t - \sinh t + C$.

21. $u = \tan^{-1} x$, $dv = dx$; $du = 1/(1+x^2)\,dx$, $v = x$;

$$\int \tan^{-1} x\,dx = x\tan^{-1} x - \int \frac{x}{1+x^2}\,dx.$$

For $\int [x/(1+x^2)]\,dx$ substitute $u = 1+x^2$, so that $du = 2x\,dx$. Then

$$\int \frac{x}{1+x^2}\,dx = \int \frac{1}{u}\frac{1}{2}\,du = \frac{1}{2}\int \frac{1}{u}\,du = \frac{1}{2}\ln|u| + C = \frac{1}{2}\ln(1+x^2) + C.$$

Thus $\int \tan^{-1} x\,dx = x\tan^{-1} x - \frac{1}{2}\ln(1+x^2) + C$.

23. $u = \cos^{-1}(-7x)$, $dv = dx$; $du = -\left(-7/\sqrt{1-(-7x)^2}\right)dx = \left(7/\sqrt{1-49x^2}\right)dx$, $v = x$;

$$\int \cos^{-1}(-7x)\,dx = x\cos^{-1}(-7x) - \int \frac{7x}{\sqrt{1-49x^2}}\,dx = x\cos^{-1}(-7x) - 7\int \frac{x}{\sqrt{1-49x^2}}\,dx.$$

For $\int \left(x/\sqrt{1-49x^2}\right)dx$ substitute $u = 1-49x^2$, so that $du = -98x\,dx$. Then

$$\int \frac{x}{\sqrt{1-49x^2}}\,dx = \int \frac{1}{\sqrt{u}}\left(-\frac{1}{98}\right)du = -\frac{1}{49}\sqrt{u} + C_1 = -\frac{1}{49}\sqrt{1-49x^2} + C_1.$$

Thus $\int \cos^{-1}(-7x)\,dx = x\cos^{-1}(-7x) + \frac{1}{7}\sqrt{1-49x^2} + C$.

25. By Exercise 24,

$$x^n \ln x\,dx = \frac{1}{n+1}x^{n+1}\ln x - \frac{1}{(n+1)^2}x^{n+1} + C_1.$$

Thus

$$\int x^n \ln x^m\,dx = m\int x^n \ln x\,dx = \frac{m}{n+1}x^{n+1}\ln x - \frac{m}{(n+1)^2}x^{n+1} + C.$$

27. By the solution of Exercise 26, we have $\int \cos(\ln x)\,dx = x\cos(\ln x) + \int \sin(\ln x)\,dx$ and $\int \sin(\ln x)\,dx = x\sin(\ln x) - \int \cos(\ln x)\,dx$, so that $\int \cos(\ln x)\,dx = \frac{1}{2}[x\cos(\ln x) + x\sin(\ln x)] + C$.

29. $u = x$, $dv = e^{5x}\,dx$; $du = dx$, $v = \frac{1}{5}e^{5x}$;

$$\int_0^1 xe^{5x}\,dx = \frac{1}{5}xe^{5x}\Big|_0^1 - \int_0^1 \frac{1}{5}e^{5x}\,dx = \frac{1}{5}e^5 - \frac{1}{5}\int_0^1 e^{5x}\,dx$$

$$= \frac{1}{5}e^{5x} - \left(\frac{1}{25}e^{5x}\right)\Big|_0^1 = \frac{1}{5}e^5 - \frac{1}{25}e^5 + \frac{1}{25}e^0 = \frac{4}{25}e^5 + \frac{1}{25}.$$

31. $u = t^2$, $dv = \cos t\,dt$; $du = 2t\,dt$, $v = \sin t$; $\int_0^\pi t^2 \cos t\,dt = t^2 \sin t\big|_0^\pi - \int_0^\pi 2t \sin t\,dt = -2\int_0^\pi t \sin t\,dt$. By the solution of Exercise 1, $-2\int_0^\pi t\sin t\,dt = -2(-t\cos t + \sin t)\big|_0^\pi = -2(-\pi\cos\pi + \sin\pi) + 2\cdot 0 = -2\pi$. Thus $\int_0^\pi t^2 \cos t\,dt = -2\pi$.

33. $u = x$, $dv = \sec^2 x\,dx$; $du = dx$, $v = \tan x$;

$$\int_{-\pi/3}^{\pi/4} x\sec^2 x\,dx = x\tan x\big|_{-\pi/3}^{\pi/4} - \int_{-\pi/3}^{\pi/4} \tan x\,dx = \frac{\pi}{4} - \frac{\pi\sqrt{3}}{3} + (\ln|\cos x|)\big|_{-\pi/3}^{\pi/4}$$

$$= \frac{\pi}{4} - \frac{\pi\sqrt{3}}{3} + \ln\frac{1}{\sqrt{2}} - \ln\frac{1}{2} = \frac{\pi}{4} - \frac{\pi\sqrt{3}}{3} + \frac{1}{2}\ln 2.$$

35. Substitute $u = x + 1$, so that $du = dx$. If $x = 0$ then $u = 1$, and if $x = 1$ then $u = 2$. Thus by the solution of Example 3,

$$\int_0^1 \ln(x+1)\,dx = \int_1^2 \ln u\,du = (u\ln u - u)\big|_1^2 = (2\ln 2 - 2) + 1 = 2\ln 2 - 1.$$

37. Substitute $u = ax$, so that $du = a\,dx$ and $x = u/a$. Then

$$\int x\sin ax\,dx = \int \frac{u}{a}(\sin u)\frac{1}{a}\,du = \frac{1}{a^2}\int u\sin u\,du.$$

By the solution of Exercise 1, $\int u\sin u\,du = -u\cos u + \sin u + C_1$. Thus

$$\int x\sin ax\,dx = \frac{1}{a^2}(-ax\cos ax + \sin ax + C_1) = -\frac{x}{a}\cos ax + \frac{1}{a^2}\sin ax + C.$$

39. Substitute $u = \cos x$, so that $du = -\sin x\,dx$. Then $\int \sin x\,\tan^{-1}(\cos x)\,dx = -\int \tan^{-1} u\,du$. By the solution of Exercise 21, $\int \tan^{-1} u\,du = u\tan^{-1} u - \frac{1}{2}\ln(1+u^2) + C_1$. Thus $\int \sin x\,\tan^{-1}(\cos x)\,dx = -\cos x\,\tan^{-1}(\cos x) + \frac{1}{2}\ln(1+\cos^2 x) + C$.

41. Substitute $s = \sqrt{t}$, so that $ds = [1/2\sqrt{t}]\,dt$. Then $\int \cos\sqrt{t}\,dt = \int(\cos s)(2s)\,ds = 2\int s\cos s\,ds$. By the solution of Example 1, $\int s\cos s\,ds = s\sin s + \cos s + C_1$. Thus $\int\cos\sqrt{t}\,dt = 2\int s\cos s\,ds = 2s\sin s + 2\cos s + C = 2\sqrt{t}\sin\sqrt{t} + 2\cos\sqrt{t} + C$.

43. Substitute $u = x/2$, so that $du = \frac{1}{2}\,dx$. If $x = 0$ then $u = 0$, and if $x = \pi/2$ then $u = \pi/4$. Thus $\int_0^{\pi/2}\cos^3(x/2)\,dx = 2\int_0^{\pi/4}\cos^3 u\,du$. By (11) with $n = 3$ we have

$$\int_0^{\pi/4} \cos^3 u\,du = \frac{1}{3}\cos^2 u \sin u\Big|_0^{\pi/4} + \frac{2}{3}\int_0^{\pi/4} \cos u\,du$$

$$= \left(\frac{1}{3}\left(\frac{\sqrt{2}}{2}\right)^3 - 0\right) + \left(\frac{2}{3}\sin u\right)\Big|_0^{\pi/4} = \frac{\sqrt{2}}{12} + \frac{\sqrt{2}}{3} = \frac{5}{12}\sqrt{2}.$$

Therefore

$$\int_0^{\pi/2} \cos^3 \frac{x}{2}\,dx = 2\int_0^{\pi/4} \cos^3 u\,du = 2\left(\frac{5}{12}\sqrt{2}\right) = \frac{5\sqrt{2}}{6}.$$

45. Using the formula for $\int \sin^n x\,dx$ in (10) and then the one for $\int \sin^2 x\,dx$, we have

$$\int \sin^4 x\,dx = \frac{-1}{4}\sin^3 x \cos x + \frac{3}{4}\int \sin^2 x\,dx = \frac{-1}{4}\sin^3 x \cos x - \frac{3}{16}\sin 2x + \frac{3}{8}x + C$$

$$= \frac{-1}{4}\sin^3 x \cos x - \frac{3}{8}\sin x \cos x + \frac{3}{8}x + C.$$

47. $u = \cos^{n-1} x$, $dv = \cos x\,dx$; $du = -(n-1)\cos^{n-2} x \sin x\,dx$, $v = \sin x$;

$$\int \cos^n\,dx = \int \cos^{n-1} x \cos x\,dx = (\cos^{n-1} x)(\sin x) - \int [\sin x][-(n-1)\cos^{n-2} x \sin x]\,dx$$

$$= \cos^{n-1} x \sin x + (n-1)\int \cos^{n-2} x \sin^2 x\,dx = \cos^{n-1} x \sin x + (n-1)\int (\cos^{n-2} x - \cos^n x)\,dx.$$

Thus $n\int \cos^n x\,dx = \cos^{n-1} x \sin x + (n-1)\int \cos^{n-2} x\,dx$, so that

$$\int \cos^n x\,dx = \frac{1}{n}\cos^{n-1} x \sin x + \frac{n-1}{n}\int \cos^{n-2} x\,dx.$$

49. $u = \dfrac{1}{(x^2+a^2)^n}$, $dv = dx$; $du = \dfrac{-2nx}{(x^2+a^2)^{n+1}}\,dx$, $v = x$;

$$\int \frac{1}{(x^2+a^2)^n}\,dx = \frac{x}{(x^2+a^2)^n} + \int \frac{2nx^2}{(x^2+a^2)^{n+1}}\,dx = \frac{x}{(x^2+a^2)^n} + \int \frac{2n(x^2+a^2-a^2)}{(x^2+a^2)^{n+1}}\,dx$$

$$= \frac{x}{(x^2+a^2)^n} + \int \frac{2n}{(x^2+a^2)^n}\,dx - \int \frac{2na^2}{(x^2+a^2)^{n+1}}\,dx.$$

Thus

$$(1-2n)\int \frac{1}{(x^2+a^2)^n}\,dx = \frac{x}{(x^2+a^2)^n} - \int \frac{2na^2}{(x^2+a^2)^{n+1}}\,dx$$

so

$$\int \frac{1}{(x^2+a^2)^{n+1}}\,dx = \frac{x}{2na^2(x^2+a^2)^n} + \frac{2n-1}{2na^2}\int \frac{1}{(x^2+a^2)^n}\,dx.$$

51. By Exercise 48 with $n = 3$, $\int (\ln x)^3\,dx = x(\ln x)^3 - 3\int (\ln x)^2\,dx$.

By Exercise 48 with $n = 2$, $\int (\ln x)^2\,dx = x(\ln x)^2 - 2\int \ln x\,dx$.

By Example 3, $\int \ln x\,dx = x\ln x - x + C_1$. Thus

$$\int (\ln x)^3\,dx = x(\ln x)^3 - 3x(\ln x)^2 + 6x\ln x - 6x + C.$$

53. For the integral $\int e^{-x}\sin x\,dx$ let $u = \sin x$ and $dv = e^{-x}\,dx$. Then $du = \cos x\,dx$ and $v = -e^{-x}$. Therefore $\int e^{-x}\sin x\,dx = (\sin x)(-e^{-x}) - \int(-e^{-x})\cos x\,dx = -e^{-x}\sin x + \int e^{-x}\cos x\,dx$. Together with the information from the first integration by parts, this yields $\int e^{-x}\cos x\,dx = e^{-x}\sin x + \left(-e^{-x}\sin x + \int e^{-x}\cos x\,dx\right) = \int e^{-x}\cos x\,dx$.

55. Using the solution of Example 3, we find that the area A is given by

$$A = \int_1^2 \ln x\,dx = (x\ln x - x)\Big|_1^2 = (2\ln 2 - 2) + 1 = 2\ln 2 - 1.$$

57. Using the solution of Exercise 1, we find that the area A is given by

$$A = \int_0^{\pi/2} x\sin x\,dx = (-x\cos x + \sin x)\Big|_0^{\pi/2} = (0+1) - (0+0) = 1.$$

59. The graphs intersect at (x, y) if $3x\ln x = y = x^2\ln x$, or $x = 1$ or 3. Since $3x\ln x \geq x^2\ln x$ on $[1, 3]$, we have $A = \int_1^3(3x\ln x - x^2\ln x)\,dx = \int_1^3(3x - x^2)\ln x\,dx$. Let $u = \ln x$, $dv = (3x - x^2)\,dx$, $du = (1/x)\,dx$, $v = \frac{3}{2}x^2 - \frac{1}{3}x^3$. Then

$$\int_1^3 (3x - x^2)\ln x\,dx = \left(\frac{3}{2}x^2 - \frac{1}{3}x^3\right)\ln x\Big|_1^3 - \int_1^3\left(\frac{3}{2}x - \frac{1}{3}x^2\right)dx$$

$$= \left(\frac{27}{2} - 9\right)\ln 3 - \left(\frac{3}{4}x^2 - \frac{1}{9}x^3\right)\Big|_1^3 = \frac{9}{2}\ln 3 - \left(\frac{27}{4} - 3\right) + \left(\frac{3}{4} - \frac{1}{9}\right) = \frac{9}{2}\ln 3 - \frac{28}{9}.$$

Thus $A = \frac{9}{2}\ln 3 - \frac{28}{9}$.

61. $$V = \int_0^\pi \pi(\sin^{3/2} x)^2\,dx = \int_0^\pi \pi\sin^3 x\,dx \overset{(10)}{=} -\frac{\pi}{3}\sin^2 x\cos x\Big|_0^\pi + \frac{2\pi}{3}\int_0^\pi \sin x\,dx$$

$$= 0 - \frac{2\pi}{3}\cos x\Big|_0^\pi = \frac{4\pi}{3}$$

63. $$M_x = \int_0^{\pi/4} \frac{1}{2}\left(\cos^2 x - \sin^2 x\right)\,dx = \frac{1}{2}\int_0^{\pi/4}\cos 2x\,dx = \frac{1}{4}\sin 2x\Big|_0^{\pi/4} = \frac{1}{4}$$

$$M_y = \int_0^{\pi/4} x(\cos x - \sin x)\,dx \overset{\text{parts}}{=} x(\sin x + \cos x)\Big|_0^{\pi/4} - \int_0^{\pi/4}(\sin x + \cos x)\,dx$$

$$= \frac{\pi}{4}\sqrt{2} - (-\cos x + \sin x)\Big|_0^{\pi/4} = \frac{\pi\sqrt{2}}{4} - 1$$

$$A = \int_0^{\pi/4}(\cos x - \sin x)\,dx = (\sin x + \cos x)\Big|_0^{\pi/4} = \sqrt{2} - 1$$

$$\bar{x} = \frac{M_y}{A} = \frac{\frac{\pi\sqrt{2}}{4} - 1}{\sqrt{2} - 1} = \frac{\pi\sqrt{2} - 4}{4\sqrt{2} - 4};\ \bar{y} = \frac{M_x}{A} = \frac{1/4}{\sqrt{2} - 1} = \frac{1}{4\sqrt{2} - 4};\ (\bar{x}, \bar{y}) = \left(\frac{\pi\sqrt{2} - 4}{4\sqrt{2} - 4}, \frac{1}{4\sqrt{2} - 4}\right)$$

65. a. By Example 1, the moment M is given by

$$M = \int_0^{\pi/2} x\cos x\,dx = (x\sin x + \cos x)\Big|_0^{\pi/2} = \frac{\pi}{2} - 1.$$

b. By Example 1, the moment M is given by

$$M = \int_{-\pi/2}^{\pi/2} x\cos x\,dx = (x\sin x + \cos x)\Big|_{-\pi/2}^{\pi/2} = 0.$$

c. By Exercise 3, the moment M is given by

$$M = \int_1^2 x\ln x\,dx = \left(\frac{1}{2}x^2\ln x - \frac{x^2}{4}\right)\Bigg|_1^2 = 2\ln 2 - \frac{3}{4}.$$

67. a. Notice that f is linear on $[-L/2, 0]$ and passes through the points $(-L/2, 0)$ and $(0, H)$. Similarly, f is linear on $[0, L/2]$ and passes through the points $(0, H)$ and $(L/2, 0)$. Thus

$$f(x) = \begin{cases} 2Hx/L + H & \text{for } -L/2 \le x \le 0 \\ -2Hx/L + H & \text{for } 0 < x \le L/2 \end{cases}$$

b. Since f and $\cos x$ are even functions, it follows that $f(x)\cos(\pi x/L)$ is also an even function. Thus

$$\int_{-L/2}^{L/2} f(x)\cos\frac{\pi}{L}x\,dx = 2\int_0^{L/2}\left(-\frac{2H}{L}x + H\right)\cos\frac{\pi}{L}x\,dx$$

$$= -\frac{4H}{L}\int_0^{L/2} x\cos\frac{\pi}{L}x\,dx + 2H\int_0^{L/2}\cos\frac{\pi}{L}x\,dx.$$

To evaluate the first integral, let $u = \pi x/L$, so $du = (\pi/L)\,dx$. If $x = 0$, then $u = 0$; if $x = L/2$, then $u = \pi/2$. Thus by Example 1,

$$\int_0^{L/2} x\cos\frac{\pi}{L}x\,dx = \int_0^{\pi/2}\frac{L}{\pi}u\cos u\frac{L}{\pi}\,du = \frac{L^2}{\pi^2}(u\sin u + \cos u)\Big|_0^{\pi/2} = \frac{L^2}{\pi^2}\left(\frac{\pi}{2} - 1\right).$$

For the second integral we have

$$\int_0^{L/2}\left(\cos\frac{\pi}{L}x\right)dx = \int_0^{\pi/2}(\cos u)\frac{L}{\pi}\,du = \frac{L}{\pi}\sin u\Big|_0^{\pi/2} = \frac{L}{\pi}.$$

Therefore

$$\int_{-L/2}^{L/2} f(x)\cos\frac{\pi}{L}x\,dx = -\frac{4HL^2}{L\pi^2}\left(\frac{\pi}{2} - 1\right) + 2H\left(\frac{L}{\pi}\right) = \frac{4HL}{\pi^2}.$$

8.2 Trigonometric Integrals

1. $u = \cos x$, $du = -\sin x\,dx$; $\int \sin^3 x \cos^2 x\,dx = -\int(-\sin x)(1-\cos^2 x)\cos^2 x\,dx = -\int(1-u^2)u^2\,du =$ $\int(-u^2+u^4)\,du = \frac{-1}{3}u^3 + \frac{1}{5}u^5 + C = \frac{-1}{3}\cos^3 x + \frac{1}{5}\cos^5 x + C$.

3. $u = \sin 3x$, $du = 3\cos 3x\,dx$; $\int \sin^3 3x \cos 3x\,dx = \frac{1}{3}\int u^3\,du = \frac{1}{12}u^4 + C = \frac{1}{12}\sin^4 3x + C$.

5. First let $u = 1/x$, so that $du = -(1/x^2)\,dx$. Then

$$\int \frac{1}{x^2}\sin^5\frac{1}{x}\cos^2\frac{1}{x}\,dx = \int -\sin^5 u\cos^2 u\,du = -\int \sin^5 u \cos^2 u\,du.$$

For $\int \sin^5 u\cos^2 u\,du$, let $v = \cos u$, so that $dv = -\sin u\,du$. Then

$$\int \sin^5 u\cos^2 u\,du = \int(1-\cos^2 u)^2(\sin u)\cos^2 u\,du = \int -(1-v^2)^2 v^2\,dv = -\int(v^2 - 2v^4 + v^6)\,dv$$

$$= -\left(\frac{1}{3}v^3 - \frac{2}{5}v^5 + \frac{1}{7}v^7\right) + C_1 = -\frac{1}{3}\cos^3 u + \frac{2}{5}\cos^5 u - \frac{1}{7}\cos^7 u + C_1.$$

Thus

$$\int \frac{1}{x^2}\sin^5\frac{1}{x}\cos^2\frac{1}{x}\,dx = \frac{1}{3}\cos^3\frac{1}{x} - \frac{2}{5}\cos^5\frac{1}{x} + \frac{1}{7}\cos^7\frac{1}{x} + C.$$

7. By (1), $\int \sin^2 y\cos^2 y\,dy = \int(\sin y\cos y)^2\,dy = \int(\frac{1}{2}\sin 2y)^2\,dy = \frac{1}{4}\int\sin^2 2y\,dy$. Next, let $u = 2y$, so that $du = 2\,dy$. Then $\int \sin^2 2y\,dy = \int(\sin^2 u)\frac{1}{2}\,du = \frac{1}{2}\int\sin^2 u\,du$. By (4), $\int\sin^2 u\,du = \frac{1}{2}u - \frac{1}{4}\sin 2u +$ C_1. Thus $\int\sin^2 y\cos^2 y\,dy = \frac{1}{4}\int\sin^2 2y\,dy = \frac{1}{8}\int\sin^2 u\,du = \frac{1}{8}(\frac{1}{2}u - \frac{1}{4}\sin 2u) + C = \frac{1}{8}y - \frac{1}{32}\sin 4y + C$.

9. By (1), $\int\sin^4 x\cos^4 x\,dx = \int\frac{1}{16}(2\sin x\cos x)^4\,dx = \frac{1}{16}\int\sin^4 2x\,dx$. Next, let $u = 2x$, so that $du =$ $2\,dx$. Then $\frac{1}{16}\int\sin^4 2x\,dx = \frac{1}{16}\int(\sin^4 u)\frac{1}{2}\,du = \frac{1}{32}\int\sin^4 u\,du$. By (12) of Section 8.1, $\frac{1}{32}\int\sin^4 u\,du =$ $\frac{1}{32}\left(-\frac{1}{4}\sin^3 u\cos u - \frac{3}{8}\sin u\cos u + \frac{3}{8}u\right) + C = -\frac{1}{128}\sin^3 2x\cos 2x - \frac{3}{256}\sin 2x\cos 2x + \frac{3}{128}x + C$.

11. $u = \sin x$, $du = \cos x\,dx$; $\int\sin^{-10}x\cos^3 x\,dx = \int\sin^{-10}x(1-\sin^2 x)\cos x\,dx = \int(u^{-10} - u^{-8})\,du =$ $-\frac{1}{9}u^{-9} + \frac{1}{7}u^{-7} + C = -\frac{1}{9}\sin^{-9}x + \frac{1}{7}\sin^{-7}x + C$.

13. $$\int(1+\sin^2 x)(1+\cos^2 x)\,dx = \int\left(1 + \frac{1-\cos 2x}{2}\right)\left(1 + \frac{1+\cos 2x}{2}\right)dx$$
$$= \int\left(\frac{3-\cos 2x}{2}\right)\left(\frac{3+\cos 2x}{2}\right)dx = \frac{1}{4}\int(9 - \cos^2 2x)\,dx$$

Using (5) we find that $\frac{1}{4}\int(9-\cos^2 2x)\,dx = \frac{9}{4}x - \frac{1}{8}x - \frac{1}{32}\sin 4x + C = \frac{17}{8}x - \frac{1}{32}\sin 4x + C$.

15. $u = \cos x$, $du = -\sin x\,dx$; if $x = 0$ then $u = 1$, and if $x = \pi/4$ then $u = \sqrt{2}/2$;

$$\int_0^{\pi/4}\frac{\sin^3 x}{\cos^2 x}\,dx = \int_0^{\pi/4}\frac{(1-\cos^2 x)}{\cos^2 x}\sin x\,dx = \int_1^{\sqrt{2}/2}\frac{-(1-u^2)}{u^2}\,du$$

$$= \int_1^{\sqrt{2}/2}\left(1-\frac{1}{u^2}\right)du = \left(u+\frac{1}{u}\right)\Big|_1^{\sqrt{2}/2} = \left(\frac{\sqrt{2}}{2}+\sqrt{2}\right)-(1+1) = \frac{3}{2}\sqrt{2}-2.$$

17. $u=\tan x$, $du=\sec^2 x\,dx$; $\int \tan^5 x\sec^2 x\,dx = \int u^5\,du = \frac{1}{6}u^6+C = \frac{1}{6}\tan^6 x + C$.

19. $u=\tan t$, $du=\sec^2 t\,dt$; if $t=0$ then $u=0$, and if $t=\pi/4$ then $u=1$; $\int_0^{\pi/4}\tan^5 t\sec^4 t\,dt = \int_0^{\pi/4}\tan^5 t\,(\tan^2 t+1)\sec^2 t\,dt = \int_0^1 u^5(u^2+1)\,du = \int_0^1(u^7+u^5)\,du = (\frac{1}{8}u^8+\frac{1}{6}u^6)\big|_0^1 = \frac{7}{24}$.

21. $u=\sec x$, $du=\sec x\tan x\,dx$; if $x=5\pi/4$ then $u=-\sqrt{2}$, and if $x=4\pi/3$ then $u=-2$;

$$\int_{5\pi/4}^{4\pi/3}\tan^3 x\sec x\,dx = \int_{5\pi/4}^{4\pi/3}(\sec^2 x-1)(\sec x\tan x)\,dx$$

$$= \int_{-\sqrt{2}}^{-2}(u^2-1)\,du = \left(\frac{1}{3}u^3-u\right)\Big|_{-\sqrt{2}}^{-2} = -\frac{2}{3}-\frac{1}{3}\sqrt{2}.$$

23. $u=\sec\sqrt{x}$, $du=[1/(2\sqrt{x})]\sec\sqrt{x}\tan\sqrt{x}\,dx$;

$$\int\frac{1}{\sqrt{x}}\tan^3\sqrt{x}\sec^3\sqrt{x}\,dx = \int(\tan^2\sqrt{x})(\sec^2\sqrt{x})\left(\frac{1}{\sqrt{x}}\sec\sqrt{x}\tan\sqrt{x}\right)dx$$

$$= \int\left[(\sec^2\sqrt{x}-1)\sec^2\sqrt{x}\right]\frac{1}{\sqrt{x}}\sec\sqrt{x}\tan\sqrt{x}\,dx = \int(u^2-1)u^2(2)\,du$$

$$= 2\int(u^4-u^2)\,du = \frac{2}{5}u^5-\frac{2}{3}u^3+C = \frac{2}{5}\sec^5\sqrt{x}-\frac{2}{3}\sec^3\sqrt{x}+C.$$

25. $u=\tan x$, $du=\sec^2 x\,dx$; $\int\tan^3 x\sec^4 x\,dx = \int\tan^3 x(\tan^2 x+1)\sec^2 x\,dx = \int u^3(u^2+1)\,du = \int(u^5+u^3)\,du = \frac{1}{6}u^6+\frac{1}{4}u^4+C = \frac{1}{6}\tan^6 x+\frac{1}{4}\tan^4 x+C$.

27. $u=\sec x$, $du=\sec x\tan x\,dx$; $\int\tan x\sec^5 x\,dx = \int u^4\,du = \frac{1}{5}u^5+C = \frac{1}{5}\sec^5 x+C$.

29. $u=\cot x$, $du=-\csc^2 x\,dx$; $\int\cot^3 x\csc^2 x\,dx = -\int u^3\,du = -\frac{1}{4}u^4+C = -\frac{1}{4}\cot^4 x+C$.

31. $u=\csc x$, $du=-\csc x\cot x\,dx$; if $x=\pi/4$ then $u=\sqrt{2}$, and if $x=\pi/2$ then $u=1$;

$$\int_{\pi/4}^{\pi/2}\cot^3 x\csc^3 x\,dx = \int_{\pi/4}^{\pi/2}(\csc^2 x-1)\csc^2 x(\cot x\csc x)\,dx$$

$$= \int_{\sqrt{2}}^{1}(u^2-1)u^2(-1)\,du = -\int_{\sqrt{2}}^{1}(u^4-u^2)\,du = \left(\frac{-1}{5}u^5+\frac{1}{3}u^3\right)\Big|_{\sqrt{2}}^{1} = \frac{2}{15}(\sqrt{2}+1).$$

33. $\displaystyle\int\cot x\csc^{-2}x\,dx = \int\frac{\cos x}{\sin x}\sin^2 x\,dx = \int\cos x\sin x\,dx = \frac{1}{2}\sin^2 x+C$

35. $u=\cos x$, $du=-\sin x\,dx$;

$$\int\frac{\tan x}{\cos^3 x}\,dx = \int\frac{\sin x}{\cos^4 x}\,dx = \int\frac{1}{u^4}(-1)\,du = \frac{1}{3u^3}+C = \frac{1}{3\cos^3 x}+C.$$

37. $\displaystyle\int \frac{\tan^2 x}{\sec^5 x}\,dx = \int \frac{\sin^2 x}{\cos^2 x}\cos^5 x\,dx = \int \sin^2 x\cos^3 x\,dx = \int \sin^2 x(1-\sin^2 x)\cos x\,dx$

$\displaystyle= \int (\sin^2 x - \sin^4 x)\cos x\,dx$

If $u = \sin x$ then $du = \cos x\,dx$, so

$$\int \frac{\tan^2 x}{\sec^5 x}\,dx = \int (\sin^2 x - \sin^4 x)\cos x\,dx = \int (u^2 - u^4)\,du = \frac{1}{3}u^3 - \frac{1}{5}u^5 + C = \frac{1}{3}\sin^3 x - \frac{1}{5}\sin^5 x + C.$$

39. $\displaystyle\int \frac{\tan x}{\sec^2 x}\,dx = \int \frac{\sin x}{\cos x}\cos^2 x\,dx = \int \sin x\cos x\,dx = \frac{1}{2}\sin^2 x + C$

41. $\int \tan^2 x\,dx = \int(\sec^2 x - 1)\,dx = \tan x - x + C$

43. $\int \tan^4 x\,dx = \int \tan^2 x(\sec^2 x - 1)\,dx = \int \tan^2 x\sec^2 x\,dx - \int \tan^2 x\,dx$. For $\int \tan^2 x\sec^2 x\,dx$, let $u = \tan x$, so that $du = \sec^2 x\,dx$. Then $\int \tan^2 x\sec^2 x\,dx = \int u^2\,du = \frac{1}{3}u^3 + C_1 = \frac{1}{3}\tan^3 x + C_1$. By Exercise 41, $\int \tan^2 x\,dx = \tan x - x + C_2$. Thus $\int \tan^4 x\,dx = \frac{1}{3}\tan^3 x - \tan x + x + C$.

45. By (8) with $a = 2$ and $b = 3$, $\int \sin 2x\cos 3x\,dx = \frac{1}{2}\int(\sin(-x)+\sin 5x)\,dx = \frac{1}{2}\cos(-x) - \frac{1}{10}\cos 5x + C = \frac{1}{2}\cos x - \frac{1}{10}\cos 5x + C$.

47. By (8) with $a = -4$ and $b = -2$, $\int \sin(-4x)\cos(-2x)\,dx = \frac{1}{2}\int\big(\sin(-2x)+\sin(-6x)\big)\,dx = \frac{1}{4}\cos(-2x) + \frac{1}{12}\cos(-6x) + C = \frac{1}{4}\cos 2x + \frac{1}{12}\cos 6x + C$.

49. By (8) with $a = \frac{1}{2}$ and $b = \frac{2}{3}$, $\int \sin\frac{1}{2}x\cos\frac{2}{3}x\,dx = \frac{1}{2}\int(\sin(-\frac{1}{6}x)+\sin\frac{7}{6}x)\,dx = 3\cos(-\frac{1}{6}x) - \frac{3}{7}\cos\frac{7}{6}x + C = 3\cos\frac{1}{6}x - \frac{3}{7}\cos\frac{7}{6}x + C$.

51. $\int \sin 2x\sin 3x\,dx = \int[-\frac{1}{2}\cos(2+3)x + \frac{1}{2}\cos(2-3)x]\,dx = -\frac{1}{10}\sin 5x - \frac{1}{2}\sin(-x) + C = -\frac{1}{10}\sin 5x + \frac{1}{2}\sin x + C$

53. $\int \cos 5x\cos(-3x)\,dx = \int[\frac{1}{2}\cos(5-3)x + \frac{1}{2}\cos(5+3)x]\,dx = \frac{1}{4}\sin 2x + \frac{1}{16}\sin 8x + C$

55. $\displaystyle\int_{\pi/4}^{\pi/2} \frac{1}{1+\cos x}\,dx = \int_{\pi/4}^{\pi/2} \frac{1}{1+\cos x}\,\frac{1-\cos x}{1-\cos x}\,dx = \int_{\pi/4}^{\pi/2} \frac{1-\cos x}{\sin^2 x}\,dx$

$\displaystyle= \int_{\pi/4}^{\pi/2} (\csc^2 x - \csc x\cot x)\,dx = (-\cot x + \csc x)\Big|_{\pi/4}^{\pi/2} = 2 - \sqrt{2}$

57. $\displaystyle\int \frac{1+\cos x}{\sin x}\,dx = \int \frac{1+\cos x}{\sin x}\,\frac{1-\cos x}{1-\cos x}\,dx = \int \frac{\sin x}{1-\cos x}\,dx = \ln|1-\cos x| + C$

59. $\displaystyle\int \tan^n x\,dx = \int \tan^{n-2} x\tan^2 x\,dx = \int \tan^{n-2} x(\sec^2 x - 1)\,dx$

$\displaystyle= \int \tan^{n-2} x\sec^2 x\,dx - \int \tan^{n-2} x\,dx = \frac{1}{n-1}\tan^{n-1} x - \int \tan^{n-2} x\,dx$

61. The area is given by $A = \int_0^{\pi/2} \sin^2 x \cos^3 x\,dx$, so by the solution of Example 1, $A = (\frac{1}{3}\sin^3 x - \frac{1}{5}\sin^5 x)\Big|_0^{\pi/2} = \frac{1}{3} - \frac{1}{5} = \frac{2}{15}$.

63. The area A is given by

$$A = \int_{\pi/4}^{\pi/3} \tan^3 x \sec x\,dx = \int_{\pi/4}^{\pi/3} \tan^2 x\,(\tan x \sec x)\,dx = \int_{\pi/4}^{\pi/3} (\sec^2 x - 1)(\tan x \sec x)\,dx.$$

Let $u = \sec x$, so that $du = \tan x \sec x\,dx$; if $x = \pi/4$, then $u = \sqrt{2}$; if $x = \pi/3$, then $u = 2$. Thus

$$A = \int_{\pi/4}^{\pi/3} \tan^3 x \sec x\,dx = \int_{\sqrt{2}}^{2} (u^2 - 1)\,du = \left(\frac{1}{3}u^3 - u\right)\Big|_{\sqrt{2}}^{2}$$

$$= \left(\frac{8}{3} - 2\right) - \left(\frac{2\sqrt{2}}{3} - \sqrt{2}\right) = \frac{2}{3} + \frac{\sqrt{2}}{3} = \frac{1}{3}(2 + \sqrt{2}).$$

65. By (7), $S = \int_0^{\pi/4} 2\pi \sec t \sqrt{\sec^2 t + (\sec t \tan t)^2}\,dt = 2\pi \int_0^{\pi/4} \sec^2 t \sqrt{1 + \tan^2 t}\,dt$

$= 2\pi \int_0^{\pi/4} \sec^3 t\,dt = 2\pi(\frac{1}{2}\sec t \tan t + \frac{1}{2}\ln|\sec t + \tan t|)\Big|_0^{\pi/4}$

$= \pi[\sqrt{2} + \ln(\sqrt{2} + 1)]$.

8.3 Trigonometric Substitutions

1. $x = \frac{1}{2}\sin u$, $dx = \frac{1}{2}\cos u\,du$; if $x = 0$ then $u = 0$, and if $x = \frac{1}{2}$ then $u = \pi/2$;

$$\int_0^{1/2} \sqrt{1 - 4x^2}\,dx = \int_0^{\pi/2} \sqrt{1 - \sin^2 u}\left(\frac{1}{2}\cos u\right)du = \frac{1}{2}\int_0^{\pi/2} \cos^2 u\,du$$

$$= \frac{1}{4}\int_0^{\pi/2} (1 + \cos 2u)\,du = \frac{1}{4}\left(u + \frac{1}{2}\sin 2u\right)\Big|_0^{\pi/2} = \frac{\pi}{8}.$$

3. $x = 2\sin u$, $dx = 2\cos u\,du$; if $x = -2$ then $u = -\pi/2$, and if $x = 2$ then $u = \pi/2$;

$$\int_{-2}^{2} \sqrt{1 - \frac{x^2}{4}}\,dx = \int_{-\pi/2}^{\pi/2} \sqrt{1 - \sin^2 u}\,(2\cos u)\,du = 2\int_{-\pi/2}^{\pi/2} \cos^2 u\,du$$

$$= \int_{-\pi/2}^{\pi/2} (1 + \cos 2u)\,du = \left(u + \frac{1}{2}\sin 2u\right)\Big|_{-\pi/2}^{\pi/2} = \pi.$$

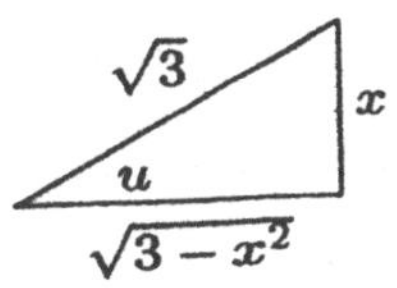

Exercise 5

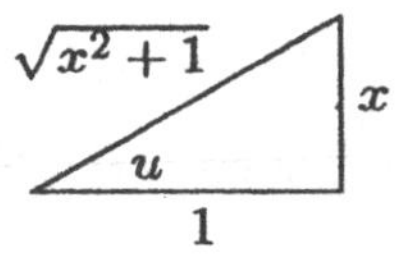

Exercise 7

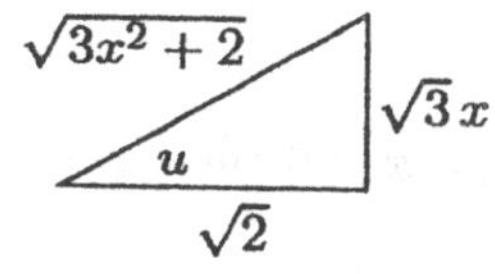

Exercise 9

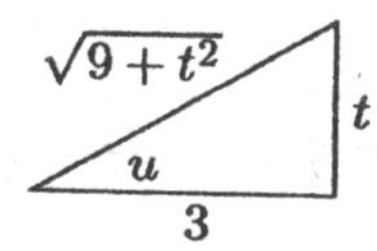

Exercise 13

5. $x = \sqrt{3}\sin u$, $dx = \sqrt{3}\cos u\,du$;

$$\int \frac{1}{(3-x^2)^{3/2}}\,dx = \int \frac{1}{(3-3\sin^2 u)^{3/2}}(\sqrt{3}\cos u)\,du = \frac{1}{3}\int \sec^2 u\,du = \frac{1}{3}\tan u + C = \frac{1}{3}\frac{x}{\sqrt{3-x^2}} + C.$$

7. $x = \tan u$, $dx = \sec^2 u\,du$; $\sin u = x/\sqrt{x^2+1}$;

$$\int \frac{1}{(x^2+1)^{3/2}}\,dx = \int \frac{1}{(\tan^2 u+1)^{3/2}}\sec^2 u\,du = \int \frac{\sec^2 u}{\sec^3 u}\,du = \int \cos u\,du = \sin u + C = \frac{x}{\sqrt{x^2+1}} + C.$$

9. $x = \sqrt{\frac{2}{3}}\tan u$, $dx = \sqrt{\frac{2}{3}}\sec^2 u\,du$; $\sin u = (\sqrt{3}\,x)/\sqrt{3x^2+2}$;

$$\int \frac{1}{(3x^2+2)^{5/2}}\,dx = \int \frac{1}{(2\tan^2 u+2)^{5/2}}\left(\sqrt{\frac{2}{3}}\sec^2 u\right)du$$

$$= \frac{1}{4\sqrt{3}}\int \frac{1}{\sec^5 u}\sec^2 u\,du = \frac{1}{4\sqrt{3}}\int \cos^3 u\,du = \frac{1}{4\sqrt{3}}\int (1-\sin^2 u)\cos u\,du$$

$$= \frac{1}{4\sqrt{3}}\left(\sin u - \frac{1}{3}\sin^3 u\right) + C = \frac{x}{4\sqrt{3x^2+2}} - \frac{x^3}{4(3x^2+2)^{3/2}} + C.$$

Thus

$$\int_0^1 \frac{1}{(3x^2+2)^{5/2}}\,dx = \left(\frac{x}{4\sqrt{3x^2+2}} - \frac{x^3}{4(3x^2+2)^{3/2}}\right)\Bigg|_0^1 = \frac{\sqrt{5}}{25}.$$

11. $x = 2\sec u$, $dx = 2\sec u\tan u\,du$; if $x = 2$ then $u = 0$, and if $x = 2\sqrt{2}$ then $u = \pi/4$;

$$\int_2^{2\sqrt{2}} \frac{\sqrt{x^2-4}}{x}\,dx = \int_0^{\pi/4} \frac{\sqrt{4\sec^2 u - 4}}{2\sec u}(2\sec u\tan u)\,du$$

$$= 2\int_0^{\pi/4} \tan^2 u\,du = 2\int_0^{\pi/4}(\sec^2 u - 1)\,du = 2(\tan u - u)\Big|_0^{\pi/4} = 2 - \frac{\pi}{2}.$$

13. $t = 3\tan u$, $dt = 3\sec^2 u\,du$;

$$\int \frac{1}{(9+t^2)^2}\,dt = \int \frac{1}{(9+9\tan^2 u)^2}3\sec^2 u\,du = \frac{1}{27}\int \frac{\sec^2 u}{\sec^4 u}\,du$$

$$= \frac{1}{27}\int \cos^2 u\,du = \frac{1}{54}\int (1+\cos 2u)\,du = \frac{1}{54}\left(u + \frac{1}{2}\sin 2u\right) + C.$$

Now $t = 3\tan u$ implies that $u = \tan^{-1} t/3$, and from the figure, $\sin u = t/\sqrt{9+t^2}$ and $\cos u = 3/\sqrt{9+t^2}$, so that $\sin 2u = 2\sin u\cos u = 6t/(9+t^2)$. Then

$$\int \frac{1}{(9+t^2)^2}\,dt = \frac{1}{54}u + \frac{1}{108}\sin 2u + C = \frac{1}{54}\tan^{-1}\frac{t}{3} + \frac{t}{18(9+t^2)} + C.$$

15. $x = \frac{5}{2}\sin u$, $dx = \frac{5}{2}\cos u\, du$; if $x = 0$ then $u = 0$, and if $x = \frac{5}{4}$ then $u = \pi/6$;

$$\int_0^{5/4} \frac{1}{\sqrt{25-4x^2}}\,dx = \int_0^{\pi/6} \frac{1}{\sqrt{25-25\sin^2 u}}\left(\frac{5}{2}\cos u\right)du = \frac{1}{2}\int_0^{\pi/6} 1\,du = \frac{1}{2}u\Big|_0^{\pi/6} = \frac{1}{12}\pi.$$

17. $$\int \frac{1}{\sqrt{4x^2+4x+2}}\,dx = \int \frac{1}{\sqrt{(4x^2+4x+1)+1}}\,dx = \int \frac{1}{\sqrt{(2x+1)^2+1}}\,dx$$

Let $2x+1 = \tan u$, so that $2\,dx = \sec^2 u\,du$. Then

$$\int \frac{1}{\sqrt{(2x+1)^2+1}}\,dx = \int \frac{1}{\sqrt{\tan^2 u+1}}\,\frac{1}{2}\sec^2 u\,du = \frac{1}{2}\int \sec u\,du = \frac{1}{2}\ln|\sec u + \tan u| + C.$$

Since $\sec u = \sqrt{\tan^2 u+1} = \sqrt{(2x+1)^2+1} = \sqrt{4x^2+4x+2}$, we have

$$\int \frac{1}{\sqrt{4x^2+4x+2}}\,dx = \frac{1}{2}\ln\left|\sqrt{4x^2+4x+2}+(2x+1)\right| + C.$$

19. $w = (1/\sqrt{2})\sin u$, $dw = (1/\sqrt{2})\cos u\,du$;

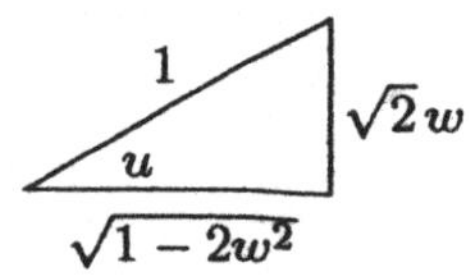

$$\int \frac{1}{(1-2w^2)^{5/2}}\,dw = \int \frac{1}{(1-\sin^2 u)^{5/2}}\left(\frac{1}{\sqrt{2}}\cos u\right)du = \frac{1}{\sqrt{2}}\int \sec^4 u\,du$$

$$= \frac{1}{\sqrt{2}}\int (\tan^2 u+1)\sec^2 u\,du = \frac{1}{\sqrt{2}}\left(\frac{1}{3}\tan^3 u + \tan u\right) + C$$

$$= \frac{1}{\sqrt{2}}\left(\frac{1}{3}\frac{2\sqrt{2}w^3}{(1-2w^2)^{3/2}} + \frac{\sqrt{2}w}{(1-2w^2)^{1/2}}\right) + C = \frac{2w^3}{3(1-2w^2)^{3/2}} + \frac{w}{(1-2w^2)^{1/2}} + C.$$

21. $$\int_0^1 \frac{1}{2x^2-2x+1}\,dx = \int_0^1 \frac{1}{2(x-\frac{1}{2})^2+\frac{1}{2}}\,dx$$

Let $x - \frac{1}{2} = \frac{1}{2}\tan u$, so that $dx = \frac{1}{2}\sec^2 u\,du$; if $x = 0$ then $u = -\pi/4$, and if $x = 1$ then $u = \pi/4$. Then

$$\int_0^1 \frac{1}{2x^2-2x+1}\,dx = \int_0^1 \frac{1}{2(x-\frac{1}{2})^2+\frac{1}{2}}\,dx = \int_{-\pi/4}^{\pi/4} \frac{1}{\frac{1}{2}\tan^2 u+\frac{1}{2}}\left(\frac{1}{2}\sec^2 u\right)du = \int_{-\pi/4}^{\pi/4} 1\,du = \frac{\pi}{2}.$$

23. $$\int \sqrt{x-x^2}\,dx = \int \sqrt{\tfrac{1}{4}-(x-\tfrac{1}{2})^2}\,dx.$$

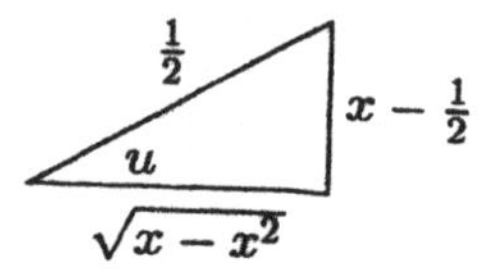

Let $x - \frac{1}{2} = \frac{1}{2}\sin u$, so that $dx = \frac{1}{2}\cos u\,du$. Then

$$\int \sqrt{\frac{1}{4}-\left(x-\frac{1}{2}\right)^2}\,dx = \int \sqrt{\frac{1}{4}-\frac{1}{4}\sin^2 u}\left(\frac{1}{2}\cos u\right)du$$

$$= \frac{1}{4}\int \cos^2 u\,du = \frac{1}{4}\int\left(\frac{1}{2}+\frac{1}{2}\cos 2u\right)du = \frac{1}{4}\left(\frac{1}{2}u + \frac{1}{4}\sin 2u\right) + C.$$

Now $x - \frac{1}{2} = \frac{1}{2}\sin u$ implies that $2x - 1 = \sin u$, so that $u = \sin^{-1}(2x-1)$. From the figure, we see that $\cos u = (\sqrt{x - x^2})/\frac{1}{2} = 2\sqrt{x-x^2}$, so that $\sin 2u = 2\sin u\,\cos u = [2(2x-1)][2\sqrt{x-x^2}] = 4(2x-1)\sqrt{x-x^2}$. Thus

$$\int \sqrt{x - x^2}\,dx = \int \sqrt{\frac{1}{4} - \left(x - \frac{1}{2}\right)^2}\,dx = \frac{1}{4}\left\{\frac{1}{2}\sin^{-1}(2x-1) + \frac{1}{4}\left[4(2x-1)(\sqrt{x-x^2})\right]\right\} + C$$

$$= \frac{1}{8}\sin^{-1}(2x-1) + \frac{1}{4}(2x-1)\sqrt{x-x^2} + C.$$

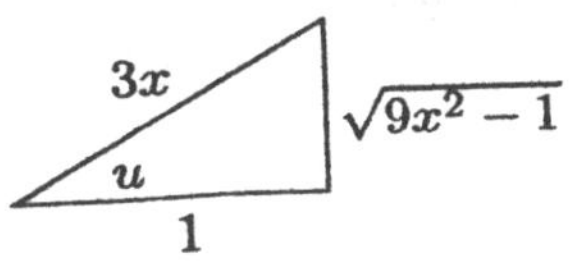

Exercise 25

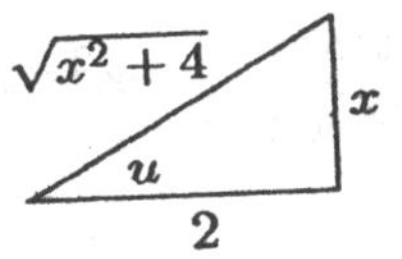

Exercise 27

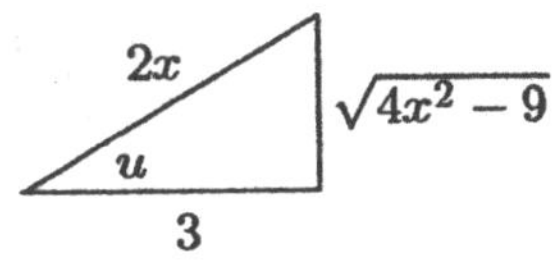

Exercise 29

25. $x = \frac{1}{3}\sec u$, $dx = \frac{1}{3}\sec u\,\tan u\,du$;

$$\int \frac{x^2}{\sqrt{9x^2-1}}\,dx = \int \frac{\frac{1}{9}\sec^2 u}{\sqrt{\sec^2 u - 1}}\,\frac{1}{3}\sec u\,\tan u\,du = \int \frac{1}{27}\sec^3 u\,du.$$

By (7) of Section 8.2, $\int \frac{1}{27}\sec^3 u\,du = \frac{1}{54}\sec u\,\tan u + \frac{1}{54}\ln|\sec u + \tan u| + C$. Now $x = \frac{1}{3}\sec u$ implies that $\sec u = 3x$ and $\tan u = \sqrt{9x^2-1}$. Thus

$$\int \frac{x^2}{\sqrt{9x^2-1}}\,dx = \frac{1}{54}\sec u\,\tan u + \frac{1}{54}\ln|\sec u + \tan u| + C = \frac{1}{18}x\sqrt{9x^2-1} + \frac{1}{54}\ln\left|3x + \sqrt{9x^2-1}\right| + C.$$

27. $x = 2\tan u$, $dx = 2\sec^2 u\,du$, $\csc u = \sqrt{x^2+4}/x$, $\cot u = 2/x$;

$$\int \frac{1}{x\sqrt{x^2+4}}\,dx = \int \frac{1}{2\tan u\sqrt{4\tan^2 u + 4}}(2\sec^2 u)\,du = \frac{1}{2}\int \frac{\sec u}{\tan u}\,du$$

$$= \frac{1}{2}\int \csc u\,du = \frac{-1}{2}\ln|\csc u + \cot u| + C = \frac{-1}{2}\ln\left|\frac{\sqrt{x^2+4}}{x} + \frac{2}{x}\right| + C.$$

29. $x = \frac{3}{2}\sec u$, $dx = \frac{3}{2}\sec u\,\tan u\,du$;

$$\int \frac{1}{x^2\sqrt{4x^2-9}}\,dx = \int \frac{1}{\frac{9}{4}\sec^2 u\sqrt{9\sec^2 u - 9}}\,\frac{3}{2}\sec u\,\tan u\,du$$

$$= \int \frac{2}{9\sec u}\,du = \frac{2}{9}\int \cos u\,du = \frac{2}{9}\sin u + C.$$

Now $x = \frac{3}{2}\sec u$ implies that $\sin u = \sqrt{4x^2-9}/(2x)$. Thus

$$\int \frac{1}{x^2\sqrt{4x^2-9}}\,dx = \frac{2}{9}\sin u + C = \frac{\sqrt{4x^2-9}}{9x} + C.$$

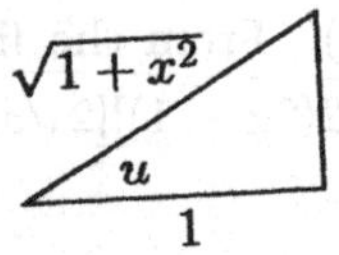

Exercise 31

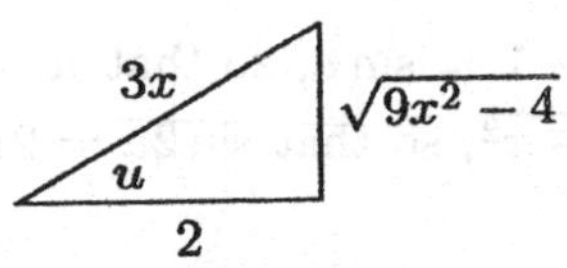

Exercise 33

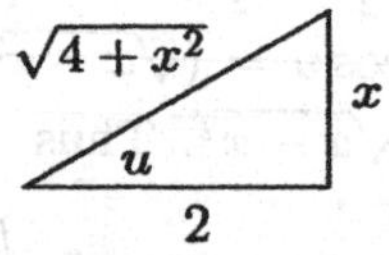

Exercise 35

31. $x = \tan u$, $dx = \sec^2 u\, du$, $\sec u = \sqrt{1+x^2}$; from (7) in Section 8.2 we find that

$$\int \frac{x^2}{\sqrt{1+x^2}}\,dx = \int \frac{\tan^2 u}{\sqrt{1+\tan^2 u}}\sec^2 u\,du = \int \tan^2 u \sec u\,du$$

$$= \int (\sec^2 u - 1)\sec u\,du = \int (\sec^3 u - \sec u)\,du$$

$$= \left(\frac{1}{2}\sec u\tan u + \frac{1}{2}\ln|\sec u + \tan u|\right) - \ln|\sec u + \tan u| + C$$

$$= \frac{1}{2}\sec u \tan u - \frac{1}{2}\ln|\sec u + \tan u| + C = \frac{1}{2}x\sqrt{1+x^2} - \frac{1}{2}\ln\left|\sqrt{1+x^2}+x\right| + C.$$

33. $x = \frac{2}{3}\sec u$, $dx = \frac{2}{3}\sec u \tan u\, du$;

$$\int \frac{1}{(9x^2-4)^{5/2}}\,dx = \int \frac{1}{(4\sec^2 u - 4)^{5/2}}\frac{2}{3}\sec u\tan u\,du = \int \frac{\sec u \tan u}{48\tan^5 u}\,du$$

$$= \frac{1}{48}\int \frac{\sec u}{\tan^4 u}\,du = \frac{1}{48}\int \frac{\cos^3 u}{\sin^4 u}\,du = \frac{1}{48}\int \frac{1-\sin^2 u}{\sin^4 u}\cos u\,du$$

$$= \frac{1}{48}\int\left(\frac{1}{\sin^4 u} - \frac{1}{\sin^2 u}\right)\cos u\,du = \frac{1}{48}\left(\frac{-1}{3\sin^3 u} + \frac{1}{\sin u}\right) + C.$$

Now $x = \frac{2}{3}\sec u$ implies that $\sin u = \sqrt{9x^2-4}/(3x)$. Thus

$$\int \frac{1}{(9x^2-4)^{5/2}}\,dx = \frac{1}{48}\left(\frac{-1}{3\sin^2 u} + \frac{1}{\sin u}\right) + C = \frac{-3x^3}{16(9x^2-4)^{3/2}} + \frac{x}{16(9x^2-4)^{1/2}} + C.$$

35. $x = 2\tan u$, $dx = 2\sec^2 u\,du$; $\sec u = \sqrt{4+x^2}/2$; from (7) in Section 8.2 we have

$$\int \sqrt{4+x^2}\,dx = \int \sqrt{4+4\tan^2 u}\,(2\sec^2 u)\,du = 4\int \sec^3 u\,du$$

$$= 2\sec u\tan u + 2\ln|\sec u + \tan u| + C = \frac{x\sqrt{4+x^2}}{2} + 2\ln\left|\frac{\sqrt{4+x^2}}{2} + \frac{x}{2}\right| + C.$$

37. $x = 3\sec u$, $dx = 3\sec u\tan u\,du$; if $x = 3\sqrt{2}$ then $u = \pi/4$, and if $x = 6$ then $u = \pi/3$. Thus

$$\int_{3\sqrt{2}}^{6} \frac{1}{x^4\sqrt{x^2-9}}\,dx = \int_{\pi/4}^{\pi/3} \frac{1}{(3\sec u)^4\sqrt{9\sec^2 u - 9}}\,3\sec u\tan u\,du = \int_{\pi/4}^{\pi/3}\frac{1}{81\sec^3 u}\,du$$

$$= \int_{\pi/4}^{\pi/3}\frac{1}{81}\cos^3 u\,du = \frac{1}{81}\int_{\pi/4}^{\pi/3}\cos u\,(1-\sin^2 u)\,du.$$

Let $v = \sin u$, so that $dv = \cos u\,du$. If $u = \pi/4$ then $v = \sqrt{2}/2$, and if $u = \pi/3$ then $v = \sqrt{3}/2$. Thus

$$\frac{1}{81}\int_{\pi/4}^{\pi/3} \cos u\,(1-\sin^2 u)\,du = \frac{1}{81}\int_{\sqrt{2}/2}^{\sqrt{3}/2} (1-v^2)\,dv = \frac{1}{81}\left(v - \frac{1}{3}v^3\right)\Bigg|_{\sqrt{2}/2}^{\sqrt{3}/2}$$

$$= \frac{1}{81}\left[\left(\frac{\sqrt{3}}{2} - \frac{1}{3}\frac{3\sqrt{3}}{8}\right) - \left(\frac{\sqrt{2}}{2} - \frac{1}{3}\frac{\sqrt{2}}{4}\right)\right] = \frac{1}{81}\left(\frac{3\sqrt{3}}{8} - \frac{5\sqrt{2}}{12}\right).$$

39. $\displaystyle\int \frac{x}{\sqrt{2x^2+12x+19}}\,dx = \int \frac{x}{\sqrt{2(x^2+6x+9)+1}}\,dx = \int \frac{x}{\sqrt{2(x+3)^2+1}}\,dx$

Let $\sqrt{2}\,(x+3) = \tan u$, so that $\sqrt{2}\,dx = \sec^2 u\,du$ and $x = (1/\sqrt{2})\tan u - 3$. Then

$$\int \frac{x}{\sqrt{2(x+3)^2+1}}\,dx = \int \frac{(1/\sqrt{2})\tan u - 3}{\sqrt{\tan^2 u + 1}}\,\frac{1}{\sqrt{2}}\sec^2 u\,du = \frac{1}{2}\int (\tan u - 3\sqrt{2})\sec u\,du$$

$$= \frac{1}{2}\int \tan u\,\sec u\,du - \frac{3}{2}\sqrt{2}\int \sec u\,du = \frac{1}{2}\sec u - \frac{3}{2}\sqrt{2}\,\ln|\sec u + \tan u| + C.$$

Since $\sqrt{2}\,(x+3) = \tan u$, we have $\sec u = \sqrt{[\sqrt{2}\,(x+3)]^2+1} = \sqrt{2x^2+12x+19}$, and thus

$$\int \frac{x}{\sqrt{2x^2+12x+19}}\,dx = \frac{1}{2}\sqrt{2x^2+12x+19} - \frac{3}{2}\sqrt{2}\,\ln\left|\sqrt{2x^2+12x+19} + \sqrt{2}\,(x+3)\right| + C.$$

41. $\displaystyle\int \sqrt{x^2+6x+5}\,dx = \int \sqrt{(x+3)^2-4}\,dx$

Let $x+3 = 2\sec u$, so that $dx = 2\sec u\,\tan u\,du$;

$$\int \sqrt{(x+3)^2-4}\,dx = \int \sqrt{4\sec^2 u - 4}\,(2\sec u\,\tan u)\,du$$

$$= \int 4\tan^2 u\,\sec u\,du = 4\int (\sec^2 u - 1)\sec u\,du = 4\int \sec^3 u\,du - 4\int \sec u\,du.$$

By (6) and (7) of Section 8.2,

$$4\int \sec^3 u\,du - 4\int \sec u\,du = 2\sec u\,\tan u + 2\ln|\sec u + \tan u| - 4\ln|\sec u + \tan u| + C$$

$$= 2\sec u\,\tan u - 2\ln|\sec u + \tan u| + C.$$

Now $x+3 = 2\sec u$ implies that $\sec u = (x+3)/2$ and $\tan u = \sqrt{x^2+6x+5}/2$. Thus

$$\int \sqrt{x^2+6x+5}\,dx = 2\sec u\,\tan u - 2\ln|\sec u + \tan u| + C$$

$$= \frac{x+3}{2}\sqrt{x^2+6x+5} - 2\ln\left|\frac{x+3}{2} + \frac{1}{2}\sqrt{x^2+6x+5}\right| + C.$$

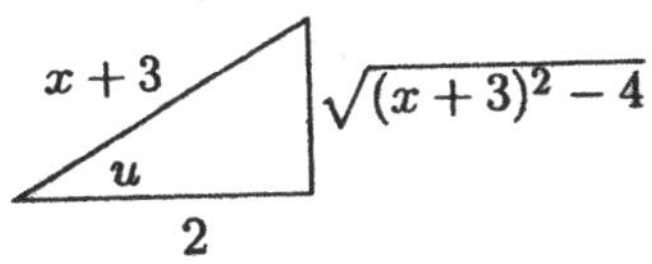

43. Let $v = e^w$, so that $dv = e^w\,dw$. Then $\int e^w\sqrt{1+e^{2w}}\,dw = \int \sqrt{1+v^2}\,dv$. Let $v = \tan u$, so that $dv = \sec^2 u\,du$. Then $\int \sqrt{1+v^2}\,dv = \int \sqrt{1+\tan^2 u}\,(\sec^2 u)\,du = \int \sec^3 u\,du$, and by (7) of Section 8.2, $\int \sec^3 u\,du = \frac{1}{2}\sec u\,\tan u + \frac{1}{2}\ln|\sec u + \tan u| + C$. By the figure, $\sec u = \sqrt{1+v^2}$, so that

$$\int e^w\sqrt{1+e^{2w}}\,dw = \int \sqrt{1+v^2}\,dv = \frac{v}{2}\sqrt{1+v^2} + \frac{1}{2}\ln\left|\sqrt{1+v^2}+v\right| + C$$

$$= \frac{1}{2}e^w\sqrt{1+e^{2w}} + \frac{1}{2}\ln\left|\sqrt{1+e^{2w}}+e^w\right| + C.$$

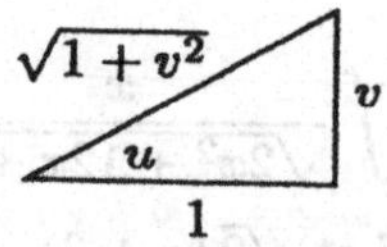

45. For integration by parts, let $u = \sin^{-1} x$, $dv = x\,dx$, so that $du = (1/\sqrt{1-x^2})\,dx$, $v = \frac{1}{2}x^2$. Then

$$\int x\sin^{-1} x\,dx = \frac{1}{2}x^2\sin^{-1} x - \int \frac{x^2/2}{\sqrt{1-x^2}}\,dx = \frac{1}{2}x^2\sin^{-1} x - \frac{1}{2}\int \frac{x^2}{\sqrt{1-x^2}}\,dx.$$

For $\int (x^2/\sqrt{1-x^2})\,dx$ let $x = \sin w$, so that $dx = \cos w\,dw$. Then

$$\int \frac{x^2}{\sqrt{1-x^2}}\,dx = \int \frac{\sin^2 w}{\sqrt{1-\sin^2 w}}\cos w\,dw = \int \sin^2 w\,dw$$

$$= \int \left(\frac{1}{2}-\frac{1}{2}\cos 2w\right)dw = \frac{1}{2}w - \frac{1}{4}\sin 2w + C_1 = \frac{1}{2}w - \frac{1}{2}\sin w\,\cos w + C_1.$$

Now $x = \sin w$, so from the figure, $\cos w = \sqrt{1-x^2}$, and thus $\int (x^2/\sqrt{1-x^2})\,dx = \frac{1}{2}w - \frac{1}{2}\sin w\,\cos w + C_1 = \frac{1}{2}\sin^{-1} x - \frac{1}{2}x\sqrt{1-x^2} + C_1$. Consequently $\int x\sin^{-1} x\,dx = \frac{1}{2}x^2\sin^{-1} x - \frac{1}{4}\sin^{-1} x + \frac{1}{4}x\sqrt{1-x^2} + C$.

47. The area A is given by $A = \int_0^1 \sqrt{1-x^2}\,dx$. Let $x = \sin u$, so that $dx = \cos u\,du$. If $x = 0$ then $u = 0$, and if $x = 1$ then $u = \pi/2$. Thus

$$A = \int_0^1 \sqrt{1-x^2}\,dx = \int_0^{\pi/2} \sqrt{1-\sin^2 u}\cos u\,du = \int_0^{\pi/2} \cos^2 u\,du$$

$$= \int_0^{\pi/2} \left(\frac{1}{2}+\frac{1}{2}\cos 2u\right)du = \left(\frac{1}{2}u + \frac{1}{4}\sin 2u\right)\Big|_0^{\pi/2} = \frac{\pi}{4}.$$

49. The area A is given by $A = \int_0^3 \sqrt{9+x^2}\,dx$. Let $x = 3\tan u$, so that $dx = 3\sec^2 u\,du$. If $x = 0$ then $u = 0$, and if $x = 3$ then $u = \pi/4$. By (7) of Section 8.2,

$$A = \int_0^3 \sqrt{9+x^2}\,dx = \int_0^{\pi/4} \sqrt{9+9\tan^2 u}\,(3\sec^2 u)\,du = 9\int_0^{\pi/4} \sqrt{1+\tan^2 u}\,\sec^2 u\,du$$

$$= 9\int_0^{\pi/4} \sec^3 u\,du = \left(\frac{9}{2}\sec u\,\tan u + \frac{9}{2}\ln|\sec u + \tan u|\right)\Big|_0^{\pi/4} = \frac{9}{2}\sqrt{2} + \frac{9}{2}\ln(\sqrt{2}+1).$$

51. By (12) of Section 6.8 and (7) of Section 8.2,

$$S = \int_0^\pi 4\pi \sin t\sqrt{(-\sqrt{2}\sin t)^2 + (2\cos t)^2}\,dt = \int_0^\pi 4\pi \sin t\sqrt{2 + 2\cos^2 t}\,dt$$

$$\overset{u=\cos t}{=} \int_{-1}^{1} 4\pi\sqrt{2+2u^2}\,du \overset{u=\tan\theta}{=} \int_{-\pi/4}^{\pi/4} 4\pi\sqrt{2}\sqrt{1+\tan^2\theta}\,\sec^2\theta\,d\theta$$

$$= 4\pi\sqrt{2}\int_{-\pi/4}^{\pi/4} \sec^3\theta\,d\theta = 4\pi\sqrt{2}\left[\frac{1}{2}\sec\theta\tan\theta + \frac{1}{2}\ln|\sec\theta + \tan\theta|\right]\Bigg|_{-\pi/4}^{\pi/4}$$

$$= 4\pi\sqrt{2}\left[\left(\frac{1}{2}\sqrt{2} + \frac{1}{2}\ln(1+\sqrt{2})\right) - \left(-\frac{1}{2}\sqrt{2} + \frac{1}{2}\ln(-1+\sqrt{2})\right)\right]$$

$$= 8\pi + 4\pi\sqrt{2}\ln(1+\sqrt{2})$$

53. $L = \int_0^1 \sqrt{1+(2x-1)^2}\,dx$. Let $2x - 1 = \tan u$. Then

$$\int_0^1 \sqrt{1+(2x-1)^2}\,dx = \int_{-\pi/4}^{\pi/4} \sqrt{1+\tan^2 u}\left(\frac{1}{2}\sec^2 u\right)du = \frac{1}{2}\int_{-\pi/4}^{\pi/4} \sec^3 u\,du.$$

By (7) of Section 8.2,

$$\frac{1}{2}\int_{-\pi/4}^{\pi/4} \sec^3 u\,du = \frac{1}{4}[\sec u\tan u + \ln|\sec u + \tan u|]\Big|_{-\pi/4}^{\pi/4}$$

$$= \frac{1}{4}\left[2\sqrt{2} + \ln(\sqrt{2}+1) - \ln(\sqrt{2}-1)\right]$$

$$= \frac{1}{2}\sqrt{2} + \frac{1}{4}\ln(3+2\sqrt{2}) = \frac{1}{2}\sqrt{2} + \frac{1}{2}\ln(\sqrt{2}+1).$$

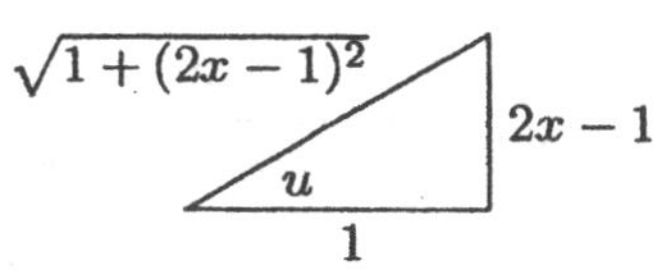

55. The volume is given by

$$V = \int_0^1 \pi\left[(1+\sqrt{1-x^2})^2 - \left(\frac{1}{2}\right)^2\right]dx = \pi\int_0^1\left(\frac{7}{4} - x^2 + 2\sqrt{1-x^2}\right)dx$$

$$= \pi\left(\frac{7}{4}x - \frac{x^3}{3}\right)\Bigg|_0^1 + 2\pi\int_0^1 \sqrt{1-x^2}\,dx \overset{x=\sin u}{=} \frac{17}{12}\pi + 2\pi\int_0^{\pi/2}\sqrt{1-\sin^2 u}\,\cos u\,du$$

$$= \frac{17}{12}\pi + 2\pi\int_0^{\pi/2}\cos^2 u\,du = \frac{17}{12}\pi + 2\pi\left(\frac{1}{2}u + \frac{1}{4}\sin 2u\right)\Bigg|_0^{\pi/2} = \frac{17\pi}{12} + \frac{\pi^2}{2}.$$

57. a. Since the ellipse is symmetric with respect to both axes, the area A is given by

$$A = 4\int_0^a \left(b\sqrt{1-x^2/a^2}\right)dx.$$

Let $x = a\sin u$, $dx = a\cos u\,du$. Then

$$A = 4\int_0^{\pi/2} b\sqrt{1-\sin^2 u}\,a\cos u\,du = 4ab\int_0^{\pi/2}\cos^2 u\,du = 4ab\int_0^{\pi/2}\left(\frac{1}{2} + \frac{1}{2}\cos 2u\right)du$$

$$= 4ab\left(\frac{1}{2}u + \frac{1}{4}\sin 2u\right)\Bigg|_0^{\pi/2} = 4ab\frac{\pi}{4} = \pi ab.$$

b. By part (a), the amount would be $\pi(750)(640) = 480{,}000\pi$ (square feet).

8.4 Partial Fractions

1. $\int \frac{x}{x+1}\,dx = \int \left(1 - \frac{1}{x+1}\right) dx = x - \ln|x+1| + C$

3. $\int \frac{x^2}{x^2-1}\,dx = \int \left(1 + \frac{1}{(x+1)(x-1)}\right) dx;\ \frac{1}{(x+1)(x-1)} = \frac{A}{x+1} + \frac{B}{x-1};$
$A(x-1) + B(x+1) = 1;\ A + B = 0,\ -A + B = 1;\ A = -\frac{1}{2},\ B = \frac{1}{2};$

$$\int \frac{x^2}{x^2-1}\,dx = \int \left(1 - \frac{1}{2(x+1)} + \frac{1}{2(x-1)}\right) dx = x - \frac{1}{2}\ln|x+1| + \frac{1}{2}\ln|x-1| + C = x + \frac{1}{2}\ln\left|\frac{x-1}{x+1}\right| + C$$

5. $\frac{x^2+4}{x(x-1)^2} = \frac{A}{x} + \frac{B}{x-1} + \frac{C}{(x-1)^2};\ A(x-1)^2 + Bx(x-1) + Cx = x^2 + 4;$
$A + B = 1,\ -2A - B + C = 0,\ A = 4;\ A = 4,\ B = -3,\ C = 5;$

$$\int \frac{x^2+4}{x(x-1)^2}\,dx = \int \left(\frac{4}{x} - \frac{3}{x-1} + \frac{5}{(x-1)^2}\right) dx$$

$$= 4\ln|x| - 3\ln|x-1| - \frac{5}{x-1} + C_1 = \ln\left|\frac{x^4}{(x-1)^3}\right| - \frac{5}{x-1} + C_1$$

7. $\frac{5}{(x-2)(x+3)} = \frac{A}{x-2} + \frac{B}{x+3}$; $A(x+3) + B(x-2) = 5;\ A + B = 0,\ 3A - 2B = 5;\ A = 1,\ B = -1;$

$$\int_3^4 \frac{5}{(x-2)(x+3)}\,dx = \int_3^4 \left(\frac{1}{x-2} - \frac{1}{x+3}\right) dx = (\ln|x-2| - \ln|x+3|)\Big|_3^4 = \ln 2 - \ln 7 + \ln 6 = \ln\frac{12}{7}$$

9. $\frac{3t}{t^2-8t+15} = \frac{A}{t-5} + \frac{B}{t-3};\ A(t-3) + B(t-5) = 3t;\ A + B = 3,\ -3A - 5B = 0;\ A = \frac{15}{2},\ B = -\frac{9}{2};$

$$\int \frac{3t}{t^2-8t+15}\,dt = \int \left(\frac{15}{2(t-5)} - \frac{9}{2(t-3)}\right) dt = \frac{15}{2}\ln|t-5| - \frac{9}{2}\ln|t-3| + C$$

11. $\int_{-1}^0 \frac{x^2+x+1}{x^2+1}\,dx = \int_{-1}^0 \left(1 + \frac{x}{x^2+1}\right) dx = \left[x + \frac{1}{2}\ln(x^2+1)\right]\Bigg|_{-1}^0 = -\left[(-1) + \frac{1}{2}\ln 2\right] = 1 - \frac{1}{2}\ln 2$

13. $\int \frac{x^2+x+1}{x^2-1}\,dx = \int \left(1 + \frac{x+2}{x^2-1}\right) dx;\ \frac{x+2}{x^2-1} = \frac{A}{x+1} + \frac{B}{x-1};\ A(x-1) + B(x+1) = x+2;\ A + B = 1,$
$-A + B = 2;\ A = -\frac{1}{2},\ B = \frac{3}{2};$

$$\int \frac{x^2+x+1}{x^2-1}\,dx = \int \left(1 - \frac{1}{2(x+1)} + \frac{3}{2(x-1)}\right) dx = x - \frac{1}{2}\ln|x+1| + \frac{3}{2}\ln|x-1| + C$$

15. $\displaystyle\int_0^1 \frac{u-1}{u^2+u+1}\,du = \frac{1}{2}\int_0^1 \frac{2u+1}{u^2+u+1}\,du - \frac{3}{2}\int_0^1 \frac{1}{(u+\frac{1}{2})^2+\frac{3}{4}}\,du;$

$\displaystyle\frac{1}{2}\int_0^1 \frac{2u+1}{u^2+u+1}\,du = \frac{1}{2}\ln(u^2+u+1)\Big|_0^1 = \frac{1}{2}\ln 3;$

$\displaystyle -\frac{3}{2}\int_0^1 \frac{1}{(u+\frac{1}{2})^2+\frac{3}{4}}\,du \overset{v=u+1/2}{=} -\frac{3}{2}\int_{1/2}^{3/2} \frac{1}{v^2+\frac{3}{4}}\,dv = -\frac{3}{2}\left(\frac{2}{\sqrt{3}}\right)\tan^{-1}\frac{2v}{\sqrt{3}}\Big|_{1/2}^{3/2} = -\frac{\sqrt{3}\pi}{6}$

Thus

$$\int_0^1 \frac{u-1}{u^2+u+1}\,du = \frac{1}{2}\ln 3 - \frac{\sqrt{3}\pi}{6}.$$

17. $\displaystyle\frac{3x}{x^2-4x+4} = \frac{3x}{(x-2)^2} = \frac{A}{x-2} + \frac{B}{(x-2)^2};\ A(x-2)+B = 3x;\ A=3,\ -2A+B=0;\ A=3,\ B=6;$

$$\int \frac{3x}{x^2-4x+4}\,dx = \int\left(\frac{3}{x-2} + \frac{6}{(x-2)^2}\right)dx = 3\ln|x-2| - \frac{6}{x-2} + C$$

19. $\displaystyle\frac{-x}{x^3-3x^2+2x} = \frac{-1}{(x-1)(x-2)} = \frac{A}{x-1} + \frac{B}{x-2};$

$A(x-2)+B(x-1) = -1;\ A+B=0,\ -2A-B=-1;\ A=1,\ B=-1;$

$$\int \frac{-x}{x^3-3x^2+2x}\,dx = \int\left(\frac{1}{x-1} - \frac{1}{x-2}\right)dx = \ln|x-1| - \ln|x-2| + C = \ln\left|\frac{x-1}{x-2}\right| + C$$

21. $\displaystyle\int \frac{u^3}{(u+1)^2}\,du = \int\left(u-2+\frac{3u+2}{(u+1)^2}\right)du$

$\displaystyle\frac{3u+2}{(u+1)^2} = \frac{A}{u+1} + \frac{B}{(u+1)^2};\ A(u+1)+B = 3u+2;\ A=3,\ A+B=2;\ A=3,\ B=-1;$

$$\int \frac{u^3}{(u+1)^2}\,du = \int\left(u-2+\frac{3}{u+1} - \frac{1}{(u+1)^2}\right)du = \frac{1}{2}u^2 - 2u + 3\ln|u+1| + \frac{1}{u+1} + C$$

23. $\displaystyle\int \frac{1}{(1-x^2)^2}\,dx = \int\left(\frac{1}{2(x+1)} - \frac{1}{2(x-1)}\right)^2 dx = \int\left(\frac{1}{4(x+1)^2} - \frac{1}{2(x+1)(x-1)} + \frac{1}{4(x-1)^2}\right)dx$

$$= \int\left[\frac{1}{4(x+1)^2} + \frac{1}{2}\left(\frac{1}{2(x+1)} - \frac{1}{2(x-1)}\right) + \frac{1}{4(x-1)^2}\right]dx$$

$$= \frac{-1}{4(x+1)} + \frac{1}{4}\ln\left|\frac{x+1}{x-1}\right| - \frac{1}{4(x-1)} + C$$

25. $\displaystyle\frac{x}{(x+1)^2(x-2)} = \frac{A}{x+1} + \frac{B}{(x+1)^2} + \frac{C}{x-2};$

$A(x+1)(x-2) + B(x-2) + C(x+1)^2 = x;\ A+C=0,\ -A+B+2C=1,\ -2A-2B+C=0;$

$A = -\frac{2}{9},\ B = \frac{1}{3},\ C = \frac{2}{9};$

$$\int \frac{x}{(x+1)^2(x-2)}\,dx = \int \left(\frac{-2}{9(x+1)} + \frac{1}{3(x+1)^2} + \frac{2}{9(x-2)}\right) dx$$

$$= \frac{-2}{9}\ln|x+1| - \frac{1}{3(x+1)} + \frac{2}{9}\ln|x-2| + C_1 = \frac{2}{9}\ln\left|\frac{x-2}{x+1}\right| - \frac{1}{3(x+1)} + C_1$$

27. $\dfrac{-x^3+x^2+x+3}{(x+1)(x^2+1)^2} = \dfrac{A}{x+1} + \dfrac{Bx+C}{x^2+1} + \dfrac{Dx+E}{(x^2+1)^2}$;

$A(x^2+1)^2 + (Bx+C)(x+1)(x^2+1) + (Dx+E)(x+1) = -x^3+x^2+x+3$;

$A+B=0,\ B+C=-1,\ 2A+B+C+D=1,\ B+C+D+E=1,\ A+C+E=3$;

$A=1,\ B=-1,\ C=0,\ D=0,\ E=2$.

With the help of the evaluation of $\int [1/(x^2+1)^2]\,dx$ in Exercise 26, we find that

$$\int \frac{-x^3+x^2+x+3}{(x+1)(x^2+1)^2}\,dx = \int \left(\frac{1}{x+1} - \frac{x}{x^2+1} + \frac{2}{(x^2+1)^2}\right) dx$$

$$= \ln|x+1| - \frac{1}{2}\ln(x^2+1) + 2\left(\frac{1}{2}\tan^{-1}x + \frac{1}{2}\frac{x}{x^2+1}\right) + C_1 = \ln|x+1| - \frac{1}{2}\ln(x^2+1) + \tan^{-1}x + \frac{x}{x^2+1} + C_1.$$

29. $\dfrac{x^2-1}{x^3+3x+4} = \dfrac{(x+1)(x-1)}{(x+1)(x^2-x+4)} = \dfrac{x-1}{x^2-x+4}$;

$$\int \frac{x^2-1}{x^3+3x+4}\,dx = \int \left(\frac{2x-1}{2(x^2-x+4)} - \frac{1}{2(x^2-x+4)}\right) dx = \frac{1}{2}\ln|x^2-x+4| - \frac{1}{2}\int \frac{1}{(x-\frac{1}{2})^2 + \frac{15}{4}}\,dx$$

$$= \frac{1}{2}\ln|x^2-x+4| - \frac{1}{2}\left(\frac{2}{\sqrt{15}}\tan^{-1}\frac{x-\frac{1}{2}}{\sqrt{15}/2}\right) + C = \frac{1}{2}\ln|x^2-x+4| - \frac{1}{\sqrt{15}}\tan^{-1}\frac{\sqrt{15}\,(2x-1)}{15} + C.$$

31. Let $u = \sqrt{x+1}$, so $du = [1/(2\sqrt{x+1})]\,dx$ and $dx = 2u\,du$;

$$\int \frac{1}{x\sqrt{x+1}}\,dx = \int \frac{2u}{(u^2-1)u}\,du = \int \frac{2}{(u+1)(u-1)}\,du = \int \left(\frac{-1}{u+1} + \frac{1}{u-1}\right) du$$

$$= -\ln|u+1| + \ln|u-1| + C = \ln\left|\frac{\sqrt{x+1}-1}{\sqrt{x+1}+1}\right| + C.$$

33. Let $u = \sqrt[6]{x}$, so $du = \frac{1}{6}x^{-5/6}\,dx$ and $dx = 6u^5\,du$;

$$\int \frac{\sqrt{x}}{1+\sqrt[3]{x}}\,dx = \int \frac{u^3}{1+u^2}(6u^5)\,du = 6\int \frac{u^8}{1+u^2}\,du = 6\int \left(u^6 - u^4 + u^2 - 1 + \frac{1}{u^2+1}\right) du$$

$$= 6\left(\frac{1}{7}u^7 - \frac{1}{5}u^5 + \frac{1}{3}u^3 - u + \tan^{-1}u\right) + C = \frac{6}{7}x^{7/6} - \frac{6}{5}x^{5/6} + 2x^{1/2} - 6x^{1/6} + 6\tan^{-1}x^{1/6} + C.$$

35. Let $u = \sqrt{\dfrac{x+1}{x-1}}$, so $dx = \dfrac{-4u}{(u^2-1)^2}\,du$ and $x = \dfrac{u^2+1}{u^2-1}$;

$$\int_{-5/3}^{-1}\sqrt{\frac{x+1}{x-1}}\,dx = \int_{1/2}^{0}\frac{-4u^2}{(u^2-1)^2}\,du = -4\int_{1/2}^{0}\left(\frac{u}{u^2-1}\right)^2 du = -4\int_{1/2}^{0}\left(\frac{1}{2(u+1)}+\frac{1}{2(u-1)}\right)^2 du$$

$$= -\int_{1/2}^{0}\left(\frac{1}{(u+1)^2}+\frac{2}{(u+1)(u-1)}+\frac{1}{(u-1)^2}\right)du$$

$$= -\int_{1/2}^{0}\left(\frac{1}{(u+1)^2}-\frac{1}{u+1}+\frac{1}{u-1}+\frac{1}{(u-1)^2}\right)du$$

$$= \left(\frac{1}{u+1}+\ln|u+1|-\ln|u-1|+\frac{1}{u-1}\right)\Bigg|_{1/2}^{0} = \frac{4}{3}-\ln 3.$$

37. Let $u = \sin x$, so that $du = \cos x\,dx$. Then

$$\int\frac{\sin^2 x\cos x}{\sin^2 x+1}\,dx = \int\frac{u^2}{u^2+1}\,du = \int\left(1-\frac{1}{u^2+1}\right)du = u-\tan^{-1}u+C = \sin x-\tan^{-1}(\sin x)+C.$$

39. Let $u = e^x$, so that $du = e^x\,dx$;

$$\int\frac{e^x}{1-e^{3x}}\,dx = \int\frac{1}{1-u^3}\,du = \int\frac{1}{(1-u)(1+u+u^2)}\,du = \int\left(\frac{A}{1-u}+\frac{Bu+C}{1+u+u^2}\right)du;$$

$A(1+u+u^2)+(Bu+C)(1-u) = 1$; $A-B=0$, $A+B-C=0$, $A+C=1$; $A=\frac{1}{3}$, $B=\frac{1}{3}$, $C=\frac{2}{3}$.
Therefore

$$\int\frac{1}{(1-u)(1+u+u^2)}\,du = \int\left[\frac{1}{3(1-u)}+\frac{u+2}{3(1+u+u^2)}\right]du$$

$$= \frac{1}{3}\int\frac{1}{1-u}\,du+\frac{1}{6}\int\frac{2u+1}{1+u+u^2}\,du+\frac{1}{2}\int\frac{1}{(u+\frac{1}{2})^2+\frac{3}{4}}\,du$$

$$= -\frac{1}{3}\ln|1-u|+\frac{1}{6}\ln(1+u+u^2)+\frac{1}{2}\,\frac{2}{\sqrt{3}}\tan^{-1}\frac{u+\frac{1}{2}}{\sqrt{3}/2}+C$$

$$= -\frac{1}{3}\ln|1-e^x|+\frac{1}{6}\ln(1+e^x+e^{2x})+\frac{1}{\sqrt{3}}\tan^{-1}\frac{2e^x+1}{\sqrt{3}}+C.$$

41. To integrate by parts, let $u = \tan^{-1}x$, $dv = x\,dx$, so that $du = [1/(1+x^2)]\,dx$, $v = \frac{1}{2}x^2$. Then

$$\int x\tan^{-1}x\,dx = \frac{1}{2}x^2\tan^{-1}x-\int\frac{1}{2}x^2\frac{1}{1+x^2}\,dx = \frac{1}{2}x^2\tan^{-1}x-\frac{1}{2}\int\frac{x^2}{1+x^2}\,dx.$$

Now

$$\int\frac{x^2}{1+x^2}\,dx = \int\left(1-\frac{1}{1+x^2}\right)dx.$$

Thus

$$\int x\tan^{-1}x\,dx = \frac{1}{2}x^2\tan^{-1}x-\frac{1}{2}\int\left(1-\frac{1}{1+x^2}\right)dx = \frac{1}{2}x^2\tan^{-1}x-\frac{1}{2}x+\frac{1}{2}\tan^{-1}x+C.$$

43. To integrate by parts, let $u = \ln(x^2+1)$, $dv = dx$, so that $du = [2x/(x^2+1)]\,dx$, $v = x$. Then

$$\int \ln(x^2+1)\,dx = x\ln(x^2+1) - \int x\frac{2x}{x^2+1}\,dx = x\ln(x^2+1) - 2\int \frac{x^2}{x^2+1}\,dx.$$

Now

$$\int \frac{x^2}{x^2+1}\,dx = \int \left(1 - \frac{1}{x^2+1}\right)dx = x - \tan^{-1}x + C_1.$$

Thus

$$\int \ln(x^2+1)\,dx = x\ln(x^2+1) - 2x + 2\tan^{-1}x + C.$$

45. Using Exercise 44 and letting $u = \sqrt{x+4}$, so that $u^2 = x+4$ and $dx = 2u\,du$, we have

$$\int \frac{\sqrt{x+4}}{x^2}\,dx = \int \frac{u}{(u^2-4)^2}(2u)\,du = 2\int \frac{u^2}{(u^2-4)^2}\,du$$

$$= \frac{1}{2}\left(\frac{-2u}{u^2-4} + \frac{1}{2}\ln\left|\frac{u-2}{u+2}\right|\right) + C = \frac{-\sqrt{x+4}}{x} + \frac{1}{4}\ln\left|\frac{\sqrt{x+4}-2}{\sqrt{x+4}+2}\right| + C.$$

47. By Exercise 46 with $r = 10$,

$$\int \frac{1}{x - x^{11}}\,dx = \frac{1}{10}\ln\left|\frac{x^{10}}{1-x^{10}}\right| + C.$$

49. $A = \displaystyle\int_0^3 \frac{x^3}{x^2+1}\,dx = \int_0^3 \left(x - \frac{x}{x^2+1}\right)dx = \left[\frac{1}{2}x^2 - \frac{1}{2}\ln(x^2+1)\right]\Big|_0^3$

$$= \left(\frac{9}{2} - \frac{1}{2}\ln 10\right) + \frac{1}{2}\ln 1 = \frac{9}{2} - \frac{1}{2}\ln 10$$

51. The graphs intersect at (x,y) if $x^2/[(x-2)(x^2+1)] = y = 1/(x-3)$, or $x^3 - 3x^2 = x^3 - 2x^2 + x - 2$, or $x^2 + x - 2 = 0$, or $x = 1$ or -2. Since $x^2/[(x-2)(x^2+1)] \ge 1/(x-3)$ on $[-2,1]$, the area is given by

$$A = \int_{-2}^1 \left(\frac{x^2}{(x-2)(x^2+1)} - \frac{1}{x-3}\right)dx.$$

Now

$$\frac{x^2}{(x-2)(x^2+1)} = \frac{B}{x-2} + \frac{Cx+D}{x^2+1};$$

$B(x^2+1) + (Cx+D)(x-2) = x^2$; $B + C = 1$, $(-2C + D) = 0$, $B - 2D = 0$; $B = \frac{4}{5}$, $C = \frac{1}{5}$, $D = \frac{2}{5}$; thus

$$A = \int_{-2}^1 \left(\frac{4}{5(x-2)} + \frac{x+2}{5(x^2+1)} - \frac{1}{x-3}\right)dx$$

$$= \int_{-2}^1 \left(\frac{4}{5}\cdot\frac{1}{x-2} + \frac{1}{10}\cdot\frac{2x}{x^2+1} + \frac{2}{5}\cdot\frac{1}{x^2+1} - \frac{1}{x-3}\right)dx$$

$$= \left(\frac{4}{5}\ln|x-2| + \frac{1}{10}\ln(x^2+1) + \frac{2}{5}\tan^{-1}x - \ln|x-3|\right)\Big|_{-2}^1$$

$$= \left(\frac{1}{10}\ln 2 + \frac{\pi}{10} - \ln 2\right) - \left(\frac{4}{5}\ln 4 + \frac{1}{10}\ln 5 + \frac{2}{5}\tan^{-1}(-2) - \ln 5\right)$$

$$= -\frac{5}{2}\ln 2 + \frac{9}{10}\ln 5 + \frac{\pi}{10} + \frac{2}{5}\tan^{-1}2.$$

53. $V = \int_0^2 2\pi y\left(\frac{1}{y+2} - \frac{1}{(y+2)^2}\right) dy = 2\pi \int_0^2 \left(1 - \frac{2}{y+2}\right) dy - 2\pi \int_0^2 \frac{y}{(y+2)^2}\, dy$

$$\overset{u=y+2}{=} 2\pi[y - 2\ln(y+2)]\big|_0^2 - 2\pi \int_2^4 \frac{u-2}{u^2}\, du$$

$$= 2\pi(2 - 2\ln 4 + 2\ln 2) - 2\pi \int_2^4 \left(\frac{1}{u} - \frac{2}{u^2}\right) du$$

$$= 2\pi(2 - 2\ln 2) - 2\pi \left(\ln|u| + \frac{2}{u}\right)\Big|_2^4 = 2\pi(\frac{5}{2} - 3\ln 2)$$

55. $L = \int_2^5 \sqrt{1 + \left(\frac{2x}{x^2-1}\right)^2}\, dx = \int_2^5 \sqrt{1 + \frac{4x^2}{(x^2-1)^2}}\, dx = \int_2^5 \sqrt{\frac{x^4 - 2x^2 + 1 + 4x^2}{(x^2-1)^2}}\, dx$

$$= \int_2^5 \sqrt{\frac{(x^2+1)^2}{(x^2-1)^2}}\, dx = \int_2^5 \frac{x^2+1}{x^2-1}\, dx = \int_2^5 \left(1 + \frac{1}{x-1} - \frac{1}{x+1}\right) dx$$

$$= (x + \ln|x-1| - \ln|x+1|)\big|_2^5 = (5 + \ln 4 - \ln 6) - (2 + \ln 1 - \ln 3) = 3 + \ln 2$$

57. $\frac{1}{y(1-y)} = \frac{A}{y} + \frac{B}{1-y}$; $A(1-y) + By = 1$; $-A + B = 0$, $A = 1$, so $A = B = 1$. Thus

$$\int \frac{1}{ay(1-y)}\, dy = \frac{1}{a}\int \left(\frac{1}{y} - \frac{1}{1-y}\right) dy = \frac{1}{a}[\ln y - \ln(1-y)] + C = \frac{1}{a}\ln\frac{y}{1-y} + C.$$

59. a. $\frac{1}{(a-y)(b-y)(c-y)} = \frac{A}{a-y} + \frac{B}{b-y} + \frac{C}{c-y}$; $A(b-y)(c-y) + B(a-y)(c-y) + C(a-y)(b-y) = 1$;
if $y = a$, then $A(b-a)(c-a) = 1$, so $A = 1/[(b-a)(c-a)]$;
if $y = b$, then $B(a-b)(c-b) = 1$, so $B = 1/[(a-b)(c-b)]$;
if $y = c$, then $C(a-c)(b-c) = 1$, so $C = 1/[(a-c)(b-c)]$.

Thus

$$t = \int \frac{1}{r(a-y)(b-y)(c-y)}\, dy$$

$$= \frac{1}{r}\int \left[\frac{1}{(a-b)(a-c)}\frac{1}{a-y} + \frac{1}{(b-c)(b-a)}\frac{1}{b-y} + \frac{1}{(c-a)(c-b)}\frac{1}{c-y}\right] dy$$

$$= \frac{1}{r(a-b)(b-c)(c-a)}\int \left(\frac{c-b}{a-y} + \frac{a-c}{b-y} + \frac{b-a}{c-y}\right) dy$$

$$= \frac{1}{r(a-b)(b-c)(c-a)}\left[-\ln(a-y)^{c-b} - \ln(b-y)^{a-c} - \ln(c-y)^{b-a}\right] + D_1.$$

Therefore

$$(a-c)(b-a)(c-b)r(t - D_1) = \ln\left[(a-y)^{c-b}(b-y)^{a-c}(c-y)^{b-a}\right]$$

so that

$$De^{(a-c)(b-a)(c-b)rt} = e^{(a-c)(b-a)(c-b)r(t-D_1)} = (a-y)^{c-b}(b-y)^{a-c}(c-y)^{b-a}$$

where $D = e^{-(a-c)(b-a)(c-b)rD_1}$.

b. If $y = 0$ when $t = 0$, then $D = a^{c-b}b^{a-c}c^{b-a}$.

8.5 Symbolic Integration

For indefinite integrals, the format of the answer given by a computer algebra system is apt to depend significantly on the particular computer algebra system used. As a result, we include here the answers provided by Mathematica for the definite integrals, but do not include computer algebra responses for the indefinite integrals.

3. 11.5304 11. −1.17689

8.6 The Trapezoidal Rule and Simpson's Rule

Let T and S be the approximations by the Trapezoidal Rule and Simpson's Rule, respectively.

1. $T = \frac{3-1}{2(6)}\left[1 + 2(\frac{1}{4/3}) + 2(\frac{1}{5/3}) + \cdots + 2(\frac{1}{8/3}) + \frac{1}{9/3}\right] \approx 1.106746032$

   $S = \frac{3-1}{3(6)}\left[1 + 4(\frac{1}{4/3}) + 2(\frac{1}{5/3}) + \cdots + 4(\frac{1}{8/3}) + \frac{1}{9/3}\right] \approx 1.098941799$

3. $T = \frac{1-0}{2(10)}\left[1 + 2e^{-(0.1)^2} + 2e^{-(0.2)^2} + \cdots + 2e^{-(0.9)^2} + e^{-1}\right] \approx 0.7462107961$

   $S = \frac{1-0}{3(10)}\left[1 + 4e^{-(0.1)^2} + 2e^{-(0.2)^2} + \cdots + 4e^{-(0.9)^2} + e^{-1}\right] \approx 0.7468249483$

5. Let $f(x) = 1/x$. Then $f'(x) = -1/x^2$ and $f''(x) = 2/x^3$. For $1 \le x \le 3$ we have $|f''(x)| \le 2/1^3 = 2$, so we can let $K_T = 2$. By (3),

$$E_{10}^T \le \frac{2}{12(10)^2}(2^3) = \frac{1}{75}.$$

7. Let $f(x) = 1/(3+x)$. Then

   $f'(x) = \dfrac{-1}{(3+x)^2}$, $f''(x) = \dfrac{2}{(3+x)^3}$, $f^{(3)}(x) = \dfrac{-6}{(3+x)^4}$, and $f^{(4)}(x) = \dfrac{24}{(3+x)^5}$

   For $-1 \le x \le 2$, we have $|f^{(4)}(x)| \le 24/(3+(-1))^5 = 24/32 = 3/4$, so we can let $K_S = 3/4$. By (8),

$$E_{10}^S \le \frac{3/4}{180(10)^4}(2-(-1))^5 = \frac{729}{720 \times 10^4} = 1.0125 \times 10^{-4}$$

9. Let $f(x) = \sqrt{1+x^2}$ for $1 \le x \le 2$. The computer yields

$$f^{(4)}(x) = \frac{-15x^4}{(1+x^2)^{7/2}} + \frac{18x^2}{(1+x^2)^{5/2}} - \frac{3}{(1+x^2)^{3/2}}.$$

By plotting the graph of $f^{(4)}$ we see that the maximum value of $|f^{(4)}(x)|$ for $1 \le x \le 2$ is $|f^{(4)}(1)|$, and $0.78 < |f^{(4)}(1)| < 0.8$. Thus we can take $K = 0.8$. By (8),

$$E_{10}^S \le \frac{0.8}{180(10)^4}(2-1)^5 = \frac{0.8}{1.8 \times 10^6} \approx 4.444444444 \times 10^{-7}.$$

11. By Simpson's Rule with $n = 10$, $A = \int_{-\pi/3}^{\pi/3} 1/(1+\cos x)\,dx \approx 1.154724372$.

13. a. By Simpson's Rule with $n = 4$, $A \approx \frac{0.9-0.1}{3(4)}[f(0.1)+4f(0.3)+2f(0.5)+4f(0.7)+f(0.9)] \approx \frac{0.8}{12}[1+4(2.4)+2(1.4)+4(1)+2.4] = 1.32$.

b. By Simpson's Rule with $n = 8$, $A \approx \frac{0.9-0.1}{3(8)}[f(0.1)+4f(0.2)+2f(0.3)+\cdots+4f(0.8)+f(0.9)] \approx \frac{0.8}{24}[1+4(2)+2(2.4)+4(2)+2(1.4)+4(1)+2(1)+4(1.4)+2.4] \approx 1.286666667$.

15. a. *i*. By the Trapezoidal Rule with $n = 10$, $\int_0^1 4/(1+x^2)\,dx \approx 3.139925989$.

ii. By the Simpson's Rule with $n = 10$, $\int_0^1 4/(1+x^2)\,dx \approx 3.141592614$.

b. For the Trapezoidal Rule the error is approximately 0.001666665; for Simpson's Rule the error is approximately 4×10^{-8}.

17. a. If $f(x) = 1/x$, then $f'(x) = -1/x^2$, $f''(x) = 2/x^3$, $f^{(3)}(x) = -6/x^4$, $f^{(4)}(x) = 24/x^5$, so $M = \max_{1\le x\le 8}|24/x^5| = 24$. Thus $E_n^S \le [(8-1)^5 24]/(180n^4)$, so $E_n^S \le 10^{-4}$ if $7^5(24)/(180n^4) \le 10^{-4}$, that is, $n^4 \ge 7^5(24)10^4/180$, or $n \ge 68.8$. Therefore, since n must be even, we take $n = 70$.

b. Since $M = \max_{1\le x\le 2} 24/x^5 = 24$, it follows that $E_n^S \le [(2-1)^5 24]/(180n^4) = 2/(15n^4)$ and thus $3E_n^S \le 10^{-4}$ if $6/(15n^4) \le 10^{-4}$, that is, $n^4 \ge (6)10^4/15$, or $n \ge 7.96$. Therefore we take $n = 8$.

19. a. If $f(x) = c_3x^3 + c_2x^2 + c_1x + c_0$, then $f^{(4)}(x) = 0$ for $a \le x \le b$, so that by (8) the error E_n^S by using Simpson's Rule satisfies

$$0 \le E_n^S \le \frac{(b-a)^5 \cdot 0}{180n^4} = 0.$$

b. By Simpson's Rule with $n = 10$, $\int_{-2}^1 (x^3 - 2x^2 + 3x - 1)\,dx = -17.25$. By direct integration, $\int_{-2}^1 (x^3-2x^2+3x-1)\,dx = (\frac{1}{4}x^4 - \frac{2}{3}x^3 + \frac{3}{2}x^2 - x)\big|_{-2}^1 = (\frac{1}{4} - \frac{2}{3} + \frac{3}{2} - 1) - (4 + \frac{16}{3} + 6 + 2) = -17.25$. Thus Simpson's Rule gives the exact value.

21. a. From the figure we see that the trapezoid above $[x_{k-1}, x_k]$ on the x axis contains the region bounded by the graph of f. Adding the corresponding areas, we obtain $T_n \ge \int_a^b f(x)\,dx$.

b. The Trapezoidal Rule approximation will decrease because the trapezoids will more closely approximate the graph of f.

c. If the graph of f is concave downward, then $T_n \le \int_a^b f(x)\,dx$, and the Trapezoidal Rule approximation would increase if n is doubled.

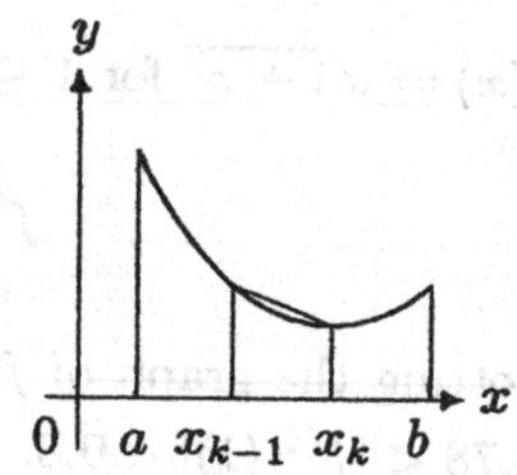

23. We have $x_1 - x_0 = x_2 - x_1 = h$ and $x_2 - x_0 = 2h$. Thus

$$\int_{x_0}^{x_2} f(x_0)\,dx = f(x_0)(x_2 - x_0) = 2hf(x_0)$$

$$\int_{x_0}^{x_2} \frac{f(x_1)-f(x_0)}{h}(x-x_0)\,dx = \frac{f(x_1)-f(x_0)}{h}\cdot\frac{1}{2}(x-x_0)^2\Big|_{x_0}^{x_2} = 2h[f(x_1)-f(x_0)].$$

By integration by parts with $u = (x - x_0)$, $dv = (x - x_1)\,dx$, $du = dx$, and $v = \frac{1}{2}(x - x_1)^2$, we have

$$\int_{x_0}^{x_2} \frac{f(x_0)-2f(x_1)+f(x_2)}{2h^2}(x-x_0)(x-x_1)\,dx$$

$$= \frac{f(x_0)-2f(x_1)+f(x_2)}{2h^2}\left[(x-x_0)\cdot\frac{1}{2}(x-x_1)^2\Big|_{x_0}^{x_2} - \int_{x_0}^{x_2}\frac{1}{2}(x-x_1)^2\,dx\right]$$

$$= \frac{f(x_0)-2f(x_1)+f(x_2)}{2h^2}\left[h^3 - \frac{1}{6}(x-x_1)^3\Big|_{x_0}^{x_2}\right] = \frac{h}{3}[f(x_0)-2f(x_1)+f(x_2)].$$

Adding these integrals, we find that

$$\int_{x_0}^{x_2} p(x)\,dx = 2hf(x_0) + 2h[f(x_1)-f(x_0)] + \frac{h}{3}[f(x_0)-2f(x_1)+f(x_2)] = \frac{h}{3}[f(x_0)+4f(x_1)+f(x_2)].$$

25. Let $v(t)$ denote the wind speed at time t, with $t = 0$ corresponding to 8 a.m. and $t = 12$ corresponding to 8 p.m. By (1) in Section 5.3, the mean speed equals $\frac{1}{12-0}\int_0^{12} v(t)\,dt$. By Simpson's Rule with $n = 6$, the mean wind speed during the 12-hour period is approximately $\frac{1}{12}\left\{\frac{12-0}{3(6)}[20 + 4(24) + 2(18) + 4(28) + 2(21) + 4(15) + 10]\right\} = \frac{376}{18} \approx 20.9$. Thus the mean wind speed is approximately 20.9 miles per hour.

27. Let $F(t)$ denote the number of flu cases that occur during the epidemic before time t (measured in days), with $t = 0$ corresponding to the outbreak of the epidemic. Then $F'(t)$ equals the rate at which the epidemic is spreading. By (5) in Section 5.4 the total number of flu cases during the 30-day period equals $\int_0^{30} F'(t)\,dt$. By Simpson's Rule with $n = 10$, the total number of cases is approximately $\frac{30-0}{3(10)}[0 + 4(30) + 2(160) + 4(370) + 2(480) + 4(490) + 2(380) + 4(200) + 2(70) + 4(20) + 0] = 6620$.

29. $L = \int_0^{2\pi} \sqrt{(-2\sin t)^2 + (3\cos t)^2}\,dt = \int_0^{2\pi} \sqrt{4\sin^2 t + 9\cos^2 t}\,dt = \int_0^{2\pi} \sqrt{4 + 5\cos^2 t}\,dt$

By Simpson's Rule with $n = 30$, $L \approx 15.86543959$.

31. Let $f(x)$ equal the width of the oil spill x miles from the left end of the spill. Then the area of the oil spill equals $\int_0^{10} f(x)\,dx$. By Simpson's Rule with $n = 10$, the area is approximately $\frac{10}{3(10)}[0 + 4(1.1) + 2(1.3) + 4(1.4) + 2(1.6) + 4(1.6) + 2(1.5) + 4(1.4) + 2(1.2) + 4(0.8) + 0] = 12.1$ (square miles).

33. By (7) in Section 5.4 the revenue equals $\int_0^5 m_r(x)\,dx$, where m_r denotes the marginal revenue function. By Simpson's Rule with $n=10$, the revenue is approximately

$$\frac{5-0}{3(10)}(100{,}000)\,[6000 + 4(5000) + 2(4000) + 4(3000) + 2(2000) + 4(1300) + 2(1000)$$

$$+4(1500) + 2(2000) + 4(1700) + 1000] = 1{,}250{,}000{,}000 \text{ (dollars)}.$$

8.7 Improper Integrals

1. converges; $\int_0^1 \frac{1}{x^{0.9}}\,dx = \lim_{c\to 0^+} \int_c^1 \frac{1}{x^{0.9}}\,dx = \lim_{c\to 0^+} 10x^{0.1}\Big|_c^1 = \lim_{c\to 0^+} 10(1 - c^{0.1}) = 10$

3. diverges; $\int_3^4 \frac{1}{(t-4)^2}\,dt = \lim_{c\to 4^-} \int_3^c \frac{1}{(t-4)^2}\,dt = \lim_{c\to 4^-} \frac{-1}{t-4}\Big|_3^c = \lim_{c\to 4^-} \left(\frac{-1}{c-4} - 1\right) = \infty$

5. converges; $\int_3^4 \frac{1}{\sqrt[3]{x-3}}\,dx = \lim_{c\to 3^+} \int_c^4 \frac{1}{\sqrt[3]{x-3}}\,dx = \lim_{c\to 3^+} \frac{3}{2}(x-3)^{2/3}\Big|_c^4 = \lim_{c\to 3^+} \left(\frac{3}{2} - \frac{3}{2}(c-3)^{2/3}\right) = \frac{3}{2}$

7. diverges $\int_0^{\pi/2} \sec^2\theta\,d\theta = \lim_{c\to \pi/2^-} \int_0^c \sec^2\theta\,d\theta = \lim_{c\to \pi/2^-} \tan\theta\Big|_0^c = \lim_{c\to \pi/2^-} \tan c = \infty$

9. converges; let $u = 1 + \cos x$, so that $du = -\sin x\,dx$; then

$$\int_0^{\pi} \frac{\sin x}{\sqrt{1+\cos x}}\,dx = \lim_{c\to\pi^-} \int_0^c \frac{\sin x}{\sqrt{1+\cos x}}\,dx = \lim_{c\to\pi^-} \int_2^{1+\cos c} \frac{-1}{u^{1/2}}\,du$$

$$= \lim_{c\to\pi^-} \left(-2u^{1/2}\right)\Big|_2^{1+\cos c} = \lim_{c\to\pi^-} \left[-2(1+\cos c)^{1/2} + 2\sqrt{2}\right] = 2\sqrt{2}.$$

11. diverges; $\int_1^2 \frac{1}{w\ln w}\,dw = \lim_{c\to 1^+} \int_c^2 \frac{1}{w\ln w}\,dw = \lim_{c\to 1^+} \ln(\ln w)\Big|_c^2 = \lim_{c\to 1^+} [\ln(\ln 2) - \ln(\ln c)] = \infty$

13. converges. Let $d = \frac{1}{2}$. Then

$$\int_0^1 \frac{3x^2-1}{\sqrt[3]{x^3-x}}\,dx = \lim_{c\to 0^+} \int_c^{1/2} \frac{3x^2-1}{\sqrt[3]{x^3-x}}\,dx + \lim_{p\to 1^-} \int_{1/2}^p \frac{3x^2-1}{\sqrt[3]{x^3-x}}\,dx$$

$$= \lim_{c\to 0^+} \frac{3}{2}(x^3-x)^{2/3}\Big|_c^{1/2} + \lim_{p\to 1^-} \frac{3}{2}(x^3-x)^{2/3}\Big|_{1/2}^p$$

$$= \lim_{c\to 0^+}\left[\frac{3}{2}\left(\frac{-3}{8}\right)^{2/3} - \frac{3}{2}(c^3-c)^{2/3}\right] + \lim_{p\to 1^-}\left[\frac{3}{2}(p^3-p)^{2/3} - \frac{3}{2}\left(\frac{-3}{8}\right)^{2/3}\right]$$

$$= \frac{3}{2}\left(\frac{3}{8}\right)^{2/3} - \frac{3}{2}\left(\frac{3}{8}\right)^{2/3} = 0.$$

15. diverges. Let $d = \pi/2$. Then

$$\int_0^\pi \csc^2 x\,dx = \lim_{c\to 0^+}\int_c^{\pi/2}\csc^2 x\,dx + \lim_{p\to\pi^-}\int_{\pi/2}^p \csc^2 x\,dx$$

$$= \lim_{c\to 0^+}(-\cot x)\Big|_c^{\pi/2} + \lim_{p\to\pi^-}(-\cot x)\Big|_{\pi/2}^p = \lim_{c\to 0^+}(0+\cot c) + \lim_{p\to\pi^-}(-\cot p + 0).$$

Neither one-sided limit exists, so the integral diverges.

17. diverges; let $u = e^t$, so that $du = e^t\,dt$ and $(1/u)\,du = dt$; then

$$\int_0^1 \frac{1}{e^t - e^{-t}}\,dt = \lim_{c\to 0^+}\int_c^1 \frac{1}{e^t-e^{-t}}\,dt = \lim_{c\to 0^+}\int_{e^c}^e \frac{1}{u - 1/u}\,\frac{1}{u}\,du$$

$$= \lim_{c\to 0^+}\int_{e^c}^e \frac{1}{u^2-1}\,du = \lim_{c\to 0^+}\int_{e^c}^e\left(\frac{1}{2}\,\frac{1}{u-1} - \frac{1}{2}\,\frac{1}{u+1}\right)du$$

$$= \lim_{c\to 0^+}\left(\frac{1}{2}\ln|u-1| - \frac{1}{2}\ln|u+1|\right)\Big|_{e^c}^e = \lim_{c\to 0^+}\left(\frac{1}{2}\ln\left|\frac{u-1}{u+1}\right|\right)\Big|_{e^c}^e$$

$$= \lim_{c\to 0^+}\left(\frac{1}{2}\ln\left|\frac{e-1}{e+1}\right| - \frac{1}{2}\ln\left|\frac{e^c-1}{e^c+1}\right|\right).$$

Since $\lim_{c\to 0^+}(e^c-1)/(e^c+1) = 0$, it follows that

$$\lim_{c\to 0^+}\left(\frac{1}{2}\ln\left|\frac{e-1}{e+1}\right| - \frac{1}{2}\ln\left|\frac{e^c-1}{e^c+1}\right|\right) = \infty$$

so the integral diverges.

19. diverges; $$\int_{-1}^2\left(\frac{1}{x}+\frac{1}{x^2}\right)dx = \lim_{c\to 0^-}\int_{-1}^c\left(\frac{1}{x}+\frac{1}{x^2}\right)dx + \lim_{p\to 0^+}\int_p^2\left(\frac{1}{x}+\frac{1}{x^2}\right)dx$$

$$= \lim_{c\to 0^-}\left(\ln|x| - \frac{1}{x}\right)\Big|_{-1}^c + \lim_{p\to 0^+}\left(\ln|x| - \frac{1}{x}\right)\Big|_p^2$$

$$= \lim_{c\to 0^-}\left(\ln|c| - \frac{1}{c} - 1\right) + \lim_{p\to 0^+}\left(\ln 2 - \frac{1}{2} - \ln p + \frac{1}{p}\right)$$

Since $\lim_{p\to 0^+}(\ln 2 - \frac{1}{2} - \ln p + 1/p) = \infty$, the integral diverges.

21. converges; $$\int_0^2 \frac{1}{(x-1)^{1/3}}\,dx = \lim_{c\to 1^-}\int_0^c \frac{1}{(x-1)^{1/3}}\,dx + \lim_{p\to 1^+}\int_p^2 \frac{1}{(x-1)^{1/3}}\,dx$$

$$= \lim_{c\to 1^-}\frac{3}{2}(x-1)^{2/3}\Big|_0^c + \lim_{p\to 1^+}\frac{3}{2}(x-1)^{2/3}\Big|_p^2$$

$$= \lim_{c\to 1^-}\left(\frac{3}{2}(c-1)^{2/3} - \frac{3}{2}\right) + \lim_{p\to 1^+}\left(\frac{3}{2} - \frac{3}{2}(p-1)^{2/3}\right) = 0$$

23. converges; $\displaystyle\int_0^1 \frac{1}{\sqrt{1-t^2}}\,dt = \lim_{c\to1^-}\int_0^c \frac{1}{\sqrt{1-t^2}}\,dt = \lim_{c\to1^-} \sin^{-1} t\Big|_0^c$

$$= \lim_{c\to1^-}(\sin^{-1} c - \sin^{-1} 0) = \sin^{-1} 1 = \frac{\pi}{2}$$

25. diverges; $\displaystyle\int_0^\infty \frac{1}{x}\,dx = \lim_{a\to0^+}\int_a^1 \frac{1}{x}\,dx + \lim_{b\to\infty}\int_1^b \frac{1}{x}\,dx = \lim_{a\to0^+} \ln|x|\Big|_a^1 + \lim_{b\to\infty} \ln|x|\Big|_1^b$

$$= \lim_{a\to0^+}(0 - \ln a) + \lim_{b\to\infty}(\ln b - 0)$$

Since both limits are infinite, the integral diverges.

27. converges; $\displaystyle\int_0^\infty \frac{1}{(2+x)^\pi}\,dx = \lim_{b\to\infty}\int_0^b (2+x)^{-\pi}\,dx = \lim_{b\to\infty} \frac{1}{-\pi+1}(2+x)^{-\pi+1}\Big|_0^b$

$$= \lim_{b\to\infty} \frac{1}{-\pi+1}\left((2+b)^{-\pi+1} - 2^{-\pi+1}\right) = \frac{1}{\pi-1}2^{-\pi+1}$$

29. diverges; $\displaystyle\int_0^\infty \sin y\,dy = \lim_{b\to\infty}\int_0^b \sin y\,dy = \lim_{b\to\infty}(-\cos y)\Big|_0^b = \lim_{b\to\infty}(1-\cos b)$

The limit does not exist, so the integral diverges.

31. converges; $\displaystyle\int_0^\infty \frac{1}{(1+x)^3}\,dx = \lim_{b\to\infty}\int_0^b \frac{1}{(1+x)^3}\,dx = \lim_{b\to\infty}\left(\frac{-1}{2}\right)\frac{1}{(1+x)^2}\Big|_0^b$

$$= \lim_{b\to\infty}\left(\frac{-1}{2}\,\frac{1}{(1+b)^2} + \frac{1}{2}\right) = \frac{1}{2}$$

33. diverges; $\displaystyle\int_0^\infty \frac{x}{1+x^2}\,dx = \lim_{b\to\infty}\int_0^b \frac{x}{1+x^2}\,dx = \lim_{b\to\infty} \frac{1}{2}\ln(1+x^2)\Big|_0^b = \lim_{b\to\infty}\frac{1}{2}\ln(1+b^2) = \infty$

35. diverges; $\displaystyle\int_3^\infty \ln x\,dx = \lim_{b\to\infty}\int_3^b \ln x\,dx = \lim_{b\to\infty}(x\ln x - x)\Big|_3^b = \lim_{b\to\infty}(b\ln b - b - 3\ln 3 + 3)$

Since $\lim_{b\to\infty}(b\ln b - b) = \lim_{b\to\infty} b(\ln b - 1) = \infty$, the integral diverges.

37. converges; let $u = \ln x$, so that $du = (1/x)\,dx$; then

$$\int_2^\infty \frac{1}{x(\ln x)^3}\,dx = \lim_{b\to\infty}\int_2^b \frac{1}{x(\ln x)^3}\,dx = \lim_{b\to\infty}\int_{\ln 2}^{\ln b} \frac{1}{u^3}\,du$$

$$= \lim_{b\to\infty} \frac{-1}{2u^2}\Big|_{\ln 2}^{\ln b} = \lim_{b\to\infty}\left[-\frac{1}{2(\ln b)^2} + \frac{1}{2(\ln 2)^2}\right] = \frac{1}{2(\ln 2)^2}.$$

39. diverges; let $d = -4$; then

$$\int_{-\infty}^0 \frac{1}{(x+3)^2}\,dx = \lim_{a\to-\infty}\int_a^{-4} \frac{1}{(x+3)^2}\,dx + \lim_{c\to-3^-}\int_{-4}^c \frac{1}{(x+3)^2}\,dx + \lim_{p\to-3^+}\int_p^0 \frac{1}{(x+3)^2}\,dx$$

$$= \lim_{a\to-\infty} \frac{-1}{x+3}\Big|_a^{-4} + \lim_{c\to-3^-} \frac{-1}{x+3}\Big|_{-4}^c + \lim_{p\to-3^+} \frac{-1}{x+3}\Big|_p^0$$

$$= \lim_{a\to-\infty}\left(1+\frac{1}{a+3}\right)+\lim_{c\to-3^-}\left(\frac{-1}{c+3}-1\right)+\lim_{p\to-3^+}\left(-\frac{1}{3}+\frac{1}{p+3}\right).$$

The latter two one-sided limits do not exist, so the integral diverges.

41. diverges; let $u=\sqrt{x}+1$, so that $du = 1/(2\sqrt{x})\,dx$; then

$$\int_1^\infty \frac{1}{\sqrt{x}\,(\sqrt{x}+1)}\,dx = \lim_{b\to\infty}\int_1^b \frac{1}{\sqrt{x}\,(\sqrt{x}+1)}\,dx = \lim_{b\to\infty} 2\int_2^{\sqrt{b}+1}\frac{1}{u}\,du$$

$$= \lim_{b\to\infty} 2\ln|u|\Big|_2^{\sqrt{b}+1} = \lim_{b\to\infty}\left(2\ln(\sqrt{b}+1)-2\ln 2\right)=\infty.$$

43. diverges; $\displaystyle\int_0^\infty e^{4x}\,dx = \lim_{b\to\infty}\int_0^b e^{4x}\,dx = \lim_{b\to\infty}\frac{1}{4}e^{4x}\Big|_0^b = \lim_{b\to\infty}\left(\frac{1}{4}e^{4b}-\frac{1}{4}\right)=\infty$

45. converges; let $u=x$, $dv=e^{-x}\,dx$, so $du=dx$, $v=-e^{-x}$;

$$\int_0^\infty xe^{-x}\,dx = \lim_{b\to\infty}\int_0^b xe^{-x}\,dx = \lim_{b\to\infty}\left(-xe^{-x}\Big|_0^b+\int_0^b e^{-x}\,dx\right) = \lim_{b\to\infty}\left(-xe^{-x}\Big|_0^b - e^{-x}\Big|_0^b\right)$$

$$= \lim_{b\to\infty}\left[(-be^{-b}+0)-(e^{-b}-1)\right] = \left(-\lim_{b\to\infty} be^{-b}\right)-\lim_{b\to\infty}e^{-b}+1 = \left(-\lim_{b\to\infty}be^{-b}\right)+1.$$

Since $\lim_{b\to\infty} b = \infty = \lim_{b\to\infty} e^b$, l'Hôpital's Rule implies that

$$\lim_{b\to\infty} be^{-b} = \lim_{b\to\infty}\frac{b}{e^b} = \lim_{b\to\infty}\frac{1}{e^b}=0.$$

47. diverges; for $\int(1/\sqrt{x^2-1})\,dx$ let $x=\sec u$, so that $dx = \sec u\,\tan u\,du$; thus

$$\int\frac{1}{\sqrt{x^2-1}}\,dx = \int\frac{1}{\sqrt{\sec^2 u-1}}\sec u\,\tan u\,du = \int \sec u\,du = \ln|\sec u+\tan u|+C.$$

Now $x=\sec u$ implies that $\tan u = \sqrt{x^2-1}$. Thus

$$\int\frac{1}{\sqrt{x^2-1}}\,dx = \ln|\sec u+\tan u|+C = \ln|x+\sqrt{x^2-1}|+C.$$

Since

$$\int_1^\infty\frac{1}{\sqrt{x^2-1}}\,dx = \lim_{a\to1^+}\int_a^2\frac{1}{\sqrt{x^2-1}}\,dx+\lim_{b\to\infty}\int_2^b\frac{1}{\sqrt{x^2-1}}\,dx$$

and since

$$\lim_{b\to\infty}\int_2^b\frac{1}{\sqrt{x^2-1}}\,dx = \lim_{b\to\infty}\ln|x+\sqrt{x^2-1}|\Big|_2^b = \lim_{b\to\infty}\left[\ln|b+\sqrt{b^2-1}|-\ln(2+\sqrt{3})\right]=\infty$$

it follows that $\int_1^\infty(1/\sqrt{x^2-1})\,dx$ diverges.

49. converges; $\displaystyle\int_1^\infty \frac{1}{t\sqrt{t^2-1}}\,dt = \lim_{a\to 1^+}\int_a^2 \frac{1}{t\sqrt{t^2-1}}\,dt + \lim_{b\to\infty}\int_2^b \frac{1}{t\sqrt{t^2-1}}\,dt$

$$= \lim_{a\to 1^+} \sec^{-1} t\Big|_a^2 + \lim_{b\to\infty} \sec^{-1} t\Big|_2^b$$

$$= \lim_{a\to 1^+} (\sec^{-1} 2 - \sec^{-1} a) + \lim_{b\to\infty} (\sec^{-1} b - \sec^{-1} 2)$$

$$= \left(\frac{\pi}{3} - 0\right) + \left(\frac{\pi}{2} - \frac{\pi}{3}\right) = \frac{1}{2}\pi$$

51. converges; since $t^2 + 4t + 8 = (t+2)^2 + 4$, let $u = t + 2$, so that $du = dt$;

$$\int_{-2}^\infty \frac{1}{t^2+4t+8}\,dt = \lim_{b\to\infty}\int_{-2}^b \frac{1}{(t+2)^2+4}\,dt = \lim_{b\to\infty}\int_0^{b+2} \frac{1}{u^2+4}\,dt$$

$$= \lim_{b\to\infty} \frac{1}{2}\tan^{-1}\frac{u}{2}\Big|_0^{b+2} = \lim_{b\to\infty} \frac{1}{2}\tan^{-1}\frac{b+2}{2} = \frac{1}{4}\pi.$$

53. diverges; $\displaystyle\int_{-\infty}^\infty x\,dx = \lim_{a\to-\infty}\int_a^0 x\,dx + \lim_{b\to\infty}\int_0^b x\,dx = \lim_{a\to-\infty}\frac{x^2}{2}\Big|_a^0 + \lim_{b\to\infty}\frac{x^2}{2}\Big|_0^b$

$$= \lim_{a\to-\infty}\frac{-1}{2}a^2 + \lim_{b\to\infty}\frac{1}{2}b^2$$

Neither limit exists, so the integral diverges.

55. diverges; let $u = x$, $dv = \sin x\,dx$, so $du = dx$, $v = -\cos x$; then

$$\int_{-\infty}^\infty x\sin x\,dx = \lim_{a\to-\infty}\int_a^0 x\sin x\,dx + \lim_{b\to\infty}\int_0^b x\sin x\,dx$$

$$= \lim_{a\to-\infty}\left[-x\cos x\Big|_a^0 + \int_a^0 \cos x\,dx\right] + \lim_{b\to\infty}\left[-x\cos x\Big|_0^b + \int_0^b \cos x\,dx\right]$$

$$= \lim_{a\to-\infty}\left[(0 + a\cos a) + (\sin x)\Big|_a^0\right] + \lim_{b\to\infty}\left[(-b\cos b + 0) + (\sin x)\Big|_0^b\right]$$

$$= \lim_{a\to-\infty}[a\cos a + 0 - \sin a] + \lim_{b\to\infty}[-b\cos b + \sin b - 0].$$

Neither limit exists, so the integral diverges.

57. converges; let $u = x^4 + 1$, so that $du = 4x^3\,dx$; then

$$\int_{-\infty}^\infty \frac{x^3}{(x^4+1)^2}\,dx = \lim_{a\to-\infty}\int_a^0 \frac{x^3}{(x^4+1)^2}\,dx + \lim_{b\to\infty}\int_0^b \frac{x^3}{(x^4+1)^2}\,dx$$

$$= \lim_{a\to-\infty}\frac{1}{4}\int_{a^4+1}^1 \frac{1}{u^2}\,du + \lim_{b\to\infty}\frac{1}{4}\int_1^{b^4+1} \frac{1}{u^2}\,du = \lim_{a\to-\infty}\frac{-1}{4u}\Big|_{a^4+1}^1 + \lim_{b\to\infty}\frac{-1}{4u}\Big|_1^{b^4+1}$$

$$= \lim_{a\to-\infty}\left[-\frac{1}{4} + \frac{1}{4(a^4+1)}\right] + \lim_{b\to\infty}\left[-\frac{1}{4(b^4+1)} + \frac{1}{4}\right] = 0.$$

59. converges; since $x^2 - 6x + 10 = (x-3)^2 + 1$, let $u = x - 3$, so that $du = dx$; then

$$\int_{-\infty}^{\infty} \frac{1}{x^2 - 6x + 10}\,dx = \lim_{a\to-\infty} \int_a^3 \frac{1}{(x-3)^2+1}\,dx + \lim_{b\to\infty} \int_3^b \frac{1}{(x-3)^2+1}\,dx$$

$$= \lim_{a\to-\infty} \int_{a-3}^{0} \frac{1}{u^2+1}\,du + \lim_{b\to\infty} \int_0^{b-3} \frac{1}{u^2+1}\,du = \lim_{a\to-\infty} \tan^{-1} u\Big|_{a-3}^{0} + \lim_{b\to\infty} \tan^{-1} u\Big|_0^{b-3}$$

$$= \lim_{a\to-\infty} -(\tan^{-1}(a-3)) + \lim_{b\to\infty} \tan^{-1}(b-3) = -\left(-\frac{\pi}{2}\right) + \frac{\pi}{2} = \pi.$$

61. a. $\dfrac{1}{x(x+1)} = \dfrac{A}{x} + \dfrac{B}{x+1} = \dfrac{A(x+1)+Bx}{x(x+1)}$; $A + B = 0$, $A = 1$; thus $B = -1$, so that

$$\frac{1}{x(x+1)} = \frac{1}{x} - \frac{1}{x+1}.$$

b. No, since both $\int_1^\infty (1/x)\,dx$ and $\int_1^\infty [1/(x+1)]\,dx$ diverge, whereas

$$\int_1^\infty \frac{1}{x(x+1)}\,dx = \lim_{b\to\infty} \int_1^b \frac{1}{x(x+1)}\,dx = \lim_{b\to\infty} \int_1^b \left(\frac{1}{x} - \frac{1}{x+1}\right) dx$$

$$= \lim_{b\to\infty} (\ln|x| - \ln|x+1|)\Big|_1^b = \lim_{b\to\infty} [\ln b - \ln(b+1) + \ln 2] = \lim_{b\to\infty} \left(\ln \frac{b}{b+1} + \ln 2\right) = \ln 2.$$

63. $A = \displaystyle\int_{-\infty}^{0} \frac{1}{(x-3)^2}\,dx = \lim_{a\to-\infty} \int_a^0 \frac{1}{(x-3)^2}\,dx = \lim_{a\to-\infty} \frac{-1}{x-3}\Big|_a^0 = \lim_{a\to-\infty} \left(\frac{1}{3} + \frac{1}{a-3}\right) = \frac{1}{3}$

so the region has finite area.

65. $A = \displaystyle\int_2^\infty \frac{\ln x}{x}\,dx = \lim_{b\to\infty} \int_2^b \frac{\ln x}{x}\,dx = \lim_{b\to\infty} \frac{1}{2}(\ln x)^2\Big|_2^b = \lim_{b\to\infty} \left[\frac{1}{2}(\ln b)^2 - \frac{1}{2}(\ln 2)^2\right] = \infty,$

so the region has infinite area.

67. $A = \displaystyle\int_2^\infty \frac{1}{\sqrt{x+1}}\,dx = \lim_{b\to\infty} \int_2^b \frac{1}{\sqrt{x+1}}\,dx = \lim_{b\to\infty} 2\sqrt{x+1}\Big|_2^b = \lim_{b\to\infty} (2\sqrt{b+1} - 2\sqrt{3}) = \infty$

so the region has infinite area.

69. Since $1/(1+x^4) \le 1/(1+x^2)$ and since

$$\int_0^\infty \frac{1}{1+x^2}\,dx = \lim_{b\to\infty} \int_0^b \frac{1}{1+x^2}\,dx = \lim_{b\to\infty} \tan^{-1} x\Big|_0^b = \lim_{b\to\infty} \tan^{-1} b = \frac{\pi}{2}$$

it follows from the Comparison Property that $\int_1^\infty 1/(1+x^4)\,dx$ converges.

71. Since $(\sin^2 x)/\sqrt{1+x^3} \le 1/\sqrt{x^3} = 1/x^{3/2}$, and since $\int_1^\infty (1/x^{3/2})\,dx$ converges by Exercise 62(b), it follows from the Comparison Property that $\int_1^\infty [(\sin^2 x)/\sqrt{1+x^3}]\,dx$ converges.

73. Since $(1/x)\sqrt{1+1/x^4} \ge 1/x$ and since $\int_1^\infty (1/x)\,dx$ diverges by Exercise 62(b), it follows from the Comparison Property that $\int_1^\infty (1/x)\sqrt{1+1/x^4}\,dx$ diverges.

75. a. If $\lim_{b\to\infty}\int_a^b f(x)\,dx$ and $\lim_{b\to\infty}\int_a^b g(x)\,dx$ both exist, then $\lim_{b\to\infty}\left(\int_a^b f(x)\,dx+\int_a^b g(x)\,dx\right)=\lim_{b\to\infty}\int_a^b(f(x)+g(x))\,dx$ exists, so $\int_a^\infty(f(x)+g(x))\,dx$ converges.

b. If $\lim_{b\to\infty}\int_a^b f(x)\,dx$ exists, then $\lim_{b\to\infty}c\int_a^b f(x)\,dx=\lim_{b\to\infty}\int_a^b cf(x)\,dx$ exists, so $\int_a^\infty cf(x)\,dx$ converges.

c. If $f(x)=1/x$ and $g(x)=-1/x$, then $0=\int_1^\infty(f(x)+g(x))\,dx$, so the integral converges; however, $\int_1^\infty(1/x)\,dx$ and $\int_1^\infty(-1/x)\,dx$ diverge since $\int_1^\infty(1/x)\,dx=\lim_{b\to\infty}\int_1^b(1/x)\,dx=\lim_{b\to\infty}\ln|x|\Big|_1^b=\lim_{b\to\infty}\ln b=\infty$.

77. a. $$\int_a^\infty f(t)\,dt=\lim_{b\to\infty}\int_a^b f(t)\,dt=\lim_{b\to\infty}\left(\int_a^x f(t)\,dt+\int_x^b f(t)\,dt\right)$$
$$=\int_a^x f(t)\,dt+\lim_{b\to\infty}\int_x^b f(t)\,dt=\int_a^x f(t)\,dt+\int_x^\infty f(t)\,dt$$

b. By part (a),
$$\frac{d}{dx}\int_a^\infty f(t)\,dt=\frac{d}{dx}\int_a^x f(t)\,dt+\frac{d}{dx}\int_x^\infty f(t)\,dt.$$
Now $\int_a^\infty f(t)\,dt$ is independent of x, so $(d/dx)\int_a^\infty f(t)\,dt=0$. By Theorem 5.12 in Section 5.4, $(d/dx)\int_a^x f(t)\,dt=f(x)$. Thus we obtain
$$0=f(x)+\frac{d}{dx}\int_x^\infty f(t)\,dt,\quad\text{and thus}\quad\frac{d}{dx}\int_x^\infty f(t)\,dt=-f(x).$$

79. a. Since $1/(x^{3/2}+1)\le 1/x^{3/2}$ for $x\ge 1$, and since $\int_1^\infty(1/x^{3/2})\,dx$ converges by Exercise 62(b), it follows from the Comparison Property that $\int_1^\infty[1/(x^{3/2}+1)]\,dx$ converges.

b. If $y=1/(x^{3/2}+1)$, then $x^{3/2}+1=1/y$, so that $x=((1/y)-1)^{2/3}$. If $1\le x$, then $0<y\le\frac{1}{2}$. Thus the area of the red region in Figure 8.30, which is $\int_0^{1/2}((1/x)-1)^{2/3}\,dx$, equals the area of the blue region, which is $\int_1^\infty[1/(x^{3/2}+1)]\,dx$.

81. a. By Simpson's Rule with $n=100$, $\int_{-5}^5(1/\sqrt{2\pi})\,e^{-x^2/2}\,dx\approx 0.9999994265$.

b. If $b=7$, Simpson's Rule with $n=100$ yields $\int_{-7}^7(1/\sqrt{2\pi})\,e^{-x^2/2}\,dx=1$.

83. To integrate by parts, let $u=ct$ and $dv=e^{-ct}\,dt$; then $du=c\,dt$ and $v=(-1/c)e^{-ct}$. Therefore
$$\int cte^{-ct}\,dt=ct\left(-\frac{1}{c}e^{-ct}\right)-\int\left(-\frac{1}{c}e^{-ct}\right)c\,dt=-te^{-ct}+\int e^{-ct}\,dt=-te^{-ct}-\frac{1}{c}e^{-ct}+C.$$
Therefore
$$\int_0^\infty cte^{-ct}\,dt=\lim_{b\to\infty}\int_0^b cte^{-ct}\,dt=\lim_{b\to\infty}\left(-te^{-ct}-\frac{1}{c}e^{-ct}\right)\Big|_0^b=\lim_{b\to\infty}\left(-be^{-cb}-\frac{1}{c}e^{-cb}+\frac{1}{c}\right)=\frac{1}{c}$$
where we have used the fact that $\lim_{b\to\infty}be^{-cb}=0$ by l'Hôpital's Rule.

85. $M = -\dfrac{1}{A}\displaystyle\int_0^\infty tkf(t)\,dt = -\dfrac{1}{A}\int_0^\infty tkAe^{kt}\,dt = -\int_0^\infty tke^{kt}\,dt$

For integration by parts, let $u = t$, $dv = ke^{kt}\,dt$, so that $du = dt$, $v = e^{kt}$. Then since $k < 0$,

$$M = \lim_{b\to\infty} -\int_0^b tke^{kt}\,dt = \lim_{b\to\infty} -\left(te^{kt}\Big|_0^b - \int_0^b e^{kt}\,dt\right)$$

$$= \lim_{b\to\infty} -\left(be^{kb} - \frac{1}{k}e^{kt}\Big|_0^b\right) = \lim_{b\to\infty} -\left(be^{kb} - \frac{1}{b}e^{kb} + \frac{1}{k}\right) = -\frac{1}{k}.$$

a. If $k = -1.24 \times 10^{-4}$, then $M = 1/1.24 \times 10^4 \approx 8060$ years.

b. If $k = -4.36 \times 10^{-4}$, then $M = 1/4.36 \times 10^4 \approx 2090$ years.

87. By Exercise 77(b),

$$-g(x) = \frac{d}{dx}\int_x^\infty g(t)\,dt = \frac{d}{dx}cx^{-1.5} = -1.5cx^{-2.5}.$$

Thus $g(x) = 1.5cx^{-2.5}$ for $x \geq s$.

89. a. We have $\mu = 7.5$ and $\sigma = 1$. Then $8 = \mu + 0.5\sigma$ and $12 = \mu + 4.5\sigma$. Thus we need to approximate

$$\int_{\mu+0.5\sigma}^{\mu+4.5\sigma} \frac{1}{\sigma\sqrt{2\pi}}e^{-(x-\mu)^2/2\sigma^2}\,dx,$$

which by Exercise 82 equals $\int_{0.5}^{4.5}(1/\sqrt{2\pi})e^{-x^2/2}\,dx$. By Simpson's Rule with $n = 100$,

$$\int_{0.5}^{4.5}\frac{1}{\sqrt{2\pi}}e^{-x^2/2}\,dx \approx 0.3085341341.$$

Therefore approximately 31% of the babies, that is, approximately 62 babies, born in the Easy-Birth Hospital weigh between 8 and 12 pounds.

b. Here $\mu = 7.5$ and $\sigma = 1$, and thus $6.5 = \mu - \sigma$. We need to approximate

$$\int_{-\infty}^{\mu-\sigma}\frac{1}{\sigma\sqrt{2\pi}}e^{-(x-\mu)^2/2\sigma^2}\,dx,$$

which by Exercise 82 equals $\int_{-\infty}^{-1}(1/\sqrt{2\pi})e^{-x^2/2}\,dx$. Notice that

$$1 = \int_{-\infty}^{\infty}\frac{1}{\sqrt{2\pi}}e^{-x^2/2}\,dx = 2\int_{-\infty}^{0}\frac{1}{\sqrt{2\pi}}e^{-x^2/2}\,dx$$

because the graph of $e^{-x^2/2}/\sqrt{2\pi}$ is symmetric with respect to the y axis. Thus

$$\int_{-\infty}^{0}\frac{1}{\sqrt{2\pi}}e^{-x^2/2}\,dx = \frac{1}{2}.$$

Then by a result analogous to Exercise 77(a),

$$\int_{-\infty}^{-1} \frac{1}{\sqrt{2\pi}} e^{-x^2/2}\,dx = \int_{-\infty}^{0} \frac{1}{\sqrt{2\pi}} e^{-x^2/2}\,dx - \int_{-1}^{0} \frac{1}{\sqrt{2\pi}} e^{-x^2/2}\,dx = \frac{1}{2} - \int_{-1}^{0} \frac{1}{\sqrt{2\pi}} e^{-x^2/2}\,dx.$$

By Simpson's Rule with $n = 100$,

$$\int_{-1}^{0} \frac{1}{\sqrt{2\pi}} e^{-x^2/2}\,dx \approx 0.3413447461.$$

Consequently

$$\int_{-\infty}^{-1} \frac{1}{\sqrt{2\pi}} e^{-x^2/2}\,dx \approx \frac{1}{2} - 0.3413447461 = 0.1586552539.$$

Thus approximately 16% of the babies, that is, approximately 32 babies, born in the hospital weighed under 6.5 pounds at birth.

Chapter 8 Review

1. $u = \ln(x^2+9)$, $dv = dx$; $du = [2x/(x^2+9)]\,dx$, $v = x$;

$$\int \ln(x^2+9)\,dx = x\ln(x^2+9) - \int \frac{2x^2}{x^2+9}\,dx = x\ln(x^2+9) - 2\int \left(1 - \frac{9}{x^2+9}\right)dx$$

$$= x\ln(x^2+9) - 2x + 18\left(\frac{1}{3}\right)\tan^{-1}\frac{x}{3} + C = x\ln(x^2+9) - 2x + 6\tan^{-1}\frac{x}{3} + C.$$

3. $u = x$, $dv = \csc^2 x\,dx$; $du = dx$, $v = -\cot x$; $\int x\csc^2 x\,dx = -x\cot x + \int \cot x\,dx = -x\cot x + \ln|\sin x| + C$.

5. $u = x$, $dv = \cosh x\,dx$; $du = dx$, $v = \sinh x$; $\int x\cosh x\,dx = x\sinh x - \int \sinh x\,dx = x\sinh x - \cosh x + C$.

7. $\int x\cos^2 x\,dx = \int x(\frac{1}{2}+\frac{1}{2}\cos 2x)\,dx = \frac{1}{2}\int x\,dx + \frac{1}{2}\int x\cos 2x\,dx = \frac{1}{4}x^2 + \frac{1}{2}\int x\cos 2x\,dx$. For $\int x\cos 2x\,dx$, let $u = x$, $dv = \cos 2x\,dx$; $du = dx$, $v = \frac{1}{2}\sin 2x$; $\int x\cos 2x\,dx = \frac{1}{2}x\sin 2x - \frac{1}{2}\int \sin 2x\,dx = \frac{1}{2}x\sin 2x + \frac{1}{4}\cos 2x + C_1$. Thus $\int x\cos^2 x\,dx = \frac{1}{4}x^2 + \frac{1}{4}x\sin 2x + \frac{1}{8}\cos 2x + C$.

9. $u = \sin x^3$, $du = 3x^2\cos x^3\,dx$; $\int x^2 \sin x^3 \cos x^3\,dx = \frac{1}{3}\int u\,du = \frac{1}{6}u^2 + C = \frac{1}{6}\sin^2 x^3 + C$.

11.
$$\begin{aligned}\int \tan^5 x\,dx &= \int \tan^3 x\,(\sec^2 x - 1)\,dx = \int \tan^3 x\,\sec^2 x\,dx - \int \tan^3 x\,dx\\ &= \int \tan^3 x\,\sec^2 x\,dx - \int \tan x\,(\sec^2 x - 1)\,dx\\ &= \int \tan^3 x\,\sec^2 x\,dx - \int \tan x\,\sec^2 x\,dx + \int \tan x\,dx\end{aligned}$$

For the first two integrals let $u = \tan x$, so that $du = \sec^2 x\,dx$. Then

$$\int \tan^3 x \sec^2 x\,dx - \int \tan x \sec^2 x\,dx = \int u^3\,du - \int u\,du = \frac{1}{4}u^4 - \frac{1}{2}u^2 + C_1 = \frac{1}{4}\tan^4 x - \frac{1}{2}\tan^2 x + C_1.$$

Thus $\int \tan^5 x\,dx = \frac{1}{4}\tan^4 x - \frac{1}{2}\tan^2 x - \ln|\cos x| + C$.

13. $u = x^2$, $dv = x\cos x^2\,dx$; $du = 2x\,dx$, $v = \frac{1}{2}\sin x^2$; $\int x^3\cos x^2\,dx = \frac{1}{2}x^2\sin x^2 - \int \frac{1}{2}(2x)\sin x^2\,dx = \frac{1}{2}x^2\sin x^2 + \frac{1}{2}\cos x^2 + C$.

15. $u = 1 - 3t$, $du = -3\,dt$, so $t = \frac{1}{3}(1-u)$;

$$\int t^2\sqrt{1-3t}\,dt = \int \left[\frac{1}{3}(1-u)\right]^2 \sqrt{u}\left(-\frac{1}{3}\right)du = -\frac{1}{27}\int \left(u^{1/2} - 2u^{3/2} + u^{5/2}\right)du$$

$$= -\frac{1}{27}\left(\frac{2}{3}u^{3/2} - \frac{4}{5}u^{5/2} + \frac{2}{7}u^{7/2}\right) + C = -\frac{2}{81}(1-3t)^{3/2} + \frac{4}{135}(1-3t)^{5/2} - \frac{2}{189}(1-3t)^{7/2} + C.$$

17. $$\int \frac{\cos x}{1+\cos x}\,dx = \int \frac{\cos x}{1+\cos x}\,\frac{1-\cos x}{1-\cos x}\,dx = \int \frac{\cos x - (1-\sin^2 x)}{\sin^2 x}\,dx$$

$$= \int \left(\frac{\cos x}{\sin^2 x} - \csc^2 x + 1\right)dx = \frac{-1}{\sin x} + \cot x + x + C = -\csc x + \cot x + x + C$$

19. $x + 1 = 3\tan u$, $dx = 3\sec^2 u\,du$;

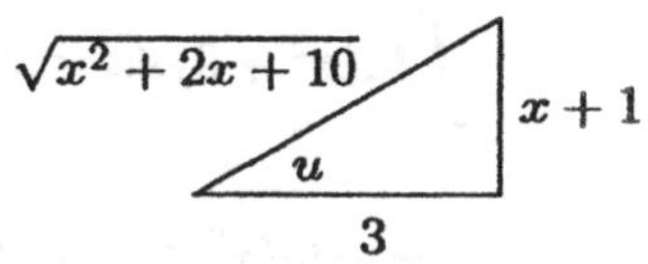

$$\int \frac{x^2}{(x^2+2x+10)^{5/2}}\,dx = \int \frac{x^2}{((x+1)^2+9)^{5/2}}\,dx$$

$$= \int \frac{(3\tan u - 1)^2}{(9\tan^2 u + 9)^{5/2}}\,3\sec^2 u\,du$$

$$= \frac{1}{81}\int \frac{9\tan^2 u - 6\tan u + 1}{\sec^5 u}\,\sec^2 u\,du$$

$$= \frac{1}{9}\int \sin^2 u\,\cos u\,du - \frac{2}{27}\int \sin u\,\cos^2 u\,du + \frac{1}{81}\int (1-\sin^2 u)\cos u\,du$$

$$= \frac{1}{27}\sin^3 u + \frac{2}{81}\cos^3 u + \frac{1}{81}\sin u - \frac{1}{243}\sin^3 u + C$$

$$= \frac{8}{243}\frac{(x+1)^3}{(x^2+2x+10)^{3/2}} + \frac{2}{3}\frac{1}{(x^2+2x+10)^{3/2}} + \frac{1}{81}\frac{x+1}{(x^2+2x+10)^{1/2}} + C.$$

21. $x = \tan u$, $dx = \sec^2 u\,du$;

$$\int \frac{x^4}{(x^2+1)^2}\,dx = \int \frac{\tan^4 u}{(\tan^2 u + 1)^2}\sec^2 u\,du = \int \frac{\tan^4 u}{\sec^2 u}\,du = \int \frac{\sin^4 u}{\cos^4 u}\cos^2 u\,du = \int \frac{\sin^4 u}{\cos^2 u}\,du$$

$$= \int \frac{(1-\cos^2 u)^2}{\cos^2 u}\,du = \int \frac{1 - 2\cos^2 u + \cos^4 u}{\cos^2 u}\,du = \int (\sec^2 u - 2 + \cos^2 u)\,du$$

$$= \tan u - 2u + \int \left(\frac{1}{2} + \frac{1}{2}\cos 2u\right)du = \tan u - 2u + \frac{1}{2}u + \frac{1}{4}\sin 2u + C$$

$$= \tan u - \frac{3}{2}u + \frac{1}{2}\sin u\,\cos u + C$$

Now $u = \tan^{-1} x$, and by the figure, $\sin u = x/\sqrt{x^2+1}$ and $\cos u = 1/\sqrt{x^2+1}$, so that

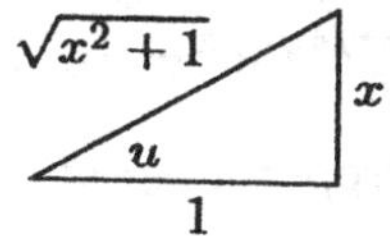

$$\int \frac{x^4}{(x^2+1)^2}\,dx = x - \frac{3}{2}\tan^{-1} x + \frac{1}{2}\frac{x}{x^2+1} + C.$$

23. $\displaystyle\int \frac{x}{x^2+3x-18}\,dx = \int \frac{x}{(x+6)(x-3)}\,dx;\ \frac{x}{(x+6)(x-3)} = \frac{A}{x+6} + \frac{B}{x-3};$
$A(x-3) + B(x+6) = x$; $A + B = 1$, $-3A + 6B = 0$; $A = \frac{2}{3}$, $B = \frac{1}{3}$;

$$\int \frac{x}{x^2+3x-18}\,dx = \int \left(\frac{2}{3}\frac{1}{x+6} + \frac{1}{3}\frac{1}{x-3}\right) dx$$

$$= \frac{2}{3}\int \frac{1}{x+6}\,dx + \frac{1}{3}\int \frac{1}{x-3}\,dx = \frac{2}{3}\ln|x+6| + \frac{1}{3}\ln|x-3| + C$$

25. **Improper integral;**

$$\int_{-2}^{1} \frac{1}{3x+4}\,dx = \lim_{c\to -4/3^-} \int_{-2}^{c} \frac{1}{3x+4}\,dx + \lim_{p\to -4/3^+} \int_{p}^{1} \frac{1}{3x+4}\,dx$$

$$= \lim_{c\to -4/3^-} \frac{1}{3}\ln|3x+4|\Big|_{-2}^{c} + \lim_{p\to -4/3^+} \frac{1}{3}\ln|3x+4|\Big|_{p}^{1}$$

$$= \lim_{c\to -4/3^-}\left(\frac{1}{3}\ln|3c+4| - \frac{1}{3}\ln 2\right) + \lim_{p\to -4/3^+}\left(\frac{1}{3}\ln 7 - \frac{1}{3}\ln|3p+4|\right).$$

Neither limit exists, so the integral diverges.

27. **Proper integral;** $u = x$, $dv = \sec x \tan x\,dx$, $du = dx$, $v = \sec x$; $\int_0^{\pi/4} x \sec x \tan x\,dx = x\sec x\big|_0^{\pi/4} - \int_0^{\pi/4} \sec x\,dx = x\sec x\big|_0^{\pi/4} - \ln|\sec x + \tan x|\big|_0^{\pi/4} = \frac{\pi\sqrt{2}}{4} - \ln(\sqrt{2}+1)$.

29. **Proper integral;** $u = 1 + \sqrt{x}$, $du = \frac{1}{2}x^{-1/2}\,dx$, so $dx = 2(u-1)\,du$;

$$\int_1^4 \frac{1}{1+\sqrt{x}}\,dx = \int_2^3 \frac{2(u-1)}{u}\,du = \int_2^3 \left(2 - \frac{2}{u}\right) du$$

$$= (2u - 2\ln|u|)\big|_2^3 = 6 - 2\ln 3 - 4 + 2\ln 2 = 2 + \ln\frac{4}{9}.$$

31. **Proper integral;** $u = \sin x$, $du = \cos x\,dx$;

$$\int_0^{\pi/4} \sin^4 x \cos^3 x\,dx = \int_0^{\pi/4} \sin^4 x\,(1 - \sin^2 x)\cos x\,dx = \int_0^{\sqrt{2}/2} (u^4 - u^6)\,du$$

$$= \left(\frac{1}{5}u^5 - \frac{1}{7}u^7\right)\Big|_0^{\sqrt{2}/2} = \frac{\sqrt{2}}{40} - \frac{\sqrt{2}}{112} = \frac{9\sqrt{2}}{560}.$$

33. Proper integral; $u = \tan x$, $du = \sec^2 x\,dx$; $\int_0^{\pi/4}(\tan^3 x + \tan^5 x)\,dx = \int_0^{\pi/4}\tan^3 x\,(1+\tan^2 x)\,dx = \int_0^{\pi/4}\tan^3 x \sec^2 x\,dx = \int_0^1 u^3\,du = \frac{u^4}{4}\Big|_0^1 = \frac{1}{4}$.

35. Proper integral; $x = \sec u$, $dx = \sec u \tan u\,du$;

$$\int_1^{\sqrt{2}} \frac{\sqrt{x^2-1}}{x^2}\,dx = \int_0^{\pi/4} \frac{\sqrt{\sec^2 u - 1}}{\sec^2 u}\sec u \tan u\,du = \int_0^{\pi/4}\frac{\tan^2 u}{\sec u}\,du = \int_0^{\pi/4}\frac{\sec^2 u - 1}{\sec u}\,du$$

$$= \int_0^{\pi/4}(\sec u - \cos u)\,du = (\ln|\sec u + \tan u| - \sin u)\Big|_0^{\pi/4} = \ln(\sqrt{2}+1) - \frac{\sqrt{2}}{2}$$

37. Proper integral; $u = x^{3/2}$, $du = \frac{3}{2}x^{1/2}\,dx$; then $u = \sin t$, $du = \cos t\,dt$;

$$\int_0^{\sqrt[3]{2}/2}\frac{\sqrt{x}}{\sqrt{1-x^3}}\,dx = \frac{2}{3}\int_0^{1/2}\frac{1}{\sqrt{1-u^2}}\,du = \frac{2}{3}\int_0^{\pi/6}\frac{1}{\sqrt{1-\sin^2 t}}\cos t\,dt = \frac{2}{3}\int_0^{\pi/6}1\,dt = \frac{2}{3}t\Big|_0^{\pi/6} = \frac{1}{9}\pi$$

39. Proper integral; $x = \tan u$, $dx = \sec^2 u\,du$;

$$\int_0^{\sqrt{3}}\sqrt{x^2+1}\,dx = \int_0^{\pi/3}\sqrt{\tan^2 u + 1}\sec^2 u\,du = \int_0^{\pi/3}\sec^3 u\,du \overset{\text{(7) of Section 7.2}}{=}$$

$$\left(\frac{1}{2}\sec u \tan u + \frac{1}{2}\ln|\sec u + \tan u|\right)\Big|_0^{\pi/3} = \frac{1}{2}(2)(\sqrt{3}) + \frac{1}{2}\ln|2+\sqrt{3}| = \sqrt{3} + \frac{1}{2}\ln(2+\sqrt{3})$$

41. $\displaystyle\int_{-5}^0 \frac{x}{x^2+4x-5}\,dx = \int_{-5}^0\frac{x}{(x+5)(x-1)}\,dx$, so the integral is improper. Next,

$$\frac{x}{(x+5)(x-1)} = \frac{A}{x+5} + \frac{B}{x-1} = \frac{A(x-1)+B(x+5)}{(x+5)(x-1)}$$

$$A(x-1)+B(x+5) = x; A+B=1, -A+5B=0; A = \frac{5}{6}, B = \frac{1}{6}.$$

Thus

$$\int_{-5}^0\frac{x}{x^2+4x-5}\,dx = \lim_{c\to -5^+}\int_c^0\left(\frac{5}{6}\frac{1}{x+5}+\frac{1}{6}\frac{1}{x-1}\right)dx = \lim_{c\to -5^+}\left(\frac{5}{6}\int_c^0\frac{1}{x+5}\,dx + \frac{1}{6}\int_c^0\frac{1}{x-1}\,dx\right)$$

$$= \lim_{c\to -5^+}\left(\frac{5}{6}\ln|x+5|\Big|_c^0 + \frac{1}{6}\ln|x-1|\Big|_c^0\right) = \lim_{c\to -5^+}\left(\frac{5}{6}\ln 5 - \frac{5}{6}\ln(c+5) - \frac{1}{6}\ln|c-1|\right).$$

Since $\lim_{c\to -5^+}\ln(c+5) = -\infty$, the integral diverges.

43. Improper integral;

$$\int_0^{\pi/2}\frac{1}{1-\sin x}\,dx = \lim_{c\to\pi/2^-}\int_0^c\frac{1}{1-\sin x}\,dx = \lim_{c\to\pi/2^-}\int_0^c\frac{1}{1-\sin x}\,\frac{1+\sin x}{1+\sin x}\,dx$$

$$= \lim_{c\to\pi/2^-}\int_0^c \frac{1+\sin x}{\cos^2 x}\,dx = \lim_{c\to\pi/2^-}\int_0^c \left(\sec^2 x + \frac{\sin x}{\cos^2 x}\right)dx = \lim_{c\to\pi/2^-}\left(\tan x\Big|_0^c + \int_0^c \frac{\sin x}{\cos^2 x}\,dx\right)$$

For $\int_0^c (\sin x)/(\cos^2 x)\,dx$ let $u = \cos x$, so $du = -\sin x\,dx$. Then

$$\int_0^c \frac{\sin x}{\cos^2 x}\,dx = \int_1^{\cos c} -\frac{1}{u^2}\,du = \frac{1}{u}\Big|_1^{\cos c} = \frac{1}{\cos c} - 1 = \sec c - 1.$$

Thus

$$\int_0^{\pi/2} \frac{1}{1-\sin x}\,dx = \lim_{c\to\pi/2^-}(\tan c + \sec c - 1).$$

Since $\lim_{c\to\pi/2^-}\tan c = \infty = \lim_{c\to\pi/2^-}\sec c$, the integral diverges.

45. Improper integral; $u = \ln x$, $dv = x\,dx$, $du = (1/x)\,dx$, $v = \frac{1}{2}x^2$; $\int x\ln x\,dx = \frac{1}{2}x^2\ln x - \int \frac{1}{2}x\,dx = \frac{1}{2}x^2\ln x - \frac{1}{4}x^2 + C$. Thus

$$\int_0^1 x\ln x\,dx = \lim_{c\to 0^+}\int_c^1 x\ln x\,dx = \lim_{c\to 0^+}\left(\frac{1}{2}x^2\ln x - \frac{1}{4}x^2\right)\Big|_c^1 = \lim_{c\to 0^+}\left(-\frac{1}{4} - \frac{1}{2}c^2\ln c + \frac{1}{4}c^2\right) = -\frac{1}{4}$$

since $\lim_{c\to 0^+} c\ln c = 0$ by Example 7 of Section 7.6.

47. Improper integral;

$$\int_1^\infty \frac{1}{x(\ln x)^2}\,dx = \lim_{c\to 1^+}\int_c^2 \frac{1}{x(\ln x)^2}\,dx + \lim_{b\to\infty}\int_2^b \frac{1}{x(\ln x)^2}\,dx$$

$$= \lim_{c\to 1^+}\frac{-1}{\ln x}\Big|_c^2 + \lim_{b\to\infty}\frac{-1}{\ln x}\Big|_2^b = \lim_{c\to 1^+}\left(\frac{1}{\ln c} - \frac{1}{\ln 2}\right) + \lim_{b\to\infty}\left(\frac{1}{\ln 2} - \frac{1}{\ln b}\right).$$

Since $\lim_{c\to 1^+}(1/(\ln c) - 1/(\ln 2)) = \infty$, the integral diverges.

49. Improper integral; by Exercise 54(b) of Section 3.1, with $a = -1$ and $b = 1$, $\int e^{-x}\cos x\,dx = \frac{1}{2}e^{-x}(-\cos x + \sin x) + C$. Thus

$$\int_0^\infty e^{-x}\cos x\,dx = \lim_{b\to\infty}\int_0^b e^{-x}\cos x\,dx = \lim_{b\to\infty}\frac{1}{2}e^{-x}(\sin x - \cos x)\Big|_0^b$$

$$= \lim_{b\to\infty}\left[\frac{1}{2}e^{-b}(\sin b - \cos b) + \frac{1}{2}\right] = \frac{1}{2}.$$

51. Improper integral;

$$\int_1^\infty \frac{1}{x(x^2+1)}\,dx = \lim_{b\to\infty}\int_1^b \frac{1}{x(x^2+1)}\,dx \overset{x=\tan u}{=} \lim_{b\to\infty}\int_{\pi/4}^{\tan^{-1}b} \frac{1}{\tan u\,(\sec^2 u)}\sec^2 u\,du$$

$$= \lim_{b\to\infty}\int_{\pi/4}^{\tan^{-1}b} \cot u\,du = \lim_{b\to\infty}\ln|\sin u|\Big|_{\pi/4}^{\tan^{-1}b}.$$

By the figure, if $u = \tan^{-1} b$, then $\sin u = b/\sqrt{b^2+1}$, so that

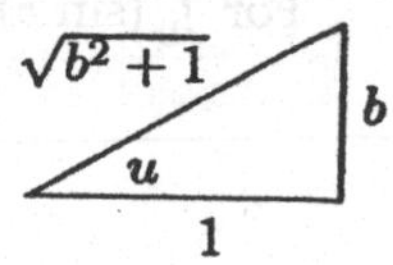

$$\lim_{b\to\infty} \ln|\sin u|\Big|_{\pi/4}^{\tan^{-1} b} = \lim_{b\to\infty}\left(\ln\frac{b}{\sqrt{b^2+1}} - \ln\left(\sin\frac{\pi}{4}\right)\right)$$
$$= \lim_{b\to\infty}\left(\ln 1 - \ln\frac{\sqrt{2}}{2}\right) = \ln\sqrt{2} = \frac{1}{2}\ln 2.$$

Therefore the integral converges, and $\int_1^\infty 1/[x(x^2+1)]\,dx = \frac{1}{2}\ln 2$.

53. $u = \ln x$, $dv = (1/x)\,dx$, $du = (1/x)\,dx$, $v = \ln x$;

$$\int \frac{\ln x}{x}\,dx = (\ln x)^2 - \int \frac{\ln x}{x}\,dx, \quad \text{so that} \quad 2\int \frac{\ln x}{x}\,dx = (\ln x)^2 + C_1$$

and thus $\int (\ln x)/x\,dx = \frac{1}{2}(\ln x)^2 + C$.

55. a. $x = \dfrac{u}{1-u}$, $dx = \dfrac{1}{(1-u)^2}\,du$, $u = \dfrac{x}{1+x}$ and $1 + x = \dfrac{1}{1-u}$;

$$\int_0^b \frac{x^{m-1}}{(1+x)^{m+n}}\,dx = \int_0^b \left(\frac{x}{1+x}\right)^{m-1}\frac{1}{(1+x)^{n+1}}\,dx$$
$$= \int_0^{b/(1+b)} u^{m-1}(1-u)^{n+1}\frac{1}{(1-u)^2}\,du = \int_0^{b/(1+b)} u^{m-1}(1-u)^{n-1}\,du.$$

b. Using (a) with $m = 4$ and $n = 1$, we obtain

$$\int_0^\infty \frac{x^3}{(1+x)^5}\,dx = \lim_{b\to\infty}\int_0^b \frac{x^3}{(1+x)^5}\,dx = \lim_{b\to\infty}\int_0^{b/(1+b)} u^3(1-u)^0\,du$$
$$= \lim_{b\to\infty}\int_0^{b/(1+b)} u^3\,du = \lim_{b\to\infty}\frac{1}{4}u^4\Big|_0^{b/(1+b)} = \lim_{b\to\infty}\frac{1}{4}\frac{b^4}{(1+b)^4} = \frac{1}{4}.$$

57. Using the given identities, we have

$$\int \frac{\tan(\pi/4 + x/2)}{\sec^2(x/2)}\,dx = \int \frac{\sin(\pi/2+x)}{1+\cos(\pi/2+x)}\,\frac{1+\cos x}{2}\,dx$$
$$= \int \left(\frac{\cos x}{1-\sin x}\right)\left(\frac{1+\cos x}{2}\right)dx = \frac{1}{2}\int \frac{(\cos x + \cos^2 x)(1+\sin x)}{(1-\sin x)(1+\sin x)}\,dx$$
$$= \frac{1}{2}\int \frac{(\cos x + \cos^2 x)(1+\sin x)}{\cos^2 x}\,dx = \frac{1}{2}\int (\sec x + 1 + \tan x + \sin x)\,dx$$
$$= \frac{1}{2}(\ln|\sec x + \tan x| + \ln|\sec x| - \cos x + x) + C = \frac{1}{2}\ln\left|\sec^2 x + \sec x\,\tan x\right| - \frac{1}{2}\cos x + \frac{x}{2} + C.$$

59. a. With $n = 10$, the Trapezoidal Rule yields $\int_0^2 \sqrt{2x - x^2}\,dx \approx 1.518524414$.

b. With $n = 10$, Simpson's Rule yields $\int_0^2 \sqrt{2x - x^2}\,dx \approx 1.55008698$.

61. a. By the solution of Exercise 60 it suffices to use the Trapezoidal Rule with $n = 5$. We obtain

$$\int_2^{2.5} \sqrt{x^2-1}\,dx \approx \frac{1}{20}\left(\sqrt{3}+2\sqrt{3.41}+2\sqrt{3.84}+2\sqrt{4.29}+2\sqrt{4.76}+\sqrt{5.25}\right) \approx 1.007085359.$$

b. By the solution of Exercise 60 it suffices to use Simpson's Rule with $n = 2$. We obtain

$$\int_2^{2.5} \sqrt{x^2-1}\,dx \approx \frac{1}{12}\left(\sqrt{3}+4\sqrt{4.0625}+\sqrt{5.25}\right) \approx 1.007133034.$$

63. $A = \int_{-3}^{0} \sqrt{9-x^2}\,dx$. Let $x = 3\sin u$, so that $dx = 3\cos u\,du$. Then

$$\int_{-3}^{0} \sqrt{9-x^2}\,dx = \int_{-\pi/2}^{0} \sqrt{9-9\sin^2 u}\,3\cos u\,du$$

$$= 9\int_{-\pi/2}^{0} \cos^2 u\,du = 9\left(\frac{1}{2}u+\frac{1}{4}\sin 2u\right)\Big|_{-\pi/2}^{0} = 0 - 9\left(-\frac{\pi}{4}\right) = \frac{9}{4}\pi.$$

65. $$A = \int_{\pi}^{3\pi/2} \left|\frac{\cos x}{1+\sin x}\right| dx = \int_{\pi}^{3\pi/2} \frac{-\cos x}{1+\sin x}\,dx = -\lim_{c\to 3\pi/2^-}\int_{\pi}^{c} \frac{\cos x}{1+\sin x}\,dx$$

$$= -\lim_{c\to 3\pi/2^-} \ln(1+\sin x)\Big|_{\pi}^{c} = -\lim_{c\to 3\pi/2^-} \ln(1+\sin c) = \infty$$

so the region has infinite area.

67. $$A = \int_{-\infty}^{\infty} \left|\frac{x^3}{2+x^4}\right| dx = \lim_{a\to-\infty}\int_a^0 \frac{-x^3}{2+x^4}\,dx + \lim_{b\to\infty}\int_0^b \frac{x^3}{2+x^4}\,dx$$

$$= \lim_{a\to-\infty}\left(\frac{-1}{4}\ln(2+x^4)\right)\Big|_a^0 + \lim_{b\to\infty}\frac{1}{4}\ln(2+x^4)\Big|_0^b$$

$$= \lim_{a\to-\infty}\left[\frac{1}{4}\ln(2+a^4) - \frac{1}{4}\ln 2\right] + \lim_{b\to\infty}\left[\frac{1}{4}\ln(2+b^4) - \frac{1}{4}\ln 2\right]$$

Neither limit exists, so the area is infinite.

69. $$A = \int_{-\infty}^{\infty} \left|\frac{x^3}{(2+x^4)^2}\right| dx = \lim_{a\to-\infty}\int_a^0 \frac{-x^3}{(2+x^4)^2}\,dx + \lim_{b\to\infty}\int_0^b \frac{x^3}{(2+x^4)^2}\,dx$$

$$= \lim_{a\to-\infty}\frac{1}{4(2+x^4)}\Big|_a^0 + \lim_{b\to\infty}\frac{-1}{4(2+x^4)}\Big|_0^b$$

$$= \lim_{a\to-\infty}\left[\frac{1}{8} - \frac{1}{4(2+a^4)}\right] + \lim_{b\to\infty}\left[\frac{1}{8} - \frac{1}{4(2+b^4)}\right] = \frac{1}{4}$$

71. Let $c = m/(2kT)$. We are to evaluate $\int_0^\infty v^3 e^{-cv^2}\,dv$ and $\int_0^\infty v^2 e^{-cv^2}\,dv$. For the first integral we integrate by parts, letting $t = v^2$ and $du = ve^{-cv^2}\,dv$; then $dt = 2v\,dv$ and $u = -(1/2c)e^{-cv^2}$. Therefore

$$\int_0^b v^3 e^{-cv^2}\,dv = \int_0^b v^2(ve^{-cv^2})\,dv = v^2\left(-\frac{1}{2c}e^{-cv^2}\right)\Big|_0^b + \int_0^b \frac{v}{c}e^{-cv^2}\,dv$$

$$= \frac{-b^2}{2c} e^{-cb^2} - \frac{1}{2c^2} e^{-cv^2}\Big|_0^b = -\frac{b^2}{2c} e^{-cb^2} - \frac{1}{2c^2} e^{-cb^2} + \frac{1}{2c^2}.$$

Since $\lim_{b\to\infty} b^2 e^{-cb^2} = 0$ by l'Hôpital's Rule, and also $\lim_{b\to\infty} e^{-cb^2} = 0$, we conclude that

$$A = \int_0^\infty v^3 e^{-cv^2}\, dv = \lim_{b\to\infty} \int_0^b v^3 e^{-cv^2}\, dv = \lim_{b\to\infty} \left[-\frac{b^2}{2c} e^{-cb^2} - \frac{1}{2c^2} e^{-cb^2} + \frac{1}{2c^2} \right] = \frac{1}{2c^2}.$$

For the second integral, we use integration by parts, letting $t = v$ and $du = ve^{-cv^2}\, dv$; thus $dt = dv$ and $u = -(1/2c)e^{-cv^2}$. Therefore

$$\int_0^b v^2 e^{-cv^2}\, dv = \int_0^b v(ve^{-cv^2})\, dv = v\left(-\frac{1}{2c} e^{-cv^2}\right)\Big|_0^b + \int_0^b \frac{1}{2c} e^{-cv^2}\, dv = -\frac{b}{2c} e^{-cb^2} + \int_0^b \frac{1}{2c} e^{-cv^2}\, dv.$$

Since $\lim_{b\to\infty} be^{-cb^2} = 0$ by l'Hôpital's Rule, it follows that

$$B = \int_0^\infty v^2 e^{-cv^2}\, dv = \lim_{b\to\infty} \left(-\frac{b}{2c} e^{-cb^2} + \int_0^b \frac{1}{2c} e^{-cv^2}\, dv \right) = \lim_{b\to\infty} \frac{1}{2c} \int_0^b e^{-cv^2}\, dv = \frac{1}{2c} \int_0^\infty e^{-cv^2}\, dv.$$

Now let $w = \sqrt{2c}\, v$, so that $dw = \sqrt{2c}\, dv$. If $v = 0$, then $w = 0$; if v approaches ∞, then so does w. Therefore by the hint,

$$\frac{1}{2c} \int_0^\infty e^{-cv^2}\, dv = \frac{1}{2c} \int_0^\infty e^{-w^2/2} \frac{1}{\sqrt{2c}}\, dw = \frac{1}{(2c)^{3/2}} \int_0^\infty e^{-w^2/2}\, dw = \frac{1}{(2c)^{3/2}} \sqrt{\frac{\pi}{2}} = \frac{\sqrt{\pi}}{4c^{3/2}}.$$

As a result, $B = \sqrt{\pi}/(4c^{3/2})$, so that

$$\frac{A}{B} = \frac{1/(2c^2)}{\sqrt{\pi}/(4c^{3/2})} = \frac{2}{\sqrt{\pi c}} = \sqrt{\frac{8kT}{\pi m}}.$$

Cumulative Review(Chapters 1–7)

1. The conditions for applying l'Hôpital's Rule three times are met;

$$\lim_{x\to 0} \frac{\sin x - \sin(\sin x)}{x^3} = \lim_{x\to 0} \frac{\cos x - \cos(\sin x)\cos x}{3x^2} = \lim_{x\to 0} \cos x \lim_{x\to 0} \frac{1 - \cos(\sin x)}{3x^2}$$
$$= \lim_{x\to 0} \frac{1 - \cos(\sin x)}{3x^2} = \lim_{x\to 0} \frac{\sin(\sin x)\cos x}{6x} = \lim_{x\to 0} \frac{\sin(\sin x)}{6x} \lim_{x\to 0} \cos x$$
$$= \lim_{x\to 0} \frac{\sin(\sin x)}{6x} = \lim_{x\to 0} \frac{\cos(\sin x)\cos x}{6}.$$

Let $y = \sin x$, so that y approaches 0 as x approaches 0. Then

$$\lim_{x\to 0} \frac{\cos(\sin x)\cos x}{6} = \lim_{x\to 0} \cos(\sin x) \lim_{x\to 0} \frac{\cos x}{6} = \lim_{y\to 0} \cos y \lim_{x\to 0} \frac{\cos x}{6} = 1 \cdot \frac{1}{6} = \frac{1}{6}.$$

Thus the given limit equals $\frac{1}{6}$.

3. $\sqrt{cx+1}-\sqrt{x}=\left(\sqrt{cx+1}-\sqrt{x}\right)\dfrac{\left(\sqrt{cx+1}+\sqrt{x}\right)}{\sqrt{cx+1}+\sqrt{x}}=\dfrac{(c-1)x+1}{\sqrt{cx+1}+\sqrt{x}}$

so that if $c=1$, then

$$\lim_{x\to\infty}\left(\sqrt{cx+1}-\sqrt{x}\right)=\lim_{x\to\infty}\frac{1}{\sqrt{x+1}+\sqrt{x}}=0.$$

If $0\le c<1$, then

$$\lim_{x\to\infty}\left(\sqrt{cx+1}-\sqrt{x}\right)=\lim_{x\to\infty}\frac{(c-1)x+1}{\sqrt{cx+1}+\sqrt{x}}=-\infty$$

and if $c>1$, then

$$\lim_{x\to\infty}\left(\sqrt{cx+1}-\sqrt{x}\right)=\lim_{x\to\infty}\frac{(c-1)x+1}{\sqrt{cx+1}+\sqrt{x}}=\infty.$$

Finally, if $c<0$, then $\sqrt{cx+1}$ is not defined for $x>1/|c|$, so $\lim_{x\to\infty}\left(\sqrt{cx+1}-\sqrt{x}\right)$ is meaningless. Thus $\lim_{x\to\infty}\left(\sqrt{cx+1}-\sqrt{x}\right)$ exists only for $c=1$.

5. $f'(x)=\dfrac{2e^{2x}(e^x+1)-e^{2x}e^x}{(e^x+1)^2}=\dfrac{e^{3x}+2e^{2x}}{(e^x+1)^2}$

7. Differentiating the equation $a^2-a=2v^2-6v$ implicitly, we have

$$2a\frac{da}{dt}-\frac{da}{dt}=4v\frac{dv}{dt}-6\frac{dv}{dt},\quad\text{so that}\quad(2a-1)\frac{da}{dt}=(4v-6)\frac{dv}{dt}.$$

At the instant at which $da/dt=6(dv/dt)$, this becomes

$$(2a-1)6\frac{dv}{dt}=(4v-6)\frac{dv}{dt}.$$

Since $dv/dt=a>0$ by hypothesis, it follows that $(2a-1)6=4v-6$, so that $v=3a$.

9. $A(x)=\int_0^x(e^t-1)\,dt=(e^t-t)\big|_0^x=e^x-x-1$. Thus finding the value of $x>0$ for which $A(x)=1$ is equivalent to solving the equation $e^x-x-1=1$, or $e^x-x-2=0$. To that end we let $f(x)=e^x-x-2$, so that $f'(x)=e^x-1$, and apply the Newton-Raphson method with initial value of c equal to 1. By computer we obtain 1.146193221 as the desired approximate zero of f, and hence the approximate value of x for which $A(x)=1$.

11. $f'(x)=\dfrac{(1)(x-3)^2-2x(x-3)}{(x-3)^4}=-\dfrac{x+3}{(x-3)^3}$; $f''(x)=-\dfrac{(1)(x-3)^3-3(x+3)(x-3)^2}{(x-3)^6}=\dfrac{2(x+6)}{(x-3)^4}$;

relative minimum value is $f(-3)=-\frac{1}{12}$; increasing on $[-3,3)$ and decreasing on $(-\infty,-3]$ and $(3,\infty)$; concave upward on $(-6,3)$ and $(3,\infty)$, and concave downward on $(-\infty,-6)$; inflection point is $(-6,-\frac{2}{27})$; vertical asymptote is $x=3$; horizontal asymptote is $y=0$.

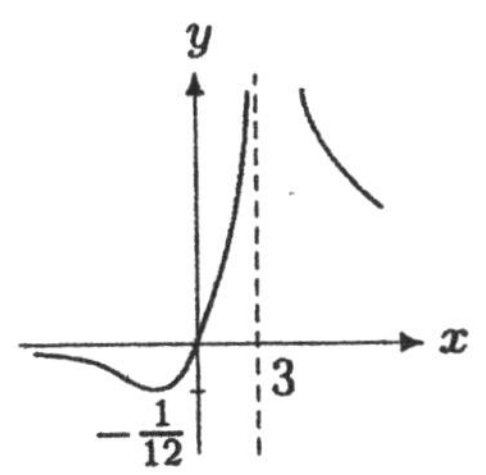

13. Using the notation in the figure, we find that the area A and circumference C are given by $A = \frac{1}{2}xy = \frac{1}{2}\sin\theta\cos\theta$ and $C = 1 + x + y = 1 + \sin\theta + \cos\theta$. Thus the ratio R is given by

$$R = \frac{C}{A} = \frac{1+\sin\theta+\cos\theta}{\frac{1}{2}\sin\theta\cos\theta} = 2(\csc\theta\sec\theta + \sec\theta + \csc\theta) \quad \text{for } 0 < \theta < \frac{\pi}{2}.$$

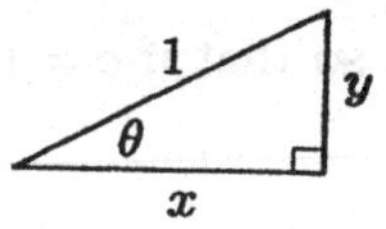

We are to minimize R. Notice that

$$R'(\theta) = 2(-\csc\theta\cot\theta\sec\theta + \csc\theta\sec\theta\tan\theta + \sec\theta\tan\theta - \csc\theta\cot\theta)$$

$$= 2\left(\frac{-1}{\sin^2\theta} + \frac{1}{\cos^2\theta} + \frac{\sin\theta}{\cos^2\theta} - \frac{\cos\theta}{\sin^2\theta}\right) = \frac{2(-\cos^2\theta + \sin^2\theta + \sin^3\theta - \cos^2\theta)}{\sin^2\theta\cos^2\theta}$$

$$= \frac{2(\sin^2\theta + \sin^3\theta - \cos^2\theta - \cos^3\theta)}{\sin^2\theta\cos^2\theta}.$$

If $\theta = \pi/4$, then $R'(\theta) = 0$ since $\sin(\pi/4) = \cos(\pi/4)$. Moreover, since $\sin\theta < \cos\theta$ on $(0, \pi/4)$ and $\sin\theta > \cos\theta$ on $(\pi/4, \pi/2)$, it follows that $R'(\theta) < 0$ for $0 < \theta < \pi/4$ and $R'(\theta) > 0$ for $\pi/4 < \theta < \pi/2$. Consequently (1) of Section 4.6 and the First Derivative Test imply that the ratio R is minimum for $\theta = \pi/4$, that is, when the right triangle is isosceles.

15. $\dfrac{2x+3}{x^2+4x-5} = \dfrac{2x+3}{(x+5)(x-1)} = \dfrac{A}{x+5} + \dfrac{B}{x-1}$; $2x+3 = A(x-1) + B(x+5)$;
$A + B = 2$ and $-A + 5B = 3$, so $A = \frac{7}{6}$ and $B = \frac{5}{6}$;

$$\int \frac{2x+3}{x^2+4x-5}\,dx = \int \left(\frac{7}{6}\frac{1}{x+5} + \frac{5}{6}\frac{1}{x-1}\right)dx = \frac{7}{6}\ln|x+5| + \frac{5}{6}\ln|x-1| + C = \frac{1}{6}\ln|x+5|^7|x-1|^5 + C$$

17. Notice that the graphs of $y = e^{2x}$ and $y = e^{-x}$ intersect at $x = 0$, and that $e^{2x} \geq e^{-x}$ for $0 \leq x \leq 1/2$. Thus

$$A = \int_0^{1/2} (e^{2x} - e^{-x})\,dx = \left(\frac{1}{2}e^{2x} + e^{-x}\right)\Big|_0^{1/2} = \frac{1}{2}e + e^{-1/2} - \frac{3}{2}$$

19. $$L = \int_{-27}^{-1} \sqrt{1 + (x^{-1/3})^2}\,dx = \int_{-27}^{-1} \sqrt{1 + \frac{1}{x^{2/3}}}\,dx = \int_{-27}^{-1} \sqrt{\frac{x^{2/3}+1}{x^{2/3}}}\,dx$$

$$= \int_{-27}^{-1} \frac{-1}{x^{1/3}}\sqrt{x^{2/3}+1}\,dx \overset{u=x^{2/3}+1}{=} -\int_{10}^{2} \frac{3}{2}\sqrt{u}\,du = -\frac{3}{2}\left(\frac{2}{3}u^{3/2}\right)\Big|_{10}^{2}$$

$$= 10\sqrt{10} - 2\sqrt{2}$$

21. Let $u = cx$, so that $du = c\,dx$. If $x = a$, then $u = ca$; if $x = b$, then $u = cb$. Therefore

$$\int_a^b f(cx)\,dx = \int_{ca}^{cb} f(u)\frac{1}{c}\,du = \frac{1}{c}\int_{ca}^{cb} f(x)\,dx.$$

Chapter 9

Sequences and Series

9.1 Polynomial Approximation

1. $f(x) = \sin x$, $f'(x) = \cos x$, $f''(x) = -\sin x$, and $f^{(3)}(x) = -\cos x$, so that $f(0) = 0$, $f'(0) = 1$, $f''(0) = 0$, and $f^{(3)}(0) = -1$. Then $p_0(x) = 0$, $p_1(x) = 0 + 1 \cdot x = x$, $p_2(x) = 0 + 1 \cdot x + (0/2!)x^2 = x$, and $p_3(x) = 0 + 1 \cdot x + (0/2!)x^2 - (1/3!)x^3 = x - x^3/3! = x - x^3/6$.

3. $f(x) = e^{-2x}$, $f'(x) = -2e^{-2x}$, $f''(x) = (-2)^2 e^{-2x}$, and $f^{(3)}(x) = (-2)^3 e^{-2x}$, so that $f(0) = 1$, $f'(0) = -2$, $f''(0) = (-2)^2 = 4$, and $f^{(3)}(0) = (-2)^3 = -8$. Then $p_0(x) = 1$, $p_1(x) = 1 - 2x$, $p_2(x) = 1 - 2x + (4/2!)x^2 = 1 - 2x + 2x^2$ and $p_3(x) = 1 - 2x + (4/2!)x^2 - (8/3!)x^3 = 1 - 2x + 2x^2 - \frac{4}{3}x^3$.

5. $f(x) = \sin(2x)$, $f'(x) = 2\cos(2x)$, $f''(x) = -(2^2)\sin(2x)$, and $f^{(3)}(x) = -(2^3)\cos(2x)$, so that $f(0) = 0$, $f'(0) = 2$, $f''(0) = 0$, and $f^{(3)}(0) = -2^3$. Then $p_0(x) = 0$, $p_1(x) = 0 + 2x = 2x$, $p_2(x) = 0 + 2x + (0/2!)x^2 = 2x$, and $p_3(x) = 0 + 2x + (0/2!)x^2 - (2^3/3!)x^3 = 2x - \frac{4}{3}x^3$.

7. $f(x) = x^2 - x - 2$, $f'(x) = 2x - 1$, $f''(x) = 2$, and $f^{(n)}(x) = 0$ for $n \geq 3$. Since $f(0) = -2$, $f'(0) = -1$, $f''(0) = 2$, and $f^{(n)}(0) = 0$ for $n \geq 3$, it follows that $p_n(x) = -2 - 1 \cdot x + (2/2!)x^2 = -2 - x + x^2$ for $n \geq 2$.

9. $f(x) = \dfrac{1}{1+x}$, $f'(x) = \dfrac{-1}{(1+x)^2}$, $f''(x) = \dfrac{2}{(1+x)^3}$, $f^{(3)}(x) = \dfrac{-6}{(1+x)^4}$,

   $f^{(k)}(x) = \dfrac{(-1)^k k!}{(1+x)^{k+1}}$; $f^{(k)}(0) = (-1)^k k!$;

   $$p_n(x) = 1 - x + \frac{2!}{2!}x^2 - \frac{3!}{3!}x^3 + \cdots + \frac{(-1)^n n!}{n!}x^n = 1 - x + x^2 - x^3 + \cdots + (-1)^n x^n$$

11. $f(x) = e^{-x}$, $f'(x) = -e^{-x}$, $f''(x) = e^{-x}$, $f^{(3)}(x) = -e^{-x}$, $f^{(k)}(x) = (-1)^k e^{-x}$; $f^{(k)}(0) = (-1)^k$

    $p_n(x) = 1 - x + \dfrac{x^2}{2!} - \dfrac{x^3}{3!} + \cdots + \dfrac{(-1)^n}{n!}x^n$

13. $f(x) = \cosh x$, $f'(x) = \sinh x$, $f''(x) = \cosh x$, $f^{(2k)}(x) = \cosh x$, $f^{(2k+1)}(x) = \sinh x$;

    $f^{(2k)}(0) = 1$, $f^{(2k+1)}(0) = 0$; $p_{2n+1}(x) = p_{2n}(x) = 1 + \dfrac{x^2}{2!} + \dfrac{x^4}{4!} + \cdots + \dfrac{x^{2n}}{(2n)!}$

15. $f(x) = \sin x$, $f'(x) = \cos x$, $f''(x) = -\sin x$, $f^{(3)}(x) = -\cos x$, $f^{(4)}(x) = \sin x$

$f^{(2k)}(x) = (-1)^k \sin x$, $f^{(2k+1)}(x) = (-1)^k \cos x$; $f^{(2k)}(0) = 0$, $f^{(2k+1)}(0) = (-1)^k$;

$$p_{2n+2}(x) = p_{2n+1}(x) = x - \frac{x^3}{3!} + \frac{x^5}{5!} - \frac{x^7}{7!} + \cdots + \frac{(-1)^n}{(2n+1)!}x^{2n+1}$$

17. $p_2(x) = 0$ for all x

19. $f(x) = e^{-(x^2)}$, $f'(x) = -2xe^{-(x^2)}$, $f''(x) = -2e^{-(x^2)} + 4x^2e^{-(x^2)}$,

$f^{(3)}(x) = 12xe^{-(x^2)} - 8x^3e^{-(x^2)}$; $f(0) = 1$, $f'(0) = 0$, $f''(0) = -2$, $f^{(3)}(0) = 0$;

$p_3(x) = 1 + (-2/2!)x^2 = 1 - x^2$

21. $f(x) = \ln(\cos x)$, $f'(x) = (-\sin x)/(\cos x) = -\tan x$, $f''(x) = -\sec^2 x$;

$f(0) = 0$, $f'(0) = 0$, $f''(0) = -1$; $p_2(x) = (-1/2!)x^2 = -\frac{1}{2}x^2$

23. $f(x) = \sec x$, $f'(x) = \sec x \tan x$, $f''(x) = \sec x \tan^2 x + \sec^3 x$, $f^{(3)}(x) = \sec x \tan^3 x + 5\sec^3 x \tan x$;

$f(0) = 1$, $f'(0) = 0$, $f''(0) = 1$, $f^{(3)}(0) = 0$; $p_3(x) = p_2(x) = (1/2!)x^2 = 1 + \frac{1}{2}x^2$

25. $f(x) = e^{-1/x^2}$ for $x \neq 0$, $f(0) = 0$; $f'(x) = (2/x^3)e^{-1/x^2}$ for $x \neq 0$; by l'Hôpital's Rule,

$$f'(0) = \lim_{x\to 0} \frac{f(x) - f(0)}{x - 0} = \lim_{x\to 0} \frac{e^{-1/x^2} - 0}{x - 0} = \lim_{x\to 0} \frac{1/x}{e^{1/x^2}} = \lim_{x\to 0} \frac{-1/x^2}{(-2/x^3)e^{1/x^2}} = \lim_{x\to 0} \frac{x}{2}e^{-1/x^2} = 0;$$

by l'Hôpital's Rule,

$$f''(0) = \lim_{x\to 0} \frac{f'(x) - f'(0)}{x - 0} = \lim_{x\to 0} \frac{(2/x^3)e^{-1/x^2} - 0}{x - 0} = \lim_{x\to 0} \frac{2/x^4}{e^{1/x^2}} = \lim_{x\to 0} \frac{-8/x^5}{(-2/x^3)e^{1/x^2}}$$

$$= \lim_{x\to 0} \frac{4/x^2}{e^{1/x^2}} = \lim_{x\to 0} \frac{-8/x^3}{(-2/x^3)e^{1/x^2}} = \lim_{x\to 0} 4e^{-1/x^2} = 0;$$

$p_2(x) = 0$.

27. a. $f(x) = \sqrt{1+x}$, $f'(x) = \dfrac{1}{2(1+x)^{1/2}}$, $f''(x) = -\dfrac{1}{4(1+x)^{3/2}}$;

$f(0) = 1$, $f'(0) = \frac{1}{2}$, $f''(0) = -\frac{1}{4}$; $p_2(x) = 1 + \frac{1}{2}x - \frac{1}{8}x^2$

b. $f(1) = \sqrt{2}$, and $p_2(1) = 1 + \frac{1}{2} - \frac{1}{8} = 1.375$ is the desired approximation to $\sqrt{2}$, whose value is 1.41421 (accurate to 6 digits).

c. $f(0.1) = \sqrt{1.1}$, and $p_2(0.1) = 1 + \frac{1}{2}(0.1) - \frac{1}{8}(0.01)^2 = 1.04875$ is the desired approximation to $\sqrt{1.1}$, whose value is 1.04881 (accurate to 6 digits).

9.2 Sequences

1. $\frac{1}{3}, \frac{1}{4}, \frac{1}{5}, \frac{1}{6}$

3. $0, \frac{1}{3}, \frac{1}{2}, \frac{3}{5}$

5. Let $\varepsilon > 0$, and let N be any integer. If $n \geq N$, then $|(-2) - (-2)| = 0 < \varepsilon$. Thus $\lim_{n\to\infty}(-2) = -2$.

7. Let $\varepsilon > 0$, and let N be any integer greater than $1/\varepsilon$. If $n \geq N$, then $1/n \leq 1/N$, so that

$$\left|\frac{3n+1}{n} - 3\right| = \left|3 + \frac{1}{n} - 3\right| = \frac{1}{n} \leq \frac{1}{N} < \varepsilon.$$

Thus $\lim_{n\to\infty}(3n+1)/n = 3$.

9. Let M be any number, and let N be any positive integer such that $N > M^2$. If $n \geq N$, then $\sqrt{n} \geq \sqrt{N} > \sqrt{M^2} = |M| \geq M$. Thus $\lim_{n\to\infty}\sqrt{n} = \infty$.

11. Let M be any number, and let N be any positive integer such that $N > \ln|M|$. If $k \geq N$, then $e^k \geq e^N > e^{\ln|M|} = |M| \geq M$, so that $\lim_{k\to\infty} e^k = \infty$.

13. Let $f(x) = \pi + 1/x$ for $x \geq 1$. Then $f(n) = \pi + 1/n$ for $n \geq 1$. Since $\lim_{x\to\infty}(\pi + 1/x) = \pi$, Theorem 9.4 implies that $\lim_{n\to\infty}(\pi + 1/n) = \pi$.

15. Since $0 < 0.8 < 1$, it follows from (4) with $r = 0.8$ that $\lim_{j\to\infty}(0.8)^j = 0$.

17. Since $e^{-n} = (1/e)^n$ and $0 < 1/e < 1$, it follows from (4) with $r = 1/e$ that $\lim_{n\to\infty} e^{-n} = 0$.

19. Let $f(x) = (x+3)/(x^2-2)$ for $x \geq 2$. Then $f(n) = (n+3)/(n^2-2)$ for $n \geq 2$. Since

$$\lim_{x\to\infty}\frac{x+3}{x^2-2} = \lim_{x\to\infty}\frac{1/x + 3/x^2}{1 - 2/x^2} = \frac{0+0}{1-0} = 0$$

Theorem 9.4 implies that $\lim_{n\to\infty}(n+3)/(n^2-2) = 0$.

21. Let $f(x) = (2x^2-4)/(-x-5)$ for $x \geq 1$. Then $f(n) = (2n^2-4)/(-n-5)$ for $n \geq 1$. Since

$$\lim_{x\to\infty}\frac{2x^2-4}{-x-5} = \lim_{x\to\infty}\frac{2x - 4/x}{-1 - 5/x} = -\infty$$

Theorem 9.4 implies that $\lim_{n\to\infty}(2n^2-4)/(-n-5) = -\infty$.

23. Let $f(x) = x\sin\pi/x$ for $x \geq 1$. Then $f(n) = n\sin\pi/n$ for $n \geq 1$. By l'Hôpital's Rule,

$$\lim_{x\to\infty} x\sin\frac{\pi}{x} = \lim_{x\to\infty}\frac{\sin\pi/x}{1/x} = \lim_{x\to\infty}\frac{(\cos\pi/x)(-\pi/x^2)}{-1/x^2} = \lim_{x\to\infty}\left(\pi\cos\frac{\pi}{x}\right) = \pi$$

so Theorem 9.4 implies that $\lim_{n\to\infty} n\sin\pi/n = \pi$.

25. Let $f(x) = (1+0.05/x)^x$ for $x \geq 1$. Then $f(n) = (1+0.05/n)^n$ for $n \geq 1$. Since $\ln f(x) = x\ln(1+0.05/x)$ for $x \geq 1$ and $\lim_{x\to\infty}\ln(1+0.05/x) = 0 = \lim_{x\to\infty} 1/x$, l'Hôpital's Rule implies that

$$\lim_{x\to\infty} \ln f(x) = \lim_{x\to\infty} x \ln\left(1 + \frac{0.05}{x}\right) = \lim_{x\to\infty} \frac{\ln(1+0.05/x)}{1/x}$$

$$= \lim_{x\to\infty} \frac{\dfrac{1}{1+0.05/x}(-0.05/x^2)}{-1/x^2} = \lim_{x\to\infty} \frac{0.05}{1+0.05/x} = 0.05.$$

Therefore $\lim_{x\to\infty} f(x) = e^{0.05}$, so that by Theorem 9.4, $\lim_{n\to\infty}(1+0.05/n)^n = e^{0.05}$.

27. Let $f(x) = (1+x)^{1/(2x)}$ for $x \geq 1$.Then $f(k) = (1+k)^{1/(2k)}$ for $k \geq 1$. Since $\ln f(x) = [1/(2x)]\ln(1+x)$ for $x \geq 1$, and $\lim_{x\to\infty} \ln(1+x) = \infty = \lim_{x\to\infty} 2x$, l'Hôpital's Rule implies that

$$\lim_{x\to\infty} \ln f(x) = \lim_{x\to\infty} \frac{\ln(1+x)}{2x} = \lim_{x\to\infty} \frac{1/(1+x)}{2} = \lim_{x\to\infty} \frac{1}{2(1+x)} = 0.$$

Therefore $\lim_{x\to\infty} f(x) = e^0 = 1$, so that by Theorem 9.4, $\lim_{k\to\infty}(1+k)^{1/(2k)} = 1$.

29. Let $f(x) = (1/\sqrt{2})\cos 1/x$ for $x \geq 1$. Then $f(k) = (1/\sqrt{2})\cos 1/k$ for $k \geq 1$. Since the cosine function is continuous at 0, we have $\lim_{x\to\infty}(1/\sqrt{2})\cos 1/x = (1/\sqrt{2})\cos 0 = 1/\sqrt{2}$. Next, let $g(x) = \sin^{-1} x$. Since the inverse sine function is continuous on $[-1, 1]$, it follows from the Substitution Theorem with $y = (1/\sqrt{2})\cos 1/x$ that $\lim_{x\to\infty} g(f(x)) = \lim_{x\to\infty} \sin^{-1}[(1/\sqrt{2})\cos 1/x] = \lim_{y\to 1/\sqrt{2}} \sin^{-1} y = \sin^{-1} 1/\sqrt{2} = \pi/4$. Therefore by Theorem 9.4, $\lim_{k\to\infty} \sin^{-1}[(1/\sqrt{2})\cos 1/k] = \pi/4$.

31. Observe that $\int_{-1/n}^{1/n} e^x\,dx = e^x\big|_{-1/n}^{1/n} = e^{1/n} - e^{-1/n}$. Since $\lim_{n\to\infty} e^{1/n} = 1$ by Exercise 18, we find that

$$\lim_{n\to\infty} \int_{-1/n}^{1/n} e^x\,dx = \lim_{n\to\infty}(e^{1/n} - e^{-1/n}) = \lim_{n\to\infty}\left(e^{1/n} - \frac{1}{e^{1/n}}\right) = 1 - \frac{1}{1} = 0.$$

33. Since $\lim_{n\to\infty}(-4n) = -\infty$, the sequence diverges.

35. Since $\lim_{n\to\infty} 1/(n^2-1) = 0$, the sequence converges to 0.

37. Since $\lim_{n\to\infty}(-\frac{1}{3})^n = 0$ by Example 7, the sequence converges to 0.

39. $(.25)^n < 10^{-5}$ if $n\log(.25) < -5$, that is, if $n > -5/\log(.25) \approx 8.3$. Thus $n = 9$ is the smallest positive integer such that $(.25)^n < 10^{-5}$.

41. Since $\sqrt[n]{1.2} > 1$, $|1 - \sqrt[n]{1.2}| < 10^{-3}$ if $1.2^{1/n} < 1.001$, that is, if $(1/n)\ln 1.2 < \ln 1.001$, or $n > (\ln 1.2)/(\ln 1.001) \approx 182.4$. Thus $n = 183$ is the smallest positive integer such that $|1 - \sqrt[n]{1.2}| < 10^{-3}$.

43. $\lim_{n\to\infty} a_n = \lim_{x\to 0^+}(1/x)\sinh(e^{-1/x})$, which appears to be 0.

45. Suppose $\lim_{n\to\infty} a_n = L_1$ and $\lim_{n\to\infty} a_n = L_2$. Then for any $\varepsilon > 0$ there are integers N_1 and N_2 such that if $n \geq N_1$, then $|a_n - L_1| < \varepsilon/2$, and if $n \geq N_2$, then $|a_n - L_2| < \varepsilon/2$. Let n be an integer greater than both N_1 and N_2. Then $|L_2 - L_1| = |(a_n - L_1) - (a_n - L_2)| \leq |a_n - L_1| + |a_n - L_2| < \varepsilon/2 + \varepsilon/2 = \varepsilon$. Since $|L_2 - L_1| < \varepsilon$ for any $\varepsilon > 0$, we have $L_2 = L_1$. Thus the limit of a convergent sequence is unique.

47. By hypothesis, $M > 0$ and $r > 0$. Suppose that $p_n > M$. Then $M - p_n < 0$ and hence

$$\frac{r(M - p_n)}{M} p_n < 0.$$

Therefore (8) implies that

$$p_{n+1} = p_n + \frac{r(M - p_n)}{M} p_n < p_n.$$

Similarly, if $p_n < M$, then $M - p_n > 0$ and hence

$$\frac{r(M - p_n)}{M} p_n > 0.$$

Thus $p_{n+1} > p_n$ by (8).

49. Let P_0, P_1, and P_2 denote the population in 1900, 1910, and 1920, respectively. Using (6) and the values of P_0 and P_1 given in Example 11, we have

$$r = \frac{P_1}{P_0} \approx \frac{92.2 \times 10^6}{76.2 \times 10^6} \approx 1.21.$$

By (7), $P_2 = r^2 P_0 \approx (1.21)^2(76.2 \times 10^6) \approx 112 \times 10^6$, which means that the population in 1920 would have been approximately 112 million. This is not as close to the actual 1920 census figure of 106,021,537 as the number obtained in the solution of Example 11. Thus the prediction is better when we use a Verhulst sequence.

51. In general, if the amount of money in the account at the beginning of a year is P, then the interest earned during the year is $(.01)rP$, so that the amount of money in the account at the end of the year is $P + (.01)rP = P(1 + (.01)r)$. Let a_n be the amount of money in the account at the end of the nth year. Then $a_1 = 1000(1 + (.01)r)$ and $a_n = a_{n-1}(1 + (.01)r)$ for $n > 1$. By induction it follows that $a_n = 1000(1 + (.01)r)^n$.

53. a. If the interest is compounded n times a year, then each interest period lasts $1/n$ years and the interest rate for each interest period is r/n percent. Thus the amount of money in the account at the end of $1/n$ years is $P(1 + (.01)r/n)$, the amount after $2/n$ years is $P(1 + (.01)r/n)(1 + (.01)r/n) = P(1 + (.01)r/n)^2$, the amount after $3/n$ years is $P(1 + (.01)r/n)^2(1 + (.01)r/n) = P(1 + (.01)r/n)^3$, and by induction the amount at the end of one year (after n interest periods) is $P(1 + (.01)r/n)^n$.

b. By Theorem 9.4,

$$\lim_{n\to\infty} P\left(1 + \frac{(.01)r}{n}\right)^n = \lim_{x\to\infty} P\left(1 + \frac{(.01)r}{x}\right)^x = P \lim_{x\to\infty}\left(1 + \frac{(.01)r}{x}\right)^x.$$

Let $f(x) = (1 + (.01)r/x)^x$. Then $\ln f(x) = x \ln(1 + (.01)r/x)$. By l'Hôpital's Rule,

$$\lim_{x\to\infty} \ln f(x) = \lim_{x\to\infty} \frac{\ln(1 + (.01)(r/x))}{1/x}$$

$$= \lim_{x\to\infty} \frac{\left(\dfrac{1}{1+(.01)r/x}\right)\left(\dfrac{-(.01)r}{x^2}\right)}{-1/x^2} = \lim_{x\to\infty} \frac{(.01)r}{1+(.01)r/x} = (.01)r.$$

Thus $\lim_{x\to\infty} f(x) = e^{(.01)r}$, so that $\lim_{n\to\infty} P(1+(.01)r/n)^n = Pe^{(.01)r}$.

c. If interest is compounded continuously, then by part (b) the amount after 1 year is $1000e^{.05} \approx 1051.27$. If interest is compounded quarterly, then by part (a) the amount after 1 year is $1000(1+(.01)5/4)^4 = 1000(1.0125)^4 \approx 1050.95$. Thus the difference is approximately 32 cents.

9.3 Convergence Properties of Sequences

1. Since $1 \le 1+2/n \le 1+2/1 = 3$ for $n \ge 1$, the sequence is bounded.

3. Since $n + 1/(2n) \ge n$ for $n \ge 1$, the sequence is unbounded.

5. Since $\cosh k = \frac{1}{2}(e^k + e^{-k}) \ge \frac{1}{2}e^k$ for $k \ge 10$, and since $\lim_{x\to\infty} \frac{1}{2}e^x = \infty$, the sequence is unbounded.

7. By (1), $\lim_{n\to\infty}(2+1/n) = \lim_{n\to\infty} 2 + \lim_{n\to\infty} 1/n = 2+0 = 2$.

9. By (1) and (2),

$$\lim_{n\to\infty}\left(\frac{1}{n} - \frac{1}{n+1}\right) = \lim_{n\to\infty}\frac{1}{n} + \lim_{n\to\infty}\frac{-1}{n+1} = \lim_{n\to\infty}\frac{1}{n} - \lim_{n\to\infty}\frac{1}{n+1} = 0 - 0 = 0.$$

11. By (2) and Example 7 of Section 9.2,

$$\lim_{n\to\infty}\frac{2^{n+1}}{5^{n+2}} = \lim_{n\to\infty}\frac{2\cdot 2^n}{5^2\,5^n} = \lim_{n\to\infty}\frac{2}{25}\left(\frac{2}{5}\right)^n = \frac{2}{25}\lim_{n\to\infty}\left(\frac{2}{5}\right)^n = \frac{2}{25}\cdot 0 = 0.$$

13. By (1), $\lim_{k\to\infty}(k+1)/k = \lim_{k\to\infty}(1+1/k) = \lim_{k\to\infty} 1 + \lim_{k\to\infty} 1/k = 1+0 = 1$.

15. By (1) – (4),

$$\lim_{n\to\infty}\frac{4n^3-5}{6n^3+3} = \lim_{n\to\infty}\frac{4-5/n^3}{6+3/n^3} = \frac{4-5(0)^3}{6+3(0)^3} = \frac{2}{3}.$$

17. By Examples 8 and 9 of Section 9.2, $\lim_{n\to\infty}\sqrt[n]{3} = 1$ and $\lim_{n\to\infty}\sqrt[n]{n} = 1$. Thus by (3) we have $\lim_{n\to\infty}\sqrt[n]{3n} = \lim_{n\to\infty}\sqrt[n]{3}\lim_{n\to\infty}\sqrt[n]{n} = 1\cdot 1 = 1$.

19. By (1), (2), and (4),

$$\lim_{n\to\infty}\frac{\sqrt{n}+1}{\sqrt{n}-1} = \lim_{n\to\infty}\frac{1+1/\sqrt{n}}{1-1/\sqrt{n}} = \frac{1+0}{1-0} = 1.$$

21. By (1) and (4),

$$\lim_{n\to\infty}\sqrt{n}(\sqrt{n+1}-\sqrt{n})=\lim_{n\to\infty}\sqrt{n}(\sqrt{n+1}-\sqrt{n})\frac{\sqrt{n+1}+\sqrt{n}}{\sqrt{n+1}+\sqrt{n}}$$

$$=\lim_{n\to\infty}\frac{\sqrt{n}\,[(n+1)-n]}{\sqrt{n+1}+\sqrt{n}}=\lim_{n\to\infty}\frac{\sqrt{n}}{\sqrt{n+1}+\sqrt{n}}=\lim_{n\to\infty}\frac{1}{\sqrt{1+1/n}+1}=\frac{1}{1+1}=\frac{1}{2}.$$

23. Observe that $-1/n\le(-1)^n/n\le 1/n$ for $n\ge 1$. Since $\lim_{n\to\infty}-1/n=0=\lim_{n\to\infty}1/n$, it follows from (5) that $\lim_{n\to\infty}(-1)^n/n=0$.

25. Since $\lim_{x\to\infty}(x/e^x)=\lim_{x\to\infty}(1/e^x)=0$ by l'Hôpital's Rule, it follows from Theorem 9.4 that $\lim_{n\to\infty}(n/e^n)=0$.

27. Since $0\le[\ln(1+1/n)]/n\le(\ln 2)/n$ for $n\ge 1$, and since $\lim_{n\to\infty}(\ln 2)/n=0$, it follows from (5) that $\lim_{n\to\infty}[\ln(1+1/n)]/n=0$.

29. Observe that $1\le(n+1/n)^{1/n}\le(2n)^{1/n}$, $\lim_{n\to\infty}1=1$, and

$$\lim_{n\to\infty}(2n)^{1/n}=\lim_{n\to\infty}\sqrt[n]{2}\lim_{n\to\infty}\sqrt[n]{n}=1.$$

Thus it follows from (5) that $\lim_{n\to\infty}(n+1/n)^{1/n}=1$.

31. Since $0<e^{-2n}<e^{-n}$ for $n\ge 1$, and since $\lim_{n\to\infty}e^{-n}=0=\lim_{n\to\infty}0$ by Exercise 17 of Section 9.2, it follows from (5) that $\lim_{n\to\infty}e^{-2n}=0$. It follows from (1), (2), and (4) that

$$\lim_{n\to\infty}\tanh n=\lim_{n\to\infty}\frac{1-e^{-2n}}{1+e^{-2n}}=\frac{1-0}{1+0}=1.$$

33. Conjecture: $L\approx 0.7390851343$

35. We will first show by induction that $1<a_n\le 2$ for all n. Since $a_0=2$, this is true for $n=0$. Observe that if $1<a_n\le 2$, then

$$\frac{1}{2}\le\frac{1}{a_n}<1,\quad\text{so that}\quad -1<-\frac{1}{a_n}\le-\frac{1}{2}\quad\text{and hence}\quad 1<a_{n+1}=2-\frac{1}{a_n}\le\frac{3}{2}\le 2.$$

Thus the sequence is bounded. Moreover, for any n we have $a_n>1$, so that $a_n^2-2a_n+1=(a_n-1)^2>0$, and thus $a_n^2>2a_n-1$. This means that $a_n>2-1/a_n=a_{n+1}$. Consequently the sequence is decreasing. By Theorem 9.6, the given sequence must converge to a limit L. We find that

$$L=\lim_{n\to\infty}a_n=\lim_{n\to\infty}a_{n+1}=\lim_{n\to\infty}\left(2-\frac{1}{a_n}\right)=2-\frac{1}{\lim_{n\to\infty}a_n}=2-\frac{1}{L}.$$

This means that $L=2-1/L$, so that $L^2-2L+1=0$, and thus $L=1$. As a result, $\lim_{n\to\infty}a_n=1$.

37. a. Let $\varepsilon > 0$. Since $\{b_n\}_{n=m}^{\infty}$ is bounded, there is an $M > 0$ such that $|b_n| \le M$ for all $n \ge m$. Since $\lim_{n\to\infty} a_n = 0$ there is an integer $N \ge m$ such that if $n \ge N$, then $|a_n| < \varepsilon/M$. Thus if $n \ge N$, then $|a_n b_n| = |a_n|\,|b_n| < (\varepsilon/M)M = \varepsilon$. Therefore $\lim_{n\to\infty} a_n b_n = 0$.

b. Let

$$a_n = |c_n| \quad \text{and} \quad b_n = \begin{cases} 1 & \text{if } c_n \ge 0 \\ -1 & \text{if } c_n < 0. \end{cases}$$

Then $\{b_n\}_{n=1}^{\infty}$ is bounded and $c_n = a_n b_n$. If $\lim_{n\to\infty} |c_n| = 0$, then $\lim_{n\to\infty} a_n = 0$, so by part (a), $\lim_{n\to\infty} c_n = \lim_{n\to\infty} a_n b_n = 0$.

c. (i) Let $a_n = 1/n$ and $b_n = \sin n$ for $n \ge 1$. Then $\lim_{n\to\infty} a_n = 0$ and $\{b_n\}_{n=1}^{\infty}$ is bounded. By part(a), $\lim_{n\to\infty}(\sin n)/n = \lim_{n\to\infty} a_n b_n = 0$.

(ii) Let $a_n = 1/n^2$ and $b_n = \ln(1+(-1)^n/n)$ for $n \ge 2$. Then $\lim_{n\to\infty} a_n = 0$ and $\ln \frac{2}{3} \le b_n \le \ln \frac{3}{2}$, so that $\{b_n\}_{n=2}^{\infty}$ is bounded. By part (a),

$$\lim_{n\to\infty} \frac{1}{n^2} \ln\left(1 + \frac{(-1)^n}{n}\right) = \lim_{n\to\infty} a_n b_n = 0.$$

(iii) Let $a_n = 1/e^n$ and $b_n = 2 + (-1)^n$ for $n \ge 1$. Then $\lim_{n\to\infty} a_n = 0$ and $1 \le b_n \le 3$, so that $\{b_n\}_{n=1}^{\infty}$ is bounded. By part (a), $\lim_{n\to\infty}[2 + (-1)^n]/e^n = \lim_{n\to\infty} a_n b_n = 0$.

(iv) Let $a_n = 1/e^n$ and $b_n = [2n + (-1)^n]/e^n$ for $n \ge 1$. Then $\lim_{n\to\infty} a_n = 0$ and $0 \le b_n \le [2n + (-1)^n]/2^n \le 2$, so that $\{b_n\}_{n=1}^{\infty}$ is bounded. By part (a), $\lim_{n\to\infty}[2n + (-1)^n]/e^{2n} = \lim_{n\to\infty} a_n b_n = 0$.

39. a. First, $a_1 = \sqrt{2} < 2$. Next, if $a_n < 2$, then $a_{n+1} = (\sqrt{2})^{a_n} < (\sqrt{2})^2 = 2$, so by induction, $a_n < 2$ for $n \ge 1$. Therefore $\{a_n\}_{n=1}^{\infty}$ is bounded. Next, we will show that $\{a_n\}_{n=1}^{\infty}$ is increasing. Since $a_n < 2$, it follows from Exercise 55(a) of Section 5.7 that $(\ln a_n)/a_n < (\ln 2)/2 = \ln\sqrt{2}$, so that $\ln a_n < a_n \ln\sqrt{2} = \ln((\sqrt{2})^{a_n}) = \ln a_{n+1}$. Since $\ln x$ is increasing, we conclude that $a_n < a_{n+1}$. By Theorem 9.6 the bounded, increasing sequence $\{a_n\}_{n=1}^{\infty}$ converges to a number L, and $L \le 2$ since $a_n < 2$ for $n \ge 1$.

b. By part (a), $\lim_{n\to\infty} a_n = L$, so that $\lim_{n\to\infty} a_{n+1} = L$. Thus $\ln L = \lim_{n\to\infty} \ln a_{n+1} = \lim_{n\to\infty} \ln((\sqrt{2})^{a_n}) = \lim_{n\to\infty}(a_n \ln\sqrt{2}) = (\ln\sqrt{2})\lim_{n\to\infty} a_n = (\ln\sqrt{2})L$. Therefore $(\ln L)/L = \ln\sqrt{2} = (\ln 2)/2$. Since $0 < L < e$ and $(\ln x)/x$ is increasing on $(0, e)$ by Exercise 55(a) in Section 5.7, it follows that $L = 2$.

9.4 Infinite Series

1. $s_4 = 1 + 1 + 1 + 1 = 4$

3. $s_4 = 1 + \frac{1}{3} + \frac{1}{9} + \frac{1}{27} = \frac{40}{27}$

5. $s_4 = \frac{1}{2} - \frac{1}{3} + \frac{1}{4} - \frac{1}{5} = \frac{13}{60}$

7. $\lim_{n\to\infty} \left(1 + \frac{1}{n}\right) = 1$; series diverges.

9. $\lim_{n\to\infty} (-1)^n \frac{1}{n^2} = 0$; series could converge.

11. $\lim_{n\to\infty} \sin\left(\frac{\pi}{2} - \frac{1}{n}\right) = \sin\frac{\pi}{2} = 1$; series diverges.

13. $\lim_{n\to\infty} n \sin\frac{1}{n} = \lim_{n\to\infty} \frac{\sin 1/n}{1/n} = 1$; series diverges.

15. $s_j = j$; $\lim_{j\to\infty} s_j = \infty$; series diverges.

17. $s_{2j} = 0$ and $s_{2j+1} = 1$ for $j \geq 1$; $\lim_{j\to\infty} s_j$ does not exist; series diverges.

19. $s_j = \left(\frac{1}{2} - \frac{1}{3}\right) + \left(\frac{1}{3} - \frac{1}{4}\right) + \cdots + \left(\frac{1}{j+1} - \frac{1}{j+2}\right) = \frac{1}{2} - \frac{1}{j+2}$;

$$\lim_{j\to\infty} s_j = \lim_{j\to\infty} \left(\frac{1}{2} - \frac{1}{j+2}\right) = \frac{1}{2};\ \sum_{n=1}^{\infty} \left(\frac{1}{n+1} - \frac{1}{n+2}\right) = \frac{1}{2}$$

21. $s_j = (1-8) + (8-27) + \cdots + (j^3 - (j+1)^3) = 1 - (j+1)^3$;

$$\lim_{j\to\infty} s_j = \lim_{j\to\infty} (1 - (j+1)^3) = -\infty;\ \text{series diverges}$$

23. $s_n = (a_1 - a_2) + (a_2 - a_3) + \cdots + (a_{n-1} - a_n) + (a_n - a_{n+1}) = a_1 - a_{n+1}$. Thus $\lim_{n\to\infty} s_n$ exists if and only if $\lim_{n\to\infty} a_{n+1}$, or equivalently, $\lim_{n\to\infty} a_n$, exists. If $\lim_{n\to\infty} a_n$ exists, then

$$\lim_{n\to\infty} s_n = \lim_{n\to\infty} (a_1 - a_{n+1}) = a_1 - \lim_{n\to\infty} a_{n+1} = a_1 - \lim_{n\to\infty} a_n.$$

25. Since the series is a geometric series with ratio $\frac{4}{7} < 1$, it converges;

$$\sum_{n=1}^{\infty} 5\left(\frac{4}{7}\right)^n = \frac{5(\frac{4}{7})}{1 - \frac{4}{7}} = \frac{20}{3}.$$

27. Since the series is a geometric series with ratio -0.3 and $|-0.3| < 1$, it converges;

$$\sum_{n=0}^{\infty} (-1)^n (0.3)^n = \sum_{n=0}^{\infty} (-0.3)^n = \frac{1}{1 - (-0.3)} = \frac{10}{13}.$$

29. Since the series is a geometric series with ratio $\frac{1}{2} < 1$, it converges:

$$\sum_{n=1}^{\infty} 5\left(\frac{1}{2}\right)^{n+1} = \sum_{n=1}^{\infty} \frac{5}{2}\left(\frac{1}{2}\right)^{n} = \frac{\frac{5}{2}(\frac{1}{2})}{1-\frac{1}{2}} = \frac{5}{2}.$$

31. Since the series is a geometric series with ratio $\frac{3}{5} < 1$, it converges;

$$\sum_{n=1}^{\infty} \frac{3^{n+3}}{5^{n-1}} = \sum_{n=1}^{\infty} (27)(5)\left(\frac{3}{5}\right)^{n} = \frac{135(\frac{3}{5})}{1-\frac{3}{5}} = \frac{405}{2}.$$

33. Since the series is a geometric series with ratio $-3/2$ and $|-3/2| > 1$, it diverges.

35. $\frac{1}{7} + \frac{1}{7^2} + \frac{1}{7^3} + \cdots = \frac{1}{7} + \frac{1}{7^2} + \frac{1}{7^3} + \sum_{n=4}^{\infty}\left(\frac{1}{7}\right)^n = \frac{57}{343} + \sum_{n=4}^{\infty}\left(\frac{1}{7}\right)^n$

37. $\sum_{n=1}^{\infty} \frac{1}{n^2+1} = \frac{1}{2} + \frac{1}{5} + \frac{1}{10} + \sum_{n=4}^{\infty} \frac{1}{n^2+1} = \frac{4}{5} + \sum_{n=4}^{\infty} \frac{1}{n^2+1}$

39. $\sum_{n=1}^{\infty} \frac{1}{n^2} = 1 + \frac{1}{4} + \frac{1}{9} + \sum_{n=4}^{\infty} \frac{1}{n^2} = \frac{49}{36} + \sum_{n=4}^{\infty} \frac{1}{n^2}$

41. $0.72727272\ldots = 72\left(\frac{1}{100}\right) + 72\left(\frac{1}{100}\right)^2 + 72\left(\frac{1}{100}\right)^3 + \cdots = \sum_{n=1}^{\infty} 72\left(\frac{1}{100}\right)^n = \frac{72(\frac{1}{100})}{1-\frac{1}{100}} = \frac{8}{11}$

43.
$$\begin{aligned} 0.232232232\ldots &= 232\left(\frac{1}{1000}\right) + 232\left(\frac{1}{1000}\right)^2 + 232\left(\frac{1}{1000}\right)^3 + \cdots \\ &= \sum_{n=1}^{\infty} 232\left(\frac{1}{1000}\right)^n = \frac{232(\frac{1}{1000})}{1-\frac{1}{1000}} = \frac{232}{999} \end{aligned}$$

45.
$$\begin{aligned} 27.56123123123\ldots &= 27.56 + 123\left(\frac{1}{10}\right)^5 + 123\left(\frac{1}{10}\right)^8 + 123\left(\frac{1}{10}\right)^{11} + \cdots \\ &= \frac{2756}{100} + \sum_{n=1}^{\infty} \frac{123}{100}\left(\frac{1}{1000}\right)^n = \frac{2756}{100} + \frac{\frac{123}{100}(\frac{1}{1000})}{1-\frac{1}{1000}} \\ &= \frac{2756}{100} + \frac{123}{99900} = \frac{917789}{33300} \end{aligned}$$

47. $0.86400000 = \frac{864}{1000} = \frac{108}{125}$

49. $\sum_{n=4}^{\infty} (-1)^{n+1}\frac{1}{n} = \sum_{n=1}^{\infty} (-1)^{n+1}\frac{1}{n} - \left(1 - \frac{1}{2} + \frac{1}{3}\right) = \ln 2 - \frac{5}{6}$

51. a. 11

b. 31

53. If $r > 0$, then either $r > 1$ or $1/r \geq 1$, so that either $\sum_{n=0}^{\infty}(1/r)^n$ or $\sum_{n=1}^{\infty} r^n$ diverges.

55. Let c_j denote the concentration of insulin just after the jth injection, for $j \geq 1$. Then $c_j = c_0 + c_0 e^{-bt_0} + c_0 e^{-2bt_0} + \cdots + c_0 e^{-(j-1)bt_0} = \sum_{n=0}^{j-1} c_0 e^{-nbt_0}$. Since the first term is c_0,

$$\lim_{j\to\infty} c_j = \sum_{n=0}^{\infty} c_0 e^{-nbt_0} = \frac{c_0}{1-e^{-bt_0}}.$$

57. If the height of the ball at its zenith is a, then until it hits the ground, the height is given by $h(t) = -4.9t^2 + a$. Thus $h(t) = 0$ if $0 = -4.9t^2 + a$, so that $t = a^{1/2}/(4.9)^{1/2}$. Therefore it takes $a^{1/2}/(4.9)^{1/2}$ seconds for the ball to hit the ground. Since the ball begins 1 meter above the surface, it takes $1/(4.9)^{1/2}$ seconds for it to drop to the surface. Between the first and second bounces the ball is in the air for $2(0.6)^{1/2}/(4.9)^{1/2}$ and likewise between the nth and $(n+1)$st bounces it takes $2(0.6)^{n/2}/(4.9)^{1/2}$ seconds. Thus the total time the ball is in the air is

$$\frac{1}{(4.9)^{1/2}} + \sum_{n=1}^{\infty} \frac{2(0.6)^{n/2}}{(4.9)^{1/2}} = \frac{1}{(4.9)^{1/2}} + \frac{2}{(4.9)^{1/2}} \sum_{n=1}^{\infty} [(0.6)^{1/2}]^n = \frac{1}{(4.9)^{1/2}} \left[1 + 2\frac{(0.6)^{1/2}}{1-(0.6)^{1/2}}\right] \approx 0.846.$$

Thus the total time the ball is in the air is approximately 0.846 seconds.

59. During the first month the person earns 2500 and spends $p2500$, so that $w_1 = 2500 - p2500 = 2500(1-p)$. If $n \geq 2$, then during the nth month the person earns 2500 and spends $p(2500 + w_{n-1})$. Thus

$$w_n = 2500 + w_{n-1} - p(2500 + w_{n-1}) = (2500 + w_{n-1})(1-p).$$

We will show by induction that $w_n = 2500(1-p) + 2500(1-p)^2 + \cdots + 2500(1-p)^n$. If

$$w_{n-1} = 2500(1-p) + 2500(1-p)^2 + \cdots + 2500(1-p)^{n-1},$$

then

$$w_n = (2500 + w_{n-1})(1-p) = 2500(1-p) + 2500(1-p)^2 + \cdots + 2500(1-p)^n$$

which is the nth partial sum of the geometric series $\sum_{n=1}^{\infty} 2500(1-p)^n$. Since $0 < p < 1$ by hypothesis, it follows that $0 < 1 - p < 1$, so that the series converges, and

$$\lim_{n\to\infty} w_n = \sum_{n=1}^{\infty} 2500(1-p)^n = \frac{2500(1-p)}{1-(1-p)} = \frac{2500(1-p)}{p}.$$

61. a. Initially there is one equilateral triangle whose sides have length 1. At each successive step in generating the Koch snowflake, each line segment of the nth step is replaced by 4 line segments, each having 1/3 the length of the line segment being replaced. Consequently after n steps there are $3 \cdot 4^n$ sides of length $1/3^n$. It follows that for the $(n+1)$st step there are $3 \cdot 4^n$ new equilateral triangles (one for each of the sides at the nth step). The side of such a triangle has length $1/3^{n+1}$, and thus the area of the triangle is given by

$$\frac{\sqrt{3}}{4}\left(\frac{1}{3^{n+1}}\right)^2 = \frac{\sqrt{3}}{4}\cdot\frac{1}{3^{2n+2}} = \frac{\sqrt{3}}{36}\cdot\frac{1}{3^{2n}}.$$

Therefore in the $(n+1)$st step the total amount of area of the $3\cdot 4^n$ new triangles is

$$3\cdot 4^n\left(\frac{\sqrt{3}}{36}\cdot\frac{1}{3^{2n}}\right) = \frac{\sqrt{3}}{12}\cdot\frac{4^n}{9^n} = \frac{\sqrt{3}}{12}\left(\frac{4}{9}\right)^n.$$

Consequently the area A of the Koch snowflake is given by

$$A = \frac{\sqrt{3}}{4} + \sum_{n=0}^{\infty}\frac{\sqrt{3}}{12}\left(\frac{4}{9}\right)^n = \frac{\sqrt{3}}{4} + \frac{\sqrt{3}}{12}\cdot\frac{1}{1-4/9} = \frac{\sqrt{3}}{4} + \frac{\sqrt{3}}{12}\cdot\frac{9}{5} = \frac{2}{5}\sqrt{3}.$$

b. As we noted in the solution of part (a), there are $3\cdot 4^n$ sides of length $1/3^n$ after n steps. Thus the length of the boundary after n steps is $(3\cdot 4^n)(1/3^n) = 3(\frac{4}{3})^n$. Since $\lim_{n\to\infty} 3(\frac{4}{3})^n = \infty$, the boundary of the snowflake has infinite length.

63. The distance run by Achilles is

$$100 + 10 + 1 + \frac{1}{10} + \cdots = \sum_{n=0}^{\infty} 100\left(\frac{1}{10}\right)^n = \frac{100}{1-\frac{1}{10}} = \frac{1000}{9}\text{ (yards)}.$$

The distance run by the tortoise during the same time is

$$10 + 1 + \frac{1}{10} + \frac{1}{100} + \cdots = \sum_{n=0}^{\infty} 10\left(\frac{1}{10}\right)^n = \frac{10}{1-\frac{1}{10}} = \frac{100}{9}\text{ (yards)}.$$

Since $\frac{1000}{9} - \frac{100}{9} = \frac{900}{9} = 100$, Achilles catches up with the tortoise after running $\frac{1000}{9}$ (yards).

65. a. By the end of the 1st minute the ant has covered 1 foot, which is $\frac{1}{3}$ of the band. The band is then stretched so it is 6 feet long. By the end of the 2nd minute the ant has covered another foot, which is $\frac{1}{6}$ of the stretched band. Thus at the end of 2 minutes the ant has covered altogether $\frac{1}{3}+\frac{1}{6}$ of the band. The band is then stretched so it is 9 feet long. In general, during the jth minute the band is $3j$ feet long, and until the ant reaches the end of the band it covers an additional foot during that minute, which means covering $1/(3j)$ of the stretched band. Thus at the end of the jth minute the ant has covered altogether $\frac{1}{3}+\frac{1}{6}+\frac{1}{9}+\cdots+1/(3j)$ of the band. So we need to find the smallest positive integer j so that $\frac{1}{3}+\frac{1}{6}+\frac{1}{9}+\cdots+1/(3j)\geq 1$, that is, $\frac{1}{3}(1+\frac{1}{2}+\frac{1}{3}+\cdots+1/j)\geq 1$, or $\sum_{n=1}^{j} 1/n \geq 3$. But by Exercise 51(a), this happens if $j = 11$. Thus the ant reaches the other end of the band, and it takes between 10 and 11 minutes to do so.

b. For $n \geq 1$ let a_n be the distance in feet from the ant to the starting point at the end of n minutes (just after the band has been stretched). During the first minute the ant crawls 1 foot and hence is $\frac{1}{20}$ of the way between the two ends. Thus after the band is stretched an additional 20 feet, the ant is still $\frac{1}{20}$ of the way between the two ends, so that $a_1 = \frac{1}{20}(40) = 2$. By an argument similar to that used in part (a), we have $a_n = [(a_{n-1}+1)/(20n)][20(n+1)]$, so that $a_n/[20(n+1)] = a_{n-1}/(20n) + 1/(20n)$. By induction it follows that $a_n/[20(n+1)] = \frac{1}{20}(1+\frac{1}{2}+\frac{1}{3}+\cdots+1/n)$. The ant will reach the other end of the band during the $(n+1)$st minute if n satisfies $a_n + 1 \geq 20(n+1)$, or $a_n/[20(n+1)] + 1/[20(n+1)] \geq 1$, or $\frac{1}{20}(1+\frac{1}{2}+\frac{1}{3}+\cdots+1/(n+1)) \geq 1$. By the comments following the solution of Example 4 the ant reaches the other end, and it takes between 272,400,599 and 272,400,600 minutes.

9.5 Positive Series: The Integral Test and the Comparison Tests

1. Since $1/(n+1)^2 \le 1/n^2$ for $n > 1$, and since $\sum_{n=1}^{\infty} 1/n^2$ is a convergent p series (with $p = 2$), $\sum_{n=1}^{\infty} 1/(n+1)^2$ converges by the Comparison Test.

3. Since $1/\sqrt{n^3+1} \le 1/n^{3/2}$ for $n \ge 1$, and since $\sum_{n=1}^{\infty} 1/n^{3/2}$ is a convergent p series (with $p = \frac{3}{2}$), $\sum_{n=1}^{\infty} 1/\sqrt{n^3+1}$ converges by the Comparison Test.

5. Since $1/\sqrt{n^2+1} \ge 1/\sqrt{n^2+2n+1} = 1/(n+1)$ for $n \ge 1$, and since $\sum_{n=1}^{\infty} 1/(n+1) = \sum_{n=2}^{\infty} 1/n$ diverges, so does $\sum_{n=1}^{\infty} 1/\sqrt{n^2+1}$ by the Comparison Test.

7. Since $1/e^{(n^2)} < 1/e^n = (1/e)^n$ for $n \ge 1$, and since $\sum_{n=1}^{\infty}(1/e)^n$ is a convergent geometric series, $\sum_{n=1}^{\infty} 1/e^{(n^2)}$ converges by the Comparison Test.

9. Since $1/(n-1)(n-2) \le 1/(n-2)^2$ for $n \ge 3$, and since $\sum_{n=3}^{\infty} 1/(n-2)^2 = \sum_{n=1}^{\infty} 1/n^2$ converges, so does $\sum_{n=3}^{\infty} 1/(n-1)(n-2)$ by the Comparison Test.

11. Since
$$\lim_{n\to\infty} \frac{(n^2-1)/(n^3-n-1)}{1/n} = \lim_{n\to\infty} \frac{n^3-n}{n^3-n-1} = \lim_{n\to\infty} \frac{1-1/n^2}{1-1/n^2-1/n^3} = 1$$
and since $\sum_{n=2}^{\infty} 1/n$ diverges, the Limit Comparison Test implies that $\sum_{n=2}^{\infty}(n^2-1)/(n^3-n-1)$ diverges.

13. Since $n/7^n < n/6^n = (n/2^n)(1/3^n) < (\frac{1}{3})^n$ for $n \ge 1$, and since $\sum_{n=1}^{\infty}(\frac{1}{3})^n$ is a convergent geometric series, the Comparison Test implies that $\sum_{n=1}^{\infty} n/7^n$ converges.

15. Since
$$\lim_{n\to\infty} \frac{\sqrt{n}/(n^2-3)}{1/n^{3/2}} = \lim_{n\to\infty} \frac{n^2}{n^2-3} = \lim_{n\to\infty} \frac{1}{1-3/n^2} = 1$$
and since $\sum_{n=4}^{\infty} 1/n^{3/2}$ is a convergent p series (with $p = \frac{3}{2}$), the Limit Comparison Test implies that $\sum_{n=4}^{\infty} \sqrt{n}/(n^2-3)$ converges.

17. Since
$$\lim_{n\to\infty} \frac{n/(n^3+1)^{3/7}}{1/n^{2/7}} = \lim_{n\to\infty} \frac{n^{9/7}}{(n^3+1)^{3/7}} = \lim_{n\to\infty} \frac{1}{(1+1/n^3)^{3/7}} = 1$$
and since $\sum_{n=1}^{\infty} 1/n^{2/7}$ is a divergent p series (with $p = \frac{2}{7}$), the Limit Comparison Test implies that $\sum_{n=1}^{\infty} n/(n^3+1)^{3/7}$ diverges.

19. Since
$$\lim_{n\to\infty} \frac{1/(n\sqrt{n^2-1})}{1/n^2} = \lim_{n\to\infty} \frac{n}{\sqrt{n^2-1}} = \lim_{n\to\infty} \frac{1}{\sqrt{1-1/n^2}} = 1$$
and since $\sum_{n=2}^{\infty} 1/n^2$ is a convergent p series (with $p = 2$), the Limit Comparison Test implies that $\sum_{n=2}^{\infty} 1/(n\sqrt{n^2-1})$ converges.

21. Since $0 \le (\tan^{-1} n)/(n^2+1) \le \pi/2n^2$ for $n \ge 1$, and since $\sum_{n=1}^{\infty} \pi/2n^2$ converges by Example 1, the Comparison Test implies that $\sum_{n=1}^{\infty} (\tan^{-1} n)/(n^2+1)$ converges.

23. Since $(\ln n)/n^2 \le 2\sqrt{n}/n^2 = 2/n^{3/2}$ for $n \ge 1$, and since $\sum_{n=1}^{\infty} 2/n^{3/2}$ converges by Example 1, the Comparison Test implies that $\sum_{n=1}^{\infty} (\ln n)/n^2$ converges.

25. Since

$$\lim_{n\to\infty} \frac{\sin 1/n}{1/n} = \lim_{x\to 0^+} \frac{\sin x}{x} = 1$$

and since $\sum_{n=1}^{\infty} 1/n$ diverges, the Limit Comparison Test implies that $\sum_{n=1}^{\infty} \sin 1/n$ diverges.

27. By (9) with $p = 3$, we have $E_j \le 1/(2j^2)$. Since $1/(2j^2) \le 0.02$ if $j^2 \ge 25$, we can let $j = 5$.

29. By (9) with $p = \frac{8}{7}$, we have

$$E_j \le \frac{1}{\frac{1}{7} j^{1/7}} = \frac{7}{j^{1/7}}.$$

Since $7/j^{1/7} \le 0.02 = 2/100$ if $j^{1/7} \ge 350$, or equivalently, $j \ge 350^7 \approx 6.43 \times 10^{17}$, we can let $j = 6.44 \times 10^{17}$.

31. a. By (6) and (7),

$$\sum_{n=1}^{j-1} a_n + \sum_{n=j+1}^{\infty} a_n \overset{(6)}{\le} \sum_{n=1}^{j-1} a_n + \int_j^{\infty} f(x)\,dx \overset{(7)}{\le} \sum_{n=1}^{j-1} a_n + \sum_{n=j}^{\infty} a_n = \sum_{n=1}^{\infty} a_n.$$

b. By part (a) and (6),

$$0 \overset{(a)}{\le} \sum_{n=1}^{\infty} a_n - \left(\sum_{n=1}^{j-1} a_n + \int_j^{\infty} f(x)\,dx\right) = \sum_{n=j}^{\infty} a_n - \int_j^{\infty} f(x)\,dx \le \sum_{n=j}^{\infty} a_n - \sum_{n=j+1}^{\infty} a_n = a_j.$$

33. Observe that if $p \ne 1$,

$$\int_2^{\infty} \frac{1}{x(\ln x)^p}\,dx = \lim_{b\to\infty} \int_2^b \frac{1}{x(\ln x)^p}\,dx = \lim_{b\to\infty} \left.\frac{-1}{(p-1)(\ln x)^{p-1}}\right|_2^b$$

$$= \lim_{b\to\infty} \left[\frac{1}{(p-1)(\ln 2)^{p-1}} - \frac{1}{(p-1)(\ln b)^{p-1}}\right].$$

Since

$$\lim_{b\to\infty} \frac{1}{(\ln b)^{p-1}} = \lim_{b\to\infty} (\ln b)^{1-p} = \begin{cases} \infty & \text{if } p < 1 \\ 0 & \text{if } p > 1 \end{cases}$$

we conclude from the Integral Test that $\sum_{n=2}^{\infty} 1/[n(\ln n)^p]$ converges provided that $p > 1$, and diverges for $p < 1$. For $p = 1$, the series diverges by Example 3.

35. Since $\sum_{n=1}^{\infty} a_n$ converges, we have $\lim_{n\to\infty} a_n = 0$. Thus there is an integer N such that $a_n < 1$ for $n \ge N$. Then $a_n^2 < a_n$ for $n \ge N$. Since $\sum_{n=N}^{\infty} a_n$ converges, the Comparison Test implies that $\sum_{n=N}^{\infty} a_n^2$ converges. Thus $\sum_{n=1}^{\infty} a_n^2$ converges also.

37. Taking $f(x) = 1/x$ and $a_n = 1/n$ in (2) and (3), we have

$$\frac{1}{2} + \frac{1}{3} + \cdots + \frac{1}{j} \leq \int_1^j \frac{1}{x}\,dx \leq 1 + \frac{1}{2} + \cdots + \frac{1}{j-1}.$$

Since $\int_1^j (1/x)\,dx = \ln j$, we have

$$\frac{1}{2} + \frac{1}{3} + \cdots + \frac{1}{j} \leq \ln j \leq 1 + \frac{1}{2} + \cdots + \frac{1}{j-1}$$

so that

$$\frac{1}{j} + \ln j \leq 1 + \frac{1}{2} + \cdots + \frac{1}{j} \leq 1 + \ln j.$$

Thus

$$\frac{1}{j \ln j} + 1 \leq \frac{\sum_{n=1}^{j} 1/n}{\ln j} \leq \frac{1}{\ln j} + 1.$$

Since

$$\lim_{j\to\infty} \left(\frac{1}{j\ln j} + 1\right) = 1 = \lim_{j\to\infty} \left(\frac{1}{\ln j} + 1\right),$$

(5) of Section 9.3 implies that

$$\lim_{j\to\infty} \frac{\sum_{n=1}^{j} 1/n}{\ln j} = 1.$$

39. a. Let $n \geq 2$. For the nth floor the circumference is $2\pi[10/(n \ln n)] = 20\pi/(n \ln n)$. Thus the total circumference of all the floors except the first is $\sum_{n=2}^{\infty} 20\pi/(n \ln n) = 20\pi \sum_{n=2}^{\infty} 1/(n \ln n)$. By Example 3, the series diverges. Therefore the total circumference is infinite.

b. For the nth floor the area is $\pi[10/(n \ln n)]^2 = 100\pi/[n^2(\ln n)^2]$. Therefore the total area of all the floors except the first is $\sum_{n=2}^{\infty} 100\pi/[n^2(\ln n)^2]$. Since $1/[n^2(\ln n)^2] \leq 1/n^2$ for $n \geq 3$, and since $\sum_{n=2}^{\infty} 1/n^2$ converges, it follows from the Comparison Test that $\sum_{n=3}^{\infty} 1/[n^2(\ln n)^2]$, and hence $\sum_{n=2}^{\infty} 100\pi/[n^2(\ln n)^2]$, converge. Consequently the total area of all the floors is finite.

9.6 Positive Series: The Ratio Test and the Root Test

1. $\displaystyle\lim_{n\to\infty} \frac{(n+1)!/2^{n+1}}{n!/2^n} = \lim_{n\to\infty} \frac{n+1}{2} = \infty$; the series diverges.

3. $\displaystyle\lim_{n\to\infty} \frac{(n+1)!\,3^{n+1}/10^{n+1}}{n!\,3^n/10^n} = \lim_{n\to\infty} \frac{3(n+1)}{10} = \infty$; the series diverges.

5. $\displaystyle\lim_{n\to\infty} \sqrt[n]{\left(\frac{n}{2n+5}\right)^n} = \lim_{n\to\infty} \frac{n}{2n+5} = \frac{1}{2} < 1$; the series converges.

7. $\displaystyle\lim_{n\to\infty} \frac{(2n+2)!/[(n+1)!]^2}{(2n)!/(n!)^2} = \lim_{n\to\infty} \frac{(2n+2)(2n+1)}{(n+1)^2} = 4 > 1$; the series diverges.

9. $\lim_{n\to\infty} \sqrt[n]{n^{100}e^{-n}} = \lim_{n\to\infty} (\sqrt[n]{n})^{100} e^{-1} = e^{-1} < 1$; the series converges.

11. $\lim_{n\to\infty} \frac{(n+1)^{1.7}/(1.7)^{n+1}}{n^{1.7}/(1.7)^n} = \lim_{n\to\infty} \frac{(n+1)^{1.7}}{(1.7)n^{1.7}} = \lim_{n\to\infty} \frac{1}{(1.7)}\left(1+\frac{1}{n}\right)^{1.7} = \frac{1}{1.7} < 1$; the series converges.

13. $\lim_{n\to\infty} \frac{(n+1)!/e^{n+1}}{n!/e^n} = \lim_{n\to\infty} \frac{n+1}{e} = \infty$; the series diverges.

15. $\lim_{n\to\infty} \sqrt[n]{\frac{1}{(\ln n)^n}} = \lim_{n\to\infty} \frac{1}{\ln n} = 0$; the series converges.

17. Since

$$\frac{1\cdot 3\cdot 5\cdots(2n-1)}{2\cdot 4\cdot 6\cdots(2n)} = \frac{3}{2}\cdot\frac{5}{4}\cdots\frac{2n-1}{2n-2}\cdot\frac{1}{2n} > \frac{1}{2n}$$

and since $\sum_{n=1}^{\infty} 1/2n$ diverges, the given series diverges by the Comparison Test.

19. $$\lim_{n\to\infty} \frac{\dfrac{(2n+2)!}{(n+1)!\,(2n+2)^{n+1}}}{\dfrac{(2n)!}{n!\,(2n)^n}} = \lim_{n\to\infty} \frac{(2n+2)(2n+1)(2n)^n}{(n+1)(2n+2)^{n+1}} = \lim_{n\to\infty}\left(\frac{2n+1}{n+1}\right)\left(\frac{n}{n+1}\right)^n$$

$$= \lim_{n\to\infty} \frac{2+1/n}{(1+1/n)(1+1/n)^n} = \frac{2}{e} < 1;$$

the series converges.

21. Since $\sin 1/n! \le 1/n!$ and $\cos 1/n! \ge \cos 1$ for $n \ge 1$, we have

$$\frac{\sin 1/n!}{\cos 1/n!} \le \frac{1}{n!\,\cos 1}.$$

It follows from Example 1 that $\sum_{n=1}^{\infty} 1/(n!\,\cos 1)$ converges. By the Comparison Test

$$\sum_{n=1}^{\infty} \frac{\sin(1/n!)}{\cos(1/n!)}$$

converges.

23. Let $b_n = (n/(2n+1))^n$ for $n \ge 1$. Then

$$\lim_{n\to\infty} \sqrt[n]{b_n} = \lim_{n\to\infty} \frac{n}{2n+1} = \lim_{n\to\infty} \frac{1}{2+1/n} = \frac{1}{2} < 1.$$

Thus $\sum_{n=1}^{\infty} b_n$ converges by the Root Test. Since $0 \le a_n \le b_n$ for $n \ge 1$, it follows from the Comparison Test that $\sum_{n=1}^{\infty} a_n$ converges.

25. a. Let n be even. Then

$$\frac{a_{n+1}}{a_n} = \frac{1/(n+1)^{n+1}}{1/(2n)^{2n}} = \frac{(2n)^{2n}}{(n+1)^{n+1}} = \frac{(2n)^{n+1}(2n)^{n-1}}{(n+1)^{n+1}} = \left(\frac{2n}{n+1}\right)^{n+1}(2n)^{n-1} \ge (2n)^{n-1}$$

which grows without bound as n increases. Next, let n be odd. Then

$$\frac{a_{n+1}}{a_n} = \frac{1/(2n+2)^{2n+2}}{1/n^n} = \frac{n^n}{(2n+2)^{2n+2}} = \frac{n^n}{(2n+2)^n(2n+2)^{n+2}}$$
$$= \left(\frac{n}{2n+2}\right)^n \frac{1}{(2n+2)^{n+2}} \leq \frac{1}{(2n+2)^{n+2}}$$

which decreases and approaches 0 as n increases. Thus $\lim_{n\to\infty} a_{n+1}/a_n$ does not exist, so the Ratio Test is inconclusive.

b. For odd n, $\sqrt[n]{a_n} = 1/n$; for even n, $\sqrt[n]{a_n} = 1/(2n)^2$. Thus $\lim_{n\to\infty} \sqrt[n]{a_n} = 0$. By the Root Test, the series converges.

27. a. Let $n = 20$. Then

$$\frac{2^n}{n!} = \frac{2^{20}}{20!} = \frac{\overbrace{2\cdot 2\cdots 2}^{\text{20 of these}}}{(20)(19)\cdots(2)(1)} < 1 = \left(\frac{2}{21}\right)^{n-20}.$$

Let $n > 20$. Then

$$\frac{2^n}{n!} < \left(\frac{2}{21}\right)^{n-20} \quad \text{is equivalent to} \quad \frac{1}{n!} < \frac{2^{n-20}}{2^n(21)^{n-20}} = \frac{1}{2^{20}(21)^{n-20}}.$$

Since $2^{20} < 20!$ and $(21)^{n-20} < n(n-1)(n-2)\cdots(22)(21)$, it follows that

$$\frac{1}{2^{20}(21)^{n-20}} > \frac{1}{(20!)[n(n-1)(n-2)\cdots(22)(21)]} = \frac{1}{n!}.$$

b. By part (a),

$$E_{20} = \sum_{n=21}^{\infty} \frac{2^n}{n!} < \sum_{n=21}^{\infty} \left(\frac{2}{21}\right)^{n-20} = \sum_{n=1}^{\infty} \left(\frac{2}{21}\right)^n = \frac{\frac{2}{21}}{1-\frac{2}{21}} = \frac{2}{19}.$$

29. a. $$\sqrt[n]{a_n} = \begin{cases} 0 & \text{for } n \text{ even} \\ \dfrac{n}{2n+1} & \text{for } n \text{ odd} \end{cases}$$

Since $n/(2n+1) \geq \frac{1}{3}$ for $n \geq 1$, it follows that $\lim_{n\to\infty} \sqrt[n]{a_n}$ does not exist. The solution of Exercise 23 shows that $\sum_{n=1}^{\infty} a_n$ converges.

b. Let $a_n = n^n$ for $n \geq 1$. Then $\sqrt[n]{a_n} = n$, so that $\lim_{n\to\infty} \sqrt[n]{a_n}$ does not exist. Since $\lim_{n\to\infty} a_n = \lim_{n\to\infty} n^n = \infty$, $\sum_{n=1}^{\infty}$ does not converge.

9.7 Alternating Series and Absolute Convergence

1. Since $\{1/(2n+1)\}_{n=1}^{\infty}$ is a decreasing, positive sequence with $\lim_{n\to\infty} 1/(2n+1) = 0$, the series converges.

3. Since $\lim_{n\to\infty}(2n+1)/(5n+1) = \frac{2}{5} \neq 0$, the nth term $(-1)^n(2n+1)/(5n+1)$ does not converge to 0, so the series diverges.

5. Let $f(x) = (x+2)/(x^2+3x+5)$. Then $f'(x) = (-x^2-4x-1)/(x^2+3x+5)^2 < 0$ for $x > 0$, so f is decreasing on $[1,\infty)$. Thus $\{(n+2)/(n^2+3n+5)\}_{n=1}^{\infty}$ is a positive, decreasing sequence, and $\lim_{n\to\infty}(n+2)/(n^2+3n+5) = 0$. Therefore the series converges.

7. Let $f(x) = (\ln x)/x$. Since $f'(x) = (1-\ln x)/x^2 < 0$ for $x > e$, the function f is decreasing on $[e,\infty)$. Thus $\{(\ln n)/n\}_{n=3}^{\infty}$ is a positive, decreasing sequence, and $\lim_{n\to\infty}(\ln n)/n = 0$. Therefore $\sum_{n=3}^{\infty}(-1)^n[(\ln n)/n]$ converges, and hence $\sum_{n=1}^{\infty}(-1)^n[(\ln n)/n]$ converges.

9. Let $f(x) = (\ln x)^p/x$. Then

$$f'(x) = \frac{p(\ln x)^{p-1} - (\ln x)^p}{x^2} = \frac{(\ln x)^{p-1}(p - \ln x)}{x^2} < 0 \quad \text{for } x > e^p.$$

Moreover, by l'Hôpital's Rule,

$$\lim_{x\to\infty}\frac{(\ln x)^p}{x} = \lim_{x\to\infty}\frac{p(\ln x)^{p-1}}{x} = \cdots = \lim_{x\to\infty}\frac{p!}{x} = 0.$$

Thus $\{(\ln n)^p/n\}_{n=3^p}^{\infty}$ is a positive, decreasing sequence with $\lim_{n\to\infty}(\ln n)^p/n = 0$. Thus

$$\sum_{n=3^p}^{\infty}(-1)^n\frac{(\ln n)^p}{n}$$

converges, and hence $\sum_{n=1}^{\infty}(-1)^n[(\ln n)^p/n]$ converges.

11. Since

$$\lim_{n\to\infty}\left|\frac{(-1)^{n+2}(n+1)!/100^{n+1}}{(-1)^{n+1}n!/100^n}\right| = \lim_{n\to\infty}\frac{n+1}{100} = \infty,$$

the series diverges by the Generalized Ratio Test.

13. Since $\lim_{n\to\infty} n^2/(2n+1) = \infty$, the nth term $(-1)^{n+1}n^2/(2n+1)$ does not converge to 0, so the series diverges.

15. For any $p > 0$, $\{1/n^p\}_{n=1}^{\infty}$ is a positive, decreasing sequence with $\lim_{n\to\infty} 1/n^p = 0$. Thus

$$\sum_{n=1}^{\infty}(-1)^n\frac{1}{n^p}$$

converges for any $p > 0$.

17. $E_{10} < a_{11} = \dfrac{1}{121}$

19. $E_{11} < \dfrac{1}{\sqrt{15}}$

21. Since $\frac{1}{n^3} < \frac{1}{1000}$ if $n \geq 11$, it follows from (1) that

$$\sum_{n=1}^{10}(-1)^{n+1}\frac{1}{n^3} \approx 0.9011164764$$

approximates the sum of the series with an error less than 0.001.

23. $\{1/(3n+4)\}_{n=1}^{\infty}$ is a positive, decreasing sequence and $\lim_{n\to\infty} 1/(3n+4) = 0$. Thus

$$\sum_{n=1}^{\infty}(-1)^{n+1}\frac{1}{3n+4}$$

converges. But notice that $\sum_{n=1}^{\infty} 1/(3n+4)$ diverges because $1/(3n+1) \geq 1/7n$ for $n \geq 1$ and $\sum_{n=1}^{\infty} 1/7n$ diverges. Thus $\sum_{n=1}^{\infty}(-1)^{n+1}[1/(3n+4)]$ converges conditionally.

25. $\lim_{n\to\infty} \dfrac{(n+1)^{n+1}/(n+1)!}{n^n/n!} = \lim_{n\to\infty}\left(\dfrac{n+1}{n}\right)^n = \lim_{n\to\infty}\left(1+\dfrac{1}{n}\right)^n = e > 1$; the series diverges.

27. Since $\lim_{n\to\infty} 1/n^{1/n} = 1$, the nth term $(-1)^{n+1}/n^{1/n}$ does not converge to 0, so the series diverges.

29. Since

$$\int_2^{\infty}\frac{1}{x(\ln x)^2}\,dx = \lim_{b\to\infty}\int_2^b \frac{1}{x(\ln x)^2}\,dx = \lim_{b\to\infty}\left(\frac{-1}{\ln x}\right)\bigg|_2^b = \lim_{b\to\infty}\left(\frac{-1}{\ln b}+\frac{1}{\ln 2}\right) = \frac{1}{\ln 2}$$

the Integral Test implies that $\sum_{n=2}^{\infty} 1/[n(\ln n)^2]$ converges. Thus $\sum_{n=2}^{\infty}(-1)^{n+1}\left\{1/[n(\ln n)^2]\right\}$ converges absolutely.

31. $\lim_{n\to\infty}\sqrt[n]{\left(\dfrac{1}{\ln n}\right)^n} = \lim_{n\to\infty}\dfrac{1}{\ln n} = 0$ so $\sum_{n=2}^{\infty}(-1)^{n+1}[1/(\ln n)^n]$ converges absolutely.

33. $$\lim_{n\to\infty}\frac{\dfrac{1\cdot 3\cdot 5\cdots(2n+3)}{2\cdot 5\cdot 8\cdots(3n+5)}}{\dfrac{1\cdot 3\cdot 5\cdots(2n+1)}{2\cdot 5\cdot 8\cdots(3n+2)}} = \lim_{n\to\infty}\frac{2n+3}{3n+5} = \lim_{n\to\infty}\frac{2+\dfrac{3}{n}}{3+\dfrac{5}{n}} = \frac{2}{3}$$

The series converges absolutely.

35. $\lim_{n\to\infty}\dfrac{(n+2)^2/(n+1)!}{(n+1)^2/n!} = \lim_{n\to\infty}\dfrac{(n+2)^2}{(n+1)^3} = 0$, so $\lim_{n\to\infty}\dfrac{(n+1)^2}{n!} = 0$.

37. $\lim_{n\to\infty}\sqrt[n]{\dfrac{x^{2n}}{n}} = \lim_{n\to\infty}\dfrac{x^2}{\sqrt[n]{n}} = x^2$

If $|x| < 1$, then $x^2 < 1$, so $\lim_{n\to\infty} x^{2n}/n = 0$.

39. The series $\sum_{n=1}^{\infty}(-1)^n(1/\sqrt{n})$ converges (conditionally), but $\left((-1)^n(1/\sqrt{n})\right)^2 = 1/n$ and $\sum_{n=1}^{\infty} 1/n$ diverges.

41. Since $\lim_{n\to\infty} a_n = a \neq 0$, there is an integer N such that if $n \geq N$, then $|a_n - a| < |a|/2$, so that $|a|/2 < |a_n| < 3|a|/2$, and thus $2/(3|a|) < 1/|a_n| < 2/|a|$. Notice that $|1/a_{n+1} - 1/a_n| = |(a_{n+1} - a_n)/a_{n+1}a_n|$, and that for $n \geq N$, we have

$$\frac{4}{9a^2}|a_{n+1} - a_n| < \left|\frac{a_{n+1} - a_n}{a_{n+1}a_n}\right| < \frac{4}{a^2}|a_{n+1} - a_n|.$$

Since $\sum_{n=N}^{\infty}(4/(9a^2))|a_{n+1} - a_n|$, $\sum_{n=N}^{\infty}(4/a^2)|a_{n+1} - a_n|$, and $\sum_{n=N}^{\infty}|a_{n+1} - a_n|$ all converge or all diverge, the same is true of $\sum_{n=1}^{\infty}|a_{n+1} - a_n|$ and $\sum_{n=1}^{\infty}|1/a_{n+1} - 1/a_n|$, and thus both converge or both diverge.

9.8 Power Series

1. $\displaystyle\lim_{n\to\infty}\sqrt[n]{\frac{|x|^n}{n^2}} = \lim_{n\to\infty}\frac{|x|}{(\sqrt[n]{n})^2} = |x|;$

the series converges for $|x| < 1$ and diverges for $|x| > 1$. Since $\sum_{n=1}^{\infty} 1/n^2$ converges and hence $\sum_{n=1}^{\infty}(-1)^n/n^2$ converges, the interval of convergence is $[-1, 1]$.

3. $\displaystyle\lim_{n\to\infty}\sqrt[n]{\frac{1}{\sqrt{n}\,3^n}|x|^n} = \lim_{n\to\infty}\frac{|x|}{3(\sqrt[n]{n})^{1/2}} = \frac{|x|}{3};$

the series converges for $|x| < 3$ and diverges for $|x| > 3$. Since $\sum_{n=1}^{\infty} 1/\sqrt{n}$ diverges and $\sum_{n=1}^{\infty}(-1)^n/\sqrt{n}$ converges, the interval of convergence is $[-3, 3)$.

5. $\displaystyle\lim_{n\to\infty}\frac{|x|^{2n+2}/(n+2)}{|x|^{2n}/(n+1)} = \lim_{n\to\infty}\frac{n+1}{n+2}x^2 = x^2;$

the series converges for $|x| < 1$ and diverges for $|x| > 1$. Since $\sum_{n=0}^{\infty}(-1)^n/(n+1)$ converges, the interval of convergence is $[-1, 1]$.

7. $\displaystyle\lim_{n\to\infty}\frac{|x|^{n+2}/(2n+1)}{|x|^{n+1}/(2n-1)} = \lim_{n\to\infty}\frac{2n-1}{2n+1}|x| = |x|;$

the series converges for $|x| < 1$ and diverges for $|x| > 1$. The alternating series

$$\sum_{n=1}^{\infty}\frac{(-1)^n}{2n-1} = \sum_{n=1}^{\infty}\frac{(-1)^n 1^{n+1}}{2n-1}$$

converges, and

$$\sum_{n=1}^{\infty}\frac{-1}{2n-1} = \sum_{n=1}^{\infty}\frac{(-1)^n(-1)^{n+1}}{2n-1}$$

diverges, so the interval of convergence is $(-1, 1]$.

9. $\lim_{n\to\infty} \sqrt[n]{\frac{|x|^n}{n^n}} = \lim_{n\to\infty} \frac{|x|}{n} = 0$; the interval of convergence is $(-\infty, \infty)$.

11. $\lim_{n\to\infty} \frac{(n+1)!\,|x|^{n+1}/(2n+2)!}{n!\,|x|^n/(2n)!} = \lim_{n\to\infty} \frac{n+1}{(2n+1)(2n+2)}|x| = 0$; the interval of convergence is $(-\infty, \infty)$.

13. $\lim_{n\to\infty} \frac{2^{n+1}|x|^{n+2}/(n+1)3^{n+3}}{2^n|x|^{n+1}/n3^{n+2}} = \lim_{n\to\infty} \frac{2n}{3(n+1)}|x| = \frac{2}{3}|x|$;
the series converges for $|x| < \frac{3}{2}$ and diverges for $|x| > \frac{3}{2}$. Since $\sum_{n=1}^{\infty} 1/(6n)$ diverges and $\sum_{n=1}^{\infty}(-1)^{n+1}/(6n)$ converges, the interval of convergence is $[-\frac{3}{2}, \frac{3}{2})$.

15. By l'Hôpital's Rule,

$$\lim_{x\to\infty} \frac{\ln(x+1)}{\ln x} = \lim_{x\to\infty} \frac{1/(x+1)}{1/x} = \lim_{x\to\infty} \frac{x}{x+1} = 1$$

so we have

$$\lim_{n\to\infty} \frac{|x|^{n+1}\ln(n+1)}{|x|^n \ln n} = \lim_{n\to\infty} \frac{\ln(n+1)}{\ln n}|x| = |x|$$

the series converges for $|x| < 1$ and diverges for $|x| > 1$. Since $\sum_{n=2}^{\infty} \ln n$ and $\sum_{n=2}^{\infty}(-1)^n \ln n$ both diverge, the interval of convergence is $(-1, 1)$.

17. By l'Hôpital's Rule, $\lim_{x\to\infty} \ln(x+1)/\ln x = 1$, so

$$\lim_{n\to\infty} \frac{|x|^{n+1}\ln(n+1)/(n+1)^2}{|x|^n(\ln n)/n^2} = \lim_{n\to\infty} \left(\frac{n}{n+1}\right)^2 \frac{\ln(n+1)}{\ln n}|x| = |x|$$

the series converges for $|x| < 1$ and diverges for $|x| > 1$. By the solution of Exercise 23 of Section 9.5, $\sum_{n=2}^{\infty}(\ln n)/n^2$ converges and hence $\sum_{n=2}^{\infty}(-1)^n(\ln n)/n^2$ converges. The interval of convergence is $[-1, 1]$.

19. $\lim_{n\to\infty} \frac{|x|^{(n+1)^2}}{|x|^{n^2}} = \lim_{n\to\infty} |x|^{2n+1}$, and $\lim_{n\to\infty} |x|^{2n+1} = 0$ if $|x| < 1$ whereas $\lim_{n\to\infty} |x|^{2n+1} = \infty$ if $|x| > 1$.
Thus the series converges for $|x| < 1$ and diverges for $|x| > 1$. Since $\sum_{n=0}^{\infty} 1^{(n^2)}$ and $\sum_{n=0}^{\infty}(-1)^{(n^2)}$ both diverge, the interval of convergence is $(-1, 1)$.

21. $\lim_{n\to\infty} \frac{|x|^{n+1}(n+1)!/(n+1)^{n+1}}{|x|^n n!/n^n} = \lim_{n\to\infty} \frac{(n+1)!}{n!}\frac{n^n}{(n+1)^{n+1}}|x| = \lim_{n\to\infty} \frac{1}{(1+1/n)^n}|x| = \frac{|x|}{e}$, so $R = e$.

23. $\lim_{n\to\infty} \frac{1^2 \cdot 3^2 \cdot 5^2 \cdots (2n+1)^2|x|^{2n+2}/[2^2 \cdot 4^2 \cdot 6^2 \cdots (2n)^2(2n+2)^2]}{1^2 \cdot 3^2 \cdot 5^2 \cdots (2n-1)^2|x|^{2n}/[2^2 \cdot 4^2 \cdot 6^2 \cdots (2n)^2]} = \lim_{n\to\infty} \frac{(2n+1)^2}{(2n+2)^2}x^2 = x^2$, so $R = 1$.

25. $\lim_{n\to\infty} \frac{1 \cdot 3 \cdot 5 \cdots (2n-1)(2n+1)|x|^{n+1}/\{2^{n+1}[1 \cdot 4 \cdot 7 \cdots (3n-2)(3n+1)]\}}{1 \cdot 3 \cdot 5 \cdots (2n-1)|x|^n/\{2^n[1 \cdot 4 \cdot 7 \cdots (3n-2)]\}} = \lim_{n\to\infty} \frac{2n+1}{2(3n+1)}|x| = \frac{|x|}{3}$,
so $R = 3$.

27. $f(x) = \sum_{n=1}^{\infty}(n+1)x^n$ for x in $(-1,1)$; $f'(x) = \sum_{n=1}^{\infty} n(n+1)x^{n-1}$ for x in $(-1,1)$;

$$\int_0^x f(t)\,dt = \sum_{n=1}^{\infty} x^{n+1} \text{ for } x \text{ in } (-1,1).$$

29. $f(x) = \sum_{n=1}^{\infty} \frac{5}{n} x^{(n^2)}$ for x in $[-1,1)$; $f'(x) = \sum_{n=1}^{\infty} 5nx^{n^2-1}$ for x in $(-1,1)$;

$$\int_0^x f(t)\,dt = \sum_{n=1}^{\infty} \frac{5}{n(n^2+1)} x^{n^2+1} \text{ for } x \text{ in } (-1,1).$$

31. Since $\sin x = \sum_{n=0}^{\infty}(-1)^n x^{2n+1}/(2n+1)!$, it follows that

$$\sin x^2 = \sum_{n=0}^{\infty} \frac{(-1)^n (x^2)^{2n+1}}{(2n+1)!} = \sum_{n=0}^{\infty} \frac{(-1)^n x^{4n+2}}{(2n+1)!}.$$

By the Integration Theorem,

$$\int_0^2 \sin x^2\,dx = \int_0^2 \sum_{n=0}^{\infty} \frac{(-1)^n x^{4n+2}}{(2n+1)!}\,dx = \sum_{n=0}^{\infty} \int_0^2 \frac{(-1)^n x^{4n+2}}{(2n+1)!}\,dx$$

$$= \sum_{n=0}^{\infty} \left(\frac{(-1)^n x^{4n+3}}{(4n+3)(2n+1)!} \Big|_0^2 \right) = \sum_{n=0}^{\infty} \frac{(-1)^n 2^{4n+3}}{(4n+3)(2n+1)!}.$$

By the Alternating Series Test,

$$\left| \sum_{n=m+1}^{\infty} \frac{(-1)^n 2^{4n+3}}{(4n+3)(2n+1)!} \right| < 10^{-2} \quad \text{if} \quad \frac{2^{4(m+1)+3}}{[4(m+1)+3][2(m+1)+1]!} < 10^{-2}$$

which occurs for $m = 4$. Thus

$$\sum_{n=0}^{4} \frac{(-1)^n 2^{4n+3}}{(4n+3)(2n+1)!} = \frac{2^3}{3(1!)} - \frac{2^7}{7(3!)} + \frac{2^{11}}{11(5!)} - \frac{2^{15}}{15(7!)} + \frac{2^{19}}{19(9!)} \approx 0.8131655739$$

is the desired approximation.

33. Since $e^x = \sum_{n=0}^{\infty} x^n/n!$ by (3), it follows that

$$\frac{1-e^{-x}}{x} = \frac{1 - \sum_{n=0}^{\infty}(-1)^n x^n/n!}{x} = \sum_{n=1}^{\infty} \frac{(-1)^{n+1} x^{n-1}}{n!} \quad \text{for } x \neq 0.$$

By the Integration Theorem,

$$\int_0^1 \frac{1-e^{-x}}{x}\,dx = \lim_{c\to 0^+} \int_c^1 \frac{1-e^{-x}}{x}\,dx = \lim_{c\to 0^+} \int_c^1 \sum_{n=1}^{\infty} \frac{(-1)^{n+1} x^{n-1}}{n!}\,dx$$

$$= \lim_{c\to 0^+} \sum_{n=1}^{\infty} \int_c^1 \frac{(-1)^{n+1}x^{n-1}}{n!}\,dx = \lim_{c\to 0^+} \sum_{n=1}^{\infty} \left(\frac{(-1)^{n+1}x^n}{n(n!)} \Big|_c^1 \right)$$

$$= \lim_{c\to 0^+} \left(\sum_{n=1}^{\infty} \frac{(-1)^{n+1}}{n(n!)} - \sum_{n=1}^{\infty} \frac{(-1)^{n+1}c^n}{n(n!)} \right) = \sum_{n=1}^{\infty} \frac{(-1)^{n+1}}{n(n!)}.$$

By the Alternating Series Test,

$$\left| \sum_{n=m+1}^{\infty} \frac{(-1)^{n+1}}{n(n!)} \right| < 10^{-3} \quad \text{if} \quad \frac{1}{(m+1)[(m+1)!]} < 10^{-3}$$

which occurs for for $m = 5$. Thus

$$\sum_{n=1}^{5} \frac{(-1)^{n+1}}{n(n!)} = 1 - \frac{1}{2(2!)} + \frac{1}{3(3!)} - \frac{1}{4(4!)} + \frac{1}{5(5!)} \approx 0.7968055556$$

is the desired approximation.

35. From (1) we have $1/(1-x) = \sum_{n=0}^{\infty} x^n$ for $|x| < 1$, so that

$$\frac{x^2}{1+x} = \sum_{n=0}^{\infty} x^2(-x)^n = \sum_{n=0}^{\infty} (-1)^n x^{n+2} \quad \text{for } |x| < 1.$$

By the Integration Theorem,

$$\int_0^{1/2} \frac{x^2}{1+x}\,dx = \int_0^{1/2} \sum_{n=0}^{\infty} (-1)^n x^{n+2}\,dx$$

$$= \sum_{n=0}^{\infty} \int_0^{1/2} (-1)^n x^{n+2}\,dx = \sum_{n=0}^{\infty} \left(\frac{(-1)^n x^{n+3}}{n+3} \right)\Big|_0^{1/2} = \sum_{n=0}^{\infty} \frac{(-1)^n}{2^{n+3}(n+3)}.$$

By the Alternating Series Test,

$$\left| \sum_{n=m+1}^{\infty} \frac{(-1)^n}{2^{n+3}(n+3)} \right| < 10^{-3} \quad \text{if} \quad \frac{1}{2^{(m+1)+3}[(m+1)+3]} < 10^{-3}$$

which occurs for $m = 4$. Thus

$$\sum_{n=0}^{4} \frac{(-1)^n}{2^{n+3}(n+3)} = \frac{1}{2^3(3)} - \frac{1}{2^4(4)} + \frac{1}{2^5(5)} - \frac{1}{2^6(6)} + \frac{1}{2^7(7)} \approx 0.0308035714$$

is the desired approximation.

37. From (3) we have $e^x = \sum_{n=0}^{\infty} x^n/n!$, so that $e^{(x^2)} = \sum_{n=0}^{\infty} (x^2)^n/n! = \sum_{n=0}^{\infty} x^{2n}/n!$. By the Integration Theorem,

$$\int_{-1}^{0} e^{(x^2)}\,dx = \int_{-1}^{0} \sum_{n=0}^{\infty} \frac{x^{2n}}{n!}\,dx = \sum_{n=0}^{\infty} \int_{-1}^{0} \frac{x^{2n}}{n!}\,dx$$

$$= \sum_{n=0}^{\infty}\left(\frac{x^{2n+1}}{(2n+1)n!}\Bigg|_{-1}^{0}\right) = \sum_{n=0}^{\infty}\frac{-(-1)^{2n+1}}{(2n+1)n!} = \sum_{n=0}^{\infty}\frac{1}{(2n+1)n!}.$$

If $a_n = 1/(2n+1)n!$ for $n \geq 0$, then $a_{n+1}/a_n \leq \frac{1}{3}$, so $\sum_{n=m+1}^{\infty} a_n \leq \sum_{n=m+1}^{\infty} 1/3^n$. But

$$\sum_{n=m+1}^{\infty}\frac{1}{3^n} = \frac{(\frac{1}{3})^{m+1}}{1-\frac{1}{3}} = \frac{1}{3^m \cdot 2}$$

by the Geometric Series Theorem, and $1/(3^m \cdot 2) < 10^{-3}$, if $m = 6$. Thus

$$\sum_{n=0}^{6}\frac{1}{(2n+1)n!} = 1 + \frac{1}{3(1)} + \frac{1}{5(2!)} + \frac{1}{7(3!)} + \frac{1}{9(4!)} + \frac{1}{11(5!)} + \frac{1}{13(6!)} \approx 1.4626369$$

is the desired approximation.

39. Since $\sum_{n=0}^{\infty} x^n = 1/(1-x)$ for $|x| < 1$, the Differentiation Theorem implies that $\sum_{n=1}^{\infty} nx^{n-1} = 1/(1-x)^2$ for $|x| < 1$. Then $\sum_{n=1}^{\infty} nx^n = x\left(\sum_{n=1}^{\infty} nx^{n-1}\right) = x/(1-x)^2$ for $|x| < 1$.

41. Since $e^x = \sum_{n=0}^{\infty} x^n/n!$, we have

$$\frac{e^x - 1}{x} = \frac{1}{x}\sum_{n=1}^{\infty}\frac{x^n}{n!} = \sum_{n=1}^{\infty}\frac{x^{n-1}}{n!} = \sum_{n=0}^{\infty}\frac{x^n}{(n+1)!} \quad \text{for all } x \neq 0.$$

The latter series converges for all x and defines a continuous function of x, with domain $(-\infty, \infty)$. Thus

$$\lim_{x\to 0}\frac{e^x - 1}{x} = \lim_{x\to 0}\left(\sum_{n=0}^{\infty}\frac{x^n}{(n+1)!}\right) = 1 + \sum_{n=1}^{\infty}\frac{0^n}{(n+1)!} = 1.$$

43. From (8) we have $\ln(1+x) = \sum_{n=1}^{\infty}((-1)^{n-1}/n)x^n$ for $|x| < 1$, so that

$$\frac{\ln(1+x)}{x} = \sum_{n=1}^{\infty}\frac{(-1)^{n-1}}{n}x^{n-1} \quad \text{for } |x| < 1.$$

Thus

$$\lim_{x\to 0}\frac{\ln(1+x)}{x} = \lim_{x\to 0}\sum_{n=1}^{\infty}\frac{(-1)^{n-1}}{n}x^{n-1} = 1 + \sum_{n=2}^{\infty}\frac{(-1)^{n-1}}{n}0^{n-1} = 1.$$

45. a. $\ln\dfrac{1}{1-x} = -\ln(1-x) = -\displaystyle\sum_{n=1}^{\infty}\frac{(-1)^{n-1}}{n}(-x)^n = \sum_{n=1}^{\infty}\frac{x^n}{n}$ for $|x| < 1$

b. $\ln 2 = \ln\dfrac{1}{1-\frac{1}{2}} = \displaystyle\sum_{n=1}^{\infty}\frac{1}{n}\left(\frac{1}{2}\right)^n = \sum_{n=1}^{\infty}\frac{1}{n2^n}$

c. Since

$$\left|\ln 2 - \sum_{n=1}^{N-1}\frac{1}{n2^n}\right| = \sum_{n=N}^{\infty}\frac{1}{n2^n} \leq \sum_{n=N}^{\infty}\frac{1}{N2^n} = \frac{1}{N}\frac{1/2^N}{1-\frac{1}{2}} = \frac{1}{N2^{N-1}}$$

we need to find N such that $1/(N2^{N-1}) < 0.01$. But if $N = 6$, then $1/N2^{N-1} = 1/(6 \cdot 2^5) < 0.01$, so $\sum_{n=1}^{5} 1/(n2^n) = \frac{1}{2} + \frac{1}{8} + \frac{1}{24} + \frac{1}{64} + \frac{1}{160} \approx 0.6885416667$ is the desired estimate of $\ln 2$.

47. $4\left|\dfrac{(-1)^4(\frac{1}{5})^9}{9}\right| = \dfrac{4}{9}\left(\dfrac{1}{5}\right)^9 < 2.276 \times 10^{-7}$ and $\left|\dfrac{(-1)^1(\frac{1}{239})^3}{3}\right| = \dfrac{1}{3}\left(\dfrac{1}{239}\right)^3 < 0.245 \times 10^{-7}$

Thus the error introduced in approximating $\pi/4$ is less that $2.276 \times 10^{-7} + 0.245 \times 10^{-7} = 2.521 \times 10^{-7}$.

49. a. From (1),

$$\frac{1}{1+t^4} = \frac{1}{1-(-t^4)} = \sum_{n=0}^{\infty}(-t^4)^n = \sum_{n=0}^{\infty}(-1)^n t^{4n}.$$

Then $t^2/(1+t^4) = \sum_{n=0}^{\infty}(-1)^n t^{4n+2}$ for $|t| < 1$.

b. $$\int_0^{1/2} \frac{t^2}{1+t^4}\,dt = \int_0^{1/2}\left(\sum_{n=0}^{\infty}(-1)^n t^{4n+2}\right)dt = \sum_{n=0}^{\infty}(-1)^n\left(\int_0^{1/2} t^{4n+2}\,dt\right)$$

$$= \sum_{n=0}^{\infty}(-1)^n\left(\left.\frac{t^{4n+3}}{4n+3}\right|_0^{1/2}\right) = \sum_{n=0}^{\infty}\frac{(-1)^n}{4n+3}\left(\frac{1}{2}\right)^{4n+3}$$

51. a. By (3), $e^x = \sum_{n=0}^{\infty}\dfrac{1}{n!}x^n$, so $xe^x = \sum_{n=0}^{\infty}\dfrac{1}{n!}x^{n+1}$.

b. On the one hand, by integration by parts with $u = x$, $dv = e^x\,dx$, we obtain

$$\int_0^1 xe^x\,dx = xe^x\Big|_0^1 - \int_0^1 e^x\,dx = e - e^x\Big|_0^1 = e - e + 1 = 1.$$

On the other hand, using the power series expansion of (a), we obtain

$$\int_0^1 xe^x\,dx = \int_0^1\left(\sum_{n=0}^{\infty}\frac{1}{n!}x^{n+1}\right)dx = \sum_{n=0}^{\infty}\frac{1}{n!}\left(\int_0^1 x^{n+1}\,dx\right) = \sum_{n=0}^{\infty}\frac{1}{n!}\left(\left.\frac{x^{n+2}}{n+2}\right|_0^1\right) = \sum_{n=0}^{\infty}\frac{1}{n!\,(n+2)}.$$

Thus $\sum_{n=0}^{\infty} 1/n!\,(n+2) = 1$.

53. We will find a positive integer n such that

$$A_n = 4\left|\frac{(-1)^{n+1}}{2(n+1)+1}\left(\frac{1}{5}\right)^{2(n+1)+1}\right| = \frac{4}{2n+3}\left(\frac{1}{5^{2n+3}}\right) < 10^{-11}$$

and a possibly different positive integer m such that

$$B_m = \left|\frac{(-1)^{m+1}}{2(m+1)+1}\left(\frac{1}{239}\right)^{2(m+1)+1}\right| = \frac{1}{2m+3}\left(\frac{1}{239^{2m+3}}\right) < 10^{-11}.$$

We have $A_6 \approx 8.738133333 \times 10^{-12}$ and $B_1 \approx 2.56472314 \times 10^{-13}$. Since $n = 6$, we find that

$$4\tan^{-1}\frac{1}{5} \approx 4\sum_{k=0}^{6}\frac{(-1)^{k+1}}{2k+1}\left(\frac{1}{5}\right)^{2k+1} \approx 4\left(\frac{1}{5} - \frac{1}{3}\frac{1}{5^3} + \frac{1}{5}\frac{1}{5^5} - + \cdots + \frac{1}{13}\frac{1}{5^{13}}\right) \approx 0.7895822394$$

and since $m = 1$, we find that

$$\tan^{-1}\frac{1}{239} = \frac{1}{239} - \frac{1}{3}\left(\frac{1}{239}\right)^3 \approx 0.004184076.$$

Thus

$$\frac{\pi}{4} = 4\tan^{-1}\frac{1}{5} - \tan^{-1}\frac{1}{239} \approx 0.7853981634$$

which is accurate to within $10^{-11} + 10^{-11} = 2 \times 10^{-11}$. Consequently $\pi = 4(\pi/4) \approx 3.1415926536$, with an error at most $4(2 \times 10^{-11}) < 10^{-10}$.

55. a. $\sum_{n=0}^{\infty} c_n x^n = f(x) = f(-x) = \sum_{n=0}^{\infty} c_n(-x)^n = \sum_{n=0}^{\infty}(-1)^n c_n x^n$
By Corollary 9.27, $c_n = (-1)^n c_n$, so that $c_n = 0$ for n odd.

b. $\sum_{n=0}^{\infty} c_n x^n = f(x) = -f(-x) = -\sum_{n=0}^{\infty} c_n(-x)^n = \sum_{n=0}^{\infty}(-1)^{n+1} c_n x^n$
By Corollary 9.27, $c_n = (-1)^{n+1} c_n$, so that $c_n = 0$ for n even.

57. Consider $f(x) = \frac{2}{9}[1/(1-x)]$ for $0 < x < 1$, so that $f'(x) = \frac{2}{9}[1/(1-x)^2]$. We also have $f(x) = \frac{2}{9}\sum_{n=0}^{\infty} x^n$, so by the Differentiation Theorem, $f'(x) = \frac{2}{9}\sum_{n=1}^{\infty} nx^{n-1}$. Letting $x = \frac{7}{9}$ we find that

$$\frac{2}{9}\sum_{n=1}^{\infty} n\left(\frac{7}{9}\right)^{n-1} = f'\left(\frac{7}{9}\right) = \frac{2}{9}\frac{1}{(1-\frac{7}{9})^2} = 4.5.$$

Thus the expected value of N is 4.5 (that is, between 4 and 5) rolls.

9.9 Taylor Series

1. $f(x) = 4x^2 - 2x + 1$, $f'(x) = 8x - 2$, $f''(x) = 8$, and $f^{(k)}(x) = 0$ for $k \geq 3$; $f(-3) = 43$, $f'(-3) = -26$, $f''(-3) = 8$, and $f^{(k)}(-3) = 0$ for $k \geq 3$. Thus the Taylor series about -3 is given by

$$43 - 26(x+3) + 8 \cdot \left(\frac{1}{2!}\right)(x+3)^2 = 43 - 26(x+3) + 4(x+3)^2.$$

3. $f^{(k)}(x) = e^x$ and $f^{(k)}(2) = e^2$ for $k \geq 0$. Thus the Taylor series about 2 is given by

$$e^2 + e^2(x-2) + e^2 \cdot \frac{1}{2!}(x-2)^2 + e^2 \cdot \frac{1}{3!}(x-2)^3 + \cdots = \sum_{n=0}^{\infty}\frac{e^2}{n!}(x-2)^n.$$

5. $f(x) = \ln x$, $f^{(k)}(x) = \dfrac{(-1)^{k+1}(k-1)!}{x^k}$, $f^{(k)}(2) = \dfrac{(-1)^{k+1}(k-1)!}{2^k}$ for any nonnegative integer k; the Taylor series about 2 is

$$\ln 2 + \frac{1}{2}(x-2) - \frac{1}{2^2}\cdot\frac{1}{2!}(x-2)^2 + \frac{2!}{2^3}\cdot\frac{1}{3!}(x-2)^3 - \frac{3!}{2^4}\cdot\frac{1}{4!}(x-2)^4 + \cdots = \ln 2 + \sum_{n=1}^{\infty}\frac{(-1)^{n+1}}{n2^n}(x-2)^n$$

7. $\sin 2x = \sum_{n=0}^{\infty}\dfrac{(-1)^n}{(2n+1)!}(2x)^{2n+1} = \sum_{n=0}^{\infty}\dfrac{(-1)^n 2^{2n+1}}{(2n+1)!}x^{2n+1}$

9. $\ln 3x = \ln 3 + \ln x = \ln 3 + \sum_{n=0}^{\infty} \frac{(-1)^n}{n+1}(x-1)^{n+1}$

11. Since $\ln(1+x) = \sum_{n=0}^{\infty}[(-1)^n/(n+1)]x^{n+1}$, we have

$$x \ln(1+x^2) = x \sum_{n=0}^{\infty} \frac{(1)^n}{n+1}(x^2)^{n+1} = \sum_{n=0}^{\infty} \frac{(-1)^n}{n+1} x^{2n+3}.$$

13. $2^x = e^{x \ln 2} = \sum_{n=0}^{\infty} \frac{1}{n!}(x \ln 2)^n = \sum_{n=0}^{\infty} \frac{(\ln 2)^n}{n!} x^n$

15. $\frac{x-1}{x+1} = 1 - \frac{2}{x+1} = 1 - \frac{2}{2+(x-1)} = 1 - \frac{1}{1+\frac{1}{2}(x-1)} = 1 - \sum_{n=0}^{\infty} \frac{(-1)^n}{2^n}(x-1)^n$

$$= \sum_{n=1}^{\infty} \frac{(-1)^{n+1}}{2^n}(x-1)^n = \sum_{n=0}^{\infty} \frac{(-1)^n}{2^{n+1}}(x-1)^{n+1}$$

17. $\cos^2 x = \frac{1}{2}(1+\cos 2x) = \frac{1}{2}\left(1 + \sum_{n=0}^{\infty} \frac{(-1)^n}{(2n)!}(2x)^{2n}\right) = 1 + \sum_{n=1}^{\infty} \frac{(-1)^n 2^{2n-1}}{(2n)!} x^{2n}$

19. For $x \neq 0$,

$$\frac{\sin x}{x} = \frac{1}{x} \sum_{n=0}^{\infty} \frac{(-1)^n}{(2n+1)!} x^{2n+1} = \sum_{n=0}^{\infty} \frac{(-1)^n}{(2n+1)!} x^{2n}, \quad \text{so} \quad f(x) = \sum_{n=0}^{\infty} \frac{(-1)^n}{(2n+1)!} x^{2n}.$$

21. $\sin x = \sin\left(x - \frac{\pi}{3} + \frac{\pi}{3}\right) = \sin\left(x - \frac{\pi}{3}\right)\cos\frac{\pi}{3} + \cos\left(x - \frac{\pi}{3}\right)\sin\frac{\pi}{3}$

$$= \frac{1}{2}\sin\left(x - \frac{\pi}{3}\right) + \frac{\sqrt{3}}{2}\cos\left(x - \frac{\pi}{3}\right) = \frac{1}{2}\sum_{n=0}^{\infty}(-1)^n \frac{(x-\pi/3)^{2n+1}}{(2n+1)!} + \frac{\sqrt{3}}{2}\sum_{n=0}^{\infty}(-1)^n \frac{(x-\pi/3)^{2n}}{(2n)!}$$

$$= \frac{1}{2}\sqrt{3} + \frac{1}{2}\left(x - \frac{\pi}{3}\right) - \frac{\sqrt{3}}{2(2!)}\left(x - \frac{\pi}{3}\right)^2 - \frac{1}{2(3!)}\left(x - \frac{\pi}{3}\right)^3 + + - - \cdots$$

23. Let $f(x) = e^x$, so that $f(\frac{1}{2}) = e^{1/2}$. We need to find n such that $|r_n(\frac{1}{2})| < 0.001$. Then $p_n(\frac{1}{2})$ is the desired approximation of $e^{1/2}$. By (11),

$$r_n\left(\frac{1}{2}\right) = \frac{f^{(n+1)}(t_{1/2})}{(n+1)!}\left(\frac{1}{2}\right)^{n+1} \quad \text{for an appropriate } t_{1/2} \text{ in } (0, \tfrac{1}{2}).$$

Since $f^{(n+1)}(x) = e^x$, we have $f^{(n+1)}(t_{1/2}) = e^{t_{1/2}}$. Since e^x is increasing and $t_{1/2} < \frac{1}{2}$, it follows that $e^{t_{1/2}} < e^{1/2}$. Since $e^{1/2} < 4^{1/2} = 2$, we conclude that

$$\left|r_n\left(\frac{1}{2}\right)\right| = \left|\frac{f^{(n+1)}(t_{1/2})}{(n+1)!}\left(\frac{1}{2}\right)^{n+1}\right| = \frac{e^{t_{1/2}}}{(n+1)!}\left(\frac{1}{2}\right)^{n+1} < \frac{2}{(n+1)!}\left(\frac{1}{2}\right)^{n+1} = \frac{1}{2^n(n+1)!}.$$

Thus $|r_n(\frac{1}{2})| < 0.001$ if $n \geq 4$. By (2) this means that

$$p_4\left(\frac{1}{2}\right) = \sum_{n=0}^{4} \frac{1}{n!}\left(\frac{1}{2}\right)^n \approx 1.6484375$$

is the desired approximation of $e^{1/2}$.

25. Let $f(x) = \ln(1+x)$, so that $f(0.1) = \ln 1.1$. We need to find n such that $|r_n(0.1)| < 0.00001$. Then $p_n(0.1)$ is the desired approximation of $\ln 1.1$. By (11),

$$r_n(0.1) = \frac{f^{(n+1)}(t_{0.1})}{(n+1)!}(0.1)^{n+1} \quad \text{for an appropriate } t_{0.1} \text{ in } (0, 0.1).$$

Since $f^{(n+1)}(x) = (-1)^n n!/(1+x)^{n+1}$, we have $f^{(n+1)}(t_{0.1}) = (-1)^n n!/(1+t_{0.1})^{n+1}$. Since $1/(1+x)^{n+1}$ is decreasing on $[0, 0.1]$, it follows that $1/(1+t_{0.1})^{n+1} < 1$, so that

$$|r_n(0.1)| = \left|\frac{f^{(n+1)}(t_{0.1})}{(n+1)!}(0.1)^{n+1}\right| = \frac{1}{(n+1)(1+t_{0.1})^{n+1}}(0.1)^{n+1} < \frac{1}{n+1}(0.1)^{n+1}.$$

Thus $|r_n(0.1)| < 0.00001$ if $n \geq 4$. By (3) this means that

$$p_4(0.1) = \sum_{n=0}^{4} \frac{(-1)^n}{n+1}(0.1)^{n+1} \approx 0.0953103333$$

is the desired approximation of $\ln 1.1$.

27. Let $f(x) = \cos x$, so that $f(-\pi/7) = \cos(-\pi/7)$. We need to find n such that $|r_n(-\pi/7)| < 0.001$. Then $p_n(-\pi/7)$ is the desired approximation of $\cos(-\pi/7)$. By (11),

$$r_n\left(-\frac{\pi}{7}\right) = \frac{f^{(n+1)}(t_{-\pi/7})}{(n+1)!}\left(-\frac{\pi}{7}\right)^{n+1} \quad \text{for an appropriate } t_{-\pi/7} \text{ in } (-\pi/7, 0).$$

Since $|f^{(n+1)}(x)| \leq 1$ for all x, it follows that

$$\left|r_n\left(-\frac{\pi}{7}\right)\right| = \left|\frac{f^{(n+1)}(t_{-\pi/7})}{(n+1)!}\left(-\frac{\pi}{7}\right)^{n+1}\right| \leq \frac{1}{(n+1)!}\left(\frac{\pi}{7}\right)^{n+1}.$$

Thus $|r_n(-\pi/7)| < 0.001$ if $n \geq 4$. By the Taylor series for $\cos x$ given after Example 1 in the text, $p_4(-\pi/7) = 1 - \frac{1}{2}(-\pi/7)^2 + \frac{1}{24}(-\pi/7)^4 \approx 0.9009801767$ is the desired approximation of $\cos(-\pi/7)$.

29. Let $f(x) = \ln[(1+x)/(1-x)] = \ln(1+x) - \ln(1-x)$. Notice that $(1+x)/(1-x) = 2$ if $1+x = 2-2x$, or $x = \frac{1}{3}$. Thus $f(\frac{1}{3}) = \ln 2$. We need to find n such that $|r_n(\frac{1}{3})| < 0.01$. Then $p_n(\frac{1}{3})$ is the desired approximation of $\ln 2$. By (11),

$$r_n\left(\frac{1}{3}\right) = \frac{f^{(n+1)}(t_{1/3})}{(n+1)!}\left(\frac{1}{3}\right)^{n+1} \quad \text{for an appropriate } t_{1/3} \text{ in } (0, \tfrac{1}{3}).$$

Now

$$f^{(n+1)}(x) = \frac{(-1)^n n!}{(1+x)^{n+1}} + \frac{n!}{(1-x)^{n+1}}$$

so that

$$\left|f^{(n+1)}(t_{1/3})\right| = \left|\frac{(-1)^n n!}{(1+t_{1/3})^{n+1}} + \frac{n!}{(1-t_{1/3})^{n+1}}\right| < \frac{n!}{1} + \frac{n!}{(2/3)^{n+1}} = n!\left[1 + \left(\frac{3}{2}\right)^{n+1}\right].$$

Therefore

$$\left|r_n\left(\frac{1}{3}\right)\right| = \left|\frac{f^{(n+1)}(t_{1/3})}{(n+1)!}\left(\frac{1}{3}\right)^{n+1}\right| < \frac{n!\left[1+(\frac{3}{2})^{n+1}\right]}{(n+1)!}\left(\frac{1}{3}\right)^{n+1} = \frac{1+(\frac{3}{2})^{n+1}}{(n+1)3^{n+1}}.$$

Thus $|r_n(\frac{1}{3})| < 0.01$ if $n \geq 4$. By the solution of Exercise 10,

$$\ln\frac{1+x}{1-x} = \sum_{n=0}^{\infty}\frac{2}{2n+1}x^{2n+1},$$

so that $p_4(\frac{1}{3}) = 2(\frac{1}{3}) + \frac{2}{3}(\frac{1}{3})^3 \approx 0.6913580247$ is the desired approximation of $\ln 2$.

31. If the x window is $[-2, 2]$ and the y window is $[-8, 8]$, then $n = 4$.

33. If the x window is $[-4, 4]$ and the y window is $[-1.5, 1.5]$, then $n = 10$.

35. a. As in Example 1 we find that $f(x) = \cos x$, $f'(x) = -\sin x$, $f''(x) = -\cos x$, $f^{(3)}(x) = \sin x$, and in general, $f^{(2k)}(x) = (-1)^k \cos x$ and $f^{(2k+1)}(x) = (-1)^{k+1}\sin x$. Therefore $f^{(2k)}(0) = (-1)^k$ and $f^{(2k+1)}(0) = 0$. Consequently the Taylor series of $\cos x$ is

$$1 - \frac{1}{2!}x^2 + \frac{1}{4!}x^4 - \frac{1}{6!}x^6 + - \cdots = \sum_{n=0}^{\infty}\frac{(-1)^n}{(2n)!}x^{2n}.$$

b. Since $|f^{(n+1)}(x)| \leq 1$ for all x and all $n \geq 0$,

$$|r_n(x)| = \left|\frac{f^{(n+1)}(t_x)}{(n+1)!}x^{n+1}\right| \leq \frac{|x|^{n+1}}{(n+1)!}.$$

By Example 7 of Section 9.7,

$$\lim_{n\to\infty}\frac{|x|^{n+1}}{(n+1)!} = \lim_{n\to\infty}\frac{|x|^n}{n!} = 0 \quad \text{for all } x.$$

Therefore $\lim_{n\to\infty} r_n(x) = 0$ for all x, so by (12) we conclude that

$$\cos x = \sum_{n=0}^{\infty}\frac{(-1)^n}{(2n)!}x^{2n}.$$

That is, the Taylor series of $\cos x$ converges to $\cos x$ for all x.

37. $\dfrac{-3x+2}{2x^2-3x+1} = \dfrac{-3x+2}{(2x-1)(x-1)} = \dfrac{A}{2x-1} + \dfrac{B}{x-1} = \dfrac{A(x-1)+B(2x-1)}{(2x-1)(x-1)}$

$A+2B=-3$ and $-A-B=2$ imply that $B=-1$, so $A=-2-B=-1$. Therefore

$$\frac{-3x+2}{2x^2-3x+1} = \frac{-1}{2x-1} + \frac{-1}{x-1} = \frac{1}{1-2x} + \frac{1}{1-x} = \sum_{n=0}^{\infty}(2x)^n + \sum_{n=0}^{\infty}x^n = \sum_{n=0}^{\infty}(2^n+1)x^n$$

39. a. $f(0) = e^{(0^2)}\int_0^0 e^{(-t^2)}\,dt = 0$; $f'(x) = 2xe^{(x^2)}\int_0^x e^{-(t^2)}\,dt + e^{(x^2)}e^{-(x^2)} = 2xf(x)+1$.

b. By part (a), we obtain $f''(x) = 2f(x)+2xf'(x)$, $f^{(3)}(x) = 2f'(x)+2f'(x)+2xf''(x) = 2(2)f'(x)+2xf''(x)$. In general, $f^{(n)}(x) = 2(n-1)f^{(n-2)}(x) + 2xf^{(n-1)}(x)$ for $n \ge 2$. Therefore $f^{(n)}(0) = 2(n-1)f^{(n-2)}(0)$ for $n \ge 2$. By part (a), $f(0) = 0$ and $f'(0) = 2(0)f(0)+1 = 1$, so that for $k \ge 0$ we have $f^{(2k)}(0) = 2(2k-1)f^{(2k-2)}(0) = \cdots = 0$. In addition,

$$f^{(2n+1)}(0) = 2(2n)f^{(2n-1)}(0) = 2(2n)(2)(2n-2)f^{(2n-3)}(0)$$

$$= 2^{2\cdot 2}n(n-1)f^{(2n-3)}(0) = 2^{2\cdot 3}n(n-1)(n-2)f^{(2n-5)}(0) = \cdots = 2^{2n}n!f'(0) = 4^n n!.$$

Consequently $f(x) = \sum_{n=0}^{\infty}[4^n n!/(2n+1)!]x^{2n+1}$.

41. If $\sum_{n=0}^{\infty} c_n t^n$ were a power series representation for f on an open interval containing 0, then $c_n = f^{(n)}(0)/n!$ for $n \ge 0$. Since f is stationary for $t \le 0$, it follows that $f^{(n)}(0) = 0$ for $n > 0$, and thus $c_n = 0$ for $n > 0$, so that $f(t) = \sum_{n=0}^{\infty} c_n t^n = c_0$. But $f'(t) > 0$ for $t > 0$ by hypothesis, so $f(t)$ cannot be c_0 for all $t > 0$.This contradiction proves that f has no such power series representation.

9.10 Binomial Series

1. $\sqrt{1.05} = \sqrt{1+\frac{1}{20}}$, so by (5) with $s = \frac{1}{2}$ and $x = \frac{1}{20}$, we have

$$\left|r_N\left(\frac{1}{20}\right)\right| \le \frac{1}{2}\frac{1/20^{N+1}}{1-1/20} = \frac{1}{38\cdot 20^N}.$$

Then $|r_N(\frac{1}{20})| < 0.001$ if $N = 2$, so by taking the first 3 terms in (3) we obtain

$$\sqrt{1.05} \approx 1 + \frac{1}{2}\left(\frac{1}{20}\right) - \frac{1}{8}\left(\frac{1}{20}\right)^2 \approx 1.0246875$$

as the desired approximation.

3. $\sqrt[4]{83} = \sqrt[4]{81\left(1+\frac{2}{81}\right)} = 3\sqrt[4]{1+\frac{2}{81}}$, so by (5) with $s = \frac{1}{4}$ and $x = \frac{2}{81}$, we have

$$\left|r_N\left(\frac{2}{81}\right)\right| \le \frac{1}{4}\,\frac{\left(\frac{2}{81}\right)^{N+1}}{1-\frac{2}{81}} = \frac{2^{N-1}}{79\cdot 81^N}.$$

Then $|r_N(\frac{2}{81})| < \frac{1}{3000}$ if $N = 1$, so by taking the first two terms in (2) we obtain

$$\sqrt[4]{83} \approx 3\left(1+\frac{1}{4}\left(\frac{2}{81}\right)\right) \approx 3.018518519$$

as the desired approximation.

5. $\sqrt[6]{65} = \sqrt[6]{64\left(1+\frac{1}{64}\right)} = 2\sqrt[6]{1+\frac{1}{64}}$, so by (5) with $s = \frac{1}{6}$ and $x = \frac{1}{64}$, we have

$$\left|r_N\left(\frac{1}{64}\right)\right| \le \frac{1}{6}\,\frac{\left(\frac{1}{64}\right)^{N+1}}{1-\frac{1}{64}} = \frac{1}{378\cdot 64^N}.$$

Then $|r_N(\frac{1}{64})| < 0.0005$ if $N = 1$, so by taking the first two terms in (2) we obtain

$$\sqrt[6]{65} \approx 2\left(1+\frac{1}{6}\left(\frac{1}{64}\right)\right) \approx 2.005208333$$

as the desired approximation.

7. $\dfrac{1}{\sqrt{1+x}} = \displaystyle\sum_{n=0}^{\infty}\binom{-\frac{1}{2}}{n}x^n = 1 - \frac{1}{2}x + \frac{1}{2!}\left(\frac{-1}{2}\right)\left(\frac{-3}{2}\right)x^2 + \frac{1}{3!}\left(\frac{-1}{2}\right)\left(\frac{-3}{2}\right)\left(\frac{-5}{2}\right)x^3 + \cdots;$

$1 - \frac{1}{2}x + \frac{3}{8}x^2 - \frac{5}{16}x^3$

9. $(1+x)^{-8/5} = \displaystyle\sum_{n=0}^{\infty}\binom{-\frac{8}{5}}{n}x^n = 1 - \frac{8}{5}x + \frac{1}{2!}\left(\frac{-8}{5}\right)\left(\frac{-13}{5}\right)x^2 + \frac{1}{3!}\left(\frac{-8}{5}\right)\left(\frac{-13}{5}\right)\left(\frac{-18}{5}\right)x^3 + \cdots;$

$1 - \frac{8}{5}x + \frac{52}{25}x^2 - \frac{312}{125}x^3$

11. $\dfrac{x}{\sqrt{1-x^2}} = x\displaystyle\sum_{n=0}^{\infty}\binom{-\frac{1}{2}}{n}(-x^2)^n = \sum_{n=0}^{\infty}(-1)^n\binom{-\frac{1}{2}}{n}x^{2n+1}$

$= x - \left(\frac{-1}{2}\right)x^3 + \frac{1}{2!}\left(\frac{-1}{2}\right)\left(\frac{-3}{2}\right)x^5 - \frac{1}{3!}\left(\frac{-1}{2}\right)\left(\frac{-3}{2}\right)\left(\frac{-5}{2}\right)x^7 + \cdots;$

$x + \frac{1}{2}x^3 + \frac{3}{8}x^5 + \frac{5}{16}x^7$

13. $\sqrt{1-(x+1)^2} = \sum_{n=0}^{\infty}\binom{\frac{1}{2}}{n}(-(x+1)^2)^n = \sum_{n=0}^{\infty}(-1)^n\binom{\frac{1}{2}}{n}(x+1)^{2n}$

$$= 1 - \frac{1}{2}(x+1)^2 + \frac{1}{2!}\left(\frac{1}{2}\right)\left(\frac{-1}{2}\right)(x+1)^4 - \frac{1}{3!}\left(\frac{1}{2}\right)\left(\frac{-1}{2}\right)\left(\frac{-3}{2}\right)(x+1)^6 + \cdots;$$

$1 - \frac{1}{2}(x+1)^2 - \frac{1}{8}(x+1)^4 - \frac{1}{16}(x+1)^6$

15. $\int_0^1 \sqrt{1-x^2}\,dx = \int_0^1\left[\sum_{n=0}^{\infty}\binom{\frac{1}{2}}{n}(-1)^n x^{2n}\right]dx = \sum_{n=0}^{\infty}\binom{\frac{1}{2}}{n}\frac{(-1)^n}{2n+1}(1)^{2n} = \sum_{n=0}^{\infty}\binom{\frac{1}{2}}{n}\frac{(-1)^n}{2n+1}$

and by integration by trigonometric substitution with $x = \sin u$, we find that

$$\int_0^1 \sqrt{1-x^2}\,dx = \int_0^{\pi/2}\sqrt{1-\sin^2 u}\cos u\,du$$

$$= \int_0^{\pi/2}\cos^2 u\,du = \frac{1}{2}\int_0^{\pi/2}(1+\cos 2u)\,du = \frac{1}{2}\left(u + \frac{1}{2}\sin 2u\right)\Big|_0^{\pi/2} = \frac{\pi}{4}$$

thus

$$\sum_{n=0}^{\infty}\binom{\frac{1}{2}}{n}\frac{(-1)^n}{2n+1} = \frac{\pi}{4}.$$

17. Without loss of generality, assume in this problem that $a > 0$.

a. $\int_0^x \sqrt{a^2+t^2}\,dt = a\int_0^x \sqrt{1+\left(\frac{t}{a}\right)^2}\,dt = a\int_0^x\left(\sum_{n=0}^{\infty}\binom{\frac{1}{2}}{n}\left(\frac{t}{a}\right)^{2n}\right)dt$

$$= a\sum_{n=0}^{\infty}\binom{\frac{1}{2}}{n}\left(\int_0^x\left(\frac{t}{a}\right)^{2n}dt\right) = a\sum_{n=0}^{\infty}\binom{\frac{1}{2}}{n}\left(\frac{a}{2n+1}\left(\frac{t}{a}\right)^{2n+1}\Big|_0^x\right)$$

$$= a^2\sum_{n=0}^{\infty}\binom{\frac{1}{2}}{n}\frac{1}{2n+1}\left(\frac{x}{a}\right)^{2n+1} = \sum_{n=0}^{\infty}\binom{\frac{1}{2}}{n}\left(\frac{1}{a}\right)^{2n-1}\frac{1}{2n+1}x^{2n+1}$$

Since

$$\lim_{n\to\infty}\left|\frac{\binom{\frac{1}{2}}{n+1}\left(\frac{1}{a}\right)^{2n+1} x^{2n+3}\Big/(2n+3)}{\binom{\frac{1}{2}}{n}\left(\frac{1}{a}\right)^{2n-1} x^{2n+1}\Big/(2n+1)}\right| = \lim_{n\to\infty}\frac{|\frac{1}{2}-n|}{n+1}\frac{2n+1}{2n+3}\left|\frac{x}{a}\right|^2 = \left|\frac{x}{a}\right|^2$$

the radius of convergence is a.

b. $$\int_0^x \sqrt{a^2-t^2}\,dt = a\int_0^x \sqrt{1-\left(\frac{t}{a}\right)^2}\,dt = a\int_0^x \left(\sum_{n=0}^{\infty}(-1)^n\binom{\frac{1}{2}}{n}\left(\frac{t}{a}\right)^{2n}\right)dt$$
$$= a\sum_{n=0}^{\infty}(-1)^n\binom{\frac{1}{2}}{n}\left(\int_0^x\left(\frac{t}{a}\right)^{2n}dt\right) = a^2\sum_{n=0}^{\infty}(-1)^n\binom{\frac{1}{2}}{n}\frac{1}{2n+1}\left(\frac{x}{a}\right)^{2n+1}$$
$$= \sum_{n=0}^{\infty}(-1)^n\binom{\frac{1}{2}}{n}\left(\frac{1}{a}\right)^{2n-1}\frac{1}{2n+1}x^{2n+1}$$

As in (a), the radius of convergence is a.

19. For $s=0$ we have $\left|\binom{0}{n}\right| = 0 = \left|\binom{0}{n-1}\right|$. For $n=1$ we have $\left|\binom{s}{1}\right| = |s| \le 1 = \left|\binom{s}{0}\right|$. For $n>1$ and $0<|s|\le 1$, we have $|s-n+1|\le n$, so that

$$\left|\frac{\binom{s}{n-1}}{\binom{s}{n}}\right| = \left|\frac{s(s-1)\cdots(s-n+2)/(n-1)!}{s(s-1)\cdots(s-n+1)/n!}\right| = \frac{n}{|s-n+1|}\ge 1$$

and thus $\left|\binom{s}{n}\right| \le \left|\binom{s}{n-1}\right|$.

21. a. By (3) with x^4 substituted for x,

$$\int_0^1 \sqrt{1+x^4}\,dx = \int_0^1 (1+x^4)^{1/2}\,dx = \int_0^1\sum_{n=0}^{\infty}\binom{1/2}{n}x^{4n}\,dx$$
$$= \sum_{n=0}^{\infty}\binom{1/2}{n}\int_0^1 x^{4n}\,dx = \sum_{n=0}^{\infty}\binom{1/2}{n}\frac{1}{4n+1}x^{4n+1}\Big|_0^1 = \sum_{n=0}^{\infty}\binom{1/2}{n}\frac{1}{4n+1}.$$

b. Since

$$\sum_{n=1}^{\infty}\binom{1/2}{n}\frac{1}{4n+1}$$

is an alternating series, we need to find the minimum positive integer j such that

$$\left|\binom{1/2}{n}\frac{1}{4j+3}\right| < 10^{-3}.$$

By calculator we find that if $j=5$, then we have

$$\left|\binom{1/2}{6}\frac{1}{23}\right| = \frac{(\frac{1}{2})(\frac{1}{2})(\frac{3}{2})(\frac{5}{2})(\frac{7}{2})(\frac{9}{2})}{6!\,(23)} < 10^{-3}.$$

Thus

$$\sum_{n=0}^{5}\binom{1/2}{n}\frac{1}{4n+1} = \binom{1/2}{0}+\binom{1/2}{1}\frac{1}{5}+\binom{1/2}{2}\frac{1}{9}+\binom{1/2}{3}\frac{1}{13}+\binom{1/2}{4}\frac{1}{17}+\binom{1/2}{5}\frac{1}{21}$$
$$= 1+\frac{1}{2}\frac{1}{5}-\frac{1}{2^2}\frac{1}{2!}\frac{1}{9}+\frac{3}{2^3}\frac{1}{3!}\frac{1}{13}-\frac{(3)(5)}{2^4}\frac{1}{4!}\frac{1}{17}+\frac{(3)(5)(7)}{2^5}\frac{1}{5!}\frac{1}{21}\approx 1.089923093$$

is an approximation of $\int_0^1\sqrt{1+x^4}\,dx$ that is accurate to within 10^{-3}.

23. From the figure, maximum depth is $4000 - x$, where $x = \sqrt{4000^2 - 100^2} = 4000\sqrt{1 - (\frac{100}{4000})^2} = 4000\sqrt{1 - \frac{1}{1600}}$. By Exercise 20,

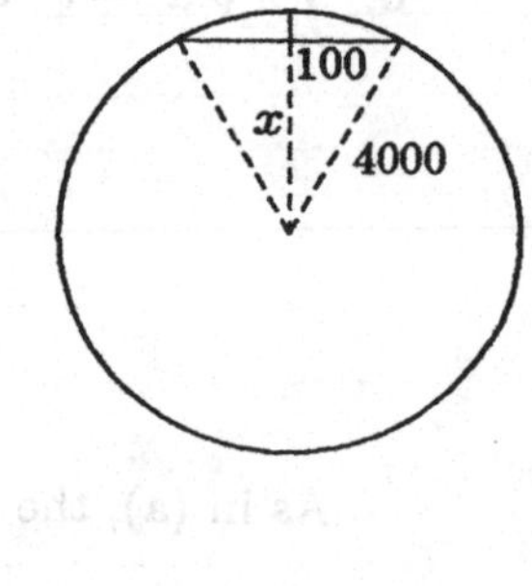

$$\sqrt{1 - \frac{1}{1600}} \approx 1 + \frac{1}{2}\left(-\frac{1}{1600}\right) = \frac{3199}{3200}$$

so that $x \approx 4000(\frac{3199}{3200}) = 3998.75$, and hence the desired approximation of the maximum depth is $4000 - 3998.75 = 1.25$ (miles). (The maximum depth is actually 1.25020, accurate to 6 places.)

25. $L = 2\int_0^{a/2} \left(1 + \frac{64b^2}{a^4}x^2\right)^{1/2} dx = 2\int_0^{a/2} \left[\sum_{n=0}^{\infty} \binom{\frac{1}{2}}{n} \left(\frac{64b^2}{a^4}\right)^n x^{2n}\right] dx$

$$= 2\sum_{n=0}^{\infty} \binom{\frac{1}{2}}{n} \left(\frac{8b}{a^2}\right)^{2n} \left(\int_0^{a/2} x^{2n}\, dx\right) = 2\sum_{n=0}^{\infty} \binom{\frac{1}{2}}{n} \left(\frac{8b}{a^2}\right)^{2n} \frac{1}{2n+1}\left(\frac{a}{2}\right)^{2n+1}$$

$$= \sum_{n=0}^{\infty} \binom{\frac{1}{2}}{n} \frac{4^{2n}a}{2n+1} \left(\frac{b}{a}\right)^{2n} \approx a\left[1 + \frac{8}{3}\left(\frac{b^2}{a^2}\right) - \frac{32}{5}\left(\frac{b^4}{a^4}\right)\right]$$

If $a = 500$ and $b = 40$, then

$$L \approx 500\left[1 + \frac{8}{3}\left(\frac{40}{500}\right)^2 - \frac{32}{5}\left(\frac{40}{500}\right)^4\right] \approx 508.402\,(\text{feet}).$$

Chapter 9 Review

1. $\lim_{n\to\infty} \frac{n^2 - \sqrt{n}}{4 - n^2} = \lim_{n\to\infty} \frac{1 - n^{-3/2}}{(4/n^2) - 1} = \frac{1 - 0}{0 - 1} = -1$

3. By replacing 0.05 by e in the solution of Exercise 25 in Section 9.2, we obtain $\lim_{n\to\infty}(1 + e/n)^n = e^e$.

5. $\lim_{n\to\infty} (\sqrt{n^2+n} - \sqrt{n^2-n}) = \lim_{n\to\infty} (\sqrt{n^2+n} - \sqrt{n^2-n}) \left(\frac{\sqrt{n^2+n} + \sqrt{n^2-n}}{\sqrt{n^2+n} + \sqrt{n^2-n}}\right)$

$$= \lim_{n\to\infty} \frac{2n}{\sqrt{n^2+n} + \sqrt{n^2-n}} = \lim_{n\to\infty} \frac{2}{\sqrt{1+1/n} + \sqrt{1-1/n}}$$

$$= \frac{2}{2} = 1$$

7. Since $1/(n + n\sqrt{n}) < 1/n^{3/2}$ for $n \geq 1$, and $\sum_{n=1}^{\infty} 1/n^{3/2}$ converges, it follows from the Comparison Test that $\sum_{n=1}^{\infty} 1/(n + n\sqrt{n})$ converges.

9. Since $\sqrt{n}/(n^2 + n + 1) \leq \sqrt{n}/n^2 = 1/n^{3/2}$ for $n \geq 1$, and $\sum_{n=1}^{\infty} 1/n^{3/2}$ converges, it follows from the Comparison Test that $\sum_{n=1}^{\infty} \sqrt{n}/(n^2 + n + 1)$ converges.

11. Since

$$\frac{6^n}{n^2(\ln n)^2} > \frac{6^n}{n^2(2n^{1/2})^2} = \frac{6^n}{4n^3}$$

and since $\lim_{n\to\infty} \sqrt[n]{6^n/4n^3} = \lim_{n\to\infty} 6/\sqrt[n]{4n^3} = 6$, it follows from the Root Test that $\sum_{n=2}^{\infty} 6^n/n^3$ diverges, and hence that $\sum_{n=2}^{\infty} 6^n/[n^2(\ln n)^2]$ diverges by the Comparison Test.

13. Let $f(x) = \sqrt{x}/(x-3)$ for $x \geq 4$. Then

$$f'(x) = \frac{(x-3)/(2\sqrt{x}) - \sqrt{x}}{(x-3)^2} = \frac{-x-3}{2\sqrt{x}\,(x-3)^2} < 0$$

so f is decreasing on $[4, \infty)$. Thus $\{\sqrt{n}/(n-3)\}_{n=4}^{\infty}$ is a nonnegative, decreasing sequence, and

$$\lim_{n\to\infty} \frac{\sqrt{n}}{n-3} = \lim_{n\to\infty} \frac{1}{\sqrt{n} - 3/\sqrt{n}} = 0.$$

Therefore $\sum_{n=4}^{\infty}(-1)^n\sqrt{n}/(n-3)$ converges by the Alternating Series Test.

15. Since

$$0 \leq \frac{\sqrt{n^2+1} - \sqrt{n^2-1}}{n} = \frac{\sqrt{n^2+1} - \sqrt{n^2-1}}{n}\,\frac{\sqrt{n^2+1} + \sqrt{n^2-1}}{\sqrt{n^2+1} + \sqrt{n^2-1}}$$

$$= \frac{(n^2+1)-(n^2-1)}{n(\sqrt{n^2+1} + \sqrt{n^2-1})} = \frac{2}{n(\sqrt{n^2+1} + \sqrt{n^2-1})} \leq \frac{2}{n\sqrt{n^2}} = \frac{2}{n^2}$$

and since $\sum_{n=1}^{\infty} 2/n^2$ converges, the given series converges by the Comparison Test.

17. $$27.1318318318\ldots = 27 + \frac{1}{10} + \frac{318}{10^4} + \frac{318}{10^7} + \cdots = 27 + \frac{1}{10} + \sum_{n=1}^{\infty} \frac{318}{10}\left(\frac{1}{1000}\right)^n$$

$$= 27 + \frac{1}{10} + \frac{318}{10}\left(\frac{1/1000}{1-1/1000}\right) = 27 + \frac{1}{10} + \frac{318}{10}\left(\frac{1}{999}\right) = \frac{90{,}349}{3330}$$

19. a. Since $\lim_{n\to 0} 0 = 0 = \lim_{n\to\infty} 1/n$, the Squeezing Theorem for sequences implies that $\lim_{n\to\infty} a_n = 0$.

    b. It is not possible to tell: if $a_n = 1/2n$, then $\sum_{n=1}^{\infty} a_n = \sum_{n=1}^{\infty} 1/2n$ diverges, whereas if $a_n = 1/2n^2$, then $\sum_{n=1}^{\infty} a_n = \sum_{n=1}^{\infty} 1/2n^2$ converges.

21. a. By hypothesis $0 < a_1 < b_1$. To use induction, assume that $0 < a_n < b_n$ for some positive integer n. Then $a_{n+1} < b_{n+1}$ is equivalent to $\sqrt{a_n b_n} < \frac{1}{2}(a_n + b_n)$, that is, $0 < a_n - 2\sqrt{a_n}\sqrt{b_n} + b_n$. Since $a_n - 2\sqrt{a_n}\sqrt{b_n} + b_n = (\sqrt{a_n} - \sqrt{b_n})^2 > 0$, it follows that $a_{n+1} < b_{n+1}$. By induction, $a_n < b_n$ for all n.

    b. By (a), $a_n < b_n$, so that $\sqrt{a_n} < \sqrt{b_n}$ and thus $a_n < \sqrt{a_n}\sqrt{b_n} = \sqrt{a_n b_n} = a_{n+1}$. Therefore $\{a_n\}_{n=1}^{\infty}$ is increasing. Likewise, by (a), $b_{n+1} = \frac{1}{2}(a_n + b_n) < \frac{1}{2}(b_n + b_n) = b_n$, so that $\{b_n\}_{n=1}^{\infty}$ is decreasing.

c. Since $0 < a_n < b_n < b_1$ for all n, $\{a_n\}_{n=1}^{\infty}$ is bounded. Since it is also increasing by part (b), Theorem 9.6 implies that $\{a_n\}_{n=1}^{\infty}$ converges to, say, L. Similarly, $b_1 > b_n > a_n > a_1$ for all n, so that $\{b_n\}_{n=1}^{\infty}$ is bounded. Since it is also decreasing by part (b), Theorem 9.6 implies that $\{b_n\}_{n=1}^{\infty}$ converges to, say, M.

d. We have $M = \lim_{n\to\infty} b_n = \lim_{n\to\infty} b_{n+1} = \lim_{n\to\infty} \frac{1}{2}(a_n + b_n) = \frac{1}{2}\left(\lim_{n\to\infty} a_n + \lim_{n\to\infty} b_n\right) = \frac{1}{2}(L+M)$, so that $2M = L + M$, or $L = M$.

e. If $a_1 = 1$ and $b_1 = 2$, then $a_2 = \sqrt{2}$ and $b_2 = 1.5$, so that $a_3 = \sqrt{\sqrt{2}(1.5)} \approx 1.456475315$ and $b_3 = \frac{1}{2}(\sqrt{2} + 1.5) \approx 1.457106781$. Thus since $a_3 < L < b_3$ and $|b_3 - a_3| \approx |1.457106781 - 1.456475315| < 0.01$, it follows that $L \approx 1.45$ to within 0.01.

23. $$\sum_{n=5}^{\infty}(-1)^n\frac{1}{2n+1} = \sum_{n=0}^{\infty}(-1)^n\frac{1}{2n+1} - \sum_{n=0}^{4}(-1)^n\frac{1}{2n+1} = \frac{\pi}{4} - \left(1 - \frac{1}{3} + \frac{1}{5} - \frac{1}{7} + \frac{1}{9}\right)$$
$$= \frac{\pi}{4} - \frac{263}{315} \approx -0.0495224715$$

25. Since $\sum_{n=1}^{\infty} x^n = x/(1-x)$ for $|x| < 1$, differentiation yields $\sum_{n=1}^{\infty} nx^{n-1} = 1/(1-x)^2$. Thus
$$(1-x)\sum_{n=1}^{\infty} nx^n = [(1-x)x]\sum_{n=1}^{\infty} nx^{n-1} = (1-x)x\frac{1}{(1-x)^2} = \frac{x}{1-x} \quad \text{for } |x| < 1.$$

27. a. If $\lim_{n\to\infty} a_n = L$ and $\lim_{n\to\infty} b_n = M$, then $\lim_{n\to\infty}(b_n - a_n) = \lim_{n\to\infty} b_n - \lim_{n\to\infty} a_n = M - L$. Let $\varepsilon > 0$. Then there is a positive integer N such that if $n \ge N$, then $|(b_n - a_n) - (M-L)| < \varepsilon$, so that $(M-L) - \varepsilon < b_n - a_n < (M-L) + \varepsilon$. By hypothesis, $a_n \le b_n$, so that $b_n - a_n \ge 0$, and thus $0 < (M-L) + \varepsilon$, or $-\varepsilon < M - L$. Since ε is arbitrary and positive, $0 \le M - L$, so that $L \le M$, which is equivalent to $\lim_{n\to\infty} a_n \le \lim_{n\to\infty} b_n$.

b. Let $s_j = \sum_{n=1}^{j} a_n$ and $s'_j = \sum_{n=1}^{j} |a_n|$ for $j \ge 1$. Since $a_n \le |a_n|$ for $n \ge 1$, we have $s_j \le s'_j$ for $j \ge 1$. Moreover, $\{s_j\}_{j=1}^{\infty}$ and $\{s'_j\}_{j=1}^{\infty}$ converge by hypothesis, and thus by part (a), $\lim_{j\to\infty} s_j \le \lim_{j\to\infty} s'_j$, so that $\sum_{n=1}^{\infty} a_n \le \sum_{n=1}^{\infty} |a_n|$. Replacing a_n by $-a_n$, we find that $-\sum_{n=1}^{\infty} a_n = \sum_{n=1}^{\infty}(-a_n) \le \sum_{n=1}^{\infty} |a_n|$. Thus $|\sum_{n=1}^{\infty} a_n| \le \sum_{n=1}^{\infty} |a_n|$.

29. $$\lim_{n\to\infty} \sqrt[n]{\frac{3^n}{5^{2n}}|x|^{3n}} = \lim_{n\to\infty} \frac{3}{25}|x|^3 = \frac{3}{25}|x|^3$$
so $\sum_{n=0}^{\infty}(3^n/5^{2n})x^{3n}$ converges for $|x| < \sqrt[3]{\frac{25}{3}}$ and diverges for $|x| > \sqrt[3]{\frac{25}{3}}$. Since $\sum_{n=0}^{\infty}(-1)^n$ and $\sum_{n=0}^{\infty} 1$ diverge, the interval of convergence is $\left(-\sqrt[3]{\frac{25}{3}}, \sqrt[3]{\frac{25}{3}}\right)$

31. $$\lim_{n\to\infty} \frac{(n+1)^{2n+2}|x|^{n+1}/(2n+2)!}{n^{2n}|x|^n/(2n)!} = \lim_{n\to\infty}\left[\left(1+\frac{1}{n}\right)^n\right]^2 \frac{n+1}{4n+2}|x| = \frac{e^2|x|}{4}$$
The radius of convergence is $4/e^2$.

33. $f(x) = \sqrt{1+x^4}$, $f'(x) = \dfrac{2x^3}{\sqrt{1+x^4}}$, $f''(x) = \dfrac{6x^2 + 2x^5}{(1+x^4)^{3/2}}$; $f(0) = 1$, $f'(0) = 0$, $f''(0) = 0$; $p_2(x) = 1$.

35. $f^{(2k)}(x) = (-1)^k \sin x$ and $f^{(2k+1)}(x) = (-1)^k \cos x$, and $f^{(2k)}(\pi/4) = (-1)^k(\sqrt{2}/2)$ and $f^{(2k+1)}(\pi/4) = (-1)^k(\sqrt{2}/2)$ for $k \geq 0$; the Taylor series about $\pi/4$ is

$$\sum_{n=0}^{\infty} \frac{f^{(n)}(\pi/4)}{n!}\left(x - \frac{\pi}{4}\right)^n = \sum_{n=0}^{\infty}(-1)^n \frac{\sqrt{2}}{2}\left[\frac{(x-\pi/4)^{2n}}{(2n)!} + \frac{(x-\pi/4)^{2n+1}}{(2n+1)!}\right]$$

37. $\dfrac{x-1}{x+1} = 1 - \dfrac{2}{x+1} = 1 - \displaystyle\sum_{n=0}^{\infty} 2(-1)^n x^n$

39. $\sqrt[4]{17} = \sqrt[4]{16\left(1+\frac{1}{16}\right)} = 2\sqrt[4]{1+\frac{1}{16}}$; by (5) in Section 9.10 with $s = \frac{1}{4}$ and $x = \frac{1}{16}$,

$$\left|r_N\left(\frac{1}{16}\right)\right| \leq \frac{1}{4}\frac{(1/16)^{N+1}}{1-(1/16)} = \frac{1}{60\cdot 16^N} < 0.0005 \quad \text{if } N = 2.$$

By (2) of Section 9.10,

$$\sqrt[4]{17} \approx 2\left[1 + \frac{1}{4}\left(\frac{1}{16}\right) + \frac{1}{2!}\left(\frac{1}{4}\right)\left(\frac{-3}{4}\right)\left(\frac{1}{16}\right)^2\right] \approx 2.030517578$$

is the desired approximation.

41. a. By (4) in Section 9.9,

$$\tan^{-1} x = \sum_{n=0}^{\infty} \frac{(-1)^n}{2n+1} x^{2n+1} = x - \frac{x^3}{3} + \frac{x^5}{5} - + \cdots.$$

Thus

$$\tan^{-1} x - x = \sum_{n=1}^{\infty} \frac{(-1)^n}{2n+1} x^{2n+1}$$

and

$$\frac{\tan^{-1} x - x}{x^3} = \sum_{n=1}^{\infty} \frac{(-1)^n}{2n+1} x^{2n-2} = -\frac{1}{3} + \frac{x^2}{5} - \frac{x^4}{7} + - \cdots.$$

Thus

$$\lim_{x\to 0} \frac{\tan^{-1} x - x}{x^3} = -\frac{1}{3}.$$

b. Since $\lim_{x\to 0}(\tan^{-1} x - x) = 0 = \lim_{x\to 0} x^3$, it follows from l'Hôpital's Rule that

$$\lim_{x\to 0} \frac{\tan^{-1} x - x}{x^3} = \lim_{x\to 0} \frac{\dfrac{1}{1+x^2} - 1}{3x^2} = \lim_{x\to 0} \frac{1-(1+x^2)}{3x^2(1+x^2)} = \lim_{x\to 0} \frac{-1}{3(1+x^2)} = -\frac{1}{3}.$$

43. a. $\sum_{n=1}^{25} a_n = s_{25} = 2 - \frac{2}{25} = \frac{48}{25}$

b. $\lim_{j\to\infty} s_j = \lim_{j\to\infty}(2 - 2/j) = 2$, so $\sum_{n=1}^{\infty} a_n$ converges and its sum is 2.

c. Since $\sum_{n=1}^{\infty} a_n$ converges, $\lim_{n\to\infty} a_n = 0$

d. $a_n = s_n - s_{n-1} = \left(2 - \frac{2}{n}\right) - \left(2 - \frac{2}{n-1}\right) = \frac{2}{n-1} - \frac{2}{n} = \frac{2}{n(n-1)}$

45. By (1) of Section 9.8 with x replaced by t^2, $\frac{1}{1-t^2} = \sum_{n=0}^{\infty} t^{2n}$, so that

$$\tanh^{-1} x = \int_0^x \frac{1}{1-t^2}\,dt = \int_0^x \left(\sum_{n=0}^{\infty} t^{2n}\right) dt = \sum_{n=0}^{\infty}\left(\int_0^x t^{2n}\,dt\right) = \sum_{n=0}^{\infty} \frac{x^{2n+1}}{2n+1}.$$

47. Since $1/(1+x^{3/2}) = \sum_{n=0}^{\infty}(-1)^n(x^{3/2})^n$, which is an alternating series, we know by Theorem 9.17 that for any $N \ge 0$,

$$\left|\frac{1}{1+x^{3/2}} - \sum_{n=0}^{N}(-1)^n x^{3n/2}\right| = \left|\sum_{n=0}^{\infty}(-1)^n x^{3n/2} - \sum_{n=0}^{N}(-1)^n x^{3n/2}\right| \le |x|^{3(N+1)/2}.$$

Thus

$$\left|\int_0^{1/4}\left(\frac{1}{1+x^{3/2}} - \sum_{n=0}^{N}(-1)^n x^{3n/2}\right)dx\right| \le \int_0^{1/4}\left|\frac{1}{1+x^{3/2}} - \sum_{n=0}^{N}(-1)^n x^{3n/2}\right| dx$$

$$\le \int_0^{1/4} x^{3(N+1)/2}\,dx = \left.\frac{2x^{(3N+5)/2}}{3N+5}\right|_0^{1/4} = \frac{2}{3N+5}\left(\frac{1}{2}\right)^{3N+5}$$

Now $2/(3N+5)(\frac{1}{2})^{3N+5} < 0.001$ if $N = 1$. Thus

$$\int_0^{1/4} \frac{1}{1+x^{3/2}}\,dx \approx \int_0^{1/4}(1 - x^{3/2})\,dx = \left.\left(x - \frac{2}{5}x^{5/2}\right)\right|_0^{1/4} = \frac{1}{4} - \frac{1}{80} = \frac{19}{80} = 0.2375.$$

Cumulative Review(Chapters 1–8)

1. The conditions for applying l'Hôpital's Rule four times are met;

$$\lim_{x\to 0^-} \frac{\cos x + \frac{1}{2}x^2 - 1}{x^5} = \lim_{x\to 0^-} \frac{-\sin x + x}{5x^4} = \lim_{x\to 0^-} \frac{-\cos x + 1}{20x^3} = \lim_{x\to 0^-} \frac{\sin x}{60x^2} = \lim_{x\to 0^-} \frac{\cos x}{120x} = -\infty.$$

3. $\lim_{x\to 0^+} 3^x = \lim_{x\to 0^+} e^{x\ln 3} = e^0 = 1$ and $\lim_{x\to 0^+} \frac{1}{x^3} = \infty$, so $\lim_{x\to 0^+} \frac{3^x}{x^3} = \infty$.

5. a. The domain of f consists of all x for which $3x - 4 > 0$ (or $x > \frac{4}{3}$) and $\ln(3x-4) > 0$. But $\ln(3x-4) > 0$ if and only if $3x - 4 > 1$, or $x > \frac{5}{3}$. Thus the domain is $(\frac{5}{3}, \infty)$.

   b. $f'(x) = \dfrac{1}{\ln(3x-4)} \cdot \dfrac{1}{3x-4} \cdot (3) = \dfrac{3}{(3x-4)\ln(3x-4)}$.

7. Differentiating the given equation implicitly, we obtain $e^x + 2e^{2x} - y^2(dy/dx) + (dy/dx) = 0$, so that $e^x + 2e^{2x} + (dy/dx)(1-y^2) = 0$. Thus if $(dy/dx)(1-y^2) = -6$ at (a,b), then $e^a + 2e^{2a} - 6 = 0$, or $2e^{2a} + e^a - 6 = 0$, or $(2e^a - 3)(e^a + 2) = 0$. Since $e^a + 2 > 0$, it follows that $e^a = \frac{3}{2}$, so that $a = \ln \frac{3}{2}$.

9. Since $A = \pi r^2$ and $dA/dt = \frac{1}{2}$, we have $\frac{1}{2} = dA/dt = 2\pi r(dr/dt)$, so that $dr/dt = 1/(4\pi r)$. Thus when $r = 1$, the radius is increasing at the rate of $1/(4\pi)$ foot per second.

11. By Exercise 54 in Section 8.1,

$$\int \frac{\cos x}{e^{3x}}\,dx = \int e^{-3x}\cos x\,dx = \frac{e^{-3x}}{10}(-3\cos x + \sin x) + C.$$

13. Let $x = 4\sin u$, so that $dx = 4\cos u\,du$. If $x = 2\sqrt{2}$, then $u = \pi/4$, and if $x = 4$, then $u = \pi/2$. Thus

$$\int_{2\sqrt{2}}^{4} \frac{\sqrt{16-x^2}}{x^3}\,dx = \int_{\pi/4}^{\pi/2} \frac{4\cos u}{64\sin^3 u} 4\cos u\,du$$

$$= \frac{1}{4}\int_{\pi/4}^{\pi/2} \frac{\cos^2 u}{\sin^3 u}\,du = \frac{1}{4}\int_{\pi/4}^{\pi/2} \frac{1-\sin^2 u}{\sin^3 u}\,du = \frac{1}{4}\int_{\pi/4}^{\pi/2} (\csc^3 u - \csc u)\,du.$$

By Exercise 28 of Section 8 2 $\int \csc^3 u\,du = -\frac{1}{2}\csc u \cot u - \frac{1}{2}\ln|\csc u + \cot u| + C$, and by Exercise 36 of Section 5.7, $\int \csc u\,du = -\ln|\csc u + \cot u| + C$. Therefore

$$\int_{2\sqrt{2}}^{4} \frac{\sqrt{16-x^2}}{x^3}\,dx = \frac{1}{4}\int_{\pi/4}^{\pi/2} (\csc^3 u - \csc u)\,du = \frac{1}{4}\left(-\frac{1}{2}\csc u \cot u + \frac{1}{2}\ln|\csc u + \cot u|\right)\Big|_{\pi/4}^{\pi/2}$$

$$= \frac{1}{4}\left(\frac{1}{2}\sqrt{2} - \frac{1}{2}\ln(\sqrt{2}+1)\right) = \frac{1}{8}(\sqrt{2} - \ln(\sqrt{2}+1)).$$

15. Let $u = 1 + e^x$, so that $du = e^x\,dx$. Then

$$\int \frac{e^x}{(1+e^x)^2}\,dx = \int \frac{1}{u^2}\,du = -\frac{1}{u} + C = \frac{-1}{1+e^x} + C.$$

Thus

$$\int_0^{\infty} \frac{e^x}{(1+e^x)^2}\,dx = \lim_{b\to\infty}\int_0^b \frac{e^x}{(1+e^x)^2}\,dx = \lim_{b\to\infty} \frac{-1}{1+e^x}\Big|_0^b = \lim_{b\to\infty}\left(\frac{-1}{1+e^b} + \frac{1}{2}\right) = \frac{1}{2}$$

and

$$\int_{-\infty}^{0} \frac{e^x}{(1+e^x)^2}\,dx = \lim_{a\to-\infty}\int_a^0 \frac{e^x}{(1+e^x)^2}\,dx$$

$$= \lim_{a \to -\infty} \frac{-1}{1+e^x}\bigg|_a^0 = \lim_{a \to -\infty}\left(-\frac{1}{2} + \frac{1}{1+e^a}\right) = -\frac{1}{2} + 1 = \frac{1}{2}.$$

Thus the given integral converges, and

$$\int_{-\infty}^{\infty} \frac{e^x}{(1+e^x)^2}\,dx = \int_{-\infty}^{0} \frac{e^x}{(1+e^x)^2}\,dx + \int_0^{\infty} \frac{e^x}{(1+e^x)^2}\,dx = \frac{1}{2} + \frac{1}{2} = 1.$$

17. $M_x = \int_0^{\ln 2} \frac{1}{2}[(1+e^x)^2 - (1-e^{-x})^2]\,dx = \int_0^{\ln 2} \frac{1}{2}[(1+2e^x+e^{2x}) - (1-2e^{-x}+e^{-2x})]\,dx$

$$= \int_0^{\ln 2}\left(e^x + \frac{1}{2}e^{2x} + e^{-x} - \frac{1}{2}e^{-2x}\right)dx = \left(e^x + \frac{1}{4}e^{2x} - e^{-x} + \frac{1}{4}e^{-2x}\right)\bigg|_0^{\ln 2}$$

$$= (2 + 1 - \tfrac{1}{2} + \tfrac{1}{16}) - (1 + \tfrac{1}{4} - 1 + \tfrac{1}{4}) = \tfrac{33}{16}$$

$$M_y = \int_0^{\ln 2} x[(1+e^x) - (1-e^{-x})]\,dx = \int_0^{\ln 2} x(e^x + e^{-x})\,dx \overset{\text{parts}}{=} x(e^x - e^{-x})\big|_0^{\ln 2} - \int_0^{\ln 2}(e^x - e^{-x})\,dx$$

$$= (\ln 2)(2 - \tfrac{1}{2}) - (e^x + e^{-x})\big|_0^{\ln 2} = \tfrac{3}{2}\ln 2 - (2 + \tfrac{1}{2}) + (1+1) = \tfrac{3}{2}\ln 2 - \tfrac{1}{2}$$

$$A = \int_0^{\ln 2} [(1+e^x) - (1-e^{-x})]\,dx = \int_0^{\ln 2}(e^x - e^{-x})\,dx = (e^x + e^{-x})\big|_0^{\ln 2} = (2 + \tfrac{1}{2}) - (1+1) = \tfrac{1}{2}$$

$$\bar{x} = \frac{M_y}{A} = \frac{\frac{3}{2}\ln 2 - \frac{1}{2}}{\frac{1}{2}} = 3\ln 2 - 1;\ \bar{y} = \frac{M_x}{A} = \frac{33/16}{1/2} = \frac{33}{8};\ (\bar{x}, \bar{y}) = (3\ln 2 - 1, \tfrac{33}{8})$$

19. If $y = x^2$ for $-2 \le x \le 2$, then $x = \sqrt{y}$ for $0 \le y \le 4$. Thus the width of the cross-section y units above the x axis is $2\sqrt{y}$, so that the cross-sectional area is given by $A(y) = 20\sqrt{y}$ for $0 \le y \le 4$. For $0 \le y \le 4$ the vertical distance the water must be pumped is $8 - y$. Therefore $W = \int_0^4 62.5(20\sqrt{y})(8-y)\,dy = 1250\int_0^4(8\sqrt{y} - y^{3/2})\,dy = 1250\left(\frac{16}{3}y^{3/2} - \frac{2}{5}y^{5/2}\right)\big|_0^4 = 1250\left(\frac{128}{3} - \frac{64}{5}\right) = \frac{112{,}000}{3}$ (foot-pounds).

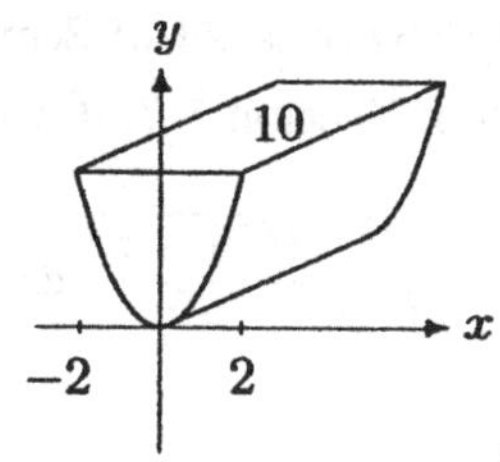

Chapter 10

Polar Coordinates and Conic Sections

10.1 Polar Coordinates

1. a. $x = 3\cos(\pi/4) = 3\sqrt{2}/2$; $y = 3\sin(\pi/4) = 3\sqrt{2}/2$; Cartesian coordinates: $(3\sqrt{2}/2, 3\sqrt{2}/2)$
   b. $x = -2\cos(-\pi/6) = -\sqrt{3}$; $y = -2\sin(-\pi/6) = 1$; Cartesian coordinates: $(-\sqrt{3}, 1)$
   c. $x = 3\cos(7\pi/3) = 3/2$; $y = 3\sin(7\pi/3) = 3\sqrt{3}/2$; Cartesian coordinates: $(3/2, 3\sqrt{3}/2)$
   d. $x = 5\cos 0 = 5$; $y = 5\sin 0 = 0$; Cartesian coordinates: $(5, 0)$
   e. $x = -2\cos(\pi/2) = 0$; $y = -2\sin(\pi/2) = -2$; Cartesian coordinates: $(0, -2)$
   f. $x = -2\cos(3\pi/2) = 0$; $y = -2\sin(3\pi/2) = 2$; Cartesian coordinates: $(0, 2)$
   g. $x = 4\cos(3\pi/4) = -2\sqrt{2}$; $y = 4\sin(3\pi/4) = 2\sqrt{2}$; Cartesian coordinates: $(-2\sqrt{2}, 2\sqrt{2})$
   h. $x = 0\cos(6\pi/7) = 0$; $y = 0\sin(6\pi/7) = 0$; Cartesian coordinates: $(0, 0)$
   i. $x = -1\cos(23\pi/3) = -1/2$; $y = -1\sin(23\pi/3) = \sqrt{3}/2$; Cartesian coordinates: $(-1/2, \sqrt{3}/2)$
   j. $x = -1\cos(-23\pi/3) = -1/2$; $y = -1\sin(-23\pi/3) = -\sqrt{3}/2$;
   Cartesian coordinates: $(-1/2, -\sqrt{3}/2)$
   k. $x = 1\cos(3\pi/2) = 0$; $y = 1\sin(3\pi/2) = -1$; Cartesian coordinates: $(0, -1)$
   l. $x = 3\cos(-5\pi/6) = -3\sqrt{3}/2$; $y = 3\sin(-5\pi/6) = -3/2$; Cartesian coordinates: $(-3\sqrt{3}/2, -3/2)$

3. If $2x + 3y = 4$, then $2(r\cos\theta) + 3(r\sin\theta) = 4$, so $r = 4/(2\cos\theta + 3\sin\theta)$.

5. If $x^2 + 9y^2 = 1$, then $r^2\cos^2\theta + 9(r^2\sin^2\theta) = 1$, so

$$r^2 = \frac{1}{\cos^2\theta + 9\sin^2\theta} = \frac{1}{8\sin^2\theta + 1}, \quad \text{or} \quad r = \frac{1}{\sqrt{8\sin^2\theta + 1}}.$$

7. If $(x^2 + y^2)^2 = x^2 - y^2$, then $(r^2)^2 = r^2\cos^2\theta - r^2\sin^2\theta$, so $r^4 = r^2\cos 2\theta$, or $r^2 = \cos 2\theta$.

9. If $x^2 + y^2 = 4y$, then it follows that $r^2 = 4r\sin\theta$, so $r = 4\sin\theta$.

11. If $y^2 = x^2(3-x)/(1+x)$, then

$$r^2 \sin^2\theta = \frac{r^2\cos^2\theta\,(3 - r\cos\theta)}{1 + r\cos\theta}$$

so $\sin^2\theta\,(1 + r\cos\theta) = \cos^2\theta\,(3 - r\cos\theta)$, or

$$r = \frac{3\cos^2\theta - \sin^2\theta}{\sin^2\theta\,\cos\theta + \cos^3\theta} = \frac{3\cos^2\theta - \sin^2\theta}{\cos\theta}.$$

13. If $r = 3\cos\theta$, then $r^2 = 3r\cos\theta$, or $x^2 + y^2 = 3x$.

15. If $\cot\theta = 3$, then $\cos\theta = 3\sin\theta$, so $r\cos\theta = 3r\sin\theta$, or $x = 3y$.

17. If $r = \sin 2\theta = 2\sin\theta\,\cos\theta$, then $r^3 = 2(r\sin\theta)(r\cos\theta)$, so $(x^2+y^2)^{3/2} = 2xy$, or $(x^2+y^2)^3 = 4x^2y^2$.

19. If $r \neq 0$, the equation $r = 1 + \cos\theta$ is equivalent to the equation $r^2 = r + r\cos\theta$ since division by r is permissible. If $r = 0$ (which corresponds to the origin), then $r^2 = r + r\cos\theta$ is satisfied by (r,θ) for any θ, whereas $r = 1 + \cos\theta$ is satisfied by (r,π). Thus the origin is on the polar graph of both equations, so the polar graphs are the same.

21. $r = 5$

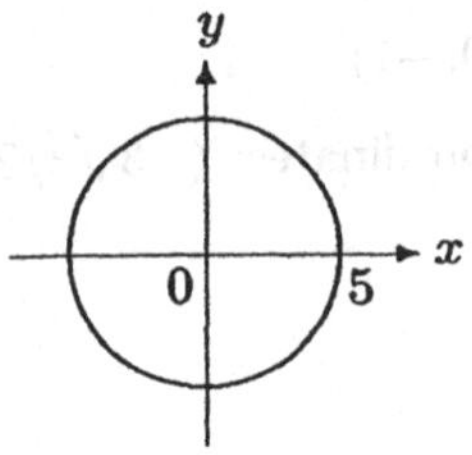

23. $r = 0$

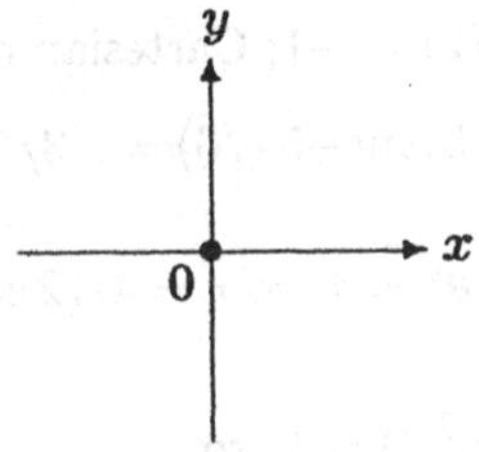

25. $\theta = -7\pi/6$

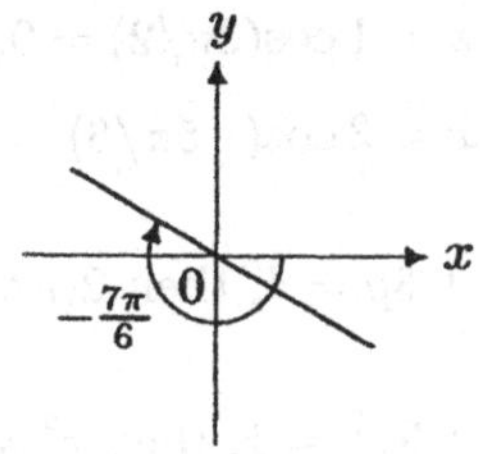

27. $r\sin\theta = 5$

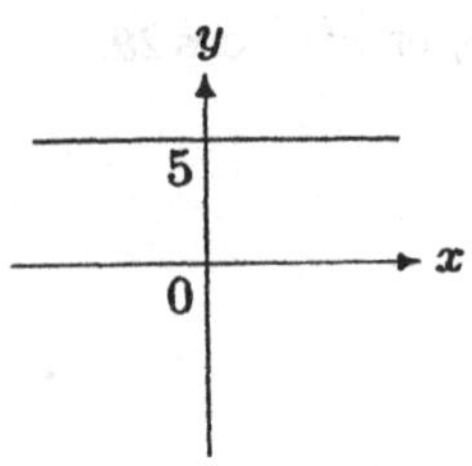

29. $r = -(3/2)\cos\theta$

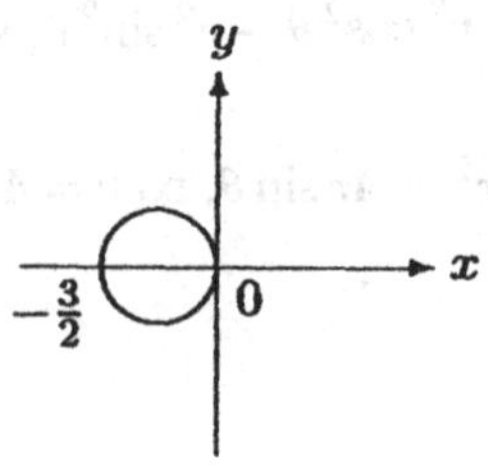

31. $r\cos(\theta - \pi/3) = 2$

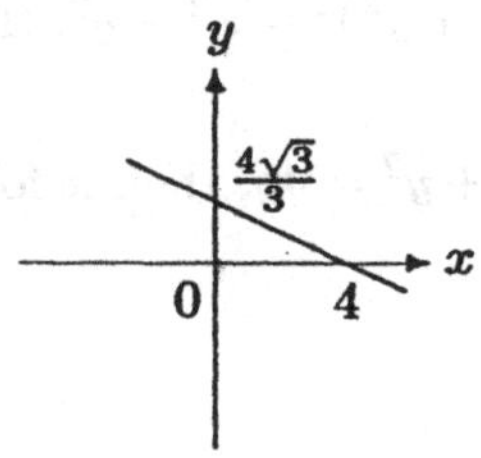

33. $r = 2\cot\theta\csc\theta$

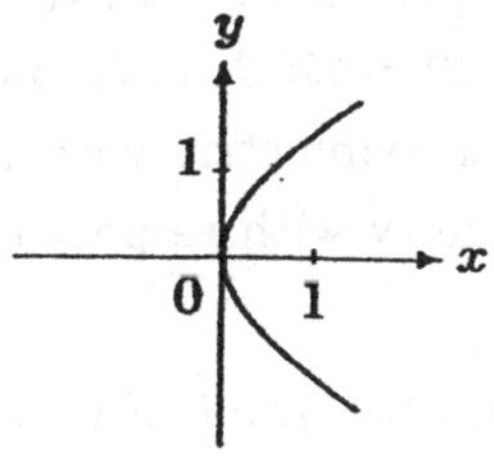

35. $r(\sin\theta + \cos\theta) = 1$

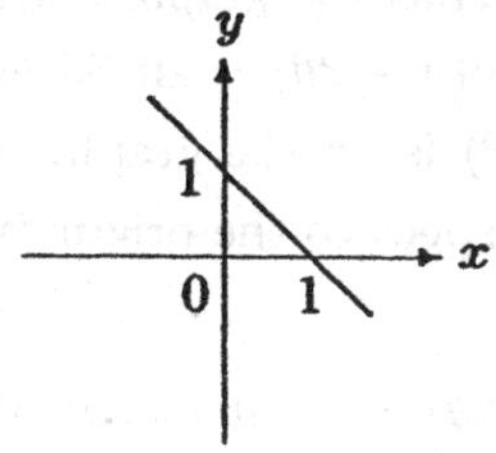

37. $r = 1 - \cos\theta$; symmetry with respect to the x axis

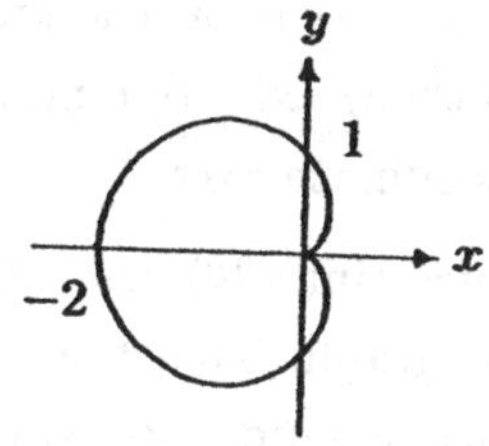

39. $r = 5(1 + \sin\theta)$; symmetry with respect to the y axis.

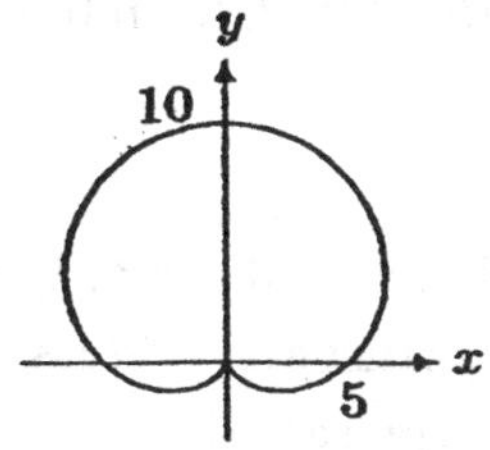

41. $r = -4\sin 3\theta$; symmetry with respect to the y axis.

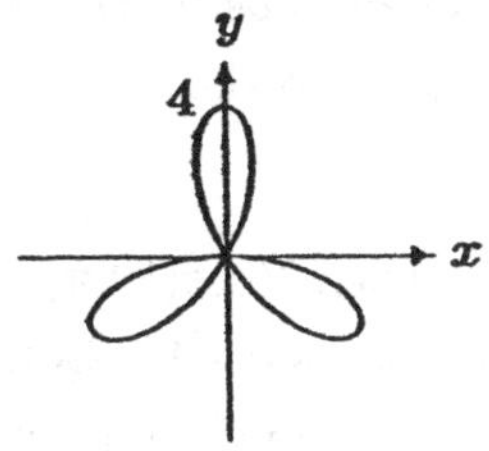

43. $r = 2\cos 6\theta$; symmetry with respect to both axes and origin.

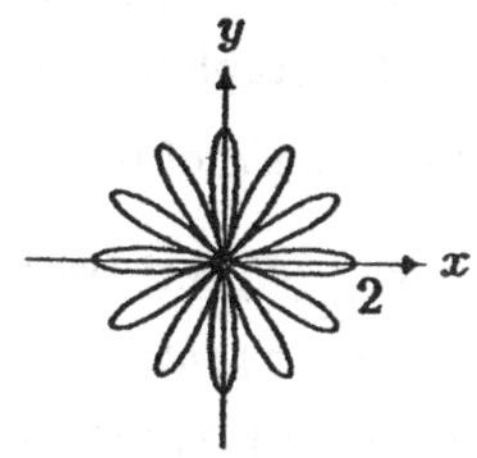

45. $r = \sin(\theta/2)$; symmetry with respect to both axes and origin (note if (r, θ) on graph then $(-r, \theta + 2\pi)$ on graph)

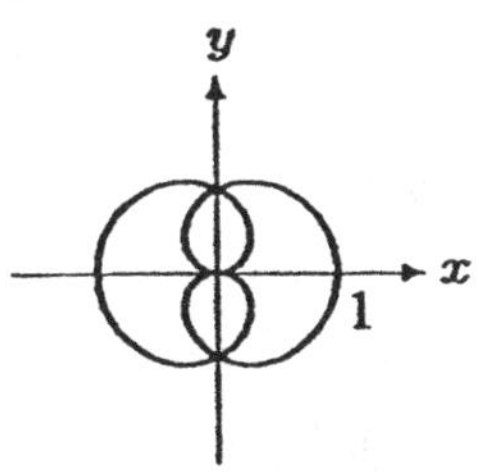

47. $r^2 = 25\cos\theta$; symmetry with respect to both axes and origin.

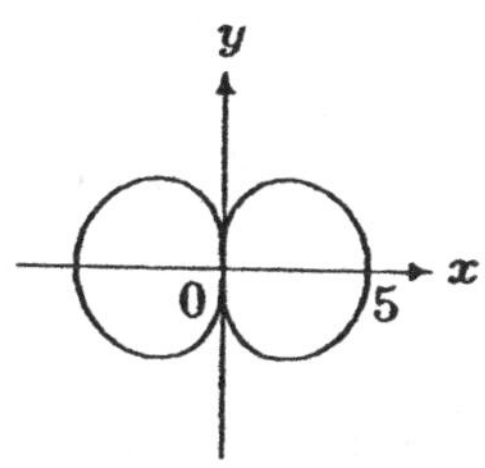

49. $r = 1 + 2\sin\theta$; symmetry with respect to the y axis.

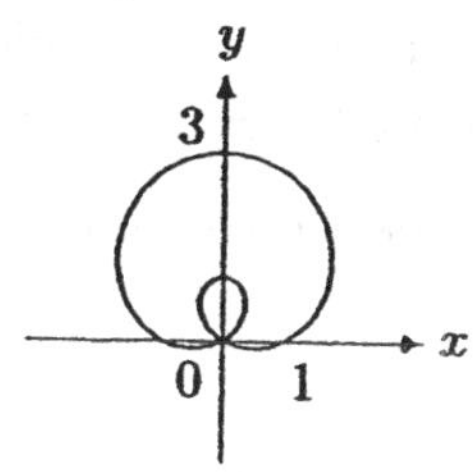

51. $r = e^{\theta/3}$; no symmetry with respect to either axis or origin.

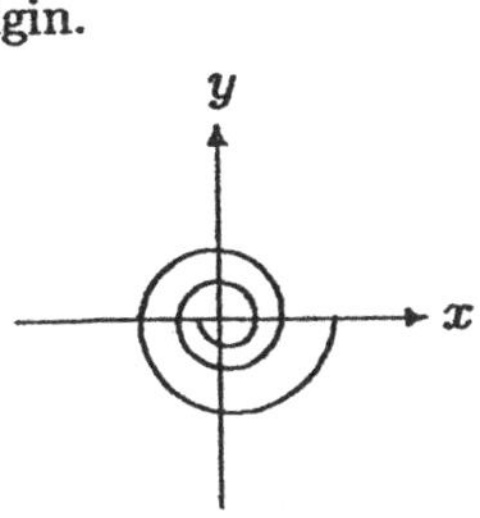

53. a. Since $\sin(-3\theta)\sin(-\theta) = -(\sin 3\theta)(-\sin\theta) = (\sin 3\theta)\sin\theta$, it follows that $(r, -\theta)$ is on the graph if (r, θ) is on the graph. Thus the graph is symmetric with respect to the x axis. Since $\sin 3(\pi - \theta) = \sin(3\pi - 3\theta) = \sin(\pi - 3\theta) = \sin 3\theta$ and since $\sin(\pi - \theta) = \sin\theta$, it follows that $(r, \pi - \theta)$ is on the graph if (r, θ) is on the graph. Thus the graph is symmetric with respect to the y axis. Symmetry with respect to the origin follows from symmetry with respect to both coordinate axes.

b. Since $\sin(-3\theta)\cos(-\theta) = -\sin(3\theta)\cos\theta$, it follows that $(-r, -\theta)$ is on the graph if (r, θ) is on the graph. Thus the graph is symmetric with respect to the y axis. Since $\sin 3(\pi - \theta) = \sin 3\theta$ by part (a) and since $\cos(\pi - \theta) = -\cos(-\theta) = -\cos\theta$, it follows that $(-r, \pi - \theta)$ is on the graph if (r, θ) is on the graph. Thus the graph is symmetric with respect to the x axis. Symmetry with respect to the origin follows from symmetry with respect to both axes.

c. Since $1 + \cos(-2\theta) + \cos(-3\theta) = 1 + \cos 2\theta + \cos 3\theta$, $(r, -\theta)$ is on the graph if (r, θ) is on the graph. Thus the graph is symmetric with respect to the x axis.

55. The graph consists of n petals.

57. a. Suppose (r, θ) satisfies $r = 3(\cos\theta + 1)$. Then $3(\cos(\theta + \pi) - 1) = 3(-\cos\theta - 1) = -3(\cos\theta + 1) = -r$, so $(-r, \theta + \pi)$ satisfies the equation $r = 3(\cos\theta - 1)$. Conversely, if (r, θ) satisfies $r = 3(\cos\theta - 1)$, then $(-r, \theta + \pi)$ satisfies $r = 3(\cos\theta + 1)$. Since (r, θ) and $(-r, \theta + \pi)$ represent the same point, the graphs are the same.

b. Suppose (r, θ) satisfies $r = 2(\sin\theta + 1)$. Then $2(\sin(\theta + \pi) - 1) = 2(-\sin\theta - 1) = -2(\sin\theta + 1) = -r$, so $(-r, \theta + \pi)$ satisfies the equation $r = 2(\sin\theta - 1)$. Conversely, if (r, θ) satisfies $r = 2(\sin\theta - 1)$, then $(-r, \theta + \pi)$ satisfies $r = 2(\sin\theta + 1)$. Since (r, θ) and $(-r, \theta + \pi)$ represent the same point, the graphs are the same.

59. Let (r, θ) be a point on the graph. Then

$$1 = \sqrt{(r\cos\theta - 1)^2 + (r\sin\theta)^2}\,\sqrt{(r\cos\theta + 1)^2 + (r\sin\theta)^2}.$$

Squaring both sides, we obtain

$$1 = (r^2 - 2r\cos\theta + 1)\cdot(r^2 + 2r\cos\theta + 1) = r^4 - 4r^2\cos^2\theta + 2r^2 + 1$$

so that $r^2[r^2 + 2(1 - 2\cos^2\theta)] = 0$. Thus either $r = 0$ or $r^2 = -2(1 - 2\cos^2\theta) = 2\cos 2\theta$. Since $(0, \pi/4)$ satisfies the equation $r^2 = 2\cos 2\theta$, we may simply write $r^2 = 2\cos 2\theta$, even when $r = 0$.

10.2 Length and Area in Polar Coordinates

1. By (2),

$$L = \int_{-\pi/2}^{\pi/2} \sqrt{(-2\sin\theta)^2 + (2\cos\theta)^2}\,d\theta = \int_{-\pi/2}^{\pi/2} \sqrt{4(\sin^2\theta + \cos^2\theta)}\,d\theta = \int_{-\pi/2}^{\pi/2} 2\,d\theta = 2\pi.$$

3. By (2),

$$L = \int_0^{2\pi} \sqrt{(2\sin\theta)^2 + (2-2\cos\theta)^2}\,d\theta = \int_0^{2\pi} \sqrt{4\sin^2\theta + 4 - 8\cos\theta + 4\cos^2\theta}\,d\theta$$

$$= \int_0^{2\pi} \sqrt{8-8\cos\theta}\,d\theta = 2\int_0^{\pi} 4\sqrt{\frac{1-\cos\theta}{2}}\,d\theta = 8\int_0^{\pi} \sin\frac{\theta}{2}\,d\theta = -16\cos\frac{\theta}{2}\Big|_0^{\pi} = 16.$$

5. By (2),

$$L = \int_0^{2\pi} \sqrt{\left(\sin^2\frac{\theta}{3}\cos\frac{\theta}{3}\right)^2 + \left(\sin^3\frac{\theta}{3}\right)^2}\,d\theta = \int_0^{2\pi} \sqrt{\sin^4\frac{\theta}{3}\left(\cos^2\frac{\theta}{3} + \sin^2\frac{\theta}{3}\right)}\,d\theta$$

$$= \int_0^{2\pi} \sin^2\frac{\theta}{3}\,d\theta = \int_0^{2\pi}\left(\frac{1}{2} - \frac{1}{2}\cos\frac{2\theta}{3}\right)d\theta = \left(\frac{1}{2}\theta - \frac{3}{4}\sin\frac{2\theta}{3}\right)\Big|_0^{2\pi} = \pi + \frac{3\sqrt{3}}{8}.$$

7. $A = \int_0^{2\pi} \left(\frac{1}{2}\right) 4^2\,d\theta = 8\theta\big|_0^{2\pi} = 16\pi$

9. The entire region is obtained from $r = 3\sin\theta$ for $0 \le \theta \le \pi$; $A = \int_0^{\pi} (\frac{1}{2}) 9\sin^2\theta\,d\theta = \frac{9}{2}(\frac{1}{2}\theta - \frac{1}{4}\sin 2\theta)\big|_0^{\pi} = \frac{9}{4}\pi$.

11. $A = \int_{\pi/2}^{3\pi/2} \frac{1}{2}(-2\cos\theta)^2\,d\theta = 2\int_{\pi/2}^{3\pi/2} \cos^2\theta\,d\theta = 2\left(\frac{1}{2}\theta + \frac{1}{4}\sin 2\theta\right)\big|_{\pi/2}^{3\pi/2} = \pi$

13. The region has the same area A as does the region described by $r = -9\cos 2\theta$ for $\pi/4 \le \theta \le \pi/2$. Now $A = \int_{\pi/4}^{\pi/2} \frac{1}{2}(-9\cos 2\theta)^2\,d\theta = \frac{81}{2}\int_{\pi/4}^{\pi/2} \cos^2 2\theta\,d\theta = \frac{81}{2}\int_{\pi/4}^{\pi/2} \left(\frac{1}{2} + \frac{1}{2}\cos 4\theta\right)d\theta = \frac{81}{2}\left(\frac{1}{2}\theta + \frac{1}{8}\sin 4\theta\right)\big|_{\pi/4}^{\pi/2} = \frac{81}{16}\pi$.

15. The three leaves have the same area. The area of one of them is given by $A_1 = \int_{-\pi/6}^{\pi/6} \frac{1}{2}\left(\frac{1}{2}\cos 3\theta\right)^2 d\theta = \frac{1}{8}\int_{-\pi/6}^{\pi/6} \cos^2 3\theta\,d\theta = \frac{1}{8}\int_{-\pi/6}^{\pi/6}\left(\frac{1}{2} + \frac{1}{2}\cos 6\theta\right)d\theta = \frac{1}{8}\left(\frac{1}{2}\theta + \frac{1}{12}\sin 6\theta\right)\big|_{-\pi/6}^{\pi/6} = \frac{1}{48}\pi$. Thus the total area $A = 3A_1 = \pi/16$.

17. $A = \int_0^{2\pi} \frac{1}{2}(4)(1-\sin\theta)^2\,d\theta = 2\int_0^{2\pi}(1 - 2\sin\theta + \sin^2\theta)\,d\theta = 2\int_0^{2\pi}\left(\frac{3}{2} - 2\sin\theta - \frac{1}{2}\cos 2\theta\right)d\theta$
$= 2\left(\frac{3}{2}\theta + 2\cos\theta - \frac{1}{4}\sin 2\theta\right)\big|_0^{2\pi} = 6\pi$

19. $A = \int_0^{2\pi} \frac{1}{2}(4 + 3\cos\theta)^2\,d\theta = \frac{1}{2}\int_0^{2\pi}(16 + 24\cos\theta + 9\cos^2\theta)\,d\theta = \frac{1}{2}\int_0^{2\pi}\left(\frac{41}{2} + 24\cos\theta + \frac{9}{2}\cos 2\theta\right)d\theta$
$= \frac{1}{2}\left(\frac{41}{2}\theta + 24\sin\theta + \frac{9}{4}\sin 2\theta\right)\big|_0^{2\pi} = \frac{41}{2}\pi$

21. The two leaves of the region have the same area. The area of one of them is given by

$$A_1 = \int_{-\pi/2}^{\pi/2} \frac{1}{2}(\sqrt{25\cos\theta})^2\,d\theta = \frac{25}{2}\int_{-\pi/2}^{\pi/2} \cos\theta\,d\theta = \frac{25}{2}\sin\theta\Big|_{-\pi/2}^{\pi/2} = 25.$$

Thus the total area $A = 2A_1 = 50$.

23. $f(\theta) = 5$, $g(\theta) = 1$; $A = \int_0^{2\pi} \frac{1}{2}(5^2 - 1^2)\,d\theta = 12\theta\big|_0^{2\pi} = 24\pi$

25. $f(\theta) = 1,\ g(\theta) = \left\{ \begin{array}{ll} \sin\theta & \text{for } 0 \le \theta \le \pi \\ 0 & \text{for } \pi \le \theta \le 2\pi \end{array} \right\};$

$A = \int_0^\pi \frac{1}{2}(1^2 - \sin^2\theta)\,d\theta + \int_\pi^{2\pi} \frac{1}{2}(1^2)\,d\theta = \frac{1}{2}\left(\frac{1}{2}\theta + \frac{1}{4}\sin 2\theta\right)\Big|_0^\pi + \frac{1}{2}\theta\Big|_\pi^{2\pi} = \frac{1}{4}\pi + \frac{1}{2}\pi = \frac{3}{4}\pi.$

27. $f(\theta) = 1,\ g(\theta) = \left\{ \begin{array}{ll} \sqrt{\cos 2\theta} & \text{for } -\pi/4 \le \theta \le \pi/4 \text{ and } 3\pi/4 \le \theta \le 5\pi/4 \\ 0 & \text{for } \pi/4 \le \theta \le 3\pi/4 \text{ and } 5\pi/4 \le \theta \le 7\pi/4 \end{array} \right\};$

$$A = \int_{-\pi/4}^{\pi/4} \tfrac{1}{2}[1^2 - (\sqrt{\cos 2\theta})^2]\,d\theta + \int_{\pi/4}^{3\pi/4} \tfrac{1}{2}(1^2)\,d\theta + \int_{3\pi/4}^{5\pi/4} \tfrac{1}{2}[1^2 - (\sqrt{\cos 2\theta})^2]\,d\theta + \int_{5\pi/4}^{7\pi/4} \tfrac{1}{2}(1^2)\,d\theta$$

$$= \tfrac{1}{2}\int_{-\pi/4}^{\pi/4}(1 - \cos 2\theta)\,d\theta + \tfrac{1}{4}\pi + \tfrac{1}{2}\int_{3\pi/4}^{5\pi/4}(1 - \cos 2\theta)\,d\theta + \tfrac{1}{4}\pi$$

$$= \tfrac{1}{2}\left(\theta - \tfrac{1}{2}\sin 2\theta\right)\Big|_{-\pi/4}^{\pi/4} + \tfrac{1}{2}\left(\theta - \tfrac{1}{2}\sin 2\theta\right)\Big|_{3\pi/4}^{5\pi/4} + \tfrac{1}{2}\pi = \left(\tfrac{1}{4}\pi - \tfrac{1}{2}\right) + \left(\tfrac{\pi}{4} - \tfrac{1}{2}\right) + \tfrac{1}{2}\pi = \pi - 1$$

29. $f(\theta) = 2 + \cos\theta,\ g(\theta) = \left\{ \begin{array}{ll} 0 & \text{for } -\pi/2 \le \theta \le \pi/2 \\ -\cos\theta & \text{for } \pi/2 \le \theta \le 3\pi/2 \end{array} \right\};$

$$A = \int_{-\pi/2}^{\pi/2} \tfrac{1}{2}(2 + \cos\theta)^2\,d\theta + \int_{\pi/2}^{3\pi/2} \tfrac{1}{2}[(2 + \cos\theta)^2 - (-\cos\theta)^2]\,d\theta$$

$$= \tfrac{1}{2}\int_{-\pi/2}^{\pi/2}(4 + 4\cos\theta + \cos^2\theta)\,d\theta + \tfrac{1}{2}\int_{\pi/2}^{3\pi/2}(4 + 4\cos\theta)\,d\theta$$

$$= \tfrac{1}{2}\left(\tfrac{9}{2}\theta + 4\sin\theta + \tfrac{1}{4}\sin 2\theta\right)\Big|_{-\pi/2}^{\pi/2} + \tfrac{1}{2}(4\theta + 4\sin\theta)\Big|_{\pi/2}^{3\pi/2}$$

$$= \left(\tfrac{9}{4}\pi + 4\right) + (2\pi - 4) = \tfrac{17}{4}\pi$$

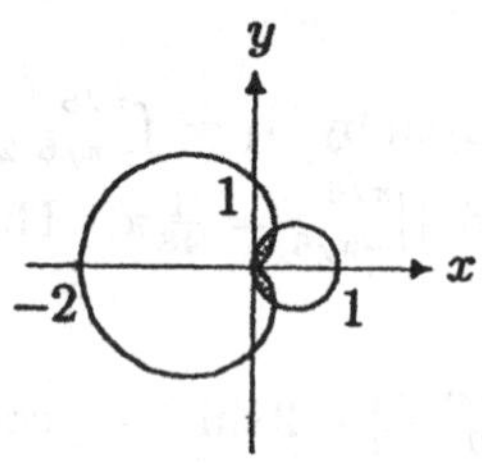

Exercise 31

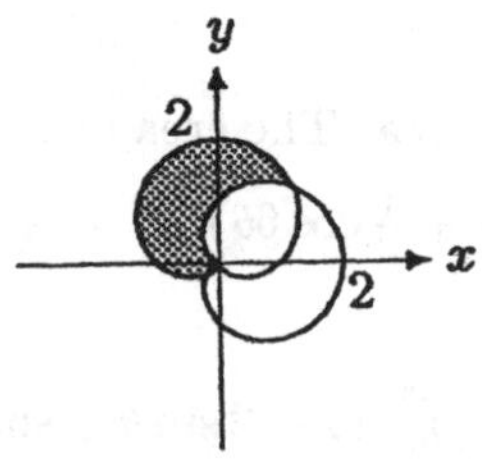

Exercise 33

31. The circle and the cardioid intersect for (r, θ) such that $r = 0$ or $\cos\theta = r = 1 - \cos\theta$, so that $\theta = -\pi/3$ or $\theta = \pi/3$. By symmetry we obtain $A = 2\left[\int_0^{\pi/3} \frac{1}{2}(1 - \cos\theta)^2\,d\theta + \int_{\pi/3}^{\pi/2} \frac{1}{2}\cos^2\theta\,d\theta\right] = \int_0^{\pi/3}(1 - 2\cos\theta + \cos^2\theta)\,d\theta + \int_{\pi/3}^{\pi/2}\cos^2\theta\,d\theta = \left(\frac{3}{2}\theta - 2\sin\theta + \frac{1}{4}\sin 2\theta\right)\Big|_0^{\pi/3} + \left(\frac{1}{2}\theta + \frac{1}{4}\sin 2\theta\right)\Big|_{\pi/3}^{\pi/2} = \left(\frac{1}{2}\pi - \frac{7}{8}\sqrt{3}\right) + \left(\frac{1}{12}\pi - \frac{1}{8}\sqrt{3}\right) = \frac{7}{12}\pi - \sqrt{3}.$

33. The two cardioids intersect for (r, θ) such that $r = 0$ or $1 + \cos\theta = r = 1 + \sin\theta$, so that $\cos\theta = \sin\theta$, and thus $\theta = \pi/4$ or $\theta = 5\pi/4$. Thus $A = \int_{\pi/4}^{5\pi/4} \frac{1}{2}[(1+\sin\theta)^2 - (1+\cos\theta)^2]\,d\theta = \frac{1}{2}\int_{\pi/4}^{5\pi/4}(2\sin\theta + \sin^2\theta - 2\cos\theta - \cos^2\theta)\,d\theta = \frac{1}{2}\int_{\pi/4}^{5\pi/4}(2\sin\theta - 2\cos\theta - \cos 2\theta)\,d\theta = \frac{1}{2}\left(-2\cos\theta - 2\sin\theta - \frac{1}{2}\sin 2\theta\right)\Big|_{\pi/4}^{5\pi/4} = 2\sqrt{2}.$

35. a. Let $g = cf$. Then $g' = cf'$. By (2) the length of the graph of g is given by

$$L = \int_\alpha^\beta \sqrt{(g'(\theta))^2 + (g(\theta))^2}\, d\theta = \int_\alpha^\beta \sqrt{c^2(f'(\theta))^2 + c^2(f(\theta))^2}\, d\theta$$

$$= c\int_\alpha^\beta \sqrt{(f'(\theta))^2 + (f(\theta))^2}\, d\theta.$$

Thus the length L of the graph of f is also multiplied by c when f is multiplied by c.

b. Let $g = cf$. By (4), the area for g is given by

$$A = \int_\alpha^\beta \frac{1}{2}[g(\theta)]^2\, d\theta = \int_\alpha^\beta \frac{1}{2}c^2[f(\theta)]^2\, d\theta = c^2\int_\alpha^\beta \frac{1}{2}[f(\theta)]^2\, d\theta.$$

Thus the area A is multiplied by c^2 when f is multiplied by c.

37. By (11) of Section 6.8 with θ replacing t, and by (1) of this section, we have

$$S = \int_\alpha^\beta 2\pi f(\theta)\sin\theta\sqrt{[f'(\theta)\cos\theta - f(\theta)\sin\theta]^2 + [f'(\theta)\sin\theta + f(\theta)\cos\theta]^2}\, d\theta$$

$$= \int_\alpha^\beta 2\pi f(\theta)\sin\theta\sqrt{(f'(\theta))^2\cos^2\theta + (f(\theta))^2\sin^2\theta + (f'(\theta))^2\sin^2\theta + (f(\theta))^2\cos^2\theta\, d\theta}$$

$$= \int_\alpha^\beta 2\pi f(\theta)\sin\theta\sqrt{(f'(\theta))^2 + (f(\theta))^2}\, d\theta.$$

39. By (6),

$$S = \int_0^{\pi/4} 2\pi\sqrt{2\cos 2\theta}\,\sin\theta\sqrt{\left(\frac{-2\sin 2\theta}{\sqrt{2\cos 2\theta}}\right)^2 + (\sqrt{2\cos 2\theta})^2}\, d\theta$$

$$= \int_0^{\pi/4} 2\pi\sin\theta\sqrt{4\sin^2 2\theta + 4\cos^2 2\theta}\, d\theta = 4\pi\int_0^{\pi/4}\sin\theta\, d\theta = -4\pi\cos\theta\Big|_0^{\pi/4} = 2\pi(2-\sqrt{2}).$$

10.3 Conic Sections

1. $c = -2$; $y^2 = -8x$

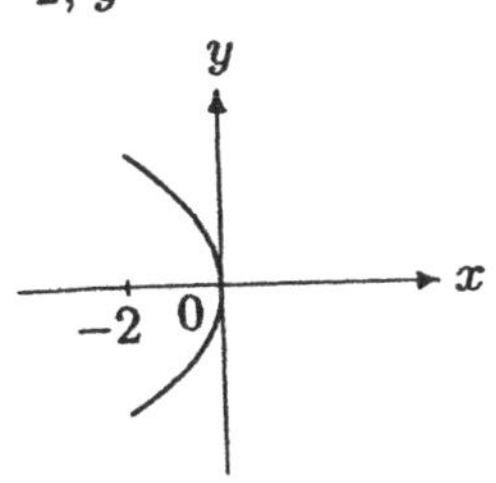

3. $c = 3$; $y^2 = 12(x-1)$

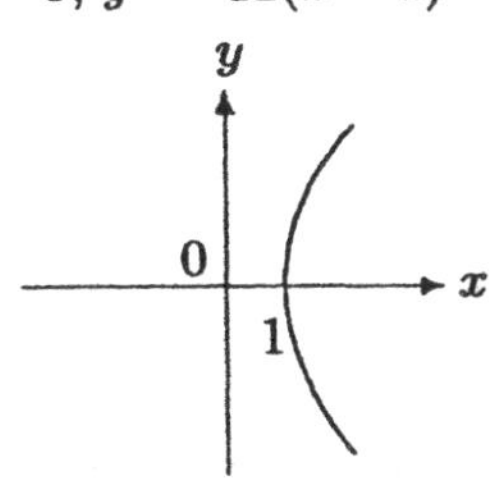

5. $y^2 = 4cx$ and $1^2 = 4c(-1)$, so $c = -\frac{1}{4}$; $y^2 = -x$.

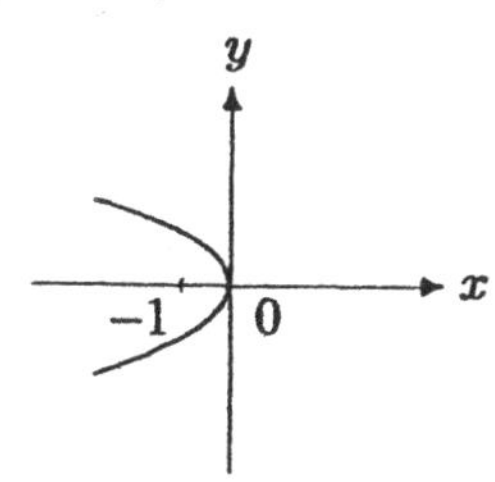

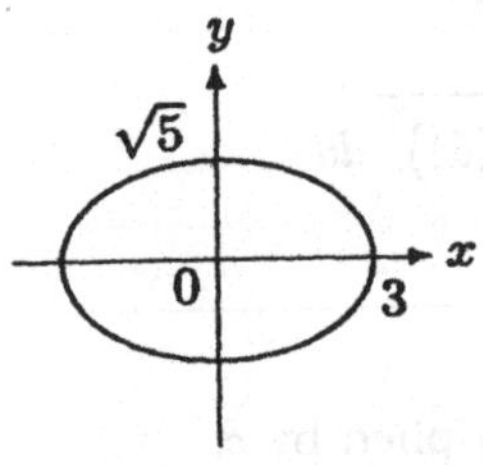

Exercise 7

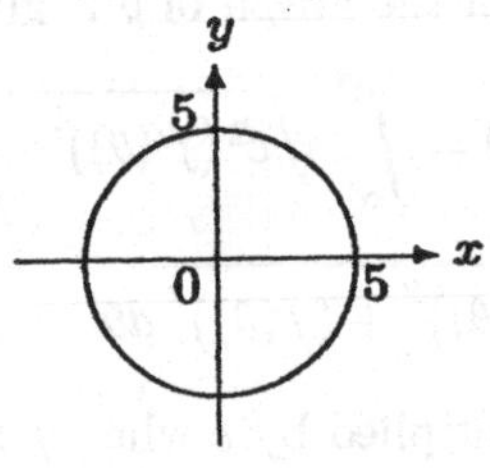

Exercise 9

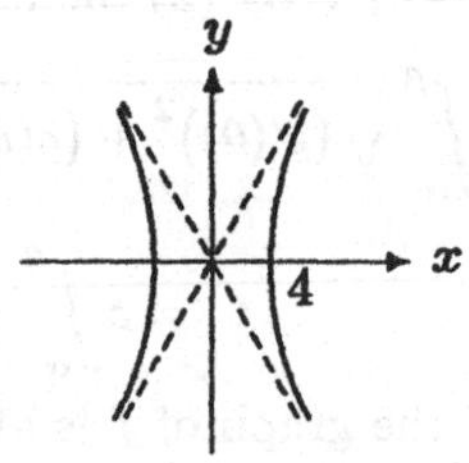

Exercise 11

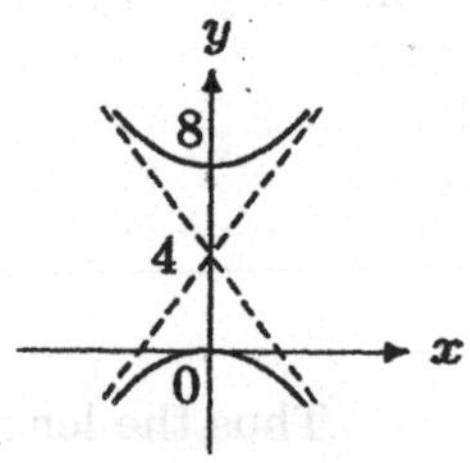

Exercise 13

7. $a = 3$, $c = 2$, $b = \sqrt{5}$; $x^2/9 + y^2/5 = 1$

9. $a = 5; x^2/25 + y^2/b^2 = 1$. Since $(3, -4)$ lies on the ellipse, $\frac{9}{25} + 16/b^2 = 1$, so $16/b^2 = \frac{16}{25}$, or $b = 5$. An equation of the ellipse is $x^2/25 + y^2/25 = 1$, or $x^2 + y^2 = 25$ (so the ellipse is a circle).

11. $a = 4$, $c = 9$, $b = \sqrt{81 - 16} = \sqrt{65}$; $x^2/16 - y^2/65 = 1$

13. The center is $(0, 4)$, $c = 5$; since $y = \frac{4}{3}x + 4$ and $y = -\frac{4}{3}x + 4$ are the asymptotes, $a/b = \frac{4}{3}$, or $a = \frac{4}{3}b$; thus $25 = (\frac{4}{3}b)^2 + b^2 = 25b^2/9$, so $b = 3$, $a = 4$; $(y-4)^2/16 - x^2/9 = 1$.

15. $c = \frac{3}{4}$, focus is $(\frac{3}{4}, 0)$; vertex is $(0, 0)$; directrix is $x = -\frac{3}{4}$; axis is $y = 0$.

17. $c = \frac{1}{4}$, focus is $(1, -\frac{7}{4})$; vertex is $(1, -2)$; directrix is $y = -\frac{9}{4}$; axis is $x = 1$.

19. $a = 5$, $b = 3$, $c = \sqrt{25 - 9} = 4$; foci are $(0, -4)$ and $(0, 4)$; vertices are $(0, -5)$ and $(0, 5)$.

21. $(x-1)^2/\frac{1}{4} + y^2/1 = 1$; $a = 1$, $b = \frac{1}{2}$, $c = \sqrt{1 - \frac{1}{4}} = \sqrt{3}/2$; center is $(1, 0)$; foci are $(1, -\sqrt{3}/2)$ and $(1, \sqrt{3}/2)$; vertices are $(1, -1)$ and $(1, 1)$.

23. $a = 3$, $b = 4$, $c = 5$; foci are $(-5, 0)$ and $(5, 0)$; vertices are $(-3, 0)$ and $(3, 0)$; asymptotes are $y = \frac{4}{3}x$ and $y = -\frac{4}{3}x$.

25. center is $(-3, -1)$; $a = 5$, $b = 12$, $c = \sqrt{25 + 144} = 13$; foci are $(-16, -1)$ and $(10, -1)$; vertices are $(-8, -1)$ and $(2, -1)$; asymptotes are $y + 1 = \frac{12}{5}(x + 3)$ and $y + 1 = -\frac{12}{5}(x + 3)$.

27. $0 = x^2 - 6x - 2y + 1 = (x^2 - 6x + 9) - 2y + 1 - 9 = (x-3)^2 - 2(y+4)$;
$X = x - 3$, $Y = y + 4$; $X^2 = 2Y$;
parabola with vertex $(3, -4)$, focus $(3, -\frac{7}{2})$, directrix $y = -\frac{9}{2}$.

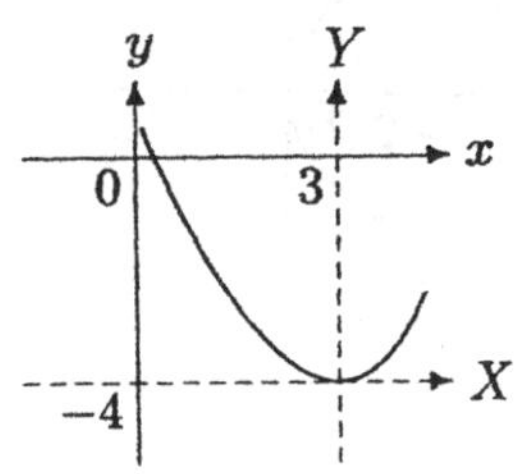

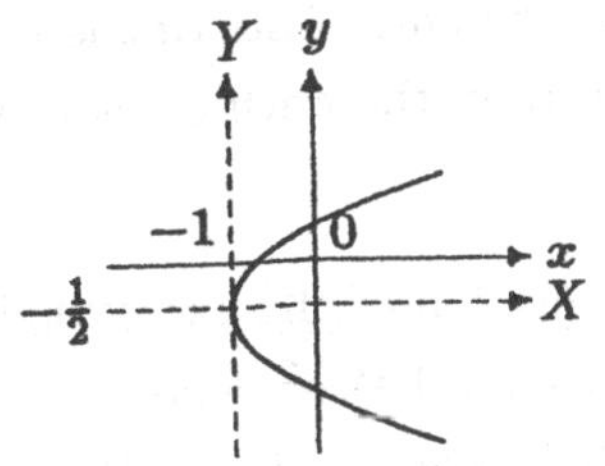

Exercise 29

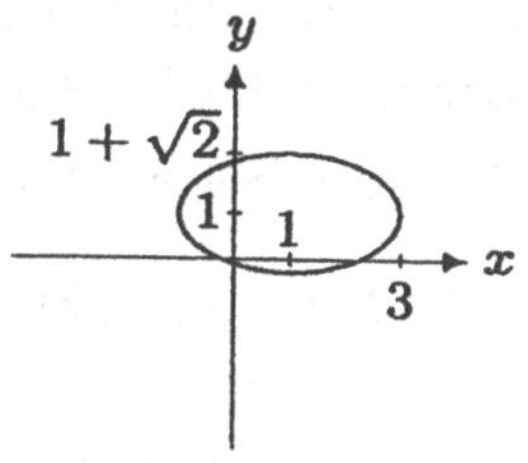

Exercise 31

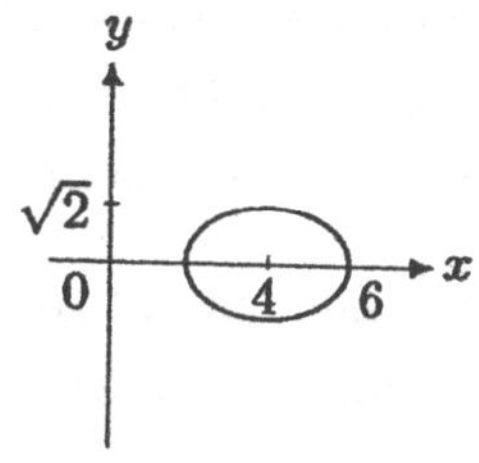

Exercise 33

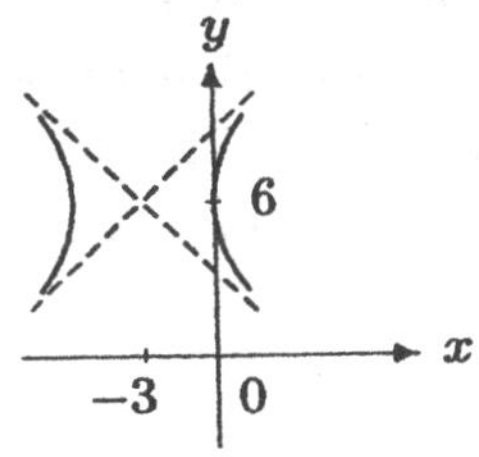

Exercise 35

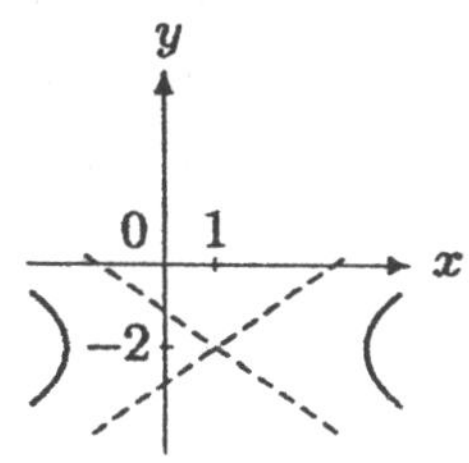

Exercise 37

29. $0 = 3y^2 - 5x + 3y - \frac{17}{4} = (3y^2 + 3y + \frac{3}{4}) - 5x - \frac{17}{4} - \frac{3}{4} = 3(y + \frac{1}{2})^2 - 5(x+1)$; $X = x + 1$, $Y = y + \frac{1}{2}$; $3Y^2 = 5X$, or $Y^2 = \frac{5}{3}X$; parabola with vertex $(-1, -\frac{1}{2})$, focus $(-\frac{7}{12}, -\frac{1}{2})$, directrix $x = -\frac{17}{12}$.

31. $x^2 - 2x + 1 + 2(y^2 - 2y + 1) = 4$, or $(x-1)^2 + 2(y-1)^2 = 4$, so that $(x-1)^2/4 + (y-1)^2/2 = 1$; $a = 2$, $b = \sqrt{2}$, $c = \sqrt{4-2} = \sqrt{2}$; center is $(1, 1)$; foci are $(1 - \sqrt{2}, 1)$ and $(1 + \sqrt{2}, 1)$; vertices are $(-1, 1)$ and $(3, 1)$.

33. $(x^2 - 8x + 16) + 2y^2 = -12 + 16 = 4$, or $(x-4)^2 + 2y^2 = 4$, so that $(x-4)^2/4 + y^2/2 = 1$; $a = 2$, $b = \sqrt{2}$, $c = \sqrt{4-2} = \sqrt{2}$; center is $(4, 0)$; foci are $(4 - \sqrt{2}, 0)$ and $(4 + \sqrt{2}, 0)$; vertices are $(2, 0)$ and $(6, 0)$.

35. $(x^2+6x+9) - (y^2-12y+36) = 9$, or $(x+3)^2 - (y-6)^2 = 9$, so that $(x+3)^2/9 - (y-6)^2/9 = 1$; $a = 3 = b$; center is $(-3, 6)$; vertices are $(-6, 6)$ and $(0, 6)$; asymptotes are $y - 6 = x + 3$ and $y - 6 = -(x+3)$.

37. $4(x^2 - 2x + 1) - 9(y^2 + 4y + 4) = 36$, or $4(x-1)^2 - 9(y+2)^2 = 36$, so that $(x-1)^2/9 - (y+2)^2/4 = 1$; center is $(1, -2)$; $a = 3$, $b = 2$; vertices are $(-2, -2)$ and $(4, -2)$; asymptotes are $y + 2 = \frac{2}{3}(x-1)$ and $y + 2 = -\frac{2}{3}(x-1)$.

39. Since the parabola has vertex $(-1, 3)$ and a vertical axis, an equation of the parabola has the form $(x+1)^2 = 4c(y-3)$, for an appropriate nonzero constant c. Differentiating implicitly with respect to x, we obtain $2(x+1) = 4c(dy/dx)$. Since the parabola has slope 2 at $x = 1$, it follows that $2(1+1) = (4c)2$, so that $c = \frac{1}{2}$. Therefore the parabola is given by $(x+1)^2 = 4(\frac{1}{2})(y-3) = 2(y-3)$.

41. If the parabola has the form $x^2 = 4cy$ with $c > 0$, then the focus is $(0, c)$ and the endpoints of the latus rectum are $(2c, c)$ and $(-2c, c)$, and its length is $4c$. If $y^2 = 4cx$ with $c > 0$, then the endpoints of the latus rectum are $(c, 2c)$ and $(c, -2c)$, and its length is again $4c$.

43. The slope of a line tangent to the parabola at a point $(a, a^2/2)$ is a, so an equation of the tangent line is $y - a^2/2 = a(x - a)$. If $(-1, -4)$ is on this line, then $-4 - a^2/2 = a(-1 - a)$, so that $a^2 + 2a - 8 = 0$, and thus $a = -4$ or $a = 2$. Equations of the two lines tangent to the parabola are $y - 8 = -4(x + 4)$ and $y - 2 = 2(x - 2)$.

45. Differentiating the equation $4x^2 + y^2 = 8$ implicitly, we obtain $8x + 2y(dy/dx) = 0$, so that at (x_0, y_0), we have $dy/dx = -4x_0/y_0$. An equation of the line tangent at (x_0, y_0) is $y - y_0 = (-4x_0/y_0)(x - x_0)$. This will be the line $2x + y = d$ if $2 = 4x_0/y_0$ and $d = y_0 + 4x_0^2/y_0$. Then $y_0 = 2x_0$. Since (x_0, y_0) is on the ellipse, $4x_0^2 + (2x_0)^2 = 8$, so that $x_0 = \pm 1$ and $y_0 = \pm 2$. Then $d = 4$ or $d = -4$, and the points of tangency are $(1, 2)$ and $(-1, -2)$.

47. Let (x, y) be a point in the collection. Then $\sqrt{(x-3)^2 + y^2} = \frac{1}{2}\sqrt{(x+3)^2 + (y-y)^2}$, or $(x-3)^2 + y^2 = \frac{1}{4}(x+3)^2$. The equation becomes $3x^2 - 30x + 27 + 4y^2 = 0$, or $3(x-5)^2 + 4y^2 = 48$, or $(x-5)^2/16 + y^2/12 = 1$.

49. Let (x, y) be the vertex of the rectangle in the first quadrant, so that $y = b\sqrt{1 - x^2/a^2}$. The area A of the rectangle is given by $A = 4xy = 4bx\sqrt{1 - x^2/a^2}$, and

$$\frac{dA}{dx} = 4b\sqrt{1 - \frac{x^2}{a^2}} - \frac{4bx^2}{a^2}\left(1 - \frac{x^2}{a^2}\right)^{-1/2} = \left(4b - \frac{8bx^2}{a^2}\right)\left(1 - \frac{x^2}{a^2}\right)^{-1/2}$$

Now $dA/dx = 0$ for $4b - 8bx^2/a^2 = 0$, or $x = (\sqrt{2}/2)a$, and $dA/dx > 0$ if $0 < x < a\sqrt{2}/2$, and $dA/dx < 0$ if $a\sqrt{2}/2 < x < a$. By (1) of Section 4.6 and the First Derivative Test, A is maximum if $x = a\sqrt{2}/2$. The vertices of the rectangle are $(\pm a\sqrt{2}/2, \pm b\sqrt{2}/2)$.

51. $$0 = Ax^2 + Cy^2 + Dx + Ey + F = A\left(x^2 + \frac{D}{A}x + \frac{D^2}{4A^2}\right) + C\left(y^2 + \frac{E}{C}y + \frac{E^2}{4C^2}\right) + F - \frac{D^2}{4A} - \frac{E^2}{4C}$$
$$= A\left(x + \frac{D}{2A}\right)^2 + C\left(y + \frac{E}{2C}\right)^2 - r, \quad \text{so that} \quad A\left(x + \frac{D}{2A}\right)^2 + C\left(y + \frac{E}{2C}\right)^2 = r.$$

a. If $r \neq 0$, then the graph of the equation is a hyperbola, since $AC < 0$.

b. If $r = 0$, then $(y + E/2C)^2 = -(A/C)(x + D/2A)^2$, which defines the two intersecting lines $y + E/2C = \pm\sqrt{-A/C}\,(x + D/2A)$.

53. Differentiating the equation $x^2/a^2 - y^2/b^2 = 1$ implicitly, we obtain $2x/a^2 - (2y/b^2)(dy/dx) = 0$, so that $dy/dx = b^2x/(a^2y)$. At (x_0, y_0) we have $dy/dx = b^2x_0/(a^2y_0)$. An equation of the tangent line is $y - y_0 = [b^2x_0/(a^2y_0)](x - x_0)$. Multiplying both sides of this equation by y_0/b^2, we obtain

$$\frac{yy_0}{b^2} - \frac{y_0^2}{b^2} = \frac{xx_0}{a^2} - \frac{x_0^2}{a^2}, \quad \text{and thus} \quad \frac{xx_0}{a^2} - \frac{yy_0}{b^2} = \frac{x_0^2}{a^2} - \frac{y_0^2}{b^2} = 1.$$

55. The path is symmetric about the vertical line which passes through its highest point, which is located at $(300, 200)$. Thus an equation of the parabola is $(x-300)^2 = 4c(y-200)$, where c is such that $(0,0)$ lies on the parabola (as does $(600,0)$). Thus $(-300)^2 = 4c(-200)$, so that $c = 90{,}000/(-800) = -225/2$. The equation of the parabola becomes $(x-300)^2 = -450(y-200)$.

57. $a \approx .1425$ and $b \approx .1205$, so that $c \approx \sqrt{(.1425)^2 - (.1205)^2} \approx .076066$. Then the distance from a vertex to the closest focus is approximately $.1425 - .076066 = .066434$ (miles).

59. If Marian is at $(4400, 0)$, Jack at $(-4400, 0)$, and Bruce midway between Marian and Jack, we find that Bruce is at the origin. Since Bruce hears the thunder 1 second after Marian does, if (x, y) represents the point at which the lightning strikes, then

$$\sqrt{x^2+y^2} - \sqrt{(x-4400)^2+y^2} = 1100$$

and this represents the equation $(x-2200)^2/a^2 - y^2/b^2 = 1$, with $x > 2200$, and $a = \frac{1100}{2} = 550$. Since Marian and Bruce are at the foci, 4400 feet apart, $2c = 4400$, so that $c = 2200$. Thus $b = \sqrt{c^2 - a^2} = \sqrt{(2200)^2 - (550)^2} = 550\sqrt{15}$. Therefore the equation of the hyperbola becomes

$$\frac{(x-2200)^2}{(550)^2} - \frac{y^2}{15(550)^2} = 1.$$

From the result of Exercise 58, along with the above equation, Marian hears the thunder 4 seconds before Jack and 1 second before Bruce only if the lightning strikes a point (x, y) satisfying

$$\frac{x^2}{(2200)^2} - \frac{y^2}{3(2200)^2} = 1 \quad \text{and} \quad \frac{(x-2200)^2}{(550)^2} - \frac{y^2}{15(550)^2} = 1.$$

Solving for y^2 in the first equation yields $y^2 = 15(x-2200)^2 - 15(550)^2$, and substituting in the first equation yields

$$\frac{x^2}{(2200)^2} - \frac{15(x-2200)^2 - 15(550)^2}{3(2200)^2} = 1.$$

This equation reduces to $4x^2 - 5(4400)x + 9(550)^2 = 0$, so by the quadratic formula,

$$x = \frac{5500 \pm \sqrt{25(16)(275)^2 - 4(91)(275)^2}}{2} = 3575 \quad \text{or} \quad 1925.$$

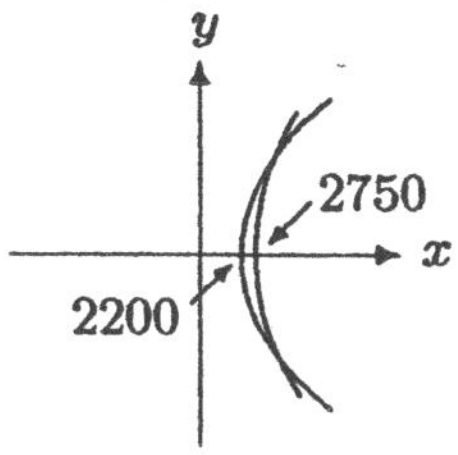

Since x must be greater than 2200, it follows that $x = 3575$, and thus $y^2 = 15(3575-2200)^2 - 15(550)^2$, so that $y = \pm 825\sqrt{35}$. Consequently lightning strikes at $(3575, 825\sqrt{35})$ or at $(3575, -825\sqrt{35})$.

61. By Exercise 61 an equation of the parabola is $x^2 = (a^2/4b)y$. If $a = 3500$ and $b = 316$, then $x^2 = [(3500)^2/4(316)]y = (765{,}625/79)y$.

10.4 Rotation of Axes

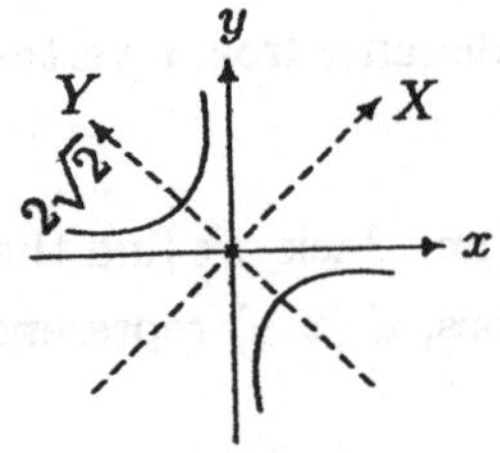

Exercise 1

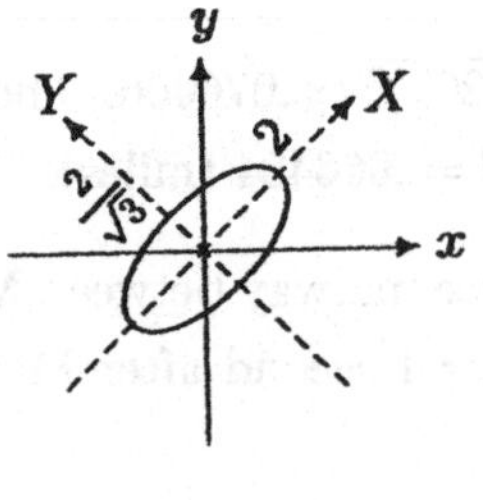

Exercise 3

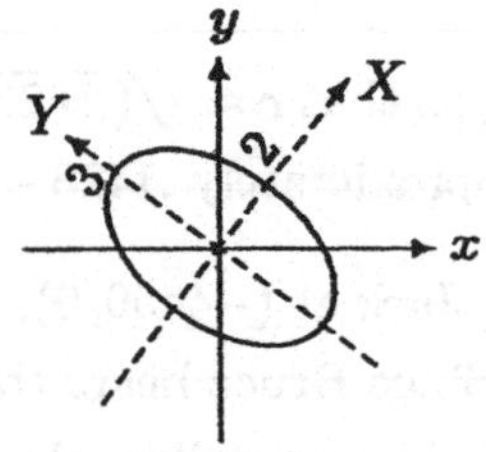

Exercise 5

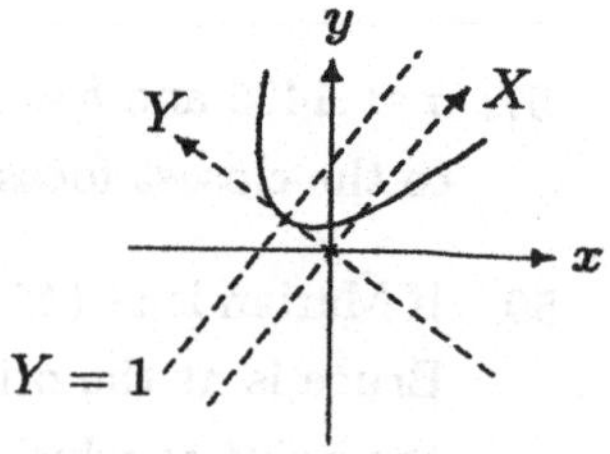

Exercise 7

1. $A = 0 = C$, so $\theta = \pi/4$; the equation becomes

$$\left(\frac{\sqrt{2}}{2}X - \frac{\sqrt{2}}{2}Y\right)\left(\frac{\sqrt{2}}{2}X + \frac{\sqrt{2}}{2}Y\right) = -4 \quad \text{or} \quad \frac{Y^2}{8} - \frac{X^2}{8} = 1.$$

The conic section is a hyperbola.

3. $A = 1 = C$, so $\theta = \pi/4$; the equation becomes

$$\left(\frac{\sqrt{2}}{2}X - \frac{\sqrt{2}}{2}Y\right)^2 - \left(\frac{\sqrt{2}}{2}X - \frac{\sqrt{2}}{2}Y\right)\left(\frac{\sqrt{2}}{2}X + \frac{\sqrt{2}}{2}Y\right) + \left(\frac{\sqrt{2}}{2}X + \frac{\sqrt{2}}{2}Y\right)^2 = 2,$$

or $X^2/4 + 3Y^2/4 = 1$. The conic section is an ellipse.

5. $A = 145$, $B = 120$, $C = 180$, $\tan 2\theta = 120/(145 - 180) = -\frac{24}{7}$; from Example 2, $\cos\theta = \frac{3}{5}$ and $\sin\theta = \frac{4}{5}$; the equation becomes

$$145\left(\frac{3}{5}X - \frac{4}{5}Y\right)^2 + 120\left(\frac{3}{5}X - \frac{4}{5}Y\right)\left(\frac{4}{5}X + \frac{3}{5}Y\right) + 180\left(\frac{4}{5}X + \frac{3}{5}Y\right)^2 = 900,$$

or $X^2/4 + Y^2/9 = 1$. The conic section is an ellipse.

7. $A = 16$, $B = -24$, $C = 9$; $\tan 2\theta = -24/(16 - 9) = -\frac{24}{7}$; from Example 2, $\cos\theta = \frac{3}{5}$ and $\sin\theta = \frac{4}{5}$; the equation becomes

$$16\left(\frac{3}{5}X - \frac{4}{5}Y\right)^2 - 24\left(\frac{3}{5}X - \frac{4}{5}Y\right)\left(\frac{4}{5}X + \frac{3}{5}Y\right) + 9\left(\frac{4}{5}X + \frac{3}{5}Y\right)^2$$

$$-5\left(\frac{3}{5}X - \frac{4}{5}Y\right) - 90\left(\frac{4}{5}X + \frac{3}{5}Y\right) + 25 = 0,$$

or $25Y^2 - 75X - 50Y + 25 = 0$, or $(Y - 1)^2 = 3X$. The conic section is a parabola.

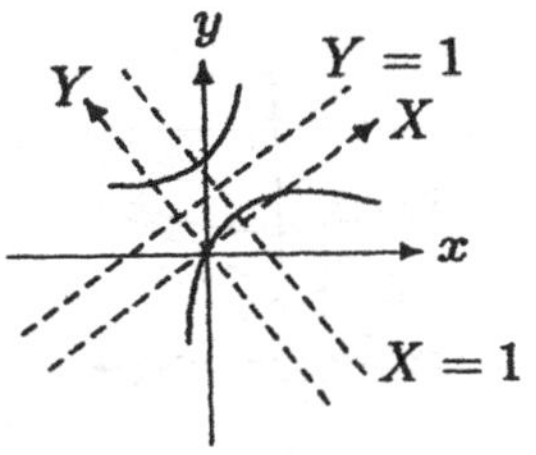

Exercise 9

9. $A = 2$, $B = -72$, $C = 23$; $\tan 2\theta = -72/(2-23) = \frac{24}{7}$; by the method of Example 2, $\cos 2\theta = \frac{7}{25}$, $\cos\theta = \frac{4}{5}$ and $\sin\theta = \frac{3}{5}$; the equation becomes

$$2\left(\frac{4}{5}X - \frac{3}{5}Y\right)^2 - 72\left(\frac{4}{5}X - \frac{3}{5}Y\right)\left(\frac{3}{5}X + \frac{4}{5}Y\right) + 23\left(\frac{3}{5}X + \frac{4}{5}Y\right)^2$$

$$+100\left(\frac{4}{5}X - \frac{3}{5}Y\right) - 50\left(\frac{3}{5}X + \frac{4}{5}Y\right) = 0,$$

or $-25X^2 + 50Y^2 + 50X - 100Y = 0$, or $2(Y-1)^2 - (X-1)^2 = 1$. The conic section is a hyperbola.

11. $A = 9$, $B = -24$, $C = 2$; $\tan 2\theta = -24/(9-2) = -\frac{24}{7}$; from Example 2, $\cos\theta = \frac{3}{5}$ and $\sin\theta = \frac{4}{5}$; the equation becomes $9(\frac{3}{5}X - \frac{4}{5}Y)^2 - 24(\frac{3}{5}X - \frac{4}{5}Y)(\frac{4}{5}X + \frac{3}{5}Y) + 2(\frac{4}{5}X + \frac{3}{5}Y)^2 = 0$, or $18Y^2 = 7X^2$; the graph consists of two intersecting lines, $Y = \sqrt{\frac{7}{18}}X$ and $Y = -\sqrt{\frac{7}{18}}X$.

13. $A = 145$, $B = 120$, $C = 180$; $\tan 2\theta = 120/(145-180) = -\frac{24}{7}$; from Example 2, $\cos\theta = \frac{3}{5}$ and $\sin\theta = \frac{4}{5}$; the equation becomes $145(\frac{3}{5}X - \frac{4}{5}Y)^2 + 120(\frac{3}{5}X - \frac{4}{5}Y)(\frac{4}{5}X + \frac{3}{5}Y) + 180(\frac{4}{5}X + \frac{3}{5}Y)^2 = -900$, or $225X^2 + 100Y^2 = -900$; no points on the graph.

15. Since $A = 0 = C$, we have $\theta = \pi/4$. The equation becomes

$$B\left(\frac{\sqrt{2}}{2}X - \frac{\sqrt{2}}{2}Y\right)\left(\frac{\sqrt{2}}{2}X + \frac{\sqrt{2}}{2}Y\right) + D\left(\frac{\sqrt{2}}{2}X - \frac{\sqrt{2}}{2}Y\right) + E\left(\frac{\sqrt{2}}{2}X + \frac{\sqrt{2}}{2}Y\right) + F = 0$$

or

$$\frac{1}{2}B(X^2 - Y^2) + \frac{\sqrt{2}}{2}(D+E)X + \frac{\sqrt{2}}{2}(E-D)Y + F = 0.$$

When the squares are completed the equation will take the form $\frac{1}{2}B(X-a)^2 - \frac{1}{2}B(Y-b)^2 = c$. If $c \neq 0$, the graph is a hyperbola; if $c = 0$, the graph is two intersecting lines.

17. Let $B' = -2A\cos\theta\sin\theta + B(\cos^2\theta - \sin^2\theta) + 2C\cos\theta\sin\theta = B(\cos^2\theta - \sin^2\theta) - 2(A-C)\sin\theta\cos\theta$. From (6) and the equation preceding (6), we find that

$$A' = A\cos^2\theta + B\sin\theta\cos\theta + C\sin^2\theta$$

$$C' = A\sin^2\theta - B\sin\theta\cos\theta + C\cos^2\theta.$$

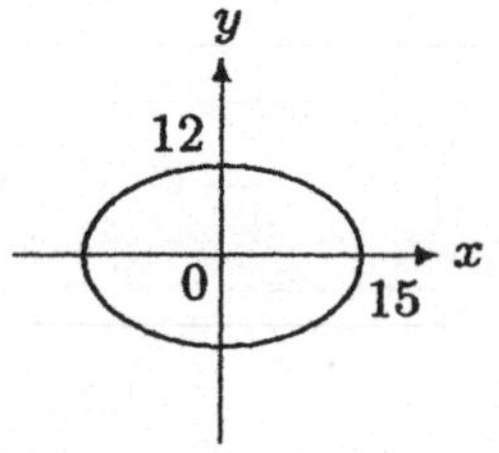

Exercise 13

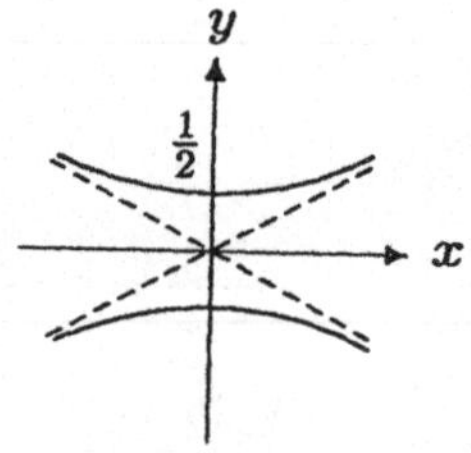

Exercise 15

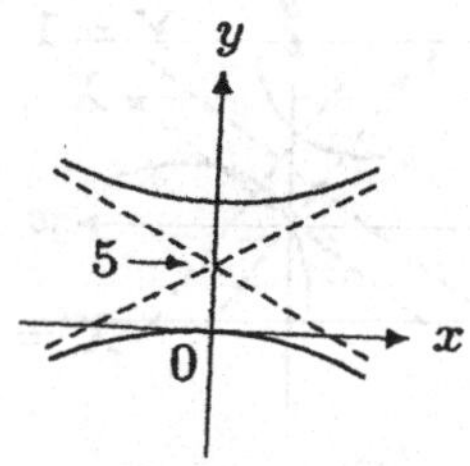

Exercise 17

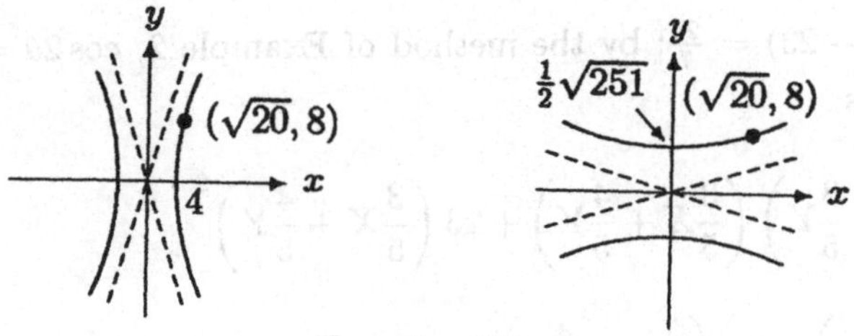

Exercise 19

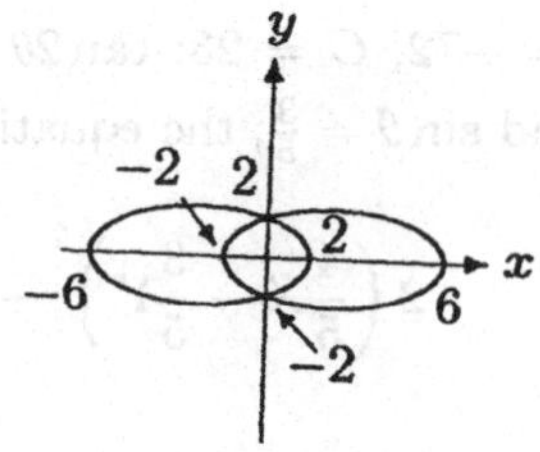

Exercise 21

13. $c = 9,\ e = \dfrac{3}{5},\ a = \dfrac{9}{3/5} = 15,\ b = \sqrt{225 - 81} = 12;\ \dfrac{x^2}{225} + \dfrac{y^2}{144} = 1$

15. $c = 1,\ e = 2,\ a = \dfrac{1}{2},\ b = \sqrt{1 - \dfrac{1}{4}} = \dfrac{\sqrt{3}}{2};\ 4y^2 - \dfrac{4x^2}{3} = 1$

17. $a = 5,\ e = \dfrac{13}{5},\ c = 5\left(\dfrac{13}{5}\right) = 13,\ b = \sqrt{169 - 25} = 12;\ \dfrac{(y-5)^2}{25} - \dfrac{x^2}{144} = 1$

19. Since $c = ae = a\sqrt{17}$, we have $b = \sqrt{17a^2 - a^2} = 4a$. Thus an equation of the hyperbola is either

$$\frac{x^2}{a^2} - \frac{y^2}{16a^2} = 1 \quad \text{or} \quad \frac{y^2}{a^2} - \frac{x^2}{16a^2} = 1.$$

Since the hyperbola passes through $(\sqrt{20}, 8)$, we find for the first of these equations that $20/a^2 - 64/16a^2 = 1$, or $a^2 = 16$, so that the equation becomes $x^2/16 - y^2/256 = 1$. For the second equation we find that $64/a^2 - 20/16a^2 = 1$, or $a^2 = \frac{251}{4}$, so that the equation becomes $4y^2/251 - x^2/1004 = 1$.

21. Since $a = 4$ and $e = \frac{1}{2}$, we have $c = 4(\frac{1}{2}) = 2$. Thus $b = \sqrt{16 - 4} = 2\sqrt{3}$, and also the center is located at either $(2, 0)$ or $(-2, 0)$. Therefore an equation of the ellipse is either $(x-2)^2/16 + y^2/12 = 1$ or $(x+2)^2/16 + y^2/12 = 1$.

23. $e = 2,\ k = \frac{1}{2}$; hyperbola with directrix $x = -\frac{1}{2}$

25. $r = \dfrac{5}{1 - \frac{3}{5}\sin\theta};\ e = \dfrac{3}{5},\ k = \dfrac{25}{3}$; ellipse with directrix $y = -\dfrac{25}{3}$

Therefore

$$\begin{aligned}(B')^2 - 4A'C' &= [B(\cos^2\theta - \sin^2\theta) - 2(A-C)\sin\theta\cos\theta]^2 \\ &\quad -4(A\cos^2\theta + B\sin\theta\cos\theta + C\sin^2\theta)(A\sin^2\theta - B\sin\theta\cos\theta + C\cos^2\theta) \\ &= B^2(\cos^4\theta - 2\sin^2\theta\cos^2\theta + \sin^4\theta + 4\sin^2\theta\cos^2\theta) \\ &\quad -4AB(\sin\theta\cos^3\theta - \sin^3\theta\cos\theta - \sin\theta\cos^3\theta + \sin^3\theta\cos\theta) \\ &\quad +4BC((\sin\theta\cos^3\theta - \sin^3\theta\cos\theta - \sin\theta\cos^3\theta + \sin^3\theta\cos\theta) \\ &\quad +4A^2(\sin^2\theta\cos^2\theta - \sin^2\theta\cos^2\theta) \\ &\quad -4AC(2\sin^2\theta\cos^2\theta + \cos^4\theta + \sin^4\theta) \\ &\quad +4C^2(\sin^2\theta\cos^2\theta - \sin^2\theta\cos^2\theta) \\ &= B^2(\cos^2\theta + \sin^2\theta)^2 - 4AC(\cos^2\theta + \sin^2\theta)^2 \\ &= B^2 - 4AC.\end{aligned}$$

But $B' = 0$ for the value of θ chosen in (8). Thus $B^2 - 4AC = -4A'C'$.

10.5 A Unified Description of Conic Sections

1. $c = \sqrt{25-9} = 4$; $e = \dfrac{4}{5}$

3. $c = \sqrt{9+25} = \sqrt{34}$; $e = \dfrac{\sqrt{34}}{3}$

5. $\dfrac{x^2}{2} + \dfrac{y^2}{8} = 1$; $c = \sqrt{8-2} = \sqrt{6}$; $e = \dfrac{\sqrt{6}}{\sqrt{8}} = \dfrac{\sqrt{3}}{2}$

7. $\dfrac{(x-3)^2}{2} + \dfrac{(y+3)^2}{8} = 1$; $c = \sqrt{8-2} = \sqrt{6}$; $e = \dfrac{\sqrt{6}}{\sqrt{8}} = \dfrac{\sqrt{3}}{2}$

9. $\dfrac{(x-1)^2}{4} + \dfrac{(y-1)^2}{2} = 1$; $c = \sqrt{4-2} = \sqrt{2}$; $e = \dfrac{\sqrt{2}}{2}$

11. $\dfrac{(x+1)^2}{9} - \dfrac{(y+2)^2}{49} = 1$; $c = \sqrt{9+49} = \sqrt{58}$; $e = \dfrac{\sqrt{58}}{3}$

27. $r = \dfrac{\frac{1}{2}}{1-\cos\theta}$; $e = 1$, $k = \dfrac{1}{2}$; parabola with directrix $x = -\dfrac{1}{2}$

29. $e = 1$, $k = 3$; parabola with directrix $y = 3$

31. Since $e = .017$ and $a = \frac{299}{2}$, we have $c = (\frac{299}{2})(.017) = 2.5415$, and thus

$$b = \sqrt{\left(\frac{299}{2}\right)^2 - (2.5415)^2} \approx 149.48.$$

An approximate equation of the orbit is $x^2/(149.5)^2 + y^2/(149.48)^2 = 1$, if we place the x axis along the major axis and the origin at the center of the orbit.

Chapter 10 Review

1. $r = \sin 5\theta$; symmetry with respect to y axis

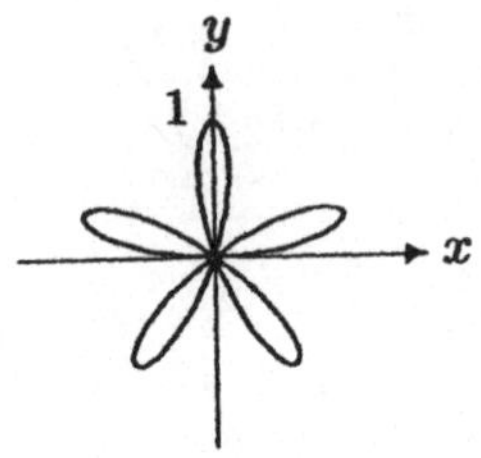

3. $r = \sqrt{3} - 2\sin\theta$; symmetry with respect to y axis

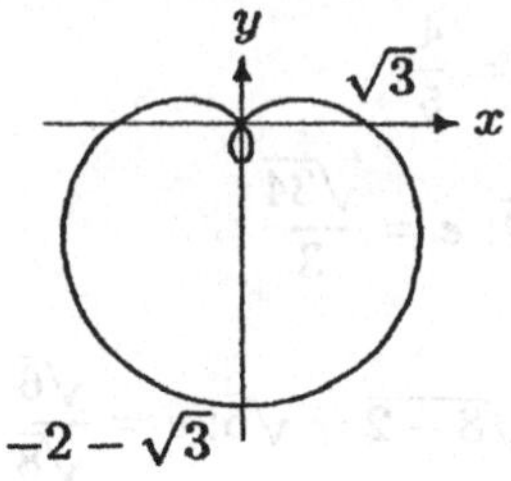

5. a. $r = 2\sin 2\theta$ and $r = 2\sin\theta$

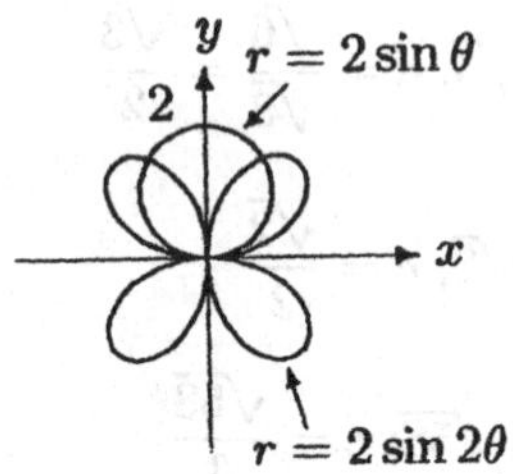

b. The graphs intersect for (r, θ) such that $2\sin 2\theta = r = 2\sin\theta$, which means $r = 0$ or $4\sin\theta\cos\theta = 2\sin\theta$, and thus $\theta = 0$, or $\theta = \pi$, or $\theta = \pi/3$ or $\theta = 5\pi/3$. In polar coordinates the points of intersection are the origin, $(\sqrt{3}, \pi/3)$ and $(-\sqrt{3}, 5\pi/3)$.

7. The region enclosed by $r = 2\cos\theta$ is obtained from $r = 2\cos\theta$ for $-\pi/2 \le \theta \le \pi/2$. The region enclosed by $r = \sin\theta + \cos\theta$ is obtained from $r = \sin\theta + \cos\theta$ for $-\pi/4 \le \theta \le 3\pi/4$. The circles intersect for (r, θ) such that $2\cos\theta = r = \sin\theta + \cos\theta$, so that $\theta = \pi/4$. Thus

$$A = \int_{-\pi/4}^{\pi/4} \frac{1}{2}(\sin\theta + \cos\theta)^2\, d\theta + \int_{\pi/4}^{\pi/2} \frac{1}{2}(2\cos\theta)^2\, d\theta$$

$$= \frac{1}{2}\int_{-\pi/4}^{\pi/4} (1 + \sin 2\theta)\, d\theta + 2\int_{\pi/4}^{\pi/2} \cos^2\theta\, d\theta$$

$$= \frac{1}{2}\left(\theta - \frac{1}{2}\cos 2\theta\right)\Bigg|_{-\pi/4}^{\pi/4} + \left(\theta + \frac{1}{2}\sin 2\theta\right)\Bigg|_{\pi/4}^{\pi/2} = \frac{\pi}{2} - \frac{1}{2}.$$

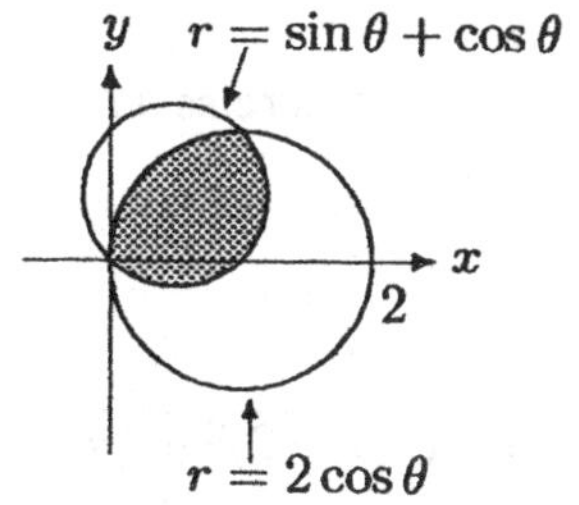

9. The vertex is $(1, 4)$; $c = -4$; $(y - 4)^2 = -16(x - 1)$

11. $a = 2\sqrt{2}$; an equation of the ellipse is $x^2/b^2 + y^2/8 = 1$. Since $(1, \sqrt{6})$ is on the ellipse, we have $1/b^2 + \frac{6}{8} = 1$, so that $b = 2$. Consequently an equation of the ellipse becomes $x^2/4 + y^2/8 = 1$.

13. The center is $(-1, 2)$; $a = 2$; an equation of the hyperbola is $(x + 1)^2/4 - (y - 2)^2/b^2 = 1$. The asymptotes have equations $y - 2 = (b/2)(x + 1)$ and $y - 2 = -(b/2)(x + 1)$. Since the asymptotes are perpendicular, $b/2 = -1/(-b/2) = 2/b$ so that $b = 2$. The equation of the hyperbola becomes $(x + 1)^2/4 - (y - 2)^2/4 = 1$.

15. The distance from the vertex to the directrix is 2. Since $e = 2$, the distance from the vertex to the focus is $2 \cdot 2 = 4$. Thus the focus corresponding to the vertex $(-2, 0)$ is $(2, 0)$. The center is $(p, 0)$, for an appropriate value of p. Then $2 = c/a = (2 - p)/(-2 - p)$, so that $p = -6$, and the center is $(-6, 0)$. Thus $a = 4$, $c = 8$, and $b = \sqrt{64 - 16} = 4\sqrt{3}$. Therefore an equation of the hyperbola is $(x + 6)^2/16 - y^2/48 = 1$.

17. $49(x^2 + 2x + 1) - 9(y^2 + 4y + 4) = 441$, or $49(x+1)^2 - 9(y+2)^2 = 441$, so that $(x+1)^2/9 - (y+2)^2/49 = 1$.

19. $9(x^2 - 4x + 4) + 5y - 15 = 0$, so that $9(x - 2)^2 = -5(y - 3)$.

21. $A = 4$, $B = -24$, $C = 11$; $\tan 2\theta = -24/(4 - 11) = \frac{24}{7}$; by the method of Example 2 in Section 10.4, $\cos 2\theta = \frac{7}{25}$, $\cos\theta = \frac{4}{5}$, and $\sin\theta = \frac{3}{5}$; the equation becomes $4(\frac{4}{5}X - \frac{3}{5}Y)^2 - 24(\frac{4}{5}X - \frac{3}{5}Y)(\frac{3}{5}X + \frac{4}{5}Y) + 11(\frac{3}{5}X + \frac{4}{5}Y)^2 + 40(\frac{4}{5}X - \frac{3}{5}Y) + 30(\frac{3}{5}X + \frac{4}{5}Y) - 45 = 0$, or $-5X^2 + 50X + 20Y^2 - 45 = 0$, or $(X - 5)^2/16 - Y^2/4 = 1$.

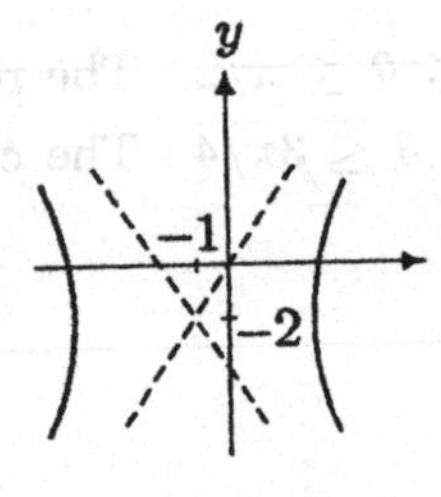

Exercise 17

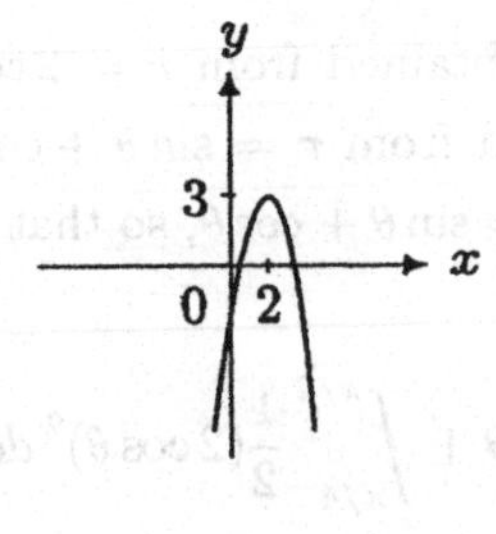

Exercise 19

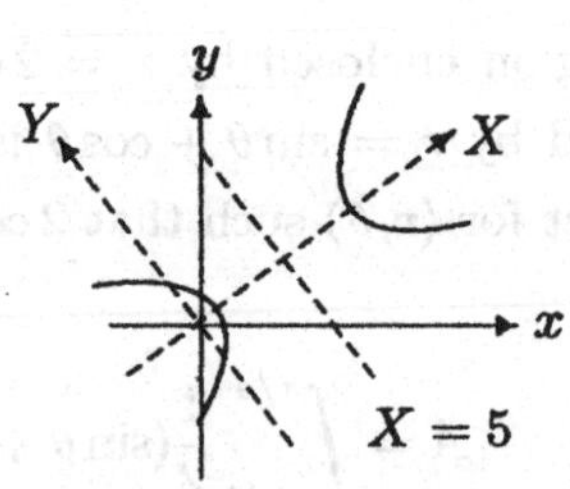

Exercise 21

23. a. $\sqrt{(x-2)^2+(y-4)^2}=3|y|$, so that $(x-2)^2+(y-4)^2=9y^2$, or $(x-2)^2-8(y^2+y+\frac{1}{4})=-18$ or $4(y+\frac{1}{2})^2/9-(x-2)^2/18=1$.

b. $\sqrt{(x-2)^2+(y-4)^2}=|y|$, so that $(x-2)^2+(y-4)^2=y^2$, or $(x-2)^2-8(y-2)=0$, or $(x-2)^2=8(y-2)$.

c. $\sqrt{(x-2)^2+(y-4)^2}=\frac{1}{2}|y|$, so that $(x-2)^2+(y-4)^2=\frac{1}{4}y^2$, or $(x-2)^2+\frac{3}{4}(y^2-\frac{32}{3}y+\frac{256}{9})=\frac{16}{3}$, or $3(x-2)^2/16+9(y-16/3)^2/64=1$.

25. $r=\dfrac{\frac{3}{4}}{1-\frac{1}{4}\cos\theta}$, so $e=\frac{1}{4}$, and hence the conic section is a ellipse.

27. If $x^2=4cy$, then the slope of the line tangent to the parabola at $(x_0,x_0^2/4c)$ is $(dy/dx)|_{x=x_0}=x_0/2c$. An equation of the tangent line is $y-x_0^2/4c=(x_0/2c)(x-x_0)$. If the tangent line passes through (a,b), then $b-x_0^2/4c=(x_0/2c)(a-x_0)$, or $(1/4c)x_0^2-(a/2c)x_0+b=0$. Solving for x_0, we find that

$$x_0=\frac{a/2c\pm\sqrt{a^2/4c^2-b/c}}{1/2c}=a\pm\sqrt{a^2-4bc}.$$

By assumption, $a^2>|4cb|$, so that two such values of x_0 exist. Consequently equations of the tangent line are

$$y-\frac{(a+\sqrt{a^2-4bc})^2}{4c}=\frac{a+\sqrt{a^2-4bc}}{2c}(x-a-\sqrt{a^2-4bc})$$

and

$$y-\frac{(a-\sqrt{a^2-4bc})^2}{4c}=\frac{a-\sqrt{a^2-4bc}}{2c}(x-a+\sqrt{a^2-4bc}).$$

Cumulative Review(Chapters 1–9)

1. $\displaystyle\lim_{x\to\infty} \frac{x^2+1}{x\sqrt{3x^2+1}} = \lim_{x\to\infty} \frac{1+1/x^2}{\sqrt{3+1/x^2}} = \frac{1}{\sqrt{3}}$

3. The conditions for applying l'Hôpital's Rule are met:

$$\lim_{x\to 0} \frac{1-\cos 4x}{3x^2} = \lim_{x\to 0} \frac{4\sin 4x}{6x} = \lim_{x\to 0} \frac{16\cos 4x}{6} = \frac{8}{3}.$$

5. Notice that

$$f(x) = \int_{-x}^{x^2} \sin\sqrt{t^3+1}\,dt = \int_{-x}^{0} \sin\sqrt{t^3+1}\,dt + \int_{0}^{x^2} \sin\sqrt{t^3+1}\,dt.$$

Let $G(x) = \int_{-x}^{0} \sin\sqrt{t^3+1}\,dt$ and $H(x) = \int_0^{x^2} \sin\sqrt{t^3+1}\,dt$, so that $G(x) = -\int_0^{-x} \sin\sqrt{t^3+1}\,dt$. Then $G'(x) = [-\sin\sqrt{-x^3+1}](-1)$ and $H'(x) = (\sin\sqrt{x^6+1})(2x) = 2x\sin\sqrt{x^6+1}$. Therefore $f'(x) = G'(x) + H'(x) = \sin\sqrt{-x^3+1} + 2x\sin\sqrt{x^6+1}$.

7. a. Let x denote the horizontal distance between the person and the kite, and y the length of string let out. Then $x^2 + 30^2 = y^2$. We want to find dx/dt when $y = 50$. Now $(2x)(dx/dt) = (2y)(dy/dt)$. By hypothesis, $dy/dt = -4$, so that $dx/dt = (y/x)(dy/dt) = -4y/x$. Because the triangle is right-angled, when $y = 50$ we have $x = \sqrt{(50)^2 - (30)^2} = 40$. Therefore at the moment when $y = 50$, we have $dx/dt = -4(50)/40 = -5$ (feet per second), so x is decreasing at 5 feet per second.

   b. Yes, because then $x = \sqrt{(45)^2 - (30)^2} = 15\sqrt{5}$, so that when $y = 45$ we have $dx/dt = -4(45)/(15\sqrt{5}) = -12/\sqrt{5}$ (feet per second).

9. $f'(x) = \cos x\cos(x-c) - \sin x\sin(x-c) = \cos[x + (x-c)] = \cos(2x-c)$, so that $f'(x) = 0$ if $\cos(2x-c) = 0$. Therefore $2x - c = \frac{1}{2}\pi + n\pi$, so that $x = \frac{1}{2}c + \frac{1}{4}\pi + \frac{1}{2}n\pi$ for some integer n. Since $f''(x) = -2\sin(2x-c)$, it follows that $f''(\frac{1}{2}c + \frac{1}{4}\pi + \frac{1}{2}n\pi) = -2\sin(c + \frac{1}{2}\pi + n\pi - c) = -2\sin(\frac{1}{2}\pi + n\pi)$, so that $f''(\frac{1}{2}c + \frac{1}{4}\pi + \frac{1}{2}n\pi) < 0$ if n is even. In that case, $f(\frac{1}{2}c + \frac{1}{4}\pi + \frac{1}{2}n\pi)$ is a relative maximum value. Since f has period π, $f(\frac{1}{2}c + \frac{1}{4}\pi)$ is the maximum value of f.

11. a. Let $f(x) = a^x$. Then $f'(x) = (\ln a)a^x$, so that $f'(0) = \ln a$. By the definition of the derivative, $f'(0) = \lim_{h\to 0}(f(h) - f(0))/(h-0) = \lim_{h\to 0}(a^h - 1)/h$. Thus $\lim_{h\to 0}(a^h-1)/h = \ln a$.

   b. Let $h = 1/n$. Then part (a) implies that $\ln a = \lim_{n\to\infty}(a^{1/n} - 1)/(1/n) = \lim_{n\to\infty} n(\sqrt[n]{a} - 1)$.

13. Let $u = 1 + 4x$, so that $du = 4\,dx$ and $36x = 9(u-1)$. If $x = 0$, then $u = 1$; if $x = \frac{3}{4}$, then $u = 4$. Thus

$$\int_0^{3/4} 36x\sqrt{1+4x}\,dx = \int_1^4 9(u-1)u^{1/2}\frac{1}{4}\,du = \frac{9}{4}\int_1^4 (u^{3/2} - u^{1/2})\,du$$

$$= \frac{9}{4}\left(\frac{2}{5}u^{5/2} - \frac{2}{3}u^{3/2}\right)\Big|_1^4 = \frac{9}{4}\left[\left(\frac{64}{5} - \frac{16}{3}\right) - \left(\frac{2}{5} - \frac{2}{3}\right)\right] = \frac{87}{5}.$$

15. $u = x^2$, $dv = e^{2x}\,dx$; $du = 2x\,dx$, $v = \frac{1}{2}e^{2x}$; $\int x^2/e^{-2x}\,dx = \int x^2e^{2x}\,dx = \frac{1}{2}x^2e^{2x} - \int xe^{2x}\,dx$. For $\int xe^{2x}\,dx$, we let $u = x$, $dv = e^{2x}\,dx$; $du = dx$, $v = \frac{1}{2}e^{2x}$; $\int xe^{2x}\,dx = \frac{1}{2}xe^{2x} - \int \frac{1}{2}e^{2x}\,dx = \frac{1}{2}xe^{2x} - \frac{1}{4}e^{2x} + C_1$. Thus $\int x^2/e^{-2x}\,dx = \frac{1}{2}x^2e^{2x} - \frac{1}{2}xe^{2x} + \frac{1}{4}e^{2x} + C$.

17. Since

$$\lim_{n\to\infty} \frac{(n+1)!/(1.4)^{n+1}}{n!/(1.4)^n} = \lim_{n\to\infty} \frac{n+1}{1.4} = \infty$$

the series diverges by the Ratio Test.

19. Since

$$\lim_{n\to\infty} \left| \frac{5(n+1)^2x^{n+1}/[4^{(n+1)}((n+1)^3+2)]}{5n^2x^n/[4^n(n^3+2)]} \right| = \lim_{n\to\infty} \left(\frac{n+1}{n}\right)^2 \frac{1}{4}\,\frac{n^3+2}{(n+1)^3+2}\,|x| = \frac{1}{4}\,|x|$$

the series converges for $|x| < 4$ and diverges for $|x| > 4$. For $x = 4$, the series $\sum_{n=1}^{\infty} 5n^2/(n^3+2)$ diverges by comparison with the series $\sum_{n=1}^{\infty} 1/n$. For $x = -4$, the series $\sum_{n=1}^{\infty} 5n^2(-1)^n/(n^3+2)$ converges by the Alternating Series Test. Therefore the interval of convergence is $[-4, 4)$.

Chapter 11

Vectors, Lines and Planes

11.1 Cartesian Coordinates in Space

1. $|PQ| = \sqrt{(0-\sqrt{2})^2 + (1-0)^2 + (1-0)^2} = \sqrt{2+1+1} = 2$

3. $|PQ| = \sqrt{(0-(-3))^2 + (8-4)^2 + (7-(-5))^2} = \sqrt{9+16+144} = 13$

5. $|PQ| = \sqrt{(4-(-1))^2 + (2-3)^2 + (7-6)^2} = \sqrt{25+1+1} = 3\sqrt{3}$

7. $$\begin{aligned} |PQ| &= \sqrt{(\sin x - 2\sin x)^2 + (2\cos x - \cos x)^2 + (0 - \tan x)^2} \\ &= \sqrt{\sin^2 x + \cos^2 x + \tan^2 x} = \sqrt{1+\tan^2 x} = |\sec x| \end{aligned}$$

9. Let $P = (3,0,2)$, $Q = (1,-1,5)$, and $R = (5,1,-1)$. Then
$$|PQ| = \sqrt{(1-3)^2 + (-1-0)^2 + (5-2)^2} = \sqrt{14}$$
and $|PR| = \sqrt{(5-3)^2 + (1-0)^2 + (-1-2)^2} = \sqrt{14}$. Thus P is equidistant from Q and R.

11. $(x-2)^2 + (y-1)^2 + (z+7)^2 = 25$

13. Completing the squares, we have $(x^2 - 2x + 1) + (y^2 - 4y + 4) + (z^2 + 6z + 9) = -10 + 1 + 4 + 9$, or $(x-1)^2 + (y-2)^2 + (z+3)^2 = 4$, which is an equation of a sphere with center $(1,2,-3)$ and radius 2.

15. $x^2 + (y+2)^2 + (z+3)^2 \le 36$

17. $|PQ| = \sqrt{(2-1)^2 + (1-(-1))^2 + (-1-1)^2} = \sqrt{1+4+4} = 3$;
$|RP| = \sqrt{1^2 + (-1)^2 + 1^2} = \sqrt{3}$; $|RQ| = \sqrt{2^2 + 1^2 + (-1)^2} = \sqrt{6}$.
Thus $|RP|^2 + |RQ|^2 = 3 + 6 = 9 = |PQ|^2$, so by the Pythagorean Theorem, the triangle is a right triangle.

19. If (x, y, z) is equidistant from $(2, 1, 0)$ and $(4, -1, -3)$, then $(x-2)^2 + (y-1)^2 + z^2 = (x-4)^2 + (y+1)^2 + (z+3)^2$, so that $4x - 4y - 6z = 21$.

21. Let $P = (x_0, y_0, z_0)$, $Q = (x_1, y_1, z_1)$ and $R = (\frac{1}{2}(x_0 + x_1), \frac{1}{2}(y_0 + y_1), \frac{1}{2}(z_0 + z_1))$. Then

$$|PR|^2 = \left[\frac{1}{2}(x_0 + x_1) - x_0\right]^2 + \left[\frac{1}{2}(y_0 + y_1) - y_0\right]^2 + \left[\frac{1}{2}(z_0 + z_1) - z_0\right]^2$$
$$= \frac{1}{4}(x_1 - x_0)^2 + \frac{1}{4}(y_1 - y_0)^2 + \frac{1}{4}(z_1 - z_0)^2$$

and

$$|QR|^2 = \left[\frac{1}{2}(x_0 + x_1) - x_1\right]^2 + \left[\frac{1}{2}(y_0 + y_1) - y_1\right]^2 + \left[\frac{1}{2}(z_0 + z_1) - z_1\right]^2$$
$$= \frac{1}{4}(x_0 - x_1)^2 + \frac{1}{4}(y_0 - y_1)^2 + \frac{1}{4}(z_0 - z_1)^2.$$

Thus $|PR|^2 = |QR|^2$, so R is the midpoint of PQ.

23. By the result of Exercise 21, the midpoint of the line segment joining $(2, -1, 3)$ and $(4, 1, 7)$ is $(3, 0, 5)$, so the center of the sphere is $(3, 0, 5)$. The distance between $(2, -1, 3)$ and $(3, 0, 5)$ is

$$\sqrt{(3-2)^2 + (0-(-1))^2 + (5-3)^2} = \sqrt{1+1+4} = \sqrt{6},$$

so an equation of the sphere is $(x-3)^2 + y^2 + (z-5)^2 = 6$.

11.2 Vectors in Space

1. $\overrightarrow{PQ} = 3\mathbf{i} - 4\mathbf{j} + 10\mathbf{k}$

3. $\overrightarrow{PQ} = 3\mathbf{i} - 2\mathbf{j} + \sqrt{7}\mathbf{k}$

5. $\mathbf{a} + \mathbf{b} = \mathbf{i} - 3\mathbf{j} + \mathbf{k}$; $\mathbf{a} - \mathbf{b} = 3\mathbf{i} - 7\mathbf{j} + 19\mathbf{k}$; $c\mathbf{a} = 4\mathbf{i} - 10\mathbf{j} + 20\mathbf{k}$

7. $\mathbf{a} + \mathbf{b} = 2\mathbf{i} + \mathbf{j} + \mathbf{k}$; $\mathbf{a} - \mathbf{b} = 2\mathbf{i} - \mathbf{j} - \mathbf{k}$; $c\mathbf{a} = \frac{2}{3}\mathbf{i}$

9. $\|\mathbf{a}\| = \sqrt{1^2 + (-1)^2 + 1^2} = \sqrt{3}$

11. $\|\mathbf{b}\| = \sqrt{(-3)^2 + 4^2 + (-12)^2} = 13$

13. $\|\mathbf{c}\| = \sqrt{(\sqrt{2})^2 + (-1)^2 + 1^2} = 2$

15. $\dfrac{\mathbf{a}}{\|\mathbf{a}\|} = \dfrac{\mathbf{a}}{13} = -\dfrac{3}{13}\mathbf{i} + \dfrac{4}{13}\mathbf{j} - \dfrac{12}{13}\mathbf{k}$

17. $\dfrac{\mathbf{b}}{\|\mathbf{b}\|} = \dfrac{\mathbf{b}}{\sqrt{13}} = \dfrac{2}{\sqrt{13}}\mathbf{i} - \dfrac{3}{\sqrt{13}}\mathbf{j}$

19. $\|\mathbf{u}\|^2 = (\sqrt{1-b_1^2})^2 + (b_1\sqrt{1-b_2^2})^2 + (b_1b_2)^2 = 1 - b_1^2 + b_1^2(1-b_2^2) + b_1^2b_2^2 = 1$. Therefore $\mathbf{u}$ is a unit vector.

21. $\mathbf{F}_1 + \mathbf{F}_2 = (10^{-3}\mathbf{i} + 0.12\mathbf{j} + 1.2\times 10^4\mathbf{k}) + (3\times 10^{-3}\mathbf{i} + 0.39\mathbf{j} - 5\times 10^4\mathbf{k}) = 4\times 10^{-3}\mathbf{i} + 0.51\mathbf{j} - 3.8\times 10^4\mathbf{k}$

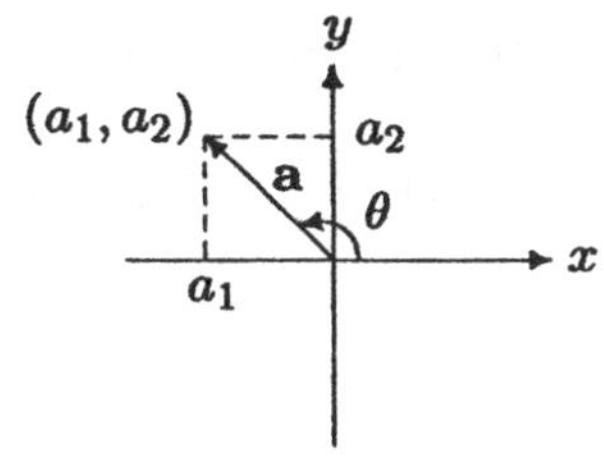

Exercise 23

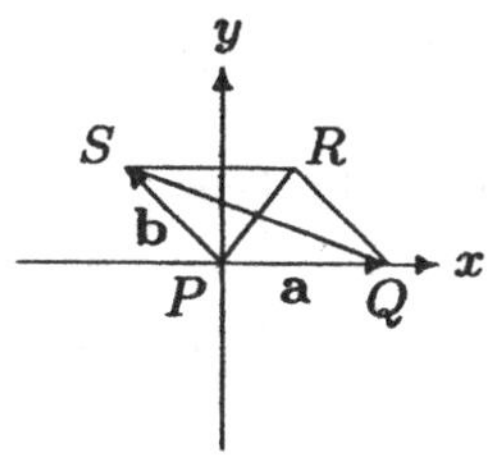

Exercise 25

23. a. If we place the vector $\mathbf{a}$ in the xy plane with its initial point at the origin and let its terminal point be (a_1, a_2), then $\mathbf{a} = a_1\mathbf{i} + a_2\mathbf{j}$. By the definition of $\cos\theta$ and $\sin\theta$, we have $\cos\theta = a_1/\|\mathbf{a}\|$ and $\sin\theta = a_2/\|\mathbf{a}\|$. Thus $\mathbf{a} = a_1\mathbf{i} + a_2\mathbf{j} = \|\mathbf{a}\|\,(\cos\theta\,\mathbf{i} + \sin\theta\,\mathbf{j})$.

b. Let θ be the angle between the positive x axis and $\mathbf{u}$. Since $\|\mathbf{u}\| = 1$, it follows from part (a) that $\mathbf{u} = \cos\theta\,\mathbf{i} + \sin\theta\,\mathbf{j}$.

25. Assume that the quadrilateral lies in the xy plane with P at the origin as in the figure, and let $\mathbf{a}$ and $\mathbf{b}$ be vectors along the two sides that contain the origin. Then $\mathbf{a} + \mathbf{b}$ and $\mathbf{a} - \mathbf{b}$ lie along the diagonals. Since the diagonals bisect each other, we have $\overrightarrow{SR} = \frac{1}{2}(\mathbf{a}-\mathbf{b}) + \frac{1}{2}(\mathbf{a}+\mathbf{b}) = \mathbf{a} = \overrightarrow{PQ}$ and $\overrightarrow{QR} = -\frac{1}{2}(\mathbf{a}-\mathbf{b}) + \frac{1}{2}(\mathbf{a}+\mathbf{b}) = \mathbf{b} = \overrightarrow{PS}$. Thus the opposite sides of the quadrilateral are parallel, so that the quadrilateral is a parallelogram.

27. The force $\mathbf{F}_1$ exerted by the first child is given by $\mathbf{F}_1 = -20\mathbf{j}$, and the force $\mathbf{F}_2$ exerted by the second child is given by

$$\mathbf{F}_2 = 100\left(\cos\frac{\pi}{3}\,\mathbf{i} + \sin\frac{\pi}{3}\,\mathbf{j}\right) = 50(\mathbf{i} + \sqrt{3}\,\mathbf{j}).$$

If $\mathbf{F}$ is the force exerted by the third child, and θ the angle $\mathbf{F}$ makes with the positive x axis, then $\mathbf{F} = \|\mathbf{F}\|\,(\cos\theta\,\mathbf{i} + \sin\theta\,\mathbf{j})$. If the total force exerted on the ball is to be $\mathbf{0}$, then $\mathbf{F}_1 + \mathbf{F}_2 + \mathbf{F} = \mathbf{0}$, so that $-20\mathbf{j} + 50(\mathbf{i} + \sqrt{3}\,\mathbf{j}) + \|\mathbf{F}\|\,(\cos\theta\,\mathbf{i} + \sin\theta\,\mathbf{j}) = \mathbf{0}$. Combining coefficients of $\mathbf{i}$ and $\mathbf{j}$, we find that

$$50 + \|\mathbf{F}\|\cos\theta = 0 \quad\text{and}\quad -20 + 50\sqrt{3} + \|\mathbf{F}\|\sin\theta = 0$$

so that

$$\|\mathbf{F}\| \cos\theta = -50 \quad \text{and} \quad \|\mathbf{F}\| \sin\theta = 20 - 50\sqrt{3}.$$

Therefore

$$\tan\theta = \frac{\sin\theta}{\cos\theta} = \frac{20 - 50\sqrt{3}}{-50} = \sqrt{3} - \frac{2}{5}$$

and

$$\mathbf{F} = \|\mathbf{F}\| \,(\cos\theta\,\mathbf{i} + \sin\theta\,\mathbf{j}) = \|\mathbf{F}\| \cos\theta\,\mathbf{i} + \|\mathbf{F}\| \sin\theta\,\mathbf{j} = -50\mathbf{i} + (20 - 50\sqrt{3})\mathbf{j}.$$

29. Let $\mathbf{v}$ be the velocity of the airplane with respect to the ground, $\mathbf{v}_1$ the velocity of the airplane with respect to the air, and $\mathbf{v}_2$ the velocity of the air with respect to the ground. By the hint,

$$\mathbf{v} = \mathbf{v}_1 + \mathbf{v}_2 = 300\left(\cos\frac{\pi}{6}\,\mathbf{i} + \sin\frac{\pi}{6}\,\mathbf{j}\right) + 20\mathbf{j} = 150\sqrt{3}\,\mathbf{i} + 170\mathbf{j}.$$

Thus $\|\mathbf{v}\| = \sqrt{(150\sqrt{3})^2 + (170)^2} = \sqrt{96{,}400} \approx 310$ (miles per hour).

31. If we let $q_1 = 1.6 \times 10^{-19}$, $\mathbf{u}_1 = (-\mathbf{j} - \mathbf{k})/\sqrt{2}$, $r_1 = 10^{-11}\sqrt{2}$, $q_2 = -1.6 \times 10^{-19}$, $\mathbf{u}_2 = (-\mathbf{i} - \mathbf{j})/\sqrt{2}$, and $r_2 = 10^{-12}\sqrt{2}$, then

$$\mathbf{F} = \frac{q_1(1)}{4\pi\varepsilon_0 r_1^2}\mathbf{u}_1 + \frac{q_2(1)}{4\pi\varepsilon_0 r_2^2}\mathbf{u}_2 = \frac{1.6 \times 10^{-19}}{4\pi\varepsilon_0(2 \times 10^{-22})}\left(\frac{-\mathbf{j} - \mathbf{k}}{\sqrt{2}}\right) + \frac{-1.6 \times 10^{-19}}{4\pi\varepsilon_0(2 \times 10^{-24})}\left(\frac{-\mathbf{i} - \mathbf{j}}{\sqrt{2}}\right)$$

$$= \frac{100\sqrt{2}}{\pi\varepsilon_0}(-\mathbf{j} - \mathbf{k} + 100\mathbf{i} + 100\mathbf{j}) = \frac{100\sqrt{2}}{\pi\varepsilon_0}(100\mathbf{i} + 99\mathbf{j} - \mathbf{k}).$$

33. Let $m = m_1 + m_2 + \cdots + m_n$. By writing the vectors in terms of the corresponding components, we find that

$$\begin{aligned}
m_1\overrightarrow{PP_1} + m_2\overrightarrow{PP_2} + \cdots + m_n\overrightarrow{PP_n} &= [m_1(x_1 - \bar{x})\mathbf{i} + m_1(y_1 - \bar{y})\mathbf{j}] \\
&\quad + [m_2(x_2 - \bar{x})\mathbf{i} + m_2(y_2 - \bar{y})\mathbf{j}] + \cdots \\
&\quad + [m_n(x_n - \bar{x})\mathbf{i} + m_n(y_n - \bar{y})\mathbf{j}] \\
&= [m_1(x_1 - \bar{x}) + m_2(x_2 - \bar{x}) + \cdots + m_n(x_n - \bar{x})]\mathbf{i} \\
&\quad + [m_1(y_1 - \bar{y}) + m_2(y_2 - \bar{y}) + \cdots + m_n(y_n - \bar{y})]\mathbf{j} \\
&= [(m_1x_1 + m_2x_2 + \cdots + m_nx_n) - (m_1 + m_2 + \cdots + m_n)\bar{x}]\mathbf{i} \\
&\quad + [(m_1y_1 + m_2y_2 + \cdots + m_ny_n) - (m_1 + m_2 + \cdots + m_n)\bar{y}]\mathbf{j} \\
&= [(m_1x_1 + m_2x_2 + \cdots + m_nx_n) - m\bar{x}]\mathbf{i} \\
&\quad + [(m_1y_1 + m_2y_2 + \cdots + m_ny_n) - m\bar{y}]\mathbf{j}
\end{aligned}$$

By (3) of Section 6.5, the coefficients of $\mathbf{i}$ and $\mathbf{j}$ are 0, so that $m_1\overrightarrow{PP_1} + m_2\overrightarrow{PP_2} + \cdots + m_n\overrightarrow{PP_n} = \mathbf{0}$.

11.3 The Dot Product

1. $\mathbf{a}\cdot\mathbf{b} = (1)(2) + (1)(-3) + (-1)(4) = -5$; $\cos\theta = \dfrac{\mathbf{a}\cdot\mathbf{b}}{\|\mathbf{a}\|\,\|\mathbf{b}\|} = \dfrac{-5}{\sqrt{3}\sqrt{29}} = \dfrac{-5}{\sqrt{87}}$

3. $\mathbf{a}\cdot\mathbf{b} = (\sqrt{2})(-\sqrt{2}) + (4)(-\sqrt{3}) + (\sqrt{3})(2) = -2 - 2\sqrt{3}$; $\cos\theta = \dfrac{\mathbf{a}\cdot\mathbf{b}}{\|\mathbf{a}\|\,\|\mathbf{b}\|} = \dfrac{-2-2\sqrt{3}}{\sqrt{21}\cdot 3} = -\dfrac{2(1+\sqrt{3})}{3\sqrt{21}}$

5. Let $\mathbf{a} = \sqrt{6}\,\mathbf{i} + \mathbf{j} - \mathbf{k}$ and $\mathbf{b} = \mathbf{i}$, and let θ be the angle between $\mathbf{a}$ and $\mathbf{b}$. Then

$$\cos\theta = \frac{\mathbf{a}\cdot\mathbf{b}}{\|\mathbf{a}\|\,\|\mathbf{b}\|} = \frac{\sqrt{6}}{\sqrt{8}\sqrt{1}} = \sqrt{\frac{3}{4}} = \frac{1}{2}\sqrt{3}, \quad \text{so} \quad \theta = \frac{\pi}{6}.$$

7. $\mathbf{a}\cdot\mathbf{b} = (\sqrt{2})(-1) + (3)(\sqrt{2}) + (1)(5) = 2\sqrt{2} + 5 \neq 0$; they are not perpendicular.

9. $\overrightarrow{PQ} = 3\mathbf{i} - 3\mathbf{j} + \mathbf{k}$ and $\overrightarrow{PR} = -2\mathbf{j} - 6\mathbf{k}$, so $\overrightarrow{PQ}\cdot\overrightarrow{PR} = (3)(0) + (-3)(-2) + (1)(-6) = 0$. Thus $\overrightarrow{PQ}$ and $\overrightarrow{PR}$ are perpendicular.

11. $\text{pr}_{\mathbf{a}}\mathbf{b} = \dfrac{\mathbf{a}\cdot\mathbf{b}}{\|\mathbf{a}\|^2}\mathbf{a} = \dfrac{2}{9}(2\mathbf{i} - \mathbf{j} + 2\mathbf{k}) = \dfrac{4}{9}\mathbf{i} - \dfrac{2}{9}\mathbf{j} + \dfrac{4}{9}\mathbf{k}$

13. $\overrightarrow{PQ} = -3\mathbf{i}$ and $\overrightarrow{PR} = -5\mathbf{i} + 4\mathbf{j}$, so $\overrightarrow{PQ}\cdot\overrightarrow{PR} = (-3)(-5) + (0)(4) = 15$ and $\|\overrightarrow{PQ}\|^2 = (-3)^2 + 0^2 = 9$. Thus

$$\text{pr}_{\overrightarrow{PQ}}\overrightarrow{PR} = \left(\frac{\overrightarrow{PQ}\cdot\overrightarrow{PR}}{\|\overrightarrow{PQ}\|^2}\right)\overrightarrow{PQ} = \frac{15}{9}(-3\mathbf{i}) = -5\mathbf{i}.$$

15. $\mathbf{a}\cdot\mathbf{a}' = 0 + 2 - 2 = 0$. Thus $\mathbf{a}$ and $\mathbf{a}'$ are perpendicular.

$$\text{pr}_{\mathbf{a}}\mathbf{b} = \frac{\mathbf{a}\cdot\mathbf{b}}{\|\mathbf{a}\|^2}\mathbf{a} = \frac{18}{6}\mathbf{a} = 3\mathbf{a} = 3(\mathbf{i} + 2\mathbf{j} - \mathbf{k})$$

$$\text{pr}_{\mathbf{a}'}\mathbf{b} = \mathbf{b} - \text{pr}_{\mathbf{a}}\mathbf{b} = 3\mathbf{i} + \mathbf{j} - 13\mathbf{k} - 3(\mathbf{i} + 2\mathbf{j} - \mathbf{k}) = -5\mathbf{j} - 10\mathbf{k} = -5\mathbf{a}'$$

Thus $\mathbf{b} = 3\mathbf{a} - 5\mathbf{a}'$.

17. $\mathbf{a}\cdot\mathbf{a}' = 4 - 24 + 20 = 0$. Thus $\mathbf{a}$ and $\mathbf{a}'$ are perpendicular.

$$\text{pr}_{\mathbf{a}}\mathbf{b} = \frac{\mathbf{a}\cdot\mathbf{b}}{\|\mathbf{a}\|^2}\mathbf{a} = \frac{-45}{45}\mathbf{a} = -\mathbf{a} = -(2\mathbf{i} - 4\mathbf{j} + 5\mathbf{k})$$

$$\text{pr}_{\mathbf{a}'}\mathbf{b} = \mathbf{b} - \text{pr}_{\mathbf{a}}\mathbf{b} = \mathbf{i} + 13\mathbf{j} + \mathbf{k} + (2\mathbf{i} - 4\mathbf{j} + 5\mathbf{k}) = 3\mathbf{i} + 9\mathbf{j} + 6\mathbf{k} = \frac{3}{2}\mathbf{a}'$$

Thus $\mathbf{b} = -\mathbf{a} + \frac{3}{2}\mathbf{a}'$.

19. Let $\mathbf{a} = \mathbf{i}$, $\mathbf{b} = \mathbf{j}$, and $\mathbf{c} = \mathbf{k}$. Then $\mathbf{a}\cdot\mathbf{b} = 0 = \mathbf{a}\cdot\mathbf{c}$, but $\mathbf{b} \neq \mathbf{c}$.

21. $\mathbf{a} \cdot \mathbf{b} = (a_1)(b_1) + (a_2)(b_2) + (a_3)(b_3) = (b_1)(a_1) + (b_2)(a_2) + (b_3)(a_3) = \mathbf{b} \cdot \mathbf{a}$.

23. a. $\mathbf{a} \cdot \mathbf{b} = \|\mathbf{a}\| \, \|\mathbf{b}\| \cos\theta = \cos\theta$ is maximum if $\theta = 0$.

b. $\mathbf{a} \cdot \mathbf{b} = \|\mathbf{a}\| \, \|\mathbf{b}\| \cos\theta = \cos\theta$ is minimum if $\theta = \pi$.

c. $|\mathbf{a} \cdot \mathbf{b}| = \|\mathbf{a}\| \, \|\mathbf{b}\| \, |\cos\theta| = |\cos\theta|$ is minimum if $\theta = \pi/2$.

25. a. $$\begin{aligned} \|\mathbf{a}+\mathbf{b}\|^2 &= (\mathbf{a}+\mathbf{b}) \cdot (\mathbf{a}+\mathbf{b}) = \mathbf{a} \cdot (\mathbf{a}+\mathbf{b}) + \mathbf{b} \cdot (\mathbf{a}+\mathbf{b}) \\ &= \mathbf{a} \cdot \mathbf{a} + \mathbf{a} \cdot \mathbf{b} + \mathbf{b} \cdot \mathbf{a} + \mathbf{b} \cdot \mathbf{b} = \|\mathbf{a}\|^2 + 2\mathbf{a} \cdot \mathbf{b} + \|\mathbf{b}\|^2 \end{aligned}$$

b. By part (a), $\|\mathbf{a}+\mathbf{b}\|^2 = \|\mathbf{a}\|^2 + \|\mathbf{b}\|^2$ if and only if $2\mathbf{a} \cdot \mathbf{b} = 0$, that is, $\mathbf{a}$ and $\mathbf{b}$ are perpendicular.

c. By Exercise 24(a), $\mathbf{a} \cdot \mathbf{b} \le |\mathbf{a} \cdot \mathbf{b}| \le \|\mathbf{a}\| \, \|\mathbf{b}\|$. Using this and part (a) of this exercise, we have $\|\mathbf{a}+\mathbf{b}\|^2 = \|\mathbf{a}\|^2 + 2\mathbf{a} \cdot \mathbf{b} + \|\mathbf{b}\|^2 \le \|\mathbf{a}\|^2 + 2\|\mathbf{a}\| \, \|\mathbf{b}\| + \|\mathbf{b}\|^2 = (\|\mathbf{a}\| + \|\mathbf{b}\|)^2$. Taking square roots, we find that $\|\mathbf{a}+\mathbf{b}\| \le \|\mathbf{a}\| + \|\mathbf{b}\|$.

27. a. $$\begin{aligned} \|\mathbf{a}+\mathbf{b}\|^2 + \|\mathbf{a}-\mathbf{b}\|^2 &= (\mathbf{a}+\mathbf{b}) \cdot (\mathbf{a}+\mathbf{b}) + (\mathbf{a}-\mathbf{b}) \cdot (\mathbf{a}-\mathbf{b}) \\ &= (\mathbf{a} \cdot \mathbf{a} + \mathbf{a} \cdot \mathbf{b} + \mathbf{b} \cdot \mathbf{a} + \mathbf{b} \cdot \mathbf{b}) + (\mathbf{a} \cdot \mathbf{a} - \mathbf{a} \cdot \mathbf{b} - \mathbf{b} \cdot \mathbf{a} + \mathbf{b} \cdot \mathbf{b} \\ &= 2(\mathbf{a} \cdot \mathbf{a}) + 2(\mathbf{b} \cdot \mathbf{b}) = 2\,\|\mathbf{a}\|^2 + 2\,\|\mathbf{b}\|^2 \end{aligned}$$

b. $$\begin{aligned} \|\mathbf{a}+\mathbf{b}\|^2 - \|\mathbf{a}-\mathbf{b}\|^2 &= (\mathbf{a}+\mathbf{b}) \cdot (\mathbf{a}+\mathbf{b}) - (\mathbf{a}-\mathbf{b}) \cdot (\mathbf{a}-\mathbf{b}) \\ &= \mathbf{a} \cdot \mathbf{a} + \mathbf{a} \cdot \mathbf{b} + \mathbf{b} \cdot \mathbf{a} + \mathbf{b} \cdot \mathbf{b} - \mathbf{a} \cdot \mathbf{a} + \mathbf{a} \cdot \mathbf{b} + \mathbf{b} \cdot \mathbf{a} - \mathbf{b} \cdot \mathbf{b} \\ &= 2(\mathbf{a} \cdot \mathbf{b}) + 2(\mathbf{b} \cdot \mathbf{a}) = 4\mathbf{a} \cdot \mathbf{b} \end{aligned}$$

c. Let $\mathbf{a}$ and $\mathbf{b}$ be vectors along two adjacent sides of the parallelogram, as in Figure 11.36. If the diagonals have equal length, then $\|\mathbf{a}+\mathbf{b}\| = \|\mathbf{a}-\mathbf{b}\|$, so that by part (b), $\mathbf{a} \cdot \mathbf{b} = 0$. Thus the sides are perpendicular, which means that the parallelogram is a rectangle. Conversely, if the parallelogram is a rectangle, then $\mathbf{a}$ and $\mathbf{b}$ are perpendicular, so that $\mathbf{a} \cdot \mathbf{b} = 0$, and thus by part (b), $\|\mathbf{a}+\mathbf{b}\|^2 - \|\mathbf{a}-\mathbf{b}\|^2 = 0$, or $\|\mathbf{a}+\mathbf{b}\|^2 = \|\mathbf{a}-\mathbf{b}\|^2$. This means that the diagonals have equal length.

29. Let the cube be placed so its center is the origin, its faces are parallel to the coordinate planes, and its edges have length $2r$. Let $\mathbf{a}$ and $\mathbf{b}$ be as in the figure, so $\mathbf{a} = r\mathbf{i} + r\mathbf{j} + r\mathbf{k}$ and $\mathbf{b} = r\mathbf{i} + r\mathbf{j} - r\mathbf{k}$. If θ is the angle between $\mathbf{a}$ and $\mathbf{b}$, then

$$\cos\theta = \frac{\mathbf{a} \cdot \mathbf{b}}{\|\mathbf{a}\| \, \|\mathbf{b}\|} = \frac{r^2 + r^2 - r^2}{(r\sqrt{3})(r\sqrt{3})} = \frac{1}{3}$$

so $\theta \approx 1.231$ (radians), or approximately $70.53°$.

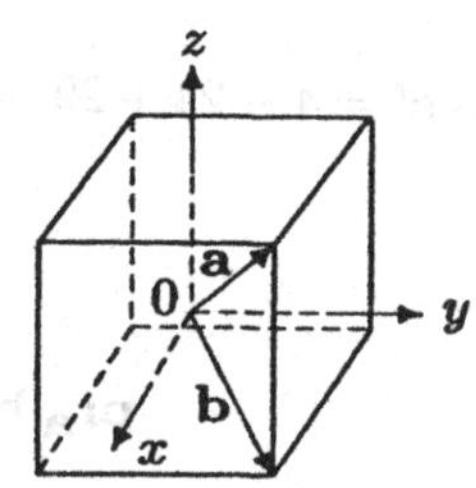

31. Let θ be the angle between **a** and **c**, and ϕ the angle between **c** and **b**. If $\mathbf{c} \neq \mathbf{0}$, then

$$\cos\theta = \frac{\mathbf{a}\cdot\mathbf{c}}{\|\mathbf{a}\|\,\|\mathbf{c}\|} = \frac{\mathbf{a}\cdot(\|\mathbf{b}\|\mathbf{a}+\|\mathbf{a}\|\mathbf{b})}{\|\mathbf{a}\|\,\|\mathbf{c}\|} = \frac{\|\mathbf{b}\|\mathbf{a}\cdot\mathbf{a}+\|\mathbf{a}\|\mathbf{a}\cdot\mathbf{b}}{\|\mathbf{a}\|\,\|\mathbf{c}\|}$$

$$= \frac{\|\mathbf{b}\|\,\|\mathbf{a}\|^2+\|\mathbf{a}\|\,\mathbf{a}\cdot\mathbf{b}}{\|\mathbf{a}\|\,\|\mathbf{c}\|} = \frac{\|\mathbf{b}\|\,\|\mathbf{a}\|+\mathbf{a}\cdot\mathbf{b}}{\|\mathbf{c}\|}$$

and

$$\cos\phi = \frac{\mathbf{c}\cdot\mathbf{b}}{\|\mathbf{c}\|\|\mathbf{b}\|} = \frac{(\|\mathbf{b}\|\,\mathbf{a}+\|\mathbf{a}\|\mathbf{b})\cdot\mathbf{b}}{\|\mathbf{c}\|\,\|\mathbf{b}\|} = \frac{\|\mathbf{b}\|\mathbf{a}\cdot\mathbf{b}+\|\mathbf{a}\|\mathbf{b}\cdot\mathbf{b}}{\|\mathbf{c}\|\,\|\mathbf{b}\|}$$

$$= \frac{\|\mathbf{b}\|\,\mathbf{a}\cdot\mathbf{b}+\|\mathbf{a}\|\,\|\mathbf{b}\|^2}{\|\mathbf{c}\|\,\|\mathbf{b}\|} = \frac{\mathbf{a}\cdot\mathbf{b}+\|\mathbf{a}\|\,\|\mathbf{b}\|}{\|\mathbf{c}\|}.$$

Thus $\cos\theta = \cos\phi$. Since $0 \le \theta \le \pi$ and $0 \le \phi \le \pi$, this means that $\theta = \phi$. Thus **c** bisects the angle formed by **a** and **b**.

33. Using the same terminology as that in Example 7, we find that

$$\overrightarrow{PQ} = 500\mathbf{i} \quad\text{and}\quad \mathbf{F} = 100\left(\cos\frac{\pi}{6}\,\mathbf{i}+\sin\frac{\pi}{6}\,\mathbf{j}\right) = 50\sqrt{3}\,\mathbf{i}+50\mathbf{j}.$$

Therefore $W = \mathbf{F}\cdot\overrightarrow{PQ} = (50\sqrt{3})500 = 25{,}000\sqrt{3}$ (foot-pounds).

11.4 The Cross Product and Triple Products

1. $\mathbf{a}\times\mathbf{b} = \begin{vmatrix} \mathbf{i} & \mathbf{j} & \mathbf{k} \\ 1 & 1 & 0 \\ 0 & 1 & 1 \end{vmatrix} = \mathbf{i}-\mathbf{j}+\mathbf{k}$

   Thus $\mathbf{c}\cdot(\mathbf{a}\times\mathbf{b}) = (-1)(1)+(-3)(-1)+(4)(1) = 6$.

3. $\mathbf{a}\times\mathbf{b} = \begin{vmatrix} \mathbf{i} & \mathbf{j} & \mathbf{k} \\ 2 & 3 & -1 \\ -1 & 4 & 5 \end{vmatrix} = 19\mathbf{i}-9\mathbf{j}+11\mathbf{k}$

   Thus $\mathbf{c}\cdot(\mathbf{a}\times\mathbf{b}) = (2)(19)+(3)(-9)+(4)(11) = 55$.

5. $\mathbf{a}\times\mathbf{b} = \begin{vmatrix} \mathbf{i} & \mathbf{j} & \mathbf{k} \\ 3 & 4 & 12 \\ 3 & 4 & 12 \end{vmatrix} = 0\mathbf{i}+0\mathbf{j}+0\mathbf{k}$

   Thus $\mathbf{c}\cdot(\mathbf{a}\times\mathbf{b}) = (1)(0)+(1)(0)+(0)(0) = 0$.

7. $$\begin{aligned}\mathbf{a}\times\mathbf{b} &= (a_2b_3-a_3b_2)\mathbf{i}+(a_3b_1-a_1b_3)\mathbf{j}+(a_1b_2-a_2b_1)\mathbf{k} \\ &= -(b_2a_3-b_3a_2)\mathbf{i}-(b_3a_1-b_1a_3)\mathbf{j}-(b_1a_2-b_2a_1)\mathbf{k} = -\mathbf{b}\times\mathbf{a}\end{aligned}$$

9. $\mathbf{0} = \mathbf{a} \times \mathbf{0} = \mathbf{a} \times (\mathbf{a} + \mathbf{b} + \mathbf{c}) = (\mathbf{a} \times \mathbf{a}) + (\mathbf{a} \times \mathbf{b}) + (\mathbf{a} \times \mathbf{c}) = (\mathbf{a} \times \mathbf{b}) + (\mathbf{a} \times \mathbf{c})$. Thus $\mathbf{a} \times \mathbf{b} = -(\mathbf{a} \times \mathbf{c}) = \mathbf{c} \times \mathbf{a}$ by Exercise 7. Similarly, $\mathbf{0} = \mathbf{b} \times \mathbf{0} = \mathbf{b} \times (\mathbf{a}+\mathbf{b}+\mathbf{c}) = (\mathbf{b} \times \mathbf{a})+(\mathbf{b} \times \mathbf{b})+(\mathbf{b} \times \mathbf{c}) = (\mathbf{b} \times \mathbf{a}) + (\mathbf{b} \times \mathbf{c})$. Thus $\mathbf{b} \times \mathbf{c} = -(\mathbf{b} \times \mathbf{a}) = \mathbf{a} \times \mathbf{b}$ by Exercise 7.

11. Let $P = (2, -3, 4)$, $Q = (1, 1, -1)$, and $R = (4, -1, -1)$. If $\mathbf{a} = \overrightarrow{OP} = 2\mathbf{i} - 3\mathbf{j} + 4\mathbf{k}$, $\mathbf{b} = \overrightarrow{OQ} = \mathbf{i} + \mathbf{j} - \mathbf{k}$, and $\mathbf{c} = \overrightarrow{OR} = 4\mathbf{i} - \mathbf{j} - \mathbf{k}$, then the volume is $|\mathbf{a} \cdot (\mathbf{b} \times \mathbf{c})|$. Since

$$\mathbf{a} \cdot (\mathbf{b} \times \mathbf{c}) = \begin{vmatrix} 2 & -3 & 4 \\ 1 & 1 & -1 \\ 4 & -1 & -1 \end{vmatrix} = (-2 + 12 - 4) - (16 + 3 + 2) = -15$$

the volume is 15.

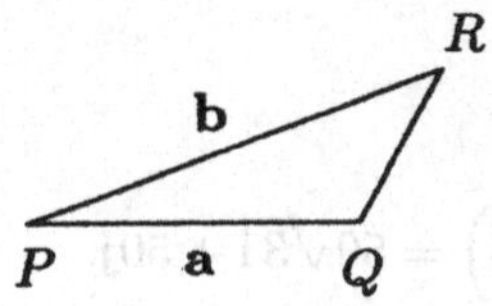

Exercise 13

13. Let $\mathbf{a} = \overrightarrow{PQ}$ and $\mathbf{b} = \overrightarrow{PR}$. As we observed in the text, the area of triangle PQR is $\frac{1}{2}\,\|\mathbf{a} \times \mathbf{b}\| = \frac{1}{2}\,\|\overrightarrow{PQ} \times \overrightarrow{PR}\|$. Similarly, the area equals $\frac{1}{2}\,\|\overrightarrow{PQ} \times \overrightarrow{QR}\|$ and $\frac{1}{2}\,\|\overrightarrow{PR} \times \overrightarrow{QR}\|$.

15. We have $\mathbf{a} \cdot (\mathbf{b} - \mathbf{c}) = \mathbf{a} \cdot \mathbf{b} - \mathbf{a} \cdot \mathbf{c} = 0$ and $\mathbf{a} \times (\mathbf{b} - \mathbf{c}) = \mathbf{a} \times \mathbf{b} - \mathbf{a} \times \mathbf{c} = \mathbf{0}$. If $\mathbf{b}$ and $\mathbf{c}$ were not equal, then $\mathbf{b} - \mathbf{c}$ would be nonzero and it would follow from the preceding calculations and the fact that $\mathbf{a} \neq \mathbf{0}$ that $\mathbf{a}$ is both perpendicular to and parallel to $\mathbf{b} - \mathbf{c}$, which is impossible. Thus it does follow that $\mathbf{b} = \mathbf{c}$.

17. If $\mathbf{u} \times \mathbf{a}$, $\mathbf{u}$ and $\mathbf{b}$ replace $\mathbf{a}$, $\mathbf{b}$, and $\mathbf{c}$ in (3), then

$$(\mathbf{u} \times \mathbf{a}) \cdot (\mathbf{u} \times \mathbf{b}) = [(\mathbf{u} \times \mathbf{a}) \times \mathbf{u}] \cdot \mathbf{b} = -[\mathbf{u} \times (\mathbf{u} \times \mathbf{a})] \cdot \mathbf{b}.$$

If $\mathbf{u}$ and $\mathbf{a}$ are perpendicular, then by (8), with $\mathbf{u}$ replacing $\mathbf{a}$ and $\mathbf{a}$ replacing $\mathbf{c}$, we find that $\mathbf{u} \times (\mathbf{u} \times \mathbf{a}) = -\|\mathbf{u}\|^2\mathbf{a}$, so $(\mathbf{u} \times \mathbf{a}) \cdot (\mathbf{u} \times \mathbf{b}) = -[-\|\mathbf{u}\|^2\mathbf{a}] \cdot \mathbf{b} = \|\mathbf{u}\|^2(\mathbf{a} \cdot \mathbf{b})$. Similarly, the formula holds if $\mathbf{u}$ and $\mathbf{b}$ are perpendicular, since $(\mathbf{u} \times \mathbf{a}) \cdot (\mathbf{u} \times \mathbf{b}) = (\mathbf{u} \times \mathbf{b}) \cdot (\mathbf{u} \times \mathbf{a})$ and $\mathbf{a} \cdot \mathbf{b} = \mathbf{b} \cdot \mathbf{a}$.

19. a. By the *bac* − *cab* rule,

$$\begin{aligned} &\mathbf{a} \times (\mathbf{b} \times \mathbf{c}) + \mathbf{b} \times (\mathbf{c} \times \mathbf{a}) + \mathbf{c} \times (\mathbf{a} \times \mathbf{b}) \\ &= [\mathbf{b}(\mathbf{a} \cdot \mathbf{c}) - \mathbf{c}(\mathbf{a} \cdot \mathbf{b})] + [\mathbf{c}(\mathbf{b} \cdot \mathbf{a}) - \mathbf{a}(\mathbf{b} \cdot \mathbf{c})] + [\mathbf{a}(\mathbf{c} \cdot \mathbf{b}) - \mathbf{b}(\mathbf{c} \cdot \mathbf{a})] \\ &= \mathbf{b}(\mathbf{a} \cdot \mathbf{c} - \mathbf{c} \cdot \mathbf{a}) + \mathbf{c}(-\mathbf{a} \cdot \mathbf{b} + \mathbf{b} \cdot \mathbf{a}) + \mathbf{a}(-\mathbf{b} \cdot \mathbf{c} + \mathbf{c} \cdot \mathbf{b}) = \mathbf{0}. \end{aligned}$$

b. $\mathbf{a} \times (\mathbf{b} \times \mathbf{c}) = (\mathbf{a} \times \mathbf{b}) \times \mathbf{c}$ if and only if $\mathbf{a} \times (\mathbf{b} \times \mathbf{c}) - (\mathbf{a} \times \mathbf{b}) \times \mathbf{c} = \mathbf{0}$ if and only if $\mathbf{a} \times (\mathbf{b} \times \mathbf{c}) + \mathbf{c} \times (\mathbf{a} \times \mathbf{b}) = \mathbf{0}$. By part (a), this equation holds if and only if $\mathbf{b} \times (\mathbf{c} \times \mathbf{a}) = \mathbf{0}$.

21. We use the coordinate system shown in Figure 11.41, with the stapler in the yz plane. Let P be the origin, and Q the point at the end of the stapler at which the force $\mathbf{F}$ is applied. Then

$$\overrightarrow{PQ} = \frac{3}{2}\left(\cos\frac{\pi}{6}\mathbf{j} + \sin\frac{\pi}{6}\mathbf{k}\right) = \frac{3}{2}\left(\frac{\sqrt{3}}{2}\mathbf{j} + \frac{1}{2}\mathbf{k}\right) = \frac{3}{4}(\sqrt{3}\mathbf{j} + \mathbf{k}) \quad \text{and} \quad \mathbf{F} = -32\mathbf{k}.$$

Thus $\mathbf{M} = \overrightarrow{PQ} \times \mathbf{F} = -24\sqrt{3}\mathbf{i}$.

11.5 Lines in Space

1. $\mathbf{r}_0 = -2\mathbf{i} + \mathbf{j}$, so that a vector equation of the line is $\mathbf{r} = (-2+3t)\mathbf{i} + (1-t)\mathbf{j} + 5t\mathbf{k}$, and thus parametric equations are $x = -2+3t$, $y = 1-t$, $z = 5t$; $a = 3$, $b = -1$, $c = 5$, so that symmetric equations of the line are $(x+2)/3 = (y-1)/(-1) = z/5$.

3. $\mathbf{r}_0 = 3\mathbf{i} + 4\mathbf{j} + 5\mathbf{k}$, so that a vector equation of the line is $\mathbf{r} = (3+\frac{1}{2}t)\mathbf{i} + (4-\frac{1}{3}t)\mathbf{j} + (5+\frac{1}{6}t)\mathbf{k}$, and thus parametric equations are $x = 3+\frac{1}{2}t$, $y = 4-\frac{1}{3}t$, $z = 5+\frac{1}{6}t$; $a = \frac{1}{2}$, $b = -\frac{1}{3}$, $c = \frac{1}{6}$, so that symmetric equations of the line are $(x-3)/\frac{1}{2} = (y-4)/(-\frac{1}{3}) = (z-5)/\frac{1}{6}$.

5. $\mathbf{r}_0 = 2\mathbf{i} + 5\mathbf{k}$, so that a vector equation of the line is $\mathbf{r} = 2\mathbf{i} + 2t\mathbf{j} + (5+3t)\mathbf{k}$, and thus parametric equations are $x = 2$, $y = 2t$, $z = 5+3t$; $a = 0$, $b = 2$, $c = 3$, so that symmetric equations of the line are $x = 2$ and $y/2 = (z-5)/3$.

7. $\mathbf{r}_0 = 4\mathbf{i} + 2\mathbf{j} - \mathbf{k}$, so that a vector equation of the line is $\mathbf{r} = 4\mathbf{i} + (2+t)\mathbf{j} - \mathbf{k}$, and thus parametric equations of the line are $x = 4$, $y = 2+t$, $z = -1$; $a = 0 = c$, $b = 1$, so that symmetric equations of the line are $x = 4$ and $z = -1$.

9. $x_0 = -1$, $y_0 = 1$, $z_0 = 0$, $a = -2-(-1) = -1$, $b = 5-1 = 4$, $c = 7-0 = 7$, so that parametric equations for the line are $x = -1-t$, $y = 1+4t$, $z = 7t$.

11. $x_0 = -1$, $y_0 = 1$, $z_0 = 0$, $a = 0$, $b = 0$, $c = 7$, so that symmetric equations for the line are $x = -1$ and $y = 1$.

13. The line through $(2,-1,3)$ and $(0,7,9)$ is parallel to $\mathbf{L}_1$, where

$$\mathbf{L}_1 = (0-2)\mathbf{i} + (7+1)\mathbf{j} + (9-3)\mathbf{k} = -2\mathbf{i} + 8\mathbf{j} + 6\mathbf{k}$$

The line through $(4,-9,-3)$ and $(7,-6,-6)$ is parallel to $\mathbf{L}_2$, where

$$\mathbf{L}_2 = (7-4)\mathbf{i} + (-6+9)\mathbf{j} + (-6+3)\mathbf{k} = 3\mathbf{i} + 3\mathbf{j} - 3\mathbf{k}$$

Since $\mathbf{L}_1 \cdot \mathbf{L}_2 = (-2)(3) + (8)(3) + (6)(-3) = 0, L_1$ and $\mathbf{L}_2$ are perpendicular. Consequently the two corresponding lines are perpendicular.

15. The line through $(7, 7, -6)$ and $(6, 8, -5)$ is parallel to $\mathbf{L}_1$, where

$$\mathbf{L}_1 = (6-7)\mathbf{i} + (8-7)\mathbf{j} + (-5+6)\mathbf{k} = -\mathbf{i} + \mathbf{j} + \mathbf{k}$$

The line with equations $x/7 = (y-3)/4 = (x+9)/3$ is parallel to $\mathbf{L}_2$, where $\mathbf{L}_2 = 7\mathbf{i} + 4\mathbf{j} + 3\mathbf{k}$. Since $\mathbf{L}_1 \cdot \mathbf{L}_2 = (-1)(7) + (1)(4) + (1)(3) = 0$, $\mathbf{L}_1$ and $\mathbf{L}_2$ are perpendicular. Consequently the two corresponding lines are perpendicular.

17. By (3), the parametric equations have the form $x = x_0 + at$, $y = y_0 + bt$, $z = z_0 + ct$. Since P_1 corresponds to $t = 0$ and P_2 corresponds to $t = 2$, we have

$$-1 = x_0 + a \cdot 0, \qquad -2 = y_0 + b \cdot 0, \qquad -3 = z_0 + c \cdot 0$$

$$2 = x_0 + 2a, \qquad -1 = y_0 + 2b, \qquad 0 = z_0 + 2c$$

Thus $x_0 = -1$, $y_0 = -2$, $z_0 = -3$, so that $a = \frac{1}{2}(2 - x_0) = \frac{3}{2}$, $b = \frac{1}{2}(-1 - y_0) = \frac{1}{2}$, $c = -\frac{1}{2}z_0 = \frac{3}{2}$. Thus the parametric equations are $x = -1 + \frac{3}{2}t$, $y = -2 + \frac{1}{2}t$, $z = -3 + \frac{3}{2}t$.

19. The point $P_0 = (1, -2, -1)$ is on the line, and the line is parallel to $\mathbf{i} - 2\mathbf{j} + 3\mathbf{k}$. If $P_1 = (5, 0, -4)$, then P_1 is not on the line, and $\overrightarrow{P_0P_1} = 4\mathbf{i} + 2\mathbf{j} - 3\mathbf{k}$, so by (5) the distance D is given by

$$D = \frac{\|(\mathbf{i} - 2\mathbf{j} + 3\mathbf{k}) \times (4\mathbf{i} + 2\mathbf{j} - 3\mathbf{k})\|}{\sqrt{1^2 + (-2)^2 + 3^2}} = \frac{\|15\mathbf{j} + 10\mathbf{k}\|}{\sqrt{14}} = 5\sqrt{\frac{13}{14}}.$$

21. The line has a vector equation $\mathbf{r} = (-3 + 2t)\mathbf{i} + (-3 - 3t)\mathbf{j} + (3 + 5t)\mathbf{k}$, and if $P_1 = (0, 0, 0)$, then P_1 is not on the line. Let $P_0 = (-3, -3, 3)$, and let $\mathbf{L} = 2\mathbf{i} - 3\mathbf{j} + 5\mathbf{k}$, so that $\mathbf{r}$ and $\mathbf{L}$ are parallel. Then $\overrightarrow{P_0P_1} = 3\mathbf{i} + 3\mathbf{j} - 3\mathbf{k}$, and by (5) the distance D from P_1 to the given line is given by

$$D = \frac{\|(2\mathbf{i} - 3\mathbf{j} + 5\mathbf{k}) \times (3\mathbf{i} + 3\mathbf{j} - 3\mathbf{k})\|}{\sqrt{2^2 + (-3)^2 + 5^2}} = \frac{3\|-2\mathbf{i} + 7\mathbf{j} + 5\mathbf{k}\|}{\sqrt{38}} = 3\sqrt{\frac{39}{19}}.$$

23. $P = (1, -1, 2)$ is on the first line; $Q = (0, 2, 3)$ is on the second line; $\mathbf{L} = 2\mathbf{i} - \mathbf{j} - 2\mathbf{k}$ is parallel to the first line. Thus

$$D = \frac{\|\mathbf{L} \times \overrightarrow{PQ}\|}{\|\mathbf{L}\|} = \frac{\|(2\mathbf{i} - \mathbf{j} - 2\mathbf{k}) \times (-\mathbf{i} + 3\mathbf{j} + \mathbf{k})\|}{\|2\mathbf{i} - \mathbf{j} - 2\mathbf{k}\|} = \frac{\|5\mathbf{i} + 5\mathbf{k}\|}{3} = \frac{5}{3}\sqrt{2}.$$

25. $\sqrt{(x-0)^2 + (y-y)^2 + (z-0)^2} = \sqrt{2}$, or $x^2 + z^2 = 2$.

27. The line through $(1, 4, 2)$ and $(4, -3, -5)$ is parallel to $3\mathbf{i} - 7\mathbf{j} - 7\mathbf{k}$, and the line through $(1, 4, 2)$ and $(-5, -10, -8)$ is parallel to $-6\mathbf{i} - 14\mathbf{j} - 10\mathbf{k}$. Since these two vectors are not parallel, the three points do not lie on the same line.

29. Notice that $\mathbf{a} - \mathbf{b}$ lies on l, so is parallel to $\mathbf{L}$. Thus $\mathbf{L} \times (\mathbf{a} - \mathbf{b}) = \mathbf{0}$. Since $\mathbf{0} = \mathbf{L} \times (\mathbf{a} - \mathbf{b}) = (\mathbf{L} \times \mathbf{a}) - (\mathbf{L} \times \mathbf{b})$, it follows that $\mathbf{L} \times \mathbf{a} = \mathbf{L} \times \mathbf{b}$.

11.6 Planes in Space

1. $x_0 = -1$, $y_0 = 2$, $z_0 = 3$, $a = -4$, $b = 15$, $c = -\frac{1}{2}$; $-4(x+1) + 15(y-2) - \frac{1}{2}(z-3) = 0$, or $8x - 30y + z = -65$.

3. $x_0 = 9$, $y_0 = 17$, $z_0 = -7$, $a = 2$, $b = 0$, $c = -3$; $2(x-9) + 0(y-17) - 3(z+7) = 0$, or $2x - 3z = 39$.

5. $x_0 = 2$, $y_0 = 3$, $z_0 = -5$, $a = 0$, $b = 1$, $c = 0$; $0(x-2) + 1(y-3) + 0(z+5) = 0$, or $y = 3$.

7. The point $P_0 = (-2, -1, -5)$ is on the line, and hence on the plane. Let $P_1 = (1, -1, 2)$, so that $\overrightarrow{P_0P_1} = 3\mathbf{i} + 7\mathbf{k}$. The vector $\mathbf{i} + \mathbf{j} + 2\mathbf{k}$ is parallel to the line but not parallel to $\overrightarrow{P_0P_1}$. For a normal to the plane we take

$$\mathbf{N} = \overrightarrow{P_0P_1} \times (\mathbf{i} + \mathbf{j} + 2\mathbf{k}) = (3\mathbf{i} + 7\mathbf{k}) \times (\mathbf{i} + \mathbf{j} + 2\mathbf{k}) = -7\mathbf{i} + \mathbf{j} + 3\mathbf{k}.$$

An equation of the plane is $-7(x+2) + 1(y+1) + 3(z+5) = 0$, or $7x - y - 3z = 2$.

9. A normal to the plane is given by $\mathbf{N} = 2\mathbf{i} + 5\mathbf{j} + 9\mathbf{k}$, so an equation of the plane is $2(x-2) + 5(y - \frac{1}{2}) + 9(z - \frac{1}{3}) = 0$, or $4x + 10y + 18z = 19$.

11. a. The line l is perpendicular to the vectors $2\mathbf{i} - 3\mathbf{j} + 4\mathbf{k}$ and $\mathbf{i} - \mathbf{k}$, which are normal to the two planes. Thus l is parallel to $(2\mathbf{i} - 3\mathbf{j} + 4\mathbf{k}) \times (\mathbf{i} - \mathbf{k}) = 3\mathbf{i} + 6\mathbf{j} + 3\mathbf{k}$ and hence to $\mathbf{i} + 2\mathbf{j} + \mathbf{k}$. Since $(1, 0, 0)$ is on the intersection of the two planes, a vector equation of l is $\mathbf{r} = \mathbf{i} + t(\mathbf{i} + 2\mathbf{j} + \mathbf{k}) = (1+t)\mathbf{i} + 2t\mathbf{j} + t\mathbf{k}$.

    b. A normal to the plane is given by $\mathbf{N} = \mathbf{i} + 2\mathbf{j} + \mathbf{k}$, so that an equation of the plane is $1(x+9) + 2(y-12) + 1(z-14) = 0$, or $x + 2y + z = 29$.

13. The point $P_1 = (3, -1, 4)$ is not on the plane, whereas $P_0 = (0, 0, 5)$ is on the plane. Then $\overrightarrow{P_0P_1} = 3\mathbf{i} - \mathbf{j} - \mathbf{k}$. If $\mathbf{N} = 2\mathbf{i} - \mathbf{j} + \mathbf{k}$, then $\mathbf{N}$ is normal to the plane, and by Theorem 11.13 the distance D from P_1 to the plane is given by

$$D = \frac{|\mathbf{N} \cdot \overrightarrow{P_0P_1}|}{\|\mathbf{N}\|} = \frac{|(2\mathbf{i} - \mathbf{j} + \mathbf{k}) \cdot (3\mathbf{i} - \mathbf{j} - \mathbf{k})|}{\sqrt{2^2 + (-1)^2 + 1^2}} = \frac{6}{\sqrt{6}} = \sqrt{6}.$$

15. If the plane passes through the origin, then $d = 0$, so that the distance from the origin to the plane and the number $|d|/\sqrt{a^2 + b^2 + c^2}$ are both 0. If the plane does not pass through the origin, let $P_1 = (0, 0, 0)$. Also assume that $c \neq 0$, and let $P_0 = (0, 0, d/c)$, so that P_0 is on the plane. By Theorem 11.13, the distance D from P_1 to the plane is given by

$$D = \frac{|(a\mathbf{i} + b\mathbf{j} + c\mathbf{k}) \cdot \overrightarrow{P_0P_1}|}{\sqrt{a^2 + b^2 + c^2}} = \frac{|(a\mathbf{i} + b\mathbf{j} + c\mathbf{k}) \cdot ((d/c)\mathbf{k})|}{\sqrt{a^2 + b^2 + c^2}} = \frac{|d|}{\sqrt{a^2 + b^2 + c^2}}.$$

The same result follows if $a \neq 0$, or if $b \neq 0$.

17. Let (x, y, z) be on the plane. Then

$$\sqrt{(x-3)^2+(y-1)^2+(z-5)^2} = \sqrt{(x-5)^2+(y+1)^2+(z-3)^2}.$$

Squaring both sides, we obtain $(x-3)^2+(y-1)^2+(z-5)^2 = (x-5)^2+(y+1)^2+(z-3)^2$. Simplifying, we obtain $x - y - z = 0$.

19.
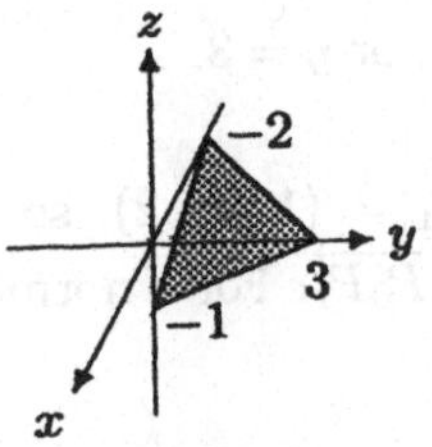

$-\frac{1}{2}x + \frac{1}{3}y - z = 1$

21.
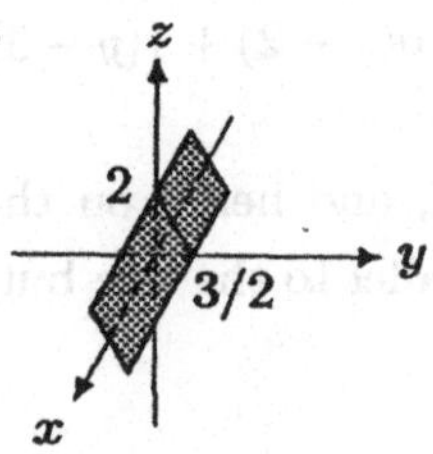

$4y + 3z = 6$

23. Let $P_0 = (2,3,2)$, $P_1 = (1,-1,-3)$, and $P_2 = (1,0,-1)$. Then $\overrightarrow{P_0P_1} = -\mathbf{i} - 4\mathbf{j} - 5\mathbf{k}$ and $\overrightarrow{P_0P_2} = -\mathbf{i} - 3\mathbf{j} - 3\mathbf{k}$. Thus a normal $\mathbf{N}$ to the plane containing the three points is given by

$$\mathbf{N} = \overrightarrow{P_0P_1} \times \overrightarrow{P_0P_2} = (-\mathbf{i} - 4\mathbf{j} - 5\mathbf{k}) \times (-\mathbf{i} - 3\mathbf{j} - 3\mathbf{k}) = -3\mathbf{i} + 2\mathbf{j} - \mathbf{k}.$$

Therefore an equation of the plane is $-3(x-2) + 2(y-3) - 1(z-2) = 0$, or $3x - 2y + z = 2$. Since $3 \cdot 5 - 2 \cdot 9 + 5 = 2$, the fourth point $(5,9,5)$ lies on the plane, so that all four given points lie on the same plane.

25. Since $(\mathbf{a} \times \mathbf{b}) \times (\mathbf{c} \times \mathbf{d})$ is perpendicular to $(\mathbf{a} \times \mathbf{b})$ and $(\mathbf{c} \times \mathbf{d})$, which are normal to $\mathcal{P}_1$ and $\mathcal{P}_2$ respectively, $(\mathbf{a} \times \mathbf{b}) \times (\mathbf{c} \times \mathbf{d})$ is parallel to all vectors that lie in both $\mathcal{P}_1$ and $\mathcal{P}_2$, and thus is parallel to the intersection of $\mathcal{P}_1$ and $\mathcal{P}_2$.

27. Normals of the planes in (a) – (d) are $\mathbf{i}+\mathbf{j}-\mathbf{k}$, $\mathbf{i}-\mathbf{j}$, $\mathbf{j}-\mathbf{k}$, and $\mathbf{i}+\mathbf{j}$. Since none of these is a multiple of another, no two planes are identical or parallel. Since $(\mathbf{i}+\mathbf{j}-\mathbf{k}) \cdot (\mathbf{i}-\mathbf{j}) = 0$ and $(\mathbf{i}-\mathbf{j}) \cdot (\mathbf{i}+\mathbf{j}) = 0$, the planes in (a) and (b) are perpendicular, as are the planes in (b) and (d).

29. a. The vector $\mathbf{i}$ is normal to the plane, so an equation of the plane is $1(x+4) = 0$, or $x = -4$.

    b. The vector $\mathbf{j}$ is normal to the plane, so an equation of the plane is $1(y+5) = 0$, or $y = -5$.

    c. The vector $\mathbf{k}$ is normal to the plane, so an equation of the plane is $1(z+3) = 0$, or $z = -3$.

31. From the first and second equations, $x = 1 - y$ and $z = 2 - y$. Substituting in the third equation, we find that $3 = x + z = (1-y) + (2-y) = 3 - 2y$, so that $y = 0$. Thus $x = 1$ and $z = 2$. The point of intersection is $(1,0,2)$.

33. Adding the first and second equations, we obtain $3x = -1$, so that $x = -\frac{1}{3}$. The second equation becomes $3y+z = -\frac{5}{3}$. Subtracting the third equation from this equation, we obtain $y = -\frac{11}{6}$. Therefore $z = \frac{23}{6}$. The point is $(-\frac{1}{3}, -\frac{11}{6}, \frac{23}{6})$.

35. The two planes are parallel. Let $P_0 = (0, 0, 2)$ and $P_1 = (0, 0, \frac{1}{3})$, which are on the first and second planes, respectively. Then $\overrightarrow{P_0P_1} = -\frac{5}{3}\mathbf{k}$, and a normal $\mathbf{N}$ to either plane is given by $\mathbf{N} = \mathbf{i} - \mathbf{j} + \mathbf{k}$. By Theorem 11.13, the distance D is given by

$$D = \frac{|\mathbf{N} \cdot \overrightarrow{P_0P_1}|}{\|\mathbf{N}\|} = \frac{\frac{5}{3}}{\sqrt{1^2 + (-1)^2 + 1^2}} = \frac{5\sqrt{3}}{9}.$$

37. The two planes are parallel. Let $P_0 = (\frac{5}{2}, 0, 0)$ and $P_1 = (-\frac{1}{4}, 0, 0)$, which are on the first and second planes, respectively. Then $\overrightarrow{P_0P_1} = -\frac{11}{4}\mathbf{i}$, and a normal $\mathbf{N}$ to either plane is given by $\mathbf{N} = 2\mathbf{i} - 3\mathbf{j} + 4\mathbf{k}$. By Theorem 11.13,

$$D = \frac{|\mathbf{N} \cdot \overrightarrow{P_0P_1}|}{\|\mathbf{N}\|} = \frac{\frac{11}{2}}{\sqrt{2^2 + (-3)^2 + 4^2}} = \frac{11}{58}\sqrt{29}.$$

Chapter 11 Review

1. $2\mathbf{a} + \mathbf{b} - 3\mathbf{c} = 5\mathbf{i} - 10\mathbf{j} + 11\mathbf{k}$; $\mathbf{a} \times \mathbf{b} = \begin{vmatrix} \mathbf{i} & \mathbf{j} & \mathbf{k} \\ 2 & -3 & 1 \\ 1 & -1 & 0 \end{vmatrix} = \mathbf{i} + \mathbf{j} + \mathbf{k}$;

$\mathbf{c} \cdot (\mathbf{a} \times \mathbf{b}) = (\mathbf{j} - 3\mathbf{k}) \cdot (\mathbf{i} + \mathbf{j} + \mathbf{k}) = -2$;

$\mathbf{a} \times (\mathbf{b} \times \mathbf{c}) = \mathbf{b}(\mathbf{a} \cdot \mathbf{c}) - \mathbf{c}(\mathbf{a} \cdot \mathbf{b}) = -6\mathbf{b} - 5\mathbf{c} = -6\mathbf{i} + \mathbf{j} + 15\mathbf{k}$

3. $2\mathbf{a} + \mathbf{b} - 3\mathbf{c} = 11\mathbf{i} - 9\mathbf{j} + 6\mathbf{k}$; $\mathbf{a} \times \mathbf{b} = \begin{vmatrix} \mathbf{i} & \mathbf{j} & \mathbf{k} \\ 3 & -2 & 1 \\ 5 & -2 & 1 \end{vmatrix} = 2\mathbf{j} + 4\mathbf{k}$;

$\mathbf{c} \cdot (\mathbf{a} \times \mathbf{b}) = (\mathbf{j} - \mathbf{k}) \cdot (2\mathbf{j} + 4\mathbf{k}) = -2$;

$\mathbf{a} \times (\mathbf{b} \times \mathbf{c}) = \mathbf{b}(\mathbf{a} \cdot \mathbf{c}) - \mathbf{c}(\mathbf{a} \cdot \mathbf{b}) = -3\mathbf{b} - 20\mathbf{c} = -15\mathbf{i} - 14\mathbf{j} + 17\mathbf{k}$

5. Let $Q = (b_1, b_2, b_3)$. Then $\overrightarrow{PQ} = (b_1 - 1)\mathbf{i} + (b_2 + 2)\mathbf{j} + (b_3 - 3)\mathbf{k}$, so that $\overrightarrow{PQ} = \mathbf{a}$ if $b_1 - 1 = 2$, $b_2 + 2 = -2$, $b_3 - 3 = 1$, that is, if $b_1 = 3$, $b_2 = -4$, $b_3 = 4$. Thus $Q = (3, -4, 4)$.

7. Let $\mathbf{b} = 2\mathbf{i} - \mathbf{j} - \mathbf{k}$, $\mathbf{a} = 2\mathbf{j} + \mathbf{k}$, and $\mathbf{a}' = -20\mathbf{i} - 2\mathbf{j} + 4\mathbf{k}$. Note that $\mathbf{a} \cdot \mathbf{a}' = -4 + 4 = 0$, so that $\mathbf{a}$ and $\mathbf{a}'$ are perpendicular. Next,

$$\mathbf{pr}_{\mathbf{a}}\mathbf{b} = \frac{\mathbf{a} \cdot \mathbf{b}}{\|\mathbf{a}\|^2}\mathbf{a} = -\frac{3}{5}\mathbf{a} = -\frac{3}{5}(2\mathbf{j} + \mathbf{k})$$

$$\mathbf{pr}_{\mathbf{a}'}\mathbf{b} = \mathbf{b} - \mathbf{pr}_{\mathbf{a}}\mathbf{b} = 2\mathbf{i} - \mathbf{j} - \mathbf{k} + \frac{3}{5}(2\mathbf{j} + \mathbf{k}) = 2\mathbf{i} + \frac{1}{5}\mathbf{j} - \frac{2}{5}\mathbf{k} = -\frac{1}{10}\mathbf{a}'.$$

Thus $\mathbf{b} = -\frac{3}{5}\mathbf{a} - \frac{1}{10}\mathbf{a}'$.

9. If $P_0 = (\frac{1}{2}, \frac{1}{3}, 0)$, $P_1 = (1, 1, -1)$, and $P_2 = (-2, -3, 5)$, then $\overrightarrow{P_0P_1} = \frac{1}{2}\mathbf{i} + \frac{2}{3}\mathbf{j} - \mathbf{k}$, whereas $\overrightarrow{P_1P_2} = -3\mathbf{i} - 4\mathbf{j} + 6\mathbf{k}$. Thus $-6\overrightarrow{P_0P_1} = \overrightarrow{P_1P_2}$, so that the three points are collinear. Symmetric equations of the line are $(x - 1)/-3 = (y - 1)/-4 = (z + 1)/6$.

11. The line is parallel to $2\mathbf{i} - 3\mathbf{j} + 4\mathbf{k}$, which is normal to the plane. Since $\mathbf{r}_0 = -3\mathbf{i} - 3\mathbf{j} + \mathbf{k}$, a vector equation of the line is $\mathbf{r} = (-3 + 2t)\mathbf{i} + (-3 - 3t)\mathbf{j} + (1 + 4t)\mathbf{k}$.

13. Let $P_0 = (-1, 1, 1)$, $P_1 = (0, 2, 1)$, and $P_2 = (0, 0, \frac{3}{2})$. Then $\overrightarrow{P_0P_1} = \mathbf{i} + \mathbf{j}$ and $\overrightarrow{P_0P_2} = \mathbf{i} - \mathbf{j} + \frac{1}{2}\mathbf{k}$. Thus a normal $\mathbf{N}$ to the plane containing the three points is given by

$$\mathbf{N} = \overrightarrow{P_0P_1} \times \overrightarrow{P_0P_2} = \frac{1}{2}\mathbf{i} - \frac{1}{2}\mathbf{j} - 2\mathbf{k}.$$

Therefore an equation of the plane is $\frac{1}{2}(x+1) - \frac{1}{2}(y-1) - 2(z-1) = 0$, or $\frac{1}{2}x - \frac{1}{2}y - 2z = -3$. Since $\frac{1}{2}(13) - \frac{1}{2}(-1) - 2(5) = -3$, the fourth point $(13, -1, 5)$ lies on the plane, so that all four points lie on the same plane.

15. Since a normal $\mathbf{N}$ of the plane is perpendicular to the z axis, $\mathbf{N} = a\mathbf{i} + b\mathbf{j}$ for appropriate choices of a and b. Thus an equation of the plane is $a(x-3) + b(y+1) + 0(z-5) = 0$, or $ax + by = 3a - b$. Since $(7, 9, 4)$ is on the plane, $7a + 9b = 3a - b$, so that $4a = -10b$, or $a = -\frac{5}{2}b$. Therefore an equation of the plane is $-\frac{5}{2}bx + by = -\frac{15}{2}b - b$, or $5x - 2y = 17$.

17. The point $P_1 = (1, -2, 5)$ is not on the plane, whereas $P_0 = (1, -2, 0)$ is on the plane. Then $\overrightarrow{P_0P_1} = 5\mathbf{k}$. If $\mathbf{N} = 3\mathbf{i} - 4\mathbf{j} + 12\mathbf{k}$, then $\mathbf{N}$ is normal to the plane, and the distance D from P_1 to the plane is given by

$$D = \frac{|\mathbf{N}\cdot\overrightarrow{P_0P_1}|}{\|\mathbf{N}\|} = \frac{|(3\mathbf{i} - 4\mathbf{j} + 12\mathbf{k})\cdot(5\mathbf{k})|}{\sqrt{3^2 + (-4)^2 + (12)^2}} = \frac{60}{13}.$$

19. a. Letting $y = 4$ and $z = 1$ in the first equation, we obtain $3x - 4 + 1 = 2$, or $x = \frac{5}{3}$. Thus the three planes have the point $(\frac{5}{3}, 4, 1)$ in common.

b. If (x, y, z) is on the first two planes, then $2x + y - 2z - 1 = 0 = 3x + y - z - 2$, so that $-z = x - 1$, or $x = 1 - z$. If (x, y, z) is on the first and third planes, then $2x + y - 2z - 1 = 0 = 2x - 2y + 2z$, so that $3y - 1 = 4z$, or $y = (4z+1)/3$. Thus if (x, y, z) is on all three planes, then $0 = x - y + z = (1 - z) - ((4z+1)/3) + z$, so that $z = \frac{1}{2}$. Thus $x = \frac{1}{2}$ and $y = 1$. Thus the planes have the point $(\frac{1}{2}, 1, \frac{1}{2})$ in common.

c. If (x, y, z) is on the first two planes, then $2x - 11y + 6z + 2 = 0 = 2x - 3y + 2z - 2$, or $z + 1 = 2y$. If (x, y, z) is on the first and third planes, then $2x - 11y + 6z + 2 = 0 = 2x - 9y + 5z + 1$, or $z + 1 = 2y$. Thus $z = 2y - 1$. Substituting $z = 2y - 1$ into the equation $2x - 3y + 2z = 2$, we obtain $2x - 3y + 2(2y - 1) = 2$, or $2x + y = 4$. Thus $x = -\frac{1}{2}y + 2$. If we let $y = t$, then $x = 2 - \frac{1}{2}t$, $y = t$, $z = -1 + 2t$ are parametric equations of the line common to the three planes.

21. An equation of the required plane is $a(x - a) + b(y - b) + c(z - c) = 0$, or $ax + by + cz = a^2 + b^2 + c^2$.

23. Using the notation in the figure, we have $\mathbf{a} = \mathbf{c} + \mathbf{d}$, $\mathbf{b} = -\mathbf{c} + \mathbf{d}$, and $\|\mathbf{c}\| = \|\mathbf{d}\| = r$, the radius of the circle. Thus

$$\mathbf{a}\cdot\mathbf{b} = (\mathbf{c} + \mathbf{d})\cdot(-\mathbf{c} + \mathbf{d}) = -\mathbf{c}\cdot\mathbf{c} + \mathbf{c}\cdot\mathbf{d} - \mathbf{c}\cdot\mathbf{d} + \mathbf{d}\cdot\mathbf{d} = \|\mathbf{d}\|^2 - \|\mathbf{c}\|^2 = 0$$

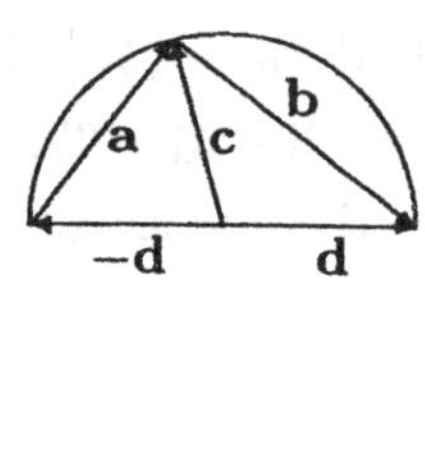

so $\mathbf{a}$ and $\mathbf{b}$ are perpendicular. Therefore every angle inscribed in a semicircle is a right angle.

25. a. Since $\mathbf{a}\cdot\mathbf{c} = \mathbf{b}\cdot\mathbf{c}$ for all $\mathbf{c}$, we let $\mathbf{c} = \mathbf{a}-\mathbf{b}$ and find that $\|\mathbf{a}-\mathbf{b}\|^2 = (\mathbf{a}-\mathbf{b})\cdot(\mathbf{a}-\mathbf{b}) = (\mathbf{a}-\mathbf{b})\cdot\mathbf{c} = \mathbf{a}\cdot\mathbf{c} - \mathbf{b}\cdot\mathbf{c} = 0$. Thus $\mathbf{a}-\mathbf{b} = \mathbf{0}$, so that $\mathbf{a} = \mathbf{b}$.

b. If $\mathbf{a} \neq \mathbf{b}$, then $\mathbf{a}-\mathbf{b} \neq \mathbf{0}$, so we let $\mathbf{c}$ be a nonzero vector perpendicular to $\mathbf{a}-\mathbf{b}$. It follows that $\mathbf{0} = (\mathbf{a}\times\mathbf{c}) - (\mathbf{b}\times\mathbf{c}) = (\mathbf{a}-\mathbf{b})\times\mathbf{c}$. Therefore $\|\mathbf{a}-\mathbf{b}\|\,\|\mathbf{c}\| = 0$, which is impossible since $\|\mathbf{a}-\mathbf{b}\| \neq 0$ and $\|\mathbf{c}\| \neq 0$. Thus $\mathbf{a} = \mathbf{b}$.

27. a.
$$\begin{aligned}\|\mathbf{a}+\mathbf{b}\|^2\,\|\mathbf{a}-\mathbf{b}\|^2 &= [(\mathbf{a}+\mathbf{b})\cdot(\mathbf{a}+\mathbf{b})][(\mathbf{a}-\mathbf{b})\cdot(\mathbf{a}-\mathbf{b})]\\ &= \left[\|\mathbf{a}\|^2 + 2\mathbf{a}\cdot\mathbf{b} + \|\mathbf{b}\|^2\right]\cdot\left[\|\mathbf{a}\|^2 - 2\mathbf{a}\cdot\mathbf{b} + \|\mathbf{b}\|^2\right]\\ &= \left(\|\mathbf{a}\|^4 + 2\|\mathbf{a}\|^2\|\mathbf{b}\|^2 + \|\mathbf{b}\|^4\right) - 4(\mathbf{a}\cdot\mathbf{b})^2\\ &= \left(\|\mathbf{a}\|^2 + \|\mathbf{b}\|^2\right)^2 - 4(\mathbf{a}\cdot\mathbf{b})^2\end{aligned}$$

b. Since $-4(\mathbf{a}\cdot\mathbf{b})^2 \leq 0$, we infer that $\|\mathbf{a}+\mathbf{b}\|^2\,\|\mathbf{a}-\mathbf{b}\|^2 \leq \left(\|\mathbf{a}\|^2 + \|\mathbf{b}\|^2\right)^2$, so that $\|\mathbf{a}+\mathbf{b}\|\,\|\mathbf{a}-\mathbf{b}\| \leq \|\mathbf{a}\|^2 + \|\mathbf{b}\|^2$.

29. Let $\mathbf{r}$, $\mathbf{s}$ be as in the figure. Then $\mathbf{s} = \cos\phi\,\mathbf{k}$ since $\mathbf{s}$ is parallel to the z axis and $\|\mathbf{u}\| = 1$. Next, $\mathbf{r}$ has length $\sin\phi$ since $\|\mathbf{u}\| = 1$. Since $\mathbf{r}$ is in the xy plane, makes the angle θ with the positive x axis, and has length $\sin\phi$, it follows that $\mathbf{r} = (\sin\phi)(\cos\theta\,\mathbf{i} + \sin\theta\,\mathbf{j}) = \sin\phi\cos\theta\,\mathbf{i} + \sin\phi\sin\theta\,\mathbf{j}$. Therefore $\mathbf{u} = \mathbf{r}+\mathbf{s} = \cos\theta\sin\phi\,\mathbf{i} + \sin\theta\sin\phi\,\mathbf{j} + \cos\phi\,\mathbf{k}$.

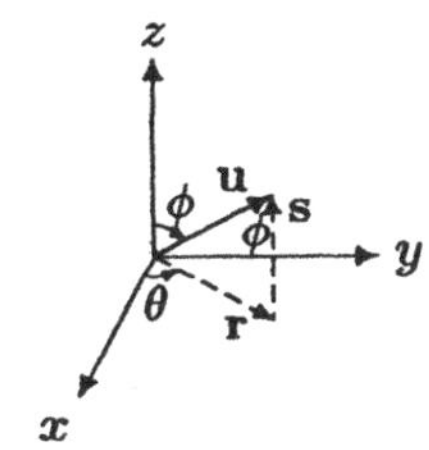

31. a. Let $\mathbf{F}_1$ and $\mathbf{F}_2$ be the forces pointing downward and upward, respectively. Then $\mathbf{F}_1 = 5(\cos\theta\,\mathbf{i} - \sin\theta\,\mathbf{j})$ and $\mathbf{F}_2 = 5(\cos\theta\,\mathbf{i} + \sin\theta\,\mathbf{j})$. We must choose θ so that $\mathbf{F}_1 + \mathbf{F}_2 = 5\mathbf{i}$. But $\mathbf{F}_1 + \mathbf{F}_2 = 5(\cos\theta\,\mathbf{i} - \sin\theta\,\mathbf{j}) + 5(\cos\theta\,\mathbf{i} + \sin\theta\,\mathbf{j}) = 10\cos\theta\,\mathbf{i}$, so that $\mathbf{F}_1 + \mathbf{F}_2 = 5\mathbf{i}$ if and only if $10\cos\theta = 5$, or $\cos\theta = \frac{1}{2}$. Thus $\theta = \pi/3$.

b. We proceed as in part (a), but now we need $\mathbf{F}_1 + \mathbf{F}_2 = 6\mathbf{i}$. Thus we need $10\cos\theta = 6$, or $\cos\theta = 0.6$. By calculator, $\theta \approx 0.927295218$ radian ($\approx 53°$)

33. Let $\mathbf{v}$ be the velocity of the jet with respect to the ground, $\mathbf{v}_1$ the velocity of the jet with respect to the air, and $\mathbf{v}_2$ the velocity of the air (jet stream). Then $\mathbf{v} = \mathbf{v}_1 + \mathbf{v}_2 = 500\mathbf{j} + 100\mathbf{i}$. The ground speed of the jet is $\|\mathbf{v}\|$ and $\|\mathbf{v}\| = \sqrt{500^2 + 100^2} = 100\sqrt{26} \approx 509.902$ miles per hour.

Cumulative Review(Chapters 1–10)

1. a. The domain consists of all x such that $(1-2x)/(1-3x) \geq 0$, or $\frac{2}{3}[(x-\frac{1}{2})/(x-\frac{1}{3})] \geq 0$. This occurs if $x \geq \frac{1}{2}$ and $x > \frac{1}{3}$, or if $x \leq \frac{1}{2}$ and $x < \frac{1}{3}$. Thus the domain is the union of $(-\infty, \frac{1}{3})$ and $[\frac{1}{2}, \infty)$.

b. $$\lim_{x\to-\infty} f(x) = \lim_{x\to-\infty} \sqrt{\frac{1-2x}{1-3x}} = \lim_{x\to-\infty} \sqrt{\frac{1/x-2}{1/x-3}} = \sqrt{\frac{0-2}{0-3}} = \sqrt{\frac{2}{3}}$$

3. $\lim_{x\to\infty} x^{\tan(1/x)} = \lim_{x\to\infty} e^{\tan(1/x)\ln x} = e^{\lim_{x\to\infty} \ln x/\cot(1/x)}$

By l'Hôpital's Rule,

$$\lim_{x\to\infty} \frac{\ln x}{\cot(1/x)} = \lim_{x\to\infty} \frac{1/x}{[-\csc^2(1/x)](-1/x^2)} = \lim_{x\to\infty} \sin\frac{1}{x}\,\frac{\sin(1/x)}{1/x} = 0 \cdot 1 = 0.$$

Thus

$$\lim_{x\to\infty} x^{\tan(1/x)} = e^{\lim_{x\to\infty} \ln x/\cot(1/x)} = e^0 = 1.$$

5. a. $f'(x) = \dfrac{1}{(\pi/2)+\tan^{-1}x}\,\dfrac{1}{x^2+1} > 0$ for all x.

Therefore f is increasing, so f^{-1} exists. The domain of f^{-1} = the range of f. Since the range of $\tan^{-1}x$ is $(-\pi/2, \pi/2)$, it follows that the range of $\pi/2+\tan^{-1}x$ is $(0,\pi)$ and the range of $\ln(\pi/2+\tan^{-1}x)$ is $(-\infty, \ln\pi)$. Thus the domain of f^{-1} is $(-\infty, \ln\pi)$. The range of f^{-1} = the domain of $f = (-\infty, \infty)$.

b. $(f^{-1})'(\ln\pi/3) = 1/f'(a)$ for the value of a in $(-\infty,\infty)$ such that $f(a) = \ln\pi/3$, or by the definition of f, $\ln(\pi/2+\tan^{-1}a) = \ln\pi/3$. This means that $\pi/2+\tan^{-1}a = \pi/3$, so that $\tan^{-1}a = -\pi/6$ and hence $a = -\sqrt{3}/3$. Thus

$$(f^{-1})'\left(\ln\frac{\pi}{3}\right) = \frac{1}{f'(-\sqrt{3}/3)} = \frac{1}{\dfrac{1}{\pi/2+\tan^{-1}(-\sqrt{3}/3)}\,\dfrac{1}{(-\sqrt{3}/3)^2+1}} = \left(\frac{\pi}{2}-\frac{\pi}{6}\right)\left(\frac{1}{3}+1\right) = \frac{4\pi}{9}.$$

c. $y = f^{-1}(x)$ if and only if $x = f(y) = \ln(\pi/2+\tan^{-1}y)$ if and only if $e^x = \pi/2+\tan^{-1}y$ if and only if $\tan^{-1}y = e^x - \pi/2$ if and only if $y = \tan(e^x-\pi/2)$. Therefore $f^{-1}(x) = \tan(e^x-\pi/2)$ for $x < \ln\pi$.

7. a. $\dfrac{dy}{dx} = \dfrac{dy}{dt}\dfrac{1}{dx/dt} = \dfrac{\sin t}{1-\cos t}$; $\cos t = 1-y$ and $0 < t < \pi$, so $\sin t = \sqrt{1-\cos^2 t} = \sqrt{1-(1-y)^2} = \sqrt{2y-y^2}$; thus

$$\frac{dy}{dx} = \frac{\sin t}{1-\cos t} = \frac{\sqrt{2y-y^2}}{y}.$$

b. Since $\cos t = 1-y$, we have $t = \cos^{-1}(1-y)$ for $0 < t < \pi$. Thus by part (a), $x = t - \sin t = \cos^{-1}(1-y) - \sqrt{2y-y^2}$.

c. By differentiating the equation for x in part (b) implicitly with respect to x, we obtain

$$1 = \frac{-1}{\sqrt{1-(1-y)^2}}\left(-\frac{dy}{dx}\right) - \frac{1}{2\sqrt{2y-y^2}}(2-2y)\frac{dy}{dx}$$

or

$$1 = \frac{1}{\sqrt{2y-y^2}}\frac{dy}{dx} - \frac{1-y}{\sqrt{2y-y^2}}\frac{dy}{dx}, \quad \text{so} \quad \frac{dy}{dx} = \frac{\sqrt{2y-y^2}}{y}.$$

9. $f'(x) = 5(x^2-5)(x^2-1)$; $f''(x) = 20x(x^2-3)$;

relative maximum values are $f(-\sqrt{5}) = 0$ and $f(1) = 16$; relative minimum values are $f(-1) = -16$ and $f(\sqrt{5}) = 0$; increasing on $(-\infty, -\sqrt{5}]$, $[-1, 1]$, and $[\sqrt{5}, \infty)$; decreasing on $[-\sqrt{5}, -1]$ and $[1, \sqrt{5}]$; concave upward on $(-\sqrt{3}, 0)$ and $(\sqrt{3}, \infty)$, and concave downward on $(-\infty, -\sqrt{3})$ and $(0, \sqrt{3})$; inflection points are $(-\sqrt{3}, -4\sqrt{3})$, $(0,0)$, and $(\sqrt{3}, 4\sqrt{3})$; symmetric with respect to the origin.

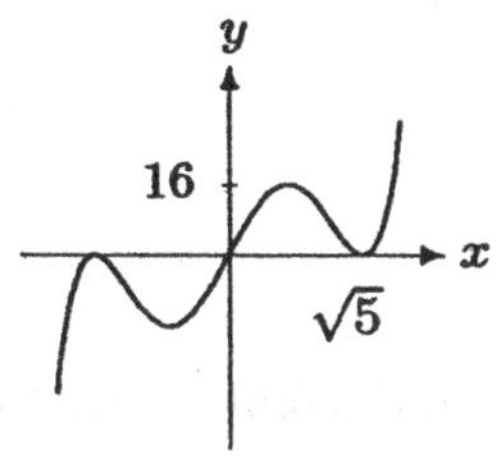

11. The graph of f and the line $y = \frac{5}{2}$ intersect at (x, y) if $x + 1/x = \frac{5}{2}$, or $2x^2 - 5x + 2 = 0$, or $(2x-1)(x-2) = 0$, or $x = \frac{1}{2}$ or $x = 2$. Since $x + 1/x \le \frac{5}{2}$ for $1 \le x \le 2$ and $x + 1/x \ge \frac{5}{2}$ for $2 \le x \le 3$, the area A is given by

$$A = \int_1^2 \left[\frac{5}{2} - \left(x + \frac{1}{x}\right)\right] dx + \int_2^3 \left[\left(x + \frac{1}{x}\right) - \frac{5}{2}\right] dx$$

$$= \left(\frac{5}{2}x - \frac{1}{2}x^2 - \ln x\right)\Big|_1^2 + \left(\frac{1}{2}x^2 + \ln x - \frac{5}{2}x\right)\Big|_2^3 = (1 - \ln 2) + (\ln 3 - \ln 2) = 1 + \ln\frac{3}{4}.$$

13. $\dfrac{x^3}{x^2-x+1} = x + 1 - \dfrac{1}{x^2-x+1} = x + 1 - \dfrac{1}{(x-\frac{1}{2})^2 + \frac{3}{4}}$

Thus

$$\int \frac{x^3}{x^2-x+1}\,dx = \int \left[x + 1 - \frac{1}{(x-\frac{1}{2})^2 + \frac{3}{4}}\right] dx$$

$$= \frac{1}{2}x^2 + x - \frac{2}{\sqrt{3}}\tan^{-1}\left[\frac{2}{\sqrt{3}}\left(x - \frac{1}{2}\right)\right] + C = \frac{1}{2}x^2 + x - \frac{2}{\sqrt{3}}\tan^{-1}\frac{2x-1}{\sqrt{3}} + C.$$

15. Let $u = -3x^2$, so that $du = -6x\,dx$. If $x = 2$, then $u = -12$, and if $x = b$, then $u = -3b^2$. Thus

$$\int_2^\infty xe^{-3x^2}\,dx = \lim_{b\to\infty}\int_2^b xe^{-3x^2}\,dx = \lim_{b\to\infty}\int_{-12}^{-3b^2} e^u\left(-\frac{1}{6}\right)du$$

$$= \lim_{b\to\infty}\left(-\frac{1}{6}e^u\right)\Big|_{-12}^{-3b^2} = \lim_{b\to\infty}\frac{1}{6}(e^{-12} - e^{-3b^2}) = \frac{1}{6}e^{-12}.$$

17. $L = \displaystyle\int_{-1}^{1}\sqrt{(-e^{-t}\sin 2t + 2e^{-t}\cos 2t)^2 + (-e^{-t}\cos 2t - 2e^{-t}\sin 2t)^2}\,dt$

$$= \int_{-1}^{1} e^{-t}\sqrt{\sin^2 2t - 4\sin 2t\cos 2t + 4\cos^2 2t + \cos^2 2t + 4\cos 2t\sin 2t + 4\sin^2 2t}\,dt$$

$$= \int_{-1}^{1} e^{-t}\sqrt{5}\,dt = -\sqrt{5}\,e^{-t}\big|_{-1}^{1} = \sqrt{5}(e - e^{-1})$$

19. For any c,

$$\lim_{n\to\infty}(\sqrt{n+c} - \sqrt{n}) = \lim_{n\to\infty}\frac{(\sqrt{n+c}-\sqrt{n})(\sqrt{n+c}+\sqrt{n})}{\sqrt{n+c}+\sqrt{n}}$$

$$= \lim_{n\to\infty}\frac{n+c-n}{\sqrt{n+c}+\sqrt{n}} = \lim_{n\to\infty}\frac{c}{\sqrt{n+c}+\sqrt{n}} = 0.$$

Thus the limit exists for all c.

21. Since

$$\lim_{n\to\infty}\frac{n^2}{\sqrt{n^4+2}} = \lim_{n\to\infty}\frac{1}{\sqrt{1+2/n^4}} = 1$$

$\lim_{n\to\infty}(-1)^n n^2/\sqrt{n^4+2}$ does not exist. By Corollary 9.9 the given series diverges.

23. By Theorem 9.28 and the uniqueness of Taylor series, the Taylor series for $\int_0^x f(t)\,dt$ is

$$\sum_{n=1}^{\infty}\int_0^x \frac{t^n}{n+2}\,dt = \sum_{n=1}^{\infty}\frac{x^{n+1}}{(n+1)(n+2)}.$$

For $x \neq 0$,

$$\lim_{n\to\infty}\frac{|x^{n+2}|/[(n+2)(n+3)]}{|x^{n+1}|/[(n+1)(n+2)]} = \lim_{n\to\infty}\frac{n+1}{n+3}|x| = |x|.$$

By the Generalized Ratio Test, the series $\sum_{n=1}^{\infty} x^{n+1}/[(n+1)(n+2)]$ converges for $|x| < 1$ and diverges for $|x| > 1$. For $x = 1$ the series becomes $\sum_{n=1}^{\infty} 1/[(n+1)(n+2)]$, which converges by comparison with the p series $\sum_{n=1}^{\infty} 1/n^2$. For $x = -1$ the series becomes $\sum_{n=1}^{\infty}(-1)^n\,1/[(n+1)(n+2)]$, which converges by the Alternating Series Test. Consequently the interval of convergence of $\sum_{n=1}^{\infty} x^{n+1}/[(n+1)(n+2)]$ is $[-1, 1]$.

Chapter 12

Vector-Valued Functions and Curves in Space

12.1 Definitions and Examples

1. domain: $(-\infty, \infty)$; $f_1(t) = t$, $f_2(t) = t^2$, $f_3(t) = t^3$

3. $\tanh t$ is defined for all t; domain: union of $(-\infty, -2)$, and $(-2, 2)$ and $(2, \infty)$; $f_1(t) = \tanh t$, $f_2(t) = 0$, $f_3(t) = -1/(t^2 - 4)$

5. $\mathbf{F}(t) = 2\sqrt{t}\,\mathbf{i} - 2t^{3/2}\mathbf{j} - (t^3 + 1)\mathbf{k}$. Thus the domain is $[0, \infty)$. Also $f_1(t) = 2\sqrt{t}$, $f_2(t) = -2t^{3/2}$, and $f_3(t) = -(t^3 + 1)$.

7. $(2\mathbf{F} - 3\mathbf{G})(t) = (2t - 3\cos t)\mathbf{i} + (2t^2 - 3\sin t)\mathbf{j} + (2t^3 - 3)\mathbf{k}$. Thus the domain is $(-\infty, \infty)$. Also $f_1(t) = 2t - 3\cos t$, $f_2(t) = 2t^2 - 3\sin t$, and $f_3(t) = 2t^3 - 3$.

9. $(\mathbf{F} \times \mathbf{G})(t) = \dfrac{1}{\sqrt{t}}(1 - \cos t)\mathbf{i} - \dfrac{1}{\sqrt{t}}(t - \sin t)\mathbf{j} - t(t - \sin t)\mathbf{k}$. Thus the domain is $(0, \infty)$. Also $f_1(t) = (1/\sqrt{t})(1 - \cos t)$, $f_2(t) = -(1/\sqrt{t})(t - \sin t)$, and $f_3(t) = -t(t - \sin t)$.

11. $(\mathbf{F} \circ g)(t) = \cos t^{1/3}\mathbf{i} + \sin t^{1/3}\mathbf{j} + \sqrt{t^{1/3} + 2}\,\mathbf{k}$. Thus the domain consists of all t for which $t^{1/3} + 2 \geq 0$, and hence is $[-8, \infty)$. Also $f_1(t) = \cos t^{1/3}$, $f_2(t) = \sin t^{1/3}$, and $f_3(t) = \sqrt{t^{1/3} + 2}$.

13.

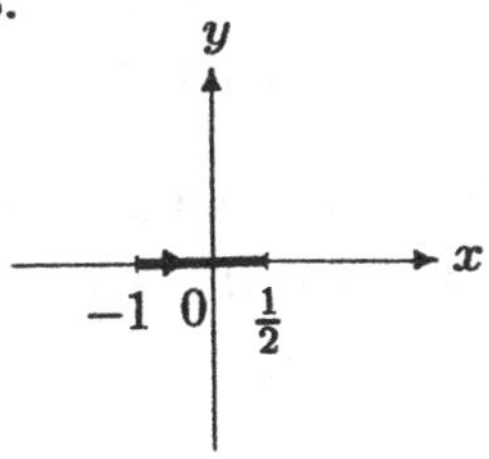

15.

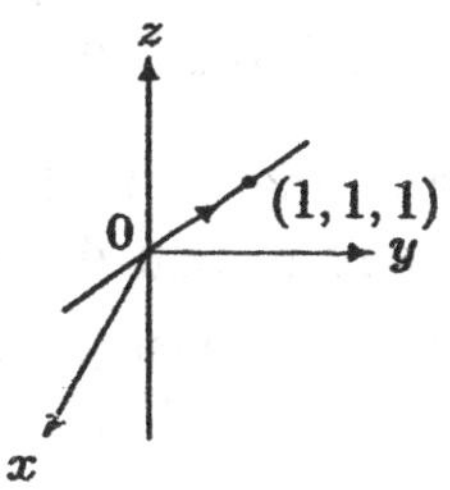

17.

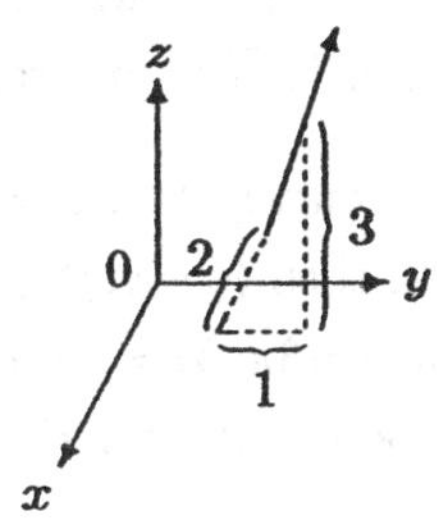

19.

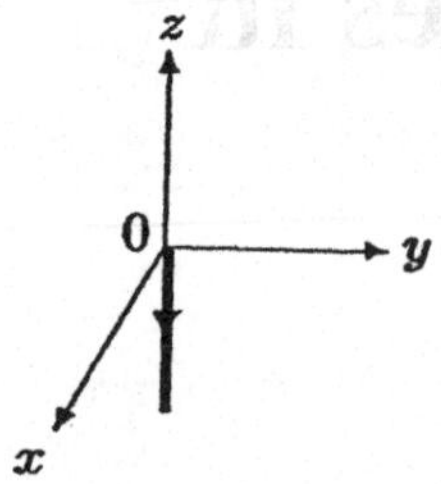

21.

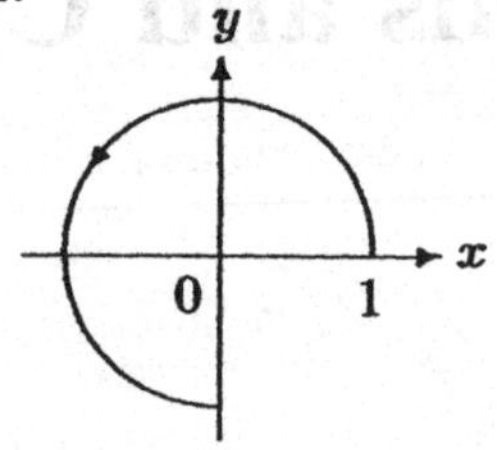

23.

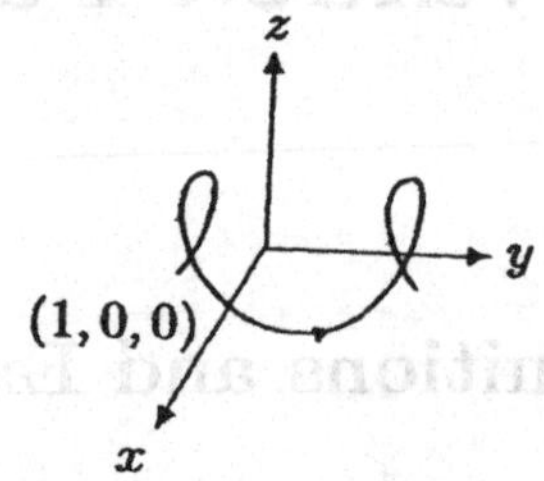

25.

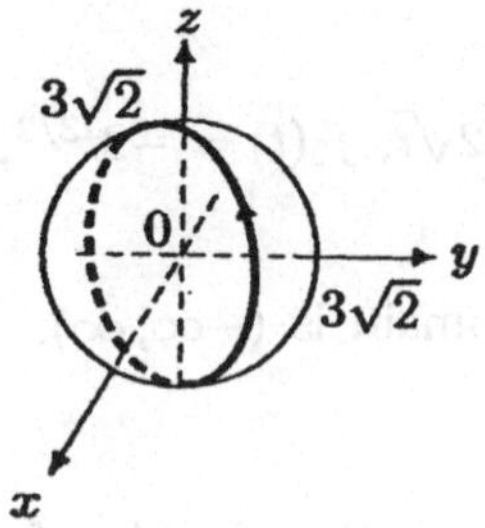

31. a. $(\mathbf{F}-\mathbf{G})(t) = (t-1)\mathbf{i} + (-2t+3)\mathbf{j} - 4t\mathbf{k}$, so parametric equations are $x = t-1$, $y = -2t+3$, and $z = -4t$.

b. $(\mathbf{F}+3\mathbf{G})(t) = (5t+3)\mathbf{i} + (10t-5)\mathbf{j}$, so parametric equations are $x = 5t+3$, $y = 10t-5$, and $z = 0$.

c. $(\mathbf{F}\circ g)(t) = 2\cos t\,\mathbf{i} + (\cos t + 1)\,\mathbf{j} - 3\cos t\,\mathbf{k}$. Then $x = 2\cos t$, $y = \cos t + 1$, and $z = -3\cos t$ for all real t, which yields the line segment parametrized by $x = 2t$, $y = t+1$, $z = -3t$ for $-1 \le t \le 1$.

33. If (x, y, z) is on the curve, then $x = \cos \pi t$, $y = \sin \pi t$, and $z = t$. If in addition (x, y, z) is on the sphere $x^2 + y^2 + z^2 = 10$, then $(\cos \pi t)^2 + (\sin \pi t)^2 + t^2 = 10$, so that $t^2 = 9$, and thus $t = \pm 3$. If $t = 3$, then $(x, y, z) = (\cos 3\pi, \sin 3\pi, 3) = (-1, 0, 3)$, and if $t = -3$, then $(x, y, z) = (\cos(-3\pi), \sin(-3\pi), -3) = (-1, 0, -3)$. The points of intersection are therefore $(-1, 0, 3)$ and $(-1, 0, -3)$.

35. $\mathbf{F}(t)$ can be written $\mathbf{a} + \mathbf{b}$, where $\mathbf{a}$ describes the motion of the center of the circle and $\mathbf{b}$ describes the motion of P around the center. As in Example 7, $\mathbf{a} = rt\mathbf{i} + r\mathbf{j}$. Also, if C is the center of the circle, then t represents the number of radians through which $\overrightarrow{CP}$ has rotated since time 0. At that time $\overrightarrow{CP}$ makes an angle of $(3\pi/2 - t)$ radians with the positive x axis, so $\mathbf{b} = \overrightarrow{CP} = b\cos(3\pi/2 - t)\mathbf{i} + b\sin(3\pi/2 - t)\mathbf{j} = -b\sin t\,\mathbf{i} - b\cos t\,\mathbf{j}$. Consequently

$$\mathbf{F}(t) = \mathbf{a} + \mathbf{b} = (rt\mathbf{i} + r\mathbf{j}) + (-b\sin t\,\mathbf{i} - b\cos t\,\mathbf{j}) = (rt - b\sin t)\mathbf{i} + (r - b\cos t)\mathbf{j}.$$

37. $\mathbf{F}(t)$ can be written $\mathbf{a}+\mathbf{b}$, where $\mathbf{a}$ describes the motion of the center C of the circle and $\mathbf{b}$ describes the motion of P around the center. First of all, $\mathbf{a} = (r+b)\cos t\,\mathbf{i} + (r+b)\sin t\,\mathbf{j}$. The number t represents the number of radians through which $\overrightarrow{OC}$ has rotated since time 0. Notice that an angle of t radians on the fixed circle corresponds to an arc of length rt, which in turn corresponds to an arc of length rt on the rolling circle. If the corresponding angle on the rolling circle has α radians, then the arc has length $b\alpha$, so that $b\alpha = rt$, or $\alpha = (r/b)t$. Thus by time t the angle OCP has $(r/b)t$ radians, so that $\overrightarrow{CP}$ makes an angle of $(r/b)t + t$ radians with the line parallel to the negative x axis, and hence an angle of $\pi + (r/b)t + t$ radians with a line parallel to the positive x axis. Therefore

$$\mathbf{b} = \overrightarrow{CP} = b\cos\left(\pi + \frac{r}{b}t + t\right)\mathbf{i} + b\sin\left(\pi + \frac{r}{b}t + t\right)\mathbf{j} = -b\cos\left(\frac{r}{b}t + t\right)\mathbf{i} - b\sin\left(\frac{r}{b}t + t\right)\mathbf{j}.$$

Consequently

$$\mathbf{F}(t) = \mathbf{a} + \mathbf{b} = [(r+b)\cos t\,\mathbf{i} + (r+b)\sin t\,\mathbf{j}] + \left[-b\cos\left(\frac{r}{b}t + t\right)\mathbf{i} - b\sin\left(\frac{r}{b}t + t\right)\mathbf{j}\right]$$

$$= \left[(r+b)\cos t - b\cos\left(\frac{r+b}{b}t\right)\right]\mathbf{i} + \left[(r+b)\sin t - b\sin\left(\frac{r+b}{b}t\right)\right]\mathbf{j}.$$

12.2 Limits and Continuity of Vector-Valued Functions

1. $\lim_{t\to 4}(\mathbf{i} - \mathbf{j} + \mathbf{k}) = \mathbf{i} - \mathbf{j} + \mathbf{k}$

3. $\lim_{t\to\pi}(\tan t\,\mathbf{i} + 3t\,\mathbf{j} - 4\mathbf{k}) = \left(\lim_{t\to\pi}\tan t\right)\mathbf{i} + \left(\lim_{t\to\pi}3t\right)\mathbf{j} + \left(\lim_{t\to\pi}(-4)\right)\mathbf{k} = 0\mathbf{i} + 3\pi\mathbf{j} - 4\mathbf{k} = 3\pi\mathbf{j} - 4\mathbf{k}$

5. $$\lim_{t\to 2^-}\mathbf{F}(t) = \lim_{t\to 2^-}(5\mathbf{i} - \sqrt{2t^2+2t+4}\,\mathbf{j} + e^{-(t-2)}\mathbf{k})$$

$$= \left(\lim_{t\to 2^-}5\right)\mathbf{i} + \left(\lim_{t\to 2^-} -\sqrt{2t^2+2t+4}\right)\mathbf{j} + \left(\lim_{t\to 2^-}e^{-(t-2)}\right)\mathbf{k} = 5\mathbf{i} - 4\mathbf{j} + \mathbf{k}$$

and

$$\lim_{t\to 2^+}\mathbf{F}(t) = \lim_{t\to 2^+}(t^2+1)\mathbf{i} + (4-t^3)\mathbf{j} + \mathbf{k} = \left(\lim_{t\to 2^+}(t^2+1)\right)\mathbf{i} + \left(\lim_{t\to 2^+}(4-t^3)\right)\mathbf{j} + \left(\lim_{t\to 2^+}1\right)\mathbf{k} = 5\mathbf{i} - 4\mathbf{j} + \mathbf{k}.$$

Since $\lim_{t\to 2^-}\mathbf{F}(t) = \lim_{t\to 2^+}\mathbf{F}(t) = 5\mathbf{i} - 4\mathbf{j} + \mathbf{k}$, it follows that $\lim_{t\to 2}\mathbf{F}(t)$ exists and that $\lim_{t\to 2}\mathbf{F}(t) = 5\mathbf{i} - 4\mathbf{j} + \mathbf{k}$.

7. $$\lim_{t\to 0}(\mathbf{F} - \mathbf{G})(t) = \lim_{t\to 0}\left[(e^{-1/t^2} + \pi)\mathbf{i} + \left(\cos t - \frac{1+\cos t}{t}\right)\mathbf{j} + t^3\mathbf{k}\right]$$

Since $\lim_{t\to 0}(1+\cos t)/t$ and hence $\lim_{t\to 0}(\cos t - (1+\cos t)/t)$ does not exist, Theorem 12.4 implies that $\lim_{t\to 0}(\mathbf{F} - \mathbf{G})(t)$ does not exist.

9. $\lim_{t\to 3}\left(\frac{t^2-5t+6}{t-3}\mathbf{i}+\frac{t^2-2t-3}{t-3}\mathbf{j}+\frac{t^2+4t-21}{t-3}\mathbf{k}\right)=\lim_{t\to 3}[(t-2)\mathbf{i}+(t+1)\mathbf{j}+(t+7)\mathbf{k}]$

$=\left(\lim_{t\to 3}(t-2)\right)\mathbf{i}+\left(\lim_{t\to 3}(t+1)\right)\mathbf{j}+\left(\lim_{t\to 3}(t+7)\right)\mathbf{k}=\mathbf{i}+4\mathbf{j}+10\mathbf{k}$

11. a. Let $\mathbf{F}$ be defined at each point in some open interval (t_0, t_1). A vector $\mathbf{L}$ is the limit of $\mathbf{F}(t)$ as t approaches t_0 from the right if for every $\varepsilon > 0$ there is a number $\delta > 0$ such that

$$\text{if } \ 0 < t - t_0 < \delta, \quad \text{then } \|\mathbf{F}(t)-\mathbf{L}\| < \varepsilon.$$

In this case we write $\lim_{t\to t_0^+}\mathbf{F}(t)=\mathbf{L}$. To prove that such limits can be computed componentwise, we can use the statement and proof of Theorem 12.4, with "$\lim_{t\to t_0}$" replaced by "$\lim_{t\to t_0^+}$", and "$0 < |t-t_0| < \delta$" replaced by "$0 < t-t_0 < \delta$".

b. Let $\mathbf{F}$ be defined at each point in some open interval (t_1, t_0) A vector $\mathbf{L}$ is the limit of $\mathbf{F}(t)$ as t approaches t_0 from the left if for every $\varepsilon > 0$ there is a number $\delta > 0$ such that

$$\text{if } \ -\delta < t - t_0 < 0, \quad \text{then } \|\mathbf{F}(t)-\mathbf{L}\| < \varepsilon.$$

In this case we write $\lim_{t\to t_0^-}\mathbf{F}(t)=\mathbf{L}$. To prove that such limits can be computed componentwise, use the statement and proof of Theorem 12.4, with "$\lim_{t\to t_0}$" replaced by "$\lim_{t\to t_0^-}$", and "$0 < |t-t_0| < \delta$" replaced by "$-\delta < t-t_0 < 0$".

13. a. Let $\mathbf{F}$ be defined at each point in an interval of the form (t_0, ∞). A vector $\mathbf{L}$ is the limit of $\mathbf{F}(t)$ as t approaches ∞ if for every $\varepsilon > 0$ there is a number M such that

$$\text{if } \ t > M, \quad \text{then } \|\mathbf{F}(t)-\mathbf{L}\| < \varepsilon.$$

In this case we write $\lim_{t\to\infty}\mathbf{F}(t)=\mathbf{L}$. To prove that such limits can be computed componentwise, use the statement and proof of Theorem 12.4, with "$\lim_{t\to t_0}$" replaced by "$\lim_{t\to\infty}$", "$\delta > 0$" by "M", and "$0 < |t-t_0| < \delta$" by "$t > M$".

b. $\lim_{t\to\infty}\left(\frac{1}{t}\mathbf{i}+\frac{t-1}{t+1}\mathbf{j}+\frac{\sin t^3}{t^2}\mathbf{k}\right)=\left(\lim_{t\to\infty}\frac{1}{t}\right)\mathbf{i}+\left(\lim_{t\to\infty}\frac{t-1}{t+1}\right)\mathbf{j}+\left(\lim_{t\to\infty}\frac{\sin t^3}{t^2}\right)\mathbf{k}$

Since $|(\sin t^3)/t^2| \le 1/t^2$, we have $\lim_{t\to\infty}(\sin t^3)/t^2 = 0$. Thus

$$\left(\lim_{t\to\infty}\frac{1}{t}\right)\mathbf{i}+\left(\lim_{t\to\infty}\frac{t-1}{t+1}\right)\mathbf{j}+\left(\lim_{t\to\infty}\frac{\sin t^3}{t^2}\right)\mathbf{k}=0\mathbf{i}+\mathbf{j}+0\mathbf{k}=\mathbf{j}.$$

12.3 Derivatives and Integrals of Vector-Valued Functions

1. $\mathbf{F}'(t) = \mathbf{j} + 5t^4\mathbf{k}$

3. $\mathbf{F}'(t) = \frac{3}{2}(1+t)^{1/2}\mathbf{i} + \frac{3}{2}(1-t)^{1/2}\mathbf{j} + \frac{3}{2}\mathbf{k}$

5. $\mathbf{F}'(t) = \sec^2 t\,\mathbf{i} + \sec t\,\tan t\,\mathbf{k}$

7. $\mathbf{F}'(t) = \sinh t\,\mathbf{i} + \cosh t\,\mathbf{j} - \dfrac{1}{2\sqrt{t}}\mathbf{k}$

9. $(4\mathbf{F} - 2\mathbf{G})(t) = (8\sec t - 6t)\mathbf{i} + (2t^2 - 12)\mathbf{j} + 12\csc t\,\mathbf{k}$, so that $(4\mathbf{F} - 2\mathbf{G})'(t) = (8\sec t\,\tan t - 6)\mathbf{i} + 4t\mathbf{j} - 12\csc t\,\cot t\,\mathbf{k}$

11. $(\mathbf{F}\times\mathbf{G})(t) = -3\ln t\,\mathbf{i} - 2\sec t\,\ln t\,\mathbf{j}$; $(\mathbf{F}\times\mathbf{G})'(t) = \dfrac{-3}{t}\mathbf{i} - \left(2\sec t\,\tan t\,\ln t + \dfrac{2}{t}\sec t\right)\mathbf{j}$

13. $(\mathbf{F}\times\mathbf{G})(t) = \left(t - \dfrac{3}{t}e^{-t}\right)\mathbf{k}$; $(\mathbf{F}\times\mathbf{G})'(t) = \left(1 + \dfrac{3}{t^2}e^{-t} + \dfrac{3}{t}e^{-t}\right)\mathbf{k}$

15. $(\mathbf{F}\circ g)'(t) = \mathbf{F}'(g(t))g'(t) = \left(\dfrac{1}{\sqrt{t}}\mathbf{i} - 8e^{2\sqrt{t}}\mathbf{j} + \dfrac{1}{(\sqrt{t})^2}\mathbf{k}\right)\dfrac{1}{2\sqrt{t}} = \dfrac{1}{2t}\mathbf{i} - \dfrac{4e^{2\sqrt{t}}}{\sqrt{t}}\mathbf{j} + \dfrac{1}{2}t^{-3/2}\mathbf{k}$

17. $\displaystyle\int\left(t^2\mathbf{i} - (3t-1)\mathbf{j} - \frac{1}{t^3}\mathbf{k}\right)dt = \frac{t^3}{3}\mathbf{i} - \left(\frac{3}{2}t^2 - t\right)\mathbf{j} + \frac{1}{2t^2}\mathbf{k} + \mathbf{C}$

19. $\int_0^1 (e^t\mathbf{i} + e^{-t}\mathbf{j} + 2t\mathbf{k})\,dt = e^t\big|_0^1\mathbf{i} - e^{-t}\big|_0^1\mathbf{j} + t^2\big|_0^1\mathbf{k} = (e-1)\mathbf{i} + (1-e^{-1})\mathbf{j} + \mathbf{k}$

21. $\int_{-1}^1 [(1+t)^{3/2}\mathbf{i} + (1-t)^{3/2}\mathbf{j}]\,dt = \frac{2}{5}(1+t)^{5/2}\big|_{-1}^1\mathbf{i} - \frac{2}{5}(1-t)^{5/2}\big|_{-1}^1\mathbf{j} = \frac{8}{5}\sqrt{2}\,(\mathbf{i}+\mathbf{j})$

23. $\mathbf{v}(t) = -\sin t\,\mathbf{i} + \cos t\,\mathbf{j} - 32\mathbf{k}$,
$\|\mathbf{v}(t)\| = \sqrt{(-\sin t)^2 + \cos^2 t + (-32t)^2} = \sqrt{1 + 1024t^2}$,
$\mathbf{a}(t) = -\cos t\,\mathbf{i} - \sin t\,\mathbf{j} - 32\mathbf{k}$

25. $\mathbf{v}(t) = 2\mathbf{i} + 2t\mathbf{j} + \dfrac{1}{t}\mathbf{k}$, $\|\mathbf{v}(t)\| = \sqrt{4 + 4t^2 + \dfrac{1}{t^2}} = \dfrac{2t^2+1}{t}$, $\mathbf{a}(t) = 2\mathbf{j} - \dfrac{1}{t^2}\mathbf{k}$

27. $\mathbf{v}(t) = (e^t\sin t + e^t\cos t)\mathbf{i} + (e^t\cos t - e^t\sin t)\mathbf{j} + e^t\mathbf{k}$,
$\|\mathbf{v}(t)\| = \sqrt{(e^{2t}\sin^2 t + 2e^{2t}\sin t\,\cos t + e^{2t}\cos^2 t) + (e^{2t}\cos^2 t - 2e^{2t}\sin t\,\cos t + e^{2t}\sin^2 t) + e^{2t}} = e^t\sqrt{3}$,
$\mathbf{a}(t) = 2e^t\cos t\,\mathbf{i} - 2e^t\sin t\,\mathbf{j} + e^t\mathbf{k}$

29. $\mathbf{v}(t) = \int \mathbf{a}(t)\,dt = \int -32\mathbf{k}\,dt = -32t\,\mathbf{k} + \mathbf{C}$;
$\mathbf{r}(t) = \int \mathbf{v}(t)\,dt = \int(-32t\,\mathbf{k} + \mathbf{C})\,dt = -16t^2\mathbf{k} + \mathbf{C}t + \mathbf{C}_1$.
Since $\mathbf{v}(0) = \mathbf{v}_0 = \mathbf{i} + \mathbf{j}$, we have $\mathbf{C} = \mathbf{i} + \mathbf{j}$, so that $\mathbf{v}(t) = \mathbf{i} + \mathbf{j} - 32t\mathbf{k}$. Since $\mathbf{r}(0) = \mathbf{r}_0 = \mathbf{0}$, we have $\mathbf{C}_1 = \mathbf{0}$, so that $\mathbf{r}(t) = t\mathbf{i} + t\mathbf{j} - 16t^2\mathbf{k}$. Finally, $\|\mathbf{v}(t)\| = \sqrt{1 + 1 + (-32t)^2} = \sqrt{2 + 1024t^2}$.

31. $\mathbf{v}(t) = \int \mathbf{a}(t)\,dt = \int(-\cos t\,\mathbf{i} - \sin t\,\mathbf{j})\,dt = -\sin t\,\mathbf{i} + \cos t\,\mathbf{j} + \mathbf{C}$;
$\mathbf{r}(t) = \int \mathbf{v}(t)\,dt = \int(-\sin t\,\mathbf{i} + \cos t\,\mathbf{j} + \mathbf{C})\,dt = \cos t\,\mathbf{i} + \sin t\,\mathbf{j} + \mathbf{C}t + \mathbf{C}_1$.
Since $\mathbf{v}(0) = \mathbf{v}_0 = \mathbf{k}$, we have $\mathbf{k} = -\sin 0\,\mathbf{i} + \cos 0\,\mathbf{j} + \mathbf{C} = \mathbf{j} + \mathbf{C}$, so that $\mathbf{C} = \mathbf{k} - \mathbf{j}$, and thus $\mathbf{v}(t) = -\sin t\,\mathbf{i} + (\cos t - 1)\mathbf{j} + \mathbf{k}$. Since $\mathbf{r}(0) = \mathbf{r}_0 = \mathbf{i}$, we have $\mathbf{i} = \cos 0\,\mathbf{i} + \sin 0\,\mathbf{j} + (\mathbf{k} - \mathbf{j})0 + \mathbf{C}_1 = \mathbf{i} + \mathbf{C}_1$, so that $\mathbf{C}_1 = \mathbf{0}$, and thus $\mathbf{r}(t) = \cos t\,\mathbf{i} + \sin t\,\mathbf{j} + t(\mathbf{k} - \mathbf{j}) = \cos t\,\mathbf{i} + (\sin t - t)\mathbf{j} + t\mathbf{k}$. Finally, $\|\mathbf{v}(t)\| = \sqrt{\sin^2 t + (\cos t - 1)^2 + 1} = \sqrt{3 - 2\cos t}$.

33. $\mathbf{F}(t) = \left(\int_0^t u \tan u^3\,du\right)\mathbf{i} + \left(\int_0^t \cos e^u\,du\right)\mathbf{j} + \left(\int_0^t e^{(u^2)}\,du\right)\mathbf{k}$, and thus $\mathbf{F}'(t) = t\tan t^3\,\mathbf{i} + \cos e^t\,\mathbf{j} + e^{(t^2)}\,\mathbf{k}$.

35. $\|\mathbf{F}(t)\| = \sqrt{\dfrac{16t^2}{(1+4t^2)^2} + \dfrac{1 - 8t^2 + 16t^4}{(1+4t^2)^2}} = \sqrt{\dfrac{(1+4t^2)^2}{(1+4t^2)^2}} = 1$, so that by Corollary 12.11, $\mathbf{F}(t)\cdot\mathbf{F}'(t) = 0$ for all t.

37. $\mathbf{F}'(t) = \cos t\,\mathbf{i} + \sin t\,\mathbf{j}$ and $\mathbf{F}''(t) = -\sin t\,\mathbf{i} + \cos t\,\mathbf{j} = -\mathbf{F}(t)$. Since $\|\mathbf{F}(t)\| = \|\mathbf{F}''(t)\| = 1$, it follows that $\mathbf{F}(t)$ and $\mathbf{F}''(t)$ are parallel but have opposite, and hence dissimilar, directions, for all t.

39. By Theorem 12.10(e), $\dfrac{d}{dt}(\mathbf{F}\times\mathbf{F}')(t) = [\mathbf{F}'(t)\times\mathbf{F}'(t)] + [\mathbf{F}(t)\times\mathbf{F}''(t)] = \mathbf{F}(t)\times\mathbf{F}''(t)$.

41. a. $\mathbf{r}(t) = \cos t\,\mathbf{i} + \sin t\,\mathbf{j} + t\,\mathbf{k}$, so that $\mathbf{v}(t) = -\sin t\,\mathbf{i} + \cos t\,\mathbf{j} + \mathbf{k}$, $\|\mathbf{r}(t)\| = \sqrt{(\cos t)^2 + (\sin t)^2 + t^2} = \sqrt{1+t^2}$, and $\|\mathbf{v}(t)\| = \sqrt{(-\sin t)^2 + (\cos t)^2 + 1^2} = \sqrt{2}$. Thus

$$\cos\theta(t) = \frac{\mathbf{r}(t)\cdot\mathbf{v}(t)}{\|\mathbf{r}(t)\|\,\|\mathbf{v}(t)\|} = \frac{(\cos t)(-\sin t) + (\sin t)(\cos t) + (t)(1)}{\sqrt{1+t^2}\,\sqrt{2}} = \frac{t}{\sqrt{2(1+t^2)}}.$$

Since $0 \le \theta(t) \le \pi$, it follows that

$$\theta(t) = \cos^{-1}\frac{t}{\sqrt{2(1+t^2)}}.$$

b. By part (a), $\theta(0) = \cos^{-1} 0 = \pi/2$.

c. By part (a),

$$\theta(\pi/2) = \cos^{-1}\frac{\pi/2}{\sqrt{2(1+\pi^2/4)}} \approx 0.9316761004 \text{ radians}.$$

43. Let $\mathbf{F}(t) = \mathbf{v}(t)$, so that $\mathbf{F}'(t) = \mathbf{a}(t)$. Since the speed is constant, we have $\|\mathbf{F}(t)\| = \|\mathbf{v}(t)\| = c$, for some constant c. By Corollary 12.11, $\mathbf{v}\cdot\mathbf{a}(t) = \mathbf{F}(t)\cdot\mathbf{F}'(t) = 0$. Therefore the acceleration vector is perpendicular to the velocity vector.

45. Let $K_1(t)$ and $K_2(t)$ be the kinetic energies of the mass at any time t during its first and second journeys, respectively. By Example 10, the first time the position is given by $\mathbf{r}_1(t) = (-16t^2 + 96)\mathbf{k}$, so that $\mathbf{r}_1(\sqrt{6}) = \mathbf{0}$, and thus the ball hits the ground when $t = \sqrt{6}$. Since $\mathbf{v}_1(t) = -32t\mathbf{k}$, and $K_1(t) = \frac{1}{2}m\|\mathbf{v}_1(t)\|^2$, we therefore have $K_1(\sqrt{6}) = \frac{1}{2}m\left|-32(\sqrt{6})\right|^2 = 3072m$. The second time the position is given by $\mathbf{r}_2(t) = (-16t^2 - 80t + 96)\mathbf{k} = -16(t^2 + 5t - 6)\mathbf{k} = -16(t+6)(t-1)\mathbf{k}$, so that $\mathbf{r}_2(1) = \mathbf{0}$, and thus the ball hits the ground when $t = 1$. Since $\mathbf{v}_2(t) = (-32t - 80)\mathbf{k}$, and $K_2(t) = \frac{1}{2}m\,\|\mathbf{v}_2(t)\|^2$, we have $K_2(1) = \frac{1}{2}m\,|-32-80|^2 = 6272m$. Consequently $K_2(1) - K_1(\sqrt{6}) = 6272m - 3072m = 3200m$, and this is how much larger the kinetic energy is at impact the second time.

47. a. Initial position: $\mathbf{r}_0 = \mathbf{0}$; velocity: $\mathbf{v}(t) = 90\sqrt{2}\,\mathbf{i} + 90\sqrt{2}\,\mathbf{j} + (64 - 32t)\mathbf{k}$; initial velocity: $\mathbf{v}_0 = 90\sqrt{2}\,\mathbf{i} + 90\sqrt{2}\,\mathbf{j} + 64\mathbf{k}$

b. $\mathbf{r}(4) = 360\sqrt{2}\,\mathbf{i} + 360\sqrt{2}\,\mathbf{j} + 0\mathbf{k}$, so the height is 0 when $t = 4$; distance from initial position: $\|\mathbf{r}(4) - \mathbf{r}(0)\| = \|360\sqrt{2}\,\mathbf{i} + 360\sqrt{2}\,\mathbf{j}\| = 720$ (feet).

49. Place the coordinate system so that the floor represents the xy plane and the ball rolls on a line directly above the negative x axis towards the edge, located at the point $(0, 0, 2.6)$. Then by Example 10, the position of the ball t seconds after it leaves the table is given by $\mathbf{r}(t) = -16t^2\mathbf{k} + t\mathbf{v}_0 + \mathbf{r}_0$. By assumption, $\mathbf{v}_0 = 2\mathbf{i}$ and $\mathbf{r}_0 = 2.6\mathbf{k}$. Thus $\mathbf{r}(t) = -16t^2\mathbf{k} + 2t\mathbf{i} + 2.6\mathbf{k} = 2t\mathbf{i} + (2.6 - 16t^2)\mathbf{k}$.

a. The $\mathbf{k}$ component of $\mathbf{r}(t)$ is 0 if $2.6 = 16t^2$, or $t = \sqrt{2.6}/4 \approx .40$. Thus the ball hits the floor after approximately .40 seconds.

b. $\mathbf{v}(t) = 2\mathbf{i} - 32t\mathbf{k}$, so that

$$\left\|\mathbf{v}\left(\frac{\sqrt{2.6}}{4}\right)\right\| = \left\|2\mathbf{i} - 32\left(\frac{\sqrt{2.6}}{4}\right)\mathbf{k}\right\| = \sqrt{4 + 166.4} = \sqrt{170.4} \approx 13 \text{ (feet per second).}$$

51. By the hint, $2\pi v_0 =$ the circumference $= 24(3600)v_0$. By (7) we have $r_0 = C/v_0^2$, so that

$$\frac{24(3600)v_0}{2\pi} = r_0 = \frac{C}{v_0^2}, \quad \text{so} \quad v_0^3 = \frac{2\pi C}{24(3600)} = \frac{2\pi(32)(3960)^2(5280)^2}{24(3600)}.$$

Therefore $v_0 \approx 10{,}057.53$ (feet per second), or approximately 6857.41 miles per hour. Then

$$r_0 = \frac{(24)(3600)v_0}{2\pi} \approx 138{,}300{,}965.1 \text{ (feet)}$$

or approximately 26,193.4 miles. (It follows that such a communications satellite circles the earth at approximately $26{,}193 - 3960 = 22{,}233$ miles above the surface of the earth.)

53. a. By (5), $\|\mathbf{a}(t)\| = \dfrac{[(2.5) \times 10^5]^2}{1} = 6.25 \times 10^{10}$ (kilometers per second per second).

b. By (5), $\|\mathbf{a}(t)\| = \dfrac{[(2.9) \times 10^5]^2}{1} = 8.41 \times 10^{10}$ (kilometers per second per second).

12.4 Space Curves and Their Lengths

1. $\mathbf{r}'(t) = \mathbf{i} + 2t\mathbf{j} + 3t^2\mathbf{k}$; $\mathbf{r}'$ is continuous and is never $\mathbf{0}$, so $\mathbf{r}$ is smooth.

3. $\mathbf{r}'(t) = \begin{cases} -\mathbf{i} + \mathbf{j} + \mathbf{k} & \text{for } t < 0 \\ \mathbf{i} + \mathbf{j} + \mathbf{k} & \text{for } t > 0 \end{cases}$

so $\mathbf{r}'$ is continuous and nonzero on $(-\infty, 0)$ and on $(0, \infty)$. Since the one-sided derivatives at 0 are $-\mathbf{i} + \mathbf{j} + \mathbf{k}$ and $\mathbf{i} + \mathbf{j} + \mathbf{k}$ respectively, we conclude that $\mathbf{r}$ is piecewise smooth but not smooth.

5. $\mathbf{r}'(t) = \frac{3}{2}(1+t)^{1/2}\mathbf{i} - \frac{3}{2}(1-t)^{1/2}\mathbf{j} + \frac{3}{2}\mathbf{k}$ for $-1 < t < 1$, and the appropriate one-sided derivatives exist at -1 and 1. Since $\mathbf{r}'$ is continuous and nonzero on $[-1, 1]$, $\mathbf{r}$ is smooth.

7. $\mathbf{r}'(t) = -2\cos t \sin t\,\mathbf{i} + 2\sin t \cos t\,\mathbf{j} + 2t\,\mathbf{k}$; $\mathbf{r}$ is continuous and $\mathbf{r}'(t) = \mathbf{0}$ only if $t = 0$. Thus $\mathbf{r}$ is piecewise smooth.

9. $\mathbf{r}'(t) = (e^t - 1)\mathbf{i} + 2t\,\mathbf{j} + 3t^2\,\mathbf{k}$; $\mathbf{r}'$ is continuous and $\mathbf{r}'(t) = \mathbf{0}$ only if $t = 0$. Thus $\mathbf{r}$ is piecewise smooth.

11. The line segment is parallel to $7\mathbf{i} - 2\mathbf{j} + 4\mathbf{k}$ and starts at $(-3, 2, 1)$. Thus a parametrization is $\mathbf{r}(t) = (-3 + 7t)\mathbf{i} + (2 - 2t)\mathbf{j} + (1 + 4t)\mathbf{k}$ for $0 \le t \le 1$, and $\mathbf{r}$ is smooth.

13. $\mathbf{r}(t) = 6\cos t\,\mathbf{i} + 6\sin t\,\mathbf{j}$ for $0 \le t \le 2\pi$ is one such smooth parametrization.

15. $\mathbf{r}(t) = \cos t\,\mathbf{i} + \sin t\,\mathbf{j}$ for $0 \le t \le \pi$ is one such smooth parametrization.

17. Notice that the quarter circle in the xy plane that extends from $(\sqrt{2}/2, \sqrt{2}/2)$ to $(\sqrt{2}/2, -\sqrt{2}/2)$ corresponds to the quarter circle from $(1, \pi/4)$ to $(1, -\pi/4)$ in polar coordinates. This leads us to $\mathbf{r}(t) = \cos t\,\mathbf{i} - \sin t\,\mathbf{j} + 4\mathbf{k}$ for $-\pi/4 \le t \le \pi/4$, which is smooth.

19. $\mathbf{r}(t) = t\,\mathbf{i} + \tan t\,\mathbf{j}$ for $0 \le t \le \pi/4$ is one such smooth parametrization.

21. $L = \int_0^{2\pi} \sqrt{[3\cos^2 t(-\sin t)]^2 + [3\sin^2 t \cos t]^2}\,dt$

$= \int_0^{2\pi} 3|\sin t \cos t|\,dt = 12\int_0^{2\pi} \sin t \cos t\,dt = 6\sin^2 t\Big|_0^{2\pi} = 6$

23. $L = \int_{-1}^1 \sqrt{\frac{1}{4}(1+t) + \frac{1}{4}(1-t) + \frac{1}{4}}\,dt = \int_{-1}^1 \frac{1}{2}\sqrt{3}\,dt = \frac{1}{2}\sqrt{3}\,t\Big|_{-1}^1 = \sqrt{3}$

25. $L = \int_0^1 \sqrt{e^{2t} + e^{-2t} + 2}\,dt = \int_0^1 \sqrt{(e^t + e^{-t})^2}\,dt = \int_0^1 (e^t + e^{-t})\,dt = (e^t - e^{-t})|_0^1 = e - e^{-1}$

27. $L = \int_1^{\sqrt{8}} \sqrt{[36t^2(t^2 - 1)] + 36t^2 + 36t^2}\,dt = \int_1^{\sqrt{8}} 6\sqrt{t^4 + t^2}\,dt$

$= \int_1^{\sqrt{8}} 6t\sqrt{t^2 + 1}\,dt = 2(t^2 + 1)^{3/2}\Big|_1^{\sqrt{8}} = 54 - 4\sqrt{2}$

29. $\dfrac{ds}{dt} = \sqrt{4\cos^2 2t + 4\sin^2 2t + t} = \sqrt{4 + t}$

31. $\dfrac{ds}{dt} = \sqrt{(\cos t - t\sin t)^2 + (\sin t + t\cos t)^2 + 1} = \sqrt{\cos^2 t + \sin^2 t + t^2(\sin^2 t + \cos^2 t) + 1} = \sqrt{2 + t^2}$

33. $\dfrac{ds}{dt} = \sqrt{(1 - \cos t)^2 + \sin^2 t + 1} = \sqrt{3 - 2\cos t}$

35. $L = \int_0^2 \sqrt{4\sin^2 2t + 4\cos^2 2t + 4t^3}\,dt = 2\int_0^2 \sqrt{1 + t^3}\,dt$. By Simpson's Rule with $n = 10$, $L \approx 6.482559337$.

37. $L = \int_0^1 \sqrt{t^2+t^4+t^6}\,dt$. By Simpson's Rule with $n = 10$, $L \approx 0.668455969$.

39. $s(t) = \int_0^t \sqrt{(x'(u))^2 + (y'(u))^2 + (z'(u))^2}\,du = \int_0^t \sqrt{2^2+\cos^2 u + \sin^2 u}\,du = \int_0^t \sqrt{5}\,du = \sqrt{5}\,t$

41. a. Assume that the radius of the circle is $r > 0$. Then $n = 8$ and $d = 2r$. Thus $\pi nd/8 = [\pi(8)(2r)]/8 = 2\pi r =$ circumference L of the circle.

b. $n = 18$, $d = \frac{1}{3}$, so $L \approx [\pi(18)(\frac{1}{3})]/8 = 3\pi/4$.

12.5 Tangents and Normals to Curves

1. $\mathbf{r}'(t) = 2t\mathbf{i} + 2\mathbf{j}$; $\|\mathbf{r}'(t)\| = \sqrt{(2t)^2+2^2} = 2\sqrt{t^2+1}$;

$$\mathbf{T}(t) = \frac{\mathbf{r}'(t)}{\|\mathbf{r}'(t)\|} = \frac{2t\mathbf{i}+2\mathbf{j}}{2\sqrt{t^2+1}} = \frac{t}{\sqrt{t^2+1}}\mathbf{i} + \frac{1}{\sqrt{t^2+1}}\mathbf{j};$$

$$\mathbf{T}'(t) = \frac{\sqrt{t^2+1} - t^2/\sqrt{t^2+1}}{t^2+1}\mathbf{i} - \frac{t}{(t^2+1)^{3/2}}\mathbf{j} = \frac{1}{(t^2+1)^{3/2}}\mathbf{i} - \frac{t}{(t^2+1)^{3/2}}\mathbf{j};$$

$$\|\mathbf{T}'(t)\| = \sqrt{\frac{1}{(t^2+1)^3} + \frac{t^2}{(t^2+1)^3}} = \frac{1}{t^2+1}; \mathbf{N}(t) = \frac{\mathbf{T}'(t)}{\|\mathbf{T}'(t)\|} = \frac{1}{\sqrt{t^2+1}}\mathbf{i} - \frac{t}{\sqrt{t^2+1}}\mathbf{j}$$

3. $\mathbf{r}'(t) = -\sin t\,\mathbf{i} - \sin t\,\mathbf{j} + \sqrt{2}\cos t\,\mathbf{k}$; $\|\mathbf{r}'(t)\| = \sqrt{(-\sin t)^2 + (-\sin t)^2 + (\sqrt{2}\cos t)^2} = \sqrt{2}$;

$$\mathbf{T}(t) = \frac{\mathbf{r}'(t)}{\|\mathbf{r}'(t)\|} = -\frac{\sin t}{\sqrt{2}}\mathbf{i} - \frac{\sin t}{\sqrt{2}}\mathbf{j} + \cos t\,\mathbf{k};\ \mathbf{T}'(t) = -\frac{\cos t}{\sqrt{2}}\mathbf{i} - \frac{\cos t}{\sqrt{2}}\mathbf{j} - \sin t\,\mathbf{k};$$

$$\|\mathbf{T}'(t)\| = \sqrt{\left(\frac{-\cos t}{\sqrt{2}}\right)^2 + \left(\frac{-\cos t}{\sqrt{2}}\right)^2 + (-\sin t)^2} = 1;$$

$$\mathbf{N}(t) = \frac{\mathbf{T}'(t)}{\|\mathbf{T}'(t)\|} = -\frac{\cos t}{\sqrt{2}}\mathbf{i} - \frac{\cos t}{\sqrt{2}}\mathbf{j} - \sin t\,\mathbf{k}$$

5. $\mathbf{r}'(t) = 2\mathbf{i} + 2t\mathbf{j} + t^2\mathbf{k}$; $\|\mathbf{r}'(t)\| = \sqrt{2^2 + (2t)^2 + (t^2)^2} = \sqrt{4+4t^2+t^4} = 2+t^2$;

$$\mathbf{T}(t) = \frac{\mathbf{r}'(t)}{\|\mathbf{r}'(t)\|} = \frac{2}{2+t^2}\mathbf{i} + \frac{2t}{2+t^2}\mathbf{j} + \frac{t^2}{2+t^2}\mathbf{k};$$

$$\mathbf{T}'(t) = \frac{-4t}{(2+t^2)^2}\mathbf{i} + \frac{2(2+t^2)-4t^2}{(2+t^2)^2}\mathbf{j} + \frac{2t(2+t^2)-2t^3}{(2+t^2)^2}\mathbf{k} = \frac{-4t}{(2+t^2)^2}\mathbf{i} + \frac{4-2t^2}{(2+t^2)^2}\mathbf{j} + \frac{4t}{(2+t^2)^2}\mathbf{k};$$

$$\|\mathbf{T}'(t)\| = \frac{1}{(2+t^2)^2}\sqrt{(-4t)^2 + (4-2t^2)^2 + (4t)^2} = \frac{1}{(2+t^2)^2}\sqrt{16+16t^2+4t^4} = \frac{2}{2+t^2};$$

$$\mathbf{N}(t) = \frac{\mathbf{T}'(t)}{\|\mathbf{T}'(t)\|} = \frac{-2t}{2+t^2}\mathbf{i} + \frac{2-t^2}{2+t^2}\mathbf{j} + \frac{2t}{2+t^2}\mathbf{k}$$

7. $\mathbf{r}'(t) = e^t\mathbf{i} - e^{-t}\mathbf{j} + \sqrt{2}\,\mathbf{k}$; $\|\mathbf{r}'(t)\| = \sqrt{e^{2t} + e^{-2t} + 2} = e^t + e^{-t}$;

$$\mathbf{T}(t) = \frac{\mathbf{r}'(t)}{\|\mathbf{r}'(t)\|} = \frac{e^t}{e^t + e^{-t}}\mathbf{i} - \frac{e^{-t}}{e^t + e^{-t}}\mathbf{j} + \frac{\sqrt{2}}{e^t + e^{-t}}\mathbf{k};$$

$$\mathbf{T}'(t) = \frac{e^t(e^t + e^{-t}) - e^t(e^t - e^{-t})}{(e^t + e^{-t})^2}\mathbf{i} - \frac{-e^{-t}(e^t + e^{-t}) - e^{-t}(e^t - e^{-t})}{(e^r + e^{-t})^2}\mathbf{j} - \frac{\sqrt{2}\,(e^t - e^{-t})}{(e^t + e^{-t})^2}\mathbf{k}$$

$$= \frac{2}{(e^t + e^{-t})^2}\mathbf{i} + \frac{2}{(e^t + e^{-t})^2}\mathbf{j} - \frac{\sqrt{2}\,(e^t - e^{-t})}{(e^t + e^{-t})^2}\mathbf{k};$$

$$\|\mathbf{T}'(t)\| = \frac{1}{(e^t + e^{-t})^2}\sqrt{2^2 + 2^2 + [\sqrt{2}\,(e^{-t} - e^t)]^2} = \frac{\sqrt{2e^{-2t} + 4 + 2e^{2t}}}{(e^t + e^{-t})^2} = \frac{\sqrt{2}}{e^t + e^{-t}};$$

$$\mathbf{N}(t) = \frac{\mathbf{T}'(t)}{\|\mathbf{T}'(t)\|} = \frac{\sqrt{2}}{e^t + e^{-t}}\mathbf{i} + \frac{\sqrt{2}}{e^t + e^{-t}}\mathbf{j} - \frac{e^t - e^{-t}}{e^t + e^{-t}}\mathbf{k}$$

9. $\mathbf{r}'(t) = 2\mathbf{i} + 2t\mathbf{j} + \frac{1}{t}\mathbf{k}$; $\|\mathbf{r}'(t)\| = \sqrt{2^2 + (2t)^2 + \left(\frac{1}{t}\right)^2} = \sqrt{4t^2 + 4 + \frac{1}{t^2}} = 2t + \frac{1}{t} = \frac{2t^2 + 1}{t}$;

$$\mathbf{T}(t) = \frac{\mathbf{r}'(t)}{\|\mathbf{r}'(t)\|} = \frac{2t}{2t^2 + 1}\mathbf{i} + \frac{2t^2}{2t^2 + 1}\mathbf{j} + \frac{1}{2t^2 + 1}\mathbf{k};$$

$$\mathbf{T}'(t) = \frac{2(2t^2 + 1) - 2t(4t)}{(2t^2 + 1)^2}\mathbf{i} + \frac{4t(2t^2 + 1) - 2t^2(4t)}{(2t^2 + 1)^2}\mathbf{j} - \frac{4t}{(2t^2 + 1)^2}\mathbf{k}$$

$$= \frac{2 - 4t^2}{(2t^2 + 1)^2}\mathbf{i} + \frac{4t}{(2t^2 + 1)^2}\mathbf{j} - \frac{4t}{(2t^2 + 1)^2}\mathbf{k};$$

$$\|\mathbf{T}'(t)\| = \frac{1}{(2t^2 + 1)^2}\sqrt{(2 - 4t^2)^2 + (4t)^2 + (-4t)^2} = \frac{\sqrt{16t^4 + 16t^2 + 4}}{(2t^2 + 1)^2} = \frac{2}{2t^2 + 1};$$

$$\mathbf{N}(t) = \frac{\mathbf{T}'(t)}{\|\mathbf{T}'(t)\|} = \frac{1 - 2t^2}{2t^2 + 1}\mathbf{i} + \frac{2t}{2t^2 + 1}\mathbf{j} - \frac{2t}{2t^2 + 1}\mathbf{k}$$

11. $\mathbf{v} = r(1 - \cos t)\mathbf{i} + r\sin t\,\mathbf{j}$; $\|\mathbf{v}\| = \sqrt{r^2(1 - \cos t)^2 + r^2\sin^2 t} = \sqrt{2r^2(1 - \cos t)}$;

$$a_{\mathbf{T}} = \frac{d\,\|\mathbf{v}\|}{dt} = \frac{r^2\sin t}{\sqrt{2r^2(1 - \cos t)}};\ \mathbf{a} = \frac{d\mathbf{v}}{dt} = r\sin t\,\mathbf{i} + r\cos t\,\mathbf{j};$$

$$\|\mathbf{a}\|^2 = r^2\sin^2 t + r^2\cos^2 t = r^2;$$

$$a_{\mathbf{N}} = \sqrt{\|\mathbf{a}\|^2 - a_{\mathbf{T}}^2} = \sqrt{r^2 - \frac{r^4\sin^2 t}{2r^2(1 - \cos t)}} = \sqrt{r^2 - \frac{r^2(1 - \cos^2 t)}{2(1 - \cos t)}} = \frac{|r|}{\sqrt{2}}\sqrt{1 - \cos t}$$

13. $\mathbf{v} = 2\mathbf{i} + 2t\mathbf{j} + t^2\mathbf{k}$; $\|\mathbf{v}\| = \sqrt{4 + 4t^2 + t^4} = 2 + t^2$; $a_{\mathbf{T}} = \frac{d\,\|\mathbf{v}\|}{dt} = 2t$;

$$\mathbf{a} = 2\mathbf{j} + 2t\mathbf{k};\ \|\mathbf{a}\|^2 = 4(1 + t^2);\ a_{\mathbf{N}} = \sqrt{\|\mathbf{a}\|^2 - a_{\mathbf{T}}^2} = \sqrt{4(1 + t^2) - 4t^2} = 2$$

15. $\mathbf{v} = e^t\mathbf{i} - e^{-t}\mathbf{j} + \sqrt{2}\,\mathbf{k}$; $\|\mathbf{v}\| = \sqrt{e^{2t} + e^{-2t} + 2} = e^t + e^{-t}$; $a_{\mathbf{T}} = \dfrac{d\,\|\mathbf{v}\|}{dt} = e^t - e^{-t}$;

$\mathbf{a} = e^t\mathbf{i} + e^{-t}\mathbf{j}$; $\|\mathbf{a}\|^2 = e^{2t} + e^{-2t}$; $a_{\mathbf{N}} = \sqrt{\|\mathbf{a}\|^2 - a_{\mathbf{T}}^2} = \sqrt{(e^{2t} + e^{-2t}) - (e^t - e^{-t})^2} = \sqrt{2}$

17. $\mathbf{r}(t) = -4\sin t\,\mathbf{j} + 4\cos t\,\mathbf{k}$ for $0 \le t \le 2\pi$

19. $$\mathbf{r}(t) = \begin{cases} t\mathbf{i} + t^2\mathbf{j} & \text{for } 0 \le t \le 1 \\ (2-t)\mathbf{i} + (2-t)\mathbf{j} & \text{for } 1 \le t \le 2 \end{cases}$$

21. $$\mathbf{r}(t) = \begin{cases} (t+1)\mathbf{i} & \text{for } 0 \le t \le 1 \\ 2\cos\frac{\pi}{2}(t-1)\,\mathbf{i} + 2\sin\frac{\pi}{2}(t-1)\,\mathbf{j} & \text{for } 1 \le t \le 2 \\ (4-t)\mathbf{j} & \text{for } 2 \le t \le 3 \\ \sin\frac{\pi}{2}(t-3)\,\mathbf{i} + \cos\frac{\pi}{2}(t-3)\,\mathbf{j} & \text{for } 3 \le t \le 4 \end{cases}$$

23. Since $\mathbf{r}_1(t) = \mathbf{i} + 2\mathbf{j} + \mathbf{k} = \mathbf{r}_2(-1)$, the two curves intersect at $(1, 2, 1)$. Since $\mathbf{r}_1'(t) = \mathbf{i} + 2\mathbf{j} + 2t\mathbf{k}$ and $\mathbf{r}_2'(t) = 2t\mathbf{i} - \mathbf{j} - 2t\mathbf{k}$, we have $\mathbf{r}_1'(1) = \mathbf{i} + 2\mathbf{j} + 2\mathbf{k}$ and $\mathbf{r}_2'(-1) = -2\mathbf{i} - \mathbf{j} + 2\mathbf{k}$. Thus $\mathbf{T}_1(1) = \frac{1}{3}(\mathbf{i} + 2\mathbf{j} + 2\mathbf{k})$ and $\mathbf{T}_2(-1) = \frac{1}{3}(-2\mathbf{i} - \mathbf{j} + 2\mathbf{k})$, so that $\mathbf{T}_1(1) \cdot \mathbf{T}_2(-1) = \frac{1}{9}(-2 - 2 + 4) = 0$. Thus the two tangent vectors are perpendicular.

25. Since $a_{\mathbf{T}} = d\,\|\mathbf{v}\|\,/dt$, it follows from Theorem 4.6 that $a_{\mathbf{T}} = 0$ only if $\|\mathbf{v}\|$ is constant.

27. Let $\mathbf{r}(t) = t\,\mathbf{i} + \sin t\,\mathbf{j}$. Then $d\mathbf{r}/dt = \mathbf{i} + \cos t\,\mathbf{j}$ and $\|d\mathbf{r}/dt\| = \sqrt{1 + \cos^2 t}$. Therefore

$$\mathbf{T}(t) = \frac{1}{\sqrt{1 + \cos^2 t}}\,\mathbf{i} + \frac{\cos t}{\sqrt{1 + \cos^2 t}}\,\mathbf{j}.$$

Thus

$$\mathbf{T}'(t) = \frac{\sin t\,\cos t}{(1 + \cos^2 t)^{3/2}}\,\mathbf{i} - \frac{\sin t}{(1 + \cos^2 t)^{3/2}}\,\mathbf{j}.$$

But $\mathbf{T}'(t) = \mathbf{0}$ if $t = n\pi$, so $\mathbf{N}(n\pi)$ fails to exist, for all integers n.

29. By (3), (4), and (7),

$$(\mathbf{v} \cdot \mathbf{v})\,\mathbf{a} - (\mathbf{v} \cdot \mathbf{a})\,\mathbf{v} = \|\mathbf{v}\|^2 \left(\frac{\mathbf{v} \cdot \mathbf{a}}{\|\mathbf{v}\|}\mathbf{T} + \frac{\|\mathbf{v} \times \mathbf{a}\|}{\|\mathbf{v}\|}\mathbf{N}\right) - (\mathbf{v} \cdot \mathbf{a})\,\|\mathbf{v}\|\,\mathbf{T} = \|\mathbf{v}\|\,\|\mathbf{v} \times \mathbf{a}\|\,\mathbf{N}.$$

Using the fact that $\mathbf{N}$ is a unit vector, and dividing each side of the previous equation by its length, we conclude that

$$\mathbf{N} = \frac{(\mathbf{v} \cdot \mathbf{v})\mathbf{a} - (\mathbf{v} \cdot \mathbf{a})\mathbf{v}}{\|(\mathbf{v} \cdot \mathbf{v})\mathbf{a} - (\mathbf{v} \cdot \mathbf{a})\mathbf{v}\|}.$$

31. a. By (3) and (4),

$$\mathbf{v} \times \mathbf{a} = \|\mathbf{v}\|\,\mathbf{T} \times \left(\frac{d\|\mathbf{v}\|}{dt}\mathbf{T} + \|\mathbf{v}\|\left\|\frac{d\mathbf{T}}{dt}\right\|\mathbf{N}\right) = \|\mathbf{v}\|^2\left\|\frac{d\mathbf{T}}{dt}\right\|\mathbf{T} \times \mathbf{N}.$$

Therefore $\mathbf{v} \times \mathbf{a}$ and $\mathbf{T} \times \mathbf{N}$ point in the same direction.

b. It follows from (a) that $\mathbf{B} = \mathbf{T} \times \mathbf{N} = c\mathbf{v} \times \mathbf{a}$, where $c > 0$. Since $\mathbf{B}$ is a unit vector, we have $1 = \|\mathbf{B}\| = c\|\mathbf{v} \times \mathbf{a}\|$, so that $c = 1/\|\mathbf{v} \times \mathbf{a}\|$. Thus

$$\mathbf{B} = c\mathbf{v} \times \mathbf{a} = \frac{\mathbf{v} \times \mathbf{a}}{\|\mathbf{v} \times \mathbf{a}\|}.$$

33. a. From (9) and (10), $\tan\theta = \dfrac{\|\mathbf{F}_n\| \sin\theta}{\|\mathbf{F}_n\| \cos\theta} = \dfrac{mv_R^2/\rho}{mg} = \dfrac{v_R^2}{\rho g}$, so that $v_R^2 = \rho g \tan\theta$.

b. $v_R = \sqrt{\rho g \tan\theta} = \sqrt{500(32)\tan(\pi/12)} \approx 65.5$ (feet per second), or 44.6 (miles per hour).

c. If ρ' and $v_{R'}$ are the new radius and rated speed, respectively, then $v_{R'} = 2v_R$ if $\sqrt{\rho' g \tan\theta} = 2\sqrt{\rho g \tan\theta} = \sqrt{4\rho g \tan\theta}$, so that $\rho' = 4\rho$. Since $\rho = 500$, it follows that if $\rho' = 2000$ (feet), then the rated speed will be doubled.

12.6 Curvature

1. From Exercise 1 in Section 12.5, $\kappa(t) = \dfrac{\|\mathbf{T}'(t)\|}{\|\mathbf{r}'(t)\|} = \dfrac{1/(t^2+1)}{2\sqrt{t^2+1}} = \dfrac{1}{2(t^2+1)^{3/2}}$.

3. From Exercise 3 in Section 12.5, $\kappa(t) = \dfrac{\|\mathbf{T}'(t)\|}{\|\mathbf{r}'(t)\|} = \dfrac{1}{\sqrt{2}}$.

5. From Exercise 5 in Section 12.5, $\kappa(t) = \dfrac{\|\mathbf{T}'(t)\|}{\|\mathbf{r}'(t)\|} = \dfrac{2/(2+t^2)}{2+t^2} = \dfrac{2}{(2+t^2)^2}$.

7. From Exercise 7 in Section 12.5, $\kappa(t) = \dfrac{\|\mathbf{T}'(t)\|}{\|\mathbf{r}'(t)\|} = \dfrac{\sqrt{2}/(e^t+e^{-t})}{e^t+e^{-t}} = \dfrac{\sqrt{2}}{(e^t+e^{-t})^2}$.

9. From Exercise 9 in Section 12.5, $\kappa(t) = \dfrac{\|\mathbf{T}'(t)\|}{\|\mathbf{r}'(t)\|} = \dfrac{2/(2t^2+1)}{(2t^2+1)/t} = \dfrac{2t}{(2t^2+1)^2}$.

11. $\mathbf{v} = \dfrac{d\mathbf{r}}{dt} = 2\mathbf{i} + 2t\mathbf{j}$; $\|\mathbf{v}\| = \sqrt{2^2+(2t)^2} = 2\sqrt{1+t^2}$; $\mathbf{a} = \dfrac{d\mathbf{v}}{dt} = 2\mathbf{j}$; $\mathbf{v} \times \mathbf{a} = \begin{vmatrix} \mathbf{i} & \mathbf{j} & \mathbf{k} \\ 2 & 2t & 0 \\ 0 & 2 & 0 \end{vmatrix} = 4\mathbf{k}$;

$\|\mathbf{v} \times \mathbf{a}\| = 4$; $\kappa = \dfrac{\|\mathbf{v} \times \mathbf{a}\|}{\|\mathbf{v}\|^3} = \dfrac{4}{(2\sqrt{1+t^2})^3} = \dfrac{1}{2(1+t^2)^{3/2}}$

13. $\mathbf{v} = \dfrac{d\mathbf{r}}{dt} = e^t(\sin t + \cos t)\mathbf{i} + e^t(\cos t - \sin t)\mathbf{j} + \mathbf{k}$;

$$\begin{aligned} \|\mathbf{v}\| &= \sqrt{e^{2t}(\sin t + \cos t)^2 + e^{2t}(\cos t - \sin t)^2 + 1} \\ &= \sqrt{e^{2t}(\sin^2 t + 2\sin t\,\cos t + \cos^2 t + \cos^2 t - 2\cos t \sin t + \sin^2 t) + 1} = \sqrt{2e^{2t}+1}; \end{aligned}$$

$$\mathbf{a}=\frac{d\mathbf{v}}{dt}=[e^t(\sin t+\cos t)+e^t(\cos t-\sin t)]\mathbf{i}+[e^t(\cos t-\sin t)+e^t(-\sin t-\cos t)]\mathbf{j}=2e^t\cos t\,\mathbf{i}-2e^t\sin t\,\mathbf{j};$$

$$\mathbf{v}\times\mathbf{a}=\begin{vmatrix}\mathbf{i}&\mathbf{j}&\mathbf{k}\\ e^t(\sin t+\cos t)&e^t(\cos t-\sin t)&1\\ 2e^t\cos t&-2e^t\sin t&0\end{vmatrix}=2e^t\sin t\,\mathbf{i}+2e^t\cos t\,\mathbf{j}-2e^{2t}\mathbf{k};$$

$$\|\mathbf{v}\times\mathbf{a}\|=2e^t\sqrt{\sin^2 t+\cos^2 t+(-e^t)^2}=2e^t\sqrt{1+e^{2t}};\ \kappa=\frac{\|\mathbf{v}\times\mathbf{a}\|}{\|\mathbf{v}\|^3}=\frac{2e^t(1+e^{2t})^{1/2}}{(2e^{2t}+1)^{3/2}}$$

15. $\mathbf{v}=\dfrac{d\mathbf{r}}{dt}=\cos t\,\mathbf{i}-\sin t\,\mathbf{j}+t^{1/2}\mathbf{k};\ \|\mathbf{v}\|=\sqrt{\cos^2 t+\sin^2 t+t}=\sqrt{1+t};$

$$\mathbf{a}=\frac{d\mathbf{v}}{dt}=-\sin t\,\mathbf{i}-\cos t\,\mathbf{j}+\tfrac{1}{2}t^{-1/2}\mathbf{k};$$

$$\mathbf{v}\times\mathbf{a}=\begin{vmatrix}\mathbf{i}&\mathbf{j}&\mathbf{k}\\ \cos t&-\sin t&t^{1/2}\\ -\sin t&-\cos t&\frac{1}{2}t^{-1/2}\end{vmatrix}=(-\tfrac{1}{2}t^{-1/2}\sin t+t^{1/2}\cos t)\,\mathbf{i}+(-t^{1/2}\sin t-\tfrac{1}{2}t^{-1/2}\cos t)\,\mathbf{j}-\mathbf{k};$$

$$\begin{aligned}\|\mathbf{v}\times\mathbf{a}\| &= \sqrt{(-\tfrac{1}{2}t^{-1/2}\sin t+t^{1/2}\cos t)^2+(-t^{1/2}\sin t-\tfrac{1}{2}t^{-1/2}\cos t)^2+1}\\ &= \sqrt{\tfrac{1}{4}t^{-1}\sin^2 t-\sin t\,\cos t+t\cos^2 t+t\sin^2 t+\sin t\,\cos t+\tfrac{1}{4}t^{-1}\cos^2 t+1}\\ &= \sqrt{\tfrac{1}{4}t^{-1}+1+t}=\sqrt{\frac{1+4t+4t^2}{4t}}=\frac{2t+1}{2\sqrt{t}};\end{aligned}$$

$$\kappa=\frac{\|\mathbf{v}\times\mathbf{a}\|}{\|\mathbf{v}\|^3}=\frac{(2t+1)/(2\sqrt{t})}{(\sqrt{1+t})^3}=\frac{2t+1}{2\sqrt{t}\,(1+t)^{3/2}}$$

17. By the solution of Exercise 16, $\kappa=6/(4+5\cos^2 t)^{3/2}$; for $t_0=\pi/2$, $\kappa=6/4^{3/2}=\frac{3}{4}$, so that $\rho=1/\kappa=\frac{4}{3}$.

19. $x=t,\ \dfrac{dx}{dt}=1,\ \dfrac{d^2x}{dt^2}=0;\ y=\dfrac{1}{3}t^3,\ \dfrac{dy}{dt}=t^2,\ \dfrac{d^2y}{dt^2}=2t;$

$$\kappa=\frac{\left|\dfrac{dx}{dt}\dfrac{d^2y}{dt^2}-\dfrac{d^2x}{dt^2}\dfrac{dy}{dt}\right|}{\left[\left(\dfrac{dx}{dt}\right)^2+\left(\dfrac{dy}{dt}\right)^2\right]^{3/2}}=\frac{2|t|}{(1+t^4)^{3/2}};\ \text{for } t_0=1,\ \kappa=\frac{2}{2^{3/2}}=\frac{1}{\sqrt{2}},\ \text{so that } \rho=\frac{1}{\kappa}=\sqrt{2}.$$

21. $\dfrac{dy}{dx}=\dfrac{1}{x},\ \dfrac{d^2y}{dx^2}=-\dfrac{1}{x^2};\ \kappa=\dfrac{|d^2y/dx^2|}{[1+(dy/dx)^2]^{3/2}}=\dfrac{1/x^2}{(1+1/x^2)^{3/2}}=\dfrac{x}{(1+x^2)^{3/2}}$

23. $\dfrac{dy}{dx}=-\dfrac{1}{x^2},\ \dfrac{d^2y}{dx^2}=\dfrac{2}{x^3};\ \kappa=\dfrac{|d^2y/dx^2|}{[1+(dy/dx)^2]^{3/2}}=\dfrac{-2/x^3}{(1+1/x^4)^{3/2}}=\dfrac{-2x^3}{(1+x^4)^{3/2}}$

25. $\dfrac{dy}{dx} = e^x = \dfrac{d^2y}{dx^2}$; $\kappa = \dfrac{|d^2y/dx^2|}{[1+(dy/dx)^2]^{3/2}} = \dfrac{e^x}{(1+e^{2x})^{3/2}}$;

$$\frac{d\kappa}{dx} = \frac{e^x(1+e^{2x})^{3/2} - e^x(\frac{3}{2})(1+e^{2x})^{1/2}(2e^{2x})}{(1+e^{2x})^3} = \frac{e^x - 2e^{3x}}{(1+e^{2x})^{5/2}} = \frac{e^x(1-2e^{2x})}{(1+e^{2x})^{5/2}}$$

$d\kappa/dx = 0$ if $1 - 2e^{2x} = 0$, or $x = -\frac{1}{2}\ln 2$. Since $d\kappa/dx > 0$ for $x < -\frac{1}{2}\ln 2$, and $d\kappa/dx < 0$ for $x > -\frac{1}{2}\ln 2$, κ is maximum at $(-\frac{1}{2}\ln 2, \sqrt{2}/2)$.

27. $\mathbf{v} = -\sin t\,\mathbf{i} + \cos t\,\mathbf{j} + \mathbf{k}$; $\|\mathbf{v}\| = \sqrt{\sin^2 t + \cos^2 t + 1} = \sqrt{2}$;

$$\mathbf{a} = -\cos t\,\mathbf{i} - \sin t\,\mathbf{j};\ \mathbf{v}\times\mathbf{a} = \begin{vmatrix} \mathbf{i} & \mathbf{j} & \mathbf{k} \\ -\sin t & \cos t & 1 \\ -\cos t & -\sin t & 0 \end{vmatrix} = \sin t\,\mathbf{i} - \cos t\,\mathbf{j} + \mathbf{k};$$

$$\|\mathbf{v}\times\mathbf{a}\| = \sqrt{\sin^2 t + \cos^2 t + 1} = \sqrt{2};\ \kappa = \frac{\|\mathbf{v}\times\mathbf{a}\|}{\|\mathbf{v}\|^3} = \frac{\sqrt{2}}{(\sqrt{2})^3} = \frac{1}{2}$$

29. By Definition 12.19, $\|d\mathbf{T}/dt\| = \kappa\|d\mathbf{r}/dt\| = \kappa\|\mathbf{v}\|$, so that by (5) of Section 12.5,

$$a_{\mathbf{N}} = \|\mathbf{v}\|\,\|d\mathbf{T}/dt\| = \kappa\,\|\mathbf{v}\|^2.$$

31. By Exercise 30,

$$\kappa(\theta) = \frac{|2(3\cos 3\theta)^2 - \sin 3\theta(-9\sin 3\theta) + \sin^2 3\theta|}{[(3\cos 3\theta)^2 + \sin^2 3\theta]^{3/2}} = \frac{18\cos^2 3\theta + 10\sin^2 3\theta}{(9\cos^2 3\theta + \sin^2 3\theta)^{3/2}} = \frac{8\cos^2 3\theta + 10}{(8\cos^2 3\theta + 1)^{3/2}}.$$

33. a. $\theta'(t) = \kappa(t)$; $\mathbf{v} = \cos\theta(t)\,\mathbf{i} + \sin\theta(t)\,\mathbf{j}$; $\|\mathbf{v}\| = \sqrt{\cos^2\theta(t) + \sin^2\theta(t)} = 1$;

$\mathbf{a} = (-\sin\theta(t))\theta'(t)\,\mathbf{i} + (\cos\theta(t))\theta'(t)\,\mathbf{j} = -\kappa(t)\sin\theta(t)\,\mathbf{i} + \kappa(t)\cos\theta(t)\,\mathbf{j}$;

$$\mathbf{v}\times\mathbf{a} = \begin{vmatrix} \mathbf{i} & \mathbf{j} & \mathbf{k} \\ \cos\theta(t) & \sin\theta(t) & 0 \\ -\kappa(t)\sin\theta(t) & \kappa(t)\cos\theta(t) & 0 \end{vmatrix} = \kappa(t)\mathbf{k};$$

$\|\mathbf{v}\times\mathbf{a}\| = \kappa(t)$. Thus the curvature is $\|\mathbf{v}\times\mathbf{a}\| / \|\mathbf{v}\|^3 = \kappa(t)$.

b. Taking $a = 0$ in part (a), we find that $\theta(t) = \int_0^t (1/\sqrt{1-u^2})\,du = \sin^{-1} t$, so that

$$\int_0^t \cos\theta(u)\,du = \int_0^t \cos(\sin^{-1} u)\,du = \int_0^t \sqrt{1-u^2}\,du$$

$$\overset{u=\sin w}{=} \int_0^{\sin^{-1} t} \sqrt{1-\sin^2 w}\,\cos w\,dw = \int_0^{\sin^{-1} t} \cos^2 w\,dw$$

$$= \int_0^{\sin^{-1} t} \left(\frac{1}{2} + \frac{1}{2}\cos 2w\right)dw = \left(\frac{1}{2}w + \frac{1}{4}\sin 2w\right)\Big|_0^{\sin^{-1} t}$$

$$= \frac{1}{2}\sin^{-1} t + \frac{1}{2}\sin(\sin^{-1} t)\cos(\sin^{-1} t) = \frac{1}{2}\sin^{-1} t + \frac{t}{2}\sqrt{1-t^2};$$

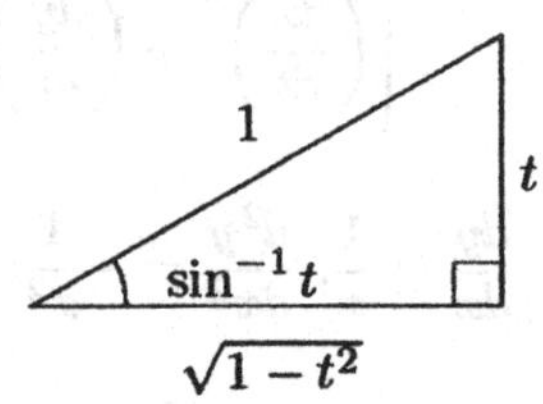

$$\int_0^t \sin\theta(u)\,du = \int_0^t \sin(\sin^{-1}u)\,du = \int_0^t u\,du = \frac{1}{2}t^2.$$

Thus the desired parametrization is

$$\mathbf{r}(t) = \left(\frac{1}{2}\sin^{-1}t + \frac{t}{2}\sqrt{1-t^2}\right)\mathbf{i} + \frac{1}{2}t^2\mathbf{j} \quad \text{for} \quad -1 < t < 1.$$

c. Taking $a = 0$ in part (a), we find that $\theta(t) = \int_0^t (1/(1+u^2))\,du = \tan^{-1}t$, so that

$$\int_0^t \cos\theta(u)\,du = \int_0^t \cos(\tan^{-1}u)\,du = \int_0^t \frac{1}{\sqrt{1+u^2}}\,du$$

$$\overset{u=\tan w}{=} \int_0^{\tan^{-1}t} \frac{1}{\sqrt{1+\tan^2 w}}\sec^2 w\,dw = \int_0^{\sin^{-1}t} \sec w\,dw$$

$$= \ln|\sec w + \tan w|\Big|_0^{\tan^{-1}t} = \ln|\sec(\tan^{-1}t) + \tan(\tan^{-1}t)|$$

$$= \ln(t + \sqrt{1+t^2});$$

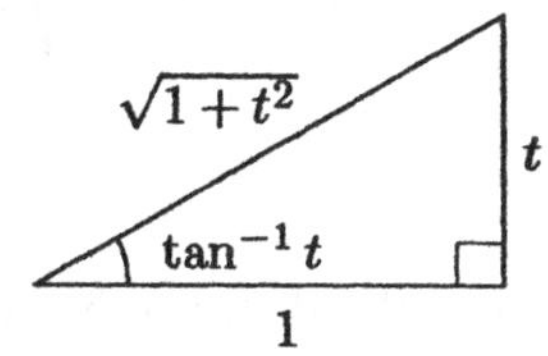

$$\int_0^t \sin\theta(u)\,du = \int_0^t \sin(\tan^{-1}u)\,du = \int_0^t \frac{u}{\sqrt{1+u^2}}\,du = \sqrt{1+u^2}\Big|_0^t = \sqrt{1+t^2} - 1.$$

Thus the desired parametrization is $\mathbf{r}(t) = \ln(t + \sqrt{1+t^2})\,\mathbf{i} + (\sqrt{1+t^2} - 1)\,\mathbf{j}$.

35. $a_\mathbf{T} = d\|\mathbf{v}\|/dt = (0-81)/(9-0) = -9$. Since $\|\mathbf{v}(0)\| = 81$ and $d\|\mathbf{v}\|/dt = -9$, we have $\|\mathbf{v}\| = 81 - 9t$. Since the radius of the circle is 729, it follows that $\kappa = 1/\rho = 1/729$, so that by Exercise 29, $a_\mathbf{N} = \kappa\|\mathbf{v}\|^2 = \frac{1}{729}(81-9t)^2 = (9-t)^2/9$. Thus $\|\mathbf{a}\| = \sqrt{a_\mathbf{T}^2 + a_\mathbf{N}^2} = \sqrt{81 + (9-t)^4/81}$.

12.7 Kepler's Laws of Motion

1. a. Since $\mathbf{u}$ is a unit vector, $\mathbf{u}$ and $d\mathbf{u}/dt$ are perpendicular by Corollary 12.11. Thus by (6),

$$p = \|p\mathbf{k}\| = \left\|r^2\left(\mathbf{u}\times\frac{d\mathbf{u}}{dt}\right)\right\| = r^2\,\|\mathbf{u}\|\left\|\frac{d\mathbf{u}}{dt}\right\|\sin\frac{\pi}{2} = r^2\left\|\frac{d\mathbf{u}}{dt}\right\|.$$

b. Since $dr/dt = 0$ when r is minimum, (5) implies that $d\mathbf{r}/dt = r(d\mathbf{u}/dt)$ when r is minimum.

c. Since r_0 is the minimum value of r, and v_0 is the corresponding speed, it follows from (a) and (b) that

$$p = r_0^2\left\|\frac{d\mathbf{u}}{dt}\right\| = r_0\left\|r_0\frac{d\mathbf{u}}{dt}\right\| = r_0\left\|\frac{d\mathbf{r}}{dt}\right\| = r_0 v_0.$$

d. For straight line motion, $\mathbf{r}$ has the form $\mathbf{r} = \mathbf{r}_0 + f(t)\mathbf{L}$ for some constant vectors $\mathbf{r}_0$ and $\mathbf{L}$, and for some scalar function f. Then $d^2\mathbf{r}/dt^2 = f''(t)\mathbf{L}$, so by (1) and (2), $f''(t)\mathbf{L} = -\frac{GM}{r^3}[\mathbf{r}_0 + f(t)\mathbf{L}]$. It follows that $\mathbf{r}_0$ is parallel to $\mathbf{L}$, so that $\mathbf{r} = \mathbf{r}_0 + f(t)\mathbf{L} = g(t)\mathbf{L}$ for some scalar function g. From this and (4) we conclude that $p\mathbf{k} = \mathbf{r}\times d\mathbf{r}/dt = g(t)\mathbf{L}\times g'(t)\mathbf{L} = \mathbf{0}$, and hence that $p = 0$.

3. If the orbit is circular, then r is constant, so that $r = r_0$. From (13) and Exercise 1(c) we have $r_0^2 = p^4/G^2M^2 = r_0^4 v_0^4/G^2M^2$. Solving for v_0, we find that $v_0 = \sqrt{GM/r_0}$.

5. By (23), $T = \sqrt{\dfrac{4\pi^2 a^3}{GM}} = \sqrt{\dfrac{4\pi^2(5000)^3}{1.237 \times 10^{12}}} \approx 1.99733$ (hours).

7. By (22),

$$c = \sqrt{a^2 - b^2} = \sqrt{\frac{p^4 G^2 M^2}{(G^2M^2 - w^2)^2} - \frac{p^4}{G^2M^2 - w^2}} = \sqrt{\frac{G^2M^2}{G^2M^2 - w^2} - 1}\,\frac{p^2}{\sqrt{G^2M^2 - w^2}}$$

$$= \frac{w}{\sqrt{G^2M^2 - w^2}}\,\frac{p^2}{\sqrt{G^2M^2 - w^2}} = \frac{wp^2}{G^2M^2 - w^2}.$$

Since c is the distance from the center of the ellipse to either focus and since the center is

$$(-wp^2/(G^2M^2 - w^2), 0)$$

by (21), it follows that one focus is the origin, where the sun is located.

9. a. Since $d\mathbf{r}/dt$ is perpendicular to $\mathbf{k}$, (9) implies that

$$\|GM\mathbf{u} + \mathbf{w}_1\| = \left\|\frac{d\mathbf{r}}{dt} \times p\mathbf{k}\right\| = p\left\|\frac{d\mathbf{r}}{dt}\right\| \|\mathbf{k}\| \sin\frac{\pi}{2} = p\left\|\frac{d\mathbf{r}}{dt}\right\|.$$

This $\|d\mathbf{r}/dt\| = (1/p)\|GM\mathbf{u} + \mathbf{w}_1\|$.

b. Since $\mathbf{w}_1 = w\mathbf{i}$ and $\mathbf{u} = \cos\theta\,\mathbf{i} + \sin\theta\,\mathbf{j}$, we have

$$\|GM\mathbf{u} + \mathbf{w}_1\| = \|(GM\cos\theta + w)\mathbf{i} + GM\sin\theta\,\mathbf{j}\|$$

$$= \sqrt{(GM\cos\theta + w)^2 + G^2M^2\sin^2\theta} = \sqrt{G^2M^2 + w^2 + 2GMw\cos\theta}.$$

Thus $\|GM\mathbf{u} + \mathbf{w}_1\|$ is maximum when $\cos\theta = 1$, or $\theta = 0$, and the maximum value is $GM + w$. By Exercises 9(a), 8, and 1(c) we find that the maximum speed is

$$\frac{GM + w}{p} = \frac{r_0 v_0^2}{p} = \frac{r_0 v_0^2}{r_0 v_0} = v_0.$$

11. a. We have $a - c = r_0 = 100 + 3960 = 4060$ and $2a = 3100 + 100 + 2(3960) = 11{,}120$, so that $a = 5560$ and $c = 5560 - 4060 = 1500$. Thus

$$b = \sqrt{a^2 - c^2} = \sqrt{5560^2 - 1500^2} \approx 5353.84.$$

By equations (20) and (23) we have

$$p = \frac{2\pi ab}{T} = 2\pi ab\sqrt{\frac{GM}{4\pi^2 a^3}} = b\sqrt{\frac{GM}{a}} \approx 5353.84\sqrt{\frac{1.237 \times 10^{12}}{5560}} \approx 7.98570 \times 10^7.$$

By Exercise 1(c),

$$v_0 = \frac{p}{r_0} \approx \frac{7.98570 \times 10^7}{4060} \approx 19{,}669.2 \text{ (miles per hour).}$$

b. By Exercise 8, $w = r_0 v_0^2 - GM \approx 4060(19{,}669.2)^2 - 1.237 \times 10^{12} \approx 3.33722 \times 10^{11}$. At aphelion $\mathbf{u} = -\mathbf{i}$, so that $\|GM\mathbf{u} + \mathbf{w}_1\| = \|-GM\mathbf{i} + w\mathbf{i}\| = GM - w$. Thus, by Exercise 9(a) the minimum velocity is given by

$$\left\|\frac{d\mathbf{r}}{dt}\right\| = \frac{1}{p}\|GM\mathbf{u} + \mathbf{w}_1\| = \frac{GM - w}{p}$$

$$\approx \frac{1.237 \times 10^{12} - 3.33722 \times 10^{11}}{7.98570 \times 10^7} \approx 11{,}311.2 \text{ (miles per hour).}$$

13. Since $GM = w$ for a parabolic orbit, it follows from Exercise 8 that $GM = w = r_0 v_0^2 - GM$, so that $v_0 = \sqrt{2GM/r_0}$.

15. Let F_e be the magnitude of the gravitational force exerted by the earth on the spacecraft, F_m the magnitude of the gravitational force exerted by the moon on the spacecraft, m the mass of the spacecraft, and M_m the mass of the moon. By Newton's Law of Gravitation,

$$\frac{F_e}{F_m} = \frac{GM_e m/(240{,}000 - 4080)^2}{GM_m m/(4080)^2} = \frac{(4080)^2}{(M_m/M_e)(235{,}920)^2} \approx \frac{(4080)^2}{(.0123)(235{,}920)^2} \approx .024316.$$

17. By (23),

$$a^3 = \frac{GMT^2}{4\pi^2} \approx \frac{(1.323 \times 10^{26})(75.6)^2}{4\pi^2},$$

so that $a \approx 2.68 \times 10^9$ (kilometers). Since $2a$ is the length of the major axis, which is the sum of the distance of the closest approach and the distance of farthest retreat of the comet, it follows that the distance of farthest retreat is $2 \times 2.68 \times 10^9 - 5.31 \times 10^7 \approx 5.31 \times 10^9$ kilometers.

19. a. When the planet makes one revolution in a circular orbit about the sun, it travels a distance of $2\pi r$ at a constant speed v, and the length of time required is T. Thus $2\pi r = vT$, or $T = 2\pi r/v$.

b. By (24), (25), and (26) in that order, $\|\mathbf{F}\| = \dfrac{mv^2}{r} = \dfrac{m}{r}\left(\dfrac{4\pi^2 r^2}{T^2}\right) = \dfrac{m}{r}\left(\dfrac{4\pi^2 r^2}{cr^3}\right) = \dfrac{4\pi^2}{c}\dfrac{m}{r^2}.$

Chapter 12 Review

1.

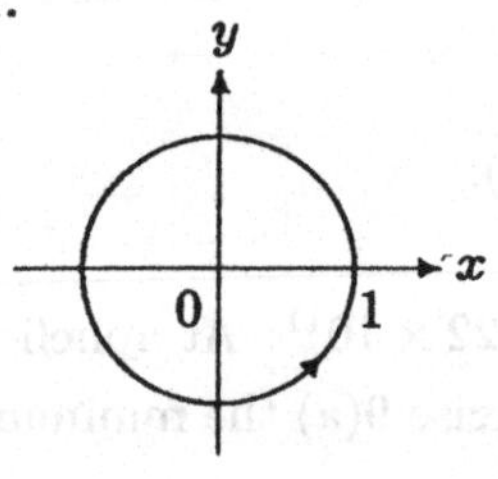

3. $(\mathbf{F}\times\mathbf{G})(t) = \begin{vmatrix} \mathbf{i} & \mathbf{j} & \mathbf{k} \\ t & 1 & 0 \\ 0 & 1 & t \end{vmatrix} = t\mathbf{i} - t^2\mathbf{j} + t\mathbf{k}$

$[(\mathbf{F}\times\mathbf{G})\times\mathbf{H}](t) = \begin{vmatrix} \mathbf{i} & \mathbf{j} & \mathbf{k} \\ t & -t^2 & t \\ 0 & t & 0 \end{vmatrix} = -t^2\mathbf{i} + t^2\mathbf{k}$

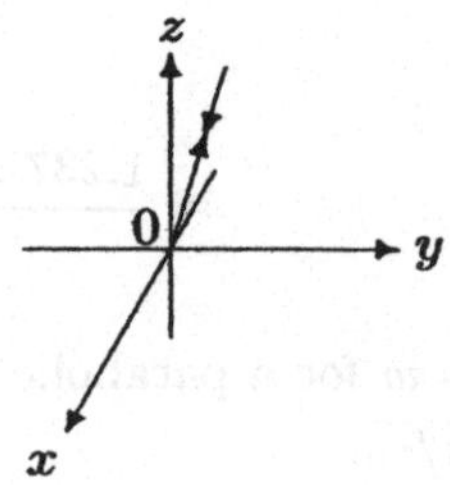

5. a. $(\mathbf{F}\cdot\mathbf{G})(t) = t^3$; $(\mathbf{F}\cdot\mathbf{G})'(t) = 3t^2$

b. $(\mathbf{F}\times\mathbf{G})(t) = \begin{vmatrix} \mathbf{i} & \mathbf{j} & \mathbf{k} \\ \frac{1}{t} & t & 0 \\ 0 & t^2 & \frac{-1}{t^2} \end{vmatrix} = \frac{-1}{t}\mathbf{i} + \frac{1}{t^3}\mathbf{j} + t\,\mathbf{k}$; $(\mathbf{F}\times\mathbf{G})'(t) = \frac{1}{t^2}\mathbf{i} - \frac{3}{t^4}\mathbf{j} + \mathbf{k}$

7. $\displaystyle\int\left(\tan 2\pi t\,\mathbf{i} + \sec^2 2\pi t\,\mathbf{j} + \frac{4}{1+t^2}\mathbf{k}\right)dt = \left(\int \tan 2\pi t\,dt\right)\mathbf{i} + \left(\int \sec^2 2\pi t\,dt\right)\mathbf{j} + \left(\int \frac{4}{1+t^2}\,dt\right)\mathbf{k}$

$\displaystyle = -\frac{1}{2\pi}\ln|\cos 2\pi t|\,\mathbf{i} + \frac{1}{2\pi}\tan 2\pi t\,\mathbf{j} + 4\tan^{-1}t\,\mathbf{k} + \mathbf{C}$

9. a. $L = \int_0^{3\pi}\sqrt{(e^t\cos t - e^t\sin t)^2 + (e^t\sin t + e^t\cos t)^2}\,dt = \int_0^{3\pi} e^t\sqrt{2}\,dt = e^t\sqrt{2}\big|_0^{3\pi} = \sqrt{2}\,(e^{3\pi} - 1)$

b. As in part (a) but with the limits of integration altered,

$$L = \int_{-2\pi}^{1} e^t\sqrt{2}\,dt = e^t\sqrt{2}\Big|_{-2\pi}^{1} = \sqrt{2}\,(e - e^{-2\pi}).$$

11. $\mathbf{r} = [(\frac{3}{2}t)^{2/3} - 1]\mathbf{i} + \frac{2}{3}[(\frac{3}{2}t)^{2/3} - 1]^{3/2}\mathbf{j}$;

$\mathbf{v} = \frac{2}{3}(\frac{3}{2}t)^{-1/3}(\frac{3}{2})\mathbf{i} + [(\frac{3}{2}t)^{2/3} - 1]^{1/2}(\frac{2}{3})(\frac{3}{2}t)^{-1/3}(\frac{3}{2})\mathbf{j} = (\frac{3}{2}t)^{-1/3}\mathbf{i} + [(\frac{3}{2}t)^{2/3} - 1]^{1/2}(\frac{3}{2}t)^{-1/3}\mathbf{j}$;

$\|\mathbf{v}\| = (\frac{3}{2}t)^{-1/3}\sqrt{1 + [(\frac{3}{2}t)^{2/3} - 1]} = (\frac{3}{2}t)^{-1/3}(\frac{3}{2}t)^{1/3} = 1$

13. $\mathbf{v} = (3 - 3t^2)\mathbf{i} + 6t\mathbf{j} + (3 + 3t^2)\mathbf{k}$;

$\|\mathbf{v}\| = \sqrt{(3 - 3t^2)^2 + (6t)^2 + (3 + 3t^2)} = \sqrt{18 + 36t^2 + 18t^4} = 3\sqrt{2}\,(1 + t^2)$;

$\mathbf{a} = -6t\mathbf{i} + 6\mathbf{j} + 6t\mathbf{k}$;

$$\mathbf{v} \times \mathbf{a} = \begin{vmatrix} \mathbf{i} & \mathbf{j} & \mathbf{k} \\ 3 - 3t^2 & 6t & 3 + 3t^2 \\ -6t & 6 & 6t \end{vmatrix} = 18(t^2 - 1)\mathbf{i} - 36t\mathbf{j} + 18(t^2 + 1)\mathbf{k}$$

$\|\mathbf{v} \times \mathbf{a}\| = 18\sqrt{(t^2 - 1)^2 + (-2t)^2 + (t^2 + 1)^2} = 18\sqrt{2t^4 + 4t^2 + 2} = 18\sqrt{2}\,(t^2 + 1)$;

$$\kappa = \frac{\|\mathbf{v} \times \mathbf{a}\|}{\|\mathbf{v}\|^3} = \frac{18\sqrt{2}\,(t^2 + 1)}{[3\sqrt{2}\,(1 + t^2)]^3} = \frac{1}{3(1 + t^2)^2}$$

15. a. $y = \dfrac{1}{x}, \dfrac{dy}{dx} = -\dfrac{1}{x^2}, \dfrac{d^2y}{dx^2} = \dfrac{2}{x^3}; \kappa = \dfrac{|d^2y/dx^2|}{[1 + (dy/dx)^2]^{3/2}} = \dfrac{2/x^3}{(1 + 1/x^4)^{1/2}} = \dfrac{2x^3}{(x^4 + 1)^{3/2}}$

b. $\dfrac{d\kappa}{dx} = \dfrac{6x^2(x^4 + 1)^{3/2} - 2x^3(\frac{3}{2})(x^4 + 1)^{1/2}(4x^3)}{(x^4 + 1)^3} = \dfrac{6x^2(1 - x^4)}{(x^4 + 1)^{5/2}}$

Since $d\kappa/dx > 0$ for $x < 1$ and $d\kappa/dx < 0$ for $x > 1$, it follows from (1) of Section 4.6 and the First Derivative Test that κ is maximum for $x = 1$. Thus the maximum value of κ is $\kappa(1) = 2/2^{3/2} = \sqrt{2}/2$.

c. Using the value of κ at $(1, 1)$ from part (b), we have $\rho(1) = 1/\kappa(1) = \sqrt{2}$. Thus the radius of curvature at $(1, 1)$ is $\sqrt{2}$.

17. $\mathbf{v} = e^t(\cos t - \sin t)\mathbf{i} + e^t(\sin t + \cos t)\mathbf{j} + e^t\mathbf{k}$;

$\mathbf{a} = e^t(\cos t - \sin t - \sin t - \cos t)\mathbf{i} + e^t(\sin t + \cos t + \cos t - \sin t)\mathbf{j} + e^t\mathbf{k} = -2e^t \sin t\,\mathbf{i} + 2e^t \cos t\,\mathbf{j} + e^t\,\mathbf{k}$;

$\|\mathbf{v}\| = \sqrt{e^{2t}(\cos t - \sin t)^2 + e^{2t}(\sin t + \cos t)^2 + e^{2t}} = e^t\sqrt{3}$

$$\mathbf{v} \times \mathbf{a} = \begin{vmatrix} \mathbf{i} & \mathbf{j} & \mathbf{k} \\ e^t(\cos t - \sin t) & e^t(\sin t + \cos t) & e^t \\ -2e^t \sin t & 2e^t \cos t & e^t \end{vmatrix} = e^{2t}(\sin t - \cos t)\mathbf{i} - e^{2t}(\sin t + \cos t)\mathbf{j} + 2e^{2t}\mathbf{k};$$

$\|\mathbf{v} \times \mathbf{a}\| = e^{2t}\sqrt{(\sin t - \cos t)^2 + (\sin t + \cos t)^2 + 2^2} = e^{2t}\sqrt{6}$

$$\kappa = \frac{\|\mathbf{v} \times \mathbf{a}\|}{\|\mathbf{v}\|^3} = \frac{e^{2t}\sqrt{6}}{(e^t\sqrt{3})^3} = \frac{\sqrt{2}}{3e^t};\ \rho = \frac{1}{\kappa} = \frac{3\sqrt{2}\,e^t}{2}$$

19. $\mathbf{r}'(t) = (1 - \cos t)\mathbf{i} + \sin t\,\mathbf{j} + 2\cos\frac{t}{2}\,\mathbf{k}$;

$$\|\mathbf{r}'(t)\| = \sqrt{(1-\cos t)^2 + \sin^2 t + 4\cos^2\frac{t}{2}} = \sqrt{1 - 2\cos t + \cos^2 t + \sin^2 t + 4\cos^2\frac{t}{2}}$$

$$= \sqrt{2 - 2\cos t + 2 + 2\cos t} = 2;$$

$$\mathbf{T}(t) = \frac{\mathbf{r}'(t)}{\|\mathbf{r}'(t)\|} = \frac{1}{2}(1-\cos t)\mathbf{i} + \frac{1}{2}\sin t\,\mathbf{j} + \cos\frac{t}{2}\,\mathbf{k};$$

$$\mathbf{T}'(t) = \frac{1}{2}\sin t\,\mathbf{i} + \frac{1}{2}\cos t\,\mathbf{j} - \frac{1}{2}\sin\frac{t}{2}\,\mathbf{k};$$

$$\|\mathbf{T}'(t)\| = \frac{1}{2}\sqrt{\sin^2 t + \cos^2 t + \sin^2\frac{t}{2}} = \frac{1}{2}\sqrt{1 + \sin^2\frac{t}{2}} = \frac{1}{2}\sqrt{\frac{3}{2} - \frac{1}{2}\cos t} = \frac{\sqrt{2}}{4}\sqrt{3 - \cos t};$$

$$\mathbf{N}(t) = \frac{\mathbf{T}'(t)}{\|\mathbf{T}'(t)\|} = \frac{\sqrt{2}\sin t}{\sqrt{3-\cos t}}\mathbf{i} + \frac{\sqrt{2}\cos t}{\sqrt{3-\cos t}}\mathbf{j} - \frac{\sqrt{2}\sin(t/2)}{\sqrt{3-\cos t}}\mathbf{k};$$

$$\kappa(t) = \frac{\|\mathbf{T}'(t)\|}{\|\mathbf{r}'(t)\|} = \frac{\sqrt{2}}{8}\sqrt{3-\cos t}$$

21. Since $\|\mathbf{v}\| = 1$ and $\mathbf{a} = d\mathbf{v}/dt$, it follows from Corollary 12.11 that $\mathbf{v}\cdot\mathbf{a} = \mathbf{v}\cdot d\mathbf{v}/dt = 0$. Since $\|\mathbf{v}\| = \|\mathbf{a}\| = 1$, we conclude that $\mathbf{v}$ and $\mathbf{a}$ are perpendicular. Thus $\|\mathbf{v}\times\mathbf{a}\| = \|\mathbf{v}\|\,\|\mathbf{a}\|\sin(\pi/2) = 1$, so that $\kappa = \|\mathbf{v}\times\mathbf{a}\|/\|\mathbf{v}\|^3 = 1$.

Cumulative Review(Chapters 1–11)

1. $\lim_{x\to -3^+} \frac{1}{|x-3|} = \frac{1}{6}$ and $\lim_{x\to -3^+} \frac{1}{x+3} = \infty$; $\lim_{x\to -3^+}\left(\frac{1}{x+3} - \frac{1}{|x-3|}\right) = \infty$.

3. For $x = 0$, $\lim_{y\to 0} \frac{x^2 - y^3}{x^2 + y^2} = \lim_{y\to 0}\left(-\frac{y^3}{y^2}\right) = \lim_{y\to 0}(-y) = 0$.

For $x \neq 0$, $\lim_{y\to 0} \frac{x^2 - y^3}{x^2 + y^2} = \frac{x^2 - 0^3}{x^2 + 0^2} = 1$.

5. $f'(x) = \frac{1}{\sqrt{1 - ((\sin x)/5)^2}}\left(\frac{\cos x}{5}\right) = \frac{\cos x}{\sqrt{25 - \sin^2 x}}$

7. Rewriting the given equation, we obtain $x^3 - 2y^3 = 6x^2 + 6y^2$. By implicit differentiation, $3x^2 - 6y^2(dy/dx) = 12x + 12y(dy/dx)$. The tangent is horizontal at (x, y) provided that $dy/dx = 0$, which means that $3x^2 = 12x$, and thus $x = 0$ or $x = 4$. If $x = 0$, then $y \neq 0$ since $(x^3 - 2y^3)/(x^2 + y^2) = 6$, and therefore the equation $x^3 - 2y^3 = 6x^2 + 6y^2$ becomes $-2y = 6$, so that $y = -3$. If $x = 4$, then the equation becomes $64 - 2y^3 = 96 + 6y^2$, or $y^3 + 3y^2 + 16 = 0$, or $(y+4)(y^2 - y + 4) = 0$, so $y = -4$. Therefore the tangent line is horizontal at $(0, -3)$ and $(4, -4)$.

9. At t hours after noon, the train traveling 60 miles per hour is $100 - 60t$ miles from the junction, and the other train is $120 - 80t$ miles from the junction. If D is the distance between the two trains, then by the Law of Cosines, $D^2 = (120 - 80t)^2 + (100 - 60t)^2 - 2(120 - 80t)(100 - 60t)\cos(\pi/3)$, so by implicit differentiation, we find that

$$2D\frac{dD}{dt} = 2(120 - 80t)(-80) + 2(100 - 60t)(-60) + 80(100 - 60t) + 60(120 - 80t).$$

At 1 p.m., $t = 1$, so $D^2 = 40^2 + 40^2 - 2(40)(40)\frac{1}{2} = 40^2$, and thus $D = 40$. Therefore $80(dD/dt) = 2(40)(-80) + 2(40)(-60) + 80(40) + 60(40)$, which yields $dD/dt = -70$. Thus at 1 p.m. the trains are approaching each other at the rate of 70 miles per hour.

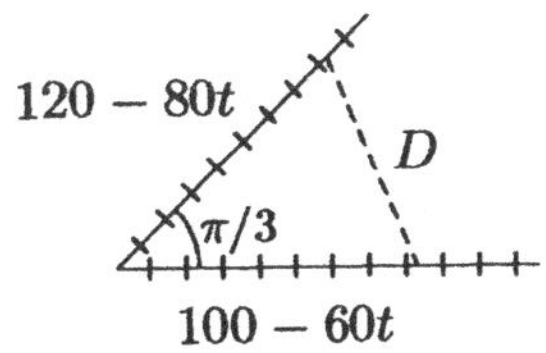

Exercise 9

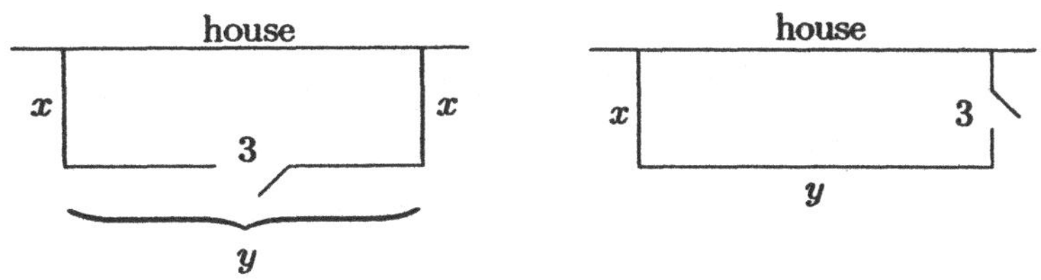

Exercise 11

11. Using the notation in the diagram, we have $2x + y - 3 = 77$, or $y = 80 - 2x$. Thus the area A is given by $A = xy = x(80 - 2x) = 80x - 2x^2$. Therefore $A'(x) = 80 - 4x$, and $A'(x) = 0$ for $x = 20$. Since $A''(x) = -4 < 0$, it follows from (1) in Section 4.6 and the Second Derivative Test that A is maximum for $x = 20$. Then $y = 80 - 2(20) = 40$. Thus the fence should be 20 feet long perpendicular to the house, and 40 feet long parallel to the house. If the gate is to be placed on one of the sides perpendicular to the house, as in the second diagram, we find that $x + (x - 3) + y = 77$, so that once again $y = 80 - 2x$, which yields the same dimensions as before.

13. Let $f(t)$ be the amount of the substance remaining after t years. Then $f(t) = f(0)e^{kt}$ for some k, and we need to find the half-life h, which is $-(1/k)\ln 2$. Now by hypothesis, $f(5) = .9f(0)$. Thus $.9f(0) = f(5) = f(0)e^{k5}$, so $e^{5k} = .9$, or $5k = \ln .9$, or $k = \frac{1}{5}\ln .9$. Then $\frac{1}{2}f(0) = f(h) = f(0)e^{kh}$, so $\frac{1}{2} = e^{kh}$, or $kh = \ln\frac{1}{2} = -\ln 2$, or $h = -(1/k)\ln 2 = -5(\ln 2/\ln .9) \approx 32.8941$. Thus the half-life is approximately 32.8941 years.

15. $\dfrac{x}{(x+2)(x^2+6)} = \dfrac{A}{x+2} + \dfrac{Bx+C}{x^2+6}$; $x = A(x^2+6)+(Bx+C)(x+2)$;

$A+B=0$, $2B+C=1$, $6A+2C=0$; $A=-\frac{1}{5}$, $B=\frac{1}{5}$, $C=\frac{3}{5}$;

$$\int \frac{x}{(x+2)(x^2+6)}\,dx = \int \left(\frac{-1}{5(x+2)} + \frac{x+3}{5(x^2+6)}\right) dx$$

$$= -\frac{1}{5}\int \frac{1}{x+2}\,dx + \frac{1}{5}\int \frac{x}{x^2+6}\,dx + \frac{3}{5}\int \frac{1}{x^2+6}\,dx$$

$$= -\frac{1}{5}\ln|x+2| + \frac{1}{10}\ln(x^2+6) + \frac{3}{5\sqrt{6}}\tan^{-1}\frac{x}{\sqrt{6}} + C$$

17. Let $u=\sqrt{x}$, so that $du = \dfrac{1}{2\sqrt{x}}\,dx$. If $x=b$, then $u=\sqrt{b}$, and if $x=\pi^2/4$, then $u=\pi/2$. Thus

$$\int_0^{\pi^2/4} \frac{\cos\sqrt{x}}{\sqrt{x}}\,dx = \lim_{b\to 0^+}\int_b^{\pi^2/4}\frac{\cos\sqrt{x}}{\sqrt{x}}\,dx = \lim_{b\to 0^+}\int_{\sqrt{b}}^{\pi/2}(\cos u)2\,du$$

$$= \lim_{b\to 0^+} 2\int_{\sqrt{b}}^{\pi/2}\cos u\,du = \lim_{b\to 0^+} 2\sin u\Big|_{\sqrt{b}}^{\pi/2} = \lim_{b\to 0^+}(2-\sin\sqrt{b}) = 2.$$

Thus the integral converges and its value is 2.

19. From the diagram, $r(x)/(10-x) = \frac{5}{10} = \frac{1}{2}$, so that $r(x) = (10-x)/2$ and therefore $A(x) = \pi[r(x)]^2 = (\pi/4)(10-x)^2$. Since a particle of water x feet from the bottom is to be raised $13-x$ feet, we find that

$$W = 62.5\int_0^4 (13-x)\left[\frac{\pi}{4}(10-x)^2\right] dx$$

$$= \frac{62.5\pi}{4}\int_0^4 (-x^3+33x^2-360x+1300)\,dx$$

$$= \frac{62.5\pi}{4}\left(-\frac{1}{4}x^4 + 11x^3 - 180x^2 + 1300x\right)\Big|_0^4$$

$$= 46{,}250\pi \text{ (foot-pounds)}.$$

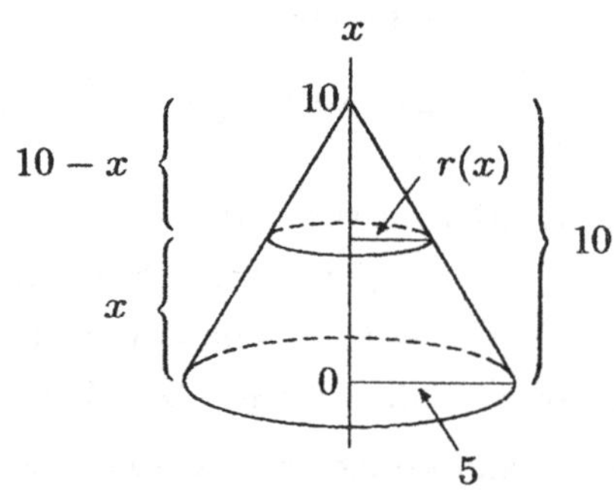

21. Since $(-1)^n n/(n+1) > 0$ for n even and $(-1)^n n/(n+1) < 0$ for n odd, the only possible limit is 0. But $\lim_{n\to\infty}|(-1)^n n/(n+1)| = \lim_{n\to\infty} n/(n+1) = 1$, so the sequence cannot approach 0. Thus $\lim_{n\to\infty}(-1)^n n/(n+1)$ does not exist, so the given sequence diverges.

23. $$\lim_{n\to\infty}\frac{\dfrac{(n+1)^{n+1}}{3^{n+1}(n+1)!}}{n^n/(3^n n!)} = \lim_{n\to\infty}\frac{(n+1)^{n+1}}{n^n}\,\frac{3^n n!}{3^{n+1}(n+1)!} = \lim_{n\to\infty}\frac{(n+1)^{n+1}}{3n^n(n+1)} = \lim_{n\to\infty}\frac{(n+1)^n}{3n^n}$$

$$= \lim_{n\to\infty}\frac{1}{3}\left(1+\frac{1}{n}\right)^n = \frac{e}{3} < 1$$

By the Ratio Test the series converges. Since the terms are positive, the series converges absolutely.

25. $f(x) = -\dfrac{2x}{(1+x^2)^2} = \dfrac{d}{dx}\left(\dfrac{1}{1+x^2}\right) = \dfrac{d}{dx}\left(\sum_{n=0}^{\infty}(-1)^n x^{2n}\right) = \sum_{n=1}^{\infty}(-1)^n(2n)x^{2n-1}$

Since

$$\lim_{n\to\infty}\left|\frac{(-1)^{n+1}(2n+2)x^{2n+1}}{(-1)^n(2n)x^{2n-1}}\right| = \lim_{n\to\infty}\frac{(2n+2)}{2n}|x^2| = |x^2|$$

the radius of convergence of the series is 1. For $x = 1$ the series becomes $\sum_{n=1}^{\infty}(-1)^n 2n$, which diverges; for $x = -1$ the series becomes $-\sum_{n=0}^{\infty}(-1)^n(2n)$, which also diverges. Thus the interval of convergence of $\sum_{n=1}^{\infty}(-1)^n(2n)x^{2n-1}$ is $(-1, 1)$.

27. Let P_1, P_2, and P_3 be the points $(1,-1,2)$, $(2,3,-1)$, and $(0,2,0)$, respectively. Then $\overrightarrow{P_1P_2} = \mathbf{i}+4\mathbf{j}-3\mathbf{k}$ and $\overrightarrow{P_1P_3} = -\mathbf{i}+3\mathbf{j}-2\mathbf{k}$, so that

$$\overrightarrow{P_1P_2}\times\overrightarrow{P_1P_3} = \begin{vmatrix} \mathbf{i} & \mathbf{j} & \mathbf{k} \\ 1 & 4 & -3 \\ -1 & 3 & -2 \end{vmatrix} = [-8-(-9)]\mathbf{i} + [3-(-2)]\mathbf{j} + [3-(-4)]\mathbf{k} = \mathbf{i}+5\mathbf{j}+7\mathbf{k}.$$

Since $\overrightarrow{P_1P_2}\times\overrightarrow{P_1P_3}$ is perpendicular to the plane, and $(1,-1,2)$ lies on the plane, an equation of the plane is $1(x-1)+5(y+1)+7(z-2) = 0$, or $x+5y+7z = 10$.

Chapter 13

Partial Derivatives

13.1 Functions of Several Variables

1. all (x, y) such that $x \geq 0$ and $y \geq 0$

3. all (x, y) such that $x \neq 0$ and $y \neq 0$

5. all (x, y) such that $x^2 + y^2 \geq 25$

7. all (x, y) such that $x + y \neq 0$

9. all (x, y, z) such that $x^2 + y^2 + z^2 \leq 1$

11. all (x, y, z) such that $x \neq 0$, $y \neq 0$, and $z \neq 0$

13.

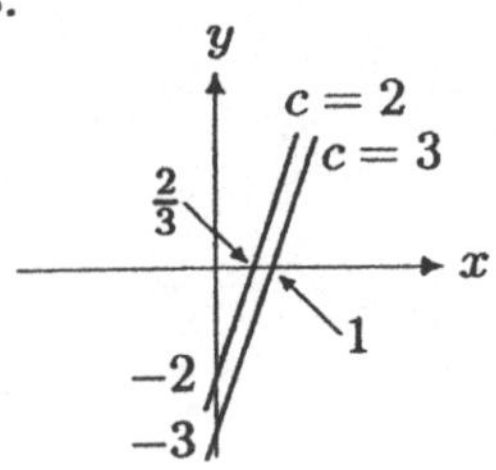

15.

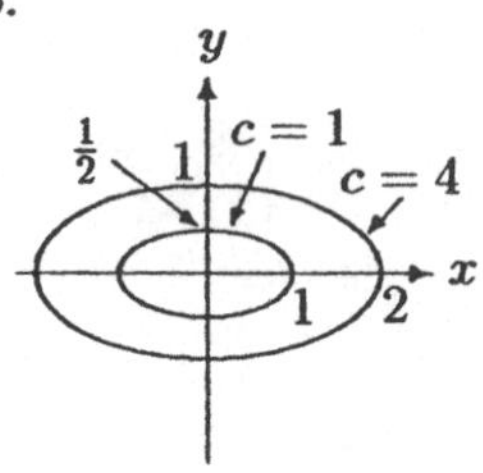

17.

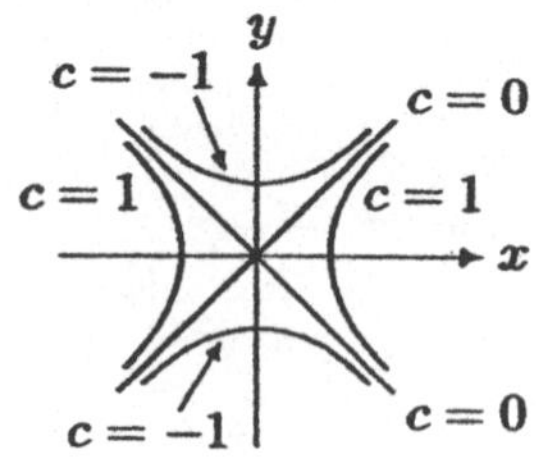

19.

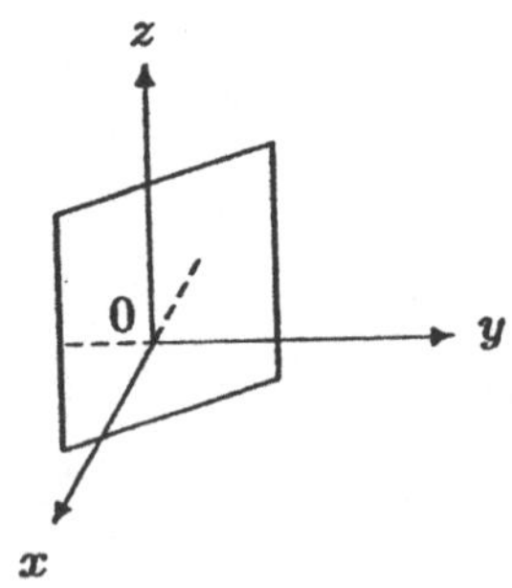

21.

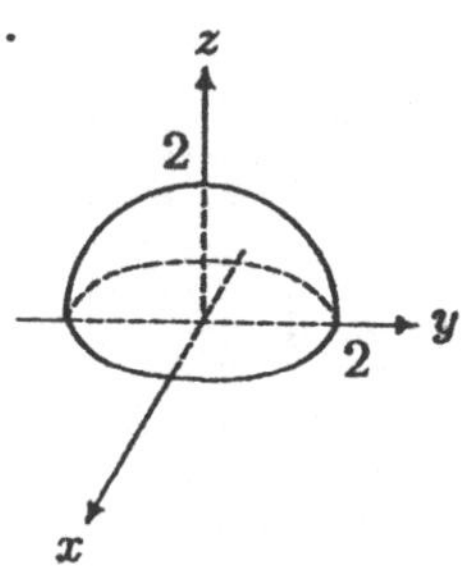

23.

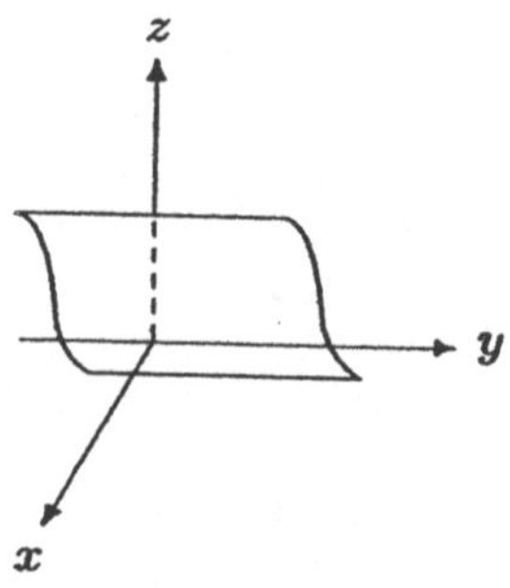

25.

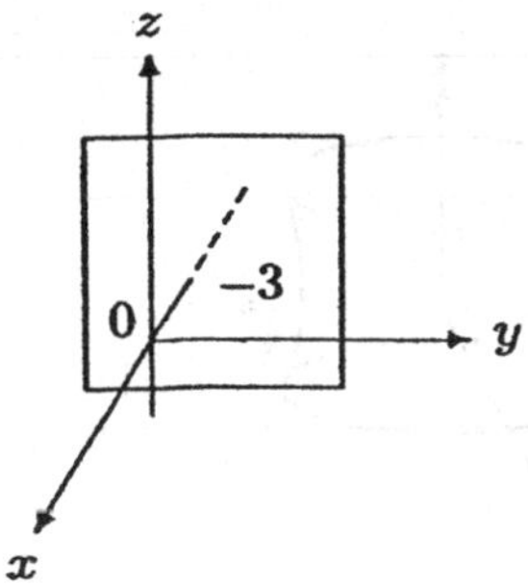

27.

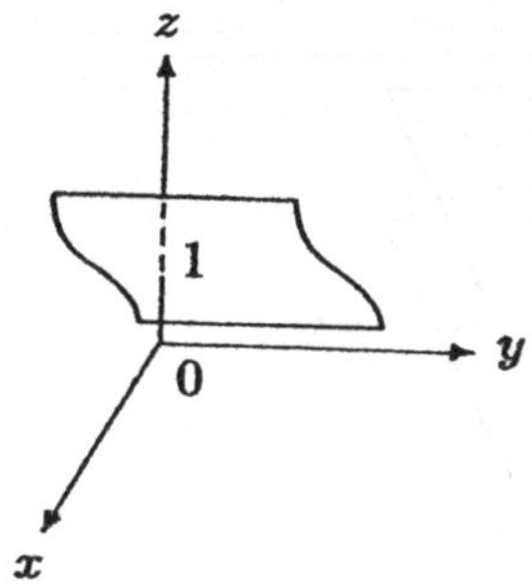

29.

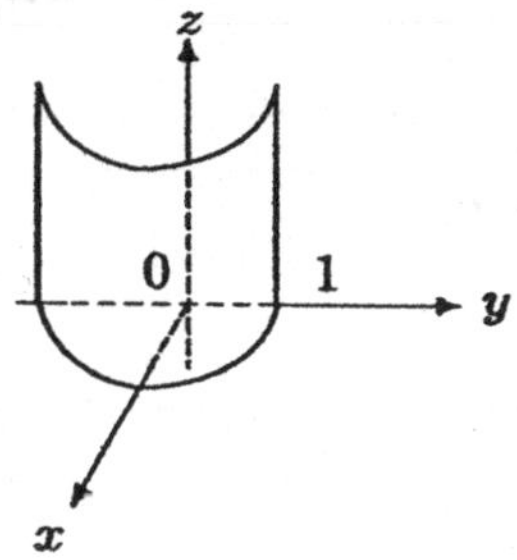

31.

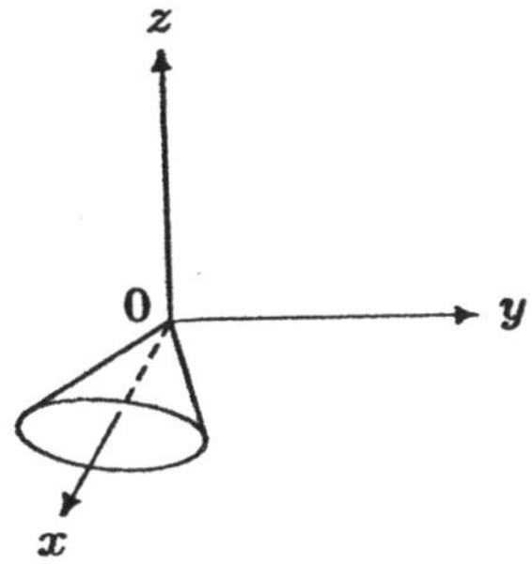

33.

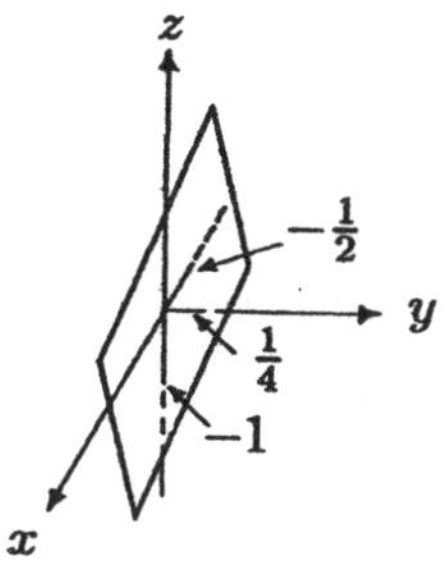

35.

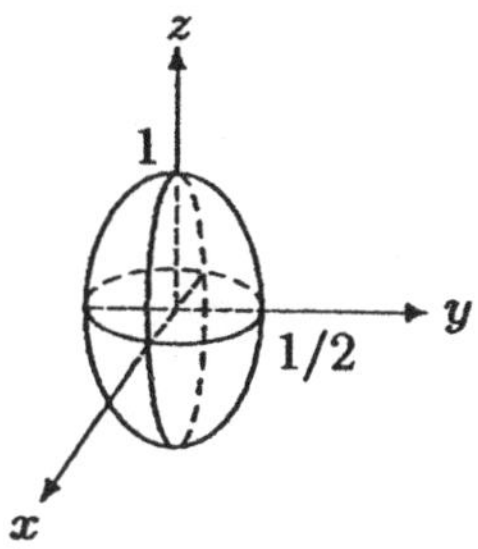

37.

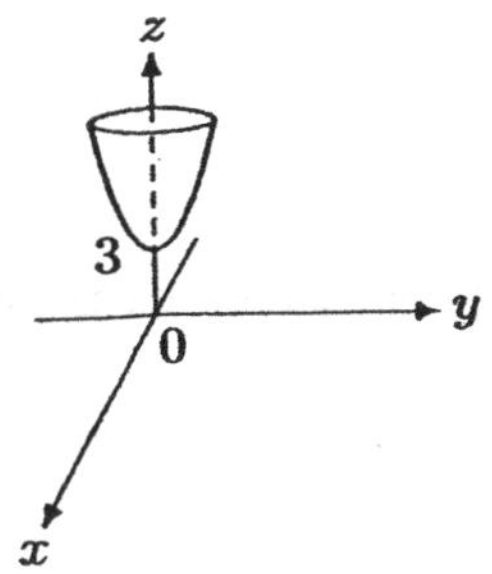

39.

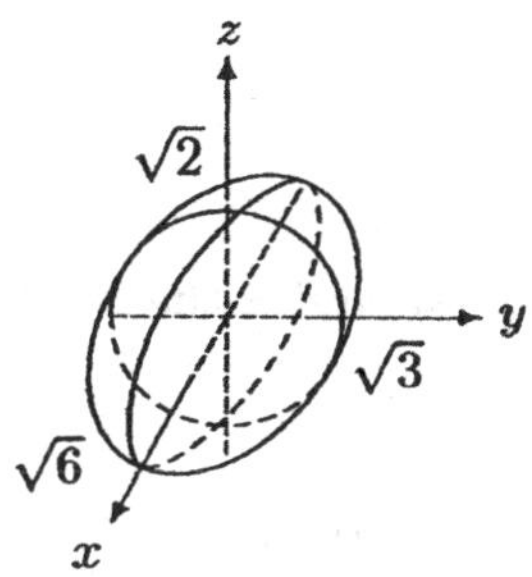

41.

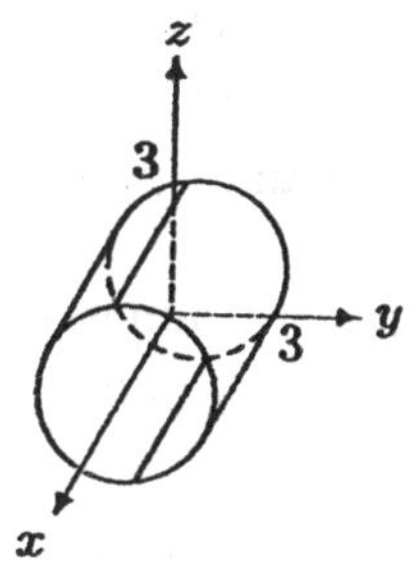

43.

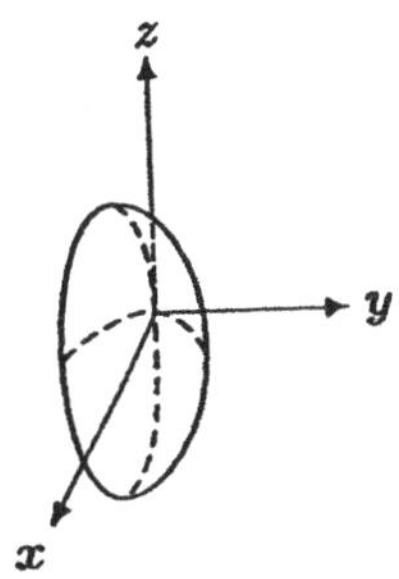

45.

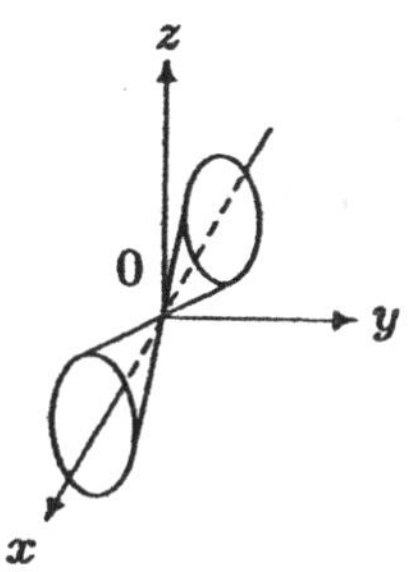

47.

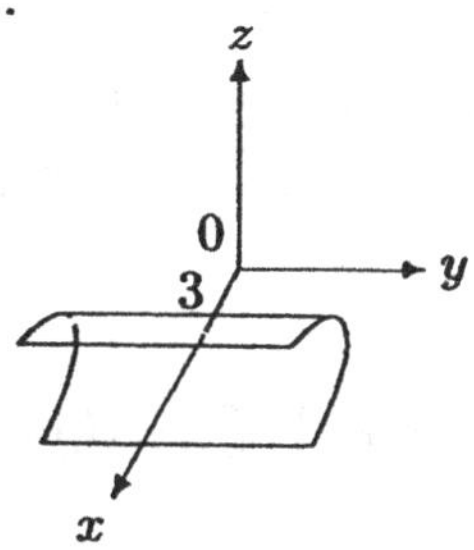

49.

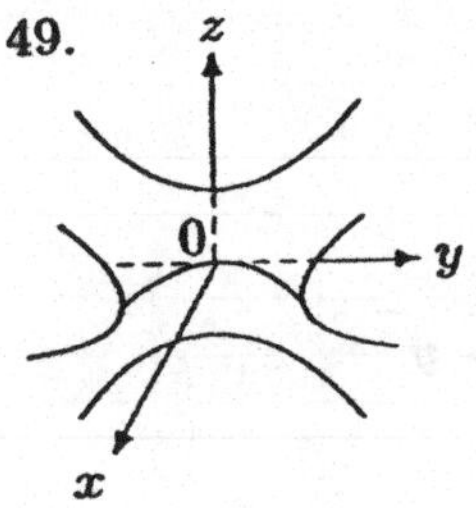

51.

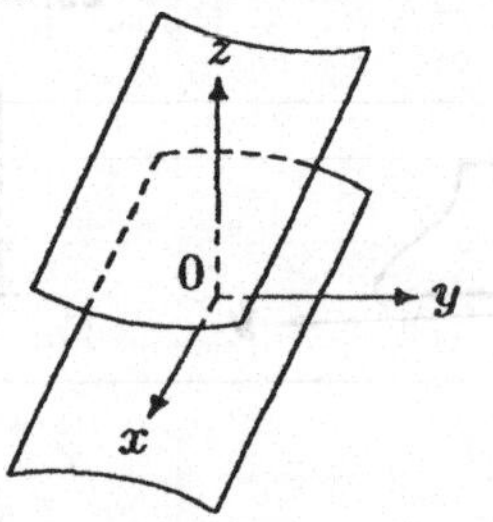

53.

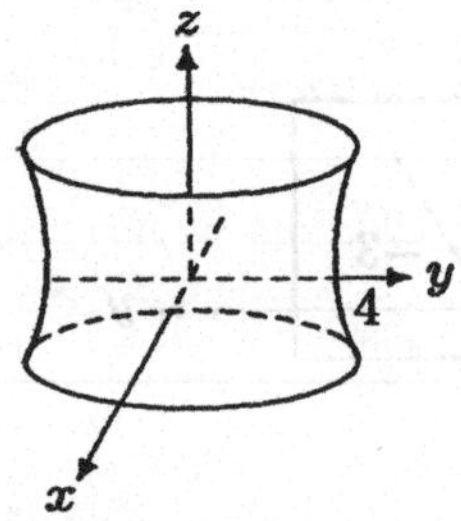

55.

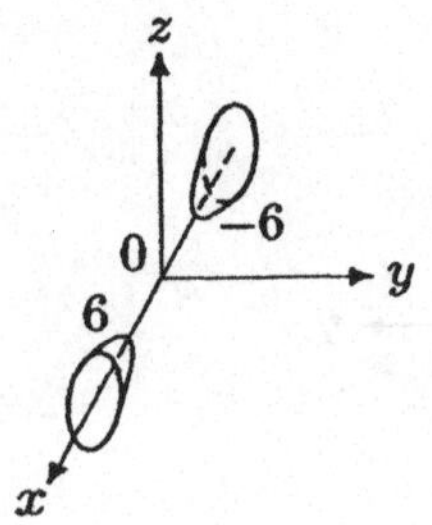

57. Surface (d); level curves (*ii*).

59. Surface (c); level curves (*iv*).

61. $h = V/\pi r^2$ for $r > 0$ and $v > 0$

63. Let x and y denote the dimensions of the base and z the height. Then the surface area is given by $S = xy + 2xz + 2yz$ for $x > 0$, $y > 0$, and $z > 0$.

65. The cost in dollars is given by $C = \frac{3}{10}xy$ for $x > 0$ and $y > 0$.

67. For any positive constant b, the level surface $E(x, y, z) = b$ consists of all (x, y, z) such that $c/\sqrt{x^2+y^2} = b$, or $x^2 + y^2 = c^2/b^2$. Thus the level surfaces are circular cylinders.

69. For any constant b, the level curve $T(x, y) = b$ consists of all (x, y) such that $x \geq 0$, $y \geq 0$, and $xy = b$. Thus the level curves are either hyperbolas or two intersecting lines.

13.2 Limits and Continuity

1. $\displaystyle\lim_{(x,y)\to(2,4)} \left(x + \frac{1}{2}\right) = 2 + \frac{1}{2} = \frac{5}{2}$

3. $\displaystyle\lim_{(x,y)\to(1,0)} \frac{x^2 - xy + 1}{x^2 + y^2} = \frac{1^2 - (1)(0) + 1}{1^2 + 0^2} = 2$

5. $$\lim_{(x,y,z)\to(2,1,-1)} \frac{2x^2y - xz^2}{y^2 - xz} = \frac{2(2)^2(1) - (2)(-1)^2}{1^2 - (2)(-1)} = 2$$

7. $$\lim_{(x,y)\to(2,1)} \frac{x^3 + 2x^2y - xy - 2y^2}{x + 2y} = \lim_{(x,y)\to(2,1)} \frac{(x + 2y)(x^2 - y)}{x + 2y} = \lim_{(x,y)\to(2,1)} (x^2 - y) = 2^2 - 1 = 3$$

9. Since $\lim_{(x,y,z)\to(\pi/2,-\pi/2,0)}(x + y + z) = \pi/2 - \pi/2 + 0 = 0$ and the cosine function is continuous, it follows from the substitution formula that

$$\lim_{(x,y,z)\to(\pi/2,-\pi/2,0)} \cos(x + y + z) = \cos 0 = 1.$$

11. Since $\lim_{(x,y)\to(0,0)}(x^2 + y^2) = 0 + 0 = 0$ and $\lim_{t\to 0}(\sin t)/t = 1$, it follows from the substitution formula that

$$\lim_{(x,y)\to(0,0)} \frac{\sin(x^2 + y^2)}{x^2 + y^2} = \lim_{t\to 0} \frac{\sin t}{t} = 1.$$

13. From (1) and (2) we have

$$\lim_{(x,y)\to(0,0)} \frac{x^3}{x^2 + y^2} = 0 = \lim_{(x,y)\to(0,0)} \frac{y^3}{x^2 + y^2}.$$

Thus

$$\lim_{(x,y)\to(0,0)} xy\frac{x^2 - y^2}{x^2 + y^2} = \lim_{(x,y)\to(0,0)} \left[y\frac{x^3}{x^2 + y^2} - x\frac{y^3}{x^2 + y^2}\right] = (0)(0) - (0)(0) = 0.$$

15. For any $\varepsilon > 0$ let $\delta = \varepsilon$. If $0 < \sqrt{x^2 + y^2 + z^2} < \delta$, then $x = \sqrt{x^2} \le \sqrt{x^2 + y^2 + z^2} < \delta = \varepsilon$, so that

$$\left|\frac{x^3}{x^2 + y^2 + z^2}\right| = \left|x\frac{x^2}{x^2 + y^2 + z^2}\right| \le |x|1 < \varepsilon.$$

Thus

$$\lim_{(x,y,z)\to(0,0,0)} \frac{x^3}{x^2 + y^2 + z^2} = 0.$$

Similarly,

$$\lim_{(x,y,z)\to(0,0,0)} \frac{y^3}{x^2 + y^2 + z^2} = 0 = \lim_{(x,y,z)\to(0,0,0)} \frac{z^3}{x^2 + y^2 + z^2}.$$

It follows that

$$\lim_{(x,y,z)\to(0,0,0)} \frac{x^3 + y^3 + z^3}{x^2 + y^2 + z^2} = \lim_{(x,y,z)\to(0,0,0)} \left[\frac{x^3}{x^2 + y^2 + z^2} + \frac{y^3}{x^2 + y^2 + z^2} + \frac{z^3}{x^2 + y^2 + z^2}\right]$$
$$= 0 + 0 + 0 = 0.$$

17. Let $x = y \ne 0$. Then

$$\frac{x^2}{x^2 + y^2} = \frac{x^2}{x^2 + y^2} = \frac{1}{2}.$$

Next, let $y = 0$. Then for $x \ne 0$, $x^2/(x^2 + y^2) = x^2/x^2 = 1$. Thus $\lim_{(x,y)\to(0,0)} x^2/(x^2 + y^2)$ does not exist.

19. Let $\varepsilon > 0$. Recall that $\lim_{x\to 0^+}(\sin x)/x = 1$, so that there is $\delta > 0$ such that if $0 < |x| < \delta$, then $|(\sin x)/x - 1| < \varepsilon$. Now let (x, y) be so close to $(0,0)$ that $|x - y| < \delta$, and assume (x, y) is in R. Then $x - y \neq 0$, and

$$\left|\frac{\sin(x-y)}{x-y} - 1\right| < \varepsilon.$$

Thus

$$\lim_{\substack{(x,y)\to(0,0)\\ R}} \frac{\sin(x-y)}{x-y} = 1.$$

21. If $y = x \neq 0$, then $y/x = 1$, whereas if $y = -x \neq 0$, then $y/x = -1$. Such pairs (x, y) exist in any disk centered at $(0,0)$ and are in R. Thus $\lim_{\substack{(x,y)\to(0,0)\\ R}} y/x$ does not exist.

23. f is continuous because it is a polynomial.

25. f is continuous because it is a rational function.

27. Let $h(x, y, z) = e^x + e^{yz}$ and $g(t) = \ln t$. Then $f = g \circ h$. Since h and g are continuous, f is also continuous.

29. a. Since $f(x, 0) = 0$ for all x and $f(0, y) = 0$ for all y, f is continuous in each variable separately at $(0,0)$.

    b. Notice that if $y = 0$, then $f(x, y) = 0$, whereas if $x = y^2 \neq 0$, then $f(x, y) = \frac{1}{2}$. By the remark preceding Example 5, it follows that $\lim_{(x,y)\to(0,0)} f(x, y)$ does not exist. Thus f is not continuous at $(0,0)$.

31. a. $(x + 3)^2 + (y - 2)^2 = 36$

    b. The boundary consists of the four line segments having the following formulas:

$$\begin{array}{lcl} y = 0 & \text{and} & 0 \le x \le 2 \\ x = 2 & \text{and} & -3 \le y \le 0 \\ y = -3 & \text{and} & 0 \le x \le 2 \\ x = 0 & \text{and} & -3 \le y \le 0 \end{array}$$

    c. The boundary consists of the three line segments having the following formulas:

$$\begin{array}{lcl} y = 1 & \text{and} & -1 \le x \le 1 \\ y = -6x - 5 & \text{and} & -1 \le x \le 0 \\ y = 6x - 5 & \text{and} & 0 \le x \le 1 \end{array}$$

    d. The x axis.

    e. $y = 4x^2$

    f. The origin.

33. Since f is constant on R, f is continuous on R.

35. If (x_0, y_0) is any point on the circle $x^2 + y^2 = 9$, then any disk around (x_0, y_0) contains points at which f takes on the value 1 and points at which f takes on the value 0. Thus $\lim_{(x,y)\to(x_0,y_0)} f(x,y)$ does not exist, so that f is not continuous at (x_0, y_0). Since (x_0, y_0) is an interior point of R, f is not continuous on R.

37. Since $-(1+x^2)/y$ and e^x are continuous, f is continuous at each (x, y) such that $y \neq 0$. Since $0 < e^{-(1+x^2)/y} \le e^{-1/y}$ and since $\lim_{y\to 0^+} e^{-1/y} = 0$, it follows that $\lim_{\substack{(x,y)\to(x_0,0)\\ R}} f(x,y) = f(x_0, 0)$ at every boundary point $(x_0, 0)$ of R. Thus f is continuous on R.

39. If $x > 0$ and $y > 0$, then $0 < 17{,}860xy/[(1.798)x + y] < 17{,}860x$. Thus $\lim_{(x,y)\to(0,0)} Q(x,y) = 0$.

13.3 Partial Derivatives

1. $f_x(x,y) = x^{1/2}$, $f_y(x,y) = 0$

3. $f_x(x,y) = 2 + 6xy^4$, $f_y(x,y) = 12x^2y^3$

5. $$g_u(u,v) = \frac{(u^2+v^2)3u^2 - (u^3+v^3)2u}{(u^2+v^2)^2} = \frac{u^4 + 3u^2v^2 - 2uv^3}{(u^2+v^2)^2},$$

$$g_v(u,v) = \frac{(u^2+v^2)3v^2 - (u^3+v^3)2v}{(u^2+v^2)^2} = \frac{v^4 + 3u^2v^2 - 2u^3v}{(u^2+v^2)^2}$$

7. $$f_x(x,y) = \frac{-x}{\sqrt{4 - x^2 - 9y^2}},\quad f_y(x,y) = \frac{-9y}{\sqrt{4 - x^2 - 9y^2}}$$

9. $$\frac{\partial z}{\partial x} = \frac{1}{2\sqrt{(1-x^{2/3})^3 - y^2}}[3(1-x^{2/3})^2]\left(\frac{-2}{3}x^{-1/3}\right) = \frac{-(1-x^{2/3})^2}{x^{1/3}\sqrt{(1-x^{2/3})^3 - y^2}},$$

$$\frac{\partial z}{\partial y} = \frac{-y}{\sqrt{(1-x^{2/3})^3 - y^2}}$$

11. $$\frac{\partial z}{\partial x} = 3(\sin x^2y)^2(\cos x^2y)(2xy) = 6xy\sin^2 x^2y\,\cos x^2y,$$

$$\frac{\partial z}{\partial y} = 3(\sin x^2y)^2(\cos x^2y)(x^2) = 3x^2\sin^2 x^2y\,\cos x^2y$$

13. $f_x(x,y,z) = 2xy^5 + z^2$, $f_y(x,y,z) = 5x^2y^4$, $f_z(x,y,z) = 2xz$

15. $f_x(x,y,z) = \dfrac{xy+yz+zx-(x+y+z)(y+z)}{(xy+yz+zx)^2} = \dfrac{-(y^2+yz+z^2)}{(xy+yz+zx)^2}$,

$f_y(x,y,z) = \dfrac{xy+yz+zx-(x+y+z)(x+z)}{(xy+yz+zx)^2} = \dfrac{-(z^2+zx+x^2)}{(xy+yz+zx)^2}$,

$f_z(x,y,z) = \dfrac{xy+yz+zx-(x+y+z)(y+x)}{(xy+yz+zx)^2} = \dfrac{-(x^2+xy+y^2)}{(xy+yz+zx)^2}$

17. $\dfrac{\partial w}{\partial x} = e^x(\cos y + \sin z)$, $\dfrac{\partial w}{\partial y} = e^x(-\sin y) = -e^x \sin y$, $\dfrac{\partial w}{\partial z} = e^x \cos z$

19. $\dfrac{\partial w}{\partial x} = \dfrac{1}{\sqrt{1-1/(1+xyz^2)^2}}\left(\dfrac{-1}{(1+xyz^2)^2}\right)yz^2 = \dfrac{-yz^2}{(1+xyz^2)^2\sqrt{1-1/(1+xyz^2)^2}}$,

$\dfrac{\partial w}{\partial y} = \dfrac{1}{\sqrt{1-1/(1+xyz^2)^2}}\left(\dfrac{-1}{(1+xyz^2)^2}\right)xz^2 = \dfrac{-xz^2}{(1+xyz^2)^2\sqrt{1-1/(1+xyz^2)^2}}$,

$\dfrac{\partial w}{\partial z} = \dfrac{1}{\sqrt{1-1/(1+xyz^2)^2}}\left(\dfrac{-1}{(1+xyz^2)^2}\right)2xyz = \dfrac{-2xyz}{(1+xyz^2)^2\sqrt{1-1/(1+xyz^2)^2}}$

21. $f_x(x,y) = \dfrac{4x}{\sqrt{4x^2+y^2}}$, $f_x(2,-3) = \dfrac{4(2)}{\sqrt{4(2)^2+(-3)^2}} = \dfrac{8}{5}$;

$f_y(x,y) = \dfrac{y}{\sqrt{4x^2+y^2}}$, $f_y(2,-3) = \dfrac{(-3)}{\sqrt{4(2)^2+(-3)^2}} = -\dfrac{3}{5}$

23. $f_x(x,y,z) = 2e^{2x-4y-z}$, $f_x(0,-1,1) = 2e^{2(0)-4(-1)-1} = 2e^3$;
$f_y(x,y,z) = -4e^{2x-4y-z}$, $f_y(0,-1,1) = -4e^{2(0)-4(-1)-1} = -4e^3$;
$f_z(x,y,z) = -e^{2x-4y-z}$, $f_z(0,-1,1) = -e^{2(0)-4(-1)-1} = -e^3$

25. $f_x(0,0) = \lim_{h\to 0}\dfrac{f(h,0)-f(0,0)}{h-0} = \lim_{h\to 0}\dfrac{0-0}{h-0} = \lim_{h\to 0} 0 = 0$

$f_y(0,0) = \lim_{h\to 0}\dfrac{f(0,h)-f(0,0)}{h-0} = \lim_{h\to 0}\dfrac{0-0}{h-0} = \lim_{h\to 0} 0 = 0$

27. Let y be fixed, and let $g(x) = f(x,y)$, $h(u) = \int_\pi^u \sin t^2\,dt$, and $k(x) = x^2+y^2$. Then

$$g(x) = f(x,y) = \int_\pi^{x^2+y^2} \sin t^2\,dt = h(k(x))$$

and $h'(u) = \sin u^2$, so that

$$f_x(x,y) = g'(x) = h'(k(x))k'(x) = [\sin(k(x))^2]2x = 2x\sin(x^2+y^2)^2.$$

Interchanging the roles of x and y in the preceding discussion, we find that $f_y(x,y) = 2y\sin(x^2+y^2)^2$.

29. $f_x(x,y) = 6x - \sqrt{2}\,y^2$, $f_{xy}(x,y) = -2\sqrt{2}\,y$; $f_y(x,y) = -2\sqrt{2}\,xy + 5y^4$, $f_{yx}(x,y) = -2\sqrt{2}\,y$

31. $f_x(x,y) = \dfrac{x}{\sqrt{x^2+y^2}}$, $f_{xy}(x,y) = \dfrac{-xy}{(x^2+y^2)^{3/2}}$;

$f_y(x,y) = \dfrac{y}{\sqrt{x^2+y^2}}$, $f_{yx}(x,y) = \dfrac{-xy}{(x^2+y^2)^{3/2}}$

33. $f_x(x,y,z) = -yz\sin xy$, $f_{xy}(x,y,z) = -z\sin xy - xyz\cos xy$;
$f_y(x,y,z) = -xz\sin xy$, $f_{yx}(x,y,z) = -z\sin xy - xyz\cos xy$

35. $f_x(x,y) = e^{x-2y}$, $f_{xx}(x,y) = e^{x-2y}$; $f_y(x,y) = -2e^{x-2y}$, $f_{yy}(x,y) = 4e^{x-2y}$

37. $f_x(x,y,z) = \dfrac{x}{\sqrt{x^2+y^2+z^2}}$, $f_{xx}(x,y,z) = \dfrac{\sqrt{x^2+y^2+z^2} - \dfrac{x^2}{\sqrt{x^2+y^2+z^2}}}{x^2+y^2+z^2} = \dfrac{y^2+z^2}{(x^2+y^2+z^2)^{3/2}}$,

$f_y(x,y,z) = \dfrac{y}{\sqrt{x^2+y^2+z^2}}$, $f_{yy}(x,y,z) = \dfrac{\sqrt{x^2+y^2+z^2} - \dfrac{y^2}{\sqrt{x^2+y^2+z^2}}}{x^2+y^2+z^2} = \dfrac{x^2+z^2}{(x^2+y^2+z^2)^{3/2}}$;

$f_z(x,y,z) = \dfrac{z}{\sqrt{x^2+y^2+z^2}}$, $f_{zz}(x,y,z) = \dfrac{\sqrt{x^2+y^2+z^2} - \dfrac{z^2}{\sqrt{x^2+y^2+z^2}}}{x^2+y^2+z^2} = \dfrac{x^2+y^2}{(x^2+y^2+z^2)^{3/2}}$

39. Let $f(x,y) = x^2 + 16y^2$. Then $f_x(x,y) = 2x$, so that $f_x(-3,1) = 2(-3) = -6$, so the line in question has equations $y = 1$ and $z - 25 = -6(x+3)$.

41. $f_x(x,y) = -1$ and $f_y(x,y) = 0$, so $\sqrt{f_x^2(x,y) + f_y^2(x,y) + 1} = \sqrt{(-1)^2 + 0^2 + 1} = \sqrt{2}$.

43. $f_x(x,y) = \dfrac{x}{\sqrt{x^2+y^2}}$ and $f_y(x,y) = \dfrac{y}{\sqrt{x^2+y^2}}$, so

$$\sqrt{f_x^2(x,y) + f_y^2(x,y) + 1} = \sqrt{\left(\frac{x}{\sqrt{x^2+y^2}}\right)^2 + \left(\frac{y}{\sqrt{x^2+y^2}}\right)^2 + 1} = \sqrt{2}.$$

45. $f_x(0,0) = \lim\limits_{h\to 0} \dfrac{f(h,0) - f(0,0)}{h-0} = \lim\limits_{h\to 0} \dfrac{0-0}{h} = \lim\limits_{h\to 0} 0 = 0$, and

$$f_y(0,0) = \lim_{h\to 0} \frac{f(0,h) - f(0,0)}{h-0} = \lim_{h\to 0} \frac{0-0}{h} = \lim_{h\to 0} 0 = 0.$$

By Exercise 29 of Section 13.2, f is not continuous at $(0,0)$.

47. $\dfrac{\partial z}{\partial x} = -ae^{-ay}\sin ax,\ \dfrac{\partial^2 z}{\partial x^2} = -a^2e^{-ay}\cos ax,\ \dfrac{\partial z}{\partial y} = -ae^{-ay}\cos ax;$

thus
$$\frac{\partial^2 z}{\partial x^2} = -a^2e^{-ay}\cos ax = a\frac{\partial z}{\partial y}.$$

49. $u_x = 2x,\ v_y = 2x$, so $u_x = v_y$; $u_y = -2y,\ v_x = 2y$, so $u_y = -v_x$.

51. $u_x = e^x\cos y,\ v_y = e^x\cos y$, so $u_x = v_y$; $u_y = -e^x\sin y,\ v_x = e^x\sin y$, so $u_y = -v_x$.

53. $\dfrac{\partial z}{\partial x} = \dfrac{2x}{x^2+y^2};\ \dfrac{\partial^2 z}{\partial x^2} = \dfrac{2y^2-2x^2}{(x^2+y^2)^2},\ \dfrac{\partial z}{\partial y} = \dfrac{2y}{x^2+y^2},\ \dfrac{\partial^2 z}{\partial y^2} = \dfrac{2x^2-2y^2}{(x^2+y^2)^2},$

so $\dfrac{\partial^2 z}{\partial x^2} + \dfrac{\partial^2 z}{\partial y^2} = \dfrac{2y^2-2x^2}{(x^2+y^2)^2} + \dfrac{2x^2-2y^2}{(x^2+y^2)^2} = 0.$

55. By Example 3, $f_x(x,y) = (x^4y + 4x^2y^3 - y^5)/(x^2+y^2)^2$ for $(x,y) \neq (0,0)$, so
$$f_{xy}(x,0) = \lim_{h\to 0}\frac{f_x(x,h) - f_x(x,0)}{h-0} = \lim_{h\to 0}\frac{x^4h + 4x^2h^3 - h^5}{h(x^2+y^2)^2} = \lim_{h\to 0}\frac{x^4 + 4x^2h^2 - h^4}{(x^2+y^2)^2} = 1 \quad \text{if } x \neq 0.$$
Thus $\lim_{h\to 0} f_{xy}(x,0) = 1$. Since $f_{xy}(0,0) = -1$ by Example 7, it follows that f_{xy} is not continuous at $(0,0)$.

57. Let y be fixed in $[c,d]$, and let $f(x) = M(x,y)$. Then $f'(x) = (\partial M/\partial x)(x,y)$ for $a \leq x \leq b$, so that by the Fundamental Theorem of Calculus,
$$\int_a^b \frac{\partial M}{\partial x}(x,y)\,dx = \int_a^b f'(x)\,dx = f(x)\Big|_a^b = f(b) - f(a) = M(b,y) - M(a,y).$$
The other formula follows in a completely analogous fashion.

59. a. $\dfrac{\partial u}{\partial x} = ae^{ax+bt},\ \dfrac{\partial^2 u}{\partial x^2} = a^2e^{ax+bt}$, and $\dfrac{\partial u}{\partial t} = be^{ax+bt}$, so that $\dfrac{b}{a^2}\dfrac{\partial^2 u}{\partial x^2} = be^{ax+bt} = \dfrac{\partial u}{\partial t}.$

b. $\dfrac{\partial u}{\partial x} = \dfrac{1}{\sqrt{t}}e^{-x^2/at}\left(\dfrac{-2x}{at}\right) = \dfrac{-2x}{at^{3/2}}e^{-x^2/at},$

$$\frac{\partial^2 u}{\partial x^2} = \frac{-2}{at^{3/2}}e^{-x^2/at} - \frac{2x}{at^{3/2}}e^{-x^2/at}\left(\frac{-2x}{at}\right) = \left(\frac{4x^2}{a^2t^{5/2}} - \frac{2}{at^{3/2}}\right)e^{-x^2/at},$$

and $\dfrac{\partial u}{\partial t} = \dfrac{-1}{2t^{3/2}}e^{-x^2/at} + \dfrac{1}{t^{1/2}}e^{-x^2/at}\left(\dfrac{x^2}{at^2}\right) = \left(\dfrac{x^2}{at^{5/2}} - \dfrac{1}{2t^{3/2}}\right)e^{-x^2/at},$

so that $\dfrac{\partial u}{\partial t} = \dfrac{a}{4}\dfrac{\partial^2 u}{\partial x^2}.$

61. $\dfrac{\partial R}{\partial R_1} = \dfrac{R_2(R_1+R_2) - R_1R_2}{(R_1+R_2)^2} = \dfrac{R_2^2}{(R_1+R_2)^2}$ and $\dfrac{\partial^2 R}{\partial R_1^2} = \dfrac{-2R_2^2}{(R_1+R_2)^3}$;

$\dfrac{\partial R}{\partial R_2} = \dfrac{R_1(R_1+R_2) - R_1R_2}{(R_1+R_2)^2} = \dfrac{R_1^2}{(R_1+R_2)^2}$ and $\dfrac{\partial^2 R}{\partial R_2^2} = \dfrac{-2R_1^2}{(R_1+R_2)^3}$.

Thus $\dfrac{\partial^2 R}{\partial R_1^2}\dfrac{\partial^2 R}{\partial R_2^2} = \dfrac{4R_1^2R_2^2}{(R_1+R_2)^6} = \dfrac{4R^2}{(R_1+R_2)^4}$.

63. $\dfrac{\partial V}{\partial T} = \dfrac{nk}{p}$, $\dfrac{\partial T}{\partial p} = \dfrac{V}{nk}$, and $\dfrac{\partial p}{\partial V} = \dfrac{-nkT}{V^2}$, so that

$$\frac{\partial V}{\partial T}\frac{\partial T}{\partial p}\frac{\partial p}{\partial V} = \left(\frac{nk}{p}\right)\left(\frac{V}{nk}\right)\left(\frac{-nkT}{V^2}\right) = \frac{-nkT}{pV} = -1.$$

65. $R = c\dfrac{2\pi r(h+2r)}{\pi r^2(h+\frac{4}{3}r)} = \dfrac{6c(h+2r)}{3hr+4r^2}$;

$$\frac{\partial R}{\partial r} = \frac{12c(3hr+4r^2) - 6c(h+2r)(3h+8r)}{(3hr+4r^2)^2} = \frac{-48chr - 18ch^2 - 48cr^2}{(3hr+4r^2)^2} < 0;$$

$$\frac{\partial R}{\partial h} = \frac{6c(3hr+4r^2) - 6c(h+2r)3r}{(3hr+4r^2)^2} = \frac{-12cr^2}{(3hr+4r^2)^2} < 0.$$

67. a. $f(tx,ty) = (tx)^\alpha(ty)^\beta = t^{\alpha+\beta}x^\alpha y^\beta = t^{\alpha+\beta}f(x,y)$

b. $\dfrac{\partial z}{\partial x} = \alpha x^{\alpha-1}y^\beta$, so $\dfrac{1}{z}\dfrac{\partial z}{\partial x} = \dfrac{1}{x^\alpha y^\beta}(\alpha x^{\alpha-1}y^\beta) = \dfrac{\alpha}{x}$; $\dfrac{\partial z}{\partial y} = \beta x^\alpha y^{\beta-1}$,

so $\dfrac{1}{z}\dfrac{\partial z}{\partial y} = \dfrac{1}{x^\alpha y^\beta}(\beta x^\alpha y^{\beta-1}) = \dfrac{\beta}{y}$;

$$x\frac{\partial z}{\partial x} + y\frac{\partial z}{\partial y} = x(\alpha x^{\alpha-1}y^\beta) + +y(\beta x^\alpha y^{\beta-1}) = (\alpha+\beta)x^\alpha y^\beta = (\alpha+\beta)z$$

69. a. Since the tax on each unit is t, the total tax on x units is tx, so the profit is given by $P(x,t) = P_0(x) - tx$.

b. By (a), $\frac{\partial P}{\partial x}(x,t) = P_0'(x) - t$ and thus $\frac{\partial^2 P}{\partial x^2}(x,t) = P_0''(x)$.

c. By (9), $0 = \frac{\partial P}{\partial x}(f(t),t)$ and by (b), $\frac{\partial P}{\partial x}(f(t),t) = P_0'(f(t)) - t$, so $P_0'(f(t)) - t = 0$.

d. From (c), $P_0''(f(t))f'(t) - 1 = 0$, so with the help of (b) and (8) we find that

$$f'(t) = \frac{1}{P_0''(f(t))} = \frac{1}{\dfrac{\partial^2 P}{\partial x^2}(f(t),t)} < 0.$$

13.4 The Chain Rule

1. $\dfrac{dz}{dt}=\dfrac{\partial z}{\partial x}\dfrac{dx}{dt}+\dfrac{\partial z}{\partial y}\dfrac{dy}{dt}=4x\left(\dfrac{1}{2\sqrt{t}}\right)-9y^2(2e^{2t})=4\sqrt{t}\left(\dfrac{1}{2\sqrt{t}}\right)-9e^{4t}(2e^{2t})=2-18e^{6t}$

3. $\dfrac{dz}{dt}=\dfrac{\partial z}{\partial x}\dfrac{dx}{dt}+\dfrac{\partial z}{\partial y}\dfrac{dy}{dt}=(\cos x-y\sin xy)(2t)+(-x\sin xy)(0)=2t(\cos t^2-\sin t^2)$

5. $\dfrac{dz}{dt}=\dfrac{\partial z}{\partial x}\dfrac{dx}{dt}+\dfrac{\partial z}{\partial y}\dfrac{dy}{dt}=\dfrac{1}{\sqrt{2x-4y}}\left(\dfrac{1}{t}\right)+\dfrac{-2}{\sqrt{2x-4y}}(-9t^2)=\dfrac{1+18t^3}{t\sqrt{2\ln t-4(1-3t^3)}}$

7. $\dfrac{\partial z}{\partial u}=\dfrac{\partial z}{\partial x}\dfrac{\partial x}{\partial u}+\dfrac{\partial z}{\partial y}\dfrac{\partial y}{\partial u}=\left(\dfrac{-4}{x^2y}-\dfrac{1}{y}\right)(2u)+\left(\dfrac{-4}{xy^2}+\dfrac{x}{y^2}\right)(v)$

$=\left(\dfrac{-4}{u^5v}-\dfrac{1}{uv}\right)(2u)+\left(\dfrac{-4}{u^4v^2}+\dfrac{1}{v^2}\right)(v)=\dfrac{-12-u^4}{u^4v};$

$\dfrac{\partial z}{\partial v}=\dfrac{\partial z}{\partial x}\dfrac{\partial x}{\partial v}+\dfrac{\partial z}{\partial y}\dfrac{\partial y}{\partial v}=\left(\dfrac{-4}{x^2y}-\dfrac{1}{y}\right)(0)+\left(\dfrac{-4}{xy^2}+\dfrac{x}{y^2}\right)(u)=\left(\dfrac{-4}{u^4v^2}+\dfrac{1}{v^2}\right)(u)=\dfrac{-4+u^4}{u^3v^2}$

9. $\dfrac{\partial z}{\partial u}=\dfrac{\partial z}{\partial x}\dfrac{\partial x}{\partial u}+\dfrac{\partial z}{\partial y}\dfrac{\partial y}{\partial u}=\dfrac{2x}{x^2-y^2}(1)-\dfrac{2y}{x^2-y^2}(2u)=\dfrac{2(u-v)-4u(u^2+v^2)}{(u-v)^2-(u^2+v^2)^2}$

$\dfrac{\partial z}{\partial v}=\dfrac{\partial z}{\partial x}\dfrac{\partial x}{\partial v}+\dfrac{\partial z}{\partial y}\dfrac{\partial y}{\partial v}=\dfrac{2x}{x^2-y^2}(-1)-\dfrac{2y}{x^2-y^2}(2v)=\dfrac{-2(u-v)-4v(u^2+v^2)}{(u-v)^2-(u^2+v^2)^2}$

11. $\dfrac{\partial z}{\partial r}=\dfrac{\partial z}{\partial u}\dfrac{\partial u}{\partial r}+\dfrac{\partial z}{\partial v}\dfrac{\partial v}{\partial r}=(2\cos 2u\cos 3v)(2(r+s))+(-3\sin 2u\sin 3v)(2(r-s))$

$=4(r+s)\cos[2(r+s)^2]\cos[3(r-s)^2]-6(r-s)\sin[2(r+s)^2]\sin[3(r-s)^2]$

$\dfrac{\partial z}{\partial s}=\dfrac{\partial z}{\partial u}\dfrac{\partial u}{\partial s}+\dfrac{\partial z}{\partial v}\dfrac{\partial v}{\partial s}=(2\cos 2u\cos 3v)(2(r+s))+(-3\sin 2u\sin 3v)(-2(r-s))$

$=4(r+s)\cos[2(r+s)^2]\cos[3(r-s)^2]+6(r-s)\sin[2(r+s)^2]\sin[3(r-s)^2]$

13. $\dfrac{\partial z}{\partial r}=\dfrac{\partial z}{\partial u}\dfrac{\partial u}{\partial r}+\dfrac{\partial z}{\partial v}\dfrac{\partial v}{\partial r}=(e^v-ve^{-u})\left(\dfrac{1}{r}\right)+(ue^v+e^{-u})\left(\dfrac{s}{r}\right)=\dfrac{1}{r}\left(r^s-\dfrac{s}{r}\ln r\right)+\dfrac{s}{r}\left(r^s\ln r+\dfrac{1}{r}\right)$

$\dfrac{\partial z}{\partial s}=\dfrac{\partial z}{\partial u}\dfrac{\partial u}{\partial s}+\dfrac{\partial z}{\partial v}\dfrac{\partial v}{\partial s}=(e^v-ve^{-u})(0)+(ue^v+e^{-u})(\ln r)=\left(r^s\ln r+\dfrac{1}{r}\right)\ln r$

15. $\dfrac{dw}{dt}=\dfrac{\partial w}{\partial x}\dfrac{dx}{dt}+\dfrac{\partial w}{\partial y}\dfrac{dy}{dt}+\dfrac{\partial w}{\partial z}\dfrac{dz}{dt}=\left(\dfrac{1}{y}+\dfrac{z}{x^2}\right)(\cos t)-\dfrac{x}{y^2}(-\sin t)-\dfrac{1}{x}(\sec^2 t)$

$=1+\csc t+\tan^2 t-\csc t\sec^2 t$

17. $\dfrac{dw}{dt}=\dfrac{\partial w}{\partial x}\dfrac{dx}{dt}+\dfrac{\partial w}{\partial y}\dfrac{dy}{dt}+\dfrac{\partial w}{\partial z}\dfrac{dz}{dt}$

$=\dfrac{x}{\sqrt{x^2+y^2+z^2}}(e^t)+\dfrac{y}{\sqrt{x^2+y^2+z^2}}(-e^{-t})+\dfrac{z}{\sqrt{x^2+y^2+z^2}}(2)=\dfrac{e^{2t}-e^{-2t}+4t}{\sqrt{e^{2t}+e^{-2t}+4t^2}}$

19. $\dfrac{dw}{dt} = \dfrac{\partial w}{\partial x}\dfrac{dx}{dt} + \dfrac{\partial w}{\partial y}\dfrac{dy}{dt} + \dfrac{\partial w}{\partial z}\dfrac{dz}{dt}$

$= (y^2z^3\cos xy^2z^3)(3) + (2xyz^3\cos xy^2z^3)\left(\dfrac{1}{2}t^{-1/2}\right) + (3xy^2z^2\cos xy^2z^3)\left(\dfrac{1}{3}t^{-2/3}\right) = 9t^2\cos 3t^3$

21. $\dfrac{\partial w}{\partial u} = \dfrac{\partial w}{\partial x}\dfrac{\partial x}{\partial u} + \dfrac{\partial w}{\partial y}\dfrac{\partial y}{\partial u} + \dfrac{\partial w}{\partial z}\dfrac{\partial z}{\partial u} = \dfrac{-yz(2x+y)}{x^2(x+y)^2}(2u) + \dfrac{\partial w}{\partial y}(0) + \dfrac{y}{x(x+y)}(2u)$

$= \dfrac{-v^2(u^2-v^2)(2u^2+v^2)2u + v^2(2u)[u^2(u^2+v^2)]}{u^4(u^2+v^2)^2} = \dfrac{2v^2(-u^4+2u^2v^2+v^4)}{u^3(u^2+v^2)^2}$

$\dfrac{\partial w}{\partial v} = \dfrac{\partial w}{\partial x}\dfrac{\partial x}{\partial v} + \dfrac{\partial w}{\partial y}\dfrac{\partial y}{\partial v} + \dfrac{\partial w}{\partial z}\dfrac{\partial z}{\partial v} = \dfrac{\partial w}{\partial x}(0) + \dfrac{(x^2+xy)z - yzx}{x^2(x+y)^2}(2v) + \dfrac{y}{x(x+y)}(-2v)$

$= \dfrac{u^4(u^2-v^2)(2v) - v^2(2v) - v^2(2v)[u^2(u^2+v^2)]}{u^4(u^2+v^2)^2} = \dfrac{2v(u^4-2u^2v^2-v^4)}{u^2(u^2+v^2)^2}$

23. $\dfrac{\partial w}{\partial u} = \dfrac{\partial w}{\partial x}\dfrac{\partial x}{\partial u} + \dfrac{\partial w}{\partial y}\dfrac{\partial y}{\partial u} + \dfrac{\partial w}{\partial z}\dfrac{\partial z}{\partial u} = \dfrac{y}{x}(ve^u) + (\ln xz)(2uv^4) + \dfrac{y}{z}e^v$

$= \dfrac{u^2v^4}{ve^u}(ve^u) + 2uv^4\ln(ve^u ue^v) + \dfrac{u^2v^4}{ue^v}e^v = uv^4[1+u+2\ln(uve^ue^v)]$

$\dfrac{\partial w}{\partial v} = \dfrac{\partial w}{\partial x}\dfrac{\partial x}{\partial v} + \dfrac{\partial w}{\partial y}\dfrac{\partial y}{\partial v} + \dfrac{\partial w}{\partial z}\dfrac{\partial z}{\partial v} = \dfrac{y}{x}(e^u) + (\ln xz)(4u^2v^3) + \dfrac{y}{z}ue^v$

$= \dfrac{u^2v^4}{ve^u}(e^u) + 4u^2v^3\ln(ve^u ue^v) + \dfrac{u^2v^4}{ue^v}(ue^v) = u^2v^3[1+v+4\ln(uve^ue^v)]$

25. Let $z = x^3 + 4x^2y - 3xy^2 + 2y^3 + 5$. Then $\dfrac{dy}{dx} = \dfrac{-\partial z/\partial x}{\partial z/\partial y} = \dfrac{-3x^2 - 8xy + 3y^2}{4x^2 - 6xy + 6y^2}$.

27. Let $z = x^2 + y^2 + \sin xy^2$. Then $\dfrac{dy}{dx} = \dfrac{-\partial z/\partial x}{\partial z/\partial y} = \dfrac{-2x - y^2\cos xy^2}{2y + 2xy\cos xy^2}$.

29. Let $z = x^2 - \dfrac{y^2}{y^2-1}$. Then $\dfrac{dy}{dx} = \dfrac{-\partial z/\partial x}{\partial z/\partial y} = \dfrac{-2x}{2y/(y^2-1)^2} = \dfrac{-x(y^2-1)^2}{y}$.

31. Let $w = x - yz + \cos xyz - 2$. Then

$$\frac{\partial z}{\partial x} = \frac{-\partial w/\partial x}{\partial w/\partial z} = \frac{-1 + yz\sin xyz}{-y - xy\sin xyz} \quad\text{and}\quad \frac{\partial z}{\partial y} = \frac{-\partial w/\partial y}{\partial w/\partial z} = \frac{z + xz\sin xyz}{-y - xy\sin xyz}.$$

33. If we let $u = x - y$, then $z = f(u)$, so that

$$\frac{\partial z}{\partial x} = \frac{dz}{du}\frac{\partial u}{\partial x} = \frac{dz}{du} \quad\text{and}\quad \frac{\partial z}{\partial y} = \frac{dz}{du}\frac{\partial u}{\partial y} = -\frac{dz}{du}.$$

Thus $dz/dx = -\partial z/\partial y$.

35. If $u = y + ax$ and $v = y - ax$, then $z = f(u) + g(v)$, so that

$$\frac{\partial z}{\partial x} = \frac{\partial z}{\partial u}\frac{\partial u}{\partial x} + \frac{\partial z}{\partial v}\frac{\partial v}{\partial x} = a\frac{df}{du} - a\frac{dg}{dv} \quad \text{and} \quad \frac{\partial z}{\partial y} = \frac{\partial z}{\partial u}\frac{\partial u}{\partial y} + \frac{\partial z}{\partial v}\frac{\partial v}{\partial y} = \frac{df}{du} + \frac{dg}{dv}.$$

It follows that

$$\frac{\partial^2 z}{\partial x^2} = \frac{\partial}{\partial x}\left(\frac{\partial z}{\partial x}\right) = \frac{\partial}{\partial x}\left(a\frac{df}{du} - a\frac{dg}{dv}\right) = \left[\frac{\partial}{\partial u}\left(a\frac{df}{du} - a\frac{dg}{dv}\right)\right]\frac{\partial u}{\partial x} + \left[\frac{\partial}{\partial v}\left(a\frac{df}{du} - a\frac{dg}{dv}\right)\right]\frac{\partial v}{\partial x}$$

$$= \left(a\frac{d^2 f}{du^2} - 0\right)(a) + \left(0 - a\frac{d^2 g}{dv^2}\right)(-a) = a^2\left(\frac{d^2 f}{du^2} + \frac{d^2 g}{dv^2}\right).$$

Similarly,

$$\frac{\partial^2 z}{\partial y^2} = \frac{\partial}{\partial y}\left(\frac{\partial z}{\partial y}\right) = \frac{\partial}{\partial y}\left(\frac{df}{du} + \frac{dg}{dv}\right) = \frac{\partial}{\partial u}\left(\frac{df}{du} + \frac{dg}{dv}\right)\frac{\partial u}{\partial y} + \frac{\partial}{\partial v}\left(\frac{df}{du} + \frac{dg}{dv}\right)\frac{\partial v}{\partial y}$$

$$= \left(\frac{d^2 f}{du^2} + 0\right)(1) + \left(0 + \frac{d^2 g}{dv^2}\right)(1) = \frac{d^2 f}{du^2} + \frac{d^2 g}{dv^2}.$$

Thus $\partial^2 z/\partial x^2 = a^2(\partial^2 z/\partial y^2)$.

37. We have

$$\frac{\partial w}{\partial s} = \frac{\partial w}{\partial x}\frac{\partial x}{\partial s} + \frac{\partial w}{\partial y}\frac{\partial y}{\partial s} = \frac{\partial w}{\partial x}e^s\cos t + \frac{\partial w}{\partial y}e^s\sin t$$

and

$$\frac{\partial w}{\partial t} = \frac{\partial w}{\partial x}\frac{\partial x}{\partial t} + \frac{\partial w}{\partial y}\frac{\partial y}{\partial t} = -\frac{\partial w}{\partial x}e^s\sin t + \frac{\partial w}{\partial y}e^s\cos t.$$

Thus

$$\frac{\partial^2 w}{\partial s^2} = \frac{\partial}{\partial s}\left(\frac{\partial w}{\partial s}\right) = \frac{\partial}{\partial s}\left(\frac{\partial w}{\partial x}e^s\cos t + \frac{\partial w}{\partial y}e^s\sin t\right)$$

$$= \left[\frac{\partial}{\partial s}\left(\frac{\partial w}{\partial x}\right)\right]e^s\cos t + \frac{\partial w}{\partial x}e^s\cos t + \left[\frac{\partial}{\partial s}\left(\frac{\partial w}{\partial y}\right)\right]e^s\sin t + \frac{\partial w}{\partial y}e^s\sin t$$

and

$$\frac{\partial^2 w}{\partial t^2} = \frac{\partial}{\partial t}\left(\frac{\partial w}{\partial t}\right) = \frac{\partial}{\partial t}\left(-\frac{\partial w}{\partial x}e^s\sin t + \frac{\partial w}{\partial y}e^s\cos t\right)$$

$$= -\left[\frac{\partial}{\partial t}\left(\frac{\partial w}{\partial x}\right)\right]e^s\sin t - \frac{\partial w}{\partial x}e^s\cos t + \left[\frac{\partial}{\partial t}\left(\frac{\partial w}{\partial y}\right)\right]e^s\cos t - \frac{\partial w}{\partial y}e^s\sin t.$$

But

$$\frac{\partial}{\partial s}\left(\frac{\partial w}{\partial x}\right) = \left[\frac{\partial}{\partial x}\left(\frac{\partial w}{\partial x}\right)\right]\frac{\partial x}{\partial s} + \left[\frac{\partial}{\partial y}\left(\frac{\partial w}{\partial x}\right)\right]\frac{\partial y}{\partial s} = \frac{\partial^2 w}{\partial x^2}e^s\cos t + \frac{\partial^2 w}{\partial y\partial x}e^s\sin t$$

$$\frac{\partial}{\partial s}\left(\frac{\partial w}{\partial y}\right) = \left[\frac{\partial}{\partial x}\left(\frac{\partial w}{\partial y}\right)\right]\frac{\partial x}{\partial s} + \left[\frac{\partial}{\partial y}\left(\frac{\partial w}{\partial y}\right)\right]\frac{\partial y}{\partial s} = \frac{\partial^2 w}{\partial x\partial y}e^s\cos t + \frac{\partial^2 w}{\partial y^2}e^s\sin t$$

$$\frac{\partial}{\partial t}\left(\frac{\partial w}{\partial x}\right) = \left[\frac{\partial}{\partial x}\left(\frac{\partial w}{\partial x}\right)\right]\frac{\partial x}{\partial t} + \left[\frac{\partial}{\partial y}\left(\frac{\partial w}{\partial x}\right)\right]\frac{\partial y}{\partial t} = -\frac{\partial^2 w}{\partial x^2}e^s\sin t + \frac{\partial^2 w}{\partial y\partial x}e^s\cos t$$

and
$$\frac{\partial}{\partial t}\left(\frac{\partial w}{\partial y}\right) = \left[\frac{\partial}{\partial x}\left(\frac{\partial w}{\partial y}\right)\right]\frac{\partial x}{\partial t} + \left[\frac{\partial}{\partial y}\left(\frac{\partial w}{\partial y}\right)\right]\frac{\partial y}{\partial t} = -\frac{\partial^2 w}{\partial x \partial y} e^s \sin t + \frac{\partial^2 w}{\partial y^2} e^s \cos t.$$

Thus
$$\begin{aligned}\frac{\partial^2 w}{\partial s^2} + \frac{\partial^2 w}{\partial t^2} &= \left[\left(\frac{\partial^2 w}{\partial x^2} e^s \cos t + \frac{\partial^2 w}{\partial y \partial x} e^s \sin t\right) e^s \cos t + \frac{\partial w}{\partial x} e^s \cos t\right. \\ &\quad \left. + \left(\frac{\partial^2 w}{\partial x \partial y} e^s \cos t + \frac{\partial^2 w}{\partial y^2} e^s \sin t\right) e^s \sin t + \frac{\partial w}{\partial y} e^s \sin t\right] \\ &\quad + \left[-\left(-\frac{\partial^2 w}{\partial x^2} e^s \sin t + \frac{\partial^2 w}{\partial y \partial x} e^s \cos t\right) e^s \sin t - \frac{\partial w}{\partial x} e^s \cos t\right. \\ &\quad \left. + \left(-\frac{\partial^2 w}{\partial x \partial y} e^s \sin t + \frac{\partial^2 w}{\partial y^2} e^s \cos t\right) e^s \cos t - \frac{\partial w}{\partial y} e^s \sin t\right] \\ &= e^{2s}\left(\frac{\partial^2 w}{\partial x^2} + \frac{\partial^2 w}{\partial y^2}\right).\end{aligned}$$

Thus
$$\frac{\partial^2 w}{\partial x^2} + \frac{\partial^2 w}{\partial y^2} = e^{-2s}\left(\frac{\partial^2 w}{\partial s^2} + \frac{\partial^2 w}{\partial t^2}\right).$$

39. Since $f(tx, ty) = \tan\{[(tx)^2 + (ty)^2]/(txty)\} = f(x, y) = t^0 f(x, y)$, we find from Exercise 38 with $n = 0$ that $xf_x(x, y) + yf_y(x, y) = 0$.

41. By the result of Exercise 16 of Section 13.2, $\lim_{(x,y)\to(0,0)} f(x, y)$ does not exist, so f is not continuous at $(0, 0)$. From Exercise 40, it follows that f is not differentiable at $(0, 0)$.

43. $\dfrac{dQ}{dt} = \dfrac{\partial Q}{\partial r}\dfrac{dr}{dt} + \dfrac{\partial Q}{\partial p}\dfrac{dp}{dt} = \left(\dfrac{4\pi p r^3}{8l\eta}\right)\left(\dfrac{1}{10}\right) + \left(\dfrac{\pi r^4}{8l\eta}\right)\left(\dfrac{-1}{5}\right) = \dfrac{\pi r^3}{40 l\eta}(2p - r)$

45. $\dfrac{dF}{dt} = \dfrac{\partial F}{\partial m}\dfrac{dm}{dt} + \dfrac{\partial F}{\partial r}\dfrac{dr}{dt} = \dfrac{GM}{r^2}\dfrac{dm}{dt} - \dfrac{2GMm}{r^3}\dfrac{dr}{dt}$

Thus if $dm/dt = -40$, $r = 6400$, and $dr/dt = 100$, we have
$$\frac{dF}{dt} = \frac{GM}{(6400)^2}(-40) - \frac{2GMm}{(6400)^3}(100) = \frac{-GM}{(6400)^2}\left(40 + \frac{m}{32}\right).$$

13.5 Directional Derivatives

1. $f_x(x, y) = 4x - 3y$, $f_y(x, y) = -3x + 2y$, and $\|\mathbf{a}\| = 1$;

$$D_{\mathbf{a}} f(1, 1) = f_x(1, 1)\left(\frac{1}{\sqrt{2}}\right) + f_y(1, 1)\left(\frac{1}{\sqrt{2}}\right) = \frac{1}{\sqrt{2}} - \frac{1}{\sqrt{2}} = 0$$

3. $f_x(x,y) = \dfrac{2x(x^2+y^2)-(x^2-y^2)2x}{(x^2+y^2)^2} = \dfrac{4xy^2}{(x^2+y^2)^2}$,

$f_y(x,y) = \dfrac{-2y(x^2+y^2)-(x^2-y^2)2y}{(x^2+y^2)^2} = \dfrac{-4x^2y}{(x^2+y^2)^2}$,

and $\|\mathbf{a}\| = 1$; $D_{\mathbf{a}}f(3,4) = f_x(3,4)\left(\dfrac{1}{2}\right) + f_y(3,4)\left(\dfrac{-\sqrt{3}}{2}\right) = \dfrac{96+72\sqrt{3}}{625}$

5. $f_x(x,y)=0$, $f_y(x,y)=4e^{4y}$, and $\|\mathbf{a}\|=4$, so that $\mathbf{u}=\mathbf{a}/\ \|\mathbf{a}\|=\mathbf{i}$;

$D_{\mathbf{u}}f(\frac{1}{2},\frac{1}{4}) = f_x(\frac{1}{2},\frac{1}{4})(1) + f_y(\frac{1}{2},\frac{1}{4})(0) = 0$

7. $f_x(x,y)=\sec^2(x+2y)$, $f_y(x,y)=2\sec^2(x+2y)$, and $\|\mathbf{a}\|=\sqrt{41}$, so that $\mathbf{u}=\dfrac{\mathbf{a}}{\|\mathbf{a}\|}=\dfrac{-4}{\sqrt{41}}\mathbf{i}+\dfrac{5}{\sqrt{41}}\mathbf{j}$;

$D_{\mathbf{u}}f\left(0,\dfrac{\pi}{6}\right) = f_x\left(0,\dfrac{\pi}{6}\right)\left(\dfrac{-4}{\sqrt{41}}\right) + f_y\left(0,\dfrac{\pi}{6}\right)\left(\dfrac{5}{\sqrt{41}}\right) = \dfrac{-16}{\sqrt{41}}+\dfrac{40}{\sqrt{41}} = \dfrac{24}{41}\sqrt{41}$

9. $f_x(x,y,z)=3x^2y^2z$, $f_y(x,y,z)=2x^3yz$, $f_z(x,y,z)=x^3y^2$, and $\|\mathbf{a}\|=3$,

so that $\mathbf{u}=\dfrac{\mathbf{a}}{\|\mathbf{a}\|}=\dfrac{2}{3}\mathbf{i}-\dfrac{1}{3}\mathbf{j}-\dfrac{2}{3}\mathbf{k}$;

$D_{\mathbf{u}}f(2,-1,2) = f_x(2,-1,2)\left(\dfrac{2}{3}\right) + f_y(2,-1,2)\left(\dfrac{-1}{3}\right) + f_z(2,-1,2)\left(\dfrac{-2}{3}\right) = 16+\dfrac{32}{3}-\dfrac{16}{3}=\dfrac{64}{3}$

11. $f_x(x,y,z) = \dfrac{(x+y+z)-(x-y-z)}{(x+y+z)^2} = \dfrac{2y+2z}{(x+y+z)^2}$,

$f_y(x,y,z) = \dfrac{-(x+y+z)-(x-y-z)}{(x+y+z)^2} = \dfrac{-2x}{(x+y+z)^2}$,

$f_z(x,y,z) = \dfrac{-(x+y+z)-(x-y-z)}{(x+y+z)^2} = \dfrac{-2x}{(x+y+z)^2}$,

and $\|\mathbf{a}\|=\sqrt{6}$, so that $\mathbf{u}=\dfrac{\mathbf{a}}{\|\mathbf{a}\|}=\dfrac{-2}{\sqrt{6}}\mathbf{i}-\dfrac{1}{\sqrt{6}}\mathbf{j}-\dfrac{1}{\sqrt{6}}\mathbf{k}$;

$$D_{\mathbf{u}}f(2,1,-1) = f_x(2,1,-1)\left(\dfrac{-2}{\sqrt{6}}\right) + f_y(2,1,-1)\left(\dfrac{-1}{\sqrt{6}}\right) + f_z(2,1,-1)\left(\dfrac{-1}{\sqrt{6}}\right)$$
$$= 0+\dfrac{1}{\sqrt{6}}+\dfrac{1}{\sqrt{6}}=\dfrac{\sqrt{6}}{3}$$

13. $f(x,y,z)=yz2^x=yze^{x\ln 2}$; $f_x(x,y,z)=(\ln 2)yze^{x\ln 2}=yz2^x\ln 2$, $f_y(x,y,z)=z2^x$, $f_z(x,y,z)=y2^x$,

$\|\mathbf{a}\|=\sqrt{5}$, so that $\mathbf{u}=\dfrac{\mathbf{a}}{\|\mathbf{a}\|}=\dfrac{2}{\sqrt{5}}\mathbf{j}-\dfrac{1}{\sqrt{5}}\mathbf{k}$;

$D_{\mathbf{u}}f(1,-1,1) = f_x(1,-1,1)(0) + f_y(1,-1,1)\left(\dfrac{2}{\sqrt{5}}\right) + f_z(1,-1,1)\left(\dfrac{-1}{\sqrt{5}}\right) = 0+\dfrac{4}{\sqrt{5}}+\dfrac{2}{\sqrt{5}}=\dfrac{6}{5}\sqrt{5}$

15. Let $\mathbf{a} = 2\mathbf{i} - \mathbf{j}$. Then $\|\mathbf{a}\| = \sqrt{4+1} = \sqrt{5}$, so we will find $D_{\mathbf{u}}f(2,-1)$, where

$$\mathbf{u} = \frac{1}{\|\mathbf{a}\|}\mathbf{a} = \frac{2}{\sqrt{5}}\mathbf{i} - \frac{1}{\sqrt{5}}\mathbf{j}.$$

Since $f_x(x,y) = (y/x)\cosh(y\ln x)$ and $f_y(x,y) = (\ln x)\cosh(y\ln x)$, it follows that

$$D_{\mathbf{u}}f(2,-1) = f_x(2,-1)\left(\frac{2}{\sqrt{5}}\right) + f_y(2,-1)\left(-\frac{1}{\sqrt{5}}\right)$$

$$= \left[-\frac{1}{2}\cosh(-\ln 2)\right]\left(\frac{2}{\sqrt{5}}\right) + [(\ln 2)\cosh(-\ln 2)]\left(-\frac{1}{\sqrt{5}}\right) = -\frac{\sqrt{5}}{4}(1+\ln 2).$$

17. a. We need to find the directional derivative of T at $(-2,4)$ in the direction tangent to the parabola $y = x^2$ at the point $(-2,4)$. Since $dy/dx = 2x$, so that $(dy/dx)\big|_{x=-2} = -4$, it follows that $\mathbf{a} = \mathbf{i} - 4\mathbf{j}$ is tangent to the parabola at $(-2,4)$. The corresponding unit vector $\mathbf{u}$ is given by $\mathbf{u} = (1/\sqrt{17})\,\mathbf{i} - (4/\sqrt{17})\,\mathbf{j}$. Thus

$$D_{\mathbf{u}}T(x,y) = \frac{-80x}{(1+x^2+y^2)^2}\frac{1}{\sqrt{17}} + \frac{-80y}{(1+x^2+y^2)^2}\left(-\frac{4}{\sqrt{17}}\right)$$

so that

$$D_{\mathbf{u}}T(-2,4) = \frac{160}{[1+(-2)^2+4^2]^2}\frac{1}{\sqrt{17}} + \frac{-320}{[1+(-2)^2+4^2]^2}\left(-\frac{4}{\sqrt{17}}\right) = \frac{160}{49\sqrt{17}}.$$

b. We have

$$\frac{dT}{dt} = \frac{\partial T}{\partial x}\frac{dx}{dt} + \frac{\partial T}{\partial y}\frac{dy}{dt} = \frac{-80x}{(1+x^2+y^2)^2}\frac{dx}{dt} + \frac{-80y}{(1+x^2+y^2)^2}\frac{dy}{dt},$$

so that

$$\left.\frac{dT}{dt}\right|_{(-2,4)} = \frac{160}{441}\frac{dx}{dt} - \frac{320}{441}\frac{dy}{dt}.$$

Since the slope of the parabola $y = x^2$ at $(-2,4)$ is $2(-2) = -4$, it follows that $dy/dt = -4(dx/dt)$. Because the speed is 5, we have

$$5 = \sqrt{\left(\frac{dx}{dt}\right)^2 + \left(\frac{dy}{dt}\right)^2} = \sqrt{\left(\frac{dx}{dt}\right)^2 + 16\left(\frac{dx}{dt}\right)^2} = \sqrt{17}\left|\frac{dx}{dt}\right|.$$

Since the particle moves from left to right, it follows that $dx/dt = 5/\sqrt{17}$ and hence $dy/dt = -4(dx/dt) = -20/\sqrt{17}$. We conclude that

$$\left.\frac{dT}{dt}\right|_{(-2,4)} = \frac{160}{441}\left(\frac{5}{\sqrt{17}}\right) - \frac{320}{441}\left(-\frac{20}{\sqrt{17}}\right) = \frac{800}{49\sqrt{17}} \text{ (meters per second).}$$

13.6 The Gradient

1. $\operatorname{grad} f(x,y) = 3\mathbf{i} - 5\mathbf{j}$

3. $\operatorname{grad} g(x,y) = -2e^{-2x}\ln(y-4)\mathbf{i} + \dfrac{e^{-2x}}{y-4}\mathbf{j}$

5. $\operatorname{grad} f(x,y,z) = 4x\mathbf{i} - 2y\mathbf{j} - 8z\mathbf{k}$

7. $$\begin{aligned}\operatorname{grad} g(x,y,z) &= \frac{-(-x+z)-(-x+y)(-1)}{(-x+z)^2}\mathbf{i} + \frac{1}{-x+z}\mathbf{j} - \frac{-x+y}{(-x+z)^2}\mathbf{k}\\ &= \frac{y-z}{(-x+z)^2}\mathbf{i} + \frac{1}{-x+z}\mathbf{j} + \frac{x-y}{(-x+z)^2}\mathbf{k}\end{aligned}$$

9. $$\begin{aligned}\operatorname{grad} f(x,y) &= \frac{5x+2y-(x+3y)(5)}{(5x+2y)^2}\mathbf{i} + \frac{3(5x+2y)-(x+3y)(2)}{(5x+2y)^2}\mathbf{j}\\ &= \frac{-13y}{(5x+2y)^2}\mathbf{i} + \frac{13x}{(5x+2y)^2}\mathbf{j}\end{aligned}$$

   $\operatorname{grad} f(-1,\frac{3}{2}) = -\frac{39}{8}\mathbf{i} - \frac{13}{4}\mathbf{j}$

11. $\operatorname{grad} g(x,y) = \left[\ln(x+y) + \dfrac{x}{x+y}\right]\mathbf{i} + \dfrac{x}{x+y}\mathbf{j}$

    $\operatorname{grad} g(-2,3) = -2\mathbf{i} - 2\mathbf{j}$

13. $\operatorname{grad} f(x,y,z) = -ze^{-x}\tan y\,\mathbf{i} + ze^{-x}\sec^2 y\,\mathbf{j} + e^{-x}\tan y\,\mathbf{k}$

    $\operatorname{grad} f(0,\pi,-2) = -2\mathbf{j}$

15. $\operatorname{grad} f(x,y) = e^x(\cos y + \sin y)\mathbf{i} + e^x(-\sin y + \cos y)\mathbf{j}$, so $\operatorname{grad} f(0,0) = \mathbf{i}+\mathbf{j}$. Consequently the direction in which f increases most rapidly at $(0,0)$ is $\mathbf{i}+\mathbf{j}$, and the maximal directional derivative is $\|\mathbf{i}+\mathbf{j}\| = \sqrt{2}$.

17. $\operatorname{grad} f(x,y) = 6x\mathbf{i} + 8y\mathbf{j}$, so that $\operatorname{grad} f(-1,1) = -6\mathbf{i} + 8\mathbf{j}$. Thus the direction in which f increases most rapidly at $(-1,1)$ is $-6\mathbf{i}+8\mathbf{j}$, and the maximal directional derivative is $\|-6\mathbf{i}+8\mathbf{j}\| = 10$.

19. $\operatorname{grad} f(x,y,z) = e^x\mathbf{i} + e^y\mathbf{j} + 2e^{2z}\mathbf{k}$, so that $\operatorname{grad} f(1,1,-1) = e\mathbf{i} + e\mathbf{j} + 2e^{-2}\mathbf{k}$. Thus the direction in which f increases most rapidly at $(1,1,-1)$ is $e\mathbf{i}+e\mathbf{j}+2e^{-2}\mathbf{k}$, and the maximal directional derivative is $\|e\mathbf{i}+e\mathbf{j}+2e^{-2}\mathbf{k})\| = e\sqrt{2+4e^{-6}}$.

21. $\operatorname{grad} f(x,y) = \pi y\cos\pi xy\,\mathbf{i} + \pi x\cos\pi xy\,\mathbf{j}$, so that $\operatorname{grad} f(\frac{1}{2},\frac{2}{3}) = (\pi/3)\mathbf{i} + (\pi/4)\mathbf{j} = (\pi/12)(4\mathbf{i}+3\mathbf{j})$. Thus the direction in which f decreases most rapidly at $(\frac{1}{2},\frac{2}{3})$ is $(\pi/12)(-4\mathbf{i}-3\mathbf{j})$.

23. $$\begin{aligned}\operatorname{grad} f(x,y,z) &= \frac{1}{y+z}\mathbf{i} - \frac{x-z}{(y+z)^2}\mathbf{j} + \frac{-(y+z)-(x-z)}{(y+z)^2}\mathbf{k} \\ &= \frac{1}{y+z}\mathbf{i} + \frac{z-x}{(y+z)^2}\mathbf{j} - \frac{y+x}{(y+z)^2}\mathbf{k}\end{aligned}$$

so that $\operatorname{grad} f(-1,1,3) = \frac{1}{4}\mathbf{i} + \frac{1}{4}\mathbf{j} = \frac{1}{4}(\mathbf{i}+\mathbf{j})$. Thus the direction in which f decreases most rapidly at $(-1,1,3)$ is $\frac{1}{4}(-\mathbf{i}-\mathbf{j})$.

25. Let $f(x,y) = \sin \pi xy$. Since the graph of the given equation is a level curve of f, Theorem 13.17 implies that $\operatorname{grad} f(\frac{1}{6},2)$ is normal to the graph at $(\frac{1}{6},2)$. Since $\operatorname{grad} f(x,y) = \pi y \cos \pi xy\mathbf{i} + \pi x \cos \pi xy\mathbf{j}$, so that $\operatorname{grad} f(\frac{1}{6},2) = \pi\mathbf{i} + (\pi/12)\mathbf{j}$, we find that $\pi\mathbf{i} + (\pi/12)\mathbf{j}$ is normal to the graph at $(\frac{1}{6},2)$.

27. By (2), $f_x(-2,1)\mathbf{i} + f_y(-2,1)\mathbf{j} - \mathbf{k}$ is normal to the graph of f at $(-2,1,16)$. Since $f_x(x,y) = 6x$ and $f_y(x,y) = 8y$, so $f_x(-2,1) = -12$ and $f_y(-2,1) = 8$, it follows that $-12\mathbf{i} + 8\mathbf{j} - \mathbf{k}$ is normal to the graph at $(-2,1,16)$.

29. By (2), $f_x(0,2)\mathbf{i} + f_y(0,2)\mathbf{j} - \mathbf{k}$ is normal to the graph of f at $(0,2,1)$. Since $f_x(x,y) = -2x$ and $f_y(x,y) = 0$, so $f_x(0,2) = 0 = f_y(0,2)$, it follows that $-\mathbf{k}$ is normal to the graph at $(0,2,1)$.

31. Since $f_x(x,y) = y-1$ and $f_y(x,y) = x+1$, so $f_x(0,2) = 1$ and $f_y(0,2) = 1$, it follows from (3) that an equation of the plane tangent at $(0,2,7)$ is $(x-0)+(y-2)-(z-7) = 0$, or $x+y-z = -5$.

33. Since $g_x(x,y) = \pi y \cos \pi xy$ and $g_y(x,y) = \pi x \cos \pi xy$, so

$$g_x(-\sqrt{2},\sqrt{2}) = \pi\sqrt{2}, \quad g_y(-\sqrt{2},\sqrt{2}) = -\pi\sqrt{2},$$

and

$$g_x\left(-\frac{1}{2},\frac{1}{3}\right) = \frac{\pi\sqrt{3}}{6}, \quad g_y\left(-\frac{1}{2},\frac{1}{3}\right) = \frac{-\pi\sqrt{3}}{4},$$

it follows from (3) that an equation of the plane tangent at $(-\sqrt{2},\sqrt{2},0)$ is

$$\pi\sqrt{2}\,(x+\sqrt{2}) - \pi\sqrt{2}\,(y-\sqrt{2}) - (z-0) = 0, \quad \text{or} \quad \pi\sqrt{2}\,x - \pi\sqrt{2}\,y - z = -4\pi$$

and that an equation of the plane tangent at $(-\frac{1}{2},\frac{1}{3},-\frac{1}{2})$ is

$$\frac{\pi\sqrt{3}}{6}\left(x+\frac{1}{2}\right) - \frac{\pi\sqrt{3}}{4}\left(y-\frac{1}{3}\right) - \left(z+\frac{1}{2}\right) = 0 \quad \text{or} \quad \frac{\pi\sqrt{3}}{6}x - \frac{\pi\sqrt{3}}{4}y - z = \frac{-\pi\sqrt{3}}{6} + \frac{1}{2}.$$

35. Since $f_x(x,y) = 2(2+x-y)$ and $f_y(x,y) = -2(2+x-y)$, so $f_x(3,-1) = 12$ and $f_y(3,-1) = -12$, it follows from (3) that an equation of the plane tangent at $(3,-1,36)$ is

$$12(x-3) - 12(y+1) - (z-36) = 0, \quad \text{or} \quad 12x - 12y - z = 12.$$

37. Since $f_x(x,y) = 2x/(x^2+y^2)$ and $f_y(x,y) = 2y/(x^2+y^2)$, so $f_x(-1,0) = -2$, $f_y(-1,0) = 0$, $f_x(-1,1) = -1$, and $f_y(-1,1) = 1$, it follows from (3) that an equation of the plane tangent at $(-1,0,0)$ is $-2(x+1) + 0(y-0) - (z-0) = 0$, or $2x + z = -2$, and that an equation of the plane tangent at $(-1,1,\ln 2)$ is $-(x+1) + (y-1) - (z - \ln 2) = 0$, or $x - y + z = \ln 2 - 2$.

39. If $f(x,y,z) = x^2 + y^2 + z^2$, then $x^2 + y^2 + z^2 = 1$ is a level surface of f. Since $f_x(x,y,z) = 2x$, $f_y(x,y,z) = 2y$, and $f_z(x,y,z) = 2z$, so

$$f_x\left(\frac{1}{2}, -\frac{1}{2}, -\frac{1}{\sqrt{2}}\right) = 1, \quad f_y\left(\frac{1}{2}, -\frac{1}{2}, -\frac{1}{\sqrt{2}}\right) = -1, \quad \text{and} \quad f_z\left(\frac{1}{2}, -\frac{1}{2}, -\frac{1}{\sqrt{2}}\right) = -\sqrt{2}$$

it follows that an equation of the plane tangent at $(\frac{1}{2}, -\frac{1}{2}, -1/\sqrt{2})$ is

$$\left(x - \frac{1}{2}\right) - \left(y + \frac{1}{2}\right) - \sqrt{2}\left(z + \frac{1}{\sqrt{2}}\right) = 0, \quad \text{or} \quad x - y - \sqrt{2}\, z = 2.$$

41. If $f(x,y,z) = xyz$, then the given surface is a level surface of f. Since $f_x(x,y,z) = yz$, $f_y(x,y,z) = xz$, and $f_z(x,y,z) = xy$, so $f_x(\frac{1}{2}, -2, -1) = 2$, $f_y(\frac{1}{2}, -2, -1) = -\frac{1}{2}$, and $f_z(\frac{1}{2}, -2, 1) = -1$, it follows than an equation of the plane tangent at $(\frac{1}{2}, -2, -1)$ is $2(x - \frac{1}{2}) - \frac{1}{2}(y+2) - (z+1) = 0$, or $2x - \frac{1}{2}y - z = 3$.

43. If $f(x,y,z) = z^2 + \sin(xy)$, then the given surface is a level surface of f. Since $f_x(x,y,z) = y\cos(xy)$, $f_y(x,y,z) = x\cos(xy)$, and $f_z(x,y,z) = 2z$, so $f_x(\pi, \frac{1}{2}, -1) = \frac{1}{2}\cos(\pi/2) = 0$, $f_y(\pi, \frac{1}{2}, -1) = \pi\cos(\pi/2) = 0$ and $f_z(\pi, \frac{1}{2}, -1) = -2$, it follows that an equation of the plane tangent at $(\pi, \frac{1}{2}, -1)$ is $0(x - \pi) + 0(y - \frac{1}{2}) - 2(z+1) = 0$, or $z = -1$.

45. If $f(x,y,z) = \ln\sqrt{x^2+1} - z = \frac{1}{2}\ln(x^2+1) - z$, then the given surface is a level surface of f. Since $f_x(x,y,z) = x/(x^2+1)$, $f_y(x,y,z) = 0$, and $f_z(x,y,z) = -1$, so $f_x(0,2,0) = 0$, $f_y(0,2,0) = 0$, and $f_z(0,2,0) = -1$, it follows that an equation of the plane tangent at $(0,2,0)$ is $0(x-0) + 0(y-2) - 1(z-0) = 0$, or $z = 0$.

47. Let $f(x,y) = 9 - 4x^2 - y^2$. Then $f_x(x,y) = -8x$ and $f_y(x,y) = -2y$. By (2), a vector normal to the tangent plane at any point (x_0, y_0, z_0) on the graph of f is $-8x_0\mathbf{i} - 2y_0\mathbf{j} - \mathbf{k}$. If the tangent plane is parallel to the plane $z = 4y$, whose normal is $-4\mathbf{j} + \mathbf{k}$, then $-8x_0\mathbf{i} - 2y_0\mathbf{j} - \mathbf{k} = c(-4\mathbf{j} + \mathbf{k})$ for some constant c. It follows that $c = -1$, so that $x_0 = 0$ and $y_0 = -2$. The corresponding point on the paraboloid is $(0, -2, 5)$.

49. Let $f(x,y) = xy - 2$. Then $f_x(x,y) = y$ and $f_y(x,y) = x$. By (2), a vector normal to the plane tangent to the graph of f at $(1,1,-1)$ is $\mathbf{i} + \mathbf{j} - \mathbf{k}$. Let $g(x,y,z) = x^2 + y^2 + z^2$. Then $\operatorname{grad} g(x,y,z) = 2x\mathbf{i} + 2y\mathbf{j} + 2z\mathbf{k}$. By Definition 13.18 a vector normal to the plane tangent to the level surface $g(x,y,z) = 3$ at $(1,1,-1)$ is $2\mathbf{i} + 2\mathbf{j} - 2\mathbf{k}$. Since the vectors normal to the two tangent planes are parallel, and since both surfaces pass through $(1,1,-1)$, the two tangent planes at $(1,1,-1)$ are identical.

51. Let $f(x,y) = x^2 + 4y^2 - 12$. Then $f_x(x,y) = 2x$ and $f_y(x,y) = 8y$, so that by (2) a vector normal to the plane tangent to the graph of f at $(-3,-1,1)$ is $-6\mathbf{i} - 8\mathbf{j} - \mathbf{k}$. Let $g(x,y) = \frac{1}{8}(4x + y^2 + 19)$. Then $g_x(x,y) = \frac{1}{2}$ and $g_y(x,y) = \frac{1}{4}y$, so that a vector normal to the plane tangent to the graph of g at $(-3,-1,1)$ is $\frac{1}{2}\mathbf{i} - \frac{1}{4}\mathbf{j} - \mathbf{k}$. Since $(-6\mathbf{i} - 8\mathbf{j} - \mathbf{k}) \cdot (\frac{1}{2}\mathbf{i} - \frac{1}{4}\mathbf{j} - \mathbf{k}) = -3 + 2 + 1 = 0$, the two normal vectors and hence the two planes tangent at $(-3,-1,1)$ are perpendicular. Thus the two surfaces are normal at $(-3,-1,1)$.

53. Let $f(x,y,z) = x^2/a^2 + y^2/b^2 + z^2/c^2$. Then $f_x(x,y,z) = 2x/a^2$, $f_y(x,y,z) = 2y/b^2$, and $f_z(x,y,z) = 2z/c^2$, so that by Definition 13.18 an equation of the plane tangent to the level surface $f(x,y,z) = 1$ at (x_0, y_0, z_0) is

$$\frac{2x_0}{a^2}(x - x_0) + \frac{2y_0}{b^2}(y - y_0) + \frac{2z_0}{c^2}(z - z_0) = 0$$

or

$$\frac{xx_0}{a^2} + \frac{yy_0}{b^2} + \frac{zz_0}{c^2} = \frac{x_0^2}{a^2} + \frac{y_0^2}{b^2} + \frac{z_0^2}{c^2} = 1.$$

55. Since $f_x(x,y) = g(y/x) - (y/x)g'(y/x)$ and $f_y(x,y) = g'(y/x)$, it follows from (3) that an equation of the plane tangent to the graph of f at $(x_0, y_0, x_0 g(y_0/x_0))$ is

$$\left[g\left(\frac{y_0}{x_0}\right) - \frac{y_0}{x_0}g'\left(\frac{y_0}{x_0}\right)\right](x - x_0) + g'\left(\frac{y_0}{x_0}\right)(y - y_0) - \left(z - x_0 g\left(\frac{y_0}{x_0}\right)\right) = 0$$

or

$$\left[g\left(\frac{y_0}{x_0}\right) - \frac{y_0}{x_0}g'\left(\frac{y_0}{x_0}\right)\right]x + g'\left(\frac{y_0}{x_0}\right)y - z = x_0 g\left(\frac{y_0}{x_0}\right) - y_0 g'\left(\frac{y_0}{x_0}\right) + y_0 g'\left(\frac{y_0}{x_0}\right) - x_0 g\left(\frac{y_0}{x_0}\right) = 0.$$

Thus the origin lies on the tangent plane.

57. Let $f(x,y,z) = z^2 - x^2 - y^2$. Then $\operatorname{grad} f(x,y,z) = -2x\mathbf{i} - 2y\mathbf{j} + 2z\mathbf{k}$. By Definition 13.18 the vector $-2x_0\mathbf{i} - 2y_0\mathbf{j} + 2z_0\mathbf{k}$ is normal to the plane tangent to the level surface $f(x,y,z) = 0$ at any point (x_0, y_0, z_0) on the level surface. Thus if x_0, y_0, and z_0 are nonzero, then equations of the normal line are

$$\frac{x - x_0}{-2x_0} = \frac{y - y_0}{-2y_0} = \frac{z - z_0}{2z_0}$$

from which it follows that $(0, 0, 2z_0)$ lies on the normal line and hence that the normal line intersects the z axis. If $x_0 = 0$, and y_0 and z_0 are nonzero, then equations of the normal line are $x_0 = 0$ and $(y - y_0)/(-2y_0) = (z - z_0)/(2z_0)$, from which it again follows that the normal line intersects the z axis at $(0, 0, 2z_0)$. All other cases are handled analogously.

59. $2x - 2z\dfrac{\partial z}{\partial x} = 0$, so that $\dfrac{\partial z}{\partial x} = \dfrac{x}{z}$; $-2y - 2z\dfrac{\partial z}{\partial y} = 0$, so that $\dfrac{\partial z}{\partial y} = -\dfrac{y}{z}$. Thus

$$\left.\frac{\partial z}{\partial x}\right|_{(\sqrt{2},0,1)} = \sqrt{2} \quad \text{and} \quad \left.\frac{\partial z}{\partial y}\right|_{(\sqrt{2},0,1)} = 0$$

so that an equation of the plane tangent to the given level surface at $(\sqrt{2}, 0, 1)$ is $\sqrt{2}(x - \sqrt{2}) + 0(y - 0) - (z - 1) = 0$, or $\sqrt{2}x - z = 1$.

61. $\dfrac{1}{x}+\dfrac{1}{z}\dfrac{\partial z}{\partial x}=0$, so that $\dfrac{\partial z}{\partial x}=-\dfrac{z}{x}$; $\dfrac{1}{y}+\dfrac{1}{z}\dfrac{\partial z}{\partial y}=0$, so that $\dfrac{\partial z}{\partial y}=-\dfrac{z}{y}$. Thus

$$\left.\frac{\partial z}{\partial x}\right|_{(1,1,e)}=-e \quad\text{and}\quad \left.\frac{\partial z}{\partial y}\right|_{(1,1,e)}=-e$$

so that an equation of the plane tangent to the given level surface at $(1,1,e)$ is $-e(x-1)-e(y-1)-(z-e)=0$, or $ex+ey+z=3e$.

63. $\operatorname{grad} T(x,y,z)=(3x^2y+6xy^2z)\mathbf{i}+(x^3+6x^2yz)\mathbf{j}+3x^2y^2\mathbf{k}$, so that $\operatorname{grad} T(1,1,-1)=-3\mathbf{i}-5\mathbf{j}+3\mathbf{k}$. Let

$$\mathbf{u}=\frac{\operatorname{grad} T(1,1,-1)}{\|\operatorname{grad} T(1,1,-1)\|}=-\frac{3}{\sqrt{43}}\mathbf{i}-\frac{5}{\sqrt{43}}\mathbf{j}+\frac{3}{\sqrt{43}}\mathbf{k}.$$

Then

$$D_{\mathbf{u}}T(1,1,-1)=T_x(1,1,-1)\left(-\frac{3}{\sqrt{43}}\right)+T_y(1,1,-1)\left(-\frac{5}{\sqrt{43}}\right)+T_z(1,1,-1)\left(\frac{3}{\sqrt{43}}\right)$$

$$=\frac{9}{\sqrt{43}}+\frac{25}{\sqrt{43}}+\frac{9}{\sqrt{43}}=\sqrt{43}.$$

65. Let $V(x,y,z)=x^2-y^2-z$. Then the electric force on a positive unit charge at the origin is parallel to $\operatorname{grad} V(0,0,0)$. But $\operatorname{grad} V(x,y,z)=2x\mathbf{i}-2y\mathbf{j}-\mathbf{k}$, so that $\operatorname{grad} V(0,0,0)=-\mathbf{k}$. Thus the electric force is parallel to $-\mathbf{k}$ and hence is perpendicular to the xy plane.

13.7 Tangent Plane Approximations and Differentials

1. Let $x_0=3$, $y_0=4$, $h=0.01$, and $k=0.03$. Since

$$f_x(x,y)=\frac{x}{\sqrt{x^2+y^2}} \quad\text{and}\quad f_y(x,y)=\frac{y}{\sqrt{x^2+y^2}}$$

we have $f_x(3,4)=\frac{3}{5}$ and $f_y(3,4)=\frac{4}{5}$. Also $f(3,4)=5$. By (6),

$$f(3.01,4.03)\approx 5+\frac{3}{5}(0.01)+\frac{4}{5}(0.03)=5.03.$$

3. Let $x_0=0$, $y_0=1$, $h=-0.03$, and $k=-0.02$. Since $f_x(x,y)=2x/(x^2+y^2)$ and $f_y(x,y)=2y/(x^2+y^2)$, we have $f_x(0,1)=0$ and $f_y(0,1)=2$. Also $f(0,1)=0$. By (6),

$$f(-0.03,0.98)\approx 0+0(-0.03)+2(-0.02)=-0.04.$$

5. Let $x_0 = \pi$, $y_0 = 0.25$, $h = -0.01\pi$, and $k = -0.01$. Since $f_x(x,y) = y\sec^2 xy$ and $f_y(x,y) = x\sec^2 xy$, we have $f_x(\pi, 0.25) = \frac{1}{4}\sec^2(\pi/4) = \frac{1}{2}$ and $f_y(\pi, 0.25) = 2\pi$. Also $f(\pi, 0.25) = \tan(\pi/4) = 1$. By (6),

$$f(0.99\pi, 0.24) \approx 1 + \frac{1}{2}(-0.01\pi) + 2\pi(-0.01) = 1 - 0.025\pi \approx 0.9214601837.$$

7. Let $x_0 = 3$, $y_0 = 4$, $z_0 = 12$, $h = 0.01$, $k = 0.02$, and $l = -0.02$. Since

$$f_x(x,y,z) = \frac{x}{\sqrt{x^2+y^2+z^2}}, \quad f_y(x,y,z) = \frac{y}{\sqrt{x^2+y^2+z^2}}, \quad \text{and} \quad f_z(x,y,z) = \frac{z}{\sqrt{x^2+y^2+z^2}},$$

we have $f_x(3,4,12) = \frac{3}{13}$, $f_y(3,4,12) = \frac{4}{13}$, and $f_z(3,4,12) = \frac{12}{13}$. Also $f(3,4,12) = 13$. By (7),

$$f(3.01, 4.02, 11.98) \approx 13 + \frac{3}{13}(0.01) + \frac{4}{13}(0.02) + \frac{12}{13}(-0.02) = 12.99.$$

9. Let $f(x,y) = \sqrt[4]{x^3+y^3}$, $x_0 = 2$, $y_0 = 2$, $h = -0.1$, and $k = 0.1$. Since

$$f_x(x,y) = \frac{3x^2}{4(x^3+y^3)^{3/4}} \quad \text{and} \quad f_y(x,y,) = \frac{3y^2}{4(x^3+y^3)^{3/4}}$$

we have $f_x(2,2) = \frac{3}{8}$ and $f_y(2,2) = \frac{3}{8}$. Also $f(2,2) = 2$. By (6),

$$\sqrt[4]{(1.9)^3 + (2.1)^3} = f(1.9, 2.1) \approx 2 + \frac{3}{8}(-0.1) + \frac{3}{8}(0.1) = 2.$$

11. Let $f(x,y) = e^x \ln y$, $x_0 = 0$, $y_0 = 1$, $h = 0.1$, and $k = -0.1$. Since $f_x(x,y) = e^x \ln y$ and $f_y(x,y) = e^x/y$, we have $f_x(0,1) = 0$ and $f_y(0,1) = 1$. Also $f(0,1) = 0$. By (6),

$$e^{0.1}\ln 0.9 = f(0.1, 0.9) \approx 0 + 0(0.1) + 1(-0.1) = -0.1.$$

13. Since $\partial f/\partial x = 9x^2 - 2xy$ and $\partial f/\partial y = -x^2 + 1$, we have $df = (9x^2 - 2xy)\,dx + (-x^2+1)\,dy$.

15. Since $\partial f/\partial x = 2x$ and $\partial f/\partial y = 2y$, we have $df = 2x\,dx + 2y\,dy$.

17. Since $\partial f/\partial x = \tan y - y\csc^2 x$ and $\partial f/\partial y = x\sec^2 y + \cot x$, we have $df = (\tan y - y\csc^2 x)\,dx + (x\sec^2 y + \cot x)\,dy$.

19. Since

$$\frac{\partial f}{\partial x} = \frac{xz^2}{\sqrt{1+x^2+y^2}}, \quad \frac{\partial f}{\partial y} = \frac{yz^2}{\sqrt{1+x^2+y^2}}, \quad \text{and} \quad \frac{\partial f}{\partial z} = 2z\sqrt{1+x^2+y^2}$$

we have

$$df = \frac{xz^2}{\sqrt{1+x^2+y^2}}\,dx + \frac{yz^2}{\sqrt{1+x^2+y^2}}\,dy + 2z\sqrt{1+x^2+y^2}\,dz.$$

21. Since $\partial f/\partial x = e^{y^2-z^2}$, $\partial f/\partial y = 2yxe^{y^2-z^2}$, and $\partial f/\partial z = -2zxe^{y^2-z^2}$, we have

$$df = e^{y^2-z^2}\,dx + 2yxe^{y^2-z^2}\,dy - 2zxe^{y^2-z^2}\,dz.$$

23. Observe that

$$\frac{\partial R}{\partial R_1} = \frac{R_2(R_1+R_2) - R_1R_2}{(R_1+R_2)^2} = \frac{R_2^2}{(R_1+R_2)^2}$$

and

$$\frac{\partial R}{\partial R_2} = \frac{R_1(R_1+R_2) - R_1R_2}{(R_1+R_2)^2} = \frac{R_1^2}{(R_1+R_2)^2}$$

Therefore

$$\frac{\partial R}{\partial R_1}(2,6) = \frac{9}{16} \quad \text{and} \quad \frac{\partial R}{\partial R_2}(2,6) = \frac{1}{16}$$

so that by (6),

$$R(2+h, 6+k) - R(2,6) \approx \frac{\partial R}{\partial R_1}h + \frac{\partial R}{\partial R_2}k = \frac{9}{16}h + \frac{1}{16}k$$

If $|h| \le 0.01$ and $|k| \le 0.02$, it follows that

$$|R(2+h, 6+k) - R(2,6)| \le \frac{9}{16}(0.01) + \frac{1}{16}(0.02) = \frac{11}{1600} = 0.006875$$

Consequently the maximum error is approximately 0.006875 ohms.

13.8 Extreme Values

1. $f_x(x,y) = 2x - 6$ and $f_y(x,y) = 4y + 8$, so $f_x(x,y) = 0 = f_y(x,y)$ if $x = 3$ and $y = -2$. Thus $(3,-2)$ is a critical point. Next, $f_{xx}(x,y) = 2$, $f_{yy}(x,y) = 4$, and $f_{xy}(x,y) = 0$. Thus $D(3,-2) = f_{xx}(3,-2)f_{yy}(3,-2) - [f_{xy}(3,-2)]^2 = (2)(4) - 0^2 = 8$. Since $D(3,-2) > 0$ and $f_{xx}(3,-2) > 0$, f has a relative minimum value at $(3,-2)$.

3. $f_x(x,y) = 2x + 6y - 6$ and $f_y(x,y) = 6x + 4y + 10$, so $f_x(x,y) = 0 = f_y(x,y)$ if $2x + 6y - 6 = 0$ and $6x + 4y + 10 = 0$. Solving for x in the first equation and substituting for it in the second equation yields $6(3 - 3y) + 4y + 10 = 0$, or $y = 2$. Then $x = -3$, so $(-3,2)$ is a critical point. Next, $f_{xx}(x,y) = 2$, $f_{yy}(x,y) = 4$, and $f_{xy}(x,y) = 6$. Thus $D(-3,2) = f_{xx}(-3,2)f_{yy}(-3,2) - [f_{xy}(-3,2)]^2 = (2)(4) - 6^2 = -28$. Since $D(-3,2) < 0$, f has a saddle point at $(-3,2)$.

5. $k_x(x,y) = -2x - 2y + 6$ and $k_y(x,y) = -2x - 4y - 10$, so $k_x(x,y) = 0 = k_y(x,y)$ if $-2x - 2y + 6 = 0$ and $-2x - 4y - 10 = 0$. Solving for y in the first equation and substituting for it in the second equation yields $-2x - 4(-x+3) - 10 = 0$, or $x = 11$. Then $y = -8$, so $(11,-8)$ is a critical point. Next, $k_{xx}(x,y) = -2$, $k_{yy}(x,y) = -4$, and $k_{xy}(x,y) = -2$. Thus $D(11,-8) = k_{xx}(11,-8)k_{yy}(11,-8) - [k_{xy}(11,-8)]^2 = (-2)(-4) - (-2)^2 = 4$. Since $D(11,-8) > 0$ and $k_{xx}(11,-8) < 0$, k has a relative maximum value at $(11,-8)$.

7. $f_x(x,y) = 2xy - 2y = 2y(x-1)$ and $f_y(x,y) = x^2 - 2x + 4y - 15$, so $f_x(x,y) = 0 = f_y(x,y)$ if $2y(x-1) = 0$ and $x^2 - 2x + 4y - 15 = 0$. From the first equation we see that $y = 0$ or $x = 1$. If $y = 0$, it follows from the second equation that $x^2 - 2x - 15 = 0$, so that $x = -3$ or $x = 5$. Thus $(-3, 0)$ and $(5, 0)$ are critical points. If $x = 1$, it follows from the second equation that $1 - 2 + 4y - 15 = 0$, or $y = 4$. Thus $(1, 4)$ is also a critical point. Next, $f_{xx}(x,y) = 2y$, $f_{yy}(x,y) = 4$, and $f_{xy}(x,y) = 2x - 2$. Thus

$$D(-3,0) = f_{xx}(-3,0)f_{yy}(-3,0) - [f_{xy}(-3,0)]^2 = (0)(4) - (-8)^2 = -64,$$

$$D(5,0) = f_{xx}(5,0)f_{yy}(5,0) - [f_{xy}(5,0)]^2 = (0)(4) - 8^2 = -64,$$

and

$$D(1,4) = f_{xx}(1,4)f_{yy}(1,4) - [f_{xy}(1,4)]^2 = (8)(4) - 0^2 = 32.$$

Since $D(1,4) > 0$ and $f_{xx}(1,4) > 0$, f has a relative minimum value at $(1, 4)$. Since $D(-3,0) < 0$ and $D(5,0) < 0$, f has saddle points at $(-3, 0)$ and $(5, 0)$.

9. $f_x(x,y) = 6x - 3y^2$ and $f_y(x,y) = -6xy + 3y^2 + 6y = 3y(-2x + y + 2)$, so $f_x(x,y) = 0 = f_y(x,y)$ if $6x - 3y^2 = 0$ and $3y(-2x + y + 2) = 0$. From the first equation we see that $x = \frac{1}{2}y^2$, so from the second equation, $y = 0$ or $0 = -2x + y + 2 = -y^2 + y + 2$, so that $y = 0$, $y = -1$, or $y = 2$. The corresponding values of x are 0, $\frac{1}{2}$, and 2. Thus $(0,0)$, $(\frac{1}{2}, -1)$, and $(2,2)$ are critical points. Next, $f_{xx}(x,y) = 6$, $f_{yy}(x,y) = -6x + 6y + 6$, and $f_{xy}(x,y) = -6y$. Thus $D(0,0) = f_{xx}(0,0)f_{yy}(0,0) - [f_{xy}(0,0)]^2 = (6)(6) - 0^2 = 36$, $D(\frac{1}{2}, -1) = f_{xx}(\frac{1}{2}, -1)f_{yy}(\frac{1}{2}, -1) - [f_{xy}(\frac{1}{2}, -1)]^2 = (6)(-3) - 6^2 = -54$, and $D(2,2) = f_{xx}(2,2)f_{yy}(2,2) - [f_{xy}(2,2)]^2 = (6)(6) - (-12)^2 = -108$. Since $D(0,0) > 0$ and $f_{xx}(0,0) > 0$, f has a relative minimum value at $(0,0)$. Since $D(\frac{1}{2}, -1) < 0$ and $D(2,2) < 0$, f has saddle points at $(\frac{1}{2}, -1)$ and $(2, 2)$.

11. $f_x(x,y) = 4y + 4xy - y^2 = y(4 + 4x - y)$ and $f_y(x,y) = 4x + 2x^2 - 2xy = 2x(2 + x - y)$, so $f_x(x,y) = 0 = f_y(x,y)$ if $y(4 + 4x - y) = 0$ and $2x(2 + x - y) = 0$. From the first equation we see that either $y = 0$ or $y = 4 + 4x$. If $y = 0$, the second equation implies that $x = 0$ or $x = -2$. Thus $(0,0)$ and $(-2,0)$ are critical points. If $y = 4 + 4x$, the second equation implies that either $x = 0$ or $0 = 2 + x - y = 2 + x - (4 + 4x)$, so that $x = 0$ or $x = -\frac{2}{3}$. The corresponding values of y are 4 and $\frac{4}{3}$. Thus $(0, 4)$ and $(-\frac{2}{3}, \frac{4}{3})$ are critical points. Next, $f_{xx}(x,y) = 4y$, $f_{yy}(x,y) = -2x$, and $f_{xy}(x,y) = 4 + 4x - 2y$. Thus

$$D(0,0) = f_{xx}(0,0)f_{yy}(0,0) - [f_{xy}(0,0)]^2 = (0)(0) - 4^2 = -16$$

$$D(-2,0) = f_{xx}(-2,0)f_{yy}(-2,0) - [f_{xy}(-2,0)]^2 = (0)(4) - (-4)^2 = -16$$

$$D(0,4) = f_{xx}(0,4)f_{yy}(0,4) - [f_{xy}(0,4)]^2 = (16)(0) - (-4)^2 = -16$$

and

$$D\left(-\frac{2}{3}, \frac{4}{3}\right) = f_{xx}\left(-\frac{2}{3}, \frac{4}{3}\right) f_{yy}\left(-\frac{2}{3}, \frac{4}{3}\right) - \left[f_{xy}\left(-\frac{2}{3}, \frac{4}{3}\right)\right]^2 = \left(\frac{16}{3}\right)\left(\frac{4}{3}\right) - \left(-\frac{4}{3}\right)^2 = \frac{16}{3}.$$

Since $D(-\frac{2}{3}, \frac{4}{3}) > 0$ and $f_{xx}(-\frac{2}{3}, \frac{4}{3}) > 0$, f has a relative minimum value at $(-\frac{2}{3}, \frac{4}{3})$. Since $D(0,0) < 0$, $D(0,4) < 0$, and $D(-2,0) < 0$, f has saddle points at $(0,0)$, $(0,4)$, and $(-2,0)$.

13. $f_x(x,y) = 2x$, $f_y(x,y) = -2ye^{y^2}$, so $f_x(x,y) = 0 = f_y(x,y)$ if $x = 0$ and $y = 0$. Thus $(0,0)$ is a critical point. Next, $f_{xx}(x,y) = 2$, $f_{yy}(x,y) = -2e^{y^2} - 4y^2e^{y^2}$, and $f_{xy}(x,y) = 0$. Thus $D(0,0) = f_{xx}(0,0)f_{yy}(0,0) - [f_{xy}(0,0)]^2 = (2)(-2) - (0)^2 = -4$. Since $D(0,0) < 0$, it follows that f has a saddle point at $(0,0)$.

15. $k_x(x,y) = e^x \sin y$ and $k_y(x,y) = e^x \cos y$, so $k_x(x,y) = 0 = k_y(x,y)$ if $e^x \sin y = 0$ and $e^x \cos y = 0$, or $\sin y = 0$ and $\cos y = 0$, which is impossible. Thus k has no critical points.

17. $f_u(u,v) = 1$ if $u > 0$, $f_u(u,v) = -1$ if $u < 0$, and $f_u(u,v)$ does not exist if $u = 0$. Similarly, $f_v(u,v) = 1$ if $v > 0$, $f_v(u,v) = -1$ if $v < 0$, and $f_v(u,v)$ does not exist if $v = 0$. Thus any point of the form $(0,v)$ or $(u,0)$ is a critical point of f. Since either f_{uu} or f_{vv} does not exist at each such point, the Second Partials Test does not apply. Notice that $f(0,0) = 0$ and $f(u,v) \geq 0$ for all u and v, so that f has a (relative) minimum value at $(0,0)$ However, at any other critical point (u,v), either $f_u(u,v) \neq 0$ or $f_v(u,v) \neq 0$, so that f does not have a relative extreme value or saddle point at any critical point except $(0,0)$.

19. $f_x(x,y) = ye^{xy}$ and $f_y(x,y) = xe^{xy}$, so $f_x(x,y) = 0 = f_y(x,y)$ if $x = 0$ and $y = 0$. Thus $(0,0)$ is a critical point. Next, $f_{xx}(x,y) = y^2e^x$, $f_{yy}(x,y) = x^2e^{xy}$, and $f_{xy}(x,y) = e^{xy} + xye^{xy}$. Thus $D(0,0) = f_{xx}(0,0)f_{yy}(0,0) - [f_{xy}(0,0)]^2 = (0)(0) - 1^2 = -1$. Since $D(0,0) < 0$, f has a saddle point at $(0,0)$.

21. $f_x(x,y) = \cos x$ and $f_y(x,y) = \cos y$, so $f_x(x,y) = 0 = f_y(x,y)$ if $\cos x = 0$ and $\cos y = 0$. Thus f has a critical point at any point of the form $(\pi/2 + m\pi, \pi/2 + n\pi)$, where m and n are integers. Next, $f_{xx}(x,y) = -\sin x$, $f_{yy}(x,y) = -\sin y$, and $f_{xy}(x,y) = 0$. Thus

$$\begin{aligned} D(\pi/2+m\pi, \pi/2+n\pi) &= f_{xx}(\pi/2+m\pi, \pi/2+n\pi)f_{yy}(\pi/2+m\pi, \pi/2+n\pi) \\ &\quad -[f_{xy}(\pi/2+m\pi, \pi/2+n\pi)]^2 \\ &= (-\sin(\pi/2+m\pi))(-\sin(\pi/2+m\pi)) - 0^2 \\ &= \sin(\pi/2+m\pi)\sin(\pi/2+m\pi). \end{aligned}$$

Since $D(\pi/2 + m\pi, \pi/2 + n\pi) = 1 > 0$ and $f_{xx}(\pi/2 + m\pi, \pi/2 + n\pi) = -1 < 0$ if m and n are both even, f has a relative maximum value at the corresponding point $(\pi/2 + m\pi, \pi/2 + n\pi)$. Since $D(\pi/2 + m\pi, \pi/2 + n\pi) = 1$ and $f_{xx}(\pi/2 + m\pi, \pi/2 + n\pi) = 1 > 0$ if m and n are both odd, f has a relative minimum value at the corresponding point $(\pi/2+m\pi, \pi/2+n\pi)$. Since $D(\pi/2+m\pi, \pi/2+n\pi) = -1 < 0$ if either m or n is odd and the other is even, f has a saddle point at the corresponding point $(\pi/2 + m\pi, \pi/2 + n\pi)$.

23. $f_x(x,y) = 2a(y + ax + b)$ and $f_y(x,y) = 2(y + ax + b)$, so $f_x(x,y) = 0 = f_y(x,y)$ if $y + ax + b = 0$, or $y = -(ax + b)$. Thus any point of the form $(x, -(ax + b))$ is a critical point. Next, $f_{xx}(x,y) = 2a^2$, $f_{yy}(x,y) = 2$, and $f_{xy}(x,y) = 2a$. Thus

$$D(x, -(ax+b)) = f_{xx}(x, -(ax+b))f_{yy}(x, -(ax+b)) - [f_{xy}(x, -(ax+b))]^2 = (2a^2)(2) - (2a)^2 = 0,$$

so the Second Partials Test yields no conclusion. However, $f(x, -(ax+b)) = 0$ and $f(x,y) \geq 0$ for all x and y. Thus f has a (relative) minimum value at any point of the form $(x, -(ax+b))$, that is, at any point on the line $y = -(ax+b) = -ax-b$.

25. $f_x(x,y) = 2x$ and $f_y(x,y) = -2y$, so $f_x(x,y) = 0 = f_y(x,y)$ if $x = 0$ and $y = 0$. Thus $(0,0)$ is a critical point in R. On the boundary $x^2+y^2 = 1$ of R we have $y^2 = 1-x^2$, so that $f(x,y) = x^2-(1-x^2) = 2x^2-1$ for $-1 \leq x \leq 1$. The maximum value of $2x^2-1$ on $[-1,1]$ is 1 and occurs for $x = 1$ and $x = -1$, and the minimum value is -1 and occurs for $x = 0$. Since $f(0,0) = 0$, it follows that the maximum value of f on R is $f(1,0) = f(-1,0) = 1$, and the minimum value is $f(0,1) = f(0,-1) = -1$.

27. $f_x(x,y) = 2\cos x$ and $f_y(x,y) = -3\sin y$, so $f_x(x,y) = 0 = f_y(x,y)$ for (x,y) in R if $x = \pi/2$ and $y = 0$. Thus $(\pi/2, 0)$ is a critical point of f in R. From the figure we observe that for (x,y) on ℓ_1 we have $y = -\pi/2$ and $\cos(-\pi/2) = 0$, so that $f(x,y) = 2\sin x$. Thus on ℓ_1 the maximum value of f is $f(\pi/2, -\pi/2) = 2\sin \pi/2 = 2$ and the minimum value is $f(0, -\pi/2) = 2\sin 0 = 0$. The same extreme values are obtained for ℓ_3. For (x,y) on ℓ_2 we have $x = \pi$, so that $f(x,y) = 3\cos y$. Thus on ℓ_2 the maximum value of f is $f(\pi, 0) = 3\cos 0 = 3$ and the minimum value is $f(\pi, -\pi/2) = 3\cos(-\pi/2) = 0$. The same extreme values are obtained for ℓ_4. Since $f(\pi/2, 0) = 2\sin \pi/2 + 3\cos 0 = 2+3 = 5$, the maximum value of f on R is $f(\pi/2, 0) = 5$ and the minimum value is 0, occurring at $(\pm\pi, \pm\pi/2)$.

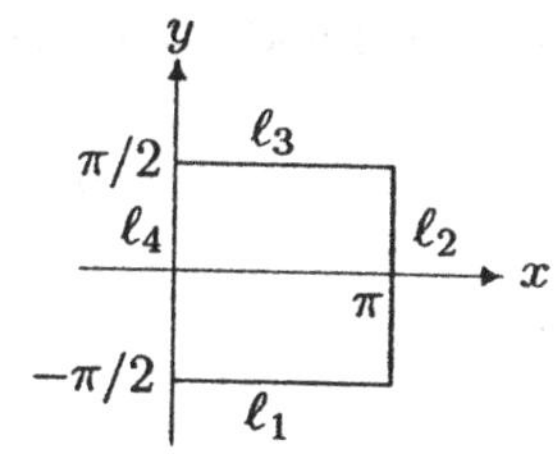

29. Let $P = xyz$ for any nonnegative numbers x, y, and z satisfying $x+y+z = 48$. Then $z = 48-x-y$, so $P = xy(48-x-y)$. We seek the maximum value of P on the triangular region R consisting of all (x,y) for which $x \geq 0$, $y \geq 0$, and $x+y \leq 48$. Such a maximum value exists by the Maximum-Minimum Theorem. Since $P = 0$ if $x = 0$, $y = 0$, or $x+y = 48$, the maximum value of P on R does not occur on the boundary of R and hence must occur at a critical point in the interior of R. $\partial P/\partial x = y(48-x-y) - xy = y(48-2x-y)$ and $\partial P/\partial y = x(48-x-y) - xy = x(48-x-2y)$, so $\partial P/\partial x = 0 = \partial P/\partial y$ if $y(48-2x-y) = 0$ and $x(48-x-2y) = 0$. From the first equation we see that $y = 0$ or $48-2x-y = 0$, and from the second equation we see that $x = 0$ or $48-x-2y = 0$. Since $x \neq 0$ and $y \neq 0$ at each interior point of R, it follows that a point (x,y) in the interior of R is a critical point of P only if $48-2x-y = 0$ and $48-x-2y = 0$. Solving for y in the first of these equations, we obtain $y = 48-2x$. Substituting for y in the second equation gives us $48-x-2(48-2x) = 0$, so that $x = 16$. Then $y = 16$, so $(16,16)$ is the only critical point of P in R, and the corresponding value of z is 16 also. The maximum product must occur at a critical point, so it is $P(16,16) = 4096$.

31. Set up a coordinate system with the origin at the center of the sphere and with the coordinate planes parallel to the sides of the box. If r is the radius of the sphere, then an equation of the sphere is $x^2+y^2+z^2 = r^2$. If (x,y,z) is the vertex of the box with $x \geq 0$, $y \geq 0$, and $z \geq 0$, then $z = \sqrt{r^2-x^2-y^2}$, and the volume V of the box is given by $V = xyz = xy\sqrt{r^2-x^2-y^2}$. We seek

the maximum value of V on the set R of points (x, y) satisfying $x \geq 0$, $y \geq 0$, and $x^2 + y^2 \leq r^2$. Such a maximum value exists by the Maximum-Minimum Theorem. Since $V = 0$ if $x = 0, y = 0$, or $x^2 + y^2 = r^2$, V does not have its maximum value on R at any boundary point of R. Thus the maximum value of V on R occurs at a critical point in the interior of R. Now

$$\frac{\partial V}{\partial x} = y\sqrt{r^2 - x^2 - y^2} - \frac{x^2 y}{\sqrt{r^2 - x^2 - y^2}} \quad \text{and} \quad \frac{\partial V}{\partial y} = x\sqrt{r^2 - x^2 - y^2} - \frac{xy^2}{\sqrt{r^2 - x^2 - y^2}}$$

so $\partial V/\partial x = 0 = \partial V/\partial y$ if

$$y\sqrt{r^2 - x^2 - y^2} - \frac{x^2 y}{\sqrt{r^2 - x^2 - y^2}} = 0 \quad \text{and} \quad x\sqrt{r^2 - x^2 - y^2} - \frac{xy^2}{\sqrt{r^2 - x^2 - y^2}} = 0.$$

Since $x \neq 0$ and $y \neq 0$ at any interior point of R, we conclude from these equations that

$$0 = \sqrt{r^2 - x^2 - y^2} - \frac{x^2}{\sqrt{r^2 - x^2 - y^2}} = \frac{r^2 - x^2 - y^2 - x^2}{\sqrt{r^2 - x^2 - y^2}}$$

and

$$0 = \sqrt{r^2 - x^2 - y^2} - \frac{y^2}{\sqrt{r^2 - x^2 - y^2}} = \frac{r^2 - x^2 - y^2 - y^2}{\sqrt{r^2 - x^2 - y^2}}.$$

Thus $r^2 - x^2 - y^2 - x^2 = 0 = r^2 - x^2 - y^2 - y^2$, or $x^2 = y^2$, or $x = y$. Therefore $0 = r^2 - x^2 - y^2 - x^2 = r^2 - 3x^2$, so that $x = r/\sqrt{3}$. Then $y = r/\sqrt{3}$, so the only critical point of V in R is $(r/\sqrt{3}, r/\sqrt{3})$. The maximum value must occur at a critical point, so it occurs for $x = r/\sqrt{3} = y$. This means that $z = \sqrt{r^2 - x^2 - y^2} = r/\sqrt{3}$ and thus the box is a cube.

33. The distance from a point to the origin will be minimum if and only if the square of the distance from the point to the origin is minimum. To simplify the calculations, we will minimize the square of the distance. Thus for points (x, y, z) satisfying $x + y + z = 48$, we seek the minimum value of $x^2 + y^2 + z^2 = x^2 + y^2 + (48 - x - y)^2$. Let $f(x, y) = x^2 + y^2 + (48 - x - y)^2$. Since $f(0, 0) = (48)^2$ and $f(x, y) > (48)^2$ if $x^2 + y^2 > (48)^2$, and since the Maximum-Minimum Theorem implies that f has a minimum value on the disk $x^2 + y^2 \leq (48)^2$, it follows that f has a minimum value. Since the domain of f is the entire xy plane, the minimum value of f occurs at a critical point. Now $f_x(x, y) = 2x - 2(48 - x - y)$ and $f_y(x, y) = 2y - 2(48 - x - y)$. Thus $f_x(x, y) = 0 = f_y(x, y)$ if $2x - 2(48 - x - y) = 0$ and $2y - 2(48 - x - y) = 0$, so $x = y$. Then $0 = 2x - 2(48 - x - y) = 2x - 2(48 - 2x)$, so that $x = 16$. Then $y = 16$, so the only critical point of f is $(16, 16)$. The minimum value of f must occur at a critical point, so it occurs at $(16, 16)$. Since $z = 16$ if $x = 16 = y$, $(16, 16, 16)$ is the point the sum of whose coordinates is 48 and whose distance to the origin is minimum.

35. a. $f_x(x, y) = e^{-y^2}(6x^2 - 6x) + e^{-y}(6x^2 - 6x) = (e^{-y^2} + e^{-y})(6x^2 - 6x)$ and $f_y(x, y) = -2ye^{-y^2}(2x^3 - 3x^2 + 1) - e^{-y}(2x^3 - 3x^2)$. To find the critical points of f, suppose that $f_x(x, y) = 0$ and $f_y(x, y) = 0$. Since $f_x(x, y) = 0$ and $e^{-y^2} + e^{-y} > 0$, it follows that $6x^2 - 6x = 0$, that is, $x = 0$ or $x = 1$. If $x = 0$, then $0 = f_y(x, y) = -2ye^{-y^2}$, so that $y = 0$. If $x = 1$, then $0 = f_y(x, y) = e^{-y}$, which is impossible. Therefore the only critical point is $(0, 0)$.

b. $f_{xx}(x,y) = (e^{-y^2}+e^{-y})(12x-6)$, $f_{yy}(x,y) = (-2e^{-y^2}+4y^2e^{-y^2})(2x^3-3x^2+1)+e^{-y}(2x^3-3x^2)$, and $f_{xy}(x,y) = (-2ye^{-y^2} - e^{-y})(6x^2-6x)$. Thus $D(0,0) = f_{xx}(0,0)f_{yy}(0,0) - [f_{xy}(0,0)]^2 = (-12)(-2) - 0^2 = 24 > 0$. Since $f_{xx}(0,0) = -12 < 0$, f has a relative maximum value at $(0,0)$.

c. Since $f(2,y) = 5e^{-y^2} + 4e^{-y} > 4e^{-y}$ and $\lim_{y\to-\infty} 4e^{-y} = \infty$, it follows that $\lim_{y\to-\infty} f(2,y) = \infty$. Thus f has no maximum value.

37. $f(m,b) = [0-(0+b)]^2+[-1-(m+b)]^2+[1-(-2m+b)]^2 = b^2+1+2(m+b)+(m+b)^2+1-2(-2m+b)+(-2m+b)^2 = 2+6m+5m^2-2mb+3b^2$. $f_m(m,b) = 6+10m-2b$ and $f_b(m,b) = -2m+6b$.

Thus $f_m(m,b) = 0 = f_b(m,b)$ if $6+10m-2b = 0$ and $-2m+6b = 0$. Solving for m in the second equation and substituting in the first equation, we obtain $0 = 6+10m-2b = 6+10(3b)-2b$, so that $b = -\frac{3}{14}$. Then $m = -\frac{9}{14}$. Thus an equation of the line of best fit is $y = -\frac{9}{14}x - \frac{3}{14}$.

39. If the dimensions of such a box are x, y, and z, then $4x+4y+4z = l$, so that $z = \frac{1}{4}l - x - y$, and the volume V of the box is given by $V = xyz = xy(\frac{1}{4}l - x - y)$. We seek the maximum value of V on the triangular region R consisting of all (x,y) such that $x \geq 0$, $y \geq 0$, and $x + y \leq \frac{1}{4}l$. Such a maximum value exists by the Maximum-Minimum Theorem. If $x = 0$, $y = 0$, or $x+y = \frac{1}{4}l$, then $V = 0$. Thus V assumes its maximum value at a critical point in the interior of R. Now $\partial V/\partial x = y(\frac{1}{4}l - x - y) - xy = y(\frac{1}{4}l - 2x - y)$ and $\partial V/\partial y = x(\frac{1}{4}l - x - y) - xy = x(\frac{1}{4}l - x - 2y)$. Thus $\partial V/\partial x = 0 = \partial V/\partial y$ if $y(\frac{1}{4}l - 2x - y) = 0$ and $x(\frac{1}{4}l - x - 2y) = 0$. Since $x \neq 0$ and $y \neq 0$ at any interior point of R, it follows that $\frac{1}{4}l - 2x - y = 0 = \frac{1}{4}l - x - 2y$, so that $x = y$. Then $0 = \frac{1}{4}l - 2x - y = \frac{1}{4}l - 3x$, so that $x = \frac{1}{12}l$. Then $y = \frac{1}{12}l$, so the only critical point of V in R is $(\frac{1}{12}l, \frac{1}{12}l)$. The corresponding value of V is given by $V = xy(\frac{1}{4}l - x - y) = (\frac{1}{12}l)(\frac{1}{12}l)(\frac{1}{4}l - \frac{1}{12}l - \frac{1}{12}l) = l^3/1728$. Thus the maximum volume is $l^3/1728$.

41. The distance the ranger travels in the thicket is $\sqrt{1+x^2}$, the distance in the marshland is $\sqrt{\frac{9}{4}+y^2}$ and the distance along the road is $10 - x - y$. Thus the time T of the walk is given by

$$T = \frac{\sqrt{1+x^2}}{3} + \frac{\sqrt{\frac{9}{4}+y^2}}{4} + \frac{10-x-y}{5}.$$

We seek the minimum value of T on the triangular region R consisting of all (x,y) satisfying $x \geq 0$, $y \geq 0$, and $x + y \leq 10$. Such a minimum exists by the Maximum-Minimum Theorem. First we find the critical points of T in the interior of R. Now

$$\frac{\partial T}{\partial x} = \frac{x}{3\sqrt{1+x^2}} - \frac{1}{5} \quad \text{and} \quad \frac{\partial T}{\partial y} = \frac{y}{4\sqrt{\frac{9}{4}+y^2}} - \frac{1}{5}.$$

Thus $\partial T/\partial x = 0 = \partial T/\partial y$ if

$$\frac{x}{3\sqrt{1+x^2}} - \frac{1}{5} = 0 \quad \text{and} \quad \frac{y}{4\sqrt{\frac{9}{4}+y^2}} - \frac{1}{5} = 0,$$

so that $x^2 = \frac{9}{25}(1+x^2)$ and $y^2 = \frac{16}{25}(\frac{9}{4}+y^2)$, or $x = \frac{3}{4}$ and $y = 2$. Thus the only critical point is $(\frac{3}{4}, 2)$, and

$$T\left(\frac{3}{4}, 2\right) = \frac{\sqrt{1+\frac{9}{16}}}{3} + \frac{\sqrt{\frac{9}{4}+4}}{4} + \frac{10-\frac{3}{4}-2}{5} = \frac{5}{12} + \frac{5}{8} + \frac{29}{20} = \frac{299}{120} < 2.5.$$

Now we find the minimum value of T on the boundary of R, which is composed of three line segments ℓ_1, ℓ_2, and ℓ_3 (see the figure). On ℓ_1, $x = 0$ and $0 \le y \le 10$. Let

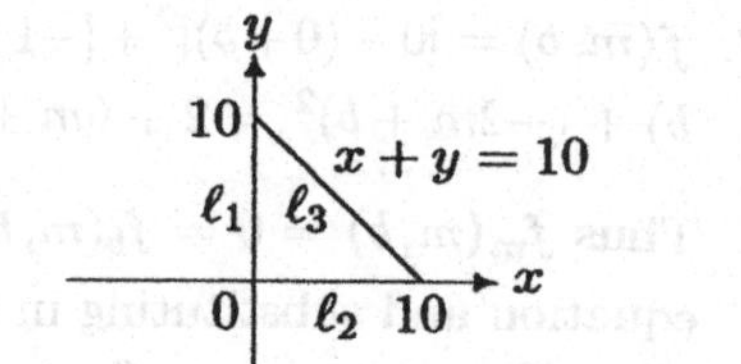

$$f(y) = T(0, y) = \frac{1}{3} + \frac{\sqrt{\frac{9}{4}+y^2}}{4} + 2 - \frac{1}{5}y$$

for $0 \le y \le 10$. Then $f'(y) = y/(4\sqrt{\frac{9}{4}+y^2}) - \frac{1}{5}$, and $f'(y) = 0$ if $y/(4\sqrt{\frac{9}{4}+y^2}) = \frac{1}{5}$, which means that $25y^2 = 16(\frac{9}{4}+y^2)$, so that $y = 2$. Thus the minimum value of f on $[0, 10]$ is $f(0)$, $f(2)$, or $f(10)$. Since $f(0) = \frac{65}{24}$, $f(2) = \frac{307}{120}$, and $f(10) = \frac{1}{3} + \sqrt{409}/8$, it follows that $T(0, y) = f(y) \ge 2.5$ for $0 \le y \le 10$. On ℓ_2, $y = 0$ and $0 \le x \le 10$. Let $g(x) = T(x, 0) = \sqrt{1+x^2}/3 + \frac{3}{8} + 2 - \frac{1}{5}x$ for $0 \le x \le 10$. Then $g'(x) = x/(3\sqrt{1+x^2}) - \frac{1}{5}$, and $g'(x) = 0$ if $x/(3\sqrt{1+x^2}) = \frac{1}{5}$, which means that $25x^2 = 9(1+x^2)$, so that $x = \frac{3}{4}$. Thus the minimum value of g on $[0, 10]$ is $g(0)$, $g(\frac{3}{4})$, or $g(10)$. Since $g(0) = \frac{65}{24}$, $g(\frac{3}{4}) = \frac{317}{120}$, and $g(10) = \sqrt{101}/3 + \frac{3}{8}$, it follows that $T(x, 0) = g(x) \ge 2.5$ for $0 \le x \le 10$. On ℓ_3, $x + y = 10$, and the ranger does not walk along the road. In that case the total time $T(x, 10-x)$ of walk is greater than it would be if the ranger walked at 4 kilometers an hour in a straight line to the ranger station, and that time is $\sqrt{(2.5)^2 + 100}/4 > 2.5$. Thus $T(x, 10-x) > 2.5$ for $0 \le x \le 10$. Consequently $T(x, y) > 2.5$ for (x, y) in ℓ_1, ℓ_2, or ℓ_3, and therefore the minimum time is attained if $x = \frac{3}{4}$ (kilometers) and $y = 2$ (kilometers).

13.9 Lagrange Multipliers

1. Let $g(x, y) = x^2 + y^2$, so the constraint is $g(x, y) = x^2 + y^2 = 4$. Next, $\operatorname{grad} f(x, y) = \mathbf{i} + 2y\mathbf{j}$ and $\operatorname{grad} g = 2x\mathbf{i} + 2y\mathbf{j}$, so that $\operatorname{grad} f(x, y) = \lambda \operatorname{grad} g(x, y)$ if

$$1 = 2x\lambda \quad \text{and} \quad 2y = 2y\lambda.$$

By the second equation, $y = 0$ or $\lambda = 1$. By the constraint,

$$\text{if} \quad y = 0, \quad \text{then } x^2 + 0^2 = 4, \quad \text{so } x = \pm 2$$

$$\text{if} \quad \lambda = 1, \quad \text{then } 1 = 2x, \quad \text{so } x = \frac{1}{2} \quad \text{and} \quad y = \pm\sqrt{4 - \frac{1}{4}} = \pm\frac{\sqrt{15}}{2}.$$

The possible extreme values of f are $f(2, 0) = 2$, $f(-2, 0) = -2$, $f(\frac{1}{2}, \sqrt{15}/2) = \frac{17}{4}$, and $f(\frac{1}{2}, -\sqrt{15}/2) = \frac{17}{4}$. The maximum value is $\frac{17}{4}$ and the minimum value is -2.

3. Let $g(x,y) = x^2 + y^2$, so the constraint is $g(x,y) = x^2 + y^2 = 1$. Next, $\operatorname{grad} f(x,y) = 3x^2\mathbf{i} + 6y^2\mathbf{j}$ and $\operatorname{grad} g(x,y) = 2x\mathbf{i} + 2y\mathbf{j}$, so that $\operatorname{grad} f(x,y) = \lambda \operatorname{grad} g(x,y)$ if

$$3x^2 = 2x\lambda \quad \text{and} \quad 6y^2 = 2y\lambda.$$

By the first equation, either $x = 0$ or $x = \frac{2}{3}\lambda$, and by the second equation, either $y = 0$ or $y = \frac{1}{3}\lambda$. Now if $x = 0$, then by the constraint, $y = \pm 1$, and if $y = 0$, then by the constraint, $x = \pm 1$. Otherwise $x = \frac{2}{3}\lambda$ and $y = \frac{1}{3}\lambda$, so the constraint becomes $(\frac{2}{3}\lambda)^2 + (\frac{1}{3}\lambda)^2 = 1$. Then $\lambda = \pm 3/\sqrt{5}$, so that $x = \pm 2/\sqrt{5}$ and $y = \pm 1/\sqrt{5}$. The possible extreme values are $f(0,1) = 2$, $f(0,-1) = -2$, $f(1,0) = 1$, and $f(-1,0) = -1$,

$$f\left(\frac{2}{\sqrt{5}}, \frac{1}{\sqrt{5}}\right) = \frac{2}{\sqrt{5}}, \quad f\left(\frac{-2}{\sqrt{5}}, \frac{-1}{\sqrt{5}}\right) = \frac{-2}{\sqrt{5}}, \quad f\left(\frac{2}{\sqrt{5}}, \frac{-1}{\sqrt{5}}\right) = \frac{6}{5\sqrt{5}}, \quad \text{and} \quad f\left(\frac{-2}{\sqrt{5}}, \frac{1}{\sqrt{5}}\right) = \frac{-6}{5\sqrt{5}}.$$

The maximum value is 2 and the minimum value is -2.

5. Let $g(x,y,z) = x^2 + y^2 + 4z^2$, so the constraint is $g(x,y,z) = x^2 + y^2 + 4z^2 = 6$. Next, $\operatorname{grad} f(x,y,z) = yz\mathbf{i} + xz\mathbf{j} + xy\mathbf{k}$ and $\operatorname{grad} g(x,y,z) = 2x\mathbf{i} + 2y\mathbf{j} + 8z\mathbf{k}$, so that $\operatorname{grad} f(x,y,z) = \lambda \operatorname{grad} g(x,y,z)$ if

$$yz = 2x\lambda, \quad xz = 2y\lambda, \quad \text{and} \quad xy = 8z\lambda.$$

If $x = 0$ or $y = 0$ or $z = 0$ or $\lambda = 0$, then $f(x,y,z) = 0$. Assume henceforth that x, y, z and λ are nonzero. Then

$$\lambda = \frac{yz}{2x} = \frac{xz}{2y} = \frac{xy}{8z}$$

so that $x^2 = y^2$ and $y^2 = 4z^2$. By the constraint, $y^2 + y^2 + y^2 = 6$, so $y = \pm\sqrt{2}$. Thus $x = \pm\sqrt{2}$ and $z = \pm\sqrt{2}/2$. Since $f(\pm\sqrt{2}, \pm\sqrt{2}, \pm\sqrt{2}/2) = \pm\sqrt{2}$, the maximum value of f is $\sqrt{2}$ and the minimum value of f is $-\sqrt{2}$.

7. Let $g(x,y) = 2x^2 + \frac{3}{2}y^2$, so the constraint is $g(x,y) = 2x^2 + \frac{3}{2}y^2 = \frac{3}{2}$. Next, $\operatorname{grad} f(x,y) = 8x\mathbf{i} + (3y^2 + 3)\mathbf{j}$, and $\operatorname{grad} g(x,y) = 4x\mathbf{i} + 3y\mathbf{j}$, so that $\operatorname{grad} f(x,y) = \lambda \operatorname{grad} g(x,y)$ if

$$8x = 4x\lambda \quad \text{and} \quad 3y^2 + 3 = 3y\lambda.$$

From the first equation, either $x = 0$ or $\lambda = 2$. If $x = 0$, then from the constraint, $2 \cdot 0^2 + \frac{3}{2}y^2 = \frac{3}{2}$, so that $y = 1$ or $y = -1$. If $\lambda = 2$, then from the second equation, $3y^2 + 3 = 6y$, so $3y^2 - 6y + 3 = 0$, and thus $y = 1$. But by the constraint, if $y = 1$, then $2x^2 + \frac{3}{2} \cdot 1^2 = \frac{3}{2}$, so $x = 0$. Since $f(0,1) = 11$ and $f(0,-1) = 3$, the minimum value of f is 3.

9. Let $g(x,y,z) = x + y + z$, so the constraint becomes $g(x,y,z) = x + y + z = \frac{11}{12}$. Next, $\operatorname{grad} f(x,y,z) = 4x^3\mathbf{i} + 32y^3\mathbf{j} + 108z^3\mathbf{k}$ and $\operatorname{grad} g(x,y,z) = \mathbf{i} + \mathbf{j} + \mathbf{k}$, so that $\operatorname{grad} f(x,y,z) = \lambda \operatorname{grad} g(x,y,z)$ if

$$4x^3 = \lambda, \quad 32y^3 = \lambda, \quad \text{and} \quad 108z^3 = \lambda.$$

Thus $4x^3 = 32y^3$, so that $y = \frac{1}{2}x$, and $4x^3 = 108z^3$, so that $z = \frac{1}{3}x$. The constraint becomes $x + \frac{1}{2}x + \frac{1}{3}x = \frac{11}{12}$, which means that $x = \frac{1}{2}$, and thus $y = \frac{1}{4}$ and $z = \frac{1}{6}$. Then $f(\frac{1}{2}, \frac{1}{4}, \frac{1}{6}) = (\frac{1}{2})^4 + 8(\frac{1}{4})^4 + 27(\frac{1}{6})^4 = \frac{11}{96}$ is the minimum value of f.

11. First we use Lagrange multipliers to find the possible extreme values of f on the circle $x^2+y^2=4$. Let $g(x,y)=x^2+y^2$, so the constraint becomes $g(x,y)=x^2+y^2=4$. Since $\text{grad}\, f(x,y)=4x\mathbf{i}+(2y+2)\mathbf{j}$ and $\text{grad}\, g(x,y)=2x\mathbf{i}+2y\mathbf{j}$, it follows that $\text{grad}\, f(x,y)=\lambda\, \text{grad}\, g(x,y)$ if

$$4x = 2x\lambda \quad \text{and} \quad 2y+2=2y\lambda.$$

From the first equation, $x=0$ or $\lambda=2$. If $x=0$, then from the constraint, $0^2+y^2=4$, so that $y=\pm 2$. If $\lambda=2$, then from the second equation, $2y+2=4y$, so that $y=1$, and then the constraint implies that $x^2+1=4$, or $x=\pm\sqrt{3}$. The possible extreme values of f on the circle $x^2+y^2=4$ are $f(0,2)=5$, $f(0,-2)=-3$, and $f(\pm\sqrt{3},1)=6$. For the interior $x^2+y^2<4$ of the disk, we find that

$$f_x(x,y)=4x \quad \text{and} \quad f_y(x,y)=2y+2$$

so that $f_x(x,y)=0=f_y(x,y)$ only if $x=0$ and $y=-1$. But $f(0,-1)=-4$. Consequently the maximum value and minimum value of f on the disk $x^2+y^2\le 4$ are 6 and -4, respectively.

13. First we use Lagrange multipliers to find the possible extreme values of f on the ellipse $2x^2+y^2=4$. Let $g(x,y)=2x^2+y^2$, so the constraint is $g(x,y)=2x^2+y^2=4$. Since $\text{grad}\, f(x,y)=y\mathbf{i}+x\mathbf{j}$ and $\text{grad}\, g(x,y)=4x\mathbf{i}+2y\mathbf{j}$, it follows that $\text{grad}\, f(x,y)=\lambda\, \text{grad}\, g(x,y)$ if

$$y=4x\lambda \quad \text{and} \quad x=2y\lambda.$$

Because of the constraint, $y\ne 0$, since if $y=0$ then the second equation would then imply that $x=0$ also. Thus $\lambda=x/2y$, so that the first equation becomes $y=4x(x/2y)$, and therefore $y^2=2x^2$. The constraint then becomes $2x^2+2x^2=4$, so $x=\pm 1$ and hence $y=\pm\sqrt{2}$. The possible extreme values of f on the ellipse are $f(1,\sqrt{2})=\sqrt{2}$, $f(1,-\sqrt{2})=-\sqrt{2}$, $f(-1,\sqrt{2})=-\sqrt{2}$, and $f(-1,-\sqrt{2})=\sqrt{2}$. For the interior $2x^2+y^2<4$ of the ellipse, we find that

$$f_x(x,y)=y \quad \text{and} \quad f_y(x,y)=x$$

so that $f_x(x,y)=0=f_y(x,y)$ only if $x=0$ and $y=0$. But $f(0,0)=0$. Consequently the maximum and minimum values of f on $2x^2+y^2\le 4$ are $\sqrt{2}$ and $-\sqrt{2}$, respectively.

15. The distance from (x,y,z) to the origin is minimum if and only if the square of the distance is minimum, so let $f(x,y,z)=x^2+y^2+z^2$. Also let $g(x,y,z)=x^2-yz$, so the constraint becomes $g(x,y,z)=x^2-yz=1$. Since $\text{grad}\, f(x,y,z)=2x\mathbf{i}+2y\mathbf{j}+2z\mathbf{k}$ and $\text{grad}\, g(x,y,z)=2x\mathbf{i}-z\mathbf{j}-y\mathbf{k}$, it follows that $\text{grad}\, f(x,y,z)=\lambda\, \text{grad}\, g(x,y,z)$ if

$$2x=2x\lambda, \quad 2y=-z\lambda, \quad \text{and} \quad 2z=-y\lambda.$$

From the first equation, $x=0$ or $\lambda=1$. If $x=0$, the constraint becomes $-yz=1$, or $z=-1/y$. Then the second equation becomes $2y=\lambda/y$, so $\lambda=2y^2$. This means that the third equation becomes $2(-1/y)=-y(2y^2)$, so $1=y^4$, and thus $y=\pm 1$, and hence $z=\mp 1$. If $x\ne 0$, then $\lambda=1$, so the third equation becomes $2z=-y$, and the second equation then becomes $2(-2z)=-z$, which means that $z=0$, and thus $y=0$ and $x=\pm 1$. The possible minimum values of f are $f(0,1,-1)=2$, $f(0,-1,1)=2$, $f(1,0,0)=1$, and $f(-1,0,0)=1$. Consequently the points on $x^2-yz=1$ closest to the origin are $(1,0,0)$ and $(-1,0,0)$.

17. Let one vertex of the triangle be (x, y), with $y > 0$. If $f(x, y) =$ the area of the triangle, then $f(x, y) = \frac{1}{2}(x+2)2y = (x+2)y$. Also, let $g(x, y) = \frac{1}{4}x^2 + y^2$, so the constraint becomes $g(x, y) = 1$. Since $\text{grad}\, f(x, y) = y\mathbf{i} + (x+2)\mathbf{j}$ and $\text{grad}\, g(x, y) = \frac{1}{2}x\mathbf{i} + 2y\mathbf{j}$, it follows that $\text{grad}\, f(x, y) = \lambda \text{grad}\, g(x, y)$ if

$$y = \frac{1}{2}\lambda x \quad \text{and} \quad x + 2 = 2\lambda y.$$

Notice that if $x = 0$, then by the first equation $y = 0$, which cannot be by the second equation. Thus $x \neq 0$. Thus we can solve for λ in the equations

$$\frac{2y}{x} = \lambda = \frac{x+2}{2y}.$$

Solving for y^2, we obtain $y^2 = \frac{1}{4}x^2 + \frac{1}{2}x$. Substituting for y^2 in the constraint, we find that

$$1 = \frac{1}{4}x^2 + y^2 = \frac{1}{4}x^2 + \left(\frac{x^2}{4} + \frac{x}{2}\right) = \frac{1}{2}(x^2 + x),$$

so that $x^2 + x = 2$. Thus $x = 1$ (or $x = -2$, which does not yield the maximum area), so that $y = \sqrt{1 - \frac{1}{4}(1^2)} = \frac{1}{2}\sqrt{3}$. Thus the maximal area of the triangle is $\frac{1}{2}(1+2)\left[2(\frac{1}{2}\sqrt{3})\right] = \frac{3}{2}\sqrt{3}$.

19. Let x, y, and z be the angles, so that $x + y + z = \pi$ and $0 \leq x, y, z \leq \pi$. Let $f(x, y, z) = \sin x \sin y \sin z$ and $g(x, y, z) = x + y + z$, so the constraint is $g(x, y, z) = x + y + z = \pi$. Since $\text{grad}\, f(x, y, z) = \cos x \sin y \sin z\, \mathbf{i} + \sin x \cos y \sin z\, \mathbf{j} + \sin x \sin y \cos z\, \mathbf{k}$ and $\text{grad}\, g(x, y, z) = \mathbf{i} + \mathbf{j} + \mathbf{k}$, it follows that $\text{grad}\, f(x, y, z) = \lambda \,\text{grad}\, g(x, y, z)$ if

$$\cos x \sin y \sin z = \lambda, \quad \sin x \cos y \sin z = \lambda, \quad \text{and} \quad \sin x \sin y \cos z = \lambda.$$

The maximum value of f is not 0, so it occurs for (x, y, z) such that $\sin x \neq 0$, $\sin y \neq 0$, and $\sin z \neq 0$. Then the first two equations yield $\sin x \cos y - \cos x \sin y = 0$, so that $\sin(x - y) = 0$, so $x = y$. Likewise the first and third equations yield $\sin(x - z) = 0$, so $x = z$. The constraint becomes $x + x + x = \pi$, so that $x = y = z = \pi/3$. The maximum value of f is

$$f\left(\frac{\pi}{3}, \frac{\pi}{3}, \frac{\pi}{3}\right) = \frac{\sqrt{3}}{2}\frac{\sqrt{3}}{2}\frac{\sqrt{3}}{2} = \frac{3\sqrt{3}}{8}.$$

21. Let (x, y, z) be on the plane $2x + y + 4z = 12$, and let $f(x, y, z) = xyz$ denote the volume of the corresponding parallelepiped with vertex at (x, y, z). Let $g(x, y, z) = 2x + y + 4z$, so that the constraint is $g(x, y, z) = 2x + y + 4z = 12$. Since $\text{grad}\, f(x, y, z) = yz\mathbf{i} + xz\mathbf{j} + xy\mathbf{k}$ and $\text{grad}\, g(x, y, z) = 2\mathbf{i} + \mathbf{j} + 4\mathbf{k}$, it follows that $\text{grad}\, f(x, y, z) = \lambda \,\text{grad}\, g(x, y, z)$ if

$$yz = 2\lambda, \quad xz = \lambda, \quad \text{and} \quad xy = 4\lambda.$$

The volume cannot be a maximum if x, y, or z is 0, so they are all positive. The first and second equations imply that $yz = 2xz$, so that $y = 2x$. The first and third equations imply that $2yz = xy$, so that $z = \frac{1}{2}x$. This means that the constraint becomes $2x + 2x + 4(\frac{1}{2}x) = 12$, so that $x = 2$, and hence $y = 4$ and $z = 1$. The maximum volume is given by $f(2, 4, 1) = 8$.

23. Let (x, y) denote a point on the parabola. The square of the distance to $(-1, 0)$ is minimized for the same point(s) as the distance is, and is given by $E(x, y) = (x+1)^2 + y^2$. Next, let $g(x, y) = x^2 + 2x - y$, so that the constraint becomes $g(x, y) = 0$. Since $\text{grad}\, E(x, y) = 2(x+1)\mathbf{i} + 2y\mathbf{j}$ and $\text{grad}\, g(x, y) = (2x+2)\mathbf{i} - \mathbf{j}$, it follows that $\text{grad}\, E(x, y) = \lambda \text{grad}\, g(x, y)$ if

$$2(x+1) = (2x+2)\lambda \quad \text{and} \quad 2y = -\lambda.$$

By the first equation, either $x = -1$ or $\lambda = 1$. If $x = -1$, then the constraint becomes $y = -1$. If $\lambda = 1$, then by the second equation $y = -\frac{1}{2}$, so the constraint becomes $x^2 + 2x + \frac{1}{2} = 0$, which means that

$$x = \frac{-2 \pm \sqrt{2^2 - 4(1)(\frac{1}{2})}}{2} = -1 \pm \frac{\sqrt{2}}{2}.$$

Thus we have three candidates for the closest point: $(-1, -1)$, $(-1 + \frac{1}{2}\sqrt{2}, -\frac{1}{2})$ and $(-1 - \frac{1}{2}\sqrt{2}, -\frac{1}{2})$. Since $E(-1, -1) = 1$, $E(-1 + \frac{1}{2}\sqrt{2}, -\frac{1}{2}) = \frac{3}{4}$, and $E(-1 - \frac{1}{2}\sqrt{2}, -\frac{1}{2}) = \frac{3}{4}$, it follows that $(-1 + \frac{1}{2}\sqrt{2}, -\frac{1}{2})$ and $(-1 - \frac{1}{2}\sqrt{2}, -\frac{1}{2})$ are the points on the parabola that are closest to $(-1, 0)$.

25. Assume that $r > 0$. Let x be half the length of the base, $y + r$ the height of the triangle, as in the figure, and A the area. Then $A(x, y) = \frac{1}{2}(2x)(y + r) = x(y + r)$. By the Pythagorean Theorem, $x^2 + y^2 = r^2$. Let $g(x, y) = x^2 + y^2$. Then the constraint becomes $g(x, y) = r^2$. Since $\text{grad}\, A(x, y) = (y + r)\mathbf{i} + x\mathbf{j}$ and $\text{grad}\, g(x, y) = 2x\mathbf{i} + 2y\mathbf{j}$, it follows that $\text{grad}\, A(x, y) = \lambda \text{grad}\, g(x, y)$ if

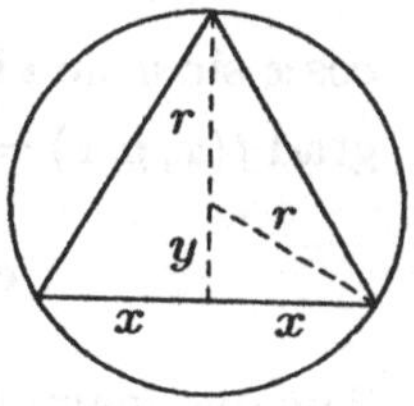

$$y + r = 2x\lambda \quad \text{and} \quad x = 2y\lambda.$$

If $x = 0$, then $y = -r$ by the first equation, which is impossible. Thus $x \neq 0$ and $y \neq 0$. Solving for λ in the equations, we obtain

$$\frac{y + r}{2x} = \lambda = \frac{x}{2y}, \quad \text{so that} \quad y^2 + ry = x^2 = r^2 - y^2.$$

Thus $2y^2 + ry - r^2 = 0$, so $(2y - r)(y + r) = 0$. Since $r > 0$, it follows that $y = \frac{1}{2}r$. Therefore $x = \sqrt{r^2 - y^2} = \sqrt{r^2 - r^2/4} = \frac{1}{2}\sqrt{3}\, r$. Consequently the maximum area of the inscribed triangle is $A(\frac{1}{2}\sqrt{3}\, r, \frac{1}{2}r) = (\frac{1}{2}\sqrt{3}\, r)(\frac{1}{2}r + r) = \frac{3}{4}\sqrt{3}\, r^2$.

27. Let x, y, and z be as in the figure, with $x > 0$, $y > 0$, and $z > 0$. Then the surface area is given by $S(x, y, z) = 3xz + 7yz + xy$. Let V denote the volume, so that $V(x, y, z) = xyz$. The constraint becomes $V(x, y, z) = xyz = 12$. Since $\text{grad}\, S(x, y, z) = (3z + y)\mathbf{i} + (7z + x)\mathbf{j} + (3x + 7y)\mathbf{k}$ and $\text{grad}\, V(x, y, z) = yz\mathbf{i} + xz\mathbf{j} + xy\mathbf{k}$, it follows that $\text{grad}\, S(x, y, z) = \lambda\, \text{grad}\, V(x, y, z)$ if

$$3z + y = yz\lambda, \quad 7z + x = xz\lambda, \quad \text{and} \quad 3x + 7y = xy\lambda.$$

Solving for λ in the three equations, and noting that $x > 0$, $y > 0$, and $z > 0$, we obtain

$$\lambda = \frac{3z + y}{yz} = \frac{3}{y} + \frac{1}{z}, \quad \lambda = \frac{7z + x}{xz} = \frac{7}{x} + \frac{1}{z}, \quad \text{and} \quad \lambda = \frac{3x + 7y}{xy} = \frac{3}{y} + \frac{7}{x}$$

so that

$$\frac{3}{y} + \frac{1}{z} = \frac{7}{x} + \frac{1}{z} \quad \text{and} \quad \frac{7}{x} + \frac{1}{z} = \frac{3}{y} + \frac{7}{x}.$$

Thus $x = \frac{7}{3}y$ and $z = \frac{1}{3}y$, so the constraint becomes $(\frac{7}{3}y)(y)(\frac{1}{3}y) = 12$, so that $y = (\frac{108}{7})^{1/3}$, and hence $x = \frac{7}{3}(\frac{108}{7})^{1/3}$, and $z = \frac{7}{3}(\frac{108}{7})^{1/3} = (\frac{4}{7})^{1/3}$. These dimensions yield the minimum surface area, and hence minimize the amount of metal required for the box.

29. The volume of the pipe is given by $V(r,l) = \pi r^2 l$, with $r > 0$ and $l > 0$. By Figure 13.61; $2r + \frac{1}{2}l = 3\sec(\pi/4) = 3\sqrt{2}$. Thus we let $g(r,l) = 2r + \frac{1}{2}l$, so that the constraint becomes $g(r,l) = 2r + \frac{1}{2}l = 3\sqrt{2}$. Since $\operatorname{grad} V(r,l) = 2\pi rl\mathbf{i} + \pi r^2\mathbf{j}$ and $\operatorname{grad} g(r,l) = 2\mathbf{i} + \frac{1}{2}\mathbf{j}$, it follows that $\operatorname{grad} V(r,l) = \lambda \operatorname{grad} g(r,l)$ if

$$2\pi rl = 2\lambda \quad \text{and} \quad \pi r^2 = \frac{1}{2}\lambda.$$

Solving for λ in the two equations, we obtain $\pi rl = \lambda = 2\pi r^2$. Since $r \neq 0$, this means that $l = 2r$. From the constraint, $2r + \frac{1}{2}(2r) = 3\sqrt{2}$, so that $r = \sqrt{2}$, and thus $l = 2\sqrt{2}$. The pipe has maximum volume if it has a radius of $\sqrt{2}$ meters and is $2\sqrt{2}$ meters long.

31. The volume of the box is given by $V(x,y,z) = xyz$, and the constraint is $V(x,y,z) = xyz = V$ for a constant V. Since $\operatorname{grad} E(x,y,z) = -h^2/(4mx^3)\mathbf{i} - h^2/(4my^3)\mathbf{j} - h^2/(4mz^3)\mathbf{k}$ and $\operatorname{grad} V(x,y,z) = yz\mathbf{i} + xz\mathbf{j} + xy\mathbf{k}$, it follows that $\operatorname{grad} E(x,y,z) = \lambda \operatorname{grad} V(x,y,z)$ if

$$\frac{-h^2}{4mx^3} = yz\lambda, \quad \frac{-h^2}{4my^3} = xz\lambda, \quad \text{and} \quad \frac{-h^2}{4mz^3} = xy\lambda.$$

Thus $-h^2/(4mxyz\lambda) = x^2 = y^2 = z^2$, so since x, y, and z are positive, $x = y = z$. The constraint becomes $x^3 = V$, so $x = y = z = V^{1/3}$, and the box is a cube whose sides are $V^{1/3}$ units long.

33. Let A be a units from the horizontal line and let B be b units from the horizontal line. Then the distance light travels in the medium containing A is $\sqrt{x^2+a^2}$ and the distance light travels in the medium containing B is $\sqrt{y^2+b^2}$. Thus the total time $t(x,y)$ required for light to travel from A to B is given by $t(x,y) = \sqrt{x^2+a^2}/v + \sqrt{y^2+b^2}/u$, and we wish to minimize t. If $D(x,y) = x + y$, then the constraint is $D(x,y) = x + y = l$. Since $\operatorname{grad} t(x,y) = x/(v\sqrt{x^2+a^2})\mathbf{i} + y/(u\sqrt{y^2+b^2})\mathbf{j}$ and $\operatorname{grad} D(x,y) = \mathbf{i} + \mathbf{j}$, it follows that $\operatorname{grad} t(x,y) = \lambda \operatorname{grad} D(x,y)$ if

$$\frac{x}{v\sqrt{x^2+a^2}} = \lambda \quad \text{and} \quad \frac{y}{u\sqrt{y^2+b^2}} = \lambda$$

which implies that

$$\frac{x}{v\sqrt{x^2+a^2}} = \frac{y}{u\sqrt{y^2+b^2}}.$$

Since $x/\sqrt{x^2+a^2} = \sin\theta$ and $y/\sqrt{y^2+b^2} = \sin\phi$, it follows that

$$\frac{\sin\theta}{v} = \frac{\sin\phi}{u}.$$

35. a. Let $g(x,y) = ax+by$, so the constraint is $g(x,y) = ax+by = c$. Since $\operatorname{grad} f(x_0,y_0) = f_x(x_0,y_0)\mathbf{i} + f_y(x_0,y_0)\mathbf{j}$ and $\operatorname{grad} g(x_0,y_0) = a\mathbf{i} + b\mathbf{j}$, it follows that $\operatorname{grad} f(x_0,y_0) = \lambda \operatorname{grad} g(x_0,y_0)$ if

$$f_x(x_0,y_0) = a\lambda \quad \text{and} \quad f_y(x_0,y_0) = b\lambda.$$

If a and b are nonzero, then

$$\frac{f_x(x_0, y_0)}{a} = \frac{f_y(x_0, y_0)}{b} = \lambda.$$

b. If $f(x, y) = x^\alpha y^\beta$, then $f_x(x, y) = \alpha x^{\alpha-1} y^\beta$ and $f_y(x, y) = \beta x^\alpha y^{\beta-1}$. By part (a) we have

$$\frac{\alpha x_0^{\alpha-1} y_0^\beta}{a} = \frac{\beta x_0^\alpha y_0^{\beta-1}}{b} = \lambda.$$

Canceling terms, we conclude that $y_0/x_0 = \beta a/(\alpha b)$.

Chapter 13 Review

1. all (x, y) such that $\frac{1}{4}x^2 + \frac{1}{25}y^2 \le 1$

3. all (x, y, z) such that $x - y + z > 0$

5. The level curves are the hyperbolas $xy = 1$ and $xy = -1$, respectively.

7. 

9. 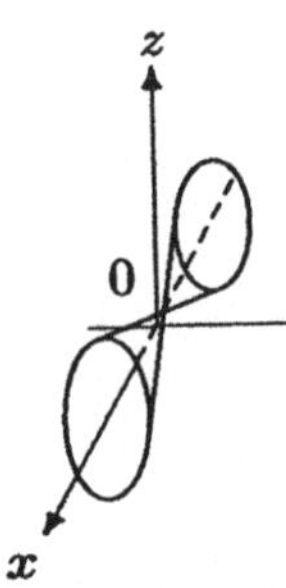

11. 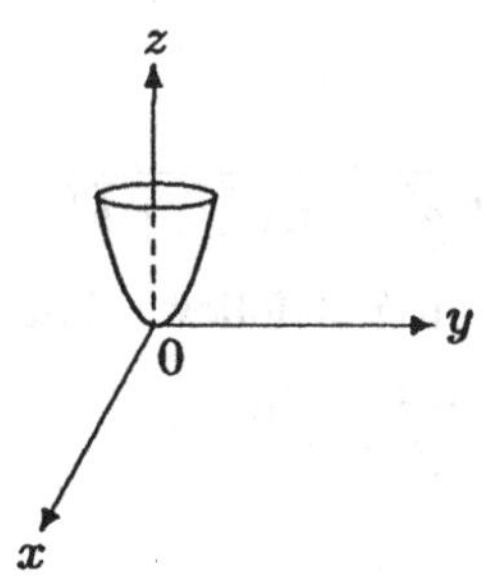

13. 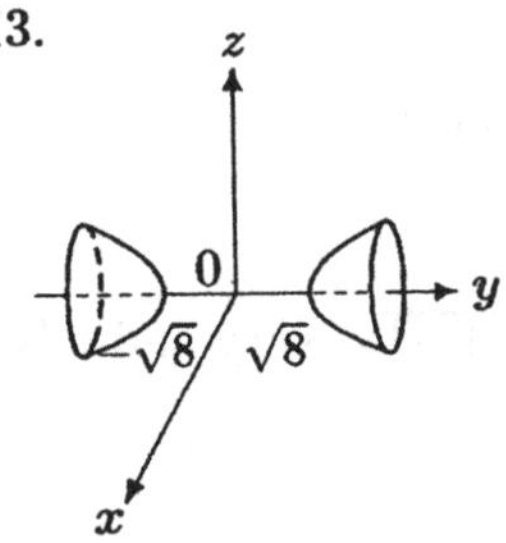

15. $$\lim_{(x,y)\to(-2,\sqrt{2})} \frac{x^4 + x^2y^2 - 6y^4}{x^2 - 2y^2} = \lim_{(x,y)\to(-2,\sqrt{2})} \frac{(x^2 + 3y^2)(x^2 - 2y^2)}{x^2 - 2y^2} = \lim_{(x,y)\to(-2,\sqrt{2})} (x^2 + 3y^2) = 10$$

17. $f_x(x,y) = 12x^2,\ f_y(x,y) = -6y$

19. $f_x(x,y,z) = 2xe^{x^2}\ln(y^2-3z),\ f_y(x,y,z) = \dfrac{2ye^{x^2}}{y^2-3z},\ f_z(x,y,z) = \dfrac{-3e^{x^2}}{y^2-3z}$

21. $k_x(x,y,z) = \frac{5}{2}[\sqrt{z}\tan(x^2+y)]^{3/2}(2x\sqrt{z}\sec^2(x^2+y)) = 5xz^{5/4}[\tan^{3/2}(x^2+y)]\sec^2(x^2+y)$

$k_y(x,y,z) = \frac{5}{2}z^{5/4}[\tan^{3/2}(x^2+y)]\sec^2(x^2+y)$

$k_z(x,y,z) = \frac{5}{2}[\sqrt{z}\tan(x^2+y)]^{3/2}[\frac{1}{2}\frac{1}{\sqrt{z}}\tan(x^2+y)] = \frac{5}{4}\sqrt[4]{z}\tan^{5/2}(x^2+y)$

23. $f_y(x,y) = 1+2xe^y$ and $g_x(x,y) = 1+2xe^y$, so $f_y = g_x$.

25. $g(u,v) = \ln\dfrac{u^2}{e^v} = \ln u^2 - \ln e^v = 2\ln u - v;\ g_u(u,v) = \dfrac{2}{u},\ g_{uu}(u,v) = \dfrac{-2}{u^2},\ g_{uv}(u,v) = 0;$

$g_v(u,v) = -1,\ g_{vv}(u,v) = g_{vu}(u,v) = 0.$

27. $\dfrac{\partial z}{\partial x} = -be^{-ay}\sin bx,\ \dfrac{\partial^2 z}{\partial x^2} = -b^2e^{-ay}\cos bx,\ \dfrac{\partial z}{\partial y} = -ae^{-ay}\cos bx,\ \dfrac{\partial^2 z}{\partial y^2} = a^2e^{-ay}\cos bx;$

thus $a^2\dfrac{\partial^2 z}{\partial x^2} + b^2\dfrac{\partial^2 z}{\partial y^2} = -a^2b^2e^{-ay}\cos bx + a^2b^2e^{-ay}\cos bx = 0.$

29. $\dfrac{dz}{dt} = \dfrac{\partial z}{\partial x}\dfrac{dx}{dt} + \dfrac{\partial z}{\partial y}\dfrac{dy}{dt} = (2xe^{x^2})(2t) + (-\dfrac{1}{2}e^{y/2})(3t^2-1) = 4t^3e^{t^4} - \dfrac{1}{2}(3t^2-1)e^{(t^3-t)/2}$

31. $\dfrac{dw}{dt} = \dfrac{\partial w}{\partial x}\dfrac{dx}{dt} + \dfrac{\partial w}{\partial y}\dfrac{dy}{dt} + \dfrac{\partial w}{\partial z}\dfrac{dz}{dt} = \dfrac{x}{\sqrt{x^2+y^2z^4}}(2) + \dfrac{yz^4}{\sqrt{x^2+y^2z^4}}(3t^2) + \dfrac{2y^2z^3}{\sqrt{x^2+y^2z^4}}\left(\dfrac{-1}{t^2}\right) = \dfrac{\sqrt{5}\,t}{|t|}$

33. $\partial z/\partial x = 2xf'(x^2+y^2)$ and $\partial z/\partial y = 2yf'(x^2+y^2)$.
Thus $y(\partial z/\partial x) - x(\partial z/\partial y) = 2xyf'(x^2+y^2) - 2xyf'(x^2+y^2) = 0.$

35. Treating z as a function of x and y and taking partial derivatives with respect to x, we find that

$$yz + xy\frac{\partial z}{\partial x} - \frac{1}{x^2yz} - \frac{1}{xyz^2}\frac{\partial z}{\partial x} = 3z^2\frac{\partial z}{\partial x}.$$

Solving for $\partial z/\partial x$, we obtain

$$\frac{\partial z}{\partial x} = \frac{yz - 1/x^2yz}{3z^2 + (1/xyz^2) - xy} = \frac{z(x^2y^2z^2-1)}{x(3xyz^4+1-x^2y^2z^2)}.$$

Interchanging the roles of x and y, we obtain

$$\frac{\partial z}{\partial y} = \frac{z(x^2y^2z^2-1)}{y(3xyz^4+1-x^2y^2z^2)}.$$

37. $f_x(x,y,z) = \dfrac{-2x}{(x^2+y^2+z^2)^2}$, $f_y(x,y,z) = \dfrac{-2y}{(x^2+y^2+z^2)^2}$, $f_z(x,y,z) = \dfrac{-2z}{(x^2+y^2+z^2)^2}$, and $\|\mathbf{a}\| = \sqrt{3}$, so that $\mathbf{u} = \dfrac{1}{\sqrt{3}}\mathbf{i} - \dfrac{1}{\sqrt{3}}\mathbf{j} - \dfrac{1}{\sqrt{3}}\mathbf{k}$;

$$D_{\mathbf{u}}f(-1,0,2) = f_x(-1,0,2)\left(\frac{1}{\sqrt{3}}\right) + f_y(-1,0,2)\left(\frac{-1}{\sqrt{3}}\right) + f_z(-1,0,2)\left(\frac{-1}{\sqrt{3}}\right)$$
$$= \frac{2}{25\sqrt{3}} + 0 + \frac{4}{25\sqrt{3}} = \frac{2}{25}\sqrt{3}$$

39. $\operatorname{grad} f(x,y) = 2e^{2x}\ln y\,\mathbf{i} + e^{2x}\dfrac{1}{y}\mathbf{j}$, so that $\operatorname{grad} f(0,1) = \mathbf{j}$.

41. $\operatorname{grad} f(x,y) = -y\sin xy\,\mathbf{i} - x\sin xy\,\mathbf{j}$, so that $\operatorname{grad} f(\frac{1}{2},\pi) = -\pi\mathbf{i} - \frac{1}{2}\mathbf{j}$. Thus the direction in which f increases most rapidly at $(\frac{1}{2},\pi)$ is $-\pi\mathbf{i} - \frac{1}{2}\mathbf{j}$.

43. Let $f(x,y) = \tan^{-1}(x^2+y)$. Since the graph of the given equation is a level curve of f, Theorem 13.16 implies that $\operatorname{grad} f(\frac{1}{2},\frac{3}{4})$ is normal to the graph at $(\frac{1}{2},\frac{3}{4})$. Since

$$\operatorname{grad} f(x,y) = \frac{2x}{1+(x^2+y)^2}\mathbf{i} + \frac{1}{1+(x^2+y)^2}\mathbf{j}$$

so that $\operatorname{grad} f(\frac{1}{2},\frac{3}{4}) = \frac{1}{2}\mathbf{i} + \frac{1}{2}\mathbf{j}$, we find that $\frac{1}{2}\mathbf{i} + \frac{1}{2}\mathbf{j}$ is normal to the graph at $(\frac{1}{2},\frac{3}{4})$.

45. Since $f_x(x,y) = -1/x^2$ and $f_y(x,y) = 1/y^2$, so that $f_x(-\frac{1}{2},\frac{1}{3}) = -4$ and $f_y(-\frac{1}{2},\frac{1}{3}) = 9$, the vector $-4\mathbf{i} + 9\mathbf{j} - \mathbf{k}$ is normal to the graph of f at $(-\frac{1}{2},\frac{1}{3},-5)$, and an equation of the plane tangent at $(-\frac{1}{2},\frac{1}{3},-5)$ is $-4(x+\frac{1}{2}) + 9(y-\frac{1}{3}) - (z+5) = 0$, or $-4x + 9y - z = 10$.

47. Since $f_x(x,y) = 3x/\sqrt{3x^2+2y^2+2}$ and $f_y(x,y) = 2y/\sqrt{3x^2+2y^2+2}$, so that $f_x(4,-5) = \frac{6}{5}$ and $f_y(4,-5) = -1$, the vector $\frac{6}{5}\mathbf{i} - \mathbf{j} - \mathbf{k}$ is normal to the graph of f at $(4,-5,10)$, and an equation of the plane tangent at $(4,-5,10)$ is $\frac{6}{5}(x-4) - (y+5) - (z-10) = 0$, or $6x - 5y - 5z = -1$.

49. If $f(x,y,z) = xe^{yz} - 2y$, then $xe^{yz} - 2y = -1$ is a level surface of f. Since $f_x(x,y,z) = e^{yz}$, $f_y(x,y,z) = xze^{yz} - 2$, and $f_z(x,y,z) = xye^{yz}$, so $f_x(1,1,0) = 1$, $f_y(1,1,0) = -2$, and $f_z(1,1,0) = 1$, it follows that an equation of the plane tangent at $(1,1,0)$ is $(x-1) - 2(y-1) + (z-0) = 0$, or $x - 2y + z = -1$.

51. Let $f(x,y,z) = 2x^3 + y - z^2$. Then $f_x(x,y,z) = 6x^2$, $f_y(x,y,z) = 1$, and $f_z(x,y,z) = -2z$, so a vector normal to the plane tangent at any point (x_0,y_0,z_0) on the level surface $2x^3 + y - z^2 = 5$ is $6x_0^2\mathbf{i} + \mathbf{j} - 2z_0\mathbf{k}$. If the tangent plane is parallel to the plane $24x + y - 6z = 3$, whose normal is $24\mathbf{i} + \mathbf{j} - 6\mathbf{k}$, then $6x_0^2\mathbf{i} + \mathbf{j} - 2z_0\mathbf{k} = c(24\mathbf{i} + \mathbf{j} - 6\mathbf{k})$ for some c. It follows that $c = 1$, so that $x_0 = \pm 2$ and $z_0 = 3$. Because (x_0,y_0,z_0) is on the level surface $2x^3 + y - z^2 = 5$ we find that if $x_0 = 2$ and $z_0 = 3$, then $y_0 = -2$, and if $x_0 = -2$ and $z_0 = 3$, then $y_0 = 30$. The required points on the level surface are $(2,-2,3)$ and $(-2,30,3)$.

53. Let $f(x,y) = x^2 + 4y^2$. Then $f_x(x,y) = 2x$ and $f_y(x,y) = 8y$, so that $f_x(2,0) = 4$ and $f_y(2,0) = 0$. Thus a vector normal to the plane tangent to the graph of f at $(2,0,4)$ is $4\mathbf{i} - \mathbf{k}$. Next, let $g(x,y) = 4x + y^2 - 4$. Then $g_x(x,y) = 4$ and $g_y(x,y) = 2y$, so that $g_x(2,0) = 4$ and $g_y(2,0) = 0$. Thus a vector normal to the plane tangent to the graph of g at $(2,0,4)$ is $4\mathbf{i} - \mathbf{k}$. Since the normals are identical and the tangent planes both pass through $(2,0,4)$, the two planes are identical.

55. Let $x_0 = 1$, $y_0 = 0$, $h = -0.03$, and $k = 0.05$. Since

$$f_x(x,y) = \frac{1}{1 + \left(\dfrac{x}{1+y}\right)^2}\left(\frac{1}{1+y}\right) = \frac{1+y}{(1+y)^2 + x^2}$$

and

$$f_y(x,y) = \frac{1}{1 + \left(\dfrac{x}{1+y}\right)^2}\left(\frac{-x}{(1+y)^2}\right) = \frac{-x}{(1+y)^2 + x^2}$$

we have $f_x(1,0) = \frac{1}{2}$ and $f_y(1,0) = -\frac{1}{2}$. Also, $f(1,0) = \pi/4$. Thus

$$f(0.97, 0.05) \approx \frac{\pi}{4} + \frac{1}{2}(-0.03) - \frac{1}{2}(0.05) = \frac{\pi}{4} - 0.04 \approx 0.7453981634.$$

57. Since $f(x,y) = \ln(x/y) = \ln x - \ln y$, we have $\partial f/\partial x = 1/x$ and $\partial f/\partial y = -1/y$, so that $df = (1/x)\,dx - (1/y)\,dy$.

59. $f_x(x,y) = 2x - 2$ and $f_y(x,y) = 2y - 4$, so $f_x(x,y) = 0 = f_y(x,y)$ if $x = 1$ and $y = 2$. Thus $(1,2)$ is a critical point. Next, $f_{xx}(x,y) = 2$, $f_{yy}(x,y) = 2$, and $f_{xy}(x,y) = 0$. Thus $D(1,2) = f_{xx}(1,2)f_{yy}(1,2) - [f_{xy}(1,2)]^2 = (2)(2) - 0^2 = 4$. Since $D(1,2) > 0$ and $f_{xx}(1,2) > 0$, f has a relative minimum value at $(1,2)$.

61. $f_x(x,y) = y - 16/x^3$ and $f_y(x,y) = x - 16/y^3$, so $f_x(x,y) = 0 = f_y(x,y)$ if $y - 16/x^3 = 0$ and $x - 16/y^3 = 0$. Solving for y in the first equation and substituting for it in the second equation yields $x - 16(x^3/16)^3 = 0$, so that $x^8 = 256$, and thus $x = \pm 2$. Then $y = \pm 2$, so that $(2,2)$ and $(-2,-2)$ are critical points. Next, $f_{xx}(x,y) = 48/x^4$, $f_{yy}(x,y) = 48/y^4$, and $f_{xy}(x,y) = 1$. Thus $D(2,2) = f_{xx}(2,2)f_{yy}(2,2) - [f_{xy}(2,2)]^2 = (3)(3) - 1^2 = 8$. Since $D(2,2) > 0$ and $f_{xx}(2,2) > 0$, f has a relative minimum value at $(2,2)$. Finally, $D(-2,-2) = f_{xx}(-2,-2)f_{yy}(-2,-2) - [f_{xy}(-2,-2)]^2 = (3)(3) - 1^2 = 8$, and since $D(-2,-2) > 0$ and $f_{xx}(-2,-2) > 0$, f has a relative minimum value at $(-2,-2)$.

63. Let $g(x,y) = 3x^2 + y^2$, so the constraint is $g(x,y) = 3x^2 + y^2 = 3$. Since $\operatorname{grad} f(x,y) = (6x - y)\mathbf{i} + (-x + 2y)\mathbf{j}$ and $\operatorname{grad} g(x,y) = 6x\mathbf{i} + 2y\mathbf{j}$, it follows that $\operatorname{grad} f(x,y) = \lambda \operatorname{grad} g(x,y)$ if

$$6x - y = 6x\lambda \quad \text{and} \quad -x + 2y = 2y\lambda.$$

If $x = 0$, then by the first equation, $y = 0$, and the constraint is contradicted. Similarly, if $y = 0$, then $x = 0$ by the second equation, and the constraint is contradicted. Thus $x \neq 0$ and $y \neq 0$, so the two equations can be solved for λ:

$$\lambda = \frac{6x-y}{6x} = 1 - \frac{y}{6x} \quad \text{and} \quad \lambda = \frac{-x+2y}{2y} = \frac{-x}{2y} + 1$$

so that $y/6x = x/2y$, and hence $y^2 = 3x^2$. The constraint becomes $3x^2 + 3x^2 = 3$, so $x = \pm 1/\sqrt{2}$, and therefore $y = \pm\sqrt{\frac{3}{2}}$. The possible extreme values of f are

$$f\left(\frac{1}{\sqrt{2}}, \sqrt{\frac{3}{2}}\right) = f\left(\frac{-1}{\sqrt{2}}, -\sqrt{\frac{3}{2}}\right) = 3 - \frac{\sqrt{3}}{2}$$

and

$$f\left(\frac{-1}{\sqrt{2}}, \sqrt{\frac{3}{2}}\right) = f\left(\frac{1}{\sqrt{2}}, -\sqrt{\frac{3}{2}}\right) = 3 + \frac{\sqrt{3}}{2}.$$

The maximum value is $3 + \sqrt{3}/2$ and the minimum value is $3 - \sqrt{3}/2$.

65. Let $g(x, y, z) = x^2 + y^2 - z$, so the constraint is $g(x, y, z) = x^2 + y^2 - z = 2$. Since

$$\operatorname{grad} f(x, y, z) = \frac{2x}{z^2+5}\mathbf{i} + \frac{2y}{z^2+5}\mathbf{j} - \frac{2z(x^2+y^2)}{(z^2+5)^2}\mathbf{k} \quad \text{and} \quad \operatorname{grad} g(x, y, z) = 2x\mathbf{i} + 2y\mathbf{j} - \mathbf{k}$$

it follows that $\operatorname{grad} f(x, y, z) = \lambda \operatorname{grad} g(x, y, z)$ if

$$\frac{2x}{z^2+5} = 2x\lambda, \quad \frac{2y}{z^2+5} = 2y\lambda, \quad \text{and} \quad \frac{-2z(x^2+y^2)}{(z^2+5)^2} = -\lambda. \qquad (*)$$

67. Let (x, y, z) be a vertex of the parallelepiped that is on the ellipsoid and in the first octant. Let $V(x, y, z)$ denote the volume of the parallelepiped, so that $V(x, y, z) = 8xyz$. Let $g(x, y, z) = x^2 + 4y^2 + 9z^2$, so the constraint is $g(x, y, z) = x^2 + 4y^2 + 9z^2 = 36$. Since

$$\operatorname{grad} V(x, y, z) = 8yz\mathbf{i} + 8xz\mathbf{j} + 8xy\mathbf{k} \quad \text{and} \quad \operatorname{grad} g(x, y, z) = 2x\mathbf{i} + 8y\mathbf{j} + 18z\mathbf{k}$$

it follows that $\operatorname{grad} V(x, y, z) = \lambda \operatorname{grad} g(x, y, z)$ if

$$8yz = 2x\lambda, \quad 8xz = 8y\lambda, \quad \text{and} \quad 8xy = 18z\lambda.$$

If x, y, or z is zero, then the volume is 0. Otherwise, $x > 0$, $y > 0$, and $z > 0$, and we can solve for λ in the three equations:

$$\lambda = \frac{4yz}{x}, \quad \lambda = \frac{xz}{y}, \quad \text{and} \quad \lambda = \frac{4xy}{9z}. \qquad (*)$$

From the first two equations in $(*)$ we obtain $4yz/x = xz/y$, so that $y = x/2$; from the first and third equations in $(*)$ we obtain $4yz/x = 4xy/9z$, so that $z = x/3$. The constraint becomes $x^2 + 4(x/2)^2 + 9(x/3)^2 = 36$, so that $x = 2\sqrt{3}$. Then $y = \sqrt{3}$ and $z = 2\sqrt{3}/3$. Consequently the dimensions of the parallelepiped with the largest volume are $4\sqrt{3}$, $2\sqrt{3}$, and $\frac{4}{3}\sqrt{3}$.

69. $\dfrac{\partial T}{\partial \dot{\theta}} = ml^2\dot{\theta}$, so that $\dfrac{d}{dt}\left(\dfrac{\partial T}{\partial \dot{\theta}}\right) = \left[\dfrac{\partial}{\partial \dot{\theta}}\left(\dfrac{\partial T}{\partial \dot{\theta}}\right)\right]\dfrac{d\dot{\theta}}{dt} = \left[\dfrac{\partial}{\partial \dot{\theta}}(ml^2\dot{\theta})\right]\dfrac{d\dot{\theta}}{dt} = ml^2\ddot{\theta}.$

Also $\partial V/\partial\theta = mgl\theta$. Thus Lagrange's equation becomes $ml^2\ddot{\theta} + mgl\theta = 0$, or rather, $\ddot{\theta} + (g/l)\theta = 0$.

From the first equation, $x = 0$ or $\lambda = 1/(z^2 + 5)$. Assume that $\lambda = 1/(z^2 + 5)$. Then the third equation is transformed into $2z(x^2 + y^2) = z^2 + 5$. With the help of the constraint this means that $2z(2 + z) = z^2 + 5$, so that $z^2 + 4z - 5 = 0$, and hence $z = 1$ or $z = -5$. If $z = 1$, then from the constraint, $x^2 + y^2 = 2 + 1 = 3$, so $f(x, y, 1) = (x^2 + y^2)/(1^2 + 5) = \frac{3}{6} = \frac{1}{2}$. If $z = -5$, then from the constraint, $x^2 + y^2 = 2 - 5 = -3$, which can't be true. Finally we assume that $\lambda \neq 1/(z^2 + 5)$. Then by the first two equations in $(*)$ we have $x = 0$ and $y = 0$, respectively. The constraint becomes $0^2 + 0^2 - z = 2$, so $z = -2$. But $f(0, 0, -2) = 0$. Consequently the maximum value of f is $\frac{1}{2}$ and the minimum value is 0.

Cumulative Review(Chapters 1–12)

1. The conditions for applying l'Hôpital's Rule are satisfied;

$$\lim_{h\to 0} \frac{\sqrt{x-h} - \sqrt{x}}{h} = \lim_{h\to 0} \frac{\dfrac{1}{2\sqrt{x-h}}(-1) - 0}{1} = -\frac{1}{2\sqrt{x}} \quad \text{for } x > 0$$

3. Since $dy/dx = 8x$, an equation of the line tangent to the parabola at $(a, 4a^2)$ is $y - 4a^2 = 8a(x - a)$. If $(0, -2)$ lies on the tangent line, then $-2 - 4a^2 = 8a(0 - a)$, so that $4a^2 = 2$, and thus $a = \sqrt{2}/2$ or $a = -\sqrt{2}/2$. Therefore equations of the tangent lines that pass through $(0, -2)$ are $y - 2 = 4\sqrt{2}\,(x - \sqrt{2}/2)$ (or equivalently $y = 4\sqrt{2}\,x - 2$) and $y - 2 = -4\sqrt{2}\,(x + \sqrt{2}/2)$ (or equivalently $y = -4\sqrt{2}\,x - 2$).

5. $$\begin{aligned} f'(x) &= n \cos nx \sin^n x + n \sin nx \sin^{n-1} x \cos x \\ &= n \sin^{n-1} x(\cos nx \sin x + \sin nx \cos x) = n(\sin^{n-1} x)(\sin(n+1)x) \end{aligned}$$

7. Differentiating implicitly, we have $1/2\sqrt{x} + (1/2\sqrt{y})(dy/dx) = 0$, so that $dy/dx = -\sqrt{y}/\sqrt{x}$. Differentiating implicitly again, and using the formula for dy/dx and the given equation, we find that

$$\frac{d^2y}{dx^2} = \frac{-(1/(2\sqrt{y}))(dy/dx)\sqrt{x} + \sqrt{y}\,(1/(2\sqrt{x}))}{x} = \frac{\frac{1}{2} + \sqrt{y}/(2\sqrt{x})}{x} = \frac{\sqrt{x} + \sqrt{y}}{2x^{3/2}} = \frac{9}{2}x^{-3/2}.$$

9. $f'(x) = \dfrac{2x}{(1 - x^2)^2}$; $f''(x) = \dfrac{2 + 6x^2}{(1 - x^2)^3}$;

relative minimum value is $f(0) = 2$; increasing on $[0, 1)$ and $(1, \infty)$, and decreasing on $(-\infty, -1)$ and $(-1, 0]$; concave upward on $(-1, 1)$, and concave downward on $(-\infty, -1)$ and $(1, \infty)$; vertical asymptotes are $x = -1$ and $x = 1$; horizontal asymptote is $y = 1$; symmetric with respect to the y axis.

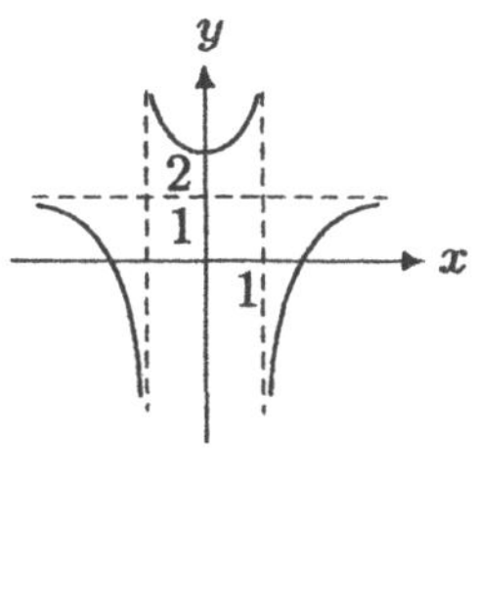

11. The curves intersect at (x, y) if $2x^3 + 2x^2 + 10x = y = x^3 - 3x^2 + 4x$, that is,

$$x^3 + 5x^2 + 6x = 0, \quad \text{or} \quad x(x+2)(x+3) = 0, \quad \text{or} \quad x = -3,\ x = -2, \text{ or } x = 0.$$

Moreover, $2x^3 + 2x^2 + 10x \geq x^3 - 3x^2 + 4x$ for $-3 \leq x \leq -2$ and $x^3 - 3x^2 + 4x \geq 2x^3 + 2x^2 + 10x$ for$-2 \leq x \leq 0$. Thus

$$A = \int_{-3}^{-2} [(2x^3 + 2x^2 + 10x) - (x^3 - 3x^2 + 4x)]\,dx + \int_{-2}^{0} [(x^3 - 3x^2 + 4x) - (2x^3 + 2x^2 + 10x)]\,dx$$

$$= \int_{-3}^{-2} (x^3 + 5x^2 + 6x)\,dx + \int_{-2}^{0} (-x^3 - 5x^2 - 6x)\,dx = \left(\frac{1}{4}x^4 + \frac{5}{3}x^3 + 3x^2\right)\Big|_{-3}^{-2} + \left(-\frac{1}{4}x^4 - \frac{5}{3}x^3 - 3x^2\right)\Big|_{-2}^{0}$$

$$= \left(4 - \frac{40}{3} + 12\right) - \left(\frac{81}{4} - 45 + 27\right) - \left(-4 + \frac{40}{3} - 12\right) = \frac{37}{12}.$$

13. $\int(\cos^2\theta - \cos^3\theta)\,d\theta = \int\left[\frac{1}{2} + \frac{1}{2}\cos 2\theta - (1 - \sin^2\theta)\cos\theta\right]d\theta = \frac{1}{2}\theta + \frac{1}{4}\sin 2\theta - \int(1 - \sin^2\theta)\cos\theta\,d\theta$.

Let $u = \sin\theta$, so that $du = \cos\theta\,d\theta$. Then

$$-\int(1 - \sin^2\theta)\cos\theta\,d\theta = -\int(1 - u^2)\,du = -u + \frac{1}{3}u^3 + C = -\sin\theta + \frac{1}{3}\sin^3\theta + C.$$

Thus

$$\int(\cos^2\theta - \cos^3\theta)\,d\theta = \frac{1}{2}\theta + \frac{1}{4}\sin 2\theta - \sin\theta + \frac{1}{3}\sin^3\theta + C.$$

15. Let $u = 3 + x^4$, so that $du = 4x^3\,dx$. If $x = 1$, then $u = 4$, and if $x = b$, then $u = 3 + b^4$. Thus

$$\int_1^\infty \frac{x^3}{(3 + x^4)^{3/2}}\,dx = \lim_{b\to\infty}\int_1^b \frac{x^3}{(3 + x^4)^{3/2}}\,dx \overset{u=3+x^4}{=} \lim_{b\to\infty}\int_4^{3+b^4} \frac{1}{u^{3/2}}\frac{1}{4}\,du$$

$$= \frac{1}{4}\lim_{b\to\infty}\int_4^{3+b^4} \frac{1}{u^{3/2}}\,du = -\frac{1}{2}\lim_{b\to\infty}\frac{1}{u^{1/2}}\Big|_4^{3+b^4} = -\frac{1}{2}\lim_{b\to\infty}\left[\frac{1}{(3 + b^4)^{1/2}} - \frac{1}{2}\right] = \frac{1}{4}.$$

It follows that the given improper integral converges, and its value is $\frac{1}{4}$.

17. The region is symmetric with respect to the y axis, so $\bar{x} = 0$ by symmetry. The graphs intersect at (x, y) if $2 - x^2 = y = 1$, or $x^2 = 1$, so $x = -1$ or $x = 1$. Thus R is the region between the graphs of $y = 1$ and $y = 2 - x^2$ on $[-1, 1]$. Thus the area of R is given by

$$A = \int_{-1}^{1} [(2 - x^2) - 1]\,dx = \int_{-1}^{1} (1 - x^2)\,dx = \left(x - \frac{1}{3}x^3\right)\Big|_{-1}^{1} = \frac{4}{3}$$

and

$$M_x = \int_{-1}^{1} \frac{1}{2}[(2 - x^2)^2 - 1^2]\,dx = \frac{1}{2}\int_{-1}^{1} (3 - 4x^2 + x^4)\,dx = \frac{1}{2}\left(3x - \frac{4}{3}x^3 + \frac{1}{5}x^5\right)\Big|_{-1}^{1} = \frac{28}{15}.$$

Therefore $\bar{y} = \frac{28/15}{4/3} = \frac{7}{5}$, so the center of gravity of R is $(0, \frac{7}{5})$.

19. The circles intersect at (r, θ) if $r = 0$ or if $\sin\theta = r = \sqrt{3}\cos\theta$, so that $\tan\theta = \sqrt{3}$, that is, $\theta = \pi/3$. If A_1 and A_2 are the areas of the subregions indicated in the figure, then

$$A = A_1 + A_2 = \frac{1}{2}\pi\left(\frac{1}{2}\right)^2 + \int_{\pi/3}^{\pi/2} \frac{1}{2}[(\sin\theta)^2 - (\sqrt{3}\cos\theta)^2]\,d\theta$$

$$= \frac{\pi}{8} + \frac{1}{2}\int_{\pi/3}^{\pi/2} (\sin^2\theta - 3\cos^2\theta)\,d\theta = \frac{\pi}{8} + \frac{1}{2}\int_{\pi/3}^{\pi/2} \left(\frac{1}{2} - \frac{1}{2}\cos 2\theta - \frac{3}{2} - \frac{3}{2}\cos 2\theta\right)d\theta$$

$$= \frac{\pi}{8} + \frac{1}{2}(-\theta - \sin 2\theta)\Big|_{\pi/3}^{\pi/2} = \frac{\pi}{8} - \frac{\pi}{12} + \frac{\sqrt{3}}{4} = \frac{\pi}{24} + \frac{\sqrt{3}}{4}.$$

21. The jth partial sum s_j is given by

$$s_j = \left(\frac{1}{\sqrt{1}} - \frac{1}{\sqrt{2}}\right) + \left(\frac{1}{\sqrt{2}} - \frac{1}{\sqrt{3}}\right) + \cdots + \left(\frac{1}{\sqrt{j}} - \frac{1}{\sqrt{j+1}}\right) = 1 - \frac{1}{\sqrt{j+1}}.$$

Therefore

$$\sum_{n=1}^{\infty}\left(\frac{1}{\sqrt{n}} - \frac{1}{\sqrt{n+1}}\right) = \lim_{j\to\infty} s_j = \lim_{j\to\infty}\left(1 - \frac{1}{\sqrt{j+1}}\right) = 1.$$

23. By (3) of Section 9.8, $e^x = \sum_{n=0}^{\infty}(1/n!)x^n$ for all x, so

$$f(x) = 2x^3e^{-x^3} = 2x^3\sum_{n=0}^{\infty}\frac{1}{n!}(-x^3)^n = \sum_{n=0}^{\infty}\frac{2}{n!}(-1)^n x^{3n+3}$$

for all x. By the Differentiation Theorem for power series,

$$f'(x) = \frac{d}{dx}\sum_{n=0}^{\infty}\frac{2}{n!}(-1)^n x^{3n+3} = \sum_{n=0}^{\infty}\frac{2(-1)^n(3n+3)}{n!}x^{3n+2} \quad \text{for all } x.$$

In particular, the radius of convergence of each series is ∞.

25. a. Since $\mathbf{T}(t)$ is parallel to $\mathbf{r}'(t) = \mathbf{i} + 2t\mathbf{j} + 2t^2\mathbf{k}$, $\mathbf{T}'(t)$ is parallel to $\mathbf{i} - \mathbf{j} + \frac{1}{2}\mathbf{k}$ if and only if $2t = -1$ and $2t^2 = \frac{1}{2}$, that is, $t = -\frac{1}{2}$.

b. By (3) in Section 12.6, $\kappa(t) = \|\mathbf{v}(t) \times \mathbf{a}(t)\|/\|\mathbf{v}(t)\|^3$. Since $\mathbf{v}(t) = \mathbf{r}'(t) = \mathbf{i} + 2t\mathbf{j} + 2t^2\mathbf{k}$ and $\mathbf{a}(t) = \mathbf{v}'(t) = 2\mathbf{j} + 4t\mathbf{k}$, we have $\|\mathbf{v}(t)\| = \sqrt{1 + (2t)^2 + (2t^2)^2} = 1 + 2t^2$ and

$$\mathbf{v}(t) \times \mathbf{a}(t) = \begin{vmatrix} \mathbf{i} & \mathbf{j} & \mathbf{k} \\ 1 & 2t & 2t^2 \\ 0 & 2 & 4t \end{vmatrix} = 4t^2\mathbf{i} - 4t\mathbf{j} + 2\mathbf{k}.$$

Therefore $\|\mathbf{v}(t) \times \mathbf{a}(t)\| = \sqrt{(4t^2)^2 + (-4t)^2 + 2^2} = 2(2t^2 + 1)$. Thus

$$\kappa(t) = \frac{\|\mathbf{v}(t) \times \mathbf{a}(t)\|}{\|\mathbf{v}(t)\|^3} = \frac{2}{(2t^2 + 1)^2}.$$

Since $\kappa(0) = 2 \geq 2/(2t^2 + 1)^2$ for all t, it follows that κ is maximum for $t = 0$.

Chapter 14

Multiple Integrals

14.1 Double Integrals

1. a. $\sum_{k=1}^{3} f(x_k, y_k)\Delta A_k = f(-2,0)\Delta A_1 + f(-2,1)\Delta A_2 + f(-1,0)\Delta A_3 = 4(1) + 3(1) + 4(1) = 11$

   b. $\sum_{k=1}^{3} f(x_k, y_k)\Delta A_k = f(-1,1)\Delta A_1 + f(-1,2)\Delta A_2 + f(0,1)\Delta A_3 = 3(1) + 2(1) + 3(1) = 8$

3. $\int_0^1 \int_{-1}^1 x\,dy\,dx = \int_0^1 xy\big|_{-1}^1\,dx = \int_0^2 2x\,dx = x^2\big|_0^1 = 1$

5. $\int_0^1 \int_0^1 e^{x+y}\,dy\,dx = \int_0^1 e^{x+y}\big|_0^1\,dx = \int_0^1 (e^{x+1} - e^x)\,dx = (e^{x+1} - e^x)\big|_0^1 = e^2 - 2e + 1$

7. $\int_0^1 \int_x^{x^2} 1\,dy\,dx = \int_0^1 y\big|_x^{x^2}\,dx = \int_0^1 (x^2 - x)\,dx = \left(\frac{1}{3}x^3 - \frac{1}{2}x^2\right)\Big|_0^1 = -\frac{1}{6}$

9. $\int_0^1 \int_0^y x\sqrt{y^2 - x^2}\,dx\,dy = \int_0^1 -\frac{1}{3}(y^2 - x^2)^{3/2}\big|_0^y\,dy = \int_0^1 \frac{1}{3}y^3\,dy = \frac{1}{12}y^4\Big|_0^1 = \frac{1}{12}$

11. $\int_0^2 \int_0^{\sqrt{4-y^2}} x\,dx\,dy = \int_0^2 \frac{1}{2}x^2\Big|_0^{\sqrt{4-y^2}}\,dy = \int_0^2 \frac{1}{2}(4 - y^2)\,dy = \left(2y - \frac{1}{6}y^3\right)\Big|_0^2 = \frac{8}{3}$

13. $\int_0^{2\pi} \int_0^1 r\sqrt{1 - r^2}\,dr\,d\theta = \int_0^{2\pi} -\frac{1}{3}(1 - r^2)^{3/2}\Big|_0^1\,d\theta = \int_0^{2\pi} \frac{1}{3}\,d\theta = \frac{1}{3}\theta\Big|_0^{2\pi} = \frac{2}{3}\pi$

15. $\int_1^3 \int_0^x \frac{2}{x^2 + y^2}\,dy\,dx = \int_1^3 \frac{2}{x}\tan^{-1}\frac{y}{x}\Big|_0^x\,dx = \int_1^3 \frac{\pi}{2x}\,dx = \frac{\pi}{2}\ln x\Big|_1^3 = \frac{\pi}{2}\ln 3$

17. $\int_0^1 \int_0^x e^{(x^2)}\,dy\,dx = \int_0^1 ye^{(x^2)}\big|_0^x\,dx = \int_0^1 xe^{(x^2)}\,dx = \frac{1}{2}e^{(x^2)}\Big|_0^1 = \frac{1}{2}(e - 1)$

19. Since $x^{15}e^{x^2y^2}$ is an odd function with respect to x and R is symmetric with respect to the y axis, $\int_{-1}^{1}\int_{0}^{2} x^{15}e^{x^2y^2}\,dy\,dx = 0$.

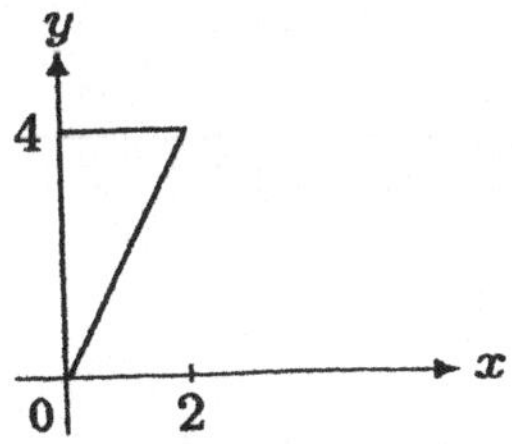

Exercise 21

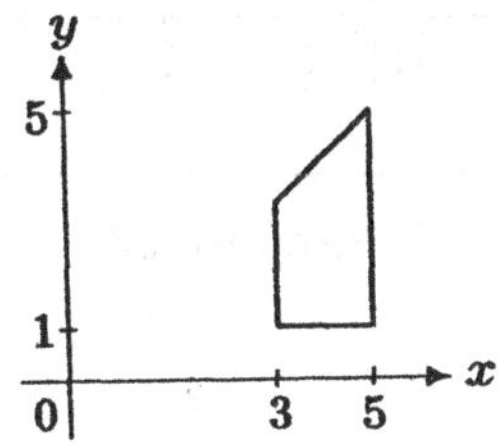

Exercise 23

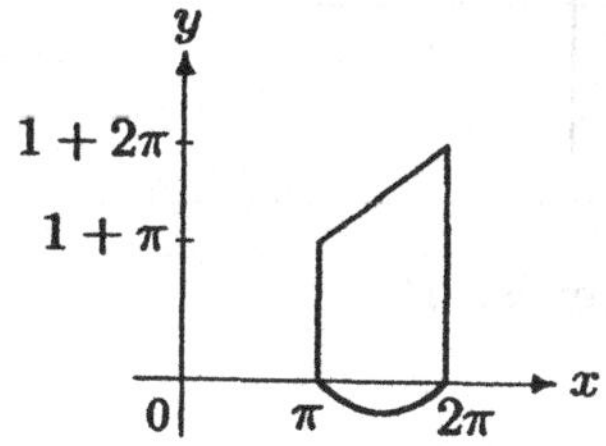

Exercise 27

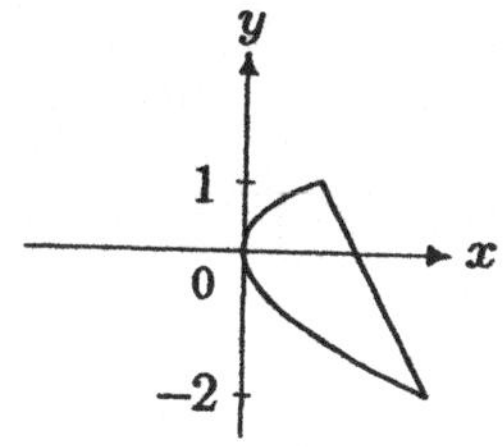

Exercise 29

21. $$\iint_R (x+y)\,dA = \int_0^2\int_{2x}^4 (x+y)\,dy\,dx = \int_0^2 \left(xy+\frac{1}{2}y^2\right)\Big|_{2x}^4 dx = \int_0^2 (4x+8-4x^2)\,dx$$
$$= \left(2x^2+8x-\frac{4}{3}x^3\right)\Big|_0^2 = \frac{40}{3}$$

23. $$\iint_R x\,dA = \int_3^5\int_1^x x\,dy\,dx = \int_3^5 xy\big|_1^x\,dx = \int_3^5 (x^2-x)\,dx = \left(\frac{1}{3}x^3-\frac{1}{2}x^2\right)\Big|_3^5 = \frac{74}{3}$$

25. The lines intersect at (x, y) such that $5+x = y = -x+7$, so that $2x = 2$, or $x = 1$. Since $5+x \geq 7-x$ for x in $[1, 10]$, we have

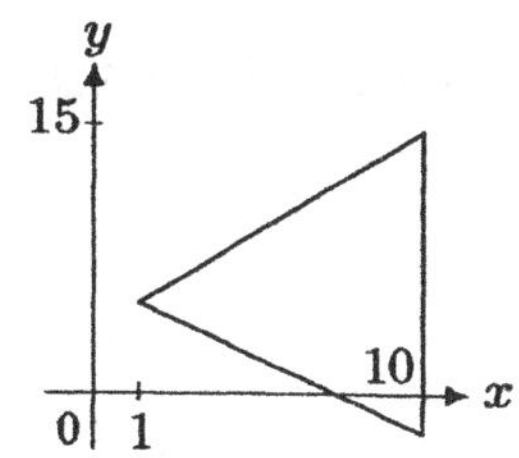

$$\iint_R (3x-5)\,dA = \int_1^{10}\int_{7-x}^{5+x} (3x-5)\,dy\,dx = \int_1^{10} (3x-5)y\big|_{7-x}^{5+x}\,dx$$
$$= \int_1^{10} (3x-5)(2x-2)\,dx = \int_1^{10} (6x^2-16x+10)\,dx = (2x^3-8x^2+10x)\big|_1^{10} = 1296$$

27. $$\iint_R 1\,dA = \int_\pi^{2\pi}\int_{\sin x}^{1+x} 1\,dy\,dx = \int_\pi^{2\pi} y\big|_{\sin x}^{1+x}\,dx = \int_\pi^{2\pi} (1+x-\sin x)\,dx$$
$$= \left(x+\frac{1}{2}x^2+\cos x\right)\Big|_\pi^{2\pi} = \frac{3}{2}\pi^2+\pi+2$$

29. The graphs intersect at (x, y) such that $y^2 = x = 2-y$, so that $y^2+y-2=0$, or $y=-2$ or $y=1$. Thus
$$\iint_R (1-y)\,dA = \int_{-2}^1\int_{y^2}^{2-y} (1-y)\,dx\,dy = \int_{-2}^1 (1-y)x\big|_{y^2}^{2-y}\,dy$$
$$= \int_{-2}^1 (1-y)(2-y-y^2)\,dy = \int_{-2}^1 (2-3y+y^3)\,dy = \left(2y-\frac{3}{2}y^2+\frac{1}{4}y^4\right)\Big|_{-2}^1 = \frac{27}{4}.$$

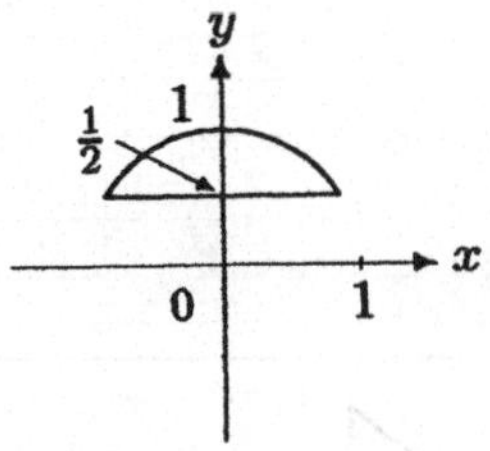

Exercise 31

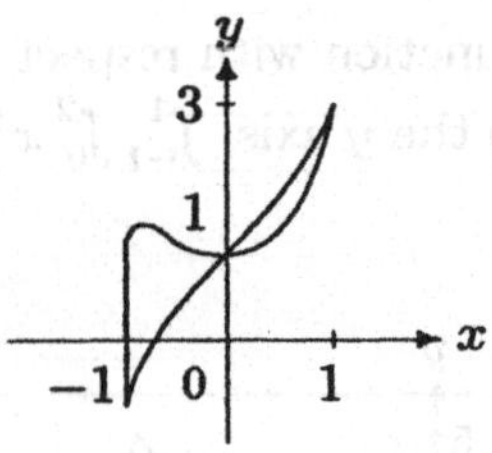

Exercise 33

31. Since $x\sqrt{y^2+1}$ is odd with respect to x, and since R is symmetric with respect to the y axis, $\iint_R x\sqrt{y^2+1}\,dA = 0$.

33. The graphs intersect at (x, y) such that $x^3+x^2+1 = y = x^3+x+1$, so that $x^2 = x$, or $x = 0$ or $x = 1$. Since $x^3+x^2+1 \geq x^3+x+1$ for $-1 \leq x \leq 0$ and $x^3+x+1 \geq x^3+x^2+1$ for $0 \leq x \leq 1$, we have

$$\iint_R x^2\,dA = \int_{-1}^{0}\int_{x^3+x+1}^{x^3+x^2+1} x^2\,dy\,dx + \int_0^1\int_{x^3+x^2+1}^{x^3+x+1} x^2\,dy\,dx$$

$$= \int_{-1}^{0} x^2y\Big|_{x^3+x+1}^{x^3+x^2+1} dx + \int_0^1 x^2y\Big|_{x^3+x^2+1}^{x^3+x+1} dx = \int_{-1}^{0}(x^4-x^3)\,dx + \int_0^1 (x^3-x^4)\,dx$$

$$= \left(\frac{1}{5}x^5 - \frac{1}{4}x^4\right)\Big|_{-1}^{0} + \left(\frac{1}{4}x^4 - \frac{1}{5}x^5\right)\Big|_0^1 = \frac{9}{20} + \frac{1}{20} = \frac{1}{2}.$$

35. The intersection of the solid region with the xy plane is the region R bounded by the lines $x + 2y = 6$ (or $y = \frac{1}{2}(6-x)$), $x = 0$, and $y = 0$. Thus

$$V = \iint_R \frac{1}{3}(6-x-2y)\,dA = \int_0^6\int_0^{(1/2)(6-x)} \frac{1}{3}(6-x-2y)\,dy\,dx$$

$$= \int_0^6 \frac{1}{3}(6y - xy - y^2)\Big|_0^{(1/2)(6-x)} dx = \int_0^6 \frac{1}{12}(6-x)^2\,dx = -\frac{1}{36}(6-x)^3\Big|_0^6 = 6.$$

37. The intersection of the solid region with the xy plane is the region R bounded by the lines $x + y = 1$ (or $y = 1 - x$), $x = 0$, and $y = 0$. Thus

$$V = \iint_R (x^2+y^2)\,dA = \int_0^1\int_0^{1-x}(x^2+y^2)\,dy\,dx = \int_0^1 \left(x^2y + \frac{1}{3}y^3\right)\Big|_0^{1-x} dx$$

$$= \int_0^1 \left[x^2 - x^3 + \frac{1}{3}(1-x)^3\right] dx = \left[\frac{1}{3}x^3 - \frac{1}{4}x^4 - \frac{1}{12}(1-x)^4\right]\Big|_0^1 = \frac{1}{6}.$$

39. The intersection of the solid region with the xy plane is the vertically simple region R between the graphs of $y=0$ and $y=\sqrt{4-x^2}$ on $[0,2]$. Thus

$$V=\iint_R y\,dA=\int_0^2\int_0^{\sqrt{4-x^2}} y\,dy\,dx=\int_0^2 \frac{1}{2}y^2\Big|_0^{\sqrt{4-x^2}}\,dx=\int_0^2\left(2-\frac{1}{2}x^2\right)dx=\left(2x-\frac{1}{6}x^3\right)\Big|_0^2=\frac{8}{3}.$$

41. The intersection of the solid region with the xy plane is the region R in the first quadrant that lies inside the circle $x^2+y^2=1$. Thus R is the vertically simple region between the graphs of $y=0$ and $y=\sqrt{1-x^2}$ on $[0,1]$, so that

$$V=\iint_R \sqrt{1-x^2}\,dA=\int_0^1\int_0^{\sqrt{1-x^2}}\sqrt{1-x^2}\,dy\,dx$$
$$=\int_0^1 y\sqrt{1-x^2}\Big|_0^{\sqrt{1-x^2}}\,dx=\int_0^1(1-x^2)\,dx=\left(x-\frac{1}{3}x^3\right)\Big|_0^1=\frac{2}{3}.$$

43. The intersection of the solid region with the xy plane is the region R between the graphs of $y=x$ and $y=x^3$. These intersect at (x,y) such that $x=y=x^3$, so that $x=-1$, $x=0$, or $x=1$. Since $x^3\geq x$ for $-1\leq x\leq 0$ and $x\geq x^3$ for $0\leq x\leq 1$, we have

$$V=\iint_R xy\,dA=\int_{-1}^0\int_x^{x^3} xy\,dy\,dx+\int_0^1\int_{x^3}^x xy\,dy\,dx=\int_{-1}^0 \frac{1}{2}xy^2\Big|_x^{x^3}\,dx+\int_0^1\frac{1}{2}xy^2\Big|_{x^3}^x\,dx$$
$$=\int_{-1}^0\frac{1}{2}(x^7-x^3)\,dx+\int_0^1\frac{1}{2}(x^3-x^7)\,dx=\frac{1}{2}\left(\frac{1}{8}x^8-\frac{1}{4}x^4\right)\Big|_{-1}^0+\frac{1}{2}\left(\frac{1}{4}x^4-\frac{1}{8}x^8\right)\Big|_0^1=\frac{1}{8}.$$

45.

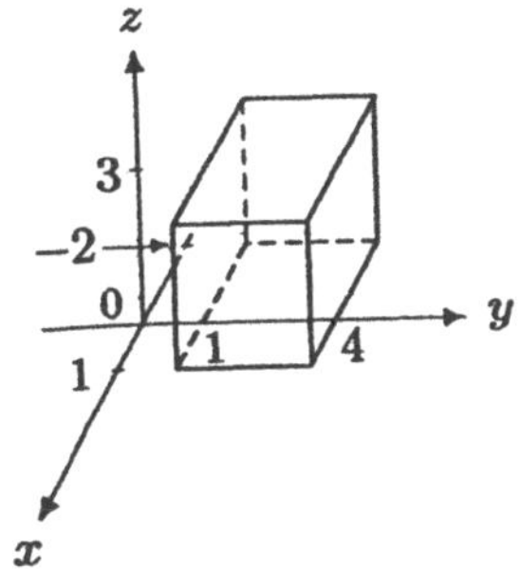

47.

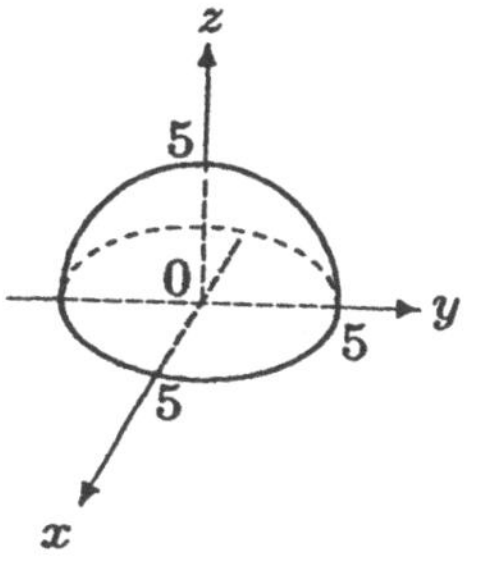

49.

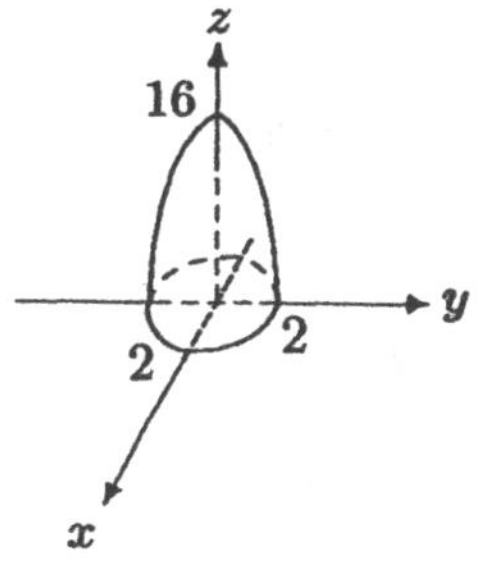

51. Since $x^2\geq\sqrt{x}$ for $1\leq x\leq 4$, we have $A=\int_1^4\int_{\sqrt{x}}^{x^2}1\,dy\,dx=\int_1^4 y\big|_{\sqrt{x}}^{x^2}dx=\int_1^4(x^2-\sqrt{x})\,dx=\left(\frac{1}{3}x^3-\frac{2}{3}x^{3/2}\right)\big|_1^4=\frac{49}{3}$.

53. The parabolas intersect at (x,y) such that $y^2=x=32-y^2$, or $y^2=16$, or $y=-4$ or $y=4$. Since $32-y^2\geq y^2$ for $-4\leq y\leq 4$, we have $A=\int_{-4}^4\int_{y^2}^{32-y^2}1\,dx\,dy=\int_{-4}^4 x\big|_{y^2}^{32-y^2}dy=\int_{-4}^4(32-2y^2)\,dy=\left(32y-\frac{2}{3}y^3\right)\big|_{-4}^4=\frac{512}{3}$.

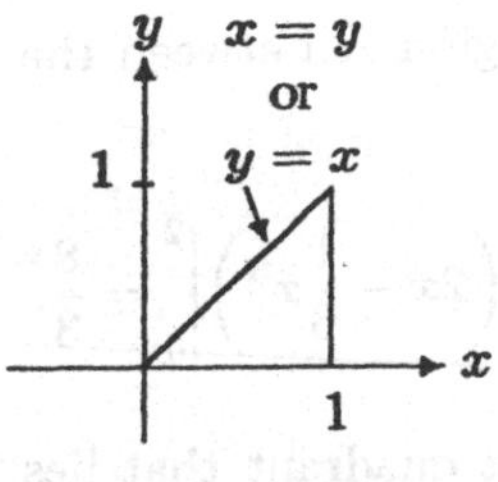

Exercise 57

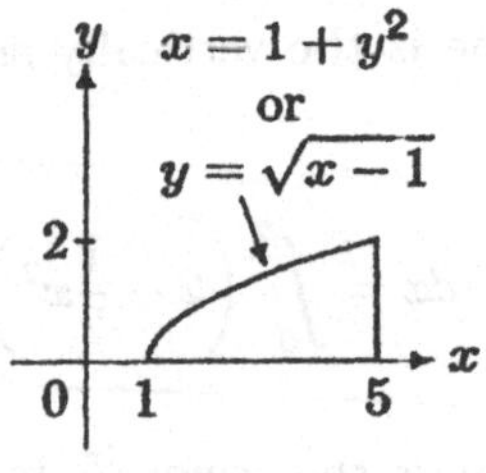

Exercise 59

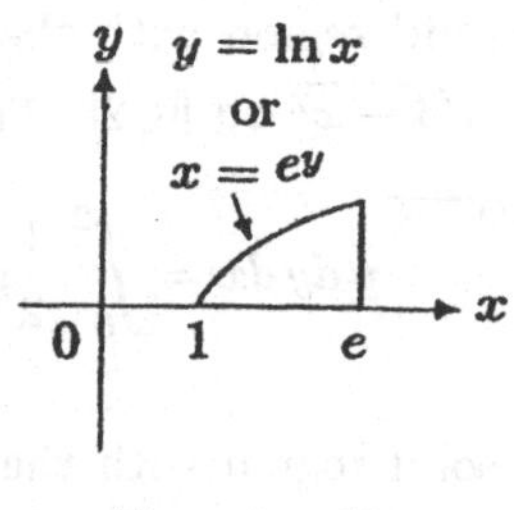

Exercise 61

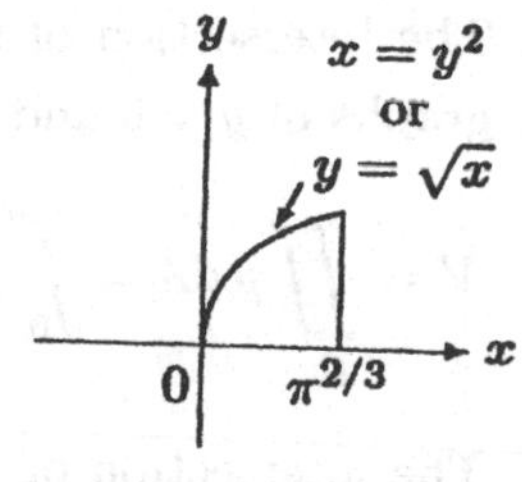

Exercise 63

55. The line $y = x$ and the parabola $x = 2 - y^2$ intersect in the first quadrant at (x, y) such that $y \geq 0$ and $y = x = 2 - y^2$, or $y^2 + y - 2 = 0$, which means that $y = 1$. Since $x = 2 - y^2$ and $y \geq 0$ imply that $y = \sqrt{2 - x}$, and since $\sqrt{2 - x} \geq x$ for $-2 \leq x \leq 1$, we have $A = \int_0^1 \int_x^{\sqrt{2-x}} 1\, dy\, dx = \int_0^1 y\big|_x^{\sqrt{2-x}} dx = \int_0^1 (\sqrt{2 - x} - x)\, dx = \left[-\frac{2}{3}(2 - x)^{3/2} - \frac{1}{2}x^2\right]\Big|_0^1 = \frac{4}{3}\sqrt{2} - \frac{7}{6}$.

57. $$\int_0^1 \int_y^1 e^{(x^2)}\, dx\, dy = \int_0^1 \int_0^x e^{(x^2)}\, dy\, dx = \int_0^1 ye^{(x^2)}\big|_0^x\, dx = \int_0^1 xe^{(x^2)}\, dx = \frac{1}{2}\, e^{(x^2)}\Big|_0^1 = \frac{1}{2}(e - 1)$$

59. $$\int_0^2 \int_{1+y^2}^5 ye^{(x-1)^2}\, dx\, dy = \int_1^5 \int_1^{\sqrt{x-1}} ye^{(x-1)^2}\, dy\, dx = \int_1^5 \frac{1}{2}\, y^2 e^{(x-1)^2}\Big|_0^{\sqrt{x-1}}\, dx$$
$$= \int_1^5 \frac{1}{2}(x - 1)e^{(x-1)^2}\, dx = \frac{1}{4}\, e^{(x-1)^2}\Big|_1^5 = \frac{1}{4}(e^{16} - 1)$$

61. $$\int_1^e \int_0^{\ln x} y\, dy\, dx = \int_0^1 \int_{e^y}^e y\, dx\, dy = \int_0^1 xy\big|_{e^y}^e\, dy = \int_0^1 (ey - ye^y)\, dy$$
$$= \left(\frac{1}{2}ey^2 - ye^y + e^y\right)\Big|_0^1 = \frac{1}{2}e - 1$$

63. $$\int_0^{\pi^{1/3}} \int_{y^2}^{\pi^{3/2}} \sin x^{3/2}\, dx\, dy = \int_0^{\pi^{2/3}} \int_0^{\sqrt{x}} \sin x^{3/2}\, dy\, dx = \int_0^{\pi^{2/3}} y \sin x^{3/2}\big|_0^{\sqrt{x}}\, dx$$
$$= \int_0^{\pi^{2/3}} \sqrt{x} \sin x^{3/2}\, dx = -\frac{2}{3} \cos x^{3/2}\Big|_0^{\pi^{2/3}} = \frac{4}{3}$$

65. Mathematica yields $\int_0^1 \int_0^1 4e^{-(x^2+y^2)}\, dy\, dx \approx 2.23099$.

67. By the symmetry with respect to the line $y = x$ of the square with vertices $(0, 0)$, $(0, 1)$, $(1, 1)$, and $(1, 0)$, and by the fact that the value of $\sqrt{x^4 + y^4}$ is unchanged when x and y are interchanged, it follows that $\int_0^1 \int_0^x \sqrt{x^4 + y^4}\, dy\, dx = \frac{1}{2} \int_0^1 \int_0^1 \sqrt{x^4 + y^4}\, dy\, dx$. Mathematica yields $\int_0^1 \int_0^1 \frac{1}{2}\sqrt{x^4 + y^4}\, dy\, dx \approx 0.272357$.

69. $\iint_R f(x, y)\, dA = 0$ since R is symmetric with respect to the y axis and f is an odd function of x.

Therefore $f_{\text{av}} = 0$.

71. The area of each square in the subdivision of the rectangle is $10 \times 10 = 100$ (square meters). We approximate the volume $V = \iint_R h(x,y)\,dA$, where h is the height of the iceberg at (x,y), by using a Riemann sum with the values shown in the table. We find that $V \approx 100(6+8+7+8+5+7+9+9+5+3+6+8+6+3+0+8+7+4+2+1) = 11{,}200$ (cubic meters).

73. Since the area of the glass surface is 1, the fraction of light absorbed by the entire surface equals $\int_0^1\int_0^1 \frac{1}{3}(1+y+x^2)\,dy\,dx = \int_0^1 \frac{1}{3}(y+\frac{1}{2}y^2+x^2y)\big|_0^1\,dx = \int_0^1 \frac{1}{3}(\frac{3}{2}+x^2)\,dx = (\frac{1}{2}x+\frac{1}{9}x^3)\big|_0^1 = \frac{1}{2}+\frac{1}{9} = \frac{11}{18}$.

14.2 Double Integrals in Polar Coordinates

1. $$\iint_R xy\,dA = \int_0^{2\pi}\int_0^5 (r\cos\theta)(r\sin\theta)r\,dr\,d\theta = \int_0^{2\pi}\int_0^5 r^3\sin\theta\cos\theta\,dr\,d\theta$$
$$= \int_0^{2\pi}\left(\frac{r^4}{4}\sin\theta\cos\theta\right)\bigg|_0^5 d\theta = \int_0^{2\pi}\frac{625}{4}\sin\theta\cos\theta\,d\theta = \frac{625}{8}\sin^2\theta\bigg|_0^{2\pi} = 0$$

3. $$\iint_R (x+y)\,dA = \int_0^{\pi/3}\int_0^2 (r\cos\theta + r\sin\theta)r\,dr\,d\theta = \int_0^{\pi/3}\int_0^2 r^2(\cos\theta+\sin\theta)\,d\theta$$
$$= \int_0^{\pi/3}\left[\frac{r^3}{3}(\cos\theta+\sin\theta)\right]\bigg|_0^2 d\theta = \frac{8}{3}\int_0^{\pi/3}(\cos\theta+\sin\theta)\,d\theta = \frac{8}{3}(\sin\theta-\cos\theta)\bigg|_0^{\pi/3}$$
$$= \frac{8}{3}\left(\frac{1}{2}\sqrt{3}+\frac{1}{2}\right) = \frac{4}{3}(\sqrt{3}+1)$$

5. $$\iint_R (x^2+y^2)\,dA = \int_0^{2\pi}\int_0^{2(1+\sin\theta)} r^3\,dr\,d\theta = \int_0^{2\pi}\frac{r^4}{4}\bigg|_0^{2(1+\sin\theta)} d\theta$$
$$= 4\int_0^{2\pi}(1+\sin\theta)^4\,d\theta = 4\int_0^{2\pi}(1+4\sin\theta+6\sin^2\theta+4\sin^3\theta+\sin^4\theta)\,d\theta$$
$$= 4\int_0^{2\pi}[1+4\sin\theta+3(1-\cos 2\theta)+4\sin\theta(1-\cos^2\theta)]\,d\theta + 4\int_0^{2\pi}\sin^4\theta\,d\theta$$
$$= 4\left(4\theta - 4\cos\theta - \frac{3}{2}\sin 2\theta - 4\cos\theta + \frac{4}{3}\cos^3\theta\right)\bigg|_0^{2\pi} + 4\left(-\frac{1}{4}\sin^3\theta\cos\theta - \frac{3}{8}\sin\theta\cos\theta + \frac{3}{8}\theta\right)\bigg|_0^{2\pi}$$
$$= 32\pi + 3\pi = 35\pi$$

with the third from the last equality coming from (12) in Section 8.1.

7. The hemisphere is bounded above by $z = \sqrt{9 - x^2 - y^2}$, and the region R over which the integral is to be taken is bounded by the circle $x^2 + y^2 = 9$. Thus

$$V = \iint\limits_R \sqrt{9 - x^2 - y^2}\, dA = \int_0^{2\pi} \int_0^3 \sqrt{9 - r^2}\, r\, dr\, d\theta$$

$$= -\frac{1}{3} \int_0^{2\pi} (9 - r^2)^{3/2} \Big|_0^3 d\theta = -\frac{1}{3} \int_0^{2\pi} -27\, d\theta = 18\pi.$$

9. The region R over which the integral is to be taken is bounded by $x^2 + y^2 = 1$. Thus

$$V = \iint\limits_R (4 + x + 2y)\, dA = \int_0^{2\pi} \int_0^1 (4 + r\cos\theta + 2r\sin\theta) r\, dr\, d\theta$$

$$= \int_0^{2\pi} \int_0^1 (4r + r^2\cos\theta + 2r^2\sin\theta)\, dr\, d\theta = \int_0^{2\pi} \left(2r^2 + \frac{r^3}{3}\cos\theta + \frac{2}{3}r^3\sin\theta\right)\Big|_0^1 d\theta$$

$$= \int_0^{2\pi} \left(2 + \frac{1}{3}\cos\theta + \frac{2}{3}\sin\theta\right) d\theta = \left(2\theta + \frac{1}{3}\sin\theta - \frac{2}{3}\cos\theta\right)\Big|_0^{2\pi} = 4\pi.$$

11. The solid region is bounded above by $z = \sqrt{4 - x^2 - y^2}$, and the region R over which the integral is to be taken is bounded by the circles $x^2 + y^2 = 1$ and $x^2 + y^2 = 4$. Thus

$$V = \iint\limits_R \sqrt{4 - x^2 - y^2}\, dA = \int_0^{2\pi} \int_1^2 \sqrt{4 - r^2}\, r\, dr\, d\theta$$

$$= \int_0^{2\pi} -\frac{1}{3}(4 - r^2)^{3/2}\Big|_1^2 d\theta = \sqrt{3} \int_0^{2\pi} 1\, d\theta = 2\sqrt{3}\,\pi.$$

13. The solid region is bounded above by $z^2 = x^2 + y^2$, whose equation can be written as $z = r$. The region R over which the integral is to be taken is bounded by $x^2 + y^2 - 4x = 0$, whose equation in polar coordinates is $r = 4\cos\theta$. Thus

$$V = \iint\limits_R \sqrt{x^2 + y^2}\, dA = \int_{-\pi/2}^{\pi/2} \int_0^{4\cos\theta} r^2\, dr\, d\theta = \int_{-\pi/2}^{\pi/2} \frac{r^3}{3}\Big|_0^{4\cos\theta} d\theta$$

$$= \int_{-\pi/2}^{\pi/2} \frac{64}{3}\cos^3\theta\, d\theta = \frac{64}{3} \int_{-\pi/2}^{\pi/2} \cos\theta(1 - \sin^2\theta)\, d\theta = \frac{64}{3}\left(\sin\theta - \frac{1}{3}\sin^3\theta\right)\Big|_{-\pi/2}^{\pi/2} = \frac{256}{9}.$$

15. $$A = \int_0^{2\pi} \int_0^{2+\sin\theta} r\, dr\, d\theta = \int_0^{2\pi} \frac{1}{2}r^2\Big|_0^{2+\sin\theta} d\theta = \int_0^{2\pi} \left(2 + 2\sin\theta + \frac{1}{2}\sin^2\theta\right) d\theta$$

$$= \int_0^{2\pi} \left[2 + 2\sin\theta + \left(\frac{1}{4} - \frac{1}{4}\cos 2\theta\right)\right] d\theta = \left(\frac{9}{4}\theta - 2\cos\theta - \frac{1}{8}\sin 2\theta\right)\Big|_0^{2\pi} = \frac{9}{2}\pi$$

17. The area A is twice the area A_1 of the portion for which $-\pi/4 \le \theta \le \pi/4$. Now

$$A_1 = \int_{-\pi/4}^{\pi/4}\int_0^{\sqrt{4\cos 2\theta}} r\,dr\,d\theta = \int_{-\pi/4}^{\pi/4} \frac{r^2}{2}\Big|_0^{\sqrt{4\cos 2\theta}} d\theta = \int_{-\pi/4}^{\pi/4} 2\cos 2\theta\,d\theta = \sin 2\theta\Big|_{-\pi/4}^{\pi/4} = 2.$$

Thus $A = 2A_1 = 4$.

19. The limaçon $r = 1 + 2\sin\theta$ intersects the origin for (r, θ) such that $1 + 2\sin\theta = 0$, so that $\sin\theta = -\frac{1}{2}$, and thus $\theta = -\pi/6$, $7\pi/6$, or $11\pi/6$. The area A_1 of the large loop is given by

$$A = \int_{-\pi/6}^{7\pi/6}\int_0^{1+2\sin\theta} r\,dr\,d\theta = \int_{-\pi/6}^{7\pi/6} \frac{r^2}{2}\Big|_0^{1+2\sin\theta} d\theta$$

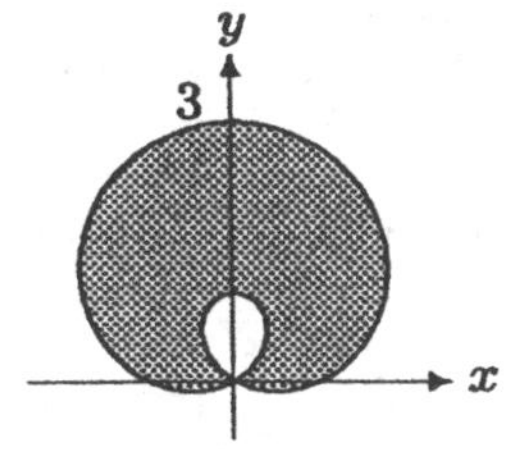

$$= \int_{-\pi/6}^{7\pi/6} \left(\frac{1}{2} + 2\sin\theta + 2\sin^2\theta\right) d\theta$$

$$= \int_{-\pi/6}^{7\pi/6} \left[\frac{1}{2} + 2\sin\theta + (1 - \cos 2\theta)\right] d\theta$$

$$= \left(\frac{3}{2}\theta - 2\cos\theta - \frac{1}{2}\sin 2\theta\right)\Big|_{-\pi/6}^{7\pi/6} = 2\pi + \frac{3}{2}\sqrt{3}.$$

For $7\pi/6 \le \theta \le 11\pi/6$, $1 + 2\sin\theta \le 0$, so the area A_2 of the small loop is given by

$$A_2 = \int_{7\pi/6}^{11\pi/6}\int_0^{-(1+2\sin\theta)} r\,dr\,d\theta = \int_{7\pi/6}^{11\pi/6} \frac{r^2}{2}\Big|_0^{-(1+2\sin\theta)} d\theta$$

$$= \int_{7\pi/6}^{11\pi/6} \left(\frac{1}{2} + 2\sin\theta + 2\sin^2\theta\right) d\theta = \left(\frac{3}{2} - 2\cos\theta - \frac{1}{2}\sin 2\theta\right)\Big|_{7\pi/6}^{11\pi/6} = \pi - \frac{3}{2}\sqrt{3}.$$

Thus the area A of the region inside the large loop and outside the small loop is given by $A = A_1 - A_2 = \pi + 3\sqrt{3}$.

21. The spirals intersect on $[0, 3\pi]$ only for $\theta = 0$. Thus the area is given by

$$A = \int_0^{3\pi}\int_{e^\theta}^{e^{2\theta}} r\,dr\,d\theta = \int_0^{3\pi} \frac{1}{2}r^2\Big|_{e^\theta}^{e^{2\theta}} d\theta = \int_0^{3\pi} \frac{1}{2}(e^{4\theta} - e^{2\theta})\,d\theta$$

$$= \left(\frac{1}{8}e^{4\theta} - \frac{1}{4}e^{2\theta}\right)\Big|_0^{3\pi} = \frac{1}{8}(e^{12\pi} - 2e^{6\pi} + 1).$$

23. $$\int_0^1\int_0^{\sqrt{2-y^2}} 1\,dx\,dy = \int_0^{\pi/4}\int_0^{\sqrt{2}} r\,dr\,d\theta = \int_0^{\pi/4} \frac{r^2}{2}\Big|_0^{\sqrt{2}} d\theta = \int_0^{\pi/4} 1\,d\theta = \frac{1}{4}\pi$$

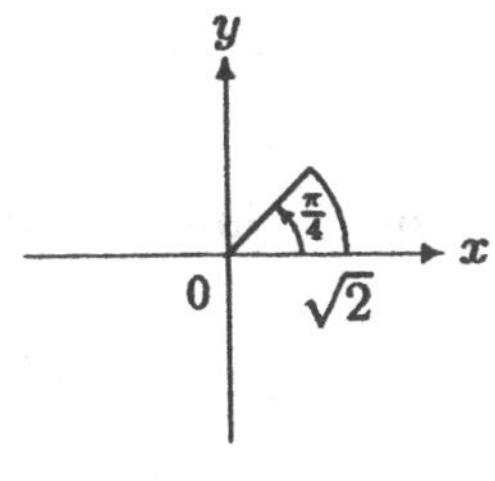

Exercise 23

25. $\displaystyle\int_0^1 \int_0^{\sqrt{1-y^2}} \sin(x^2+y^2)\,dx\,dy = \int_0^{\pi/2} \int_0^1 (\sin r^2) r\,dr\,d\theta = \int_0^{\pi/2} -\frac{1}{2}\cos r^2\Big|_0^1\,d\theta$

$\displaystyle = \int_0^{\pi/2} \left(\frac{1}{2} - \frac{1}{2}\cos 1\right) d\theta = \frac{\pi}{4}(1-\cos 1)$

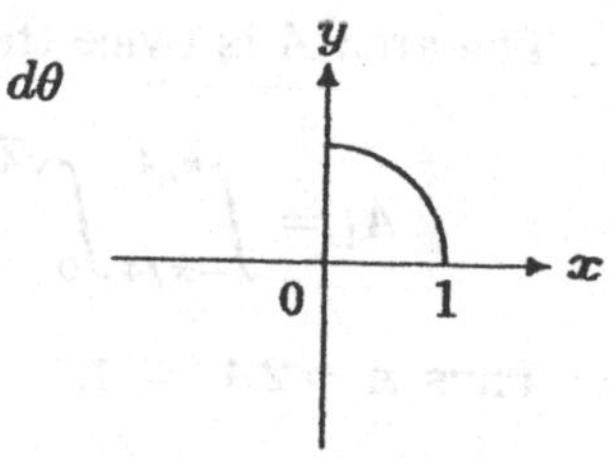

Exercise 25

27. $\displaystyle\int_0^1 \int_0^{\sqrt{1-x^2}} e^{-(x^2+y^2)}\,dy\,dx = \int_0^{\pi/2} \int_0^1 e^{-r^2} r\,dr\,d\theta = \int_0^{\pi/2} -\frac{1}{2}e^{-r^2}\Big|_0^1\,d\theta$

$\displaystyle = \int_0^{\pi/2} \frac{1}{2}(1-e^{-1})\,d\theta = \frac{\pi}{4}(1-e^{-1})$

29. **If $y = \sqrt{x-x^2}$, then $y^2 = x - x^2$, so $x^2+y^2 = x$, which becomes $r = \cos\theta$ in polar coordinates. Thus**

$$\int_0^1 \int_{-\sqrt{x-x^2}}^{\sqrt{x-x^2}} (x^2+y^2)\,dy\,dx = \int_{-\pi/2}^{\pi/2} \int_0^{\cos\theta} r^3\,dr\,d\theta = \int_{-\pi/2}^{\pi/2} \frac{r^4}{4}\Big|_0^{\cos\theta}\,d\theta$$

$$= \frac{1}{4}\int_{-\pi/2}^{\pi/2} \cos^4\theta\,d\theta = \frac{1}{4}\left(\frac{1}{4}\cos^3\theta\sin\theta + \frac{3}{8}\cos\theta\sin\theta + \frac{3}{8}\theta\right)\Big|_{-\pi/2}^{\pi/2} = \frac{3}{32}\pi$$

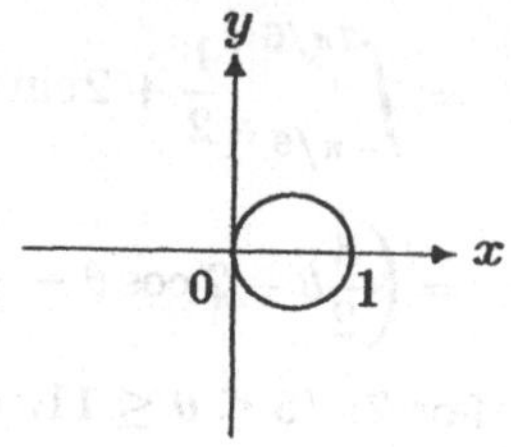

with the next to the last equality coming from (13) in Section 8.1.

31. **a. By reversing the order of integration, we find that the average value of f on $[0,1]$ equals $\frac{1}{1-0}\int_0^1 f(x)\,dx = \int_0^1 \int_x^1 e^{-t^2}\,dt\,dx = \int_0^1 \int_0^t e^{-t^2}\,dx\,dt = \int_0^1 te^{-t^2}\,dt = -\frac{1}{2}e^{-t^2}\big|_0^1 = \frac{1}{2}(1-e^{-1})$.**

b. We have $g_{av} = \frac{1}{1-0}\int_0^1 g(x)\,dx = \int_0^1 \int_0^x e^{-t^2}\,dt\,dx$. We cannot evaluate the iterated integral directly because we do not know an antiderivative of g. If we reverse the order of integration, we obtain $\int_0^1 \int_t^1 e^{-t^2}\,dx\,dt = \int_0^1 e^{-t^2}x\big|_t^1\,dt = \int_0^1 (1-t)e^{-t^2}\,dt = \int_0^1 e^{-t^2}\,dt - \int_0^1 te^{-t^2}\,dt$. Once again we are hampered in evaluating the first integral by not having an antiderivative of e^{-t^2} available. Thus we cannot find the average value of g on $[0,1]$ by integration.

14.3 Surface Area

1. **Let $f(x,y) = \frac{1}{3}(6-x-2y)$. Then $f_x(x,y) = -\frac{1}{3}$ and $f_y(x,y) = -\frac{2}{3}$. The portion of the plane $z = \frac{1}{3}(6-x-2y)$ in the first octant lies over the region R in the first quadrant of the xy plane bounded by the lines $x=0$, $y=0$, and $x+2y=6$. Thus by (1),**

$$S = \iint_R \sqrt{\left(-\frac{1}{3}\right)^2 + \left(-\frac{2}{3}\right)^2 + 1}\,dA = \int_0^6 \int_0^{(1/2)(6-x)} \frac{\sqrt{14}}{3}\,dy\,dx = \int_0^6 \frac{\sqrt{14}}{3}y\Big|_0^{(1/2)(6-x)}\,dx$$

$$= \int_0^6 \frac{\sqrt{14}}{6}(6-x)\,dx = \frac{\sqrt{14}}{6}\left(6x - \frac{1}{2}x^2\right)\Big|_0^6 = 3\sqrt{14}.$$

3. Let $f(x,y) = 9 - x^2 - y^2$. Then $f_x(x,y) = -2x$ and $f_y(x,y) = -2y$. The paraboloid $z = 9 - x^2 - y^2$ intersects the plane $z = 5$ over the circle $5 = 9 - x^2 - y^2$, or $x^2 + y^2 = 4$. Thus the portion of the paraboloid above the plane $z = 5$ lies above the region in the xy plane bounded by the circle $x^2+y^2 = 4$. Thus by (1),

$$S = \iint_R \sqrt{4x^2 + 4y^2 + 1}\, dA = \int_0^{2\pi}\int_0^2 r\sqrt{4r^2+1}\, dr\, d\theta = \int_0^{2\pi} \frac{1}{12}(4r^2+1)^{3/2}\Big|_0^2\, d\theta$$

$$= \int_0^{2\pi} \frac{1}{12}(17^{3/2} - 1)\, d\theta = \frac{\pi}{6}(17^{3/2} - 1).$$

5. The surface area of the portion of the sphere $x^2 + y^2 + z^2 = 16$ is 4 times the surface area of the portion that lies in the first octant. If $f(x,y) = \sqrt{16 - x^2 - y^2}$, then that portion of the sphere is the graph of f on the region R in the first quadrant of the xy plane bounded by the x axis and the circle $x^2 - 4x + y^2 = 0$ (whose equation in polar coordinates is $r = 4\cos\theta$). By (1), the surface area S of the graph of f on R is given as an improper integral by

$$S = \iint_R \sqrt{\frac{x^2}{16 - x^2 - y^2} + \frac{y^2}{16 - x^2 - y^2} + 1}\, dA = \iint_R \sqrt{\frac{x^2 + y^2 + 16 - x^2 - y^2}{16 - x^2 - y^2}}\, dA$$

$$= \iint_R \frac{4}{\sqrt{16 - x^2 - y^2}}\, dA = \int_0^{\pi/2}\int_0^{4\cos\theta} \frac{4r}{\sqrt{16 - r^2}}\, dr\, d\theta = \int_0^{\pi/2} -4\sqrt{16 - r^2}\Big|_0^{4\cos\theta}\, d\theta$$

$$= \int_0^{\pi/2} 16(1 - \sin\theta)\, d\theta = 16(\theta + \cos\theta)\Big|_0^{\pi/2} = 8\pi - 16.$$

Thus the surface area of the entire portion of the sphere that is inside the cylinder is $4(8\pi - 16) = 32(\pi - 2)$.

7. Let $f(x,y) = x^2$. Then $f_x(x,y) = 2x$ and $f_y(x,y) = 0$. The region R bounded by the given triangle is bounded by the lines $y = 0$, $x = 1$, and $y = x$. Thus by (1),

$$S = \iint_R \sqrt{4x^2 + 0^2 + 1}\, dA = \int_0^1\int_0^x \sqrt{4x^2+1}\, dy\, dx = \int_0^1 y\sqrt{4x^2+1}\Big|_0^x\, dx$$

$$= \int_0^1 x\sqrt{4x^2+1}\, dx = \frac{1}{12}(4x^2+1)^{3/2}\Big|_0^1 = \frac{1}{12}(5^{3/2} - 1).$$

9. If $x^2 + y^2 + z^2 = 14z$, then $x^2 + y^2 + (z-7)^2 = 49$, or $z = 7 \pm \sqrt{49 - x^2 - y^2}$. The sphere and the paraboloid intersect at (x,y,z) such that $5z = x^2 + y^2 = 14z - z^2$, so that $z^2 - 9z = 0$, or $z = 0$ or $z = 9$, and consequently $x^2 + y^2 = 0$ or $x^2 + y^2 = 45$. Thus the portion of the sphere that is inside the paraboloid lies over the region R in the xy plane bounded by the circle $x^2 + y^2 = 45$. Since $z \geq 7$ if

(x, y, z) is a point on the sphere inside the given paraboloid, we let $f(x,y) = 7 + \sqrt{49 - x^2 - y^2}$. Then $f_x(x,y) = -x/\sqrt{49 - x^2 - y^2}$ and $f_y(x,y) = -y/\sqrt{49 - x^2 - y^2}$. Thus by (1),

$$S = \iint_R \sqrt{\frac{x^2}{49 - x^2 - y^2} + \frac{y^2}{49 - x^2 - y^2} + 1}\, dA = \iint_R \frac{7}{\sqrt{49 - x^2 - y^2}}\, dA$$

$$= \int_0^{2\pi} \int_0^{\sqrt{45}} \frac{7r}{\sqrt{49 - r^2}}\, dr\, d\theta = \int_0^{2\pi} -7\sqrt{49 - r^2}\Big|_0^{\sqrt{45}}\, d\theta = \int_0^{2\pi} 35\, d\theta = 70\pi.$$

14.4 Triple Integrals

1. $$\int_0^3 \int_{-1}^1 \int_2^4 (y - xz)\, dz\, dy\, dx = \int_0^3 \int_{-1}^1 \left(yz - \frac{xz^2}{2}\right)\Big|_2^4\, dy\, dx = \int_0^3 \int_{-1}^1 (2y - 6x)\, dy\, dx$$
$$= \int_0^3 (y^2 - 6xy)\Big|_{-1}^1\, dx = \int_0^3 (-12x)\, dx = -6x^2\Big|_0^3 = -54$$

3. $$\int_{-1}^1 \int_0^x \int_{x-y}^{x+y} (z - 2x - y)\, dz\, dy\, dx = \int_{-1}^1 \int_0^x \left(\frac{z^2}{2} - 2xz - yz\right)\Big|_{x-y}^{x+y}\, dy\, dx$$
$$= \int_{-1}^1 \int_0^x \left[\frac{(x+y)^2}{2} - 2x(x+y) - y(x+y) - \frac{(x-y)^2}{2} + 2x(x-y) + y(x-y)\right] dy\, dx$$
$$= \int_{-1}^1 \int_0^x (-2xy - 2y^2)\, dy\, dx = \int_{-1}^1 \left(-xy^2 - \frac{2}{3}y^3\right)\Big|_0^x\, dx = \int_{-1}^1 -\frac{5}{3}x^3\, dx = -\frac{5}{12}x^4\Big|_{-1}^1 = 0$$

5. $$\int_0^{\ln 3} \int_0^1 \int_0^y (z^2 + 1)e^{(y^2)}\, dx\, dz\, dy = \int_0^{\ln 3} \int_0^1 (z^2 + 1)e^{(y^2)}x\Big|_0^y\, dz\, dy = \int_0^{\ln 3} \int_0^1 (z^2 + 1)ye^{(y^2)}\, dz\, dy$$
$$= \int_0^{\ln 3} \left[\left(\frac{z^3}{3} + z\right) ye^{(y^2)}\right]\Big|_0^1\, dy = \frac{4}{3}\int_0^{\ln 3} ye^{(y^2)}\, dy = \frac{2}{3}e^{(y^2)}\Big|_0^{\ln 3} = \frac{2}{3}e^{(\ln 3)^2} - \frac{2}{3}$$

7. $$\int_{-13}^{13} \int_1^e \int_1^{1/\sqrt{x}} z(\ln x)^2\, dz\, dx\, dy = \int_{-13}^{13} \int_1^e \frac{1}{2}z^2(\ln x)^2\Big|_0^{1/\sqrt{x}}\, dx\, dy = \int_{-13}^{13} \int_1^e [1/(2x)](\ln x)^2\, dx\, dy$$
$$= \int_{-13}^{13} \frac{1}{6}(\ln x)^3\Big|_1^e\, dy = \int_{-13}^{13} \frac{1}{6}[(\ln e)^3 - (\ln 1)^3]\, dy = \int_{-13}^{13} \frac{1}{6}\, dy = \frac{1}{6}(26) = \frac{13}{3}$$

9. $\displaystyle\int_0^{\pi/2}\int_0^{\pi/2}\int_0^{\sin z} x^2 \sin y\,dx\,dy\,dz = \int_0^{\pi/2}\int_0^{\pi/2} \frac{x^3}{3}\sin y\bigg|_0^{\sin z} dy\,dz = \int_0^{\pi/2}\int_0^{\pi/2} \frac{1}{3}\sin^3 z \sin y\,dy\,dz$

$\displaystyle = \frac{1}{3}\int_0^{\pi/2} (-\sin^3 z\cos y)\big|_0^{\pi/2}\,dz = \frac{1}{3}\int_0^{\pi/2} \sin^3 z\,dz = \frac{1}{3}\int_0^{\pi/2} \sin z(1-\cos^2 z)\,dz$

$\displaystyle = \frac{1}{3}\left(-\cos z + \frac{1}{3}\cos^3 z\right)\bigg|_0^{\pi/2} = \frac{2}{9}$

11. The region R in the xy plane is the horizontally simple region between the graphs of $x=-y$ and $x=y$ on $[0,1]$; D is the solid region between the graphs of $z=0$ and $z=y$ on R. Thus

$$\iiint_D e^y\,dV = \int_0^1\int_{-y}^{y}\int_0^y e^y\,dz\,dx\,dy = \int_0^1\int_{-y}^{y} ze^y\big|_0^y\,dx\,dy = \int_0^1\int_{-y}^{y} ye^y\,dx\,dy$$

$$= \int_0^1 xye^y\big|_{-y}^{y}\,dy = \int_0^1 2y^2e^y\,dy \overset{\text{parts}}{=} 2y^2e^y\big|_0^1 - \int_0^1 4ye^y\,dy$$

$$\overset{\text{parts}}{=} 2e - \left(4ye^y\big|_0^1\right) + 4\int_0^1 e^y\,dy = 2e - 4e + \left(4e^y\big|_0^1\right) = 2e-4.$$

13. The region R in the xy plane is the horizontally simple region between the graphs of $x=1$ and $x=3$ on $[0,2]$; D is the solid region between the graphs of $z=-2$ and $z=0$ on R. Thus

$$\iiint_D ye^{xy}\,dV = \int_0^2\int_1^3\int_{-2}^0 ye^{xy}\,dz\,dx\,dy = \int_0^2\int_1^3 zye^{xy}\big|_{-2}^0\,dx\,dy$$

$$= 2\int_0^2\int_1^3 ye^{xy}\,dx\,dy = 2\int_0^2 e^{xy}\big|_1^3\,dy = 2\int_0^2 (e^{3y}-e^y)\,dy = \left(\frac{2}{3}e^{3y} - 2e^y\right)\bigg|_0^2 = \frac{2}{3}e^6 - 2e^2 + \frac{4}{3}.$$

15. The region R in the xy plane is the portion of the region $x^2+y^2\le 1$ that is in the first quadrant, so is the vertically simple region between the graph of $y=\sqrt{1-x^2}$ and the x axis on $[0,1]$; D is the solid region between the graphs of $z=\sqrt{x^2+y^2}$ and $z=1$ on R. Thus

$$\iiint_D zy\,dV = \int_0^1\int_0^{\sqrt{1-x^2}}\int_{\sqrt{x^2+y^2}}^1 zy\,dz\,dy\,dx = \int_0^1\int_0^{\sqrt{1-x^2}} \frac{z^2}{2}y\bigg|_{\sqrt{x^2+y^2}}^1\,dy\,dx$$

$$= \frac{1}{2}\int_0^1\int_0^{\sqrt{1-x^2}} [y - y(x^2+y^2)]\,dy\,dx = \frac{1}{2}\int_0^1\int_0^{\sqrt{1-x^2}} (y - x^2y - y^3)\,dy\,dx$$

$$= \frac{1}{2}\int_0^1 \left(\frac{y^2}{2} - \frac{x^2y^2}{2} - \frac{y^4}{4}\right)\bigg|_0^{\sqrt{1-x^2}}\,dx = \frac{1}{2}\int_0^1\left(\frac{1-x^2}{2} - \frac{x^2(1-x^2)}{2} - \frac{(1-x^2)^2}{4}\right)dx$$

$$= \frac{1}{8}\int_0^1 (1-2x^2+x^4)\,dx = \frac{1}{8}\left(x - \frac{2}{3}x^3 + \frac{1}{5}x^5\right)\bigg|_0^1 = \frac{1}{15}.$$

17. The region R in the xy plane is a square, and is the vertically simple region between the graphs of $y=-1$ and $y=1$ on $[-1,1]$; D is the solid region between the graphs of $z=0$ and $z=\sqrt{9-x^2-y^2}$ on R. Thus

$$\iiint_D z\,dV = \int_{-1}^{1}\int_{-1}^{1}\int_0^{\sqrt{9-x^2-y^2}} z\,dz\,dy\,dx = \int_{-1}^{1}\int_{-1}^{1} \left.\frac{z^2}{2}\right|_0^{\sqrt{9-x^2-y^2}} dy\,dx$$

$$= \frac{1}{2}\int_{-1}^{1}\int_{-1}^{1}(9-x^2-y^2)\,dy\,dx = \frac{1}{2}\int_{-1}^{1}\left.\left(9y-x^2y-\frac{y^3}{3}\right)\right|_{-1}^{1} dx$$

$$= \int_{-1}^{1}\left(\frac{26}{3}-x^2\right)dx = \left.\left(\frac{26}{3}x-\frac{x^3}{3}\right)\right|_{-1}^{1} = \frac{50}{3}.$$

19. At any point (x,y,z) of intersection of the cone and cylinder, x and y must satisfy $x^2+y^2=z^2=1-x^2$, so that $y^2=1-2x^2$, and thus $y=-\sqrt{1-2x^2}$ or $y=\sqrt{1-2x^2}$ for $-1/\sqrt{2}\le x\le 1/\sqrt{2}$. Therefore the region R in the xy plane is the vertically simple region between the graphs of $y=-\sqrt{1-2x^2}$ and $y=\sqrt{1-2x^2}$ on $[-1/\sqrt{2},1/\sqrt{2}]$. Then D is the solid region between the graphs of $z=\sqrt{x^2+y^2}$ and $z=\sqrt{1-x^2}$ on R. Thus

$$\iiint_D 3xy\,dV = \int_{-1/\sqrt{2}}^{1/\sqrt{2}}\int_{-\sqrt{1-2x^2}}^{\sqrt{1-2x^2}}\int_{\sqrt{x^2+y^2}}^{\sqrt{1-x^2}} 3xy\,dz\,dy\,dx$$

$$= \int_{-1/\sqrt{2}}^{1/\sqrt{2}}\int_{-\sqrt{1-2x^2}}^{\sqrt{1-2x^2}} \left.3xyz\right|_{\sqrt{x^2+y^2}}^{\sqrt{1-x^2}} dy\,dx = \int_{-1/\sqrt{2}}^{1/\sqrt{2}}\int_{-\sqrt{1-2x^2}}^{\sqrt{1-2x^2}} 3xy\left(\sqrt{1-x^2}-\sqrt{x^2+y^2}\right)dy\,dx$$

$$= \int_{-1/\sqrt{2}}^{1/\sqrt{2}}\left.\left[\frac{3}{2}xy^2\sqrt{1-x^2}-x(x^2+y^2)^{3/2}\right]\right|_{-\sqrt{1-2x^2}}^{\sqrt{1-2x^2}} dx = \int_{-1/\sqrt{2}}^{1/\sqrt{2}} 0\,dx = 0.$$

21. Since the graphs of $y=x$ and $y=2-x$ intersect for (x,y) such that $x=y=2-x$, so that $x=1$, the region R in the xy plane is the vertically simple region between the graphs of $y=x$ and $y=2-x$ on $[0,1]$. Then D is the solid region between the graphs of $z=0$ and $z=10+x+y$ on R. Thus

$$V = \iiint_D 1\,dV = \int_0^1\int_x^{2-x}\int_0^{10+x+y} 1\,dz\,dy\,dx = \int_0^1\int_x^{2-x} \left.z\right|_0^{10+x+y} dy\,dx$$

$$\int_0^1\int_x^{2-x}(10+x+y)\,dy\,dx = \int_0^1 \left.\left(10y+xy+\frac{y^2}{2}\right)\right|_x^{2-x} dx$$

$$= \int_0^1\left[10(2-x)+x(2-x)+\frac{(2-x)^2}{2}-10x-x^2-\frac{x^2}{2}\right]dx$$

$$= \int_0^1(22-20x-2x^2)\,dx = \left.\left(22x-10x^2-\frac{2}{3}x^3\right)\right|_0^1 = \frac{34}{3}.$$

23. Since the graphs of $y = x^2$ and $y = x$ intersect for (x, y) such that $x^2 = y = x$, so that $x = 0$ or $x = 1$, the region R in the xy plane is the vertically simple region between the graphs of $y = x^2$ and $y = x$ on $[0, 1]$. Then D is the solid region between the graphs of $z = -2$ and $z = 4(x^2 + y^2)$ on R. Thus

$$V = \iiint_D 1\, dV = \int_0^1 \int_{x^2}^x \int_{-2}^{4(x^2+y^2)} 1\, dz\, dy\, dx = \int_0^1 \int_{x^2}^x z\Big|_{-2}^{4(x^2+y^2)} dy\, dx$$

$$= \int_0^1 \int_{x^2}^x (4x^2 + 4y^2 + 2)\, dy\, dx = \int_0^1 \left(4x^2y + \frac{4}{3}y^3 + 2y\right)\Big|_{x^2}^x dx$$

$$= \int_0^1 \left[4x^2(x - x^2) + \frac{4}{3}(x^3 - x^6) + 2(x - x^2)\right] dx$$

$$= \int_0^1 \left(-\frac{4}{3}x^6 - 4x^4 + \frac{16}{3}x^3 - 2x^2 + 2x\right) dx$$

$$= \left(-\frac{4}{21}x^7 - \frac{4}{5}x^5 + \frac{4}{3}x^4 - \frac{2}{3}x^3 + x^2\right)\Big|_0^1 = \frac{71}{105}.$$

25. The cone and plane intersect for (x, y, z) such that $h^2(x^2 + y^2) = z^2 = h^2$, so that $x^2 + y^2 = 1$. Thus the region R in the xy plane is circular, and is the vertically simple region between the graphs of $y = -\sqrt{1 - x^2}$ and $y = \sqrt{1 - x^2}$ on $[-1, 1]$. Then D is the solid region between the graphs of $z = h\sqrt{x^2 + y^2}$ and $z = h$ on R. Thus

$$V = \iiint_D 1\, dV = \int_{-1}^1 \int_{-\sqrt{1-x^2}}^{\sqrt{1-x^2}} \int_{h\sqrt{x^2+y^2}}^{h} 1\, dz\, dy\, dx$$

$$= \int_{-1}^1 \int_{-\sqrt{1-x^2}}^{\sqrt{1-x^2}} z\Big|_{h\sqrt{x^2+y^2}}^{h} dy\, dx = \int_{-1}^1 \int_{-\sqrt{1-x^2}}^{\sqrt{1-x^2}} (h - h\sqrt{x^2 + y^2})\, dy\, dx$$

$$= \int_{-1}^1 \int_{-\sqrt{1-x^2}}^{\sqrt{1-x^2}} h\, dy\, dx - \int_{-1}^1 \int_{-\sqrt{1-x^2}}^{\sqrt{1-x^2}} h\sqrt{x^2 + y^2}\, dy\, dx.$$

Now

$$\int_{-1}^1 \int_{-\sqrt{1-x^2}}^{\sqrt{1-x^2}} h\, dy\, dx = \int_{-1}^1 hy\Big|_{-\sqrt{1-x^2}}^{\sqrt{1-x^2}} dx = \int_{-1}^1 2h\sqrt{1 - x^2}\, dx$$

$$\overset{x = \sin u}{=} 2h \int_{-\pi/2}^{\pi/2} \sqrt{1 - \sin^2 u}\, \cos u\, du = 2h \int_{-\pi/2}^{\pi/2} \cos^2 u\, du$$

$$= 2h \int_{-\pi/2}^{\pi/2} \left(\frac{1}{2} + \frac{1}{2}\cos 2u\right) du = h\left(u + \frac{1}{2}\sin 2u\right)\Big|_{-\pi/2}^{\pi/2} = \pi h.$$

Next, by changing to polar coordinates we find that

$$\int_{-1}^1 \int_{-\sqrt{1-x^2}}^{\sqrt{1-x^2}} h\sqrt{x^2 + y^2}\, dy\, dx = \int_0^{2\pi} \int_0^1 hr^2\, dr\, d\theta = \int_0^{2\pi} \frac{hr^3}{3}\Big|_0^1 d\theta = \int_0^{2\pi} \frac{h}{3}\, d\theta = \frac{2\pi}{3}h.$$

Adding our results, we conclude that $V = \pi h - \frac{2}{3}\pi h = \frac{1}{3}\pi h$.

27. The plane $2x + y + z = 1$ intersects the xy plane for (x, y, z) such that $0 = z = 1 - 2x - y$, so that $y = 1 - 2x$. Thus the region R in the xy plane is the vertically simple region between the graphs of $y = 0$ and $y = 1 - 2x$ on $[0, \frac{1}{2}]$. Then D is the solid region between the graphs of $z = 0$ and $z = 1 - 2x - y$ on R.

a. Since $\delta(x, y, z) = z$, the total mass m is given by

$$m = \int_0^{1/2} \int_0^{1-2x} \int_0^{1-2x-y} z\, dz\, dy\, dx = \int_0^{1/2} \int_0^{1-2x} \left. \frac{z^2}{2} \right|_0^{1-2x-y} dy\, dx$$

$$= \frac{1}{2} \int_0^{1/2} \int_0^{1-2x} (1 - 2x - y)^2\, dy\, dx = \frac{1}{2} \int_0^{1/2} \left. -\frac{1}{3}(1 - 2x - y)^3 \right|_0^{1-2x} dx$$

$$= \frac{1}{6} \int_0^{1/2} (1 - 2x)^3\, dx = \left. -\frac{1}{48}(1 - 2x)^4 \right|_0^{1/2} = \frac{1}{48}.$$

b. Since $\delta(x, y, z) = 2y$, the total mass m is given by

$$m = \int_0^{1/2} \int_0^{1-2x} \int_0^{1-2x-y} 2y\, dz\, dy\, dx = \int_0^{1/2} \int_0^{1-2x} 2yz\Big|_0^{1-2x-y} dy\, dx$$

$$= \int_0^{1/2} \int_0^{1-2x} 2y(1 - 2x - y)\, dy\, dx = \int_0^{1/2} \int_0^{1-2x} [2y(1 - 2x) - 2y^2]\, dy\, dx$$

$$= \int_0^{1/2} \left[y^2(1 - 2x) - \frac{2}{3}y^3 \right] \Bigg|_0^{1-2x} dx = \int_0^{1/2} \left[(1 - 2x)^3 - \frac{2}{3}(1 - 2x)^3 \right] dx$$

$$= \int_0^{1/2} \frac{1}{3}(1 - 2x)^3\, dx = \left. -\frac{1}{24}(1 - 2x)^4 \right|_0^{1/2} = \frac{1}{24}.$$

c. Since $\delta(x, y, z) = x$, the total mass m is given by

$$m = \int_0^{1/2} \int_0^{1-2x} \int_0^{1-2x-y} x\, dz\, dy\, dx = \int_0^{1/2} \int_0^{1-2x} xz\Big|_0^{1-2x-y} dy\, dx$$

$$= \int_0^{1/2} \int_0^{1-2x} x(1 - 2x - y)\, dy\, dx = \int_0^{1/2} \int_0^{1-2x} [x(1 - 2x) - xy]\, dy\, dx$$

$$= \int_0^{1/2} \left[x(1 - 2x)y - \frac{xy^2}{2} \right] \Bigg|_0^{1-2x} dx = \int_0^{1/2} \left[x(1 - 2x)^2 - \frac{x}{2}(1 - 2x)^2 \right] dx$$

$$= \int_0^{1/2} \left(\frac{x}{2} - 2x^2 + 2x^3 \right) dx = \left. \left(\frac{x^2}{4} - \frac{2}{3}x^3 + \frac{1}{2}x^4 \right) \right|_0^{1/2} = \frac{1}{96}.$$

29. First $\rho(x,y,z) = z$. Next, R is circular, and is the vertically simple region between the graphs of $y = -\sqrt{9-x^2}$ and $y = \sqrt{9-x^2}$ on $[-3,3]$. Finally, D is the solid region between the graphs of $z = 0$ and $z = \sqrt{9-x^2-y^2}$ on R. Thus the total charge is given by

$$q = \iiint\limits_D z\,dV = \int_{-3}^{3}\int_{-\sqrt{9-x^2}}^{\sqrt{9-x^2}}\int_0^{\sqrt{9-x^2-y^2}} z\,dz\,dy\,dx = \int_{-3}^{3}\int_{-\sqrt{9-x^2}}^{\sqrt{9-x^2}} \left.\frac{z^2}{2}\right|_0^{\sqrt{9-x^2-y^2}} dy\,dx$$

$$= \frac{1}{2}\int_{-3}^{3}\int_{-\sqrt{9-x^2}}^{\sqrt{9-x^2}} (9-x^2-y^2)\,dy\,dx = \frac{1}{2}\int_{-3}^{3}\left[(9-x^2)y - \frac{y^3}{3}\right]\Bigg|_{-\sqrt{9-x^2}}^{\sqrt{9-x^2}} dx$$

$$= \frac{1}{2}\int_{-3}^{3}\frac{4}{3}(9-x^2)^{3/2}\,dx \overset{x=3\sin u}{=} \frac{2}{3}\int_{-\pi/2}^{\pi/2} 27\cos^3 u\,(3\cos u)\,du = 54\int_{-\pi/2}^{\pi/2}\cos^4 u\,du$$

$$= 54\left(\frac{1}{4}\cos^3 u\sin u + \frac{3}{8}\cos u\sin u + \frac{3}{8}u\right)\Bigg|_{-\pi/2}^{\pi/2} = \frac{81}{4}\pi$$

with the next to last equality coming from (13) in Section 7.1.

31. a. The sheet and the plane $z = x$ intersect for (x,y,z) such that $x^2 = z = x$, so that $x = 0$ or $x = 1$. Thus the region R in the xy plane is the square bounded by the lines $y = 0$ and $y = 1$ on $[0,1]$. Finally, D is the solid region between the graphs of $z = x^2$ and $z = x$ on R. Therefore the volume is given by

$$V = \iiint\limits_D 1\,dV = \int_0^1\int_0^1\int_{x^2}^{x} 1\,dz\,dy\,dx = \int_0^1\int_0^1 z\Big|_{x^2}^{x} dy\,dx = \int_0^1\int_0^1 (x-x^2)\,dy\,dx$$

$$= \int_0^1 (x-x^2)y\Big|_0^1 dx = \int_0^1 (x-x^2)\,dx = \left(\frac{x^2}{2} - \frac{x^3}{3}\right)\Bigg|_0^1 = \frac{1}{6}.$$

Consequently the average value of f on D is given by

$$\frac{1}{V}\iiint\limits_D (x+y+z)\,dV = 6\int_0^1\int_0^1\int_{x^2}^{x}(x+y+z)\,dz\,dy\,dx$$

$$= 6\int_0^1\int_0^1\left[(x+y)z + \frac{1}{2}z^2\right]\Bigg|_{x^2}^{x} dy\,dx = 6\int_0^1\int_0^1\left[-\frac{1}{2}x^4 - x^3 + \frac{3}{2}x^2 + (x-x^2)y\right] dy\,dx$$

$$= 6\int_0^1\left[\left(-\frac{1}{2}x^4 - x^3 + \frac{3}{2}x^2\right)y + (x-x^2)\left(\frac{1}{2}y^2\right)\right]\Bigg|_0^1 dx$$

$$= 6\int_0^1\left(-\frac{1}{2}x^4 - x^3 + x^2 + \frac{1}{2}x\right)dx = 6\left(-\frac{1}{10}x^5 - \frac{1}{4}x^4 + \frac{1}{3}x^3 + \frac{1}{4}x^2\right)\Bigg|_0^1 = \frac{7}{5}.$$

b. The paraboloids intersect for (x,y,z) such that $2 - x^2 - y^2 = z = x^2 + y^2$, that is, $x^2 + y^2 = 1$. Thus R is circular, and is the vertically simple region between the graph of $y = \sqrt{1-x^2}$ and the x axis on $[0,1]$. Finally, D is the solid region between the graphs of $z = x^2 + y^2$ and $z = 2 - x^2 - y^2$ on R. Therefore the volume is given by

$$V = \iiint\limits_D 1\,dV = \int_0^1 \int_0^{\sqrt{1-x^2}} \int_{x^2+y^2}^{2-x^2-y^2} 1\,dz\,dy\,dz = \int_0^1 \int_0^{\sqrt{1-x^2}} (2 - 2x^2 - 2y^2)\,dy\,dx.$$

Converting to polar coordinates, we find that

$$V = \int_0^1 \int_0^{\sqrt{1-x^2}} (2 - 2x^2 - 2y^2)\,dy\,dx = \int_0^{\pi/2} \int_0^1 (2 - 2r^2) r\,dr\,d\theta$$

$$= \int_0^{\pi/2} \left(r^2 - \frac{1}{2} r^4 \right)\Big|_0^1 d\theta = \int_0^{\pi/2} \frac{1}{2}\,d\theta = \frac{1}{2} \cdot \frac{\pi}{2} = \frac{1}{4}\pi.$$

Consequently the average value of f on D is given by

$$\frac{1}{V} \iiint\limits_D xy\,dV = \frac{4}{\pi} \int_0^1 \int_0^{\sqrt{1-x^2}} \int_{x^2+y^2}^{2-x^2-y^2} xy\,dz\,dy\,dx = \frac{4}{\pi} \int_0^1 \int_0^{\sqrt{1-x^2}} xyz\Big|_{x^2+y^2}^{2-x^2-y^2} dy\,dx$$

$$= \frac{4}{\pi} \int_0^1 \int_0^{\sqrt{1-x^2}} (2xy - 2x^3y - 2xy^3)\,dy\,dx = \frac{4}{\pi} \int_0^1 \left(xy^2 - x^3y^2 - \frac{1}{2} xy^4 \right)\Big|_0^{\sqrt{1-x^2}} dx$$

$$= \frac{4}{\pi} \int_0^1 \left[x(1-x^2) - x^3(1-x^2) - \frac{1}{2} x(1-x^2)^2 \right] dx = \frac{4}{\pi} \int_0^1 \left(\frac{1}{2} x - x^3 + \frac{1}{2} x^5 \right) dx$$

$$= \frac{4}{\pi} \left(\frac{1}{4} x^2 - \frac{1}{4} x^4 + \frac{1}{12} x^6 \right)\Big|_0^1 = \frac{1}{3\pi}.$$

14.5 Triple Integrals in Cylindrical Coordinates

1. $r \sin\theta = -4$

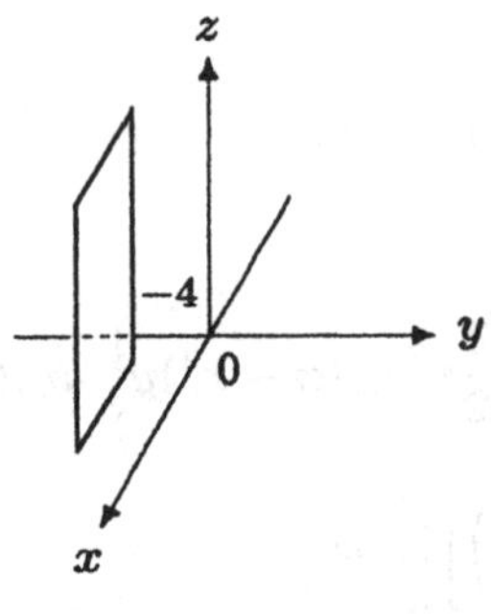

3. $r(\cos\theta + \sin\theta) + z = 3$

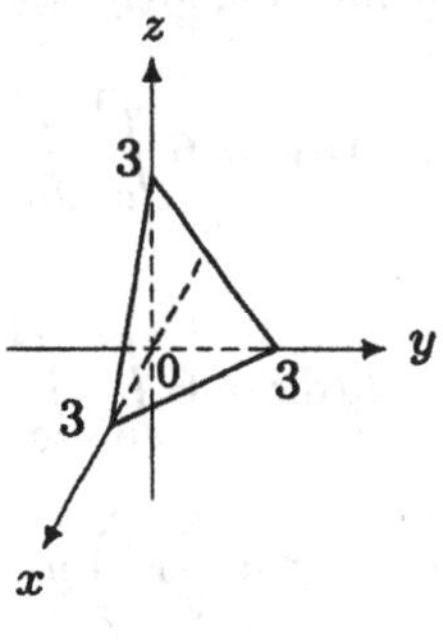

5. $r^2 + z = 1$

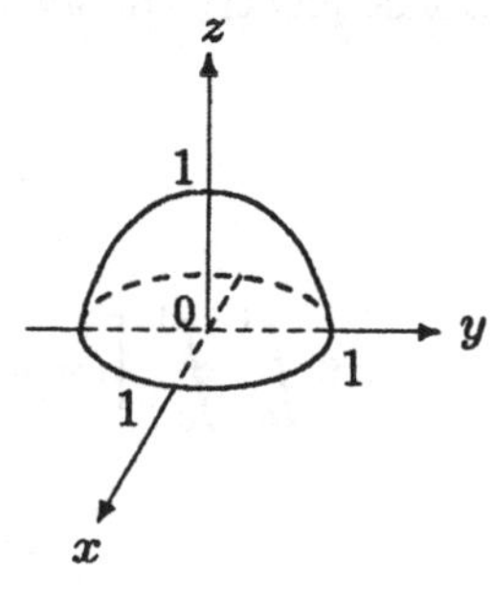

7. $4r^2 = z^2$, or $z = 2r$

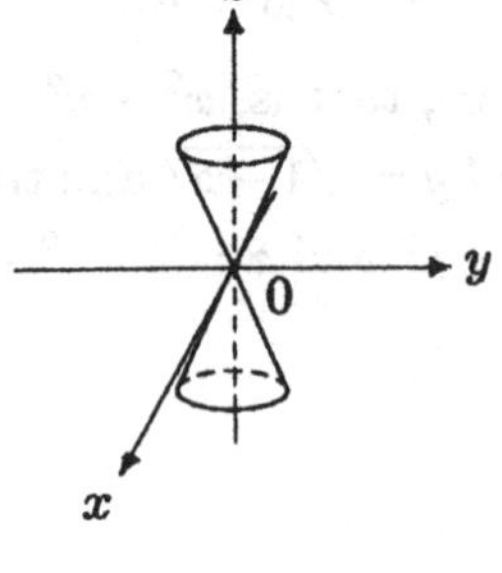

9. $\displaystyle\int_0^{2\pi}\int_1^2\int_0^5 e^z r\,dz\,dr\,d\theta = \int_0^{2\pi}\int_1^2 e^z r\Big|_0^5\,dr\,d\theta = \int_0^{2\pi}\int_1^2 (e^5-1)r\,dr\,d\theta$

$\displaystyle = \int_0^{2\pi}\frac{1}{2}(e^5-1)r^2\Big|_1^2\,d\theta = \int_0^{2\pi}\frac{3}{2}(e^5-1)\,d\theta = 3\pi(e^5-1)$

11. $\displaystyle\int_{-\pi/2}^0\int_0^{2\sin\theta}\int_0^{r^2} r^2\cos\theta\,dz\,dr\,d\theta = \int_{-\pi/2}^0\int_0^{2\sin\theta} zr^2\cos\theta\Big|_0^{r^2}\,dr\,d\theta = \int_{-\pi/2}^0\int_0^{2\sin\theta} r^4\cos\theta\,dr\,d\theta$

$\displaystyle = \int_{-\pi/2}^0 \frac{1}{5}r^5\cos\theta\Big|_0^{2\sin\theta}\,d\theta = \int_{-\pi/2}^0 \frac{32}{5}\sin^5\theta\cos\theta\,d\theta = \frac{16}{15}\sin^6\theta\Big|_{-\pi/2}^0 = -\frac{16}{15}$

13. $\displaystyle\int_{-\pi/4}^{\pi/4}\int_0^{1-2\cos^2\theta}\int_0^1 r\sin\theta\,dz\,dr\,d\theta = \int_{-\pi/4}^{\pi/4}\int_0^{1-2\cos^2\theta} zr\sin\theta\Big|_0^1\,dr\,d\theta$

$\displaystyle = \int_{-\pi/4}^{\pi/4}\int_0^{1-2\cos^2\theta} r\sin\theta\,dr\,d\theta = \int_{-\pi/4}^{\pi/4}\frac{1}{2}r^2\sin\theta\Big|_0^{1-2\cos^2\theta}\,d\theta$

$\displaystyle = \int_{-\pi/4}^{\pi/4}\left(\frac{1}{2}-2\cos^2\theta+2\cos^4\theta\right)\sin\theta\,d\theta = \left(-\frac{1}{2}\cos\theta+\frac{2}{3}\cos^3\theta-\frac{2}{3}\cos^5\theta\right)\Big|_{-\pi/4}^{\pi/4} = 0$

15. D is the solid region between the graphs of $z=0$ and $z=\sqrt{1-r^2}$ on the region bounded by the polar graphs of $r=0$ and $r=1$ for $0\le\theta\le\pi/2$. Therefore

$$\iiint_D z\,dV = \int_0^{\pi/2}\int_0^1\int_0^{\sqrt{1-r^2}} zr\,dz\,dr\,d\theta = \int_0^{\pi/2}\int_0^1 \frac{1}{2}z^2r\Big|_0^{\sqrt{1-r^2}}\,dr\,d\theta$$

$$= \int_0^{\pi/2}\int_0^1\left(\frac{1}{2}r-\frac{1}{2}r^3\right)dr\,d\theta = \int_0^{\pi/2}\left(\frac{1}{4}r^2-\frac{1}{8}r^4\right)\Big|_0^1\,d\theta = \int_0^{\pi/2}\frac{1}{8}\,d\theta = \frac{\pi}{16}.$$

17. D is the solid region between the graphs of $z=0$ and $z=\sqrt{4-r^2}$ on the quarter disk $0\le r\le 2$ with $0\le\theta\le\pi/2$. Therefore

$$\iiint_D xz\,dV = \int_0^{\pi/2}\int_0^2\int_0^{\sqrt{4-r^2}} r^2z\cos\theta\,dz\,dr\,d\theta = \int_0^{\pi/2}\int_0^2 \frac{1}{2}r^2z^2\cos\theta\Big|_0^{\sqrt{4-r^2}}\,dr\,d\theta$$

$$= \int_0^{\pi/2}\int_0^2\left(2r^2-\frac{1}{2}r^4\right)\cos\theta\,dr\,d\theta = \int_0^{\pi/2}\left(\frac{2}{3}r^3-\frac{1}{10}r^5\right)\cos\theta\Big|_0^2\,d\theta$$

$$= \int_0^{\pi/2}\frac{32}{15}\cos\theta\,d\theta = \frac{32}{15}\sin\theta\Big|_0^{\pi/2} = \frac{32}{15}.$$

19. The surface $z = \sqrt{r}$ and the plane $z = 1$ intersect over the circle $r = 1$ in the xy plane. Thus the given region D is the solid region between the graphs of $z = \sqrt{r}$ and $z = 1$ on the disk $0 \le r \le 1$. Therefore

$$V = \iiint_D 1\,dV = \int_0^{2\pi}\int_0^1\int_{\sqrt{r}}^1 r\,dz\,dr\,d\theta = \int_0^{2\pi}\int_0^1 rz\Big|_{\sqrt{r}}^1 dr\,d\theta = \int_0^{2\pi}\int_0^1 (r - r^{3/2})\,dr\,d\theta$$

$$= \int_0^{2\pi}\left(\frac{1}{2}r^2 - \frac{2}{5}r^{5/2}\right)\Big|_0^1 d\theta = \int_0^{2\pi}\frac{1}{10}\,d\theta = \frac{1}{5}\pi.$$

21. The given region D is the solid region between the graphs of $z = 0$ and $z = e^{-r^2}$ on the disk $0 \le r \le 1$. Therefore

$$V = \iiint_D 1\,dV = \int_0^{2\pi}\int_0^1\int_0^{e^{-r^2}} r\,dz\,dr\,d\theta = \int_0^{2\pi}\int_0^1 rz\Big|_0^{e^{-r^2}} dr\,d\theta$$

$$= \int_0^{2\pi}\int_0^1 re^{-r^2}\,dr\,d\theta = \int_0^{2\pi} -\frac{1}{2}e^{-r^2}\Big|_0^1 d\theta = \int_0^{2\pi}\frac{1}{2}(1 - e^{-1})\,d\theta = \pi(1 - e^{-1}).$$

23. At any point (x, y, z) of intersection of the sphere $x^2 + y^2 + z^2 = 4$ and the cone $z^2 = 3x^2 + 3y^2$, x and y must satisfy $x^2 + y^2 + 3x^2 + 3y^2 = 4$, or $r = 1$. Thus the given region D is the solid region between the graphs of $z = \sqrt{3}\,r$ and $z = \sqrt{4 - r^2}$ on the disk $0 \le r \le 1$. Therefore

$$V = \iiint_D 1\,dV = \int_0^{2\pi}\int_0^1\int_{\sqrt{3}r}^{\sqrt{4-r^2}} r\,dz\,dr\,d\theta = \int_0^{2\pi}\int_0^1 rz\Big|_{\sqrt{3}r}^{\sqrt{4-r^2}} dr\,d\theta$$

$$= \int_0^{2\pi}\int_0^1\left(r\sqrt{4 - r^2} - \sqrt{3}r^2\right)dr\,d\theta = \int_0^{2\pi}\left[\frac{-1}{3}(4 - r^2)^{3/2} - \frac{\sqrt{3}}{3}r^3\right]\Bigg|_0^1 d\theta$$

$$= \int_0^{2\pi}\left(\frac{8}{3} - \frac{4\sqrt{3}}{3}\right)d\theta = \frac{8\pi}{3}(2 - \sqrt{3}).$$

25. The cone $z = r$ intersects the planes $z = 1$ and $z = 2$ over the circles $r = 1$ and $r = 2$, respectively, in the xy plane. Thus the given solid region D consists of the regions D_1 and D_2, where D_1 is the solid region between the graphs of $z = 1$ and $z = 2$ on the disk $0 \le r \le 1$, and D_2 is the solid region between the graphs of $z = r$ and $z = 2$ on the ring $1 \le r \le 2$. Therefore

$$V = \iiint_D 1\,dV = \iiint_{D_1} 1\,dV + \iiint_{D_2} 1\,dV = \int_0^{2\pi}\int_0^1\int_1^2 r\,dz\,dr\,d\theta + \int_0^{2\pi}\int_1^2\int_r^2 r\,dz\,dr\,d\theta$$

$$= \int_0^{2\pi}\int_0^1 rz\Big|_1^2 dr\,d\theta + \int_0^{2\pi}\int_1^2 rz\Big|_r^2 dr\,d\theta = \int_0^{2\pi}\int_0^1 r\,dr\,d\theta + \int_0^{2\pi}\int_1^2 (2r - r^2)\,dr\,d\theta$$

$$= \int_0^{2\pi}\frac{1}{2}r^2\Big|_0^1 d\theta + \int_0^{2\pi}\left(r^2 - \frac{1}{3}r^3\right)\Big|_1^2 d\theta = \int_0^{2\pi}\frac{1}{2}\,d\theta + \int_0^{2\pi}\frac{2}{3}\,d\theta = \pi + \frac{4}{3}\pi = \frac{7}{3}\pi.$$

27. At any point (x, y, z) of intersection of the plane $z = y$ and the paraboloid $z = x^2 + y^2$, x, y, and z must satisfy $y = z = x^2 + y^2$, or $r\sin\theta = r^2$, or $r = \sin\theta$. Thus the given region D is the solid region between the graphs of $z = r^2$ and $z = r\sin\theta$ on the region bounded by the circle $r = \sin\theta$. Therefore

$$V = \iiint\limits_D 1\,dV = \int_0^\pi \int_0^{\sin\theta} \int_{r^2}^{r\sin\theta} r\,dz\,dr\,d\theta = \int_0^\pi \int_0^{\sin\theta} rz\Big|_{r^2}^{r\sin\theta} dr\,d\theta$$

$$= \int_0^\pi \int_0^{\sin\theta} (r^2\sin\theta - r^3)\,dr\,d\theta = \int_0^\pi \left(\frac{1}{3}r^3\sin\theta - \frac{1}{4}r^4\right)\Big|_0^{\sin\theta} d\theta$$

$$= \int_0^\pi \frac{1}{12}\sin^4\theta\,d\theta = \frac{1}{12}\left(-\frac{1}{4}\sin^3\theta\cos\theta - \frac{3}{8}\sin\theta\cos\theta + \frac{3}{8}\theta\right)\Big|_0^\pi = \frac{\pi}{32}$$

with the next to last equality coming from (12) in Section 8.1.

29. The given region D is the solid region between the graphs of $z = -\sqrt{a^2 - r^2}$ and $z = \sqrt{a^2 - r^2}$ on the region bounded by the circle $r = a\sin\theta$. Therefore

$$V = \iiint\limits_D 1\,dV = \int_0^\pi \int_0^{a\sin\theta} \int_{-\sqrt{a^2-r^2}}^{\sqrt{a^2-r^2}} r\,dz\,dr\,d\theta = \int_0^\pi \int_0^{a\sin\theta} rz\Big|_{-\sqrt{a^2-r^2}}^{\sqrt{a^2-r^2}} dr\,d\theta$$

$$= \int_0^\pi \int_0^{a\sin\theta} 2r\sqrt{a^2 - r^2}\,dr\,d\theta = \int_0^\pi -\frac{2}{3}(a^2 - r^2)^{3/2}\Big|_0^{a\sin\theta} d\theta = \int_0^\pi \frac{2}{3}a^3\left(1 - |\cos^3\theta|\right)d\theta$$

$$= \int_0^{\pi/2} \frac{2}{3}a^3(1 - \cos^3\theta)\,d\theta + \int_{\pi/2}^\pi \frac{2}{3}a^3(1 + \cos^3\theta)\,d\theta$$

$$= \frac{2}{3}a^3 \int_0^{\pi/2} [1 - (1 - \sin^2\theta)\cos\theta]\,d\theta + \frac{2}{3}a^3 \int_{\pi/2}^\pi [1 + (1 - \sin^2\theta)\cos\theta]\,d\theta$$

$$= \frac{2}{3}a^3\left(\theta - \sin\theta + \frac{1}{3}\sin^3\theta\right)\Big|_0^{\pi/2} + \frac{2}{3}a^3\left(\theta + \sin\theta - \frac{1}{3}\sin^3\theta\right)\Big|_{\pi/2}^\pi$$

$$= \frac{2}{3}a^3\left(\frac{\pi}{2} - \frac{2}{3}\right) + \frac{2}{3}a^3\left(\frac{\pi}{2} - \frac{2}{3}\right) = \frac{4}{3}a^3\left(\frac{\pi}{2} - \frac{2}{3}\right).$$

31. The given region D is the solid region between the graphs of $z = 0$ and $z = r^2$ on the region in the first quadrant bounded by the x axis and the polar graph of $r = 2\sqrt{\cos\theta}$. Therefore

$$V = \iiint\limits_D 1\,dV = \int_0^{\pi/2} \int_0^{2\sqrt{\cos\theta}} \int_0^{r^2} r\,dz\,dr\,d\theta = \int_0^{\pi/2} \int_0^{2\sqrt{\cos\theta}} rz\Big|_0^{r^2} dr\,d\theta$$

$$= \int_0^{\pi/2} \int_0^{2\sqrt{\cos\theta}} r^3\,dr\,d\theta = \int_0^{\pi/2} \frac{1}{4}r^4\Big|_0^{2\sqrt{\cos\theta}} d\theta = \int_0^{\pi/2} 4\cos^2\theta\,d\theta$$

$$= \int_0^{\pi/2} (2 + 2\cos 2\theta)\,d\theta = (2\theta + \sin 2\theta)\Big|_0^{\pi/2} = \pi.$$

33. At any point (x, y, z) of intersection of the cone $z^2 = 9x^2 + 9y^2$ and the plane $z = 9$, x, y, and z must satisfy $81 = z^2 = 9x^2 + 9y^2$, or $r = 3$. Thus the object occupies the solid region D between the graphs of $z = 3r$ and $z = 9$ on the disk $0 \le r \le 3$. Since the mass density is given by $\delta(x, y, z) = 9 - z$, the total mass m is given by

$$m = \iiint_D (9 - z)\,dV = \int_0^{2\pi}\int_0^3\int_{3r}^9 (9 - z)r\,dz\,dr\,d\theta = \int_0^{2\pi}\int_0^3 \left(9z - \frac{1}{2}z^2\right)r\Big|_{3r}^9 dr\,d\theta$$

$$= \int_0^{2\pi}\int_0^3 \left(\frac{81}{2}r - 27r^2 + \frac{9}{2}r^3\right) dr\,d\theta = \int_0^{2\pi} \left(\frac{81}{4}r^2 - 9r^3 + \frac{9}{8}r^4\right)\Big|_0^3 d\theta = \int_0^{2\pi} \frac{243}{8}\,d\theta = \frac{243\pi}{4}.$$

35. At any point (x, y, z) of intersection of the plane $z = h$ and the cone $z^2 = h^2(x^2 + y^2)$, x, y, and z must satisfy $h^2 = z^2 = h^2(x^2 + y^2)$, or $x^2 + y^2 = 1$, or $r = 1$. Thus the given region D is the solid region between the graphs of $z = hr$ and $z = h$ on the disk $0 \le r \le 1$. Therefore

$$V = \iiint_D 1\,dV = \int_0^{2\pi}\int_0^1\int_{hr}^h r\,dz\,dr\,d\theta = \int_0^{2\pi}\int_0^1 rz\Big|_{hr}^h dr\,d\theta = \int_0^{2\pi}\int_0^1 (hr - hr^2)\,dr\,d\theta$$

$$= \int_0^{2\pi} \left(\frac{1}{2}hr^2 - \frac{1}{3}hr^3\right)\Big|_0^1 d\theta = \int_0^{2\pi} \frac{1}{6}h\,d\theta = \frac{1}{3}\pi h.$$

37. Place the coordinate system so that the top of the bowl is on the xy plane, symmetrically placed with respect to the x and y axes. By (7),

$$m = \int_0^{2\pi}\int_0^8\int_{-\sqrt{64-r^2}}^0 \left(1 - \frac{z}{16}\right) r\,dz\,dr\,d\theta$$

$$= \int_0^{2\pi}\int_0^8 \left(z - \frac{z^2}{32}\right) r\Big|_{-\sqrt{64-r^2}}^0 dr\,d\theta$$

$$= \int_0^{2\pi}\int_0^8 \left[r\sqrt{64 - r^2} + \frac{r}{32}(64 - r^2)\right] dr\,d\theta$$

$$= 2\pi\left[-\frac{1}{3}(64 - r^2)^{3/2} + r^2 - \frac{r^4}{128}\right]\Big|_0^8$$

$$= = 2\pi\left[0 + 64 - \frac{64^2}{128} + \frac{1}{3}\cdot 64\cdot 8\right] = 2\pi\left[64 - 32 + \frac{512}{3}\right] = \frac{1216\pi}{3} \approx 1273\,\text{(grams)}.$$

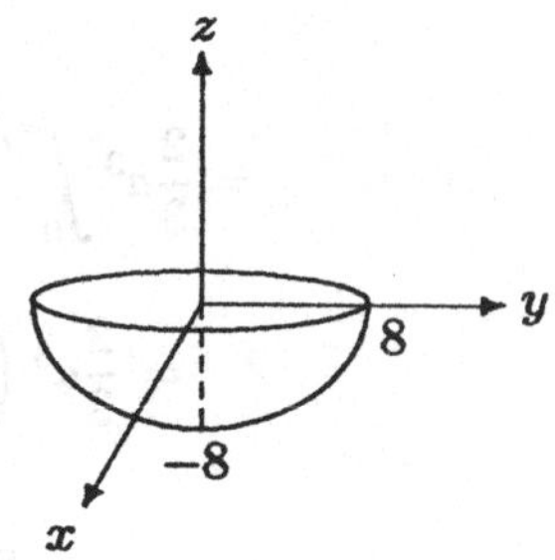

14.6 Triple Integrals in Spherical Coordinates

1. a. $x = 1(1)(\frac{\sqrt{3}}{2}) = \frac{\sqrt{3}}{2}$, $y = 1(1)(\frac{1}{2}) = \frac{1}{2}$, $z = 1(0) = 0$; $(\frac{\sqrt{3}}{2}, \frac{1}{2}, 0)$

   b. $x = 2(0)(0) = 0$, $y = 2(0)(1) = 0$, $z = 2(-1) = -2$; $(0, 0, -2)$

   c. $x = 3(\frac{\sqrt{2}}{2})(\frac{-1}{2}) = \frac{-3}{4}\sqrt{2}$, $y = 3(\frac{\sqrt{2}}{2})(\frac{-\sqrt{3}}{2}) = \frac{-3}{4}\sqrt{6}$, $z = 3(\frac{\sqrt{2}}{2}) = \frac{3}{2}\sqrt{2}$; $(\frac{-3}{4}\sqrt{2}, \frac{-3}{4}\sqrt{6}, \frac{3}{2}\sqrt{2})$

   d. $x = \frac{1}{2}(\frac{\sqrt{3}}{2})(\frac{-\sqrt{2}}{2}) = \frac{-\sqrt{6}}{8}$, $y = \frac{1}{2}(\frac{\sqrt{3}}{2})(\frac{-\sqrt{2}}{2}) = \frac{-\sqrt{6}}{8}$, $z = \frac{1}{2}(\frac{1}{2}) = \frac{1}{4}$; $(\frac{-\sqrt{6}}{8}, \frac{-\sqrt{6}}{8}, \frac{1}{4})$

   e. $x = 1(0)(\frac{-\sqrt{3}}{2}) = 0$, $y = 1(0)(\frac{-1}{2}) = 0$, $z = 1(1) = 1$; $(0, 0, 1)$

   f. $x = 5(1)(1) = 5$, $y = 5(1)(0) = 0$, $z = 5(0) = 0$; $(5, 0, 0)$

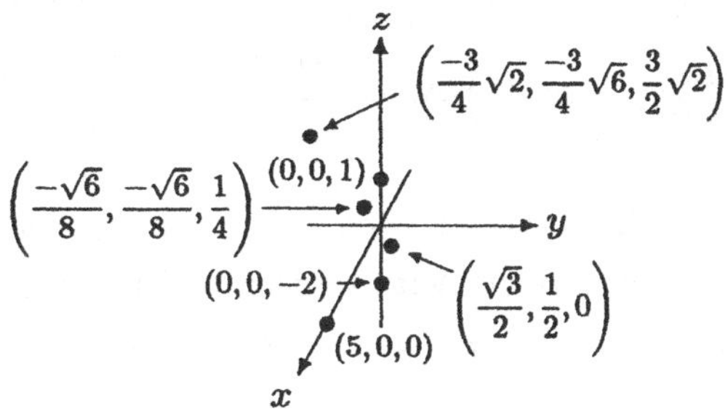

3. $$\int_0^{2\pi}\int_0^{\pi/4}\int_0^1 \rho^2 \sin\phi \, d\rho \, d\phi \, d\theta = \int_0^{2\pi}\int_0^{\pi/4} \left(\frac{\rho^3}{3}\sin\phi\right)\Big|_0^1 d\phi \, d\theta = \int_0^{2\pi}\int_0^{\pi/4} \frac{1}{3}\sin\phi \, d\phi \, d\theta$$
$$= \int_0^{2\pi} \left(\frac{-1}{3}\cos\phi\right)\Big|_0^{\pi/4} d\theta = \int_0^{2\pi} \frac{1}{6}(2-\sqrt{2})\, d\theta = \frac{\pi}{3}(2-\sqrt{2})$$

5. $$\int_0^{\pi}\int_{\pi/2}^{\pi}\int_1^2 \rho^4 \sin^2\phi\cos^2\theta \, d\rho \, d\phi \, d\theta = \int_0^{\pi}\int_{\pi/2}^{\pi} \left(\frac{\rho^5}{5}\sin^2\phi\cos^2\theta\right)\Big|_1^2 d\phi \, d\theta$$
$$= \frac{31}{5}\int_0^{\pi}\int_{\pi/2}^{\pi} \sin^2\phi\cos^2\theta \, d\phi \, d\theta = \frac{31}{5}\int_0^{\pi}\int_{\pi/2}^{\pi} \left(\frac{1}{2} - \frac{1}{2}\cos 2\phi\right)\cos^2\theta \, d\phi \, d\theta$$
$$= \frac{31}{5}\int_0^{\pi} \left[\left(\frac{1}{2}\phi - \frac{1}{4}\sin 2\phi\right)\cos^2\theta\right]\Big|_{\pi/2}^{\pi} d\theta = \frac{31}{20}\pi\int_0^{\pi}\cos^2\theta \, d\theta = \frac{31}{20}\pi\int_0^{\pi}\left(\frac{1}{2} + \frac{1}{2}\cos 2\theta\right) d\theta$$
$$= \frac{31}{20}\pi\left(\frac{1}{2}\theta + \frac{1}{4}\sin 2\theta\right)\Big|_0^{\pi} = \frac{31}{40}\pi^2$$

7. $$\int_{\pi/4}^{\pi/3}\int_0^{\theta}\int_0^{9\sec\phi} \rho\cos^2\phi\cos\theta \, d\rho \, d\phi \, d\theta = \int_{\pi/4}^{\pi/3}\int_0^{\theta} \left(\frac{\rho^2}{2}\cos^2\phi\cos\theta\right)\Big|_0^{9\sec\phi} d\phi \, d\theta$$
$$= \int_{\pi/4}^{\pi/3}\int_0^{\theta} \frac{81}{2}\cos\theta \, d\phi \, d\theta = \int_{\pi/4}^{\pi/3} \left(\phi\frac{81}{2}\cos\theta\right)\Big|_0^{\theta} d\theta = \frac{81}{2}\int_{\pi/4}^{\pi/3} \theta\cos\theta \, d\theta$$
$$\overset{\text{parts}}{=} \frac{81}{2}(\theta\sin\theta)\Big|_{\pi/4}^{\pi/3} - \frac{81}{2}\int_{\pi/4}^{\pi/3}\sin\theta \, d\theta = \frac{81}{2}(\theta\sin\theta)\Big|_{\pi/4}^{\pi/3} + \frac{81}{2}\cos\theta\Big|_{\pi/4}^{\pi/3}$$
$$= \frac{27}{4}\pi\left(\sqrt{3} - \frac{3}{4}\sqrt{2}\right) + \frac{81}{4}(1-\sqrt{2})$$

9. D is the collection of all points with spherical coordinates (ρ, ϕ, θ) such that $0 \le \theta \le 2\pi$, $0 \le \phi \le \pi$, and $2 \le \rho \le 3$. Since $x^2 = \rho^2 \sin^2 \phi \cos^2 \theta$, we find that

$$\iiint\limits_D x^2\, dV = \int_0^{2\pi} \int_0^{\pi} \int_2^3 (\rho^2 \sin^2 \phi \cos^2 \theta)\rho^2 \sin\phi \, d\rho \, d\phi \, d\theta = \int_0^{2\pi} \int_0^{\pi} \left(\frac{\rho^5}{5} \sin^3 \phi \cos^2 \theta \right)\Bigg|_2^3 d\phi \, d\theta$$

$$= \frac{211}{5} \int_0^{2\pi} \int_0^{\pi} \sin^3 \phi \cos^2 \theta \, d\phi \, d\theta = \frac{211}{5} \int_0^{2\pi} \int_0^{\pi} \sin\phi (1 - \cos^2 \phi) \cos^2 \theta \, d\phi \, d\theta$$

$$= \frac{211}{5} \int_0^{2\pi} \left(-\cos\phi + \frac{1}{3}\cos^3 \phi \right) \cos^2 \theta \Bigg|_0^{\pi} d\theta = \frac{844}{15} \int_0^{2\pi} \cos^2 \theta \, d\theta$$

$$= \frac{844}{15} \int_0^{2\pi} \left(\frac{1}{2} + \frac{1}{2}\cos 2\theta \right) d\theta = \frac{844}{15} \left(\frac{1}{2}\theta + \frac{1}{4}\sin 2\theta \right)\Bigg|_0^{2\pi} = \frac{844}{15}\pi.$$

11. In spherical coordinates the cone has equation $\rho \cos \phi = \sqrt{3}\, \rho \sin \phi$, so that $\cot \phi = \sqrt{3}$, and hence $\phi = \pi/6$. Then D is the collection of all points with spherical coordinates (ρ, ϕ, θ) such that $0 \le \theta \le 2\pi$, $0 \le \phi \le \pi/6$, and $3 \le \rho \le 9$. Since $1/(x^2 + y^2 + z^2) = 1/\rho^2$, we find that

$$\iiint\limits_D \frac{1}{x^2 + y^2 + z^2}\, dV = \int_0^{2\pi} \int_0^{\pi/6} \int_3^9 \frac{1}{\rho^2}(\rho^2 \sin\phi)\, d\rho \, d\phi \, d\theta = \int_0^{2\pi} \int_0^{\pi/6} (\rho \sin\phi)\big|_3^9 \, d\phi \, d\theta$$

$$= \int_0^{2\pi} \int_0^{\pi/6} 6 \sin\phi \, d\phi \, d\theta = 6 \int_0^{2\pi} (-\cos\phi)\big|_0^{\pi/6} d\theta = 6 \int_0^{2\pi} \left(1 - \frac{\sqrt{3}}{2} \right) d\theta = 6\pi(2 - \sqrt{3}).$$

13. In spherical coordinates the planes $x = \sqrt{3}\, y$ and $x = y$ have equations $\theta = \pi/6$ and $\theta = \pi/4$, respectively. Thus D is the collection of all points with spherical coordinates (ρ, ϕ, θ) such that $\pi/6 \le \theta \le \pi/4$, $0 \le \phi \le \pi/2$, and $0 \le \rho \le 4$. Since $\sqrt{z} = \sqrt{\rho \cos \phi}$, we find that

$$\iiint\limits_D \sqrt{z}\, dV = \int_{\pi/6}^{\pi/4} \int_0^{\pi/2} \int_0^4 \sqrt{\rho \cos \phi}\, (\rho^2 \sin\phi)\, d\rho \, d\phi \, d\theta$$

$$= \int_{\pi/6}^{\pi/4} \int_0^{\pi/2} \left(\frac{2}{7}\rho^{7/2} \cos^{1/2} \phi \sin\phi \right)\Bigg|_0^4 d\phi \, d\theta = \int_{\pi/6}^{\pi/4} \int_0^{\pi/2} \frac{256}{7} \cos^{1/2} \phi \sin \phi \, d\phi \, d\theta$$

$$= \frac{256}{7} \int_{\pi/6}^{\pi/4} \frac{-2}{3} \cos^{3/2} \phi \Bigg|_0^{\pi/2} d\theta = \frac{512}{21} \int_{\pi/6}^{\pi/4} 1 \, d\theta = \left(\frac{512}{21} \right) \left(\frac{\pi}{12} \right) = \frac{128}{63}\pi.$$

15. D is the collection of all points with spherical coordinates (ρ, ϕ, θ) such that $0 \le \theta \le 2\pi$, $0 \le \phi \le \pi/4$, and $0 \le \rho \le 2$. Thus

$$V = \iiint\limits_D 1\, dV = \int_0^{2\pi} \int_0^{\pi/4} \int_0^2 \rho^2 \sin\phi \, d\rho \, d\phi \, d\theta = \int_0^{2\pi} \int_0^{\pi/4} \left(\frac{\rho^3}{3} \sin \phi \right)\Bigg|_0^2 d\phi \, d\theta$$

$$= \int_0^{2\pi} \int_0^{\pi/4} \frac{8}{3} \sin \phi \, d\phi \, d\theta = \int_0^{2\pi} \left(-\frac{8}{3} \cos\phi \right)\Bigg|_0^{\pi/4} d\theta = \int_0^{2\pi} \left(\frac{8}{3} - \frac{4}{3}\sqrt{2} \right) d\theta = \frac{8}{3}\pi(2 - \sqrt{2}).$$

17. In spherical coordinates the cone has equation $3\rho^2\cos^2\phi = \rho^2\sin^2\phi$, so that $\cot\phi = 1/\sqrt{3}$, and thus $\phi = \pi/3$. Therefore D is the collection of all points with spherical coordinates (ρ, ϕ, θ) such that $0 \le \theta \le 2\pi$, $\pi/3 \le \phi \le \pi$, and $1 \le \rho \le 2$. Thus

$$V = \iiint\limits_D 1\,dV = \int_0^{2\pi}\int_{\pi/3}^{\pi}\int_1^2 \rho^2\sin\phi\,d\rho\,d\phi\,d\theta = \int_0^{2\pi}\int_{\pi/3}^{\pi}\left(\frac{\rho^3}{3}\sin\phi\right)\Big|_1^2\,d\phi\,d\theta$$

$$= \int_0^{2\pi}\int_{\pi/3}^{\pi}\frac{7}{3}\sin\phi\,d\phi\,d\theta = \int_0^{2\pi}\left(-\frac{7}{3}\cos\phi\right)\Big|_{\pi/3}^{\pi}\,d\theta = \int_0^{2\pi}\frac{7}{2}\,d\theta = 7\pi.$$

19. In spherical coordinates the cone has equation $\rho^2\sin^2\phi = \rho^2\cos^2\phi$, so that $\phi = \pi/4$, and the cylinder has equation $\rho^2\sin^2\phi = 4$, so that $\rho\sin\phi = 2$, and thus $\rho = 2\csc\phi$. Therefore D is the collection of all points with spherical coordinates (ρ, ϕ, θ) such that $0 \le \theta \le 2\pi$, $\pi/4 \le \phi \le \pi/2$, and $0 \le \rho \le 2\csc\phi$. Thus

$$V = \iiint\limits_D 1\,dV = \int_0^{2\pi}\int_{\pi/4}^{\pi/2}\int_0^{2\csc\phi} \rho^2\sin\phi\,d\rho\,d\phi\,d\theta = \int_0^{2\pi}\int_{\pi/4}^{\pi/2}\left(\frac{\rho^3}{3}\sin\phi\right)\Big|_0^{2\csc\phi}\,d\phi\,d\theta$$

$$= \int_0^{2\pi}\int_{\pi/4}^{\pi/2}\frac{8}{3}\csc^2\phi\,d\phi\,d\theta = \int_0^{2\pi}\left(-\frac{8}{3}\cot\phi\right)\Big|_{\pi/4}^{\pi/2}\,d\theta = \int_0^{2\pi}\frac{8}{3}\,d\theta = \frac{16}{3}\pi.$$

21. In spherical coordinates the plane $z = -4\sqrt{3}$ has equation $\rho\cos\phi = -4\sqrt{3}$, or $\rho = -4\sqrt{3}\sec\phi$. Thus the plane $z = -4\sqrt{3}$ and the sphere $x^2 + y^2 + z^2 = 64$ intersect at points having spherical coordinates (ρ, ϕ, θ) satisfying $8 = \rho = -4\sqrt{3}\sec\phi$, so that $\sec\phi = -2/\sqrt{3}$, or $\phi = 5\pi/6$. Thus the points that are inside the sphere $x^2 + y^2 + z^2 = 64$ and lie *below* the plane $z = -4\sqrt{3}$ have spherical coordinates (ρ, ϕ, θ) such that $0 \le \theta \le 2\pi$, $5\pi/6 \le \phi \le \pi$, and $-4\sqrt{3}\sec\phi \le \rho \le 8$. The volume V_1 of the set of such points is given by

$$V_1 = \int_0^{2\pi}\int_{5\pi/6}^{\pi}\int_{-4\sqrt{3}\sec\phi}^{8} \rho^2\sin\phi\,d\rho\,d\phi\,d\theta = \int_0^{2\pi}\int_{5\pi/6}^{\pi}\left(\frac{\rho^3}{3}\sin\phi\right)\Big|_{-4\sqrt{3}\sec\phi}^{8}\,d\phi\,d\theta$$

$$= \int_0^{2\pi}\int_{5\pi/6}^{\pi}\left(\frac{512}{3}\sin\phi + 64\sqrt{3}\sec^3\phi\sin\phi\right)d\phi\,d\theta = \int_0^{2\pi}\left(-\frac{512}{3}\cos\phi + \frac{32\sqrt{3}}{\cos^2\phi}\right)\Big|_{5\pi/6}^{\pi}\,d\theta$$

$$= \int_0^{2\pi}\left(\frac{512}{3} - \frac{288}{3}\sqrt{3}\right)d\theta = \frac{64\pi}{3}(16 - 9\sqrt{3}).$$

Thus $V = \frac{4}{3}\pi 8^3 - V_1 = (64\pi/3)(32 - (16 - 9\sqrt{3})) = (64\pi/3)(16 + 9\sqrt{3})$.

23. Since $\rho \ge 0$ and $0 \le \phi \le \pi$, it follows that if $\rho = \cos\phi$, then $0 \le \phi \le \pi/2$. Thus D is the collection of all points with spherical coordinates (ρ, ϕ, θ) such that $0 \le \theta \le 2\pi$, $0 \le \phi \le \pi/2$, and $0 \le \rho \le \cos\phi$. Thus

$$V = \iiint\limits_D 1\,dV = \int_0^{2\pi}\int_0^{\pi/2}\int_0^{\cos\phi} \rho^2\sin\phi\,d\rho\,d\phi\,d\theta = \int_0^{2\pi}\int_0^{\pi/2}\left(\frac{\rho^3}{3}\sin\phi\right)\Big|_0^{\cos\phi}\,d\phi\,d\theta$$

$$= \int_0^{2\pi}\int_0^{\pi/2} \frac{1}{3}\cos^3\phi\sin\phi\, d\phi\, d\theta = \int_0^{2\pi}\left(-\frac{1}{12}\cos^4\phi\right)\Bigg|_0^{\pi/2} d\theta = \int_0^{2\pi}\frac{1}{12}\, d\theta = \frac{1}{6}\pi.$$

Notice that D is a ball with radius $\frac{1}{2}$, centered at $(0, 0, \frac{1}{2})$.

25. D is the collection of all points with spherical coordinates (ρ, ϕ, θ) such that $0 \le \theta \le 2\pi$, $0 \le \phi \le \pi$, and $2 \le \rho \le 4$. Since $\delta(x, y, z) = 1/\sqrt{x^2 + y^2 + z^2}$, the total mass m is given by

$$m = \int_0^{2\pi}\int_0^{\pi}\int_2^4 \frac{1}{\rho}(\rho^2\sin\phi)\, d\rho\, d\phi\, d\theta = \int_0^{2\pi}\int_0^{\pi}\left(\frac{\rho^2}{2}\sin\phi\right)\Bigg|_2^4 d\phi\, d\theta$$

$$= \int_0^{2\pi}\int_0^{\pi} 6\sin\phi\, d\phi\, d\theta = \int_0^{2\pi}(-6\cos\phi)\big|_0^{\pi}\, d\theta = \int_0^{2\pi} 12\, d\theta = 24\pi.$$

27. Let D be the unit ball centered at the origin. Then

$$\iiint_D f(x, y, z)\, dV = \int_0^{2\pi}\int_0^{\pi}\int_0^1 (\sin\rho^3)\rho^2\sin\varphi\, d\rho\, d\varphi\, d\theta = \int_0^{2\pi}\int_0^{\pi} -\frac{1}{3}\cos\rho^3\Bigg|_0^1 \sin\varphi\, d\varphi\, d\theta$$

$$= \int_0^{2\pi}\int_0^{\pi}\frac{1}{3}(1 - \cos 1)\sin\varphi\, d\varphi\, d\theta = \int_0^{2\pi}\frac{1}{3}(1 - \cos 1)(-\cos\varphi)\Bigg|_0^{\pi} d\theta = \frac{4\pi}{3}(1 - \cos 1).$$

Since the volume of D is $\frac{4}{3}\pi$, the average value of f on D equals

$$\frac{\frac{4}{3}\pi(1 - \cos 1)}{\frac{4}{3}\pi} = 1 - \cos 1.$$

14.7 Moments and Centers of Gravity

1. By symmetry, $\bar{x} = 0$. The graphs of $y = 5$ and $y = 1 + x^2$ intersect at (x, y) such that $5 = y = 1 + x^2$, so that $x = -2$ or $x = 2$. Consequently

$$M_x = \iint_R y\, dA = \int_{-2}^{2}\int_{1+x^2}^{5} y\, dy\, dx = \int_{-2}^{2}\frac{1}{2}y^2\Bigg|_{1+x^2}^{5} dx$$

$$= \int_{-2}^{2}\left(12 - x^2 - \frac{1}{2}x^4\right) dx = \left(12x - \frac{1}{3}x^3 - \frac{1}{10}x^5\right)\Bigg|_{-2}^{2} = \frac{544}{15}.$$

Since

$$A = \int_{-2}^{2}\int_{1+x^2}^{5} 1\, dy\, dx = \int_{-2}^{2} y\big|_{1+x^2}^{5}\, dx = \int_{-2}^{2}(4 - x^2)\, dx = \left(4x - \frac{1}{2}x^3\right)\Bigg|_{-2}^{2} = \frac{32}{3}$$

it follows that $\bar{y} = M_x/A = \frac{544}{15}/\frac{32}{3} = \frac{17}{5}$. Thus $(\bar{x}, \bar{y}) = (0, \frac{17}{5})$.

3. The given region is symmetric with respect to the lines $y = x$ and $x + y = 1$. Thus $\bar{x} = \bar{y}$ and $\bar{x} + \bar{y} = 1$, so that $(\bar{x}, \bar{y}) = (\frac{1}{2}, \frac{1}{2})$.

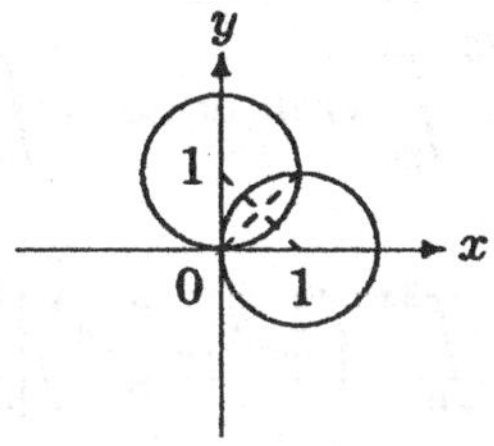

5. By symmetry, $\bar{y} = 0$. Now

$$M_y = \iint_R x\,dA = \int_0^{2\pi}\int_0^{1+\cos\theta} r^2\cos\theta\,dr\,d\theta = \int_0^{2\pi} \frac{1}{3}r^3\cos\theta\Big|_0^{1+\cos\theta}\,d\theta$$

$$= \int_0^{2\pi}\left(\frac{1}{3}\cos\theta + \cos^2\theta + \cos^3\theta + \frac{1}{3}\cos^4\theta\right)d\theta$$

$$= \int_0^{2\pi}\left[\frac{1}{3}\cos\theta + \frac{1}{2} + \frac{1}{2}\cos 2\theta + (1-\sin^2\theta)\cos\theta\right]d\theta + \int_0^{2\pi}\frac{1}{3}\cos^4\theta\,d\theta$$

$$= \left(\frac{1}{3}\sin\theta + \frac{1}{2}\theta + \frac{1}{4}\sin 2\theta + \sin\theta - \frac{1}{3}\sin^3\theta\right)\Big|_0^{2\pi} + \left(\frac{1}{12}\cos^3\theta\sin\theta + \frac{1}{8}\cos\theta\,\sin\theta + \frac{1}{8}\theta\right)\Big|_0^{2\pi} = \frac{5}{4}\pi$$

with the next to last equality coming from (13) in Section 8.1. Since

$$A = \int_0^{2\pi}\int_0^{1+\cos\theta} r\,dr\,d\theta = \int_0^{2\pi}\frac{1}{2}r^2\Big|_0^{1+\cos\theta}d\theta = \int_0^{2\pi}\left(\frac{1}{2} + \cos\theta + \frac{1}{2}\cos^2\theta\right)d\theta$$

$$= \int_0^{2\pi}\left(\frac{1}{2} + \cos\theta + \frac{1}{4} + \frac{1}{4}\cos 2\theta\right)d\theta = \left(\frac{3}{4}\theta + \sin\theta + \frac{1}{8}\sin 2\theta\right)\Big|_0^{2\pi} = \frac{3}{2}\pi$$

it follows that $\bar{x} = M_y/A = (5\pi/4)/(3\pi/2) = \frac{5}{6}$. Thus $(\bar{x}, \bar{y}) = (\frac{5}{6}, 0)$.

7. By symmetry, $\bar{x} = \bar{y} = 0$. Taking $\delta = 1$, we have

$$M_{xy} = \iiint_D z\,dV = \int_0^{2\pi}\int_0^a\int_0^{\sqrt{a^2-r^2}} zr\,dz\,dr\,d\theta = \int_0^{2\pi}\int_0^a \frac{1}{2}z^2r\Big|_0^{\sqrt{a^2-r^2}}dr\,d\theta$$

$$= \int_0^{2\pi}\int_0^a\left(\frac{1}{2}a^2r - \frac{1}{2}r^3\right)dr\,d\theta = \int_0^{2\pi}\left(\frac{1}{4}a^2r^2 - \frac{1}{8}r^4\right)\Big|_0^a d\theta = \int_0^{2\pi}\frac{1}{8}a^4\,d\theta = \frac{1}{4}\pi a^4.$$

Since $m = \frac{2}{3}\pi a^3$, it follows that $\bar{z} = M_{xy}/m = (\pi a^4/4)/(2\pi a^3/3) = \frac{3}{8}a$. Thus $(\bar{x}, \bar{y}, \bar{z}) = (0, 0, \frac{3}{8}a)$.

9. By symmetry, $\bar{x} = \bar{y} = 0$. The plane $z = 1$ and the cone $z^2 = 9x^2 + 9y^2$ intersect over the circle $x^2 + y^2 = \frac{1}{9}$. Taking $\delta = 1$, we have

$$M_{xy} = \iiint\limits_D z\,dV = \int_0^{2\pi}\int_0^{1/3}\int_{3r}^1 zr\,dz\,dr\,d\theta = \int_0^{2\pi}\int_0^{1/3} \frac{1}{2}z^2 r\Big|_{3r}^1 \,dr\,d\theta$$

$$= \int_0^{2\pi}\int_0^{1/3}\left(\frac{1}{2}r - \frac{9}{2}r^3\right)dr\,d\theta = \int_0^{2\pi}\left(\frac{1}{4}r^2 - \frac{9}{8}r^4\right)\Big|_0^{1/3} d\theta = \int_0^{2\pi}\frac{1}{72}\,d\theta = \frac{1}{36}\pi.$$

Since

$$m = \iiint\limits_D 1\,dV = \int_0^{2\pi}\int_0^{1/3}\int_{3r}^1 r\,dz\,dr\,d\theta = \int_0^{2\pi}\int_0^{1/3} rz\Big|_{3r}^1 \,dr\,d\theta$$

$$= \int_0^{2\pi}\int_0^{1/3}(r - 3r^2)\,dr\,d\theta = \int_0^{2\pi}\left(\frac{1}{2}r^2 - r^3\right)\Big|_0^{1/3} d\theta = \int_0^{2\pi}\frac{1}{54}\,d\theta = \frac{1}{27}\pi$$

it follows that $\bar{z} = M_{xy}/m = (\pi/36)/(\pi/27) = \frac{3}{4}$. Thus $(\bar{x}, \bar{y}, \bar{z}) = (0, 0, \frac{3}{4})$.

11. The paraboloids $z = 1 - x^2 - y^2$ and $z = x^2 + y^2$ intersect at (x, y, z) such that $1 - x^2 - y^2 = z = x^2 + y^2$, so that $x^2 + y^2 = \frac{1}{2}$. Taking $\delta = 1$, we have

$$M_{xy} = \iiint\limits_D z\,dV = \int_0^{\pi/2}\int_0^{1/\sqrt{2}}\int_{r^2}^{1-r^2} zr\,dz\,dr\,d\theta = \int_0^{\pi/2}\int_0^{1/\sqrt{2}} \frac{1}{2}z^2 r\Big|_{r^2}^{1-r^2} dr\,d\theta$$

$$= \int_0^{\pi/2}\int_0^{1/\sqrt{2}}\left(\frac{1}{2}r - r^3\right)dr\,d\theta = \int_0^{\pi/2}\left(\frac{1}{4}r^2 - \frac{1}{4}r^4\right)\Big|_0^{1/\sqrt{2}} d\theta = \int_0^{\pi/2}\frac{1}{16}\,d\theta = \frac{1}{32}\pi,$$

$$M_{xz} = \iiint\limits_D y\,dV = \int_0^{\pi/2}\int_0^{1/\sqrt{2}}\int_{r^2}^{1-r^2} r^2\sin\theta\,dz\,dr\,d\theta = \int_0^{\pi/2}\int_0^{1/\sqrt{2}} r^2 z\sin\theta\Big|_{r^2}^{1-r^2} dr\,d\theta$$

$$= \int_0^{\pi/2}\int_0^{1/\sqrt{2}}(r^2 - 2r^4)\sin\theta\,dr\,d\theta = \int_0^{\pi/2}\left(\frac{1}{3}r^3 - \frac{2}{5}r^5\right)\sin\theta\Big|_0^{1/\sqrt{2}} d\theta$$

$$= \int_0^{\pi/2}\frac{\sqrt{2}}{30}\sin\theta\,d\theta = \frac{-\sqrt{2}}{30}\cos\theta\Big|_0^{\pi/2} = \frac{\sqrt{2}}{30},$$

and

$$M_{yz} = \iiint\limits_D x\,dV = \int_0^{\pi/2}\int_0^{1/\sqrt{2}}\int_{r^2}^{1-r^2} r^2\cos\theta\,dz\,dr\,d\theta = \int_0^{\pi/2}\int_0^{1/\sqrt{2}} r^2 z\cos\theta\Big|_{r^2}^{1-r^2} dr\,d\theta$$

$$= \int_0^{\pi/2}\int_0^{1/\sqrt{2}}(r^2 - 2r^4)\cos\theta\,dr\,d\theta = \int_0^{\pi/2}\left(\frac{1}{3}r^3 - \frac{2}{5}r^5\right)\cos\theta\Big|_0^{1/\sqrt{2}} d\theta$$

$$= \int_0^{\pi/2}\frac{\sqrt{2}}{30}\cos\theta\,d\theta = \frac{\sqrt{2}}{30}\sin\theta\Big|_0^{\pi/2} = \frac{\sqrt{2}}{30}.$$

Since

$$m = \iiint\limits_D 1\, dV = \int_0^{\pi/2}\int_0^{1/\sqrt{2}}\int_{r^2}^{1-r^2} r\, dz\, dr\, d\theta = \int_0^{\pi/2}\int_0^{1/\sqrt{2}} rz\Big|_{r^2}^{1-r^2} dr\, d\theta$$

$$= \int_0^{\pi/2}\int_0^{1/\sqrt{2}} (r - 2r^3)\, dr\, d\theta = \int_0^{\pi/2} \left(\frac{1}{2}r^2 - \frac{1}{2}r^4\right)\Big|_0^{1/\sqrt{2}} d\theta = \int_0^{\pi/2} \frac{1}{8}\, d\theta = \frac{1}{16}\pi$$

it follows that

$$\bar{x} = \frac{M_{yz}}{m} = \frac{\sqrt{2}/30}{\pi/16} = \frac{8\sqrt{2}}{15\pi}, \quad \bar{y} = \frac{M_{xz}}{m} = \frac{\sqrt{2}/30}{\pi/16} = \frac{8\sqrt{2}}{15\pi}, \quad \text{and} \quad \bar{z} = \frac{M_{xy}}{m} = \frac{\pi/32}{\pi/16} = \frac{1}{2}.$$

Thus $(\bar{x}, \bar{y}, \bar{z}) = (8\sqrt{2}/(15\pi), 8\sqrt{2}/(15\pi), \frac{1}{2})$.

13. By symmetry, $\bar{x} = \bar{y} = 0$. An equation of the plane that contains the face of the pyramid in the first octant is $x + y + z/2 = 1$, or $z = 2(1 - x - y)$. Thus, taking $\delta = 1$, we have

$$M_{xy} = \iiint\limits_D z\, dV = 4\int_0^1\int_0^{1-x}\int_0^{2(1-x-y)} z\, dz\, dy\, dx = \int_0^1\int_0^{1-x} 2z^2\Big|_0^{2(1-x-y)} dy\, dx$$

$$= \int_0^1\int_0^{1-x} 8(1-x-y)^2\, dy\, dx = \int_0^1 -\frac{8}{3}(1-x-y)^3\Big|_0^{1-x} dx = \int_0^1 \frac{8}{3}(1-x)^3\, dx = -\frac{2}{3}(1-x)^4\Big|_0^1 = \frac{2}{3}.$$

Since

$$m = \iiint\limits_D 1\, dV = 4\int_0^1\int_0^{1-x}\int_0^{2(1-x-y)} 1\, dz\, dy\, dx = \int_0^1\int_0^{1-x} 4z\Big|_0^{2(1-x-y)} dy\, dx$$

$$= \int_0^1\int_0^{1-x} 8(1-x-y)\, dy\, dx = \int_0^1 -4(1-x-y)^2\Big|_0^{1-x} dx = \int_0^1 4(1-x)^2\, dx = -\frac{4}{3}(1-x)^3\Big|_0^1 = \frac{4}{3}$$

it follows that $\bar{z} = M_{xy}/m = \frac{2}{3}/\frac{4}{3} = \frac{1}{2}$. Thus $(\bar{x}, \bar{y}, \bar{z}) = (0, 0, \frac{1}{2})$.

15. Since $\delta(x, y, z)$ does not depend on x or y and since the given region is symmetric with respect to the yz plane, we have $\bar{x} = 0 = \bar{y}$. Next,

$$M_{xy} = \iiint\limits_D z\delta(x, y, z)\, dV = \iiint\limits_D z(z^2 + 1)\, dV = 0$$

so that $\bar{z} = M_{xy}/m = 0/m = 0$. Thus $(\bar{x}, \bar{y}, \bar{z}) = (0, 0, 0)$.

17. $M_{xy} = \iiint\limits_D z\delta(x, y, z)\, dV = \int_0^2\int_0^2\int_0^2 z(1+x)\, dz\, dy\, dx = \int_0^2\int_0^2 \frac{1}{2}z^2(1+x)\Big|_0^2 dy\, dx$

$= \int_0^2\int_0^2 2(1+x)\, dy\, dx = \int_0^2 2(1+x)y\Big|_0^2 dx = \int_0^2 4(1+x)\, dx = (4x + 2x^2)\Big|_0^2 = 16,$

$M_{xz} = \iiint\limits_D y\delta(x, y, z)\, dV = \int_0^2\int_0^2\int_0^2 y(1+x)\, dz\, dy\, dx = \int_0^2\int_0^2 y(1+x)z\Big|_0^2 dy\, dx$

$= \int_0^2\int_0^2 2y(1+x)\, dy\, dx = \int_0^2 y^2(1+x)\Big|_0^2 dx = \int_0^2 4(1+x)\, dx = (4x + 2x^2)\Big|_0^2 = 16,$

and

$$M_{yz} = \iiint\limits_D x\delta(x,y,z)\,dV = \int_0^2\int_0^2\int_0^2 x(1+x)\,dz\,dy\,dx = \int_0^2\int_0^2 x(1+x)z\Big|_0^2\,dy\,dx$$

$$= \int_0^2\int_0^2 2x(1+x)\,dy\,dx = \int_0^2 2x(1+x)y\Big|_0^2\,dx = \int_0^2 4x(1+x)\,dx = (2x^2+\tfrac{4}{3}x^3)\Big|_0^2 = \tfrac{56}{3}.$$

Since

$$m = \iiint\limits_D \delta(x,y,z)\,dV = \int_0^2\int_0^2\int_0^2 (1+x)\,dz\,dy\,dx = \int_0^2\int_0^2 (1+x)z\Big|_0^2\,dy\,dx$$

$$= \int_0^2\int_0^2 2(1+x)\,dy\,dx = \int_0^2 2(1+x)y\Big|_0^2\,dx = \int_0^2 4(1+x)\,dx = (4x+2x^2)\Big|_0^2 = 16,$$

it follows that $\bar{x} = M_{yz}/m = \frac{56}{3}/16 = \frac{7}{6}$, $\bar{y} = M_{xz}/m = \frac{16}{16} = 1$, and $\bar{z} = M_{xy}/m = \frac{16}{16} = 1$. Thus $(\bar{x},\bar{y},\bar{z}) = (\frac{7}{6},1,1)$.

19. Since $\delta(x,y,z) = \sqrt{x^2+y^2}$, we have

$$M_{xy} = \iiint\limits_D z\delta(x,y,z)\,dV = \iiint\limits_D z\sqrt{x^2+y^2}\,dV = \int_0^{2\pi}\int_0^3\int_0^{\sqrt{9-r^2}} zr^2\,dz\,dr\,d\theta$$

$$= \int_0^{2\pi}\int_0^3 \frac{1}{2}z^2r^2\Big|_0^{\sqrt{9-r^2}}\,dr\,d\theta = \int_0^{2\pi}\int_0^3 \left(\frac{9}{2}r^2 - \frac{1}{2}r^4\right)dr\,d\theta$$

$$= \int_0^{2\pi}\left(\frac{3}{2}r^3 - \frac{1}{10}r^5\right)\Big|_0^3\,d\theta = \int_0^{2\pi}\frac{81}{5}\,d\theta = \frac{162}{5}\pi.$$

By symmetry,

$$M_{xz} = \iiint\limits_D y\delta(x,y,z)\,dV = \iiint\limits_D \sqrt{x^2+y^2}\,dV = 0$$

and

$$M_{yz} = \iiint\limits_D x\delta(x,y,z)\,dV = \iiint\limits_D x\sqrt{x^2+y^2}\,dV = 0.$$

Since

$$m = \iiint\limits_D \delta(x,y,z)\,dV = \iiint\limits_D \sqrt{x^2+y^2}\,dV = \int_0^{2\pi}\int_0^3\int_0^{\sqrt{9-r^2}} r^2\,dz\,dr\,d\theta$$

$$= \int_0^{2\pi}\int_0^3 r^2z\Big|_0^{\sqrt{9-r^2}}\,dr\,d\theta = \int_0^{2\pi}\int_0^3 r^2\sqrt{9-r^2}\,dr\,d\theta$$

$$\overset{r=3\sin u}{=} \int_0^{2\pi}\int_0^{\pi/2}(9\sin^2 u)(3\cos u)(3\cos u)\,du\,d\theta = \int_0^{2\pi}\int_0^{\pi/2}\frac{81}{4}\sin^2 2u\,du\,d\theta$$

$$= \int_0^{2\pi}\int_0^{\pi/2}\frac{81}{4}\left(\frac{1}{2}-\frac{1}{2}\cos 4u\right)du\,d\theta = \int_0^{2\pi}\frac{81}{4}\left(\frac{1}{2}u - \frac{1}{8}\sin 4u\right)\Big|_0^{\pi/2}\,d\theta = \int_0^{2\pi}\frac{81}{16}\pi\,d\theta = \frac{81}{8}\pi^2$$

it follows that

$$\bar{x} = M_{yz}/m = 0/m = 0, \quad \bar{y} = M_{xz}/m = 0/m = 0,$$

25. $I_x = \iiint\limits_D (y^2 + z^2)\, dV = \int_0^{2\pi}\int_0^2\int_0^6 2(r^2\sin^2\theta + z^2)r\, dz\, dr\, d\theta$

$$= \int_0^{2\pi}\int_0^2 \left(2r^3 z\sin^2\theta + \frac{2}{3}z^3 r\right)\Big|_0^6 dr\, d\theta = \int_0^{2\pi}\int_0^2 (12r^3\sin^2\theta + 144r)\, dr\, d\theta$$

$$= \int_0^{2\pi} (3r^4\sin^2\theta + 72r^2)\big|_0^2\, d\theta = \int_0^{2\pi} (48\sin^2\theta + 288)\, d\theta$$

$$= \int_0^{2\pi} (24 - 24\cos 2\theta + 288)\, d\theta = (312\theta - 12\sin 2\theta)\big|_0^{2\pi} = 624\pi$$

By symmetry of the region and the mass density, $I_y = 624\pi$. Finally,

$$I_z = \iiint\limits_D (x^2 + y^2)2\, dV = \int_0^{2\pi}\int_0^2\int_0^6 2r^3\, dz\, dr\, d\theta = \int_0^{2\pi}\int_0^2 2r^3 z\big|_0^6\, dr\, d\theta$$

$$= \int_0^{2\pi}\int_0^2 12r^3\, dr\, d\theta = \int_0^{2\pi} 3r^4\big|_0^2\, d\theta = \int_0^{2\pi} 48\, d\theta = 96\pi.$$

27. Set up a coordinate system with the block of ice in the first octant with one corner at the origin. Then

$$I_z = \int_0^{10}\int_0^{10}\int_0^{10} (x^2 + y^2)(0.917)\, dz\, dy\, dx = 0.917\int_0^{10}\int_0^{10} 10(x^2 + y^2)\, dy\, dx$$

$$= 9.17\int_0^{10} \left(x^2 y + \frac{1}{3}y^3\right)\Big|_0^{10} dx = 9.17\int_0^{10} \left(10x^2 + \frac{1000}{3}\right) dx = 9.17\left(\frac{10}{3}x^3 + \frac{1000}{3}x\right)\Big|_0^{10}$$

$$= 9.17\left(\frac{20{,}000}{3}\right) = \frac{1.834}{3} \times 10^5 \approx 6.113 \times 10^4 \text{ (gram centimeters squared).}$$

29. Since the volume of the turntable is $\pi \cdot 15^2 \cdot 3 = 675\pi$ cubic centimeters and the mass is 2 kilograms, $\delta = 2/(675\pi)$ kilograms per cubic centimeter. Using a coordinate system with the z axis along the central axis of the turntable and using cylindrical coordinates, we have $I_z = \int_0^{2\pi}\int_0^{15}\int_0^3 (r^2\delta)r\, dz\, dr\, d\theta = \int_0^{2\pi}\int_0^{15} 3\delta r^3\, dr\, d\theta = \int_0^{2\pi} \frac{3}{4}\delta r^4\big|_0^{15}\, d\theta = 2\pi\left(\frac{3}{4}\delta(15)^4\right) = \frac{3}{2}\pi(15)^4[2/(675\pi)] = 225$ kilogram centimeters squared, or 2.25×10^5 gram centimeters squared.

14.8 Change of Variables in Multiple Integrals

1. $$\frac{\partial(x,y)}{\partial(u,v)} = \begin{vmatrix} \dfrac{\partial x}{\partial u} & \dfrac{\partial x}{\partial v} \\ \dfrac{\partial y}{\partial u} & \dfrac{\partial y}{\partial v} \end{vmatrix} = \begin{vmatrix} 3 & -4 \\ \frac{1}{2} & \frac{1}{6} \end{vmatrix} = (3)\left(\frac{1}{6}\right) - (-4)\left(\frac{1}{2}\right) = \frac{5}{2}$$

3. $$\frac{\partial(x,y)}{\partial(u,v)} = \begin{vmatrix} \dfrac{\partial x}{\partial u} & \dfrac{\partial x}{\partial v} \\ \dfrac{\partial y}{\partial u} & \dfrac{\partial y}{\partial v} \end{vmatrix} = \begin{vmatrix} v & u \\ 2u & 2v \end{vmatrix} = (v)(2v) - (u)(2u) = 2v^2 - 2u^2$$

5. $\dfrac{\partial(x,y)}{\partial(u,v)} = \begin{vmatrix} \dfrac{\partial x}{\partial u} & \dfrac{\partial x}{\partial v} \\ \dfrac{\partial y}{\partial u} & \dfrac{\partial y}{\partial v} \end{vmatrix} = \begin{vmatrix} 0 & e^v \\ e^v & ue^v \end{vmatrix} = (0)(ue^v) - (e^v)(e^v) = -e^{2v}$

7. $\dfrac{\partial(x,y,z)}{\partial(u,v,w)} = \begin{vmatrix} \dfrac{\partial x}{\partial u} & \dfrac{\partial x}{\partial v} & \dfrac{\partial x}{\partial w} \\ \dfrac{\partial y}{\partial u} & \dfrac{\partial y}{\partial v} & \dfrac{\partial y}{\partial w} \\ \dfrac{\partial z}{\partial u} & \dfrac{\partial z}{\partial v} & \dfrac{\partial z}{\partial w} \end{vmatrix} = \begin{vmatrix} a & 0 & 0 \\ 0 & b & 0 \\ 0 & 0 & 1 \end{vmatrix}$

$= a[(b)(1) - (0)(0)] + 0[(0)(0) - (0)(1)] + 0[(0)(0) - (b)(0)] = ab$

9. First,

$$\frac{\partial(x,y)}{\partial(u,v)} = \begin{vmatrix} \dfrac{\partial x}{\partial u} & \dfrac{\partial x}{\partial v} \\ \dfrac{\partial y}{\partial u} & \dfrac{\partial y}{\partial v} \end{vmatrix} = \begin{vmatrix} 3 & 1 \\ 1 & 0 \end{vmatrix} = (3)(0) - (1)(1) = -1.$$

To find S we observe that for $y = 1$ we have $1 = y = u$; for $y = \frac{1}{4}x$ we have $u = \frac{1}{4}(3u + v)$, so that $\frac{1}{4}u = \frac{1}{4}v$, or $u = v$; for $x - 3y = e$ we have $(3u + v) - 3u = e$, or $v = e$. Consequently S is the region in the uv plane bounded by the lines $u = 1$, $u = v$, and $v = e$ (see the figure). By (9),

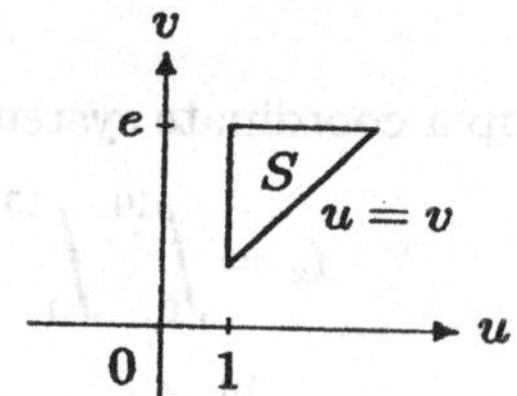

$$\iint\limits_R \frac{y}{x-3y}\,dA = \iint\limits_S \frac{u}{(3u+v)-3u}\left|\frac{\partial(x,y)}{\partial(u,v)}\right| dA = \int_1^e \int_1^v \frac{u}{v}\,|-1|\,du\,dv = \int_1^e \frac{u^2}{2}\frac{1}{v}\bigg|_1^v\,dv$$

$$= \int_1^e \left(\frac{v}{2} - \frac{1}{2v}\right) dv = \left(\frac{v^2}{4} - \frac{1}{2}\ln v\right)\bigg|_1^e = \left(\frac{e^2}{4} - \frac{1}{2}\right) - \left(\frac{1}{4} - 0\right) = \frac{1}{4}(e^2 - 3).$$

11. First,

$$\frac{\partial(x,y)}{\partial(u,v)} = \begin{vmatrix} \dfrac{\partial x}{\partial u} & \dfrac{\partial x}{\partial v} \\ \dfrac{\partial y}{\partial u} & \dfrac{\partial y}{\partial v} \end{vmatrix} = \begin{vmatrix} \frac{3}{5} & \frac{1}{5} \\ -\frac{2}{5} & \frac{1}{5} \end{vmatrix} = \left(\frac{3}{5}\right)\left(\frac{1}{5}\right) - \left(\frac{1}{5}\right)\left(-\frac{2}{5}\right) = \frac{1}{5}.$$

To find S we observe that for $x - y = 2$ we have $\frac{1}{5}(3u+v) - \frac{1}{5}(v-2u) = 2$, so that $u = 2$; for $x - y = -1$ we have $\frac{1}{5}(3u+v) - \frac{1}{5}(v-2u) = -1$, so that $u = -1$; for $2x + 3y = 1$ we have $\frac{2}{5}(3u+v) + \frac{3}{5}(v-2u) = 1$, so that $v = 1$; for $2x + 3y = 0$ we have $\frac{2}{5}(3u+v) + \frac{3}{5}(v-2u) = 0$, so that $v = 0$. Thus S is the square region bounded by the lines $u = -1$, $u = 2$, $v = 1$ and $v = 0$. By (9),

$$\iint\limits_R xy^2\,dA = \iint\limits_S \left[\frac{1}{5}(3u+v)\right]\left[\frac{1}{5}(v-2u)\right]^2 \left|\frac{\partial(x,y)}{\partial(u,v)}\right| dA = \iint\limits_S \frac{1}{125}(3u+v)(v^2 - 4uv + 4u^2)\left(\frac{1}{5}\right) dA$$

$$= \frac{1}{625}\int_0^1\int_{-1}^2 (12u^3 - 8u^2v - uv^2 + v^3)\,du\,dv = \frac{1}{625}\int_0^1 \left(3u^4 - \frac{8}{3}u^3v - \frac{1}{2}u^2v^2 + uv^3\right)\Bigg|_{-1}^2 dv$$

$$= \frac{1}{625}\int_0^1 \left(45 - 24v - \frac{3}{2}v^2 + 3v^3\right) dv = \frac{1}{625}\left(45v - 12v^2 - \frac{1}{2}v^3 + \frac{3}{4}v^4\right)\Bigg|_0^1 = \frac{133}{2500}.$$

13. First,

$$\frac{\partial(x,y)}{\partial(u,v)} = \begin{vmatrix} \dfrac{\partial x}{\partial u} & \dfrac{\partial x}{\partial v} \\[2ex] \dfrac{\partial y}{\partial u} & \dfrac{\partial y}{\partial v} \end{vmatrix} = \begin{vmatrix} \sec v & u\sec v\tan v \\ \tan v & u\sec^2 v \end{vmatrix}$$

$$= (\sec v)(u\sec^2 v) - (u\sec v\tan v)(\tan v) = u\sec v\,(\sec^2 v - \tan^2 v) = u\sec v.$$

To find S we observe that for $x^2 - y^2 = 1$ we have $1 = x^2 - y^2 = u^2\sec^2 v - u^2\tan^2 v = u^2$, and since $u > 0$ by hypothesis, $u = 1$; similarly, for $x^2 - y^2 = 4$ we have $u = 2$; for $x = 2y$ we have $u\sec v = 2u\tan v$, or $u/\cos v = (2u\sin v)/\cos v$; since $u > 0$ by hypothesis, we have $\sin v = \frac{1}{2}$, and since $0 < v < \pi/2$, $v = \pi/6$; similarly, for $x = \sqrt{2}\,y$ we have $\sin v = 1/\sqrt{2} = \frac{1}{2}\sqrt{2}$, and thus $v = \pi/4$. Consequently S is the rectangular region in the uv plane bounded by the lines $u = 1$, $u = 2$, $v = \pi/6$ and $v = \pi/4$. By (9),

$$\iint_R \frac{y}{x}e^{x^2-y^2}\,dA = \iint_S \frac{u\tan v}{u\sec v}e^{u^2\sec^2 v - u^2\tan^2 v}\left|\frac{\partial(x,y)}{\partial(u,v)}\right| dA = \iint_S \frac{\tan v}{\sec v}e^{u^2}u\sec v\,dA$$

$$= \iint_S ue^{u^2}\tan v\,dA = \int_{\pi/6}^{\pi/4}\int_1^2 ue^{u^2}\tan v\,du\,dv = \int_{\pi/6}^{\pi/4} \frac{1}{2}e^{u^2}\tan v\Bigg|_1^2 dv = \int_{\pi/6}^{\pi/4}\frac{1}{2}(e^4 - e)\tan v\,dv$$

$$= -\frac{1}{2}(e^4 - e)\ln\cos v\Bigg|_{\pi/6}^{\pi/4} = -\frac{1}{2}(e^4 - e)\left(\ln\frac{\sqrt{2}}{2} - \ln\frac{\sqrt{3}}{2}\right) = \frac{1}{2}(e^4 - e)\ln\sqrt{\frac{3}{2}} = \frac{1}{4}(e^4 - e)\ln\frac{3}{2}.$$

15. First,

$$\frac{\partial(x,y)}{\partial(u,v)} = \begin{vmatrix} \dfrac{\partial x}{\partial u} & \dfrac{\partial x}{\partial v} \\[2ex] \dfrac{\partial y}{\partial u} & \dfrac{\partial y}{\partial v} \end{vmatrix} = \begin{vmatrix} \cosh v & u\sinh v \\ \sinh v & u\cosh v \end{vmatrix}$$

$$= (\cosh v)(u\cosh v) - (u\sinh v)(\sinh v) = u(\cosh^2 v - \sinh^2 v) = u.$$

To find S we observe that for $x^2 - y^2 = 1$ we have $1 = u^2\cosh^2 v - u^2\sinh^2 v = u^2$, and since $u > 0$ by hypothesis, $u = 1$; similarly, for $x^2 - y^2 = 4$ we have $u = 2$; for $y = 0$ we have $0 = y = u\sinh v$, and since $u > 0$ by hypothesis, $v = 0$; for $y = \frac{3}{5}x$ we have $u\sinh v = \frac{3}{5}u\cosh v$, and since $u > 0$ by hypothesis, $5\sinh v = 3\cosh v$, or $5(e^v - e^{-v}) = 3(e^v + e^{-v})$, or $2e^v = 8e^{-v}$, or $e^{2v} = 4$, so $v = \ln 2$. Consequently S is the rectangular region in the uv plane bounded by the lines $u = 1$, $u = 2$, $v = 0$ and $v = \ln 2$. By (9),

$$\iint_R e^{x^2-y^2}\,dA = \iint_S e^{u^2\cosh^2 v - u^2\sinh^2 v}\left|\frac{\partial(x,y)}{\partial(u,v)}\right| dA = \iint_S e^{u^2}(u)\,dA = \int_0^{\ln 2}\int_1^2 ue^{u^2}\,du\,dv$$

$$= \int_1^3 \frac{15}{16}\frac{1}{v^3}\,dv = -\frac{15}{32}\frac{1}{v^2}\Big|_1^3 = -\frac{15}{32}\left(\frac{1}{9}-1\right) = \frac{5}{12}.$$

21. Let T_R be defined by $u = x/a$ and $v = y/b$. To find S we observe that for $x^2/a^2 + y^2/b^2 = 1$ we have $u^2 + v^2 = 1$, so that S is the unit disk $u^2 + v^2 \le 1$ in the uv plane. Next,

$$\frac{\partial(u,v)}{\partial(x,y)} = \begin{vmatrix} \dfrac{\partial u}{\partial x} & \dfrac{\partial u}{\partial y} \\ \dfrac{\partial v}{\partial x} & \dfrac{\partial v}{\partial y} \end{vmatrix} = \begin{vmatrix} 1/a & 0 \\ 0 & 1/b \end{vmatrix} = \left(\frac{1}{a}\right)\left(\frac{1}{b}\right) - (0)(0) = \frac{1}{ab}.$$

Solving for x in terms of a, we have $x = au$. Then by (9) and (14),

$$\iint_R x^2\,dA = \iint_S (au)^2 \left|\frac{\partial(x,y)}{\partial(u,v)}\right| dA = \iint_S a^2u^2(ab)\,dA = \iint_S a^3bu^2\,dA.$$

Converting to polar coordinates with $u = r\cos\theta$ and $v = r\sin\theta$, we find that

$$\iint_R x^2\,dA = \iint_S a^3bu^2\,dA = \int_0^{2\pi}\int_0^1 a^3b(r^2\cos^2\theta)r\,dr\,d\theta = \int_0^{2\pi} a^3b(\cos^2\theta)\,\frac{r^4}{4}\Big|_0^1\,d\theta$$

$$= \int_0^{2\pi} \frac{a^3b}{4}\cos^2\theta\,d\theta = \int_0^{2\pi}\frac{a^3b}{4}\left(\frac{1}{2}+\frac{1}{2}\cos 2\theta\right)d\theta = \frac{a^3b}{4}\left(\frac{\theta}{2}+\frac{1}{4}\sin 2\theta\right)\Big|_0^{2\pi} = \frac{a^3b\pi}{4}.$$

23. Let T_R be defined by $u = y - x$ and $v = y + x$. To find S we observe that for $x + y = 1$ we have $v = 1$; for $x + y = 2$ we have $v = 2$; for $x = 0$ we have $u = y$ and $v = y$, so that $u = v$; for $y = 0$ we have $u = -x$ and $v = x$, so that $u = -v$. Consequently S is the trapezoidal region in the uv plane bounded by the lines $v = 1$, $v = 2$, $u = v$ and $u = -v$ (see the figure). Next,

$$\frac{\partial(u,v)}{\partial(x,y)} = \begin{vmatrix} \dfrac{\partial u}{\partial x} & \dfrac{\partial u}{\partial y} \\ \dfrac{\partial v}{\partial x} & \dfrac{\partial v}{\partial y} \end{vmatrix} = \begin{vmatrix} -1 & 1 \\ 1 & 1 \end{vmatrix} = (-1)(1) - (1)(1) = -2.$$

By (9) and (14),

$$\iint_R \sin\left[\pi\left(\frac{y-x}{y+x}\right)\right] dA = \iint_S \sin\left(\pi\frac{u}{v}\right)\left|\frac{\partial(x,y)}{\partial(u,v)}\right| dA$$

$$= \int_1^2\int_{-v}^{v}\frac{1}{2}\sin\left(\pi\frac{u}{v}\right)du\,dv = \int_1^2 -\frac{v}{2\pi}\cos\left(\pi\frac{u}{v}\right)\Big|_{-v}^{v}\,dv$$

$$= \int_1^2 -\frac{v}{2\pi}(\cos\pi - \cos(-\pi))\,dv = \int_1^2 0\,dv = 0.$$

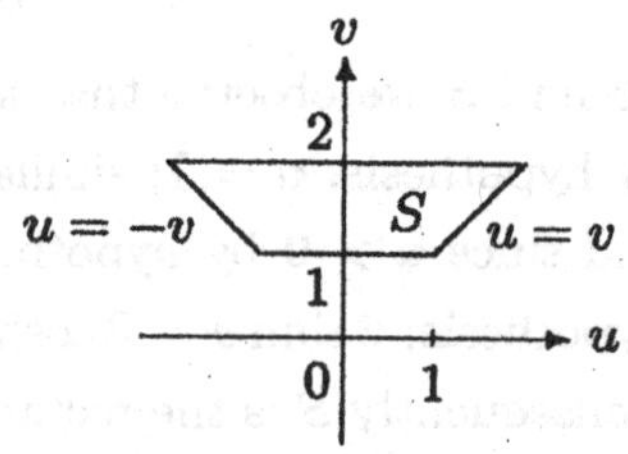

25. Let T_R be defined by $u = 2x - y$ and $v = x + y$. To find S, we observe that for $y = 2x$ we have $2x - y = 0$, so that $u = 0$; for $x+y = 1$ we have $v = 1$; for $x+y = 2$ we have $v = 2$. For $x = 2y$ we solve for x and y in terms of u and v to obtain $u + v = (2x - y) + (x + y) = 3x$, so that $x = \frac{1}{3}(u+v)$, and

$2v-u = 2(x+y)-(2x-y) = 3y$, so that $y = \frac{1}{3}(2v-u)$. Thus for $x = 2y$ we have $\frac{1}{3}(u+v) = \frac{2}{3}(2v-u)$, so that $u = v$. Consequently S is the trapezoidal region in the uv plane bounded by the lines $u = 0$, $v = 1$, $v = 2$ and $u = v$ (see the figure). Next,

$$\frac{\partial(x,y)}{\partial(u,v)} = \begin{vmatrix} \frac{\partial x}{\partial u} & \frac{\partial x}{\partial v} \\ \frac{\partial y}{\partial u} & \frac{\partial y}{\partial v} \end{vmatrix} = \begin{vmatrix} \frac{1}{3} & \frac{1}{3} \\ -\frac{1}{3} & \frac{2}{3} \end{vmatrix} = \left(\frac{1}{3}\right)\left(\frac{2}{3}\right) - \left(\frac{1}{3}\right)\left(-\frac{1}{3}\right) = \frac{1}{3}.$$

By (9),

$$\iint_R e^{(2x-y)/(x+y)}\,dA = \iint_S e^{u/v}\left|\frac{\partial(x,y)}{\partial(u,v)}\right| dA$$

$$= \int_1^2 \int_0^v e^{u/v}\frac{1}{3}\,du\,dv = \int_1^2 \frac{v}{3}e^{u/v}\Big|_0^v\,dv$$

$$= \int_1^2 \frac{e-1}{3}v\,dv = \frac{1}{6}(e-1)v^2\Big|_1^2 = \frac{1}{2}(e-1).$$

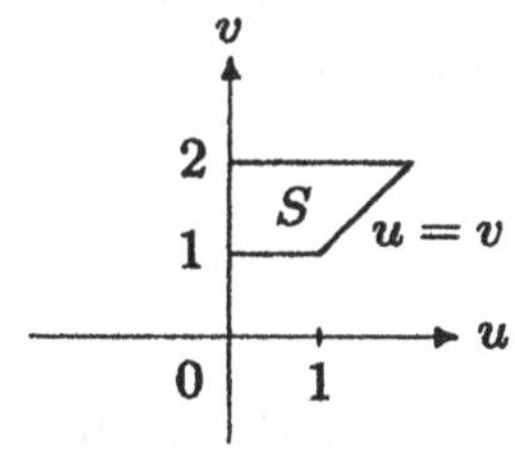

27. First, $A = \iint_R 1\,dA$. Next,

$$\frac{\partial(x,y)}{\partial(u,v)} = \begin{vmatrix} \frac{\partial x}{\partial u} & \frac{\partial x}{\partial v} \\ \frac{\partial y}{\partial u} & \frac{\partial y}{\partial v} \end{vmatrix} = \begin{vmatrix} \cosh v & u\sinh v \\ \sinh v & u\cosh v \end{vmatrix} = (\cosh v)(u\cosh v) - (u\sinh v)(\sinh v) = u.$$

Since $a > 0$, we have $x^2 \geq x^2 - y^2 \geq a^2 > 0$, so that $u > 0$. To find S we observe that for $x^2 - y^2 = a^2$ we have $a^2 = u^2\cosh^2 v - u^2\sinh^2 v = u^2$, so since $a \geq 0$ and $u \geq 0$ we have $u = a$; similarly, if $x^2 - y^2 = b^2$ we have $u = b$; for $y = 0$ we have $0 = u\sinh v$, so since $u \geq 0$ we have $v = 0$; for $y = \frac{1}{2}x$ we have $u\sinh v = \frac{1}{2}u\cosh v$, so since $u > 0$ we have $\sinh v = \frac{1}{2}\cosh v$, so that $e^v - e^{-v} = \frac{1}{2}(e^v + e^{-v})$, or $\frac{1}{2}e^v = \frac{3}{2}e^{-v}$, or $e^{2v} = 3$, or $v = \frac{1}{2}\ln 3$. Consequently S is the rectangular region in the uv plane bounded by the lines $u = a$, $u = b$, $v = 0$ and $v = \frac{1}{2}\ln 3$. By (9),

$$A = \iint_R 1\,dA = \iint_S \left|\frac{\partial(x,y)}{\partial(u,v)}\right| dA = \int_a^b \int_0^{(1/2)\ln 3} u\,dv\,du$$

$$= \int_a^b \frac{1}{2}(\ln 3)u\,du = \frac{1}{2}(\ln 3)\frac{u^2}{2}\Big|_a^b = \frac{1}{4}(b^2 - a^2)\ln 3.$$

29. Let R be the region in the xy plane bounded by the ellipse. Let T be defined by $x = au$ and $y = bv$. Then $x^2/a^2 + y^2/b^2 = u^2 + v^2$, so T maps R onto the region S in the uv plane bounded by the circle $u^2 + v^2 = 1$. Since

$$\frac{\partial(x,y)}{\partial(u,v)} = \begin{vmatrix} a & 0 \\ 0 & b \end{vmatrix} = ab$$

the area A is given by $A = \iint_R 1\,dA = \iint_S ab\,dA = ab\iint_S 1\,dA = ab \cdot (\text{area of } S) = ab \cdot \pi = \pi ab$.

31. For the transformation in Exercise 30 the equation $x^2 - xy + y^2 = 2$ becomes

$$\frac{1}{2}(u^2 - 2uv + v^2) - \frac{1}{2}(u^2 - v^2) + \frac{1}{2}(u^2 + 2uv + v^2) = 2$$

or $\frac{1}{4}u^2 + \frac{3}{4}v^2 = 1$. Also

$$\frac{\partial(x,y)}{\partial(u,v)} = \begin{vmatrix} \frac{1}{\sqrt{2}} & -\frac{1}{\sqrt{2}} \\ \frac{1}{\sqrt{2}} & \frac{1}{\sqrt{2}} \end{vmatrix} = 1.$$

Thus the area of the original ellipse is the same as the rotated ellipse, which by Exercise 29 is $\pi \cdot 2 \cdot (2/\sqrt{3}) = 4\pi/\sqrt{3}$.

14.9 Parametrized Surface

1. Let (x, y, z) be any point on the surface parametrized by $\mathbf{r}$. Then $x = u+v$, $y = u-v$ and $z = 4uv$ for some u and v. Consequently $x^2-y^2 = (u+v)^2-(u-v)^2 = u^2+2uv+v^2-(u^2-2uv+v^2) = 4uv = z$. Therefore the surface parametrized by $\mathbf{r}$ lies on the hyperbolic paraboloid $z = x^2 - y^2$. Moreover, for any x, y, and z satisfying $z = x^2 - y^2$, we can find u and v such that $x = u + v$ and $y = u - v$. Then $z = (u+v)^2 - (u-v)^2 = 4uv$, so every point on the hyperbolic paraboloid is also on the surface parametrized by $\mathbf{r}$. Thus the two surfaces are the same.

3. Assume that a, b, and c are positive. Let (x, y, z) be any point on the surface parametrized by $x = a\cos u \sinh v$, $y = b \sin u \sinh v$, and $z = c\cosh v$ for $0 \le u \le 2\pi$. Then $z > 0$, and

$$\frac{z^2}{c^2} - \frac{x^2}{a^2} - \frac{y^2}{b^2} = \cosh^2 v - \cos^2 u \sinh^2 v - \sin^2 u \sinh^2 v = \cosh^2 v - \sinh^2 v = 1$$

Therefore (x, y, z) lies on the upper sheet of the hyperboloid of the two sheets. Similarly, if $z = -c\cosh v$, then (x, y, z) lies on the lower sheet. Conversely, let (x, y, z) be any point on the hyperboloid of two sheets. If $z \ge c$, we can choose v such that $\cosh v = z/c$. If $z \le -c$, we can choose v such that $\cosh v = -z/c$. On the one hand, if $|z| = c$, then $x = y = 0$, so we choose $u = 0$. On the other hand, if $|z| > c$, then $v \ne 0$ and

$$\frac{x^2}{a^2} + \frac{y^2}{b^2} = \frac{z^2}{c^2} - 1 = \cosh^2 v - 1 = \sinh^2 v$$

so we can choose u so that $x/(a \sinh v) = \cos u$ and $y/(b \sinh v) = \sin u$. Thus $x = a \cos u \sinh v$, $y = b \sin u \sinh v$, and $z = c\cosh v$. Consequently (x, y, z) lies on the parametrized surface. Thus the two surfaces are the same.

5. The circle $x^2 + y^2 = 1$ in the xy plane is parametrized by $x = \cos\theta$ and $y = \sin\theta$ for $0 \le \theta \le 2\pi$. Therefore Σ is parametrized by $\mathbf{r}(\theta, z) = \cos\theta\mathbf{i} + \sin\theta\mathbf{j} + z\mathbf{k}$ for $0 \le \theta \le 2\pi$ and $-1 \le z \le 1$.

7. The cylinder intersects the xz plane in a circle of radius 1 that bounds the region R defined by $x = r\cos\theta$ and $z = r\sin\theta$ for $0 \le r \le 1$ and $0 \le \theta \le 2\pi$. Since $y = x + z - 3$ on Σ, it follows that a parametrization of Σ is

$$\mathbf{r}(r,\theta) = r\cos\theta\mathbf{i} + (r\cos\theta + r\sin\theta - 3)\mathbf{j} + r\sin\theta\mathbf{k} \quad \text{for } 0 \le r \le 1 \text{ and } 0 \le \theta \le 2\pi$$

9. The intersection of the (double) cone Σ with the plane parallel to the yz plane that has x intercept x is a circle of radius $|x|$ and can be parametrized by $y = x\cos\theta$ and $z = x\sin\theta$ for $0 \le \theta \le 2\pi$. Therefore Σ is parametrized by

$$\mathbf{r}(x,\theta) = x\mathbf{i} + x\cos\theta\mathbf{j} + x\sin\theta\mathbf{k} \quad \text{for arbitrary } x \text{ and } 0 \le \theta \le 2\pi$$

11. By (6) with $f(x) = \cos x$, Σ is parametrized by

$$\mathbf{r}(x,\theta) = x\mathbf{i} + \cos x\cos\theta\mathbf{j} + \cos x\sin\theta\mathbf{k} \quad \text{for } -\pi/2 \le x \le \pi/2 \text{ and } 0 \le \theta \le 2\pi$$

13. By (6) with x and z interchanged and with $f(z) = e^z$, Σ is parametrized by

$$\mathbf{r}(z,\theta) = e^z\sin\theta\mathbf{i} + e^z\cos\theta\mathbf{j} + z\mathbf{k} \quad \text{for } 0 \le z \le 1 \text{ and } 0 \le \theta \le 2\pi$$

15. By (6) with $f(x) = 1/x$, Σ is parametrized by

$$\mathbf{r}(x,\theta) = x\mathbf{i} + \frac{1}{x}\cos\theta\mathbf{j} + \frac{1}{x}\sin\theta\mathbf{k} \quad \text{for } x \ge 1 \text{ and } 0 \le \theta \le 2\pi$$

17. If $u_0 = \pi/4$ and $v_0 = -\pi/4$, then $x_0 = 2\sin u_0\cos v_0 = 1$, $y_0 = 3\sin u_0\sin v_0 = -3/2$, and $z_0 = 4\cos u_0 = 2\sqrt{2}$. Thus the point on Σ corresponding to $(u_0, v_0) = (\pi/4, -\pi/4)$ is $(1, -3/2, 2\sqrt{2})$. Next,

$$\mathbf{r}_u(u,v) \times \mathbf{r}_v(u,v) = \begin{vmatrix} \mathbf{i} & \mathbf{j} & \mathbf{k} \\ 2\cos u\cos v & 3\cos u\sin v & -4\sin u \\ -2\sin u\sin v & 3\sin u\cos v & 0 \end{vmatrix}$$

$$= 12\sin^2 u\cos v\mathbf{i} + 8\sin^2 u\sin v\mathbf{j} + 6\cos u\sin u\mathbf{k}$$

so that by (12), $\mathbf{r}_u(\pi/4, -\pi/4) \times \mathbf{r}_v(\pi/4, -\pi/4) = 3\sqrt{2}\mathbf{i} - 2\sqrt{2}\mathbf{j} + 3\mathbf{k}$ is normal to Σ at the point $(1, -3/2, 2\sqrt{2})$. Therefore an equation of the plane tangent at that point is $3\sqrt{2}(x-1) - 2\sqrt{2}(y + 3/2) + 3(z - 2\sqrt{2}) = 0$, or equivalently, $3\sqrt{2}x - 2\sqrt{2}y + 3z = 12\sqrt{2}$.

19. If $u_0 = 2$ and $v_0 = \pi$, then $x_0 = u_0^2 = 4$, $y_0 = 3u_0\cos v_0 = -6$, and $z_0 = 5u_0\sin v_0 = 0$. Thus the point on Σ corresponding to $(u_0, v_0) = (2, \pi)$ is $(4, -6, 0)$. Next,

$$\mathbf{r}_u(u,v) \times \mathbf{r}_v(u,v) = \begin{vmatrix} \mathbf{i} & \mathbf{j} & \mathbf{k} \\ 2u & 3\cos v & 5\sin v \\ 0 & -3u\sin v & 5u\cos v \end{vmatrix} = 15u\mathbf{i} - 10u^2\cos v\mathbf{j} - 6u^2\sin v\mathbf{k}$$

so that by (12), $\mathbf{r}_u(2,\pi) \times \mathbf{r}_v(2,\pi) = 30\mathbf{i} + 40\mathbf{j}$ is normal to Σ at the point $(4,-6,0)$. Therefore an equation of the plane tangent at that point is $30(x-4)+40(y+6)=0$, or equivalently, $3x+4y=-12$.

21. If $u_0 = -\pi/2$ and $v_0 = 0$, then $x_0 = (3+2\cos u_0)\sin v_0 = 0$, $y_0 = (3+2\cos u_0)\cos v_0 = 3$, and $z_0 = 2\sin u_0 = -2$. Thus the point on Σ corresponding to $(u_0, v_0) = (-\pi/2, 0)$ is $(0,3,-2)$. Next,

$$\mathbf{r}_u(u,v) \times \mathbf{r}_v(u,v) = \begin{vmatrix} \mathbf{i} & \mathbf{j} & \mathbf{k} \\ -2\sin u \sin v & -2\sin u \cos v & 2\sin u \\ (3+2\cos u)\cos v & -(3+2\cos u)\sin v & 0 \end{vmatrix}$$

$$= 2\cos u(3+2\cos u)\sin v\mathbf{i} + 2\cos u(3+2\cos u)\cos v\mathbf{j} + 2\sin u(3+2\cos u)\mathbf{k}$$

so that by (12), $\mathbf{r}_u(-\pi/2,0) \times \mathbf{r}_v(-\pi/2,0) = -6\mathbf{k}$ is normal to Σ at the point $(0,3,-2)$. Therefore an equation of the plane tangent at that point is $-6(z+2)=0$, or equivalently, $z=-2$.

23. Using the parametrization in Example 7, we have

$$\mathbf{r}_\phi \times \mathbf{r}_\theta = \begin{vmatrix} \mathbf{i} & \mathbf{j} & \mathbf{k} \\ -(b+r\cos\theta)\sin\phi & (b+r\cos\theta)\cos\phi & 0 \\ -r\sin\theta\cos\phi & -r\sin\theta\sin\phi & r\cos\theta \end{vmatrix}$$

$$= r(b+r\cos\theta)\cos\phi\cos\theta\mathbf{i} + r(b+r\cos\theta)\sin\phi\cos\theta\mathbf{j} + r(b+r\cos\theta)\sin\theta\mathbf{k}$$

and hence

$$\|\mathbf{r}_\phi \times \mathbf{r}_\theta\| = r(b+r\cos\theta)\sqrt{\cos^2\phi\cos^2\theta + \sin^2\phi\cos^2\theta + \sin^2\theta}$$
$$= r(b+r\cos\theta)\sqrt{\cos^2\theta + \sin^2\theta} = r(b+r\cos\theta)$$

Since both ϕ and θ can vary from 0 to 2π, we conclude from (14) that

$$S = \int_0^{2\pi}\int_0^{2\pi} r(b+r\cos\theta)\,d\phi\,d\theta = \int_0^{2\pi} 2\pi r(b+r\cos\theta)\,d\theta = 2\pi r(b\theta + r\sin\theta)\Big|_0^{2\pi} = 4\pi^2 br$$

25. In order to use (14), we first find that

$$\mathbf{r}_u \times \mathbf{r}_v = \begin{vmatrix} \mathbf{i} & \mathbf{j} & \mathbf{k} \\ \cos\omega v & \sin\omega v & 0 \\ -\omega u\sin\omega v & \omega u\cos\omega v & c \end{vmatrix} = c\sin\omega v\mathbf{i} - c\cos\omega v\mathbf{j} + \omega u\mathbf{k}$$

so that

$$\|\mathbf{r}_u \times \mathbf{r}_v\| = \sqrt{c^2\sin^2\omega v + c^2\cos^2\omega v + \omega^2 u^2} = \sqrt{c^2 + \omega^2 u^2}$$

Since one complete revolution is traced out when v varies from 0 to $2\pi/\omega$ and u varies from 0 to r, we conclude that

$$S = \int_0^{2\pi/\omega} \int_0^r \sqrt{c^2 + \omega^2 u^2}\, du\, dv$$

The evaluation of the integral is best obtained by using a computer algebra system.

27. a. Since Σ is parametrized by $\mathbf{r}(r,\theta) = r\cos\theta\mathbf{i} + r\sin\theta\mathbf{j} + f(r,\theta)\mathbf{k}$, we have

$$\mathbf{r}_r \times \mathbf{r}_\theta = \begin{vmatrix} \mathbf{i} & \mathbf{j} & \mathbf{k} \\ \cos\theta & \sin\theta & f_r(r,\theta) \\ -r\sin\theta & r\cos\theta & f_\theta(r,\theta) \end{vmatrix}$$

$$= [f_\theta(r,\theta)\sin\theta - f_r(r,\theta)r\cos\theta]\mathbf{i} - [f_\theta(r,\theta)\cos\theta + f_r(r,\theta)r\sin\theta]\mathbf{j} + r\mathbf{k}$$

so that

$$\begin{aligned} \|\mathbf{r}_r \times \mathbf{r}_\theta\| &= \sqrt{[f_\theta(r,\theta)\sin\theta - f_r(r,\theta)r\cos\theta]^2 + [f_\theta(r,\theta)\cos\theta + f_r(r,\theta)r\sin\theta]^2 + r^2} \\ &= \sqrt{[f_\theta(r,\theta)]^2(\sin^2\theta + \cos^2\theta) + r^2[f_r(r,\theta)]^2(\cos^2\theta + \sin^2\theta) + r^2} \\ &= \sqrt{r^2[f_r(r,\theta)]^2 + [f_\theta(r,\theta)]^2 + r^2} \end{aligned}$$

We conclude from (14) that

$$S = \iint_R \sqrt{r^2[f_r(r,\theta)]^2 + [f_\theta(r,\theta)]^2 + r^2}\, dA$$

b. The cylinder intersects the xy plane in the circle $r = 3$. Thus R is the region in the $r\theta$ plane defined by $0 \le r \le 3$ and $0 \le \theta \le 2\pi$. By part (a) with $f(r,\theta) = 5 - r^2$, we have

$$\begin{aligned} S &= \iint_R \sqrt{r^2(-2r)^2 + (0)^2 + r^2}\, dA = \int_0^3 \int_0^{2\pi} \sqrt{4r^4 + r^2}\, d\theta\, dr = \int_0^3 2\pi r \sqrt{4r^2+1}\, dr \\ &= \frac{\pi}{6}(4r^2+1)^{3/2}\Big|_0^3 = \frac{\pi}{6}(37^{3/2} - 1) = \frac{\pi}{6}(37\sqrt{37} - 1) \end{aligned}$$

29. a. $\mathbf{s}(t) = \mathbf{r}(2t, 3t) = (1 + \cos 3t)\cos 2t\mathbf{i} + (1 + \cos 3t)\sin 2t\mathbf{j} + \sin 3t\mathbf{k}$. Since each of the component functions of $\mathbf{s}$ is periodic with period 2π, it follows that $\mathbf{s}$ is also periodic with period 2π.

c. $\mathbf{w}(t) = \mathbf{r}(t, bt) = (1 + \cos bt)\cos t\mathbf{i} + (1 + \cos bt)\sin t\mathbf{j} + \sin bt\mathbf{k}$. Notice that the $\mathbf{k}$ component $\sin bt$ has period $2\pi/b$, so the only possible period for $\mathbf{w}$ is $2\pi n/b$ for some nonzero integer n. But the $\mathbf{i}$ component $(1 + \cos bt)\cos t$ does not have period $2\pi n/b$ for any nonzero integer n, because

$$\{1+\cos[b(t+2\pi n/b)]\}\cos(t+2\pi n/b) = [1+\cos(bt+2\pi n)]\cos(t+2\pi n/b) = (1+\cos bt)\cos(t+2\pi n/b)$$

which does not equal $(1 + \cos bt)\cos t$. (For example, let $t = 0$, and use the fact that for any integer n, n/b is *not* an integer.)

31. a. If the point (x, y, z) is given by $x = a\cos v \cos u, y = a\cos v \sin u$, and $z = a\sin v$, then

$$x^2 + y^2 + z^2 = a^2\cos^2 v\cos^2 u + a^2\cos^2 v\sin^2 u + a^2\sin^2 v = a^2\cos^2 v + a^2\sin^2 v = a^2$$

and hence (x, y, z) lies on the sphere Σ. Conversely, if $x^2 + y^2 + z^2 = a^2$, then $-a \le z \le a$, so that $z = a\sin v$ for some v with $-\pi/2 \le v \le \pi/2$. Then $x^2 + y^2 = a^2 - z^2 = a^2(1 - \sin^2 v) = a^2\cos^2 v$. This means that the (x, y) lies on the circle centered at the origin with radius $|a\cos v|$. Consequently there is u in $[0, 2\pi]$ such that $x = a\cos v\cos u$ and $y = a\cos v\sin u$. Thus every point on the sphere Σ is traced out by $\mathbf{r}$.

b. ii. For longitude, u radians correspond to $360 - u(180/\pi)$ degrees. Since degrees are measured east from Greenwich, 77 degrees west of Greenwich is 283 degrees east of Greenwich. Therefore in terms of u radians, $283 = 360 - u(180/\pi)$, so that $u = 77\pi/180 \approx 1.34$ radians. Similarly, for longitude, v radians correspond to $v(180/\pi)$ degrees. Thus in terms of radians, 39 degrees (north of the equator) yields $39 = v(180/\pi)$, so that $v = 39\pi/180 \approx 0.68$ radians.

c. The surface Σ under discussion is parametrized by

$$\mathbf{r}(u, v) = a\cos v\cos u\,\mathbf{i} + a\cos v\sin u\,\mathbf{j} + a\sin v\,\mathbf{k}$$

for $0 \le u \le 2\pi$ and $-\pi/4 \le v \le \pi/4$, where $a =$ the radius of the earth. The area S is twice the area for $0 \le u \le 2\pi$ and $0 \le v \le \pi/4$. Next, notice that

$$\mathbf{r}_u \times \mathbf{r}_v = \begin{vmatrix} \mathbf{i} & \mathbf{j} & \mathbf{k} \\ -a\cos v\sin u & a\cos v\cos u & 0 \\ -a\sin v\cos u & -a\sin v\sin u & a\cos v \end{vmatrix} = a^2\cos^2 v\cos u\,\mathbf{i} + a^2\cos^2 v\sin u\,\mathbf{j} + a^2\sin v\cos v\,\mathbf{k}$$

so that $\|\mathbf{r}_u \times \mathbf{r}_v\| = a^2\cos v$. Therefore

$$S = 2\int_0^{2\pi}\int_0^{\pi/4} a^2\cos v\,dv\,du = 2(2\pi)a^2\sin v\Big|_0^{\pi/4} = 2\sqrt{2}\pi a^2$$

Since the area of the total sphere is $4\pi a^2$, the portion between 45° in the southern hemisphere and 45° in the northern hemisphere is the ratio

$$\frac{2\sqrt{2}\pi a^2}{4\pi a^2} = \frac{\sqrt{2}}{2}$$

which is approximately 70.7%.

Chapter 14 Review

1. $\int_0^1 \int_x^{3x} ye^{(x^3)}\,dy\,dx = \int_0^1 \frac{y^2}{2}e^{(x^3)}\Big|_x^{3x}\,dx = \int_0^1 4x^2e^{(x^3)}\,dx = \frac{4}{3}e^{(x^3)}\Big|_0^1 = \frac{4}{3}(e-1)$

3. $\int_{-1}^1 \int_0^2 \int_{2x}^{5x} e^{xy}\,dz\,dy\,dx = \int_{-1}^1 \int_0^2 ze^{xy}\Big|_{2x}^{5x}\,dy\,dx = \int_{-1}^1 \int_0^2 3xe^{xy}\,dy\,dx$
$= \int_{-1}^1 3e^{xy}\Big|_0^2\,dx = \int_{-1}^1 3(e^{2x}-1)\,dx = \left(\frac{3}{2}e^{2x} - 3x\right)\Big|_{-1}^1 = \frac{3}{2}(e^2 - e^{-2}) - 6$

5. The region R of integration is the vertically simple region between the graphs of $y = \sqrt{x}$ and $y = 1$ on $[0,1]$. It is also the horizontally simple region between the graphs of $x = 0$ and $x = y^2$ on $[0,1]$. Thus

$$\int_0^1 \int_{\sqrt{x}}^1 e^{(y^3)}\,dy\,dx = \int_0^1 \int_0^{y^2} e^{(y^3)}\,dx\,dy = \int_0^1 xe^{(y^3)}\Big|_0^{y^2}\,dy$$

$$= \int_0^1 y^2e^{(y^3)}\,dy = \frac{1}{3}e^{(y^3)}\Big|_0^1 = \frac{1}{3}(e-1).$$

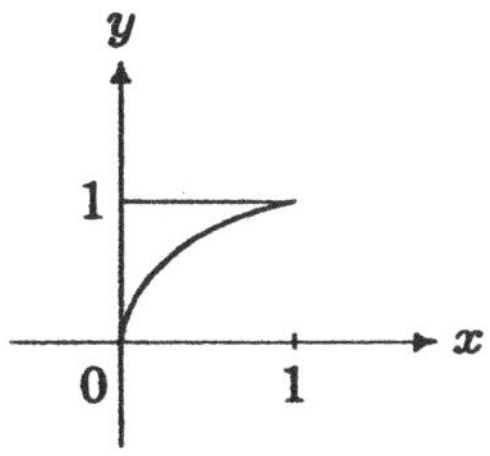

Exercise 5

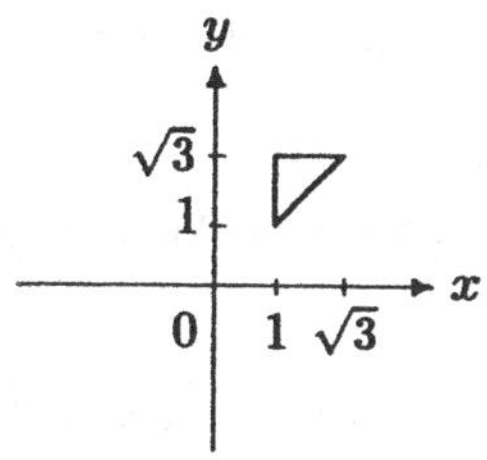

Exercise 7

7. The region R of integration is the vertically simple region between the graphs of $y = x$ and $y = \sqrt{3}$ on $[1, \sqrt{3}]$. It is also the horizontally simple region between the graphs of $x = 1$ and $x = y$ on $[1, \sqrt{3}]$. Thus

$$\int_1^{\sqrt{3}} \int_x^{\sqrt{3}} \frac{x}{(x^2+y^2)^{3/2}}\,dy\,dx = \int_1^{\sqrt{3}} \int_1^y \frac{x}{(x^2+y^2)^{3/2}}\,dx\,dy = \int_1^{\sqrt{3}} \frac{-1}{(x^2+y^2)^{1/2}}\Big|_1^y\,dy$$

$$= \int_1^{\sqrt{3}} \left[\frac{1}{(1+y^2)^{1/2}} - \frac{1}{\sqrt{2}y}\right] dy = \int_1^{\sqrt{3}} \frac{1}{(1+y^2)^{1/2}}\,dy - \frac{\sqrt{2}}{2}\int_1^{\sqrt{3}} \frac{1}{y}\,dy$$

$$\overset{y=\tan u}{=} \int_{\pi/4}^{\pi/3} \sec u\,du - \frac{\sqrt{2}}{2}\ln y\Big|_1^{\sqrt{3}} = \ln|\sec u + \tan u|\Big|_{\pi/4}^{\pi/3} - \frac{\sqrt{2}}{2}\ln\sqrt{3} = \ln\frac{2+\sqrt{3}}{1+\sqrt{2}} - \frac{\sqrt{2}}{4}\ln 3.$$

9. The graphs of $\sqrt{x}+\sqrt{y}=\sqrt{a}$ and $x+y=a$ intersect for (x,y) such that $a-x=y=(\sqrt{a}-\sqrt{x})^2=a+x-2\sqrt{a}\sqrt{x}$, so that $x=\sqrt{a}\sqrt{x}$, and thus $x=0$ or $x=a$. Since $a-x\geq(\sqrt{a}-\sqrt{x})^2$ for $0\leq x\leq a$, we have

$$A=\int_0^a\int_{(\sqrt{a}-\sqrt{x})^2}^{a-x}1\,dy\,dx=\int_0^a y\Big|_{(\sqrt{a}-\sqrt{x})^2}^{a-x}dx$$
$$=\int_0^a[(a-x)-(a+x-2\sqrt{a}\sqrt{x})]\,dx$$
$$=\int_0^a(-2x+2\sqrt{a}\sqrt{x})\,dx=\left(-x^2+\frac{4}{3}\sqrt{a}x^{3/2}\right)\Big|_0^a=\frac{1}{3}a^2.$$

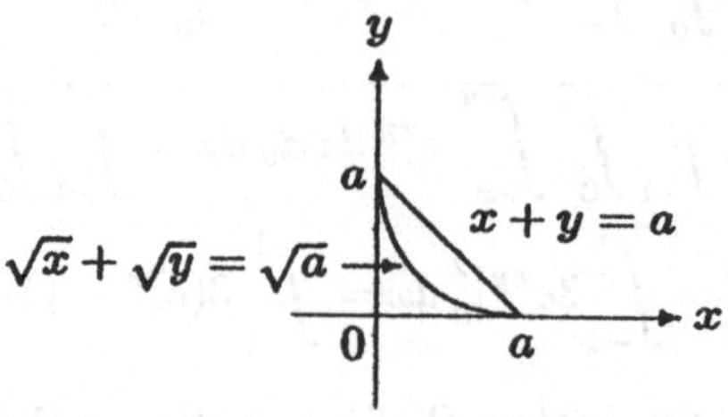

11. The limaçon and the circle intersect for (r,θ) such that $3-\sin\theta=r=5\sin\theta$, so that $\sin\theta=\frac{1}{2}$, and thus $\theta=\pi/6$ or $\theta=5\pi/6$. Therefore

$$A=\int_{\pi/6}^{5\pi/6}\int_{3-\sin\theta}^{5\sin\theta}r\,dr\,d\theta=\int_{\pi/6}^{5\pi/6}\frac{1}{2}r^2\Big|_{3-\sin\theta}^{5\sin\theta}d\theta$$
$$=\int_{\pi/6}^{5\pi/6}\left(12\sin^2\theta+3\sin\theta-\frac{9}{2}\right)d\theta$$
$$=\int_{\pi/6}^{5\pi/6}\left(6-6\cos2\theta+3\sin\theta-\frac{9}{2}\right)d\theta$$
$$=\left(\frac{3}{2}\theta-3\sin2\theta-3\cos\theta\right)\Big|_{\pi/6}^{5\pi/6}=\pi+6\sqrt{3}.$$

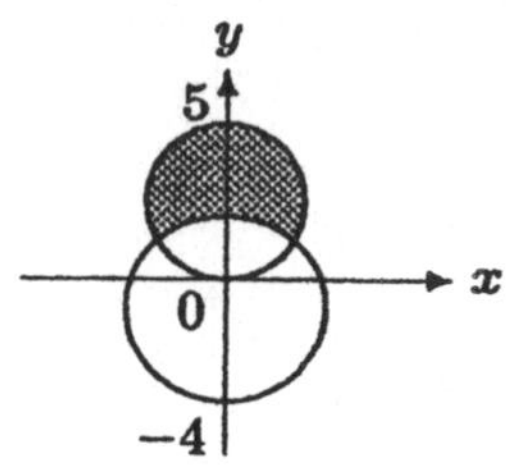

13. The lines $y=5+x$ and $y=-x+7$ intersect for (x,y) such that $5+x=y=-x+7$, so that $x=1$. Thus R is the vertically simple region between the graphs of $y=5+x$ and $y=-x+7$ on $[0,1]$. Therefore

$$\iint_R(3x-5)\,dA=\int_0^1\int_{5+x}^{-x+7}(3x-5)\,dy\,dx=\int_0^1(3x-5)y\Big|_{5+x}^{-x+7}dx$$
$$=\int_0^1(3x-5)(2-2x)\,dx=\int_0^1(-6x^2+16x-10)\,dx=(-2x^3+8x^2-10x)\Big|_0^1=-4.$$

15. In spherical coordinates the cone has equation $\rho^2\cos^2\varphi=3\rho^2\sin^2\varphi$, so that $\cot\varphi=\sqrt{3}$, and thus $\varphi=\pi/6$. Thus D is the collection of all points with spherical coordinates (ρ,φ,θ) such that $0\leq\theta\leq2\pi$, $0\leq\varphi\leq\pi/6$, and $0\leq\rho\leq2$. Since $z^2+1=\rho^2\cos^2\varphi+1$, we find that

$$\iiint_D(z^2+1)\,dV=\int_0^{2\pi}\int_0^{\pi/6}\int_0^2(\rho^2\cos^2\varphi+1)\rho^2\sin\varphi\,d\rho\,d\varphi\,d\theta$$

$$= \int_0^{2\pi}\int_0^{\pi/6}\left(\frac{\rho^5}{5}\cos^2\varphi\sin\varphi + \frac{\rho^3}{3}\sin\varphi\right)\Big|_0^2 d\varphi\, d\theta = \int_0^{2\pi}\int_0^{\pi/6}\left(\frac{32}{5}\cos^2\varphi\sin\varphi + \frac{8}{3}\sin\varphi\right) d\varphi\, d\theta$$

$$= \int_0^{2\pi}\left(\frac{-32}{15}\cos^3\varphi - \frac{8}{3}\cos\varphi\right)\Big|_0^{\pi/6} d\theta = \int_0^{2\pi}\left[\frac{32}{15}\left(1-\frac{3\sqrt{3}}{8}\right) + \frac{8}{3}\left(1-\frac{\sqrt{3}}{2}\right)\right] d\theta$$

$$= \left(\frac{72-32\sqrt{3}}{15}\right)2\pi = \frac{16\pi}{15}(9-4\sqrt{3}).$$

17. The region R in the xy plane is the vertically simple circular region in the xy plane between the graphs of $y = -\sqrt{9-x^2}$ and $y = \sqrt{9-x^2}$ on $[-3,3]$; D is the solid region between the graphs of $z = -\sqrt{9-x^2-y^2}$ and $z=0$ on R. Thus

$$\iiint_D xyz\, dV = \int_{-3}^3\int_{-\sqrt{9-x^2}}^{\sqrt{9-x^2}}\int_{-\sqrt{9-x^2-y^2}}^0 xyz\, dz\, dy\, dx = \int_{-3}^3\int_{-\sqrt{9-x^2}}^{\sqrt{9-x^2}} \frac{1}{2}xyz^2\Big|_{-\sqrt{9-x^2-y^2}}^0 dy\, dx$$

$$= \int_{-3}^3\int_{-\sqrt{9-x^2}}^{\sqrt{9-x^2}} -\frac{1}{2}xy(9-x^2-y^2)\, dy\, dx = \int_{-3}^3 \frac{1}{8}x(9-x^2-y^2)^2\Big|_{-\sqrt{9-x^2}}^{\sqrt{9-x^2}} dx = \int_{-3}^3 0\, dx = 0.$$

19. Let $f(x,y) = xy$. Then $f_x(x,y) = y$ and $f_y(x,y) = x$. The portion of the surface $z = xy$ that is inside the cylinder $x^2+y^2=1$ lies over the disk R in the xy plane bounded by the circle $r=1$. Thus

$$S = \iint_R \sqrt{y^2+x^2+1}\, dA = \int_0^{2\pi}\int_0^1 \sqrt{r^2+1}\, r\, dr\, d\theta$$

$$= \int_0^{2\pi} \frac{1}{3}(r^2+1)^{3/2}\Big|_0^1 d\theta = \int_0^{2\pi}\frac{1}{3}(2\sqrt{2}-1)\, d\theta = \frac{2\pi}{3}(2\sqrt{2}-1).$$

21. The paraboloid and cone intersect for (x,y,z) such that $z = x^2+y^2 = z^2$, so that $z=0$ or $z=1$, and thus $x^2+y^2=0$ or $x^2+y^2=1$. In polar coordinates R is the region in the xy plane between the graphs of $r=0$ and $r=1$ on $[0,2\pi]$. Then D is the solid region between the graphs of $z = x^2+y^2 = r^2$ and $z = \sqrt{x^2+y^2} = r$ on R. Therefore

$$V = \iiint_D 1\, dV = \int_0^{2\pi}\int_0^1\int_{r^2}^r r\, dz\, dr\, d\theta = \int_0^{2\pi}\int_0^1 rz\Big|_{r^2}^r dr\, d\theta$$

$$= \int_0^{2\pi}\int_0^1 (r^2-r^3)\, dr\, d\theta = \int_0^{2\pi}\left(\frac{r^3}{3}-\frac{r^4}{4}\right)\Big|_0^1 d\theta = \int_0^{2\pi}\frac{1}{12}\, d\theta = \frac{1}{6}\pi.$$

23. In polar coordinates R is the region in the xy plane between the graphs of $r=0$ and $r = 4\sin\theta$ on $[0,\pi]$; D is the solid region between the graphs of $z=0$ and $z=r$ on R. Thus

$$V = \iiint_D 1\, dV = \int_0^\pi\int_0^{4\sin\theta}\int_0^r r\, dz\, dr\, d\theta = \int_0^\pi\int_0^{4\sin\theta} rz\Big|_0^r dr\, d\theta = \int_0^\pi\int_0^{4\sin\theta} r^2\, dr\, d\theta$$

$$= \int_0^\pi \frac{r^3}{3}\Big|_0^{4\sin\theta} d\theta = \int_0^\pi \frac{64}{3}\sin^3\theta\, d\theta = \frac{64}{3}\int_0^\pi (1-\cos^2\theta)\sin\theta\, d\theta = \frac{64}{3}\left(-\cos\theta + \frac{1}{3}\cos^3\theta\right)\Big|_0^\pi = \frac{256}{9}.$$

25. The paraboloid and sheet intersect for (x, y, z) such that $4x^2 + y^2 = z = 16 - 3y^2$, so that $x^2 + y^2 = 4$. In polar coordinates R is the circular region in the xy plane between the graphs of $r = 0$ and $r = 2$ on $[0, 2\pi]$. Then D is the solid region between the graphs of $z = 4x^2 + y^2$ and $z = 16 - 3y^2$ on R. Thus

$$V = \iiint\limits_D 1\,dV = \int_0^{2\pi}\int_0^2\int_{r^2+3r^2\cos^2\theta}^{16-3r^2\sin^2\theta} r\,dz\,dr\,d\theta = \int_0^{2\pi}\int_0^2 rz\Big|_{r^2+3r^2\cos^2\theta}^{16-3r^2\sin^2\theta} dr\,d\theta$$

$$= \int_0^{2\pi}\int_0^2 r(16-4r^2)\,dr\,d\theta = \int_0^{2\pi}(8r^2 - r^4)\Big|_0^2 d\theta = \int_0^{2\pi} 16\,d\theta = 32\pi.$$

27. The region R in the xy plane is the horizontally simple region between the lines $x = 0$ and $x = 6 - 3y$ on $[0, 2]$; D is the solid region between the graphs of $z = 0$ and $z = x + 2y$ on R. Thus

$$V = \iiint\limits_D 1\,dV = \int_0^2\int_0^{6-3y}\int_0^{x+2y} 1\,dz\,dx\,dy = \int_0^2\int_0^{6-3y} z\Big|_0^{x+2y} dx\,dy$$

$$= \int_0^2\int_0^{6-3y}(x+2y)\,dx\,dy = \int_0^2\left(\frac{x^2}{2} + 2yx\right)\Big|_0^{6-3y} dy$$

$$= \int_0^2\left[\frac{1}{2}(6-3y)^2 + 12y - 6y^2\right] dy = \left[\frac{-1}{18}(6-3y)^3 + 6y^2 - 2y^3\right]\Big|_0^2 = 20.$$

29. The sphere and paraboloid intersect for (x, y, z) such that $3z + 21 = 49 - z^2$, so that $z^2 + 3z - 28 = 0$, and thus $z = -7$ or $z = 4$. If $z = -7$, then $x^2 + y^2 = 0$, so that $r = 0$, and if $z = 4$, then $x^2 + y^2 = 33$, so that $r = \sqrt{33}$. In polar coordinates R is the circular region in the xy plane between the graphs of $r = 0$ and $r = \sqrt{33}$ on $[0, 2\pi]$. Then D is the solid region between the graphs of $z = \frac{1}{3}r^2 - 7$ and $z = \sqrt{49 - r^2}$ on R. Thus

$$V = \iiint\limits_D 1\,dV = \int_0^{2\pi}\int_0^{\sqrt{33}}\int_{(r^2/3)-7}^{\sqrt{49-r^2}} r\,dz\,dr\,d\theta = \int_0^{2\pi}\int_0^{\sqrt{33}} rz\Big|_{(r^2/3)-7}^{\sqrt{49-r^2}} dr\,d\theta$$

$$= \int_0^{2\pi}\int_0^{\sqrt{33}}\left(r\sqrt{49-r^2} - \frac{1}{3}r^3 + 7r\right) dr\,d\theta = \int_0^{2\pi}\left[-\frac{1}{3}(49-r^2)^{3/2} - \frac{1}{12}r^4 + \frac{7}{2}r^2\right]\Big|_0^{\sqrt{33}} d\theta$$

$$= \int_0^{2\pi}\frac{471}{4}\,d\theta = \frac{471}{2}\pi.$$

31. Since the region D occupied by the solid is symmetric with respect to the yz plane and the xz plane, and since $\delta(x, y, z) = z$, so that $\delta(x, y, z)$ is independent of x and y, we have $\bar{x} = \bar{y} = 0$. The cone $z = \sqrt{x^2 + y^2}$ and the plane $z = 3$ intersect for (x, y, z) such that $\sqrt{x^2 + y^2} = z = 3$, and thus $r = 3$. Consequently D is the solid region between the graphs of $z = r$ and $z = 3$ on the disk $0 \le r \le 3$. Therefore

$$M_{xy} = \iiint\limits_D z\delta(x, y, z)\,dV = \int_0^{2\pi}\int_0^3\int_r^3 z^2 r\,dz\,dr\,d\theta = \int_0^{2\pi}\int_0^3 \frac{1}{3}z^3 r\Big|_r^3 dr\,d\theta$$

$$= \int_0^{2\pi}\int_0^3 \left(9r - \frac{1}{3}r^4\right) dr\, d\theta = \int_0^{2\pi} \left(\frac{9}{2}r^2 - \frac{1}{15}r^5\right)\Big|_0^3 d\theta = \int_0^{2\pi} \frac{243}{10}\, d\theta = \frac{243}{5}\pi.$$

Since

$$m = \iiint_D \delta(x,y,z)\, dV = \int_0^{2\pi}\int_0^3\int_r^3 zr\, dz\, dr\, d\theta = \int_0^{2\pi}\int_0^3 \frac{1}{2}z^2 r\Big|_r^3 dr\, d\theta$$

$$= \int_0^{2\pi}\int_0^3 \left(\frac{9}{2}r - \frac{1}{2}r^3\right) dr\, d\theta = \int_0^{2\pi} \left(\frac{9}{4}r^2 - \frac{1}{8}r^4\right)\Big|_0^3 d\theta = \int_0^{2\pi} \frac{81}{8}\, d\theta = \frac{81}{4}\pi$$

we have $\bar{z} = M_{xy}/m = (243\pi/5)/(81\pi/4) = \frac{12}{5}$. Thus $(\bar{x}, \bar{y}, \bar{z}) = (0, 0, \frac{12}{5})$.

33. By symmetry, $\bar{x} = \bar{y} = 0$. The paraboloid $z = x^2 + y^2$ and the upper nappe of the cone $z^2 = x^2 + y^2$ intersect for (r, θ, z) such that $z = r^2 = z^2$, so that $z = 0$ or $z = 1$, and thus $r = 0$ or $r = 1$. The given region D is the solid region between the graphs of $z = r^2$ and $z = r$ on the disk $0 \le r \le 1$. Taking $\delta = 1$, we have

$$M_{xy} = \iiint_D z\, dV = \int_0^{2\pi}\int_0^1\int_{r^2}^r zr\, dz\, dr\, d\theta = \int_0^{2\pi}\int_0^1 \frac{1}{2}z^2 r\Big|_{r^2}^r dr\, d\theta$$

$$= \int_0^{2\pi}\int_0^1 \frac{1}{2}(r^3 - r^5)\, dr\, d\theta = \int_0^{2\pi} \frac{1}{2}\left(\frac{1}{4}r^4 - \frac{1}{6}r^6\right)\Big|_0^1 d\theta = \int_0^{2\pi} \frac{1}{24}\, d\theta = \frac{1}{12}\pi.$$

Since $\delta = 1$, we have

$$m = \iiint_D 1\, dV = \int_0^{2\pi}\int_0^1\int_{r^2}^r r\, dz\, dr\, d\theta = \int_0^{2\pi}\int_0^1 rz\Big|_{r^2}^r dr\, d\theta = \int_0^{2\pi}\int_0^1 (r^2 - r^3)\, dr\, d\theta$$

$$= \int_0^{2\pi} \left(\frac{1}{3}r^3 - \frac{1}{4}r^4\right)\Big|_0^1 d\theta = \int_0^{2\pi} \frac{1}{12}\, d\theta = \frac{1}{6}\pi$$

and thus $\bar{z} = M_{xy}/m = (\pi/12)/(\pi/6) = \frac{1}{2}$. Therefore $(\bar{x}, \bar{y}, \bar{z}) = (0, 0, \frac{1}{2})$.

35. First,

$$\frac{\partial(x,y)}{\partial(u,v)} = \begin{vmatrix} \dfrac{\partial x}{\partial u} & \dfrac{\partial x}{\partial v} \\ \dfrac{\partial y}{\partial u} & \dfrac{\partial y}{\partial v} \end{vmatrix} = \begin{vmatrix} 1 & 1 \\ \dfrac{-v}{(u+v)^2} & \dfrac{u}{(u+v)^2} \end{vmatrix} = (1)\left(\frac{u}{(u+v)^2}\right) - (1)\left(\frac{-v}{(u+v)^2}\right) = \frac{1}{u+v}.$$

To find S we observe that for $xy = 1$ we have $(u+v)(v/(u+v)) = 1$, or $v = 1$; similarly, for $xy = 2$ we have $v = 2$. For $x(1-y) = 1$ we have $(u+v)(1 - v/(u+v)) = 1$, or $u + v - v = 1$, or $u = 1$; similarly, for $x(1-y) = 2$ we have $u = 2$. Thus S is the square region in the uv plane bounded by the lines $u = 1$, $u = 2$, $v = 1$, and $v = 2$. By (9) in Section 14.8,

$$\iint_R x\, dA = \iint_S (u+v)\left|\frac{\partial(x,y)}{\partial(u,v)}\right| dA = \int_1^2\int_1^2 (u+v)\frac{1}{u+v}\, du\, dv = \int_1^2\int_1^2 1\, du\, dv = 1.$$

37. Let T_R be defined by $u = x - y$ and $v = x + y$. To find S we observe that for $x - y = 1$ we have $u = 1$; for $x - y = 3$ we have $u = 3$; for $x + y = 2$ we have $v = 2$; for $x + y = 4$ we have $v = 4$. Consequently S is the square region in the uv plane bounded by the lines $u = 1$, $u = 3$, $v = 2$, and $v = 4$. Solving for x and y in terms of u and v, we find that $u + v = (x - y) + (x + y) = 2x$, so that $x = \frac{1}{2}(u + v)$, and $v - u = (x + y) - (x - y) = 2y$, so that $y = \frac{1}{2}(v - u)$. Then T is defined by $x = \frac{1}{2}(u + v)$ and $y = \frac{1}{2}(v - u)$. Thus

$$\frac{\partial(x, y)}{\partial(u, v)} = \begin{vmatrix} \frac{\partial x}{\partial u} & \frac{\partial x}{\partial v} \\ \frac{\partial y}{\partial u} & \frac{\partial y}{\partial v} \end{vmatrix} = \begin{vmatrix} \frac{1}{2} & \frac{1}{2} \\ -\frac{1}{2} & \frac{1}{2} \end{vmatrix} = \left(\frac{1}{2}\right)\left(\frac{1}{2}\right) - \left(\frac{1}{2}\right)\left(-\frac{1}{2}\right) = \frac{1}{2}.$$

By (9) in Section 14.8,

$$\iint_R (2x - y^2)\, dA = \iint_S \left[(u + v) - \frac{1}{4}(v - u)^2\right] \left|\frac{\partial(x, y)}{\partial(u, v)}\right| dA$$

$$= \int_2^4 \int_1^3 \left(u + v - \frac{1}{4}v^2 + \frac{1}{2}uv - \frac{1}{4}u^2\right) \frac{1}{2}\, du\, dv = \int_2^4 \left(\frac{1}{4}u^2 + \frac{1}{2}uv - \frac{1}{8}uv^2 + \frac{1}{8}u^2 v - \frac{1}{24}u^3\right)\Big|_1^3 dv$$

$$= \int_2^4 \left(\frac{11}{12} + 2v - \frac{1}{4}v^2\right) dv = \left(\frac{11}{12}v + v^2 - \frac{1}{12}v^3\right)\Big|_2^4 = \frac{55}{6}.$$

39. The cylinder intersects the xz plane in the circle $y^2 + z^2 = 4$, which bounds the region R defined by $y = r\cos\theta$ and $z = r\sin\theta$ for $0 \le r \le 2$ and $0 \le \theta \le 2\pi$. Since $x = 2 - y - z$ on the given plane, a parametrization of the part of the plane inside the cylinder is

$$\mathbf{r}(r, \theta) = (2 - r\cos\theta - r\sin\theta)\mathbf{i} + r\cos\theta\mathbf{j} + r\sin\theta\mathbf{k} \quad \text{for } 0 \le r \le 2 \text{ and } 0 \le \theta \le 2\pi$$

41. a. We need to find values of u and v such that $u^2 + v^2 = 2$, $u^2 - v^2 = 0$, and $u - v = -2$. Since $u^2 - v^2 = 0$, it follows that $u = v$ or $u = -v$. Then $u^2 + v^2 = 2$ implies that $u = 1$ or $u = -1$. Because $u - v = -2$, we conclude that $u = -1$ and $v = 1$. Thus $(2, 0, -2)$ is on Σ.

b. By hypothesis, Σ is parametrized by $\mathbf{r}(u, v) = (u^2 + v^2)\mathbf{i} + (u^2 - v^2)\mathbf{j} + (u - v)\mathbf{k}$. By part (a), $(-1, 1)$ in the uv plane corresponds to the point $(2, 0, -2)$ on Σ. Next,

$$\mathbf{r}_u(u, v) \times \mathbf{r}_v(u, v) = \begin{vmatrix} \mathbf{i} & \mathbf{j} & \mathbf{k} \\ 2u & 2u & 1 \\ 2v & -2v & -1 \end{vmatrix} = 2(-u + v)\mathbf{i} + 2(u + v)\mathbf{j} - 8uv\mathbf{k}$$

so that by (12) in Section 14.9, $\mathbf{r}_u \times \mathbf{r}_v(-1, 1) = 4\mathbf{i} + 8\mathbf{k}$ is normal to Σ at $(2, 0, -2)$. Therefore an equation of the plane tangent at $(2, 0, -2)$ is $4(x - 2) + 8(z + 2) = 0$, or equivalently, $x + 2z = -2$.

43. First we calculate that

$$\mathbf{r}_u(u, v) \times \mathbf{r}_v(u, v) = \begin{vmatrix} \mathbf{i} & \mathbf{j} & \mathbf{k} \\ -\sin u \sinh v & \cos u \sinh v & 0 \\ \cos u \cosh v & \sin u \cosh v & \sinh v \end{vmatrix} = \cos u \sinh^2 v\mathbf{i} + \sin u \sinh^2 v\mathbf{j} - \sinh v \cosh v\mathbf{k}$$

Thus

$$||\mathbf{r}_u \times \mathbf{r}_v|| = \sqrt{\cos^2 u \sinh^4 v + \sin^2 u \sinh^4 v + \sinh^2 v \cosh^2 v} = \sqrt{\sinh^4 v + \sinh^2 v \cosh^2 v}$$

$$= \sinh v \sqrt{\sinh^2 v + \cosh^2 v} = \sinh v \sqrt{2\cosh^2 v - 1}$$

By (14) in Section 14.9,

$$S = \int_0^1 \int_0^{2\pi} \sinh v \sqrt{2\cosh^2 v - 1}\, du\, dv = \int_0^1 2\pi \sinh v \sqrt{2\cosh^2 v - 1}\, dv$$

$$\overset{w=\cosh v}{=} \int_1^{\cosh(1)} 2\pi \sqrt{2w^2 - 1}\, dw$$

If we make the trigonometric substitution $\sec t = \sqrt{2}w$, then

$$\int 2\pi\sqrt{2w^2-1}\, dw = \int 2\pi \sqrt{\sec^2 t - 1}\, \frac{1}{\sqrt{2}} \sec t \tan t\, dt = \int \sqrt{2}\pi \tan^2 t \sec t\, dt$$

By Example 6 in Section 8.2,

$$\int \sqrt{2}\pi \tan^2 t \sec t\, dt = \frac{\pi}{\sqrt{2}} \sec t \tan t - \frac{\pi}{\sqrt{2}} \ln|\sec t + \tan t| + C$$

Thus

$$\int_1^{\cosh(1)} 2\pi\sqrt{2w^2-1}\, dw = \left(\pi w \sqrt{2w^2-1} - \frac{\pi}{\sqrt{2}} \ln|\sqrt{2}w + \sqrt{2w^2-1}|\right)\Big|_1^{\cosh(1)))} \approx 5.072919875$$

45. The intersection of the cylinder with the xy plane is the circle with polar equation $r = \sin\theta$. It bounds the region R defined in polar coordinates by $0 \le r \le \sin\theta$ and $0 \le \theta \le \pi$. For any point (x, y, z) that lies on the sphere and inside the cylinder, we have $x = r\cos\theta$, $y = r\sin\theta$, and $z = \sqrt{1 - x^2 - y^2} = \sqrt{1 - r^2}$. Thus the part of the sphere inside the cylinder is parametrized by $\mathbf{r}(r,\theta) = r\cos\theta\mathbf{i} + r\sin\theta\mathbf{j} + \sqrt{1-r^2}\mathbf{k}$. Then

$$\mathbf{r}_r(r,\theta) \times \mathbf{r}_\theta(r,\theta) = \begin{vmatrix} \mathbf{i} & \mathbf{j} & \mathbf{k} \\ \cos\theta & \sin\theta & -r/\sqrt{1-r^2} \\ -r\sin\theta & r\cos\theta & 0 \end{vmatrix} = \frac{r^2\cos\theta}{\sqrt{1-r^2}}\mathbf{i} + \frac{r^2\sin\theta}{\sqrt{1-r^2}}\mathbf{j} + r\mathbf{k}$$

Thus

$$||\mathbf{r}_r \times \mathbf{r}_\theta|| = \sqrt{\frac{r^4\cos^2\theta}{1-r^2} + \frac{r^4\sin^2\theta}{1-r^2} + r^2} = \sqrt{\frac{r^4}{1-r^2} + r^2} = \sqrt{\frac{r^2}{1-r^2}} = \frac{r}{\sqrt{1-r^2}}$$

The surface area S is twice the part of the surface area for $0 \le \theta \le \pi/2$. Therefore by (14) in Section 14.9,

$$S = 2\int_0^{\pi/2} \int_0^{\sin\theta} \frac{r}{\sqrt{1-r^2}}\, dr\, d\theta = 2\int_0^{\pi/2} -\sqrt{1-r^2}\Big|_0^{\sin\theta} d\theta$$

$$= 2\int_0^{\pi/2} (-\sqrt{1-\sin^2\theta} + 1)\, d\theta = 2\int_0^{\pi/2} (1 - \cos\theta)\, d\theta = 2(\theta - \sin\theta)|_0^{\pi/2} = \pi - 2$$

Cumulative Review(Chapters 1–13)

1. If $x \neq 0$,

$$\lim_{h\to 0} \frac{2x^2h - 5xh^2 - 6h^3}{3xh + 2h^2} = \lim_{h\to 0} \frac{2x^2 - 5xh - 6h^2}{3x + 2h} = \frac{2x^2}{3x} = \frac{2}{3}x.$$

If $x = 0$,

$$\lim_{h\to 0} \frac{2x^2h - 5xh^2 - 6h^3}{3xh + 2h^2} = \lim_{h\to 0} \frac{-6h^3}{2h^2} = \lim_{h\to 0}(-3h) = 0.$$

Thus

$$\lim_{h\to 0} \frac{2x^2h - 5xh^2 - 6h^3}{3xh + 2h^2} = \frac{2}{3}x \quad \text{for all } x.$$

3. a. The domain of f consists of all x for which $x^2 - 4 \geq 0$ (that is, $|x| \geq 2$) and $x + \sqrt{x^2 - 4} \geq 0$. Since $x + \sqrt{x^2-4} < 0$ for $x \leq -2$, and since $x + \sqrt{x^2-4} > 0$ for $x \geq 2$, the domain of f is $[2, \infty)$.

   b. $f'(x) = \dfrac{1}{2\sqrt{x + \sqrt{x^2-4}}}\left(1 + \dfrac{x}{\sqrt{x^2-4}}\right)$

5. Differentiating the given equation implicitly, we obtain $2x - y - x(dy/dx) + 2y(dy/dx) = 0$, so if $dy/dx = 1$, then $2x - y - x + 2y = 0$, or $y = -x$. Then the original equation becomes $x^2 - x(-x) + (-x)^2 = 4$, or $3x^2 = 4$, so that $x = -\frac{2}{3}\sqrt{3}$ or $x = \frac{2}{3}\sqrt{3}$. Thus the two points at which the slope of the tangent line is 1 are $(-\frac{2}{3}\sqrt{3}, \frac{2}{3}\sqrt{3})$ and $(\frac{2}{3}\sqrt{3}, -\frac{2}{3}\sqrt{3})$.

7. Let x and y be as in the figure. Since the volume of the toolshed is to be 1512 cubic feet, we have $7xy = 1512$, so that $y = 216/x$. Let k be the cost per square foot of the sides. Then the cost C of the toolshed is given by

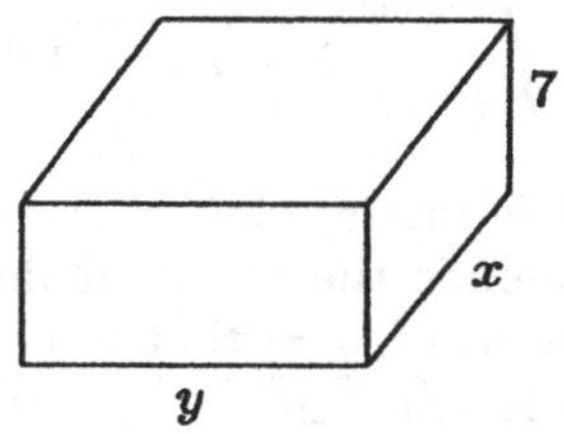

$$C = \underset{\text{sides}}{14xk} + \underset{\text{front and back}}{14y\left(\frac{3}{2}k\right)} + \underset{\text{top}}{xy(2k)} = k\left[14x + 21\left(\frac{216}{x}\right) + 432\right].$$

We are to minimize C. Now $C'(x) = k[14 - (21)(216)/x^2]$, so that $C'(x) = 0$ if $14x^2 = (21)(216)$, or $x = 18$. Since $C''(x) = k(42)(216)/x^3 > 0$, (1) of Section 4.6 and the Second Derivative Test imply that C is minimized for $x = 18$. Then $y = \frac{216}{18} = 12$. Thus for minimal cost the front should be 12 feet wide and the sides should be 18 feet wide.

9. $f'(x) = \dfrac{\ln x - 1}{(\ln x)^2}$; $f''(x) = \dfrac{2 - \ln x}{x(\ln x)^3}$

relative minimum value is $f(e) = e$; increasing on $[e, \infty)$, and decreasing on $(0, 1)$ and $(1, e]$; concave upward on $(1, e^2)$, and concave downward on $(0, 1)$ and (e^2, ∞); inflection point is $(e^2, \frac{1}{2}e^2)$; vertical asymptote is $x = 1$; $\lim_{x\to 0^+}(x/\ln x) = 0$ by l'Hôpital's Rule.

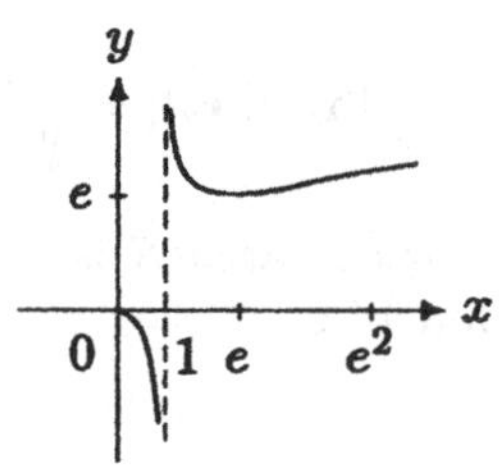

11. $\int \tan x \sin^2 x \cos^5 x\,dx = \int \frac{\sin x}{\cos x} \sin^2 x \cos^5 x\,dx = \int \sin^3 x \cos^4 x\,dx = \int (1-\cos^2 x)\cos^4 x \sin x\,dx$

Let $u = \cos x$, so that $du = -\sin x\,dx$. Then $\int \tan x \sin^2 x \cos^5 x\,dx = \int (1-\cos^2 x)\cos^4 x \sin x\,dx = \int (1-u^2)u^4(-1)\,du = -\frac{1}{5}u^5 + \frac{1}{7}u^7 + C = -\frac{1}{5}\cos^5 x + \frac{1}{7}\cos^7 x + C$.

13. $\int_{1/2}^{1} \frac{1}{\sqrt{1-x^2}}\,dx = \lim_{b\to 1^-} \int_{1/2}^{b} \frac{1}{\sqrt{1-x^2}}\,dx = \lim_{b\to 1^-} \left(\sin^{-1} x\Big|_{1/2}^{b}\right) = \lim_{b\to 1^-} \left(\sin^{-1} b - \sin^{-1}\frac{1}{2}\right)$

$= \frac{1}{2}\pi - \frac{1}{6}\pi = \frac{1}{3}\pi$

Thus the region has finite area, and the area of the region is $\pi/3$.

15. $L = \int_0^{\pi/2} \sqrt{\left(\frac{dx}{dt}\right)^2 + \left(\frac{dy}{dt}\right)^2}\,dt = \int_0^{\pi/2} \sqrt{(\sin^{1/2} t \cos t)^2 + (\cos t)^2}\,dt$

$= \int_0^{\pi/2} \sqrt{\sin t \cos^2 t + \cos^2 t}\,dt = \int_0^{\pi/2} \sqrt{\sin t + 1}\,\cos t\,dt$

Let $u = \sin t + 1$, so that $du = \cos t\,dt$. If $t = 0$, then $u = 1$, and if $t = \pi/2$, then $u = 2$. Thus

$$L = \int_0^{\pi/2} \sqrt{\sin t + 1}\,\cos t\,dt = \int_1^2 \sqrt{u}\,du = \frac{2}{3}u^{3/2}\Big|_1^2 = \frac{2}{3}(2^{3/2} - 1).$$

17. $\frac{a_{k+1}}{a_k} = \frac{\dfrac{[(k+1)!]^3(27)^{k+1}}{(3k+3)!}}{\dfrac{(k!)^3(27)^k}{(3k)!}} = 27\left(\frac{(k+1)!}{k!}\right)^3 \frac{(3k)!}{(3k+3)!} = \frac{27(k+1)^3}{(3k+1)(3k+2)(3k+3)}$

$= \frac{9(k+1)^2}{(3k+1)(3k+2)} = \frac{3k+3}{3k+1}\cdot\frac{3k+3}{3k+2} > 1\cdot 1 = 1$

Thus $a_{k+1} > a_k$, so the given sequence is increasing.

19. $\lim_{n\to\infty} \left|\frac{[(-1)^{n+1}/(n+2)]x^{2n+2}}{[(-1)^n/(n+1)]x^{2n}}\right| = \lim_{n\to\infty} \frac{n+1}{n+2}|x|^2 = |x|^2$

By the Generalized Ratio Test, the power series converges for $|x| < 1$ and diverges for $|x| > 1$. For $x = -1$ and for $x = 1$ the series becomes $\sum_{n=1}^{\infty}(-1)^n/(n+1)$, which converges by the Alternating Series Test. Thus the interval of convergence of the given power series is $[-1, 1]$.

21. We set up a coordinate system so that the sack moves from the origin O to the point $Q = (0, 6)$. Then the force $\mathbf{F}$ is given by

$$\mathbf{F} = 100\left(\cos\frac{\pi}{4}\mathbf{i} + \sin\frac{\pi}{4}\mathbf{j}\right) = 50\sqrt{2}\,(\mathbf{i}+\mathbf{j}).$$

The work is given by $W = \mathbf{F}\cdot\overrightarrow{OQ} = 50\sqrt{2}\,(\mathbf{i}+\mathbf{j})\cdot 6\mathbf{j} = (50\sqrt{2})(6) = 300\sqrt{2}$ (foot-pounds).

23. a. $\mathbf{r}'(t) = 3t^2\mathbf{i} + 6\mathbf{j} + 6t\mathbf{k}$; $\|\mathbf{r}'(t)\| = \sqrt{(3t^2)^2 + 6^2 + (6t)^2} = \sqrt{9t^4 + 36t^2 + 36} = 3(t^2+2)$;

$$\mathbf{T}(t) = \frac{\mathbf{r}'(t)}{\|\mathbf{r}'(t)\|} = \frac{3t^2\mathbf{i} + 6\mathbf{j} + 6t\mathbf{k}}{3(t^2+2)} = \frac{t^2}{t^2+2}\mathbf{i} + \frac{2}{t^2+2}\mathbf{j} + \frac{2t}{t^2+2}\mathbf{k}.$$

Therefore $\mathbf{T}(t)$ is parallel to the y axis if $t = 0$.

b. $\mathbf{T}'(t) = \dfrac{2t(t^2+2) - 2t(t^2)}{(t^2+2)^2}\mathbf{i} - \dfrac{2(2t)}{(t^2+2)^2}\mathbf{j} + \dfrac{2(t^2+2) - 2t(2t)}{(t^2+2)^2}\mathbf{k}$

$= \dfrac{4t}{(t^2+2)^2}\mathbf{i} - \dfrac{4t}{(t^2+2)^2}\mathbf{j} + \dfrac{4-2t^2}{(t^2+2)^2}\mathbf{k};$

$\|\mathbf{T}'(t)\| = \dfrac{1}{t^2+2)^2}\sqrt{(4t)^2+(-4)^2+(4-2t)^2} = \dfrac{\sqrt{4t^4+16t^2+16}}{(t^2+2)^2} = \dfrac{2}{t^2+2};$

$\mathbf{N}(t) = \dfrac{\mathbf{T}'(t)}{\|\mathbf{T}'(t)\|} = \dfrac{2t}{t^2+2}\mathbf{i} - \dfrac{2t}{t^2+2}\mathbf{j} + \dfrac{2-t^2}{t^2+2}\mathbf{k}.$

c. $\mathbf{T}(1) = \frac{1}{3}\mathbf{i} + \frac{2}{3}\mathbf{j} + \frac{2}{3}\mathbf{k}$; $\mathbf{N}(1) = \frac{2}{3}\mathbf{i} - \frac{2}{3}\mathbf{j} + \frac{1}{3}\mathbf{k}$; $\mathbf{T}(1) \times \mathbf{N}(1) = \begin{vmatrix} \mathbf{i} & \mathbf{j} & \mathbf{k} \\ \frac{1}{3} & \frac{2}{3} & \frac{2}{3} \\ \frac{2}{3} & -\frac{2}{3} & \frac{1}{3} \end{vmatrix} = \frac{2}{3}\mathbf{i} + \frac{1}{3}\mathbf{j} - \frac{2}{3}\mathbf{k}.$

$\mathbf{T}(1) \times \mathbf{N}(1)$ is perpendicular to both $\mathbf{T}(1)$ and $\mathbf{N}(1)$, but has positive $\mathbf{j}$ component. Also, $\|\mathbf{T}(1) \times \mathbf{N}(1)\| = \sqrt{(\frac{2}{3})^2 + (\frac{1}{3})^2 + (-\frac{2}{3})^2} = 1$. Therefore $-\frac{2}{3}\mathbf{i} - \frac{1}{3}\mathbf{j} + \frac{2}{3}\mathbf{k}$ is the unit vector with negative $\mathbf{j}$ component that is perpendicular to both $\mathbf{T}(1)$ and $\mathbf{N}(1)$.

25. a.

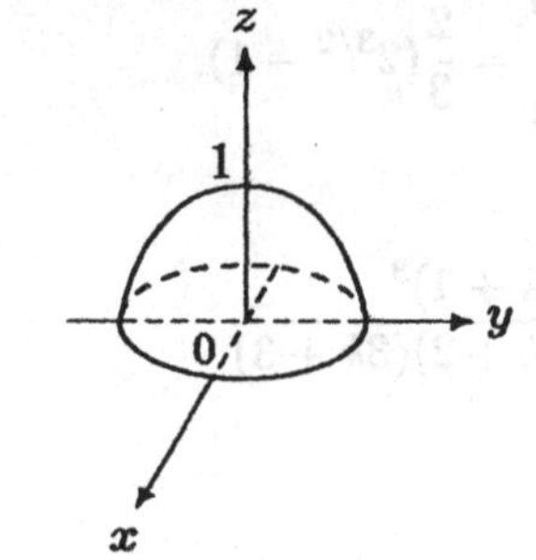

b. Let $f(x, y) = 1 - x^2 - y^2$. Notice that $f_x(x, y) = -2x$, so that $f_x(-1, 1) = 2$; also $f_y(x, y) = -2y$, so that $f_y(-1, 1) = -2$. Therefore by (2) in Section 13.6, the vector $f_x(-1, 1)\mathbf{i} + f_y(-1, 1)\mathbf{j} - \mathbf{k} = 2\mathbf{i} - 2\mathbf{j} - \mathbf{k}$ is normal to the paraboloid at $(-1, 1, -1)$. Consequently symmetric equations for the normal line ℓ through $(-1, 1, -1)$ are

$$\frac{x+1}{2} = \frac{y-1}{-2} = \frac{z+1}{-1}.$$

Chapter 15

Calculus of Vector Fields

15.1 Vector Fields

1. $\mathbf{F}(x,y) = \mathbf{i} + y\mathbf{j}$

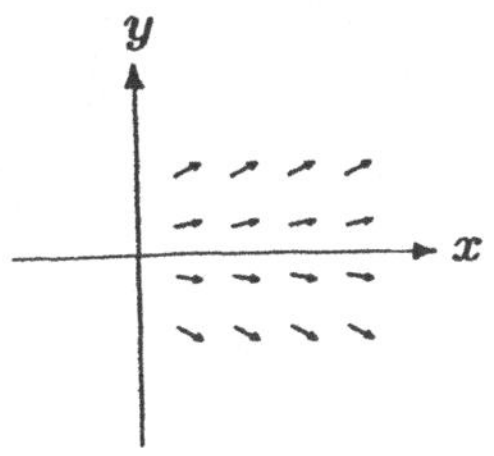

3. $\text{curl}\,\mathbf{F}(x,y) = \left(\dfrac{\partial N}{\partial x} - \dfrac{\partial M}{\partial y}\right)\mathbf{k} = (0-0)\mathbf{k} = \mathbf{0}$; $\text{div}\,\mathbf{F}(x,y) = \dfrac{\partial M}{\partial x} + \dfrac{\partial N}{\partial y} = 1+1 = 2$

5. $$\text{curl}\,\mathbf{F}(x,y,z) = \begin{vmatrix} \mathbf{i} & \mathbf{j} & \mathbf{k} \\ \dfrac{\partial}{\partial x} & \dfrac{\partial}{\partial y} & \dfrac{\partial}{\partial z} \\ x^2 & y^2 & z^2 \end{vmatrix} = 0\mathbf{i} + 0\mathbf{j} + 0\mathbf{k} = \mathbf{0}$$

$$\text{div}\,\mathbf{F}(x,y,z) = \frac{\partial M}{\partial x} + \frac{\partial N}{\partial y} + \frac{\partial P}{\partial z} = 2x + 2y + 2z = 2(x+y+z)$$

7. $$\text{curl}\,\mathbf{F}(x,y,z) = \begin{vmatrix} \mathbf{i} & \mathbf{j} & \mathbf{k} \\ \dfrac{\partial}{\partial x} & \dfrac{\partial}{\partial y} & \dfrac{\partial}{\partial z} \\ \dfrac{-x}{z} & \dfrac{-y}{z} & \dfrac{1}{z} \end{vmatrix} = \frac{-y}{z^2}\mathbf{i} + \frac{x}{z^2}\mathbf{j} + 0\mathbf{k} = \frac{-y}{z^2}\mathbf{i} + \frac{x}{z^2}\mathbf{j}$$

$$\text{div}\,\mathbf{F}(x,y,z) = \frac{\partial M}{\partial x} + \frac{\partial N}{\partial y} + \frac{\partial P}{\partial z} = -\frac{1}{z} - \frac{1}{z} - \frac{1}{z^2} = -\frac{2}{z} - \frac{1}{z^2}$$

9. $\text{curl}\,\mathbf{F}(x,y,z) = \begin{vmatrix} \mathbf{i} & \mathbf{j} & \mathbf{k} \\ \dfrac{\partial}{\partial x} & \dfrac{\partial}{\partial y} & \dfrac{\partial}{\partial z} \\ e^x \cos y & e^x \sin y & z \end{vmatrix} = 0\mathbf{i} + 0\mathbf{j} + (e^x \sin y + e^x \sin y)\mathbf{k} = 2e^x \sin y\mathbf{k}$

   $\text{div}\,\mathbf{F}(x,y,z) = \dfrac{\partial M}{\partial x} + \dfrac{\partial N}{\partial y} + \dfrac{\partial P}{\partial z} = e^x \cos y + e^x \cos y + 1 = 2e^x \cos y + 1$

11. Since $\partial M/\partial x > 0$ and $\partial N/\partial y > 0$ in Figure 15.7, it follows that $\text{div}\,\mathbf{F} > 0$.

13. $\dfrac{\partial^2 f}{\partial x^2} + \dfrac{\partial^2 f}{\partial y^2} = 2 - 2 = 0$

15. $\dfrac{\partial^2 f}{\partial x^2} + \dfrac{\partial^2 f}{\partial y^2} + \dfrac{\partial^2 f}{\partial z^2} = 2 + 2 - 4 = 0.$

17. $\partial N/\partial x = e^y = \partial M/\partial y$, so $\mathbf{F}$ is the gradient of some function f, and

$$\frac{\partial f}{\partial x} = e^y \quad \text{and} \quad \frac{\partial f}{\partial y} = xe^y + y. \qquad (*)$$

   Integrating both sides of the first equation in $(*)$ with respect to x, we obtain $f(x,y) = xe^y + g(y)$. Taking partial derivatives with respect to y, we find that $\partial f/\partial y = xe^y + dg/dy$. Comparing this with the second equation in $(*)$, we conclude that $dg/dy = y$, so that $g(y) = y^2/2 + C$. Therefore $f(x,y) = xe^y + y^2/2 + C$.

19. By Example 6, $\mathbf{F}$ is the gradient of some function f, and

$$\frac{\partial f}{\partial x} = 2xyz, \quad \frac{\partial f}{\partial y} = x^2 z, \quad \text{and} \quad \frac{\partial f}{\partial z} = x^2 y + 1. \qquad (*)$$

   Integrating both sides of the first equation in $(*)$ with respect to x, we obtain $f(x,y,z) = x^2yz + g(y,z)$. Taking partial derivatives with respect to y, we find that $\partial f/\partial y = x^2 z + \partial g/\partial y$. Comparing this with the second equation in $(*)$, we deduce that $\partial g/\partial y = 0$, so that g is constant with respect to y. Thus $f(x,y,z) = x^2yz + h(z)$. Taking partial derivatives with respect to z, we obtain $\partial f/\partial z = x^2 y + dh/dz$. Comparing this with the third equation in $(*)$, we conclude that $dh/dz = 1$, so that $h(z) = z + C$. Therefore $f(x,y,z) = x^2yz + z + C$.

21. Since $\partial P/\partial y = 0$ and $\partial N/\partial z = y$, $\mathbf{F}$ is not a gradient.

23. Since $\partial P/\partial y = 0$ and $\partial N/\partial z = 2z$, $\mathbf{F}$ is not a gradient.

25. a. $\text{grad}\,(fg)$ is a vector field.

    b. $\text{grad}\,\mathbf{F}$ is meaningless.

    c. $\text{curl}\,(\text{grad} f)$ is a vector field.

    d. $\text{grad}\,(\text{div}\,\mathbf{F})$ is a vector field.

    e. $\text{curl}\,(\text{curl}\,\mathbf{F})$ is a vector field.

    f. $\text{div}\,(\text{grad}\,f)$ is a function of several variables.

    g. $(\text{grad}\,f) \times (\text{curl}\,\mathbf{F})$ is a vector field.

h. div (curl (grad f)) is a function of several variables.

i. curl (div (grad f)) is meaningless.

27. Let $\mathbf{F} = M\mathbf{i} + N\mathbf{j} + P\mathbf{k}$. Then

$$\text{curl}\,\mathbf{F} = \left(\frac{\partial P}{\partial y} - \frac{\partial N}{\partial z}\right)\mathbf{i} + \left(\frac{\partial M}{\partial z} - \frac{\partial P}{\partial x}\right)\mathbf{j} + \left(\frac{\partial N}{\partial x} - \frac{\partial M}{\partial y}\right)\mathbf{k}$$

so that

$$\text{div}\,(\text{curl}\,\mathbf{F}) = \left(\frac{\partial^2 P}{\partial x \partial y} - \frac{\partial^2 N}{\partial x \partial z}\right) + \left(\frac{\partial^2 M}{\partial y \partial z} - \frac{\partial^2 P}{\partial y \partial x}\right) + \left(\frac{\partial^2 N}{\partial z \partial x} - \frac{\partial^2 M}{\partial z \partial y}\right)$$

Since all second partials are continuous by assumption,

$$\frac{\partial^2 M}{\partial y \partial z} = \frac{\partial^2 M}{\partial z \partial y}, \quad \frac{\partial^2 N}{\partial x \partial z} = \frac{\partial^2 N}{\partial z \partial x}, \quad \text{and} \quad \frac{\partial^2 P}{\partial x \partial y} = \frac{\partial^2 P}{\partial y \partial x}.$$

Thus $\text{div}\,(\text{curl}\,\mathbf{F}) = 0$.

29. Let $\mathbf{F} = M\mathbf{i} + N\mathbf{j} + P\mathbf{k}$. Then $f\mathbf{F} = fM\mathbf{i} + fN\mathbf{j} + fP\mathbf{k}$, so that

$$\text{div}\,(f\mathbf{F}) = \left(f\frac{\partial M}{\partial x} + \frac{\partial f}{\partial x}M\right) + \left(f\frac{\partial N}{\partial y} + \frac{\partial f}{\partial y}N\right) + \left(f\frac{\partial P}{\partial z} + \frac{\partial f}{\partial z}P\right)$$

$$= f\left(\frac{\partial M}{\partial x} + \frac{\partial N}{\partial y} + \frac{\partial P}{\partial z}\right) + \left(\frac{\partial f}{\partial x}M + \frac{\partial f}{\partial y}N + \frac{\partial f}{\partial z}P\right) = f\,\text{div}\,\mathbf{F} + (\text{grad}\,f)\cdot\mathbf{F}.$$

31. Let $\mathbf{F} = M\mathbf{i} + N\mathbf{j} + P\mathbf{k}$. Then $f\mathbf{F} = fM\mathbf{i} + fN\mathbf{j} + fP\mathbf{k}$, so that

$$\text{curl}\,f\mathbf{F} = \begin{vmatrix} \mathbf{i} & \mathbf{j} & \mathbf{k} \\ \dfrac{\partial}{\partial x} & \dfrac{\partial}{\partial y} & \dfrac{\partial}{\partial z} \\ fM & fN & fP \end{vmatrix} = \left(\frac{\partial f}{\partial y}P + f\frac{\partial P}{\partial y} - \frac{\partial f}{\partial z}N - f\frac{\partial N}{\partial z}\right)\mathbf{i}$$

$$+ \left(\frac{\partial f}{\partial z}M + f\frac{\partial M}{\partial z} - \frac{\partial f}{\partial x}P - f\frac{\partial P}{\partial x}\right)\mathbf{j} + \left(\frac{\partial f}{\partial x}N + f\frac{\partial N}{\partial x} - \frac{\partial f}{\partial y}M - f\frac{\partial M}{\partial y}\right)\mathbf{k}$$

$$= f\left[\left(\frac{\partial P}{\partial y} - \frac{\partial N}{\partial z}\right)\mathbf{i} + \left(\frac{\partial M}{\partial z} - \frac{\partial P}{\partial x}\right)\mathbf{j} + \left(\frac{\partial N}{\partial x} - \frac{\partial M}{\partial y}\right)\mathbf{k}\right]$$

$$+ \left[\left(\frac{\partial f}{\partial y}P - \frac{\partial f}{\partial z}N\right)\mathbf{i} + \left(\frac{\partial f}{\partial z}M - \frac{\partial f}{\partial x}P\right)\mathbf{j} + \left(\frac{\partial f}{\partial x}N - \frac{\partial f}{\partial y}M\right)\mathbf{k}\right]$$

$$= f\,(\text{curl}\,\mathbf{F}) + (\text{grad}\,f)\times\mathbf{F}.$$

33. Let $\mathbf{F} = a\mathbf{i} + b\mathbf{j} + c\mathbf{k}$. Then

$$(\mathbf{F} \times \mathbf{G})(x, y, z) = \begin{vmatrix} \mathbf{i} & \mathbf{j} & \mathbf{k} \\ a & b & c \\ x & y & z \end{vmatrix} = (bz - cy)\mathbf{i} + (cx - az)\mathbf{j} + (ay - bx)\mathbf{k}$$

and thus

$$[\operatorname{curl}(\mathbf{F} \times \mathbf{G})](x, y, z) = \begin{vmatrix} \mathbf{i} & \mathbf{j} & \mathbf{k} \\ \dfrac{\partial}{\partial x} & \dfrac{\partial}{\partial y} & \dfrac{\partial}{\partial z} \\ bz - cy & cx - az & ay - bx \end{vmatrix}$$

$$= [a - (-a)]\mathbf{i} + [b - (-b)]\mathbf{j} + [c - (-c)]\mathbf{k} = 2\mathbf{F}(x, y, z).$$

35. Let $\mathbf{F} = \operatorname{grad} f$ and $\mathbf{G} = \operatorname{grad} g$. Then

$$\begin{aligned} \mathbf{F}+\mathbf{G} &= \left(\frac{\partial f}{\partial x}\mathbf{i} + \frac{\partial f}{\partial y}\mathbf{j} + \frac{\partial f}{\partial z}\mathbf{k}\right) + \left(\frac{\partial g}{\partial x}\mathbf{i} + \frac{\partial g}{\partial y}\mathbf{j} + \frac{\partial g}{\partial z}\mathbf{k}\right) \\ &= \left(\frac{\partial f}{\partial x} + \frac{\partial g}{\partial x}\right)\mathbf{i} + \left(\frac{\partial f}{\partial y} + \frac{\partial g}{\partial y}\right)\mathbf{j} + \left(\frac{\partial f}{\partial z} + \frac{\partial g}{\partial z}\right)\mathbf{k} \\ &= \frac{\partial(f+g)}{\partial x}\mathbf{i} + \frac{\partial(f+g)}{\partial y}\mathbf{j} + \frac{\partial(f+g)}{\partial z}\mathbf{k} \\ &= \operatorname{grad}(f+g). \end{aligned}$$

Thus $\mathbf{F} + \mathbf{G}$ is conservative.

37. $\operatorname{grad} f(x, y, z) = (m\omega^2/2)(2x\mathbf{i} + 2y\mathbf{j} + 2z\mathbf{k}) = m\omega^2(x\mathbf{i} + y\mathbf{j} + z\mathbf{k}) = \mathbf{F}(x, y, z)$. Therefore f is a potential function for $\mathbf{F}$.

39. If $\mathbf{F} = \operatorname{curl} \mathbf{G}$, then by Exercise 27, $\operatorname{div} \mathbf{F} = \operatorname{div}(\operatorname{curl} \mathbf{G}) = 0$, so that $\mathbf{F}$ is solenoidal.

41. Let $\mathbf{G} = M\mathbf{i} + N\mathbf{j} + P\mathbf{k}$, and assume that $\mathbf{F}(x, y, z) = 8\mathbf{i} = \operatorname{curl} \mathbf{G}(x, y, z)$. Then

$$\frac{\partial P}{\partial y} - \frac{\partial N}{\partial z} = 8, \quad \frac{\partial M}{\partial z} - \frac{\partial P}{\partial x} = 0, \quad \text{and} \quad \frac{\partial N}{\partial x} - \frac{\partial M}{\partial y} = 0.$$

Notice that if $P(x, y, z) = 8y$ and $M(x, y, z) = 0 = N(x, y, z)$, then the three equations are satisfied. Consequently if we let $\mathbf{G}(x, y, z) = 8y\mathbf{k}$, then $\operatorname{curl} \mathbf{G}(x, y, z) = 8\mathbf{i} = \mathbf{F}(x, y, z)$. Thus $\mathbf{G}$ is a vector potential of $\mathbf{F}$. (Other solutions are also possible.)

43. a. $\operatorname{grad} \mathbf{T}(x, y, z) = -4x\mathbf{i} - 2y\mathbf{j} - 8z\mathbf{k}$. Since each component is continuous, so is $\operatorname{grad} \mathbf{T}$.

b. $\| \operatorname{grad} \mathbf{T}(x, y, z) \| = \sqrt{16x^2 + 4y^2 + 64z^2}$. Then $\| \operatorname{grad} \mathbf{T}(1, 0, 0) \| = 4$ and $\| \operatorname{grad} \mathbf{T}(0, 1, 0) \| = 2$. Since $(1, 0, 0)$ and $(0, 1, 0)$ are the same distance from the origin, and since the magnitudes of $\operatorname{grad} \mathbf{T}$ at these points are distinct, $\operatorname{grad} \mathbf{T}$ is not a central force field.

15.2 Line Integrals

1. Since $x(t) = 2t^{3/2}$, $y(t) = t^2$, $z(t) = 0$, and $\|d\mathbf{r}/dt\| = \sqrt{(3t^{1/2})^2 + (2t)^2 + 0^2} = \sqrt{9t + 4t^2}$, (4) implies that

$$\int_C (9 + 8y^{1/2})\,ds = \int_0^1 (9 + 8t)\sqrt{9t + 4t^2}\,dt = \frac{2}{3}(9t + 4t^2)^{3/2}\Big|_0^1 = \frac{26}{3}\sqrt{13}.$$

3. Since $x(t) = t$, $y(t) = t^3$, $z(t) = 0$, and $\|d\mathbf{r}/dt\| = \sqrt{1^2 + (3t^2)^2 + 0^2} = \sqrt{1 + 9t^4}$, (4) implies that

$$\int_C y\,ds = \int_{-1}^0 t^3\sqrt{1 + 9t^4}\,dt = \frac{1}{54}(1 + 9t^4)^{3/2}\Big|_{-1}^0 = \frac{1}{54}(1 - 10^{3/2}).$$

5. Since $x(t) = e^t$, $y(t) = e^{-t}$, $z(t) = \sqrt{2}\,t$, and $\|d\mathbf{r}/dt\| = \sqrt{(e^t)^2 + (-e^{-t})^2 + (\sqrt{2})^2} = \sqrt{e^{2t} + 2 + e^{-2t}} = \sqrt{(e^t + e^{-t})^2} = e^t + e^{-t}$, (4) implies that

$$\int_C 2xyz\,ds = \int_0^1 2(e^t)(e^{-t})(\sqrt{2}\,t)(e^t + e^{-t})\,dt = \int_0^1 2\sqrt{2}\,t(e^t + e^{-t})\,dt$$

$$\overset{\text{parts}}{=} 2\sqrt{2}\,t(e^t - e^{-t})\Big|_0^1 - \int_0^1 2\sqrt{2}\,(e^t - e^{-t})\,dt$$

$$= 2\sqrt{2}\,(e - e^{-1}) - [2\sqrt{2}\,(e^t + e^{-t})]\Big|_0^1 = 4\sqrt{2}\,(1 - e^{-1}).$$

7. Since $x(t) = \cos t$, $y(t) = \sin t$, $z(t) = t^{3/2}$, and $\|d\mathbf{r}/dt\| = \sqrt{(-\sin t)^2 + (\cos t)^2 + (\frac{3}{2}t^{1/2})^2} = \sqrt{1 + \frac{9}{4}t}$, (4) implies that

$$\int_C (1 + \frac{9}{4}z^{2/3})^{1/4}\,ds = \int_0^{20/3}\left(1 + \frac{9}{4}t\right)^{1/4}\sqrt{1 + \frac{9}{4}t}\,dt$$

$$= \int_0^{20/3}\left(1 + \frac{9}{4}t\right)^{3/4}dt = \frac{16}{63}\left(1 + \frac{9}{4}t\right)^{7/4}\Big|_0^{20/3} = \frac{2032}{63}.$$

9. The triangular path is composed of the three line segments C_1, C_2, and C_3 parametrized respectively by $\mathbf{r}_1(t) = (-1 + t)\mathbf{i} + t\mathbf{j}$ for $0 \le t \le 1$, $\mathbf{r}_2(t) = (1 - t)\mathbf{i} + t\mathbf{k}$ for $0 \le t \le 1$, and $\mathbf{r}_3(t) = -t\mathbf{i} + (1 - t)\mathbf{k}$ for $0 \le t \le 1$. Since $x(t) = -1 + t$, $y(t) = t$, $z(t) = 0$, and $\|d\mathbf{r}_1/dt\| = \sqrt{1^2 + 1^2 + 0^2} = \sqrt{2}$ for C_1, (4) implies that

$$\int_{C_1} (y + 2z)\,ds = \int_0^1 [t + 2(0)]\sqrt{2}\,dt = \int_0^1 \sqrt{2}\,t\,dt = \frac{\sqrt{2}}{2}t^2\Big|_0^1 = \frac{\sqrt{2}}{2}.$$

Since $x(t) = 0$, $y(t) = 1 - t$, $z(t) = t$, and $\|d\mathbf{r}_2/dt\| = \sqrt{0^2 + (-1)^2 + 1^2} = \sqrt{2}$ for C_2, (4) implies that

$$\int_{C_2} (y + 2z)\,ds = \int_0^1 (1 - t + 2t)\sqrt{2}\,dt = \int_0^1 \sqrt{2}\,(1 + t)\,dt = \sqrt{2}\,(t + \frac{1}{2}t^2)\Big|_0^1 = \frac{3}{2}\sqrt{2}.$$

Since $x(t) = -t$, $y(t) = 0$, $z(t) = 1 - t$, and $\|d\mathbf{r}_3/dt\| = \sqrt{(-1)^2 + 0^2 + (-1)^2} = \sqrt{2}$ for C_3, (4) implies that

$$\int_{C_3} (y + 2z)\,ds = \int_0^1 [0 + 2(1 - t)]\sqrt{2}\,dt = \int_0^1 2\sqrt{2}\,(1 - t)\,dt = 2\sqrt{2}\,(t - \frac{1}{2}t^2)\Big|_0^1 = \sqrt{2}.$$

It follows that

$$\int_C (y+2z)\,ds = \int_{C_1} (y+2z)\,ds + \int_{C_2} (y+2z)\,ds + \int_{C_3} (y+2z)\,ds = \frac{\sqrt{2}}{2} + \frac{3}{2}\sqrt{2} + \sqrt{2} = 3\sqrt{2}.$$

11. Since $x(t) = 5$, $y(t) = -\sin t$, $z(t) = -\cos t$, and $d\mathbf{r}/dt = -\cos t\mathbf{j} + \sin t\mathbf{k}$, (7) implies that

$$\int_C \mathbf{F}\cdot d\mathbf{r} = \int_0^{\pi/4} (-\cos t\mathbf{i} + \sin t\mathbf{j} - 5\mathbf{k})\cdot(-\cos t\mathbf{j} + \sin t\mathbf{k})\,dt$$

$$= \int_0^{\pi/4} (-\sin t\cos t - 5\sin t)\,dt = \left(-\frac{1}{2}\sin^2 t + 5\cos t\right)\Bigg|_0^{\pi/4} = \frac{5}{2}\sqrt{2} - \frac{21}{4}.$$

13. Since $x(t) = \cos t$, $y(t) = \sin t$, $z(t) = 2t$, and $d\mathbf{r}/dt = -\sin t\mathbf{i} + \cos t\mathbf{j} + 2\mathbf{k}$, (7) implies that

$$\int_C \mathbf{F}\cdot d\mathbf{r} = \int_0^{\pi/2} (\sin t\mathbf{i} + \cos t\sin t\mathbf{j} + 8t^3\mathbf{k})\cdot(-\sin t\mathbf{i} + \cos t\mathbf{j} + 2\mathbf{k})\,dt$$

$$= \int_0^{\pi/2} (-\sin^2 t + \cos^2 t\sin t + 16t^3)\,dt = \int_0^{\pi/2} \left(-\frac{1}{2} + \frac{1}{2}\cos 2t + \cos^2 t\sin t + 16t^3\right) dt$$

$$= \left(-\frac{1}{2}t + \frac{1}{4}\sin 2t - \frac{1}{3}\cos^3 t + 4t^4\right)\Bigg|_0^{\pi/2} = -\frac{1}{4}\pi + \frac{1}{4}\pi^4 + \frac{1}{3}.$$

15. $\int_{-C} \mathbf{F}\cdot d\mathbf{r} = -\int_C \mathbf{F}\cdot d\mathbf{r} = -\pi$, by the solution of Exercise 14.

17. Since $x(t) = \frac{1}{2}$, $y(t) = 2$, $z(t) = -\ln(\cosh t)$, and $d\mathbf{r}/dt = -(\sinh t/\cosh t)\mathbf{k}$, we have

$$\mathbf{F}(x(t), y(t), z(t))\cdot d\mathbf{r}/dt = 0,$$

so that by (7), $\int_C \mathbf{F}\cdot d\mathbf{r} = \int_0^{\pi/6} 0\,dt = 0$.

19. Since $x(t) = e^{-t}$, $y(t) = e^t$, $z(t) = t$, $dx/dt = -e^{-t}$, $dy/dt = e^t$ and $dz/dt = 1$, (9) implies that

$$\int_C y\,dx - x\,dy + xyz^2\,dz = \int_0^1 [(e^t)(-e^{-t}) - (e^{-t})(e^t) + (e^{-t}e^t t^2)(1)]\,dt$$

$$= \int_0^1 (-2 + t^2)\,dt = \left(-2t + \frac{1}{3}t^3\right)\Bigg|_0^1 = -\frac{5}{3}.$$

21. $\int_C e^x\,dx + xy\,dy + xyz\,dz = -\int_{-C} e^x\,dx + xy\,dy + xyz\,dz = e^{-1} - e - \frac{2}{3}$ by the solution of Exercise 20.

23. Since $x(t) = t+1$, $y(t) = t-1$, $z(t) = t^2$, $dx/dt = 1$, $dy/dt = 1$, and $dz/dt = 2t$, (9) implies that

$$\int_C xy\,dx + (x+z)\,dy + z^2\,dz = \int_{-1}^2 [(t+1)(t-1)(1) + (t+1+t^2)(1) + (t^4)(2t)]\,dt$$

$$= \int_{-1}^2 (2t^5 + 2t^2 + t)\,dt = \left(\frac{1}{3}t^6 + \frac{2}{3}t^3 + \frac{1}{2}t^2\right)\Bigg|_{-1}^2 = \frac{57}{2}.$$

25. Since C is parametrized by $\mathbf{r}(t) = \cos t\mathbf{i} + \sin t\mathbf{j}$ for $0 \le t \le \pi/2$, and since $x(t) = \cos t$, $y(t) = \sin t$, $dx/dt = -\sin t$, and $dy/dt = \cos t$, the two-dimensional version of (9) implies that

$$\int_C \frac{1}{1+x^2}\,dx + \frac{2}{1+y^2}\,dy = \int_0^{\pi/2} \left[\frac{1}{1+\cos^2 t}(-\sin t) + \frac{2}{1+\sin^2 t}(\cos t)\right] dt$$

$$= [\tan^{-1}(\cos t) + 2\tan^{-1}(\sin t)]\Big|_0^{\pi/2} = \frac{1}{4}\pi.$$

27. Since $x(t) = t$, $y(t) = t^2$, $z(t) = t^3$, $dx/dt = 1$, $dy/dt = 2t$, and $dz/dt = 3t^2$, (9) implies that

$$\int_c x\ln\left(\frac{xz}{y}\right) dx + \cos\left(\frac{\pi xy}{z}\right) dy = \int_1^2 [(t\ln t^2)(1) + (\cos\pi)(2t)]\,dt$$

$$= \int_1^2 (2t\ln t - 2)\,dt \overset{\text{Parts}}{=} t^2\ln t\Big|_1^2 - \int_1^2 t\,dt - t^2\Big|_1^2 = 4\ln 2 - 3 - \frac{1}{2}t^2\Big|_1^2 = 4\ln 2 - \frac{9}{2}.$$

29. a. The curve C is composed of the line segments C_1 and C_2 parametrized respectively by $\mathbf{r}_1(t) = -5t\mathbf{j}$ for $0 \le t \le 1$, and $\mathbf{r}_2(t) = (-5+6t)\mathbf{j} + t\mathbf{k}$ for $0 \le t \le 1$. By (9),

$$\int_C y\,dx + z\,dy + x\,dz = \int_{C_1} y\,dx + z\,dy + x\,dz + \int_{C_2} y\,dx + z\,dy + x\,dz$$

$$= \int_0^1 [(-5t)(0) + (0)(-5) + (0)(0)]\,dt + \int_0^1 [(-5+6t)(0) + (t)(6) + (0)(1)]\,dt$$

$$= \int_0^1 0\,dt + \int_0^1 6t\,dt = 3t^2\Big|_0^1 = 3.$$

b. The curve C is composed of the line segments C_1 and C_2 parametrized respectively by $\mathbf{r}_1(t) = t\mathbf{i}$ for $0 \le t \le 1$, and $\mathbf{r}_2(t) = (1-t)\mathbf{i} + t\mathbf{j} + t\mathbf{k}$ for $0 \le t \le 1$. By (9),

$$\int_C y\,dx + z\,dy + x\,dz = \int_{C_1} y\,dx + z\,dy + x\,dz + \int_{C_2} y\,dx + z\,dy + x\,dz$$

$$= \int_0^1 [(0)(1) + (0)(0) + (t)(0)]\,dt + \int_0^1 [(t)(-1) + (t)(1) + (1-t)(1)]\,dt$$

$$= \int_0^1 0\,dt + \int_0^1 (1-t)\,dt = \left(t - \frac{1}{2}t^2\right)\Big|_0^1 = \frac{1}{2}.$$

31. If the first component of a parametrization is constant, then $dx/dt = 0$, so that by the first formula in (10),

$$\int_C M(x,y,z)\,dx = \int_a^b M(x(t), y(t), z(t))\,\frac{dx}{dt}\,dt = \int_a^b 0\,dt = 0.$$

33. Since the density at (x, y, z) is given by $f(x, y, z) = x^2 + y^2 + z^2$,

$$m = \int_C (x^2 + y^2 + z^2)\, ds = \int_{\pi}^{2\pi} (\sin^2 t + \cos^2 t + 16t^2)\sqrt{17}\, dt$$

$$= \sqrt{17}\int_{\pi}^{2\pi} (1 + 16t^2)\, dt = \sqrt{17}\left(t + \frac{16}{3}t^3\right)\Big|_{\pi}^{2\pi} = \sqrt{17}\left(\pi + \frac{112}{3}\pi^3\right).$$

35. a. The line segment C from $(0, 0, 0)$ to $(1, 1, 1)$ is parametrized by $\mathbf{r}(t) = t\mathbf{i} + t\mathbf{j} + t\mathbf{k}$ for $0 \le t \le 1$. By (6), $W = \int_C \mathbf{F} \cdot d\mathbf{r}$. Since $x(t) = t$, $y(t) = t$, $z(t) = t$, and $d\mathbf{r}/dt = \mathbf{i} + \mathbf{j} + \mathbf{k}$, (7) implies that

$$W = \int_C \mathbf{F} \cdot d\mathbf{r} = \int_0^1 (t\mathbf{i} + 2t\mathbf{j} + 0\mathbf{k}) \cdot (\mathbf{i} + \mathbf{j} + \mathbf{k})\, dt = \int_0^1 (t + 2t)\, dt = \frac{3}{2}t^2\Big|_0^1 = \frac{3}{2}.$$

b. Let C be the curve parametrized by $\mathbf{r}$. By (6), $W = \int_C \mathbf{F} \cdot d\mathbf{r}$. Since $x(t) = \sin(\pi t/2)$, $y(t) = \sin(\pi t/2)$, $z(t) = t$, and $d\mathbf{r}/dt = (\pi/2)\cos(\pi t/2)\mathbf{i} + (\pi/2)\cos(\pi t/2)\mathbf{j} + \mathbf{k}$, (7) implies that

$$W = \int_C \mathbf{F} \cdot d\mathbf{r} = \int_0^1 \left[\sin\frac{\pi t}{2}\mathbf{i} + 2t\mathbf{j} + \left(\sin\frac{\pi t}{2} - t\right)\mathbf{k}\right] \cdot \left[\frac{\pi}{2}\cos\frac{\pi t}{2}\mathbf{i} + \frac{\pi}{2}\cos\frac{\pi t}{2}\mathbf{j} + \mathbf{k}\right] dt$$

$$= \int_0^1 \left(\frac{\pi}{2}\sin\frac{\pi t}{2}\cos\frac{\pi t}{2} + \pi t\cos\frac{\pi t}{2} + \sin\frac{\pi t}{2} - t\right) dt$$

$$= \left[\frac{1}{2}\sin^2\left(\frac{\pi t}{2}\right) - \frac{2}{\pi}\cos\frac{\pi t}{2} - \frac{1}{2}t^2\right]\Big|_0^1 + \int_0^1 \pi t\cos\frac{\pi t}{2}\, dt \overset{\text{parts}}{=} \frac{2}{\pi} + 2t\sin\frac{\pi t}{2}\Big|_0^1 - \int_0^1 2\sin\frac{\pi t}{2}\, dt$$

$$= \frac{2}{\pi} + 2 + \frac{4}{\pi}\cos\frac{\pi t}{2}\Big|_0^1 = 2 - \frac{2}{\pi}.$$

37. When the height of the painter is z, the weight of the painter and pail is $120 + (30 - \frac{1}{10}z) = 150 - \frac{1}{10}z$. Thus $\mathbf{F}(x, y, z) = -(150 - \frac{1}{10}z)\mathbf{k} = (\frac{1}{10}z - 150)\mathbf{k}$. By (6) and (7),

$$W = \int_C \mathbf{F} \cdot d\mathbf{r} = \int_0^{8\pi} \left[\frac{1}{10}\left(\frac{25t}{\pi}\right) - 150\right]\mathbf{k} \cdot \left[-50\sin t\,\mathbf{i} + 50\cos t\,\mathbf{j} + \frac{25}{\pi}\mathbf{k}\right] dt$$

$$= \int_0^{8\pi} \frac{25}{\pi}\left[\frac{1}{10}\left(\frac{25t}{\pi}\right) - 150\right] dt = \frac{25}{\pi}\left(\frac{5}{4\pi}t^2 - 150t\right)\Big|_0^{8\pi} = -28{,}000 \text{(foot-pounds)}.$$

39. a. If the force acts in a direction normal to the path, then $\mathbf{F}(x, y, z) = f(x, y, z)\mathbf{N}(x, y, z)$, where $\mathbf{N}$ is the normal vector at (x, y, z) and f is a function of three variables. Then by (5),

$$W = \int_C \mathbf{F}(x, y, z) \cdot \mathbf{T}(x, y, z)\, ds = \int_C f(x, y, z)\mathbf{N}(x, y, z) \cdot \mathbf{T}(x, y, z)\, ds = \int_C 0\, ds = 0.$$

b. Since the gravitational field of the earth is a central force field and since the normal vector at any point on the circular orbit also points toward the origin, it follows that the gravitational force acts in a direction normal to the path of the object. Consequently by part (a), the work done by the force is 0.

15.3 The Fundamental Theorem of Line Integrals

Note: In Exercises 1–8 of this section, the domain of $\mathbf{F}$ is a region without holes, so conditions 1–4 are equivalent. In particular, if $\operatorname{curl}\mathbf{F} = \mathbf{0}$, then $\int_C \mathbf{F}\cdot d\mathbf{r}$ is independent of path.

1. Let $\mathbf{F}(x,y) = (e^x + y)\mathbf{i} + (x+2y)\mathbf{j} = M(x,y)\mathbf{i} + N(x,y)\mathbf{j}$. Since $\partial N/\partial x = 1 = \partial M/\partial y$, so $\operatorname{curl}\mathbf{F} = \mathbf{0}$, the line integral is independent of path. Let $\operatorname{grad} f(x,y) = \mathbf{F}(x,y)$. Then

$$\frac{\partial f}{\partial x} = e^x + y \quad \text{and} \quad \frac{\partial f}{\partial y} = x + 2y. \qquad (*)$$

Integrating both sides of the first equation in $(*)$ with respect to x, we obtain $f(x,y) = e^x + xy + g(y)$. Taking partial derivatives with respect to y, we find that $\partial f/\partial y = x + dg/dy$. Comparing this with the second equation in $(*)$, we conclude that $dg/dy = 2y$, so that $g(y) = y^2 + c$. If $c = 0$, then $f(x,y) = e^x + xy + y^2$. Thus

$$\int_C (e^x + y)\,dy + (x+2y)\,dy = f(2,3) - f(0,1) = (e^2 + 15) - 2 = e^2 + 13.$$

3. Let $\mathbf{F}(x,y,z) = y\mathbf{i} + (x+z)\mathbf{j} + y\mathbf{k} = M(x,y,z)\mathbf{i} + N(x,y,z)\mathbf{j} + P(x,y,z)\mathbf{k}$. Since $\partial P/\partial y = 1 = \partial N/\partial z$, $\partial M/\partial z = 0 = \partial P/\partial x$, and $\partial N/\partial x = 1 = \partial M/\partial y$, so $\operatorname{curl}\mathbf{F} = \mathbf{0}$, the line integral is independent of path. Let $\operatorname{grad} f(x,y,z) = \mathbf{F}(x,y,z)$. Then

$$\frac{\partial f}{\partial x} = y, \quad \frac{\partial f}{\partial y} = x + z, \quad \frac{\partial f}{\partial z} = y. \qquad (*)$$

Integrating both sides of the first equation in $(*)$ with respect to x, we obtain $f(x,y,z) = xy + g(y,z)$. Taking partial derivatives with respect to y, we find that $\partial f/\partial y = x + \partial g/\partial y$. Comparing this with the second equation in $(*)$, we deduce that $\partial g/\partial y = z$, so that by integrating with respect to y we obtain $g(y,z) = yz + h(z)$. Thus $f(x,y,z) = xy + yz + h(z)$. Taking partial derivatives with respect to z, we find that $\partial f/\partial z = y + dh/dz$. Comparing this with the third equation in $(*)$, we conclude that $dh/dz = 0$, so that $h(z) = c$. If $c = 0$, then $f(x,y,z) = xy + yz$. Since $\mathbf{r}(0) = -\mathbf{i} + \mathbf{j}$ and $\mathbf{r}(\frac{1}{2}) = -\frac{5}{3}\mathbf{i} + \mathbf{k}$, it follows that the initial point of C is $(-1,1,0)$ and the terminal point is $(-\frac{5}{3},0,1)$. Thus

$$\int_C y\,dx + (x+z)\,dy + y\,dz = f(-\tfrac{5}{3},0,1) - f(-1,1,0) = 0 - (-1) = 1.$$

5. Let

$$\begin{aligned} \mathbf{F}(x,y,z) &= \frac{x}{1+x^2+y^2+z^2}\mathbf{i} + \frac{y}{1+x^2+y^2+z^2}\mathbf{j} + \frac{z}{1+x^2+y^2+z^2}\mathbf{k} \\ &= M(x,y,z)\mathbf{i} + N(x,y,z)\mathbf{j} + P(x,y,z)\mathbf{k}. \end{aligned}$$

Since

$$\frac{\partial P}{\partial y} = \frac{-2yz}{(1+x^2+y^2+z^2)^2} = \frac{\partial N}{\partial z}, \quad \frac{\partial M}{\partial z} = \frac{-2xz}{(1+x^2+y^2+z^2)^2} = \frac{\partial P}{\partial x},$$

and

$$\frac{\partial N}{\partial x} = \frac{-2xy}{(1+x^2+y^2+z^2)^2} = \frac{\partial M}{\partial y},$$

so $\operatorname{curl}\mathbf{F} = \mathbf{0}$, the line integral is independent of path. Let $\operatorname{grad} f(x,y,z) = \mathbf{F}(x,y,z)$. Then

$$\frac{\partial f}{\partial x} = \frac{x}{1+x^2+y^2+z^2}, \quad \frac{\partial f}{\partial y} = \frac{y}{1+x^2+y^2+z^2}, \quad \text{and} \quad \frac{\partial f}{\partial z} = \frac{z}{1+x^2+y^2+z^2}. \qquad (*)$$

Integrating both sides of the first equation in (∗) with respect to x, we obtain $f(x,y,z) = \frac{1}{2}\ln(1 + x^2 + y^2 + z^2) + g(y,z)$. Taking partial derivatives with respect to y, we find that $\partial f/\partial y = y/(1+x^2+y^2+z^2) + \partial g/\partial y$. Comparing this with the second equation in (∗), we deduce that $\partial g/\partial y = 0$, so that integrating with respect to y yields $g(y,z) = h(z)$. Thus $f(x,y,z) = \frac{1}{2}\ln(1+x^2+y^2+z^2) + h(z)$. Taking partial derivatives with respect to z, we find that $\partial f/\partial z = z/(1+x^2+y^2+z^2) + dh/dz$. Comparing this with the third equation in (∗), we conclude that $dh/dz = 0$, so that integration yields $h(z) = c$. If $c = 0$, then $f(x,y,z) = \frac{1}{2}\ln(1+x^2+y^2+z^2)$. Since $\mathbf{r}(0) = \mathbf{0}$ and $\mathbf{r}(1) = \mathbf{i}+\mathbf{j}+\mathbf{k}$, it follows that the initial and terminal points of C are $(0,0,0)$ and $(1,1,1)$, respectively. Then

$$\int_C \frac{x}{1+x^2+y^2+z^2}\,dx + \frac{y}{1+x^2+y^2+z^2}\,dy + \frac{z}{1+x^2+y^2+z^2}\,dz$$

$$= f(1,1,1) - f(0,0,0) = \frac{1}{2}\ln 4 - 0 = \ln 2.$$

7. Let $\mathbf{F}(x,y,z) = e^{-x}\ln y\mathbf{i} - (e^{-x}/y)\mathbf{j} + z\mathbf{k} = M(x,y,z)\mathbf{i} + N(x,y,z)\mathbf{j} + P(x,y,z)\mathbf{k}$. Since $\partial P/\partial y = 0 = \partial N/\partial z$, $\partial M/\partial z = 0 = \partial P/\partial x$, and $\partial N/\partial x = e^{-x}/y = \partial M/\partial y$, so $\operatorname{curl}\mathbf{F} = \mathbf{0}$, the line integral is independent of path. Let $\operatorname{grad} f(x,y,z) = \mathbf{F}(x,y,z)$. Then

$$\frac{\partial f}{\partial x} = e^{-x}\ln y, \quad \frac{\partial f}{\partial y} = \frac{-e^{-x}}{y}, \quad \text{and} \quad \frac{\partial f}{\partial z} = z. \tag{$*$}$$

Integrating both sides of the first equation in (∗) with respect to x, we obtain $f(x,y,z) = -e^{-x}\ln y + g(y,z)$. Taking partial derivatives with respect to y, we find that $\partial f/\partial y = -e^{-x}/y + \partial g/\partial y$. Comparing this with the second equation in (∗), we deduce that $\partial g/\partial y = 0$, so integration yields $g(y,z) = h(z)$. Then $f(x,y,z) = -e^{-x}\ln y + h(z)$. Taking partial derivatives with respect to z, we find that $\partial f/\partial z = dh/dz$. Comparing this with the third equation in (∗), we conclude that $dh/dz = z$, so integration yields $h(z) = \frac{1}{2}z^2 + c$. If $c = 0$, then we find that $f(x,y,z) = -e^{-x}\ln y + \frac{1}{2}z^2$. Since $\mathbf{r}(0) = -\mathbf{i}+\mathbf{j}+\mathbf{k}$ and $\mathbf{r}(1) = e\mathbf{j} + 2\mathbf{k}$, it follows that the initial and terminal points of C are $(-1,1,1)$ and $(0,e,2)$, respectively. Then

$$\int_C e^{-x}\ln y\,dx - \frac{e^{-x}}{y}\,dy + z\,dz = f(0,e,2) - f(-1,1,1) = 1 - \frac{1}{2} = \frac{1}{2}.$$

9. Let $\mathbf{F}(x,y,z) = f(x)\mathbf{i} + g(y)\mathbf{j} + h(z)\mathbf{k} = M(x,y,z)\mathbf{i} + N(x,y,z)\mathbf{j} + P(x,y,z)\mathbf{k}$. Since $\partial P/\partial y = 0 = \partial N/\partial z$, $\partial M/\partial z = 0 = \partial P/\partial x$, and $\partial N/\partial x = 0 = \partial M/\partial y$, it follows that $\operatorname{curl}\mathbf{F} = \mathbf{0}$, so that $\int_C \mathbf{F}\cdot d\mathbf{r} = \int_C f(x)\,dx + g(y)\,dy + h(z)\,dz$ is independent of path.

11. a. Let $h(u) = \int g(u)\,du$ and $f(x,y,z) = \frac{1}{2}h(x^2+y^2+z^2)$. Then

$$\operatorname{grad} f = \frac{\partial f}{\partial x}\mathbf{i} + \frac{\partial f}{\partial y}\mathbf{j} + \frac{\partial f}{\partial z}\mathbf{k}$$

$$= xh'(x^2+y^2+z^2)\mathbf{i} + yh'(x^2+y^2+z^2)\mathbf{j} + zh'(x^2+y^2+z^2)\mathbf{k}$$

$$= g(x^2+y^2+z^2)(x\mathbf{i}+y\mathbf{j}+z\mathbf{k}) = \mathbf{F}(x,y,z).$$

Thus $\mathbf{F}$ is conservative.

b. Since curl $\mathbf{F}$ = curl (grad f) from (a), and since curl (grad f) = $\mathbf{0}$ from (5) in Section 15.1, it follows that curl $\mathbf{F} = \mathbf{0}$, so $\mathbf{F}$ is irrotational.

13. Since the electric field $\mathbf{E}$ is conservative, the work is independent of path. Thus we may assume that the charge moves in a straight line. We set up a coordinate system so that the electron is at the origin and the charge moves from $(0, 0, 10^{-11})$ to $(0, 0, 10^{-12})$. At this time we recall from Section 15.1 that

$$\mathbf{E}(x, y, z) = \frac{(-1.6) \times 10^{-19}}{4\pi\varepsilon_0(x^2 + y^2 + z^2)^{3/2}}(x\mathbf{i} + y\mathbf{j} + z\mathbf{k}).$$

Now if

$$f(x, y, z) = \frac{(1.6) \times 10^{-19}}{4\pi\varepsilon_0(x^2 + y^2 + z^2)^{1/2}}$$

then grad $f(x, y, z) = \mathbf{E}(x, y, z)$, so

$$W = \int_C \mathbf{E} \cdot d\mathbf{r} = f(0, 0, 10^{-12}) - f(0, 0, 10^{-11}) = \frac{(1.6) \times 10^{-19}}{4\pi\varepsilon_0}(10^{12} - 10^{11}) \approx 1294\text{(joules)}.$$

15. Let $t = a$ when the speed of the object is 50 meters per second, and $t = b$ when the speed of the object is 10 meters per second. The change in kinetic energy (in joules) is

$$\frac{m}{2}\,\|\mathbf{r}'(b)\|^2 - \frac{m}{2}\,\|\mathbf{r}'(a)\|^2 = \frac{5}{2}(10)^2 - \frac{5}{2}(50)^2 = -6000.$$

It follows from the Law of Conservation of Energy that the change in kinetic energy is the negative of the change in potential energy, so this latter change is 6000 joules.

15.4 Green's Theorem

In Exercises 1–18, R denotes the region enclosed by C.

1. By Green's Theorem,

$$\int_C M(x, y)\,dx + N(x, y)\,dy = \iint_R \left(\frac{\partial N}{\partial x} - \frac{\partial M}{\partial y}\right) dA = \iint_R (0 - 1)\,dA = -(\text{area of } R) = -\pi.$$

3. By Green's Theorem,

$$\int_C M(x, y)\,dx + N(x, y)\,dy = \iint_R \left(\frac{\partial N}{\partial x} - \frac{\partial M}{\partial y}\right) dA = \iint_R \left(\frac{3}{2}x^{1/2} - x\right) dA$$

$$= \int_0^1 \int_0^1 \left(\frac{3}{2}x^{1/2} - x\right) dx\,dy = \int_0^1 \left(x^{3/2} - \frac{1}{2}x^2\right)\Big|_0^1 dy = \int_0^1 \frac{1}{2}\,dy = \frac{1}{2}.$$

5. By Green's Theorem,

$$\int_C M(x, y)\,dx + N(x, y)\,dy = \iint_R \left(\frac{\partial N}{\partial x} - \frac{\partial M}{\partial y}\right) dA$$

$$= \iint_R [3x(x^2 + y^2)^{1/2} - 3y(x^2 + y^2)^{1/2}]\,dA = \int_0^{2\pi} \int_0^1 3r^3(\cos\theta - \sin\theta)\,dr\,d\theta$$

$$= \int_0^{2\pi} \frac{3}{4}r^4(\cos\theta - \sin\theta)\Big|_0^1 d\theta = \int_0^{2\pi} \frac{3}{4}(\cos\theta - \sin\theta)\,d\theta = \frac{3}{4}(\sin\theta + \cos\theta)\Big|_0^{2\pi} = 0.$$

7. By Green's Theorem,

$$\int_C y\,dx - x\,dy = \iint_R \left[\frac{\partial}{\partial x}(-x) - \frac{\partial y}{\partial y}\right] dA = \iint_R -2\,dA$$

$$= \int_0^{2\pi}\int_0^{1-\cos\theta} -2r\,dr\,d\theta = \int_0^{2\pi} -r^2\Big|_0^{1-\cos\theta} d\theta = \int_0^{2\pi} (-1 + 2\cos\theta - \cos^2\theta)\,d\theta$$

$$= \int_0^{2\pi}\left(-1 + 2\cos\theta - \frac{1}{2} - \frac{1}{2}\cos 2\theta\right) d\theta = \left(-\frac{3}{2}\theta + 2\sin\theta - \frac{1}{4}\sin 2\theta\right)\Big|_0^{2\pi} = -3\pi.$$

9. By Green's Theorem,

$$\int_C e^x \sin y\,dx + e^x\cos y\,dy = \iint_R \left[\frac{\partial}{\partial x}(e^x\cos y) - \frac{\partial}{\partial y}(e^x\sin y)\right] dA$$

$$= \iint_R (e^x\cos y - e^x\cos y)\,dA = \iint_R 0\,dA = 0.$$

11. By Green's Theorem,

$$\int_C xy\,dx + \left(\frac{1}{2}x^2 + xy\right) dy = \iint_R \left[\frac{\partial}{\partial x}\left(\frac{1}{2}x^2 + xy\right) - \frac{\partial}{\partial y}(xy)\right] dA = \iint_R (x + y - x)\,dA$$

$$= \iint_R y\,dA = \int_{-1}^{1}\int_0^{(1/2)\sqrt{1-x^2}} y\,dy\,dx = \int_{-1}^{1} \frac{1}{2}y^2\Big|_0^{(1/2)\sqrt{1-x^2}} dx$$

$$= \int_{-1}^{1}\frac{1}{8}(1-x^2)\,dx = \frac{1}{8}\left(x - \frac{1}{3}x^3\right)\Big|_{-1}^{1} = \frac{1}{6}.$$

13. By Green's Theorem,

$$\int_C (\cos^3 x + e^x)\,dx + e^y\,dy = \iint_R \left[\frac{\partial}{\partial x}(e^y) - \frac{\partial}{\partial y}(\cos^3 x + e^x)\right] dA = \iint_R 0\,dA = 0.$$

15. By Green's Theorem,

$$\int_C \mathbf{F}\cdot d\mathbf{r} = \int_C y\,dx + 3x\,dy = \iint_R \left[\frac{\partial}{\partial x}(3x) - \frac{\partial y}{\partial y}\right] dA = \iint_R (3-1)\,dA = 2(\text{area of } R) = 8\pi.$$

17. By Green's Theorem,

$$\int_C \mathbf{F}\cdot d\mathbf{r} = \int_C y\sin x\,dx - \cos x\,dy = \iint_R \left[\frac{\partial}{\partial x}(-\cos x) - \frac{\partial}{\partial y}(y\sin x)\right] dA$$

$$= \iint_R (\sin x - \sin x)\,dA = \iint_R 0\,dA = 0.$$

19. Let C be the boundary of the given region R, with C oriented counterclockwise. Then C is composed of C_1 and $-C_2$, where C_1 is the interval $[0, 2\pi]$, parametrized by $\mathbf{r}_1(t) = t\mathbf{i}$ for $0 \le t \le 2\pi$, and C_2 is the cycloid parametrized by the given vector-valued function $\mathbf{r}$. Thus by (4),

$$A = \frac{1}{2}\int_C x\,dy - y\,dx = \frac{1}{2}\int_{C_1} x\,dy - y\,dx - \frac{1}{2}\int_{C_2} x\,dy - y\,dx$$

$$= \frac{1}{2}\int_0^{2\pi} [(t)(0) - (0)(1)]\,dt - \frac{1}{2}\int_0^{2\pi} [(t - \sin t)(\sin t) - (1 - \cos t)(1 - \cos t)]\,dt$$

$$= -\frac{1}{2}\int_0^{2\pi} (t\sin t - \sin^2 t - 1 + 2\cos t - \cos^2 t)\,dt = -\frac{1}{2}\int_0^{2\pi} (-2 + 2\cos t)\,dt - \frac{1}{2}\int_0^{2\pi} t\sin t\,dt$$

$$\overset{\text{parts}}{=} -\frac{1}{2}(-2t + 2\sin t)\Big|_0^{2\pi} + \frac{1}{2}t\cos t\Big|_0^{2\pi} - \frac{1}{2}\int_0^{2\pi} \cos t\,dt = 2\pi + \pi - \frac{1}{2}\sin t\Big|_0^{2\pi} = 3\pi.$$

21. Let C be the boundary of the given region, with C oriented in the counterclockwise direction. Then C consists of C_1, C_2, and C_3, where C_1 is the curve parametrized by the given vector-valued function $\mathbf{r}$, C_2 is the curve parametrized by $\mathbf{r}_2(t) = (1 - t)\mathbf{i} + \frac{1}{4}\mathbf{j}$ for $0 \le t \le 1$, and C_3 is the curve parametrized by $\mathbf{r}_3(t) = (\frac{1}{4} - t)\mathbf{j}$ for $0 \le t \le \frac{1}{4}$. Thus by (4),

$$A = \int_C x\,dy = \int_{C_1} x\,dy + \int_{C_2} x\,dy + \int_{C_3} x\,dy$$

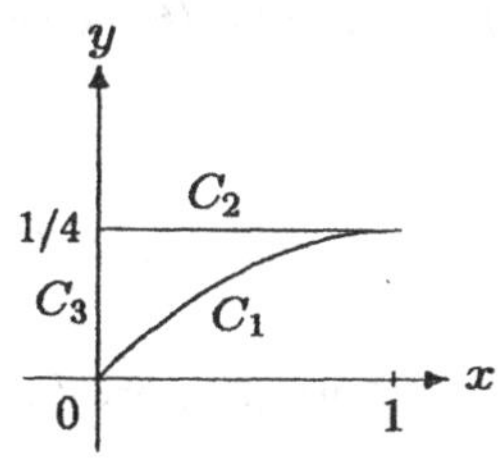

$$= \int_0^{1/2} (\sin \pi t)(1 - 2t)\,dt + \int_0^1 (1 - t)(0)\,dt + \int_0^{1/4} (0)(-1)\,dt$$

$$= \int_0^{1/2} \sin \pi t\,dt - 2\int_0^{1/2} t\sin \pi t\,dt$$

$$\overset{\text{parts}}{=} -\frac{1}{\pi}\cos \pi t\Big|_0^{1/2} + \frac{2}{\pi}t\cos \pi t\Big|_0^{1/2} - 2\int_0^{1/2} \frac{1}{\pi}\cos \pi t\,dt = \frac{1}{\pi} - \frac{2}{\pi^2}\sin \pi t\Big|_0^{1/2} = \frac{1}{\pi} - \frac{2}{\pi^2}.$$

23. By (4),

$$A = \int_C x\,dy = \int_0^{2\pi} (2\cos t - \sin 2t)(2\cos t)\,dt = \int_0^{2\pi} (4\cos^2 t - 2\sin 2t\,\cos t)\,dt$$

$$= \int_0^{2\pi} [2(1 + \cos 2t) - 4\sin t\,\cos^2 t)]\,dt = \left(2t + \sin 2t + \frac{4}{3}\cos^3 t\right)\Big|_0^{2\pi} = 4\pi.$$

Next, notice that if $\mathbf{r}(t) = a\mathbf{i} + b\mathbf{j}$, then $\mathbf{r}(\pi - t) = -a\mathbf{i} + b\mathbf{j}$. Thus C and hence R are symmetric with respect to the y axis, which means that $\bar{x} = 0$. In addition, by Exercise 22,

$$\bar{y} = -\frac{1}{2A}\int_C y^2\,dx = -\frac{1}{2A}\int_0^{2\pi} (4\sin^2 t)(-2\sin t - 2\cos 2t)\,dt$$

$$= \frac{4}{A}\int_0^{2\pi} (\sin^3 t + \sin^2 t\,\cos 2t)\,dt = \frac{4}{A}\int_0^{2\pi} \left[\sin t\,(1 - \cos^2 t) + \sin^2 t\,(\cos^2 t - \sin^2 t)\right]\,dt$$

$$= \frac{4}{A}\int_0^{2\pi} (\sin t - \sin t\,\cos^2 t + \sin^2 t\,\cos^2 t - \sin^4 t)\,dt$$

$$= \frac{4}{A}\int_0^{2\pi} (\sin t - \sin t\,\cos^2 t + \sin^2 t - 2\sin^4 t)\,dt.$$

With the help of (12) of Section 8.1, we find that

$$\bar{y} = \frac{4}{A}\left[-\cos t + \frac{1}{3}\cos^3 t + \left(\frac{1}{2}t - \frac{1}{4}\sin 2t\right) - 2\left(-\frac{1}{4}\sin^3 t\cos t - \frac{3}{8}\sin t\cos t + \frac{3}{8}t\right)\right]\Bigg|_0^{2\pi}$$

$$= -\frac{2\pi}{A} = -\frac{2\pi}{4\pi} = -\frac{1}{2}.$$

Consequently $(\bar{x}, \bar{y}) = (0, -\frac{1}{2})$.

25. Let R_1 and R_2 be the regions enclosed by C_1 and C_2, respectively. Then by Green's Theorem,

$$\int_{C_1} M(x,y)\,dx + N(x,y)\,dy = \iint_{R_1}\left(\frac{\partial N}{\partial x} - \frac{\partial M}{\partial y}\right)dA = \iint_{R_1} 0\,dA = 0$$

and

$$\int_{C_2} M(x,y)\,dx + N(x,y)\,dy = \iint_{R_2}\left(\frac{\partial N}{\partial x} - \frac{\partial M}{\partial y}\right)dA = \iint_{R_2} 0\,dA = 0.$$

Thus

$$\int_{C_1} M(x,y)\,dx + N(x,y)\,dy = \int_{C_2} M(x,y)\,dx + N(x,y)\,dy.$$

27. a. Applying (6) with $\mathbf{F}$ replaced by $g\mathbf{F}$, we obtain

$$\int_C g\mathbf{F}\cdot\mathbf{n}\,ds = \iint_R \operatorname{div}(g\mathbf{F})\,dA = \iint_R (g\operatorname{div}\mathbf{F} + (\operatorname{grad} g)\cdot\mathbf{F})\,dA.$$

b. Applying part (a) with $\mathbf{F}$ replaced by $\operatorname{grad} f$, and using the fact that $\operatorname{div}(\operatorname{grad} f) = \nabla^2 f$, we obtain

$$\int_C g(\operatorname{grad} f)\cdot\mathbf{n}\,ds = \iint_R [g\operatorname{div}(\operatorname{grad} f) + (\operatorname{grad} g)\cdot(\operatorname{grad} f)]\,dA$$

$$= \iint_R [g\nabla^2 f + (\operatorname{grad} g)\cdot(\operatorname{grad} f)]\,dA.$$

c. From (8) we obtain

$$\int_C (g\operatorname{grad} f - f\operatorname{grad} g)\cdot\mathbf{n}\,ds = \int_C g(\operatorname{grad} f)\cdot\mathbf{n}\,ds - \int_C f(\operatorname{grad} g)\cdot\mathbf{n}\,ds$$

$$= \iint_R [g\nabla^2 f + (\operatorname{grad} g)\cdot(\operatorname{grad} f)]\,dA - \iint_R [f\nabla^2 g + (\operatorname{grad} f)\cdot(\operatorname{grad} g)]\,dA = \iint_R (g\nabla^2 f - f\nabla^2 g)\,dA.$$

29. If we apply Green's Theorem to a closed curve C that is the boundary of the region R, and then use the fact that f is harmonic, we find that

$$\int_C -f_y\,dx + f_x\,dy = \iint_R [f_{xx} - (-f_{yy})]\,dA = \iint_R (f_{xx} + f_{yy})\,dA = \iint_R 0\,dA = 0.$$

By the results of Section 15.3, the given line integral is independent of path.

15.5 Surface Integrals

1. Let R be the region in the xy plane between the graphs of $y = 0$ and $y = (6 - 2x)/3$ on $[0,3]$. If $f(x,y) = 6 - 2x - 3y$ for (x,y) in R, then Σ is the graph of f on R. Since

$$\sqrt{f_x^2(x,y) + f_y^2(x,y) + 1} = \sqrt{(-2)^2 + (-3)^2 + 1} = \sqrt{14}$$

it follows from (4) that

$$\iint_\Sigma x\,dS = \iint_R x\sqrt{14}\,dA = \sqrt{14}\int_0^3 \int_0^{(6-2x)/3} x\,dy\,dx$$

$$= \sqrt{14}\int_0^3 xy\Big|_0^{(6-2x)/3} dx = \sqrt{14}\int_0^3 \left(2x - \frac{2}{3}x^2\right) dx = \sqrt{14}\left(x^2 - \frac{2}{9}x^3\right)\Big|_0^3 = 3\sqrt{14}.$$

3. Let R be the circular region in the xy plane bounded by the circle $r = 2$ in polar coordinates. If $f(x,y) = 3x - 2$ for (x,y) in R, then Σ is the graph of f on R. Since

$$\sqrt{f_x^2(x,y) + f_y^2(x,y) + 1} = \sqrt{3^2 + 1} = \sqrt{10}$$

it follows from (4) that

$$\iint_\Sigma (2x^2 + 1)\,dS = \iint_R (2x^2 + 1)\sqrt{10}\,dA = \int_0^{2\pi} \int_0^2 (2r^2\cos^2\theta + 1)\sqrt{10}\,r\,dr\,d\theta$$

$$= \sqrt{10}\int_0^{2\pi} \left(\frac{1}{2}r^4\cos^2\theta + \frac{1}{2}r^2\right)\Big|_0^2 d\theta = \sqrt{10}\int_0^{2\pi} (8\cos^2\theta + 2)\,d\theta$$

$$= \sqrt{10}\int_0^{2\pi} (4 + 4\cos 2\theta + 2)\,d\theta = \sqrt{10}\,(6\theta + 2\sin 2\theta)\Big|_0^{2\pi} = 12\pi\sqrt{10}.$$

5. The paraboloid and plane intersect for (x, y, z) such that $x^2 + y^2 = z = y$, which in polar coordinates means that $r^2 = r\sin\theta$, so that $r = \sin\theta$ for $0 \le \theta \le \pi$. Let R be the region in the xy plane bounded by $r = \sin\theta$ on $[0, \pi]$ in polar coordinates. If $f(x, y) = x^2 + y^2$ for (x, y) in R, then Σ is the graph of f on R. Since

$$\sqrt{f_x^2(x,y) + f_y^2(x,y) + 1} = \sqrt{(2x)^2 + (2y)^2 + 1} = \sqrt{4x^2 + 4y^2 + 1}$$

it follows from (4) that

$$\iint_\Sigma \sqrt{4x^2 + 4y^2 + 1}\, dS = \iint_R \sqrt{4x^2 + 4y^2 + 1}\sqrt{4x^2 + 4y^2 + 1}\, dA$$

$$= \int_0^\pi \int_0^{\sin\theta} (4r^2 + 1) r\, dr\, d\theta = \int_0^\pi \left(r^4 + \frac{1}{2}r^2\right)\Big|_0^{\sin\theta} = \int_0^\pi \left(\sin^4\theta + \frac{1}{2}\sin^2\theta\right) d\theta$$

$$= \int_0^\pi \left(\sin^4\theta + \frac{1}{4} - \frac{1}{4}\cos 2\theta\right) d\theta = \left(-\frac{1}{4}\sin^3\theta\cos\theta - \frac{3}{8}\sin\theta\cos\theta + \frac{3}{8}\theta + \frac{1}{4}\theta - \frac{1}{8}\sin 2\theta\right)\Big|_0^\pi = \frac{5}{8}\pi$$

with the next to the last equality coming from (12) in Section 8.1.

7. Let R be the region in the xy plane between the graphs of $y = 0$ and $y = 2$ on $[0, 3]$. If $f(x, y) = 4 - y^2$ for (x, y) in R, then Σ is the graph of f on R. Since

$$\sqrt{f_x^2(x,y) + f_y^2(x,y) + 1} = \sqrt{(-2y)^2 + 1} = \sqrt{4y^2 + 1}$$

it follows from (4) that

$$\iint_\Sigma y\, dS = \iint_R y\sqrt{4y^2 + 1}\, dA = \int_0^3 \int_0^2 y\sqrt{4y^2 + 1}\, dy\, dx$$

$$= \int_0^3 \frac{1}{12}(4y^2 + 1)^{3/2}\Big|_0^2 dx = \int_0^3 \frac{1}{12}(17^{3/2} - 1)\, dx = \frac{1}{4}(17^{3/2} - 1).$$

9. As in Example 1 (with $a = 2$), Σ is parametrized by

$$\mathbf{r}(\phi, \theta) = 2\sin\phi\cos\theta\,\mathbf{i} + 2\sin\phi\sin\theta\,\mathbf{j} + 2\cos\phi\,\mathbf{k} \quad \text{for } 0 \le \phi \le \pi/2 \text{ and } 0 \le \theta \le 2\pi$$

Thus $\|\mathbf{r}_\phi(\phi, \theta) \times \mathbf{r}_\theta(\phi, \theta)\| = 4\sin\phi$. Therefore (2) implies that

$$\iint_\Sigma z(x^2 + y^2)\, dS = \int_0^{2\pi} \int_0^{\pi/2} 2\cos\phi\,(4\sin^2\phi\cos^2\theta + 4\sin^2\phi\sin^2\theta) 4\sin\phi\, d\phi\, d\theta$$

$$= \int_0^{2\pi} \int_0^{\pi/2} 32\sin^3\phi\cos\phi\, d\phi\, d\theta = \int_0^{2\pi} 8\sin^4\phi\Big|_0^{\pi/2} d\theta = \int_0^{2\pi} 8\, d\theta = 16\pi$$

11. Let Σ be composed of the six surfaces Σ_1, Σ_2, Σ_3, Σ_4, Σ_5, and Σ_6, as in the figure. For Σ_1 and Σ_2, we use the region R_1 in the xy plane bounded by $y = 0$ and $y = 1$ on $[0,1]$. If $f(x,y) = 0$, then $\sqrt{f_x^2(x,y) + f_y^2(x,y) + 1} = 1$, and

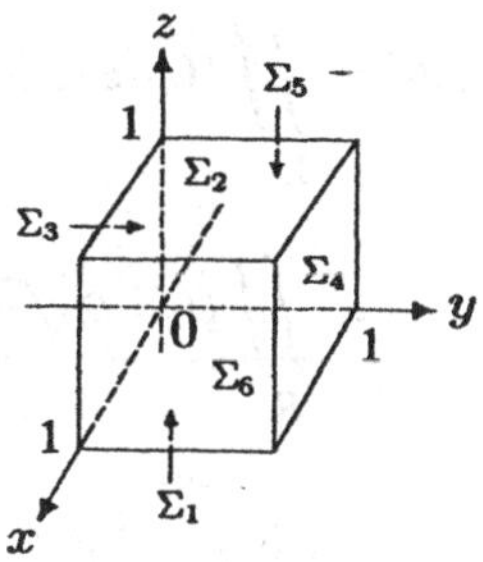

$$\iint_{\Sigma_1} (x+y)\,dS = \iint_{R_1} (x+y)\,dA$$

$$= \int_0^1 \int_0^1 (x+y)\,dy\,dx = \int_0^1 \left(xy + \frac{y^2}{2}\right)\bigg|_0^1 dx$$

$$= \int_0^1 \left(x + \frac{1}{2}\right) dx = \left(\frac{1}{2}x^2 + \frac{1}{2}x\right)\bigg|_0^1 = 1.$$

If $f(x,y) = 1$, then $\sqrt{f_x^2(x,y) + f_y^2(x,y) + 1} = 1$, and as above, $\iint_{\Sigma_2} (x+y)\,dS = \iint_{R_1} (x+y)\,dA = 1$. For Σ_3 and Σ_4, we use the region R_3 in the xz plane bounded by $z = 0$ and $z = 1$ on $[0,1]$. If $f(x,z) = 0$, then $\sqrt{f_x^2(x,z) + f_z^2(x,z) + 1} = 1$, and

$$\iint_{\Sigma_3} (x+y)\,dS = \iint_{R_3} x\,dA = \int_0^1 \int_0^1 x\,dz\,dx = \int_0^1 xz\big|_0^1 dx = \int_0^1 x\,dx = \frac{1}{2}x^2\bigg|_0^1 = \frac{1}{2}.$$

If $f(x,z) = 1$, then $\sqrt{f_x^2(x,z) + f_z^2(x,z) + 1} = 1$, and

$$\iint_{\Sigma_4} (x+y)\,dS = \iint_{R_3} (x+1)\,dA = \int_0^1 \int_0^1 (x+1)\,dz\,dx$$

$$= \int_0^1 (xz+z)\big|_0^1 dx = \int_0^1 (x+1)\,dx = \left(\frac{1}{2}x^2 + x\right)\bigg|_0^1 = \frac{3}{2}.$$

For Σ_5 and Σ_6, we use the region R_5 in the yz plane bounded by $z = 0$ and $z = 1$ on $[0,1]$. If $f(y,z) = 0$, then $\sqrt{f_y^2(y,z) + f_z^2(y,z) + 1} = 1$, and

$$\iint_{\Sigma_5} (x+y)\,dS = \iint_{R_5} y\,dA = \int_0^1 \int_0^1 y\,dz\,dy = \int_0^1 yz\big|_0^1 dy = \int_0^1 y\,dy = \frac{1}{2}y^2\bigg|_0^1 = \frac{1}{2}.$$

Finally, if $f(y,z) = 1$, then $\sqrt{f_y^2(y,z) + f_z^2(y,z) + 1} = 1$, and

$$\iint_{\Sigma_6} (x+y)\,dS = \iint_{R_5} (1+y)\,dA = \int_0^1 \int_0^1 (1+y)\,dz\,dy$$

$$= \int_0^1 (z+yz)\big|_0^1 dy = \int_0^1 (1+y)\,dy = \left(y + \frac{1}{2}y^2\right)\bigg|_0^1 = \frac{3}{2}.$$

Consequently

$$\iint_{\Sigma} (x+y)\,dS = \sum_{k=1}^{6} \iint_{\Sigma_k} (x+y)\,dS = 1 + 1 + \frac{1}{2} + \frac{3}{2} + \frac{1}{2} + \frac{3}{2} = 6.$$

13. Let R be the region in the xy plane bounded by $r = \frac{1}{4}$ and $r = 2$ in polar coordinates and let $f(x, y) = 2\sqrt{x^2 + y^2}$ for (x, y) in R. Then the funnel occupies the region Σ which is the graph of f on R. The mass of the funnel is given by

$$m = \iint_{\Sigma} \delta(x, y, z)\, dS = \iint_{\Sigma} (6 - z)\, dS = \iint_{R} (6 - 2\sqrt{x^2 + y^2}) \sqrt{\left(\frac{2x}{\sqrt{x^2 + y^2}}\right)^2 + \left(\frac{2y}{\sqrt{x^2 + y^2}}\right)^2 + 1}\, dA$$

$$= \iint_{R} (6 - 2\sqrt{x^2 + y^2})\sqrt{5}\, dA = \int_0^{2\pi} \int_{1/4}^{2} (6 - 2r)\sqrt{5}\, r\, dr\, d\theta$$

$$= \sqrt{5} \int_0^{2\pi} \left(3r^2 - \frac{2}{3}r^3\right)\Big|_{1/4}^{2} d\theta = \sqrt{5} \int_0^{2\pi} \frac{623}{96}\, d\theta = \frac{623}{48}\sqrt{5}\,\pi.$$

15. The hydrostatic force F_1 on the hemispherical part Σ_1 of the tank is given by

$$F_1 = \iint_{\Sigma_1} 62.5(z_0 - z)\, dS = \iint_{\Sigma_1} 62.5(10 - z)\, dS$$

From Example 1 we obtain a parametrization of the hemispherical part of the tank:

$$\mathbf{r}(\phi, \theta) = 10 \sin\phi \cos\theta\, \mathbf{i} + 10 \sin\phi \sin\theta\, \mathbf{j} + 10 \cos\phi\, \mathbf{k} \quad \text{for } 0 \le \phi \le \pi/2 \text{ and } 0 \le \theta \le 2\pi$$

Thus $\|\mathbf{r}_\phi(\phi, \theta) \times \mathbf{r}_\theta(\phi, \theta)\| = 100 \sin\phi$. Therefore (2) implies that

$$F_1 = \int_0^{2\pi} \int_0^{\pi/2} 62.5(10 - 10\cos\phi)\, 100 \sin\phi\, d\phi\, d\theta = \int_0^{2\pi} 62,500(-\cos\phi + \frac{1}{2}\cos^2\phi)\Big|_0^{\pi/2} d\theta$$

$$= \int_0^{2\pi} 62,500\left(\frac{1}{2}\right) d\theta = 62,500\pi \text{ (foot-pounds)}$$

The hydrostatic force F_2 on the bottom Σ_2 of the tank is given by

$$F_2 = \iint_{\Sigma_2} (10 - z)\, dz$$

Since Σ_2 is the region in the xy plane bounded by the circle $r = 10$, we have $z = 0$ on Σ_2 and hence

$$F_2 = \iint_{\Sigma_2} 62.5(10)\, dz = 625\, (\text{area of } \Sigma_2) = 625(100\pi) = 62,500\pi \text{ foot-pounds}$$

Thus the total force F on the tank is given by

$$F = F_1 + F_2 = 62,500\pi + 62,500\pi = 125,000\pi \text{ (foot-pounds)}$$

17. The hydrostatic force F is given by

$$F = \iint_{\Sigma} 62.5\,(10 - z)\, dS$$

As in Example 1 the sphere Σ is parametrized by

$$\mathbf{r}(\phi, \theta) = 10 \sin\phi \cos\theta\,\mathbf{i} + 10 \sin\phi \sin\theta\,\mathbf{j} + 10 \cos\phi\,\mathbf{k} \quad \text{for } 0 \le \phi \le \pi \text{ and } 0 \le \theta \le 2\pi$$

Thus $\|\mathbf{r}_\phi(\phi, \theta) \times \mathbf{r}_\theta(\phi, \theta)\| = 100 \sin\phi$. Therefore

$$F = \int_0^{2\pi} \int_0^{\pi} 62.5(10 - 10\cos\phi)\, 100 \sin\phi\, d\phi\, d\theta = \int_0^{2\pi} 62,500\left(-\cos\phi + \frac{1}{2}\cos^2\phi\right)\Big|_0^{\pi}\, d\theta$$

$$= \int_0^{2\pi} 125,000\, d\theta = 250,000\pi \text{ (pounds)}$$

15.6 Integrals over Oriented Surfaces

1.

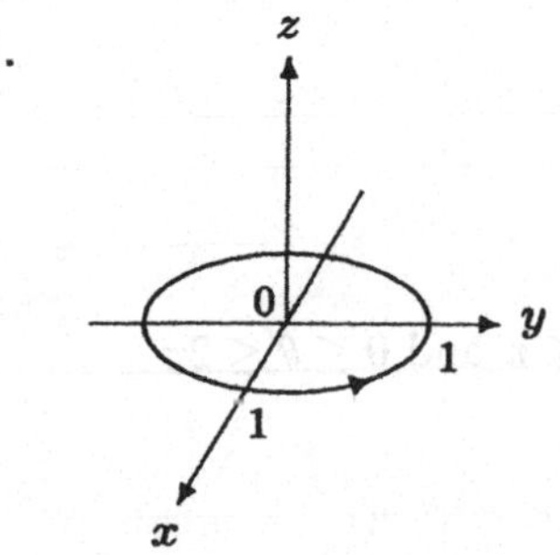

3.

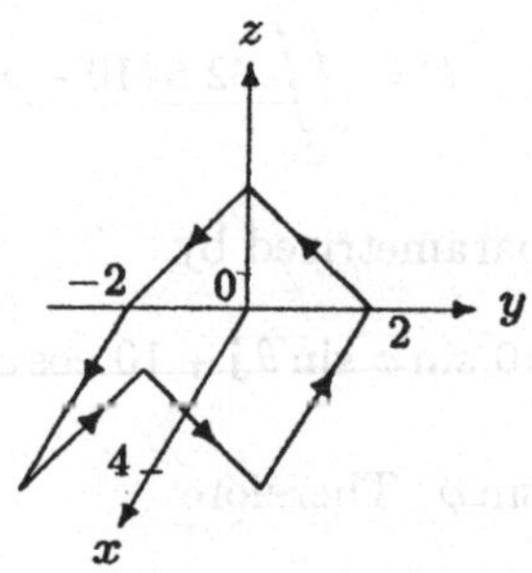

5. As in Example 9 of Section 14.9, Σ can be parametrized by

$$\mathbf{r}(\phi, \theta) = 2 \sin\phi \cos\theta\,\mathbf{i} + 2 \sin\phi \sin\theta\,\mathbf{j} + 2 \cos\phi\,\mathbf{k} \quad \text{for } 0 \le \phi \le \pi \text{ and } 0 \le \theta \le 2\pi$$

Thus $\mathbf{r}_\phi(\phi, \theta) \times \mathbf{r}_\theta(\phi, \theta) = 4 \sin^2\phi \cos\theta\,\mathbf{i} + 4 \sin^2\phi \sin\theta\,\mathbf{j} + 4 \cos\phi \sin\phi\,\mathbf{k}$. Therefore $\mathbf{r}_\phi \times \mathbf{r}_\theta$ points in the same direction as the normal for the given orientation of Σ_1, and hence by (4) or (5) with the plus sign,

$$\iint_\Sigma \mathbf{F}\cdot\mathbf{n}\,dS = \int_0^\pi \int_0^{2\pi} [(-2 \sin\phi \sin\theta)(4 \sin^2\phi \cos\theta) + (2 \sin\phi \cos\theta)(4 \sin^2\phi \sin\theta)$$

$$+(2\cos\phi)^4(4 \cos\phi \sin\theta)]\,d\theta\,d\phi = \int_0^\pi \int_0^{2\pi} 64 \cos^5\phi \sin\phi\,d\theta\,d\phi = \int_0^\pi 128\pi \cos^5\phi \sin\phi\,d\phi$$

$$= -\frac{64}{3}\cos^6\phi \Big|_0^\pi = 0$$

7. You can check that Σ, which the lower half of the ellipsoid $4x^2+4y^2+z^2 = 4$, has the parametrization

$$\mathbf{r}(\phi, \theta) = \sin\phi \cos\theta\,\mathbf{i} + \sin\phi \sin\theta\,\mathbf{j} + 2 \cos\phi\,\mathbf{k} \quad \text{for } \pi/2 \le \phi \le \pi \text{ and } 0 \le \theta \le 2\pi$$

It follows that

$$\mathbf{r}_\phi(\phi, \theta) \times \mathbf{r}_\theta(\phi, \theta) = \begin{vmatrix} \mathbf{i} & \mathbf{j} & \mathbf{k} \\ \cos\phi \cos\theta & \cos\phi \sin\theta & -2 \sin\phi \\ -\sin\phi \sin\theta & \sin\phi \cos\theta & 0 \end{vmatrix}$$

$$= 2 \sin^2\phi \cos\theta\,\mathbf{i} + 2 \sin^2\phi \sin\theta\,\mathbf{j} + \cos\phi \sin\phi\,\mathbf{k}$$

9. Σ is composed of the bottom Σ_1, the top Σ_2, and the side portion Σ_3. For Σ_1, let R_1 be the region in the xy plane bounded by the circle $r = \sqrt{\pi/2}$, and let $f(x, y) = 0$. By (6) with the plus sign,

$$\iint_{\Sigma_1} \mathbf{F}\cdot\mathbf{n}\,dS = \iint_{R_1} [-(\sin(x^2+y^2))(0) - (y \sin(x^2+y^2))(0) + (0)]\,dA = 0$$

For Σ_2 we let $R_2 = R_1$ and $f(x, y) = 1$. Then we use (6) with the minus sign to find that

$$\iint_{\Sigma_2} \mathbf{F}\cdot\mathbf{n}\,dS = -\iint_{R_1} [-(\sin(x^2+y^2))(0) - (y \sin(x^2+y^2))(0) + (0)]\,dA = 0$$

For Σ_3 we use the parametrization

$$\mathbf{r}(\theta, z) = \sqrt{\frac{\pi}{2}}\cos\theta\,\mathbf{i} + \sqrt{\frac{\pi}{2}}\sin\theta\,\mathbf{j} + z\,\mathbf{k} \quad \text{for } 0 \le \theta \le 2\pi \text{ and } 0 \le z \le 1$$

Then

$$\mathbf{r}_\theta(\theta, z) \times \mathbf{r}_z(\theta, z) = \begin{vmatrix} \mathbf{i} & \mathbf{j} & \mathbf{k} \\ -\sqrt{\pi/2}\sin\theta & \sqrt{\pi/2}\cos\theta & 0 \\ 0 & 0 & 1 \end{vmatrix} = \sqrt{\frac{\pi}{2}}\cos\theta\,\mathbf{i} + \sqrt{\frac{\pi}{2}}\sin\theta\,\mathbf{j}$$

Therefore by (4) or (5) with the plus sign,

$$\iint_{\Sigma_3} \mathbf{F}\cdot\mathbf{n}\,dS = \int_0^{2\pi}\int_0^1 \left[(\sin\frac{\pi}{2})(\sqrt{\frac{\pi}{2}}\cos\theta) + (\sqrt{\frac{\pi}{2}}\sin\theta\,\sin\frac{\pi}{2})(\sqrt{\frac{\pi}{2}}\sin\theta) + (0)(0)\right] dz\,d\theta$$

$$= \int_0^{2\pi}\left[\sqrt{\frac{\pi}{2}}\cos\theta + \frac{\pi}{2}\sin^2\theta\right] d\theta = \int_0^{2\pi}\left[\sqrt{\frac{\pi}{2}}\cos\theta + \frac{\pi}{2}(\frac{1}{2} - \cos 2\theta)\right] d\theta$$

$$= \left(\sqrt{\frac{\pi}{2}}\sin\theta + \frac{\pi}{4}\theta - \frac{\pi}{4}\sin 2\theta\right)\Big|_0^{2\pi} = \frac{1}{2}\pi^2$$

Therefore

$$\iint_\Sigma \mathbf{F}\cdot\mathbf{n}\,dS = \sum_{k=1}^{3}\iint_{\Sigma_k}\mathbf{F}\cdot\mathbf{n}\,dS = 0 + 0 + \frac{1}{2}\pi^2 = \frac{1}{2}\pi^2$$

11. The parabolas $x = 1 - y^2$ and $x = y^2 - 1$ intersect for (x, y) such that $y = -1$ or $y = 1$. Let R be the region in the xy plane between the graphs of $x = y^2 - 1$ and $x = 1 - y^2$ on $[-1, 1]$, and let $f(x, y) = x^2 + y^2$ for (x, y) in R. We use (6) with the minus sign, and with $f_x(x, y) = 2x$, $f_y(x, y) = 2y$, $M(x, y, f(x, y)) = x$, $N(x, y, f(x, y)) = -1$, and $P(x, y, f(x, y)) = 2x^2$, and find that

$$\iint_\Sigma \mathbf{F}\cdot\mathbf{n}\,dS = \iint_R [x(2x) - 2y - 2x^2]\,dA = \int_{-1}^{1}\int_{y^2-1}^{1-y^2}(-2y)\,dx\,dy$$

$$= \int_{-1}^{1}(-2yx)\Big|_{y^2-1}^{1-y^2}dy = \int_{-1}^{1} -2y(2 - 2y^2)\,dy = 4\int_{-1}^{1}(y^3 - y)\,dy = 4\left(\frac{1}{4}y^4 - \frac{1}{2}y^2\right)\Big|_{-1}^{1} = 0.$$

13. The rate of mass flow is

$$\iint_{\Sigma} \delta[\mathbf{v}(x,y,z)\cdot\mathbf{n}(x,y,z)]\,dS$$

where Σ is the given sphere, and where $\mathbf{v}(x,y,z) = x\mathbf{i}+y\mathbf{j}+z\mathbf{k}$ and $\delta(x,y,z) = 50$. By Example 9 in Section 14.9, Σ is parametrized by

$$\mathbf{r}(\phi,\theta) = \sqrt{10}\sin\phi\cos\theta\,\mathbf{i} + \sqrt{10}\sin\phi\sin\theta\,\mathbf{j} + \sqrt{10}\cos\phi\,\mathbf{k} \quad \text{for } 0 \le \phi \le \pi \text{ and } 0 \le \theta \le 2\pi$$

Then $\mathbf{r}_\phi(\phi,\theta)\times\mathbf{r}_\theta(\phi,\theta) = 10\sin^2\phi\cos\theta\,\mathbf{i}+10\sin^2\phi\sin\theta\,\mathbf{j}+10\cos\phi\sin\phi\mathbf{k}$. Thus $\mathbf{r}_\phi(\phi,\theta)\times\mathbf{r}_\theta(\phi,\theta)$ points in the direction of the outward normal, so we use (4) or (5) with the plus sign to find that the rate of mass flow is

$$\int_0^{2\pi}\int_0^{\pi} 50\Big[(\sqrt{10}\sin\phi\cos\theta)(10\sin^2\phi\cos\theta) + (\sqrt{10}\sin\phi\sin\theta)(10\sin^2\phi\sin\theta)$$
$$+(\sqrt{10}\cos\phi)(10\cos\phi\sin\phi)\Big]\,d\phi\,d\theta$$

$$= \int_0^{2\pi}\int_0^{\pi} 500\sqrt{10}[\sin^3\phi\cos^2\theta + \sin^3\phi\sin^2\theta + \cos^2\phi\sin\phi]\,d\phi\,d\theta$$

$$= \int_0^{2\pi}\int_0^{\pi} 500\sqrt{10}(\sin^3\phi + \cos^2\phi\sin\phi)\,d\phi\,d\theta = \int_0^{2\pi}\int_0^{\pi} 500\sqrt{10}\sin\phi\,d\phi\,d\theta$$

$$= \int_0^{2\pi} -500\sqrt{10}\cos\phi\Big|_0^{\pi}\,d\theta = \int_0^{2\pi} 1000\sqrt{10}\,d\theta = 2000\pi\sqrt{10}$$

15. Σ is parametrized by

$$\mathbf{r}(r,\theta) = r\cos\theta\,\mathbf{i} + r\,\mathbf{j} + r\sin\theta\,\mathbf{k} \quad \text{for } 0 \le \theta \le \pi \text{ and } 0 \le r \le 1$$

Thus

$$\mathbf{r}_r(r,\theta)\times\mathbf{r}_\theta(r,\theta) = \begin{vmatrix} \mathbf{i} & \mathbf{j} & \mathbf{k} \\ \cos\theta & 1 & \sin\theta \\ -r\sin\theta & 0 & r\cos\theta \end{vmatrix} = r\cos\theta\,\mathbf{i} - r\,\mathbf{j} + r\sin\theta\,\mathbf{k}$$

Therefore $\mathbf{r}_r(r,\theta)\times\mathbf{r}_\theta(r,\theta)$ points upward. Thus we may use (4) or (5) with the plus sign to find that the flux of $\mathbf{E}$ is

$$\iint_{\Sigma} \mathbf{E}\cdot\mathbf{n}\,dS = \int_0^{\pi}\int_0^{1} [(r\cos\theta)(r\cos\theta) + (r)(-r) + (0)(r\sin\theta)]\,dr\,d\theta$$

$$= \int_0^{\pi}\int_0^{1} r^2(\cos^2\theta - 1)\,dr\,d\theta = \int_0^{\pi} \frac{1}{3}r^3(\cos^2\theta - 1)\Big|_0^1\,d\theta$$

$$= \int_0^{\pi} -\frac{1}{3}\sin^2\theta\,d\theta = \int_0^{\pi}\left(-\frac{1}{6} + \frac{1}{6}\cos 2\theta\right)d\theta = \left(-\frac{\theta}{6} + \frac{1}{12}\sin 2\theta\right)\Big|_0^{\pi} = -\frac{1}{6}\pi$$

17. The sphere Σ is the graph of $x^2+y^2+z^2=1$, so $\mathbf{n}=x\mathbf{i}+y\mathbf{j}+z\mathbf{k}$. Therefore

$$\mathbf{E}\cdot\mathbf{n}=\frac{q}{4\pi\epsilon_0}[(x)(x)+(y)(y)+(z)(z)]=\frac{q}{4\pi\epsilon_0}$$

Therefore

$$\iint_{\Sigma}\mathbf{E}\cdot\mathbf{n}\,dS=\iint_{\Sigma}\frac{q}{4\pi\epsilon_0}\,dS=\frac{q}{4\pi\epsilon_0}\,(\text{area of}\,\Sigma)=\frac{q}{4\pi\epsilon_0}(4\pi\cdot 1^2)=\frac{q}{\epsilon_0}$$

19. By symmetry $\mathbf{E}$ has constant magnitude on spheres centered at the origin, and is normal to the spheres, so there is a function f of a single variable such that

$$\mathbf{E}(x,y,z)=f(\sqrt{x^2+y^2+z^2})\,\mathbf{n}=f(\sqrt{x^2+y^2+z^2})\frac{x\mathbf{i}+y\mathbf{j}+z\mathbf{k}}{\sqrt{x^2+y^2+z^2}}\quad\text{for } x^2+y^2+z^2>0.$$

Since no charge is contained in any such sphere with radius less than a, $f(\sqrt{x^2+y^2+z^2})=0$ if $0\le x^2+y^2+z^2<a^2$. If $x_0^2+y_0^2+z_0^2\ge a^2$, then let Σ be the sphere centered at the origin and passing through (x_0,y_0,z_0). Then

$$\iint_{\Sigma}\mathbf{E}\cdot\mathbf{n}\,dS=\iint_{\Sigma}f(\sqrt{x_0^2+y_0^2+z_0^2})\,dS$$

$$=f(\sqrt{x_0^2+y_0^2+z_0^2})\iint_{\Sigma}1\,dS=f(\sqrt{x_0^2+y_0^2+z_0^2})\,4\pi(x_0^2+y_0^2+z_0^2).$$

The total charge q inside the sphere is $4\pi a^2\sigma$, so by Gauss's Law this means that

$$\frac{4\pi a^2\sigma}{\varepsilon_0}=\frac{q}{\varepsilon_0}=\iint_{\Sigma}\mathbf{E}\cdot\mathbf{n}\,dS=f(\sqrt{x_0^2+y_0^2+z_0^2})\,4\pi(x_0^2+y_0^2+z_0^2).$$

Thus

$$f(\sqrt{x_0^2+y_0^2+z_0^2})=\frac{a^2\sigma}{\varepsilon_0(x_0^2+y_0^2+z_0^2)}.$$

Consequently the electric field is given by

$$\mathbf{E}(x,y,z)=\begin{cases}\mathbf{0} & \text{for } 0\le x^2+y^2+z^2<a^2\\[2ex] \dfrac{a^2\sigma}{\varepsilon_0(x^2+y^2+z^2)^{3/2}}(x\mathbf{i}+y\mathbf{j}+z\mathbf{k}) & \text{for } x^2+y^2+z^2\ge a^2\end{cases}$$

15.7 Stokes's Theorem

1. $\operatorname{curl}\mathbf{F}(x,y,z)=\begin{vmatrix}\mathbf{i} & \mathbf{j} & \mathbf{k}\\ \dfrac{\partial}{\partial x} & \dfrac{\partial}{\partial y} & \dfrac{\partial}{\partial z}\\ z & x & y\end{vmatrix}=\mathbf{i}+\mathbf{j}+\mathbf{k}$

Let $f(x,y) = 1 - x^2 - y^2$, and let R be the region in the first quadrant of the xy plane bounded by the lines $x = 0$, $y = 0$, and the circle $x^2 + y^2 = 1$. Then Σ is the graph of f on R. Thus by Stokes's Theorem and (6) of Section 15.6 with the minus sign,

$$\int_C \mathbf{F}\cdot d\mathbf{r} = \iint_\Sigma (\operatorname{curl}\mathbf{F})\cdot\mathbf{n}\,dS = \iint_R [(1)(-2x) + (1)(-2y) - 1]\,dA$$

$$= \int_0^{\pi/2}\int_0^1 (-2r\cos\theta - 2r\sin\theta - 1)r\,dr\,d\theta = \int_0^{\pi/2}\left(-\frac{2}{3}r^3\cos\theta - \frac{2}{3}r^3\sin\theta - \frac{1}{2}r^2\right)\Big|_0^1 d\theta$$

$$= \int_0^{\pi/2}\left(-\frac{2}{3}\cos\theta - \frac{2}{3}\sin\theta - \frac{1}{2}\right)d\theta = \left(-\frac{2}{3}\sin\theta + \frac{2}{3}\cos\theta - \frac{1}{2}\theta\right)\Big|_0^{\pi/2} = -\frac{4}{3} - \frac{\pi}{4}.$$

3. $$\operatorname{curl}\mathbf{F}(x,y,z) = \begin{vmatrix} \mathbf{i} & \mathbf{j} & \mathbf{k} \\ \dfrac{\partial}{\partial x} & \dfrac{\partial}{\partial y} & \dfrac{\partial}{\partial z} \\ y & -x & z \end{vmatrix} = -2\mathbf{k}$$

Let Σ_1 be the part of Σ on the plane $z = 1$, and Σ_2 the part of Σ on the cylinder $x^2 + y^2 = 1$. On Σ_1, $\mathbf{n} = \mathbf{k}$, so that $(\operatorname{curl}\mathbf{F})\cdot\mathbf{n} = (-2\mathbf{k})\cdot\mathbf{k} = -2$. On Σ_2, $\mathbf{n}$ is perpendicular to the z axis, so that $(\operatorname{curl}\mathbf{F})\cdot\mathbf{n} = 0$. Thus by Stokes's Theorem,

$$\int_C \mathbf{F}\cdot d\mathbf{r} = \iint_\Sigma (\operatorname{curl}\mathbf{F})\cdot\mathbf{n}\,dS$$

$$= \iint_{\Sigma_1} (\operatorname{curl}\mathbf{F})\cdot\mathbf{n}\,dS + \iint_{\Sigma_2} (\operatorname{curl}\mathbf{F})\cdot\mathbf{n}\,dS = \iint_{\Sigma_1} -2\,dS + 0 = -2\,(\text{area of }\Sigma_1) = -2\pi.$$

5. $$\operatorname{curl}\mathbf{F}(x,y,z) = \begin{vmatrix} \mathbf{i} & \mathbf{j} & \mathbf{k} \\ \dfrac{\partial}{\partial x} & \dfrac{\partial}{\partial y} & \dfrac{\partial}{\partial z} \\ 2y & 3z & -2x \end{vmatrix} = -3\mathbf{i} + 2\mathbf{j} - 2\mathbf{k}$$

By Example 9 of Section 14.9, Σ is parametrized by

$$\mathbf{r}(\phi,\theta) = \sin\phi\,\cos\theta\,\mathbf{i} + \sin\phi\,\sin\theta\,\mathbf{j} + \cos\phi\,\mathbf{k} \quad \text{for } 0 \le \phi \le \pi/2 \text{ and } 0 \le \theta \le \pi/2$$

and

$$\mathbf{r}_\phi \times \mathbf{r}_\theta = \sin^2\phi\,\cos\theta\mathbf{i} + \sin^2\phi\,\sin\theta\mathbf{j} + \cos\phi\,\sin\phi\mathbf{k}$$

It follows that $\mathbf{r}_\phi \times \mathbf{r}_\theta$ points upward, as does the normal for the given orientation of Σ. Thus by Stokes's Theorem and (4) or (5) of Section 15.6 with the plus sign,

$$\int_C \mathbf{F}\cdot d\mathbf{r} = \iint_\Sigma (\operatorname{curl}\mathbf{F})\cdot\mathbf{n}\,dS$$

$$= \int_0^{\pi/2}\int_0^{\pi/2} [(-3)(\sin^2\phi\,\cos\theta) + (2)(\sin^2\phi\,\sin\theta) + (-2)(\cos\phi\,\sin\phi)]\,d\theta\,d\phi$$

$$= \int_0^{\pi/2} (-3\sin^2\phi\,\sin\theta - 2\sin^2\phi\cos\theta - 2\theta\cos\phi\,\sin\phi)\Big|_0^{\pi/2} d\phi$$

$$= \int_0^{\pi/2} (-3\sin^2\phi + 2\sin^2\phi - \pi\cos\phi\sin\phi)\,d\phi$$

$$= \int_0^{\pi/2} \left(-\frac{1}{2} + \frac{1}{2}\cos 2\phi - \pi\cos\phi\sin\phi\right) d\phi$$

$$= \left(-\frac{1}{2}\phi + \frac{1}{4}\sin 2\phi - \frac{1}{2}\pi\sin^2\phi\right)\bigg|_0^{\pi/2} = -\frac{3}{4}\pi$$

7. $\operatorname{curl} \mathbf{F}(x,y,z) = \begin{vmatrix} \mathbf{i} & \mathbf{j} & \mathbf{k} \\ \dfrac{\partial}{\partial x} & \dfrac{\partial}{\partial y} & \dfrac{\partial}{\partial z} \\ xz & y^2 & x^2 \end{vmatrix} = -x\mathbf{j}$

Let Σ be the part of the plane $x+y+z=5$ that lies inside the cylinder $x^2 + \frac{1}{4}y^2 = 1$, and let Σ be oriented with normal $\mathbf{n}$ directed upward. If $f(x,y) = 5 - x - y$, then Σ is the graph of f on the region R in the xy plane bounded by the ellipse $x^2 + \frac{1}{4}y^2 = 1$. Thus by Stokes's Theorem and (6) of Section 15.6 with the plus sign,

$$\int_C \mathbf{F}\cdot d\mathbf{r} = \iint_\Sigma (\operatorname{curl}\mathbf{F})\cdot\mathbf{n}\,dS = \iint_R [-(0)(-1) - (-x)(-1) + 0]\,dA$$

$$= \int_{-2}^{2}\int_{-\sqrt{1-y^2/4}}^{\sqrt{1-y^2/4}} -x\,dx\,dy = \int_{-2}^{2} -\frac{1}{2}x^2\bigg|_{-\sqrt{1-y^2/4}}^{\sqrt{1-y^2/4}} dy = \int_{-2}^{2} 0\,dy = 0.$$

9. $\operatorname{curl} \mathbf{F}(x,y,z) = \begin{vmatrix} \mathbf{i} & \mathbf{j} & \mathbf{k} \\ \dfrac{\partial}{\partial x} & \dfrac{\partial}{\partial y} & \dfrac{\partial}{\partial z} \\ y(x^2+y^2) & -x(x^2+y^2) & 0 \end{vmatrix} = -4(x^2+y^2)\mathbf{k}$

Notice that in each of the points $(0,0,0), (1,0,0), (1,1,1)$, and $(0,1,1)$ the y and z coordinates are equal. Thus the rectangle C lies in the plane $z = y$. Let Σ be the part of the plane $z = y$ bounded by C, and let $f(x,y) = y$. Then Σ is the graph of f on the region R in the xy plane bounded by the rectangle with vertices $(0,0), (1,0), (1,1)$, and $(0,1)$. By Stokes's Theorem and (6) of Section 15.6 with the plus sign,

$$\int_C \mathbf{F}\cdot d\mathbf{r} = \iint_\Sigma (\operatorname{curl}\mathbf{F})\cdot\mathbf{n}\,dS = \iint_R [-(0)(0) - (0)(1) - 4(x^2+y^2)]\,dA$$

$$= \int_0^1\int_0^1 -4(x^2+y^2)\,dy\,dx = \int_0^1 -4\left(x^2y + \frac{1}{3}y^3\right)\bigg|_0^1 dx$$

$$= \int_0^1 -4\left(x^2 + \frac{1}{3}\right) dx = -4\left(\frac{1}{3}x^3 + \frac{1}{3}x\right)\bigg|_0^1 = -\frac{8}{3}.$$

11. $\operatorname{curl} \mathbf{F}(x,y,z) = \begin{vmatrix} \mathbf{i} & \mathbf{j} & \mathbf{k} \\ \dfrac{\partial}{\partial x} & \dfrac{\partial}{\partial y} & \dfrac{\partial}{\partial z} \\ z-y & y & x \end{vmatrix} = \mathbf{k}$

Let Σ be the part of the sphere $x^2+y^2+z^2 = 1$ that is contained in the cylinder $r = \cos\theta$ and lies above the xy plane, and let Σ be oriented with normal $\mathbf{n}$ directed upward. If $f(x,y) = \sqrt{1-x^2-y^2}$, then Σ is the graph of f on the region R in the xy plane bounded by the circle $r = \cos\theta$. Thus by Stokes's Theorem and (6) of Section 15.6 with the plus sign,

$$\int_C \mathbf{F}\cdot d\mathbf{r} = \iint_\Sigma (\operatorname{curl}\mathbf{F})\cdot\mathbf{n}\,dS = \iint_R \left[-(0)\frac{-x}{\sqrt{1-x^2-y^2}} - (0)\frac{-y}{\sqrt{1-x^2-y^2}} + 1\right] dA$$

$$= \iint_R 1\,dA = \text{area of } R = \pi\left(\frac{1}{2}\right)^2 = \frac{\pi}{4}.$$

13. $\operatorname{curl}\mathbf{F}(x,y,z) = \begin{vmatrix} \mathbf{i} & \mathbf{j} & \mathbf{k} \\ \dfrac{\partial}{\partial x} & \dfrac{\partial}{\partial y} & \dfrac{\partial}{\partial z} \\ x^2+z & y^2+x & z^2+y \end{vmatrix} = \mathbf{i}+\mathbf{j}+\mathbf{k}$

The sphere $x^2+y^2+z^2=1$ and the cone $z=\sqrt{x^2+y^2}$ intersect at (x,y,z) such that $z \ge 0$ and $1-x^2-y^2 = z^2 = x^2+y^2$, so that $x^2+y^2=\frac{1}{2}$ and thus $z=\sqrt{2}/2$. Making use of (9), we let Σ be the part of the plane $z=\sqrt{2}/2$ that lies inside the cone $z=\sqrt{x^2+y^2}$, and let Σ be oriented with normal $\mathbf{n}$ directed upward. If $f(x,y)=\sqrt{2}/2$, then Σ is the graph of f on the region R in the xy plane bounded by the circle $x^2+y^2=\frac{1}{2}$. Thus by Stokes's Theorem and (6) of Section 15.6 with the plus sign,

$$\int_C \mathbf{F}\cdot d\mathbf{r} = \iint_\Sigma (\operatorname{curl}\mathbf{F})\cdot\mathbf{n}\,dS = \iint_R [-(1)(0)-(1)(0)+1]\,dA = \text{area of } R = \frac{1}{2}\pi.$$

15. $\operatorname{curl}\mathbf{F}(x,y,z) = \begin{vmatrix} \mathbf{i} & \mathbf{j} & \mathbf{k} \\ \dfrac{\partial}{\partial x} & \dfrac{\partial}{\partial y} & \dfrac{\partial}{\partial z} \\ x & x^2+y^2+z^2 & z(y^4-1) \end{vmatrix} = (4zy^3-2z)\mathbf{i}+2x\mathbf{k}$

Let Σ_1 be the rectangular region in the xy plane bounded by lines $x=-1$, $x=1$, $y=-1$, and $y=1$, and orient Σ_1 with normal $\mathbf{n}=\mathbf{k}$. Then Σ and Σ_1 induce the same orientation on their common boundary. Thus by (9),

$$\iint_\Sigma (\operatorname{curl}\mathbf{F})\cdot\mathbf{n}\,dS = \iint_{\Sigma_1} (\operatorname{curl}\mathbf{F})\cdot\mathbf{n}\,dS = \iint_{\Sigma_1} [(4zy^3-2z)\mathbf{j}+2x\mathbf{k}]\cdot\mathbf{k}\,dS = \iint_{\Sigma_1} 2x\,dS$$

$$= \int_{-1}^{1}\int_{-1}^{1} 2x\,dy\,dx = \int_{-1}^{1} 2xy\Big|_{-1}^{1}\,dx = \int_{-1}^{1} 4x\,dx = 2x^2\Big|_{-1}^{1} = 0.$$

17. $\operatorname{curl}\mathbf{F}(x,y,z) = \begin{vmatrix} \mathbf{i} & \mathbf{j} & \mathbf{k} \\ \dfrac{\partial}{\partial x} & \dfrac{\partial}{\partial y} & \dfrac{\partial}{\partial z} \\ x\sin z & xy & yz \end{vmatrix} = z\mathbf{i}+x\cos z\mathbf{j}+y\mathbf{k}$

Let Σ_1 be the face of the cube in the plane $z=1$, and let Σ_1 be oriented with normal $\mathbf{n}=-\mathbf{k}$. Then Σ and Σ_1 induce the same orientation on their common boundary. Thus by (9),

$$\iint_\Sigma (\operatorname{curl}\mathbf{F})\cdot\mathbf{n}\,dS = \iint_{\Sigma_1} (\operatorname{curl}\mathbf{F})\cdot\mathbf{n}\,dS = \iint_{\Sigma_1} [(z\mathbf{i}+x\cos z\mathbf{j}+y\mathbf{k})\cdot(-\mathbf{k})]\,dS$$

$$= \iint_{\Sigma_1} -y\,dS = \int_0^1\int_0^1 -y\,dy\,dx = \int_0^1 -\frac{1}{2}y^2\Big|_0^1\,dx = \int_0^1 -\frac{1}{2}\,dx = -\frac{1}{2}.$$

19. $\operatorname{curl}\mathbf{F}(x,y,z) = \begin{vmatrix} \mathbf{i} & \mathbf{j} & \mathbf{k} \\ \dfrac{\partial}{\partial x} & \dfrac{\partial}{\partial y} & \dfrac{\partial}{\partial z} \\ xz^2 & x^3 & \cos xz \end{vmatrix} = (2xz + z\sin xz)\mathbf{j} + 3x^2\mathbf{k}$

Let Σ_1 be the disk in the xy plane bounded by the circle $x^2+y^2=1$, and let Σ_1 be oriented with normal $\mathbf{n} = -\mathbf{k}$. Then Σ and Σ_1 induce the same orientation on their common boundary. Thus by (9),

$$\iint_{\Sigma} (\operatorname{curl}\mathbf{F})\cdot\mathbf{n}\,dS = \iint_{\Sigma_1} (\operatorname{curl}\mathbf{F})\cdot\mathbf{n}\,dS = \iint_{\Sigma_1} [(2xz + z\sin xz)\mathbf{j} + 3x^2\mathbf{k}]\cdot(-\mathbf{k})\,dS$$

$$= \iint_{\Sigma_1} -3x^2\,dS = \int_0^{2\pi}\int_0^1 -3r^3\cos^2\theta\,dr\,d\theta = \int_0^{2\pi} -\frac{3}{4}r^4\cos^2\theta\Big|_0^1\,d\theta$$

$$= \int_0^{2\pi} -\frac{3}{4}\cos^2\theta\,d\theta = \int_0^{2\pi} -\left(\frac{3}{8}+\frac{3}{8}\cos 2\theta\right)d\theta = -\left(\frac{3}{8}\theta + \frac{3}{16}\sin 2\theta\right)\Big|_0^{2\pi} = -\frac{3}{4}\pi.$$

21. Since $\mathbf{F}$ is a constant vector field, $\operatorname{curl}\mathbf{F}(x,y,z) = \mathbf{0}$. Thus by Stokes's Theorem,

$$\int_C \mathbf{F}\cdot d\mathbf{r} = \iint_{\Sigma} (\operatorname{curl}\mathbf{F})\cdot\mathbf{n}\,dS = \iint_{\Sigma} \mathbf{0}\cdot\mathbf{n}\,dS = \iint_{\Sigma} 0\,dS = 0.$$

23. $\operatorname{curl}\mathbf{v}(x,y,z) = \begin{vmatrix} \mathbf{i} & \mathbf{j} & \mathbf{k} \\ \dfrac{\partial}{\partial x} & \dfrac{\partial}{\partial y} & \dfrac{\partial}{\partial z} \\ x^3 & -zy & x \end{vmatrix} = y\mathbf{i} - \mathbf{j}$

The sphere $x^2+y^2+z^2=1$ and the plane $z=y$ intersect at (x,y,z) such that $1 = x^2+y^2+z^2 = x^2+2y^2$. Let Σ_1 be the part of the plane $z=y$ that is enclosed by the boundary C of Σ, and orient Σ_1 with normal $\mathbf{n}$ directed upward. Then Σ and Σ_1 induce the same orientation on C. Let $f(x,y)=y$, and let R be the region in the xy plane bounded by the ellipse $x^2+2y^2=1$. Then Σ_1 is the graph of f on R. By the definition of circulation, by Stokes's Theorem, and by (6) in Section 15.6 with the plus sign, the circulation of the fluid is given by

$$\int_C \mathbf{v}\cdot d\mathbf{r} = \iint_{\Sigma_1} (\operatorname{curl}\mathbf{v})\cdot\mathbf{n}\,dS = \iint_R [-(y)(0) - (-1)(1) + 0]\,dA = \iint_R 1\,dA = \text{area of } R = \frac{\pi}{\sqrt{2}}.$$

15.8 The Divergence Theorem

1. Yes

3. Yes

5. No

7. No

9. $\operatorname{div} \mathbf{F}(x,y,z) = 2x + x - 2x = x$

$$\iint_\Sigma \mathbf{F}\cdot\mathbf{n}\,dS = \iiint_D \operatorname{div}\mathbf{F}(x,y,z)\,dV = \iiint_D x\,dV = \int_0^1\int_0^{1-x}\int_0^{1-x-y} x\,dz\,dy\,dx$$

$$= \int_0^1\int_0^{1-x} xz\Big|_0^{1-x-y} dy\,dx = \int_0^1\int_0^{1-x}(x - x^2 - xy)\,dy\,dx = \int_0^1 \left[(x-x^2)y - \frac{1}{2}xy^2\right]\Bigg|_0^{1-x} dx$$

$$= \int_0^1 \left[x(1-x)^2 - \frac{1}{2}x(1-x)^2\right] dx = \frac{1}{2}\int_0^1 (x - 2x^2 + x^3)\,dx = \frac{1}{2}\left(\frac{1}{2}x^2 - \frac{2}{3}x^3 + \frac{1}{4}x^4\right)\Bigg|_0^1 = \frac{1}{24}$$

11. $\operatorname{div} \mathbf{F}(x,y,z) = 1 + 1 + 1 = 3$

$$\iint_\Sigma \mathbf{F}\cdot\mathbf{n}\,dS = \iiint_D \operatorname{div}\mathbf{F}(x,y,z)\,dV = \iiint_D 3\,dV = 3(\text{volume of } D) = \frac{3\pi}{4}$$

13. $\operatorname{div} \mathbf{F}(x,y,z) = 1 + 1 + 1 = 3$

$$\iint_\Sigma \mathbf{F}\cdot\mathbf{n}\,dS = \iiint_D \operatorname{div}\mathbf{F}(x,y,z)\,dV = \iiint_D 3\,dV = 3(\text{volume of } D) = 2\pi$$

15. $\operatorname{div} \mathbf{F}(x,y,z) = 2x + 2y + 2z$

$$\iint_\Sigma \mathbf{F}\cdot\mathbf{n}\,dS = \iiint_D \operatorname{div}\mathbf{F}(x,y,z)\,dV = \iiint_D (2x + 2y + 2z)\,dV$$

$$= \int_0^{2\pi}\int_0^2\int_0^2 [2r^2(\cos\theta + \sin\theta) + 2zr]\,dz\,dr\,d\theta = \int_0^{2\pi}\int_0^2 [2r^2 z(\cos\theta + \sin\theta) + z^2 r]\big|_0^2\,dr\,d\theta$$

$$= \int_0^{2\pi}\int_0^2 [4r^2(\cos\theta + \sin\theta) + 4r]\,dr\,d\theta = \int_0^{2\pi}\left[\frac{4}{3}r^3(\cos\theta + \sin\theta) + 2r^2\right]\Bigg|_0^2 d\theta$$

$$= \int_0^{2\pi}\left[\frac{32}{3}(\cos\theta + \sin\theta) + 8\right] d\theta = \left[\frac{32}{3}(\sin\theta - \cos\theta) + 8\theta\right]\Bigg|_0^{2\pi} = 16\pi$$

17. At any point (x,y,z) of intersection of the plane $z = 2x$ and the paraboloid $z = x^2 + y^2$, x, y, and z must satisfy $2x = z = x^2 + y^2$, or in polar coordinates, $2r\cos\theta = r^2$, or $r = 2\cos\theta$. Thus the given region D is the solid region between the graphs of $z = r^2$ and $z = 2r\cos\theta$ on the region in the xy plane bounded by $r = 2\cos\theta$ for $-\pi/2 \le \theta \le \pi/2$. Since $\operatorname{div}\mathbf{F}(x,y,z) = 3xy(x^2+y^2)^{1/2} - 3xy(x^2+y^2)^{1/2} + 1 = 1$, we have

$$\iint_\Sigma \mathbf{F}\cdot\mathbf{n}\,dS = \iiint_D \operatorname{div}\mathbf{F}(x,y,z)\,dV = \iiint_D 1\,dV = \int_{-\pi/2}^{\pi/2}\int_0^{2\cos\theta}\int_{r^2}^{2r\cos\theta} r\,dz\,dr\,d\theta$$

$$= \int_{-\pi/2}^{\pi/2}\int_0^{2\cos\theta} rz\Big|_{r^2}^{2r\cos\theta} dr\,d\theta = \int_{-\pi/2}^{\pi/2}\int_0^{2\cos\theta}(2r^2\cos\theta - r^3)dr\,d\theta$$

$$= \int_{-\pi/2}^{\pi/2}\left(\frac{2}{3}r^3\cos\theta - \frac{1}{4}r^4\right)\Bigg|_0^{2\cos\theta} d\theta = \int_{-\pi/2}^{\pi/2}\left(\frac{16}{3}\cos^4\theta - 4\cos^4\theta\right) d\theta$$

$$= \int_{-\pi/2}^{\pi/2} \frac{4}{3}\cos^4\theta\, d\theta = \frac{4}{3}\left(\frac{1}{4}\cos^3\theta\sin\theta + \frac{3}{8}\cos\theta\sin\theta + \frac{3}{8}\theta\right)\Big|_{-\pi/2}^{\pi/2} = \frac{1}{2}\pi$$

with the next to the last equality coming from (13) in Section 8.1.

19. $\operatorname{div}\mathbf{F}(x,y,z) = -2+4-7=-5$

$$\iint_\Sigma \mathbf{F}\cdot\mathbf{n}\,dS = \iiint_D \operatorname{div}\mathbf{F}(x,y,z)\,dV = \iiint_D -5\,dV = \int_0^{2\pi}\int_1^2\int_{-\sqrt{4-r^2}}^{\sqrt{4-r^2}} -5r\,dz\,dr\,d\theta$$

$$= \int_0^{2\pi}\int_1^2 -5rz\Big|_{-\sqrt{4-r^2}}^{\sqrt{4-r^2}} dr\,d\theta = \int_0^{2\pi}\int_1^2 -10r\sqrt{4-r^2}\,dr\,d\theta = \int_0^{2\pi}\frac{10}{3}(4-r^2)^{3/2}\Big|_1^2 d\theta$$

$$= \int_0^{2\pi} -\frac{10}{3}3^{3/2}\,d\theta = -20\pi\sqrt{3}$$

21. $\operatorname{div}\mathbf{F}(x,y,z) = 2x+1-4z$

$$\iint_\Sigma \mathbf{F}\cdot\mathbf{n}\,dS = \iiint_D \operatorname{div}\mathbf{F}(x,y,z)\,dV = \iiint_D (2x+1-4z)\,dV$$

$$= \int_{-\sqrt{2}}^{\sqrt{2}}\int_0^{2-y^2}\int_0^x (2x+1-4z)\,dz\,dx\,dy = \int_{-\sqrt{2}}^{\sqrt{2}}\int_0^{2-y^2}[(2x+1)z-2z^2]\big|_0^x\,dx\,dy$$

$$= \int_{-\sqrt{2}}^{\sqrt{2}}\int_0^{2-y^2}[(2x+1)x-2x^2]\,dx\,dy = \int_{-\sqrt{2}}^{\sqrt{2}}\int_0^{2-y^2} x\,dx\,dy = \int_{-\sqrt{2}}^{\sqrt{2}}\frac{1}{2}x^2\Big|_0^{2-y^2} dy$$

$$= \int_{-\sqrt{2}}^{\sqrt{2}}\frac{1}{2}(4-4y^2+y^4)\,dy = \frac{1}{2}\left(4y-\frac{4}{3}y^3+\frac{1}{5}y^5\right)\Big|_{-\sqrt{2}}^{\sqrt{2}} = \frac{32}{15}\sqrt{2}$$

23. $\operatorname{div}\mathbf{F}(x,y,z) = (3x^2+y^2+z^2)+(x^2+3y^2+z^2)+0 = 4x^2+4y^2+2z^2$

$$\iint_\Sigma \mathbf{F}\cdot\mathbf{n}\,dS = \iiint_D \operatorname{div}\mathbf{F}(x,y,z)\,dV = \iiint_D (4x^2+4y^2+2z^2)\,dV =$$

$$= \int_0^{2\pi}\int_0^\pi\int_0^3 (2\rho^2+2\rho^2\sin^2\phi)\rho^2\sin\phi\,d\rho\,d\phi\,d\theta = \int_0^{2\pi}\int_0^\pi \frac{2}{5}\rho^5(\sin\phi+\sin^3\phi)\Big|_0^3 d\phi\,d\theta$$

$$= \frac{486}{5}\int_0^{2\pi}\int_0^\pi(\sin\phi+\sin^3\phi)\,d\phi\,d\theta = \frac{486}{5}\int_0^{2\pi}\int_0^\pi[\sin\phi+\sin\phi(1-\cos^2\phi)]\,d\phi\,d\theta$$

$$= \frac{486}{5}\int_0^{2\pi}\left(-2\cos\phi+\frac{1}{3}\cos^3\phi\right)\Big|_0^\pi d\theta = \frac{486}{5}\int_0^{2\pi}\frac{10}{3}\,d\theta = 648\pi$$

25. $\iint_\Sigma \mathbf{F}\cdot\mathbf{n}\,dS = \iiint_D \operatorname{div}\mathbf{F}(x,y,z)\,dV = \iiint_D \operatorname{div}(\operatorname{curl}\mathbf{G})(x,y,z)\,dV = \iiint_D 0\,dV = 0$ by (4) in Section 15.1.

27. Let R be the region in the xy plane bounded by the circle $r=a$. Let Σ_1 denote the graph of $z=(h/a)\sqrt{x^2+y^2}$ for (x,y) in R, and Σ_2 the graph of $z=h$ on R. Let Σ consist of Σ_1 and Σ_2. By Exercise 26, the volume of the conical region is given by

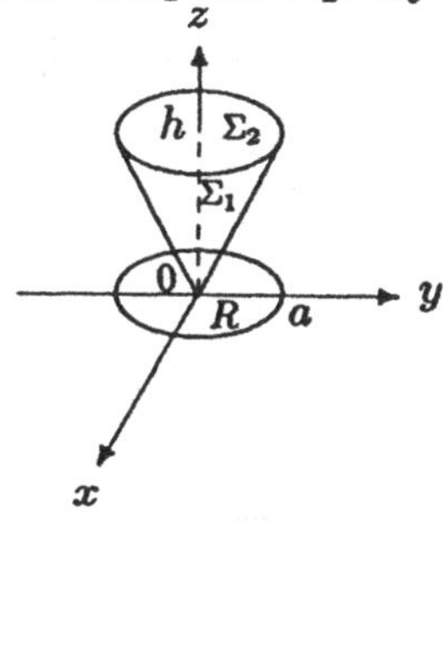

$$V = \frac{1}{3}\iint_\Sigma \mathbf{F}\cdot\mathbf{n}\,dS = \frac{1}{3}\iint_{\Sigma_1}\mathbf{F}\cdot\mathbf{n}\,dS + \frac{1}{3}\iint_{\Sigma_2}\mathbf{F}\cdot\mathbf{n}\,dS$$

with $\mathbf{F}(x,y,z) = x\mathbf{i}+y\mathbf{j}+z\mathbf{k}$. To evaluate $\iint_{\Sigma_1}\mathbf{F}\cdot\mathbf{n}\,dS$ we use the parametrization $\mathbf{r}(\theta, z) = \frac{a}{h}z\cos\theta\mathbf{i} + \frac{a}{h}z\sin\theta\mathbf{j} + z\mathbf{k}$ for $0 \le \theta \le 2\pi$ and $0 \le z \le h$. Then

$$\mathbf{r}_\theta(\theta, z) \times \mathbf{r}_z(\theta, z) = \begin{vmatrix} \mathbf{i} & \mathbf{j} & \mathbf{k} \\ -\frac{a}{h}z\sin\theta & \frac{a}{h}z\cos\theta & 0 \\ \frac{a}{h}\cos\theta & \frac{a}{h}\sin\theta & 1 \end{vmatrix} = \frac{a}{h}z\cos\theta\,\mathbf{i} + \frac{a}{h}z\sin\theta\,\mathbf{j} - \frac{a^2}{h^2}z\,\mathbf{k}$$

so that $\mathbf{r}_\theta \times \mathbf{r}_z$ points in the direction of the normal for the given orientation. Consequently we may use (4) or (5) of Section 15.6 with the plus sign to find that

$$\iint_{\Sigma_1} \mathbf{F}\cdot\mathbf{n}\,dS = \int_0^{2\pi}\int_0^h \left[\left(\frac{a}{h}z\cos\theta\right)\left(\frac{a}{h}z\cos\theta\right) + \left(\frac{a}{h}z\sin\theta\right)\left(\frac{a}{h}z\sin\theta\right) + (z)\left(-\frac{a^2}{h^2}z\right)\right] dz\,d\theta$$

$$= \int_0^{2\pi}\int_0^h 0\,dz\,d\theta = 0$$

Next,

$$\iint_{\Sigma_2} \mathbf{F}\cdot\mathbf{n}\,dS = \iint_{\Sigma_2} (x\mathbf{i} + y\mathbf{j} + z\mathbf{k})\cdot\mathbf{k}\,dS = \iint_{\Sigma_2} z\,dS = \iint_R h\,dA = \pi a^2 h.$$

Therefore

$$V = \frac{1}{3}\iint_\Sigma \mathbf{F}\cdot\mathbf{n}\,dS = \frac{1}{3}\iint_{\Sigma_1} \mathbf{F}\cdot\mathbf{n}\,dS + \frac{1}{3}\iint_{\Sigma_2} \mathbf{F}\cdot\mathbf{n}\,dS = 0 + \frac{1}{3}\pi a^2 h = \frac{1}{3}\pi a^2 h.$$

29. If $\mathbf{F}$ is constant, then $\operatorname{div}\mathbf{F} = 0$, so that by the Divergence Theorem,

$$\iint_\Sigma \mathbf{F}\cdot\mathbf{n}\,dS = \iiint_D \operatorname{div}\mathbf{F}(x, y, z)\,dV = \iiint_D 0\,dV = 0.$$

31. By the Divergence Theorem, (4) of the present section, and (8) of Section 14.4, we have

$$\iint_\Sigma \mathbf{E}\cdot\mathbf{n}\,dS = \iiint_D \operatorname{div}\mathbf{E}\,dV = \iiint_D \frac{1}{\varepsilon_0}\rho\,dV = \frac{1}{\varepsilon_0}\iiint_D \rho\,dV = \frac{1}{\varepsilon_0}q.$$

33. Notice that $\partial E/\partial y = -(a\omega/c)\sin\omega(y/c - t)$ and $-\partial B/\partial t = -b\omega\sin\omega(y/c - t)$. Since a, b, c, and $\omega > 0$, it follows that $\partial E/\partial y = -\partial B/\partial t$ if and only if $a/c = b$, that is, $a = bc$. Similarly, $\partial B/\partial y = -(b\omega/c)\sin\omega(y/c - t)$ and $\partial E/\partial t = a\omega\sin\omega(y/c - t)$. Since a, b, c and $\omega > 0$, we conclude that $\partial B/\partial y = -(1/c^2)\partial E/\partial t$ if and only if $b/c = a/c^2$, or $a = bc$.

35. In order to apply Stokes's Theorem in step (3), we must have C be the boundary of Σ. But Σ is the boundary of a simple solid region D, and hence Σ has no boundary of its own.

Taking partial derivatives with respect to z, we find that $\partial f/\partial z = dh/dz$. Comparing this with the third equation in $(*)$, we conclude that $dh/dz = z^3$, so that $h(z) = \frac{1}{4}z^4 + c$. If $c = 0$, then $f(x, y, z) = xy + \frac{1}{4}z^4$.

Cumulative Review (Chapters 1–14)

1. $$\lim_{x\to-\infty} \frac{1}{x^2+2}\sqrt{\frac{x^5-1}{2x+1}} = \lim_{x\to-\infty} \frac{1}{x^2+2}\sqrt{\frac{x^5(1-1/x^5)}{x(2+1/x)}} = \lim_{x\to-\infty} \frac{x^2}{x^2+2}\sqrt{\frac{1-1/x^5}{2+1/x}}$$
$$= \lim_{x\to-\infty} \frac{1}{1+2/x^2}\sqrt{\frac{1-1/x^5}{2+1/x}} = \sqrt{\frac{1}{2}} = \frac{\sqrt{2}}{2}$$

3. $f'(x) = 3x^2\cos(x^3)$, so $f'(0) = 0$; $f''(x) = 6x\cos(x^3) - [3x^2\sin(x^3)](3x^2) = 6x\cos(x^3) - 9x^4\sin(x^3)$, so $f''(0) = 0$; $f^{(3)}(x) = 6\cos(x^3) - [6x\sin(x^3)](3x^2) - 36x^3\sin(x^3) - [9x^4\cos(x^3)](3x^2)$, so $f^{(3)}(0) = 6$. Thus 3 is the smallest positive integer n such that $f^{(n)}(0) \neq 0$.

5. Using the notation in the figure, we find by similar triangles that $r(x)/x = (1/2)/1 = \frac{1}{2}$, that is, $r(x) = \frac{1}{2}x$. Therefore the volume V of the part of the drill that has penetrated the plank is given by $V = \frac{1}{3}\pi\left(r(x)\right)^2 x = (\pi/3)\left(\frac{1}{2}x\right)^2 x = \pi x^3/12$. By implicit differentiation, $dV/dt = (\pi x^2/4)dx/dt$. Since $dx/dt = \frac{1}{12}$, we have

$$\left.\frac{dV}{dt}\right|_{x=2/3} = \frac{\pi(2/3)^2}{4}\cdot\frac{1}{12} = \frac{\pi}{108}.$$

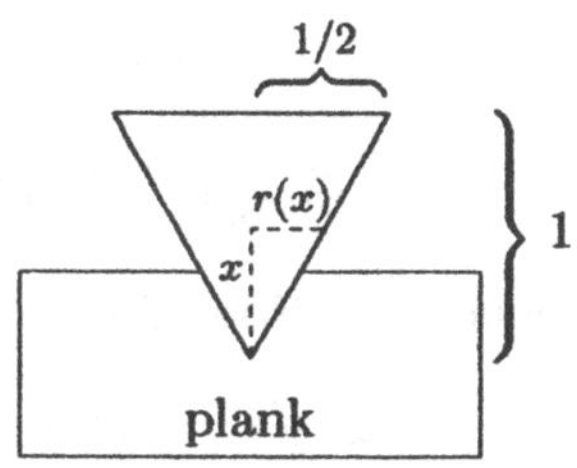

Thus the volume of the drilled hole increases at the rate of $\pi/108$ cubic inches per second when the drill tip is $\frac{2}{3}$ inch in the wood.

7. $f'(x) = \dfrac{2-x^2}{(x^2+2x+2)^2}$; $f''(x) = \dfrac{2(x^3-6x-4)}{(x^2+2x+2)^3} = \dfrac{2(x+2)(x-1-\sqrt{3})(x-1+\sqrt{3})}{(x^2+2x+2)^3}$;
relative maximum value is $f(\sqrt{2}) = \sqrt{2}/(4+2\sqrt{2}) = \frac{1}{2}(\sqrt{2}-1)$; relative minimum value is $f(-\sqrt{2}) = -\sqrt{2}/(4-2\sqrt{2}) = -\frac{1}{2}(\sqrt{2}+1)$; increasing on $[-\sqrt{2}, \sqrt{2}]$, and decreasing on $(-\infty, -\sqrt{2}]$ and $[\sqrt{2}, \infty)$; concave upward on $(-2, 1-\sqrt{3})$ and $(1+\sqrt{3}, \infty)$, and concave downward on $(-\infty, -2)$ and $(1-\sqrt{3}, 1+\sqrt{3})$; inflection points are $(-2, -1)$, $(1-\sqrt{3}, (-1-\sqrt{3})/4)$ and $(1+\sqrt{3}, (-1+\sqrt{3})/4)$; horizontal asymptote is $y = 0$.

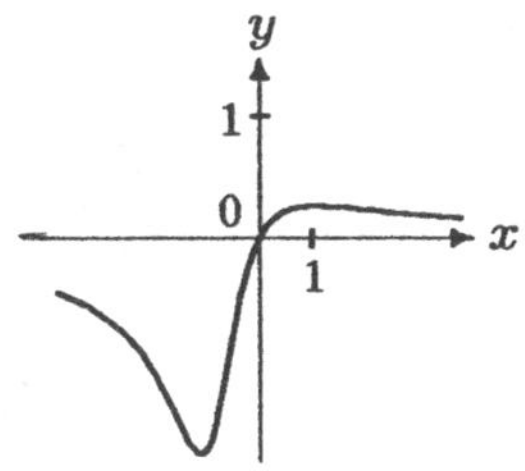

9. To integrate by parts, let $u = x^2$ and $dv = \sec^2 x \tan x\, dx$, so that $du = 2x\, dx$ and $v = \frac{1}{2}\tan^2 x$. Then

$$\int x^2\sec^2 x\tan x\, dx = \frac{1}{2}x^2\tan^2 x - \int 2x\left(\frac{1}{2}\tan^2 x\right)dx = \frac{1}{2}x^2\tan^2 x - \int x(\sec^2 x - 1)\, dx.$$

11. First we will find $\int(x^2/\sqrt{4-x^2})\, dx$. Let $x = 2\sin u$, so that $dx = 2\cos u\, du$;

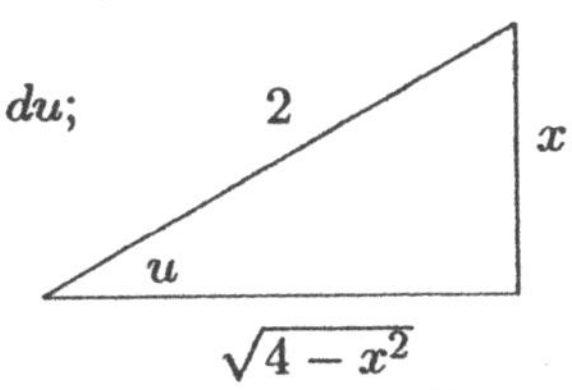

$$\int \frac{x^2}{\sqrt{4-x^2}}\, dx = \int \frac{4\sin^2 u}{\sqrt{4-4\sin^2 u}}(2\cos u)\, du$$
$$= \int \frac{8\sin^2 u\cos u}{2\cos u}\, du = 4\int \sin^2 u\, du = 4\int\left(\frac{1}{2} - \frac{1}{2}\cos 2u\right)du$$

$$= 2u - \sin 2u + C = 2u - 2\sin u \cos u + C = 2\sin^{-1}\frac{x}{2} - \frac{x}{2}\sqrt{4-x^2} + C.$$

Therefore

$$\int_0^2 \frac{x^2}{\sqrt{4-x^2}}\,dx = \lim_{b\to 2^-}\int_0^b \frac{x^2}{\sqrt{4-x^2}}\,dx = \lim_{b\to 2^-}\left(2\sin^{-1}\frac{x}{2} - \frac{x}{2}\sqrt{4-x^2}\right)\Big|_0^b$$

$$= \lim_{b\to 2^-}\left(2\sin^{-1}\frac{b}{2} - \frac{b}{2}\sqrt{4-b^2}\right) = 2\sin^{-1}1 = 2\left(\frac{\pi}{2}\right) = \pi.$$

Thus the given integral converges, and its value is π.

13. a. Since $x = r\cos\theta$, $y = r\sin\theta$, and $x^2 + y^2 = r^2$, the given equation is equivalent to $(r^2)^3 = 3(r\cos\theta)^2(r\sin\theta)^2 = 3r^4\cos^2\theta\sin^2\theta$ or $r^2 = 3\cos^2\theta\sin^2\theta = \frac{3}{4}\sin^2 2\theta$, or $r = (\sqrt{3}/2)\sin 2\theta$. The graph is a four-leaved rose, sketched in the figure.

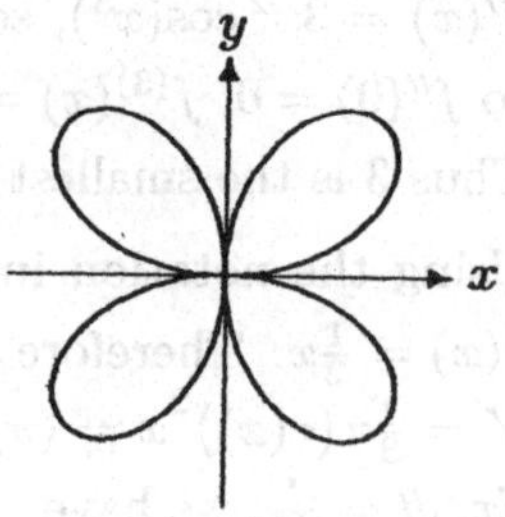

b. From part (a), $|r| \le \sqrt{3}/2$ and $r = \sqrt{3}/2$ for $\theta = \pi/4$. Therefore the maximum distance between points on the graph of the equation and the origin is $\sqrt{3}/2$.

15. Since

$$\lim_{n\to\infty}\sqrt[n]{\left|(-1)^n\left(1+\frac{1}{n}\right)^{-n^2}\right|} = \lim_{n\to\infty}\left(1+\frac{1}{n}\right)^{-n} = \lim_{n\to\infty}\frac{1}{(1+1/n)^n} = \frac{1}{e} < 1$$

the Generalized Root Test implies that the given series converges absolutely.

17. Since $e^x = \sum_{n=0}^{\infty}(1/n!)x^n$ for all x, we have

$$e^{-t^2/2} = \sum_{n=0}^{\infty}\frac{1}{n!}\left(-\frac{t^2}{2}\right)^n = \sum_{n=0}^{\infty}\frac{(-1)^n}{n!\,2^n}t^{2n} \quad \text{for all } t.$$

By the Integration Theorem for Power Series,

$$\int_0^1 e^{-t^2/2}\,dt = \int_0^1 \sum_{n=0}^{\infty}\frac{(-1)^n}{n!\,2^n}t^{2n}\,dt = \sum_{n=0}^{\infty}\int_0^1 \frac{(-1)^n}{n!\,2^n}t^{2n}\,dt$$

$$= \sum_{n=0}^{\infty}\left(\frac{(-1)^n}{n!\,2^n(2n+1)}\right)t^{2n+1}\Big|_0^1 = \sum_{n=0}^{\infty}\frac{(-1)^n}{n!\,2^n(2n+1)}.$$

Since this series is alternating, it follows from the Alternating Series Test Theorem (Theorem 9.17) that

$$\frac{5}{6} = s_2 \le \sum_{n=0}^{\infty}\frac{(-1)^n}{n!\,2^n(2n+1)} \le s_3 = \frac{103}{120}.$$

19. a. $L = \int_0^b \sqrt{\left(\frac{dx}{dt}\right)^2 + \left(\frac{dy}{dt}\right)^2 + \left(\frac{dz}{dt}\right)^2}\, dt = \int_0^b \sqrt{1^2 + (2t^{1/2})^2 + (2t)^2}\, dt.$

$$= \int_0^b \sqrt{1 + 4t + 4t^2}\, dt = \int_0^b (1 + 2t)\, dt = (t + t^2)\Big|_0^b = b + b^2$$

Thus $L = 30$ if $b + b^2 = 30$, that is $(b+6)(b-5) = 0$, so 5 is the positive value of b that makes $L = 30$.

b. $\mathbf{v}(t) = \mathbf{r}'(t) = \mathbf{i} + 2t^{1/2}\mathbf{j} + 2t\mathbf{k}$; $\|\mathbf{v}(t)\| = \sqrt{1^2 + (2t^{1/2})^2 + (2t)^2} = 1 + 2t$; $a_{\mathrm{T}} = \frac{d}{dt}\|\mathbf{v}(t)\| = 2.$

c. $\mathbf{a}(t) = \mathbf{v}'(t) = t^{-1/2}\mathbf{j} + 2\mathbf{k}$; $\|\mathbf{a}(t)\| = \sqrt{(t^{-1/2})^2 + 2^2} = \sqrt{t^{-1} + 4}$;

$a_{\mathrm{N}} = \sqrt{\|\mathbf{a}\|^2 - a_{\mathrm{T}}^2} = \sqrt{(t^{-1} + 4) - 2^2} = t^{-1/2}$, so that a_{N} is decreasing on $(0, \infty)$.

21. Let $f(x, y, z) = x^2 - 2y^2 + z^2$. Then it follows that $\operatorname{grad} f(x, y, z) = 2x\mathbf{i} - 4y\mathbf{j} + 2z\mathbf{k}$ is normal to the level surface $x^2 - 2y^2 + z^2 = 0$ of f. Similarly, let $g(x, y, z) = xyz$, so that $\operatorname{grad} g(x, y, z) = yz\mathbf{i} + xz\mathbf{j} + xy\mathbf{k}$ is normal to the surface $xyz = 1$. Since

$$\operatorname{grad} f(x, y, z) \cdot \operatorname{grad} g(x, y, z) = 2xyz - 4xyz + 2xyz = 0$$

it follows that the given surfaces are perpendicular at each point of intersection.

23. Let R be the region in the first quadrant bounded by the curve $x = y^{1/3}$ (or equivalently, $y = x^3$) and the lines $y = 1$ and $x = 2$. Reversing the order of integration, we find that

$$\int_1^8 \int_{y^{1/3}}^2 \cos\left(\frac{x^4}{4} - x\right) dx\, dy = \iint_R \cos\left(\frac{x^4}{4} - x\right) dA$$

$$= \int_1^2 \int_1^{x^3} \cos\left(\frac{x^4}{4} - x\right) dy\, dx = \int_1^2 y\cos\left(\frac{x^4}{4} - x\right)\Big|_1^{x^3} dx$$

$$= \int_1^2 (x^3 - 1)\cos\left(\frac{x^4}{4} - x\right) dx = \sin\left(\frac{x^4}{4} - x\right)\Big|_1^2$$

$$= \sin 2 - \sin\left(-\frac{3}{4}\right) = \sin 2 + \sin\frac{3}{4}.$$

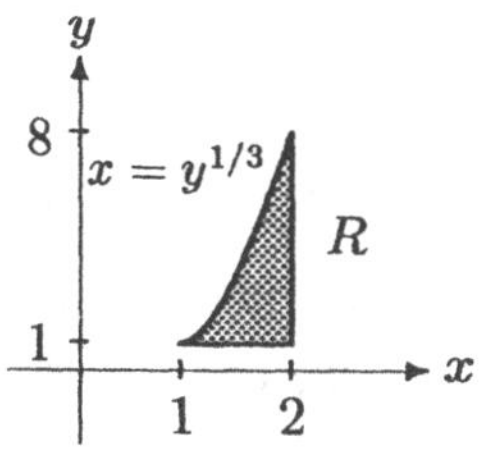

25. The paraboloid and the sphere intersect at (x, y, z) if $2z = x^2 + y^2$ and $x^2 + y^2 + z^2 = 8$, so that $2z + z^2 = 8$, or $(z+4)(z-2) = 0$. Since $z = \frac{1}{2}(x^2 + y^2) \ge 0$, it follows that $z = 2$ and hence that $x^2 + y^2 = 4$. Thus D is the solid region that lies over the region R in the xy plane bounded by the circle $x^2 + y^2 = 4$. In cylindrical coordinates the equation $2z = x^2 + y^2$ becomes $z = \frac{1}{2}r^2$, and the equation $x^2 + y^2 + z^2 = 8$ becomes $r^2 + z^2 = 8$, or $z = \sqrt{8 - r^2}$. Therefore

$$\iiint_D z\, dV = \int_0^{2\pi} \int_0^2 \int_{r^2/2}^{\sqrt{8-r^2}} zr\, dz\, dr\, d\theta = \int_0^{2\pi} \int_0^2 \frac{1}{2}z^2 r\Big|_{r^2/2}^{\sqrt{8-r^2}} dr\, d\theta$$

$$= \int_0^{2\pi} \int_0^2 \frac{1}{2}r\left(8 - r^2 - \frac{1}{4}r^4\right) dr\, d\theta = \int_0^{2\pi} \int_0^2 \left(4r - \frac{1}{2}r^3 - \frac{1}{8}r^5\right) dr\, d\theta$$

$$= \int_0^{2\pi} \left(2r^2 - \frac{1}{8}r^4 - \frac{1}{48}r^6\right)\Big|_0^2 = \int_0^{2\pi} \frac{14}{3}\, d\theta = \frac{28\pi}{3}.$$

Chapter 15 Review

1. Let $\mathbf{a} = a_1\mathbf{i} + a_2\mathbf{j} + a_3\mathbf{k}$ and $\mathbf{b} = b_1\mathbf{i} + b_2\mathbf{j} + b_3\mathbf{k}$. Then

$$\mathbf{b}\times\mathbf{r} = \begin{vmatrix} \mathbf{i} & \mathbf{j} & \mathbf{k} \\ b_1 & b_2 & b_3 \\ x & y & z \end{vmatrix} = (b_2z - b_3y)\mathbf{i} + (b_3x - b_1z)\mathbf{j} + (b_1y - b_2x)\mathbf{k}$$

so that $\mathbf{a}\cdot(\mathbf{b}\times\mathbf{r}) = a_1(b_2z - b_3y) + a_2(b_3x - b_1z) + a_3(b_1y - b_2x)$. Thus $\operatorname{grad}[\mathbf{a}\cdot(\mathbf{b}\times\mathbf{r})] = (a_2b_3 - a_3b_2)\mathbf{i} + (a_3b_1 - a_1b_3)\mathbf{j} + (a_1b_2 - a_2b_1)\mathbf{k} = \mathbf{a}\times\mathbf{b}$.

3. To begin with,
$$\frac{\partial f}{\partial x} = y^2 - y\sin xy \quad \text{and} \quad \frac{\partial f}{\partial y} = 2xy - x\sin xy. \qquad (*)$$

Integrating both sides of the first equation in (*) with respect to x, we obtain $f(x,y) = xy^2 + \cos xy + g(y)$.Taking partial derivatives with respect to y, we find that $\partial f/\partial y = 2xy - x\sin xy + dg/dy$. Comparing this with the second equation in (*), we conclude that $dg/dy = 0$, so that $g(y) = C$. Therefore $f(x,y) = xy^2 + \cos xy + C$.

5. Since $x(t) = \cos t$, $y(t) = \sin t$, $z(t) = t$, and $\|d\mathbf{r}/dt\| = \sqrt{(-\sin t)^2 + \cos^t + 1^2} = \sqrt{2}$, (4) in Section 15.2 implies that

$$\int_C (xy + z^2)\,ds = \int_{\pi/4}^{3\pi/4} (\cos t\sin t + t^2)\sqrt{2}\,dt = \left(\frac{1}{2}\sin^2 t + \frac{1}{3}t^3\right)\sqrt{2}\bigg|_{\pi/4}^{3\pi/4} = \frac{13\sqrt{2}}{96}\pi^3.$$

7. Since $x(t) = t$, $y(t) = \cos t$, $z(t) = \sin t$, $dx/dt = 1$, $dy/dt = -\sin t$, and $dz/dt = \cos t$, (9) of Section 15.2 implies that

$$\int_C xy\,dx + z\cos x\,dy + z\,dz = \int_0^{\pi/2} [(t\cos t)(1) + (\sin t\cos t)(-\sin t) + (\sin t)(\cos t)]\,dt$$

$$= \int_0^{\pi/2} t\cos t\,dt + \int_0^{\pi/2} (-\sin^2 t\cos t + \sin t\cos t)\,dt$$

$$\overset{\text{parts}}{=} t\sin t\Big|_0^{\pi/2} - \int_0^{\pi/2}\sin t\,dt + \left(-\frac{1}{3}\sin^3 t + \frac{1}{2}\sin^2 t\right)\bigg|_0^{\pi/2} = \frac{\pi}{2} + \cos t\Big|_0^{\pi/2} + \frac{1}{6} = \frac{\pi}{2} - \frac{5}{6}.$$

9. The curve C is parametrized by $\mathbf{r}(t) = t\mathbf{i} + t^2\mathbf{j}$ for $0 \le t \le 1$. Since $x(t) = t$, $dx/dt = 1$, $y(t) = t^2$, $dy/dt = 2t$, and $\|d\mathbf{r}/dt\| = \sqrt{1^2 + (2t)^2} = \sqrt{1 + 4t^2}$, it follows from (4) in Section 15.2 that

$$\int_C x\,ds = \int_0^1 t\sqrt{1 + 4t^2}\,dt = \frac{1}{12}(1 + 4t^2)^{3/2}\bigg|_0^1 = \frac{1}{12}(5^{3/2} - 1).$$

11. Let $\mathbf{F}(x,y,z) = e^x\cos z\mathbf{i} + y\mathbf{j} - e^x\sin z\mathbf{k} = M(x,y,z)\mathbf{i} + N(x,y,z)\mathbf{j} + P(x,y,z)\mathbf{k}$. Since $\partial P/\partial y = 0 = \partial N/\partial z$, $\partial M/\partial z = -e^x\sin z = \partial P/\partial x$, and $\partial N/\partial x = 0 = \partial M/\partial y$, the line integral is independent of path. Let $\operatorname{grad} f(x,y,z) = \mathbf{F}(x,y,z)$. Then

$$\frac{\partial f}{\partial x} = e^x\cos z, \quad \frac{\partial f}{\partial y} = y, \quad \text{and} \quad \frac{\partial f}{\partial z} = -e^x\sin z. \qquad (*)$$

Integrating both sides of the first equation in (*) with respect to x, we obtain $f(x,y,z) = e^x\cos z + g(y,z)$. Taking partial derivatives with respect to y, we find that $\partial f/\partial y = \partial g/\partial y$. Comparing this with the second equation in (*), we deduce that $\partial g/\partial y = y$, so that $g(y,z) = \frac{1}{2}y^2 + h(z)$. Thus $f(x,y,z) = e^x\cos z + \frac{1}{2}y^2 + h(z)$. Taking partial derivatives with respect to z, we find that $\partial f/\partial z = -e^x\sin z + dh/dz$. Comparing this with the third equation in (*), we conclude that $dh/dz = 0$, so that $h(z) = c$. If $c = 0$, then $f(x,y,z) = e^x\cos z + \frac{1}{2}y^2$. Since $\mathbf{r}(0) = \mathbf{i}$ and $\mathbf{r}(1) = \mathbf{i} + \mathbf{j} + \mathbf{k}$, it follows that the initial and terminal points of C are $(1,0,0)$ and $(1,1,1)$, respectively. Thus

$$\int_C e^x\cos z\,dx + y\,dy - e^x\sin z\,dz = f(1,1,1) - f(1,0,0) = \left(e\cos 1 + \frac{1}{2}\right) - e = \frac{1}{2} + e(\cos 1 - 1).$$

13. Since $\operatorname{div}\mathbf{F}(x,y,z) = 1 + x + 1 = x + 2$, the Divergence Theorem implies that

$$\iint\limits_{\Sigma} \mathbf{F}\cdot\mathbf{n}\,dS = \iiint\limits_{D} \operatorname{div}\mathbf{F}(x,y,z)\,dV = \iiint\limits_{D} (x+2)\,dV = \int_0^1\int_0^{1-y^2}\int_0^{1+x} (x+2)\,dz\,dx\,dy$$

$$= \int_0^1\int_0^{1-y^2} (x+2)z\Big|_0^{1+x}\,dx\,dy = \int_0^1\int_0^{1-y^2} (x^2+3x+2)\,dx\,dy$$

$$= \int_0^1 \left(\frac{1}{3}x^3 + \frac{3}{2}x^2 + 2x\right)\Big|_0^{1-y^2} dy = \int_0^1 \left[\frac{1}{3}(1-y^2)^3 + \frac{3}{2}(1-y^2)^2 + 2(1-y^2)\right] dy$$

$$= \int_0^1 \left(-\frac{1}{3}y^6 + \frac{5}{2}y^4 - 6y^2 + \frac{23}{6}\right) dy = \left(-\frac{1}{21}y^7 + \frac{1}{2}y^5 - 2y^3 + \frac{23}{6}y\right)\Big|_0^1 = \frac{16}{7}.$$

15. Since $\operatorname{div}\mathbf{F}(x,y,z) = yz + 2z$, the Divergence Theorem implies that

$$\iint\limits_{\Sigma} \mathbf{F}\cdot\mathbf{n}\,dS = \iiint\limits_{D} \operatorname{div}\mathbf{F}(x,y,z)\,dV = \iiint\limits_{D} (yz+2z)\,dA$$

$$= \int_0^{2\pi}\int_0^2\int_{-2}^3 (r\sin\theta + 2)zr\,dz\,dr\,d\theta = \int_0^{2\pi}\int_0^2 (r^2\sin\theta + 2r)\frac{z^2}{2}\Big|_{-2}^3\,dr\,d\theta$$

$$= \int_0^{2\pi}\int_0^2 \frac{5}{2}(r^2\sin\theta + 2r)\,dr\,d\theta = \int_0^{2\pi} \frac{5}{2}\left(\frac{1}{3}r^3\sin\theta + r^2\right)\Big|_0^2 d\theta$$

$$= \int_0^{2\pi} \frac{5}{2}\left(\frac{8}{3}\sin\theta + 4\right) d\theta = \frac{5}{2}\left(-\frac{8}{3}\cos\theta + 4\theta\right)\Big|_0^{2\pi} = 20\pi.$$

17. The paraboloid $z = -1 + x^2 + y^2$ and the plane $z = 1$ intersect at (x,y,z) such that $-1 + x^2 + y^2 = z = 1$, which means that $x^2 + y^2 = 2$, or in polar coordinates, $r = \sqrt{2}$. Let R be the region in the xy plane bounded by $r = \sqrt{2}$, and let $f(x,y) = -1 + x^2 + y^2$ for (x,y) in R. We use (6) in Section 15.6 with $f_x(x,y) = 2x$, $f_y(x,y) = 2y$, $M(x,y,f(x,y)) = y$, $N(x,y,f(x,y)) = -x$, and $P(x,y,f(x,y)) = -1 + x^2 + y^2$, and with the minus sign, and find that

$$\iint\limits_{\Sigma} \mathbf{F}\cdot\mathbf{n}\,dS = \iint\limits_{R} [(y)(2x) + (-x)(2y) - (-1 + x^2 + y^2)]\,dA = \iint\limits_{R} (1 - x^2 - y^2)\,dA$$

$$= \int_0^{2\pi}\int_0^{\sqrt{2}} (1 - r^2)r\,dr\,d\theta = \int_0^{2\pi} \left(\frac{1}{2}r^2 - \frac{1}{4}r^4\right)\Big|_0^{\sqrt{2}} d\theta = \int_0^{2\pi} 0\,d\theta = 0.$$

19. Let Σ_1 be the part of the plane $z = 1$ inside the sphere $x^2 + y^2 + z^2 = 2$, and let Σ_1 be oriented with normal $\mathbf{n} = -\mathbf{k}$. Let R be the region in the xy plane bounded by the circle $x^2 + y^2 = 1$, or in polar coordinates $r = 1$. Since Σ and Σ_1 induce the same orientation on their common boundary, (9) in Section 15.7 implies that

$$\iint\limits_{\Sigma} (\operatorname{curl}\mathbf{F})\cdot\mathbf{n} = \iint\limits_{\Sigma_1} (\operatorname{curl}\mathbf{F})\cdot\mathbf{n} = \iint\limits_{\Sigma_1} \left\{\left[\frac{\partial}{\partial x}(-y^3) - \frac{\partial}{\partial y}(x^3y)\right]\mathbf{k}\right\}\cdot(-\mathbf{k})\,dS$$

$$= \iint_{\Sigma_1} x^3\, dS = \iint_R x^3\, dA = \int_0^{2\pi}\int_0^1 r^4 \cos^3\theta\, dr\, d\theta = \int_0^{2\pi} \frac{1}{5} r^5 \cos^3\theta \Big|_0^1 d\theta$$

$$= \int_0^{2\pi} \frac{1}{5}\cos^3\theta\, d\theta = \int_0^{2\pi} \frac{1}{5}(1-\sin^2\theta)\cos\theta\, d\theta = \frac{1}{5}\left(\sin\theta - \frac{1}{3}\sin^3\theta\right)\Big|_0^{2\pi} = 0.$$

21. Since $\operatorname{div}\mathbf{F}(x,y,z) = \sec^2 x - (1+\tan^2 x) - 6 = -6$, the Divergence Theorem implies that

$$\iint_\Sigma \mathbf{F}\cdot\mathbf{n}\, dS = \iiint_D \operatorname{div}\mathbf{F}(x,y,z)\, dV = \iiint_D -6\, dV$$

$$= -6\,(\text{volume of the hemisphere}) = -6\left(\frac{2}{3}\pi 2^3\right) = -32\pi.$$

23. Let $\mathbf{F}(x,y,z) = y\mathbf{i} + y\mathbf{j} + x^2\mathbf{k}$. Then

$$\operatorname{curl}\mathbf{F}(x,y,z) = \begin{vmatrix} \mathbf{i} & \mathbf{j} & \mathbf{k} \\ \dfrac{\partial}{\partial x} & \dfrac{\partial}{\partial y} & \dfrac{\partial}{\partial z} \\ y & y & x^2 \end{vmatrix} = -2x\mathbf{j} - \mathbf{k}.$$

The surfaces $z = x^2 + y^2$ and $z = 1 - y^2$ intersect at (x,y,z) such that $x^2 + y^2 = z = 1 - y^2$, so that $x^2 + 2y^2 = 1$. Let $f(x,y) = 1 - y^2$, let Σ be the graph of f on the region R in the xy plane bounded by the ellipse $x^2 + 2y^2 = 1$, and let Σ be oriented with normal $\mathbf{n}$ directed upward. Then by Stokes's Theorem and (6) of Section 15.6,

$$\int_C y\,dx + y\,dy + x^2\,dz = \int_C \mathbf{F}\cdot d\mathbf{r} = \iint_\Sigma (\operatorname{curl}\mathbf{F})\cdot\mathbf{n}\, dS = \iint_R [-(0)(0) - (-2x)(-2y) - 1]\, dA$$

$$= \int_{-1}^1 \int_{-\sqrt{(1-x^2)/2}}^{\sqrt{(1-x^2)/2}} (-4xy - 1)\, dy\, dx = \int_{-1}^1 (-2xy^2 - y)\Big|_{-\sqrt{(1-x^2)/2}}^{\sqrt{(1-x^2)/2}} dx$$

$$= -\sqrt{2}\int_{-1}^1 \sqrt{1-x^2}\, dx = -\sqrt{2}\,(\text{area of semicircle of radius } 1) = -\frac{\sqrt{2}}{2}\pi.$$

25. Since $x(t) = t$, $y(t) = 0$, $z(t) = -t^3$, and $d\mathbf{r}/dt = \mathbf{i} - 3t^2\mathbf{k}$, (7) in Section 15.2 implies that

$$\int_C \mathbf{F}\cdot d\mathbf{r} = \int_{-1}^1 (t^4\mathbf{i} - t^3\mathbf{k})\cdot(\mathbf{i} - 3t^2\mathbf{k})\, dt = \int_{-1}^1 (t^4 + 3t^5)\, dt = \left(\frac{1}{5}t^5 + \frac{1}{2}t^6\right)\Big|_{-1}^1 = \frac{2}{5}.$$

27. Let $\mathbf{F}(x,y,z) = y\mathbf{i} + x\mathbf{j} + z^3\mathbf{k} = M(x,y,z)\mathbf{i} + N(x,y,z)\mathbf{j} + P(x,y,z)\mathbf{k}$. Since $\partial P/\partial y = 0 = \partial N/\partial z$, $\partial M/\partial z = 0 = \partial P/\partial x$, and $\partial N/\partial x = 1 = \partial M/\partial y$, the line integral $\int_C \mathbf{F}\cdot d\mathbf{r}$ is independent of path. Let $\operatorname{grad} f(x,y,z) = \mathbf{F}(x,y,z)$. Then

$$\frac{\partial f}{\partial x} = y, \quad \frac{\partial f}{\partial y} = x, \quad \text{and} \quad \frac{\partial f}{\partial z} = z^3. \qquad (*)$$

Integrating both sides of the first equation in $(*)$ with respect to x, we obtain $f(x,y,z) = xy + g(y,z)$. Taking partial derivatives with respect to y, we find that $\partial f/\partial y = x + \partial g/\partial y$. Comparing this with the second equation in $(*)$, we deduce that $\partial g/\partial y = 0$, so that $g(y,z) = h(z)$. Thus $f(x,y,z) = xy + h(z)$.

Appendix

1. The least upper bound is 1; the greatest lower bound is -1.

3. The least upper bound is π; the greatest lower bound is 0.

5. The least upper bound is 5; the greatest lower bound is 0.

7. The least upper bound is 1; the greatest lower bound is 0.

9. Assume that the set S of positive integers has an upper bound. By the Least Upper Bound Axiom S would then have a least upper bound M. For any number n in S, $n+1$ is in S, so that by the definition of M, we have $n+1 \leq M$. Thus $n \leq M-1$, and consequently $M-1$ is an upper bound of S. This contradicts the property of M that M is the least upper bound. Therefore S has no upper bound.

11. Let S be a set that is bounded below, and let T be the set of all numbers of the form $-s$, for s in S. Since S is bounded below, T is bounded above, so by the Least Upper Bound Axiom, there is a least upper bound M for T. If s is in S, then $-s$ is in T, so $-s \leq M$, and thus $s \geq -M$. Consequently $-M$ is a lower bound of S. If N is any lower bound of S, then $-N$ is an upper bound of T, so $-N \geq M$ since M is the least upper bound of T. Therefore $N \leq -M$. Thus $-M$ is the greatest lower bound of S.

13. Let $\varepsilon = 1$, and δ be any positive number less than 1. If $x = \delta$ and $y = \frac{1}{2}\delta$, then

$$|x-y| = \frac{1}{2}\delta < \delta, \quad \text{and} \quad \left|\frac{1}{x} - \frac{1}{y}\right| = \left|\frac{1}{\delta} - \frac{2}{\delta}\right| = \frac{1}{\delta} > 1 = \varepsilon.$$

Thus $1/x$ is not uniformly continuous on $(0,1)$.